Bringing Chemistry to Life: From Matter to Man

R. J. P. WILLIAMS
Emeritus Research Professor, University of Oxford

and

J. J. R. FRAÚSTO da SILVA
Professor of Analytical Chemistry, Instituto Superior Técnico,
Universidade Técnica de Lisboa

OXFORD
UNIVERSITY PRESS

OXFORD

UNIVERSITY PRESS

Great Clarendon Street, Oxford OX2 6DP

Oxford University Press is a department of the University of Oxford.
It furthers the University's objective of excellence in research, scholarship,
and education by publishing worldwide in

Oxford New York

Athens Auckland Bangkok Bogotá Buenos Aires Calcutta
Cape Town Chennai Dar es Salaam Delhi Florence Hong Kong Istanbul
Karachi Kuala Lumpur Madrid Melbourne Mexico City Mumbai
Nairobi Paris São Paulo Singapore Taipei Tokyo Toronto Warsaw

with associated companies in Berlin Ibadan

Oxford is a registered trade mark of Oxford University Press
in the UK and in certain other countries

Published in the United States
by Oxford University Press Inc., New York

A catalogue record for this book is available from the British Library

Library of Congress Cataloging in Publication Data
Williams, R. J. P. (Robert Joseph Paton)
    Bringing chemistry to life: from matter to man/R. J. P. Williams
and J. J. R. Fraústo da Silva.
    Includes bibliographical references and index.
    1. Chemistry.   I. Silva, J. J. R. Fraústo da.   II. Title.
QD33.W718   1999   540–dc21   99–24780
    ISBN 0 19 850546 9 (acid-free paper)

ISBN 0 19 850546 9

Typeset by EXPO Holdings, Malaysia

Printed in Great Britain
on acid-free paper by
The Bath Press, Avon

Preface

The aim of this book is to provide graduate students and teachers with knowledge of the connection between physical and biological sciences. It gives them the background logical structure to our two previous books, *The biological chemistry of the elements* and *The natural selection of the chemical elements*. In those two books we discussed the elementary content of living organisms, the physical chemistry of their use and how their functions have evolved. Against the convincing knowledge that all organisms use some 20–30 chemical elements we here explore the logic of the evolutionary progression from the primordial inanimate matter to the highest form of life we know about—modern man. By matter we mean, in this context, the original material of the early universe and its condition.

The progression, as we see it, is a gradual development of systems of chemicals, where a system is any self-contained ensemble of atoms or molecules going all the way from a contained volume of a gas to a human. Since the universe started as a limited volume of gaseous matter plus radiation how has it managed to generate man? The project demands that we analyse everything we see around us, or believe to have existed, into the units of which matter is made and their properties. The properties arise since the creation of matter was, at the stage of the Big Bang, associated with huge radiation energy (high temperature) and outward momentum, expansion. The initial system was homogeneous, but this homogeneity was soon broken. We need to follow, in as logical a manner as we can, the stages of development of the units of matter and their associations together with those of energy and momentum. Obvious stages are the breakdown into separate systems after the association of masses and the appearance of flow patterns, which produced apparently random distributions of astronomical bodies, including our planet, in systems up to the level of galaxies. During these processes various bodies formed at very different temperatures. Those at low temperatures contain atoms associated in chemicals. Our major interest centres on one such body—the cool Earth. Immediately we ask: why does Earth have so many different chemicals in its solid, liquid and gaseous compartments? The analysis requires a definition of the atomic constituents in all these materials, i.e. their composition. It continues with analysis of why particular compositions arise. We can call this *chemical speciation*. The topics here are the forces between particles, atoms, and molecules, and their time-independent, on average, statistical dispersion over space and with energy. In this description, we are led from consideration of single contained units or pairs of units, described approximately by simple fundamental space–time variables and links between them, to the examination of bulk phases and chemical equilibria. We find that this treatment requires us to use complex derived variables of statistical thermodynamics of bulk systems, whether we use

Newtonian or quantum mechanics. While we are so engaged in tackling time-independent static or dynamic systems we ask: what limits the substances we observe both in chemical composition and physical state? For example, why does liquid water exist on Earth? We shall find that there is a balance of forces holding units, e.g. water molecules, together, and statistical probabilities leading them to remain separated. In this way phases containing specific chemicals arise.

The fact that the universe has broken up into small dispersed zones separated by boundaries forces us to look at such zones, asking about their shapes and the forces (fields) they exert upon one another. After all 'shape' is used to characterise everything from molecules to crystals, to planetary systems and nebulae. Moreover, fields of force control much of what we see, and the radiation field of the sun generates much of the objects on the surface of Earth, including all life.

When we have satisfied ourselves that an appreciation of unchanging objects has been achieved in terms of specified units and their time-independent, though dynamic (time-averaged), spatial arrangements, many of which are energised, we turn to the analysis of changing systems, especially living organisms. Here we are confronted with time and the derived variable flow, or more precisely, flux. Material and energy flows become particularly intriguing because they can form patterns (dynamic shapes) as we see in living organisms. What are the units (chemicals) in these organisms, in what ratios and why? The logic here must relate not to 'stability' but to 'survival strength' (of species) and we note that survival strength of a given living organism includes its ability to reproduce. How can we appreciate the variables of composition, energy flow and so on that contribute to survival? How does 'design' arise within living things? How and why have the sophisticated designs of organisms evolved and yet the most primitive remain abundant? Does evolution depend on changes in the variables or available units? This discussion will lead us from our starting point, matter, all the way to a description of human beings, which we will refer to generally by the name 'man'.

In the course of answering these questions we have to appreciate self-organisation, dependent upon feed-forward and feedback messages from sources of information. These considerations introduce still further levels of complexity. One major source of information is the DNA in genes, but we must be careful when we examine DNA. First, it is not a static independent biomolecule, and secondly, its sequence does not have a linear relationship to the properties of an organism. As life developed, new coding possibilities evolved in the brain and then, externally, in man-made machines, such as computers.

We then turn to the self-conscious activity of man and his efforts to improve his own survival and well-being using a quite different approach to that of 'blind' biological evolution. How should we look upon man's own nature, his purposeful organisation of units and his application of variables to the creation of a man-made world for man's own benefit? Here we need to appreciate the peculiarities of the brain in individuals and the effect of mankind's activities on environmental conditions and then on local and global ecosystems.

A final question is whether we now have enough knowledge of the logic of such complex systems, not only involving physics, chemistry and biology, but also reaching to mankind's organisations, to be able to answer securely how we should proceed in the next millennium. We are forced to look closely at man to ensure that what he senses, thinks and desires, i.e. his biological reactions (inevitably linked to biological survival strength), do not conflict with the objective, scientific application of knowledge gained by using the rational power of the brain. While we can provide no answers we can indicate where care is needed to reconcile subjective aspirations and objective advantages. Only in this way can humankind preserve the present network of life, of which it is but a (vulnerable) part, rather than assist the development of some other in which its presence is not assured. The linchpin of this aim is education, particularly in chemistry. It is our hope that this book will assist in arousing awareness of the connection between organisms and this science in particular—hence its title *Bringing chemistry to life*.

Oxford and Lisbon   R. J. P. W.
November 1998   J. J. R. F.dS.

Acknowledgements

Our special thanks go to Mrs S. Compton who typed the manuscript and to Oxford University Press for its assistance in its final preparation.

The work reviewed here reflects discussion with a large number of colleagues, mainly research scientists from all over the world. We wish to thank them for all that they have brought to the book. We trust that the integrated view in it will be seen as the result of an international effort to examine a further part of the wonders of chemical and biological investigation. Mr T. D. Wess was of considerable assistance with the clarification of concepts in early chapters.

J. J. R. Fraústo da Silva acknowledges the Fundação para a Ciência e Tecnologia (Ministério da Ciência e da Tecnologia), Portugal, for general support of research activities, and Mrs Teresa Maria Carreiras da Silva and Miss Cristina Sequeira da Silva for secretarial assistance and for typing parts of the preliminary versions of the manuscript.

R. J. P. Williams acknowledges the generous support of Wadham College, Oxford, Oxford University, two British research councils (the Medical Research Council and the Science and Engineering Research Council) and The Royal Society over some 40 years. He is grateful to the Leverhulme Trust for a Research Award, which assisted the production of this book.

Contents

Units of energy and work and the values of some physical constants

The joule, SI unit of energy
$$1 \text{ J} = 1 \text{ kg m}^2 \text{ s}^{-2}$$
$$= 1 \text{ N m (newton meter)}$$
$$= 1 \text{ W s (watt second)}$$
$$= 1 \text{ C V (coulomb volt)}$$
$$= 0.24 \text{ cal (thermochemical calorie)}$$
$$= 6.242 \times 10^{18} \text{ eV (electron-volt)}$$

Thermochemical calorie
$$1 \text{ cal} = 4.184 \text{ J}$$

Large calorie
$$1 \text{ Cal} = 1 \text{ kcal} = 4.184 \text{ kJ}$$

Electron-volt
$$1 \text{ eV} = 1.602 \times 10^{-19} \text{ J (joule)}$$

Work required to raise 1 kg 1 m on earth
(at sea level) = 9.807 J

Free energy of hydrolysis of 1 mol of ATP
at pH 7, millimolar concentrations = −12.48 kcal = −52.2 kJ

Work required to concentrate 1 mole of a substance
1000-fold, e.g., from 10^{-6} to 10^{-3} M
$$= 4.09 \text{ kcal} = 17.1 \text{ kJ}$$

Avogadro's number, the number of particles in a mole
$$N = 6.0220 \times 10^{23}$$

Faraday $\quad$ 1 F = 96 485 C mol^{-1} (coulombs per mole)

Coulomb $\quad$ 1 C = 1 A s (ampere second)
$$= 6.241 \times 10^{18} \text{ electronic charges}$$

Electronic charge 1 e = 1.602×10^{-19} C (coulomb)

The Planck constant $h = 6.626 \times 10^{-34}$ J s (Joule second)

The Boltzmann constant
$$k_B = 1.3807 \times 10^{-23} \text{ J deg}^{-1}$$

The gas constant, $R = N k_B$
$$R = 8.3144 \text{ J deg}^{-1} \text{ mol}^{-1}$$
$$= 1.9872 \text{ cal deg}^{-1} \text{ mol}^{-1}$$
$$= 0.08206 \text{ 1 atm deg}^{-1} \text{ mol}^{-1}$$
and at 25° $RT = 2.479$ kJ mol^{-1}

The unit of temperature is °K (or simply K); 0 °C = 273.16 °K

Speed of light in vacuum
$$c = 3.0 \times 10^8 \text{ m s}^{-1} \text{ (metres per second)}$$

Mass of the electron
$$m_e = 9.10956 \times 10^{-31} \text{ kg}$$
$$= 0.511 \text{ MeV (mega electron-volt)}$$

Mass of the proton
$$m_p = 1.67261 \times 10^{-27} \text{ kg}$$
$$= 939.5 \text{ MeV (mega electron-volt)}$$

Mass of the neutron
$$m_n = 1.675 \times 10^{-27} \text{ kg}$$
$$= 939.5 \text{ MeV (mega electron-volt)}$$

Wave number unit, cm^{-1} (waves per centimetre)
$$1 \text{ cm}^{-1} = 1.9862 \times 10^{-23} \text{ J molecule}^{-1} \text{ (joules per molecule)}$$
$$= 2.8593 \text{ cal mol}^{-1} \text{ (thermochemical calories per mole)}$$

Frequency unit, s^{-1} (cycles per second)

1

The development of man's ideas concerning nature

The Intellect: 'Apparently there is colour, apparently sweetness, apparently bitterness, actually there are only atoms and the void'.
The Sense: 'Poor Intellect, do you hope to defeat us, while from us you borrow your very evidence?'

Democritus (*c.* 420 BC)

1.1 The early views

Since man became self-conscious he has tried to build up knowledge about himself and his relationship with his surroundings, even with the universe. One approach through chemistry and the associated physics of chemicals, which we will use in this book, has existed for thousands of years in developing form. It is based on the concept, which we call reductionist,* that all we observe can be systematised first in well-defined species, mineral or living, and then broken down into basic primary units and variables, of which there are a limited number. The variables are those 'qualities' that affect the primary units and cause them to change. In these early views, such as those of the most prominent Greek philosophers, e.g. Aristotle (384–322 BC), the primary units, which will be analysed in this chapter, were not thought to be a variety of material atoms† in different conditions,

---

*The *reductionist* approach implies that by dividing an observable system into parts the whole can be understood. The *holistic* approach assumes that the whole is greater than the sum of the parts. Scientists lean towards the reductionist method and it has been extremely successful, but in this book it will be seen to have unsatisfactory features.
†In fact 'atoms' were suggested as the ultimate indivisible pieces of all matter by other early philosophers, such as Democritus (*c.* 470–380 BC), but their ideas did not gain wide acceptance, although they were revived later by Epicurus (*c.* 342–270 BC) and his followers and much praised and elaborated by the poet Lucretius (*c.* 95–55 BC) in his long poem *De rerum natura*.

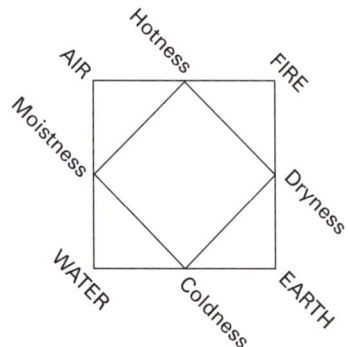

Fig. 1.1 The four primary manifestations of underlying 'form', as four units, 'elements' (in upper case) connected by 'qualifiers', variables: as proposed by Greek philosophers.

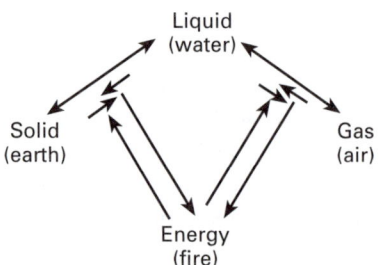

Fig. 1.2 A modern view of the four 'elements'. Notice that each physical state can now be transformed in an understandable way by the use of a fourth factor, energy, which, however, becomes a variable and which we must analyse later.

but were described by a few vague concepts such as 'earth' or 'fire' (Fig. 1.1). One of the underlying difficulties, then, was the attempt to include the description of all observables, including inanimate and animate matter, and their changes in one system of very few separate units and variables (only four), an approach which lasted in various guises for nearly 2000 years and is worth describing briefly, not only to compare it with the approach we all adopt today, but also because it reveals some of the fundamental difficulties facing our description of the natural world to which we shall return in the last chapter.

According to Aristotle, the basis of the material world was a primitive matter which had, however, only potential existence until it was imposed upon by 'form'. By 'form' he did not mean shape only, but all that conferred upon a body its specific properties in a variety of ways by given 'forces'. Thus, underlying this thinking of a division into units is a second notion of some deeper holistic system. In its simple manifestations, it was form that gave rise to the four primary units (or 'elements'), distinguished by their internal 'qualities', i.e. the variables. The examination of the world led Empedocles, and then Aristotle and his followers, to suppose that these primary units were air, fire, water and earth, each with some pair of general 'qualities' (variables), e.g. hotness and dryness (FIRE), coldness and fluidness (moistness) (WATER), one of which was predominant over the other: in earth, dryness; in water, coldness; in air, moistness; in fire, hotness (see Fig. 1.1). None of the four units was considered to be unchangeable; they might pass into one another through the application of that 'quality' (variable) that they possess in common.* (To account for the substance and the brightness of the bodies 'floating' in the sky, Aristotle admitted a fifth element, pure, eternal and incorruptible, which Plato had called ether,† a subtler kind of air). The history of these ideas is readily accessible (see references in 'Further reading') and the reasoning behind them, as well as the choices of primary units and variables, is clear enough and certainly not naïve given the state of knowledge at the time. We notice, however, that the descriptions are based upon examination using the human senses, by seeing and touching objects.

We stress now only one point. The description in these terms contains the three physical states of matter as we list them today—gas, liquid and solid, i.e. air, water and earth, which represent the general ways *matter* can be arranged in *space* and can be changed into one another—and what we recognise now as energy (fire). It is worth noting immediately that this division of the observable units of substances in the environment is stated in terms of physical features without reference to today's chemical analysis of composition, about which the Greek philosophers knew nothing, and without the concept of energy. We can rewrite the diagram as in Fig. 1.2.

---

*It is clear that the Greeks were trying to classify objects by units of composition (air, water, earth and fire) and their variables (coldness, hotness, moistness and dryness) which would convert one composition into the other. It is exactly this procedure that we follow, but with much greater knowledge we have been able to find the true units of chemical composition and the real variables.

†From this fifth element derives the expression 'quinta essentia', referring to the highest possible quality or purity.

Fig. 1.3 The *Ta-chi* symbol of balance between opposite tendencies. Consider, for example, order and disorder (Section 5.1), attraction and repulsion (Section 2.2), but even light and darkness, good and evil, love and hate.

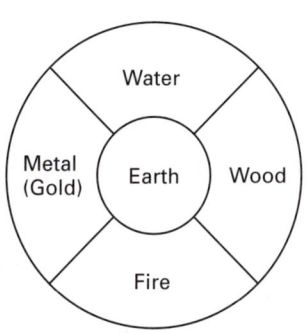

Fig. 1.4 The five 'elements', 'agents' or 'movers' (*wu-hsing*), which in Chinese philosophy are fundamental stages of any process in space–time, i.e., they are, in our sense of the word, not material, and have even been associated with spatial directions (north, south, east and west – with Earth in the centre), seasons of the year, musical notes, organs of the body and even five human senses or attitudes, as well as states of materials. As in the Greek scheme, they can all change into one another. The idea of a permanent material appears to have been foreign to Chinese thinking.

Written in this way it is an extremely attractive division of states of matter, even today (see Chapter 5). It so happens that if we still believed that under the holistic influence of 'form', as described above, matter is continuously adjustable within the limitation of three physical states (solid/liquid/gas) and not further separable into inviolate chemical units, then the above division could be said to be a correct description of materials and their changes (in physical though not in chemical terms). What is more, it makes clear that the physical states, of water for example, can be changed into one another by a supply of energy, which in a general sense is to be likened to 'fire' (see Chapters 2 and 5). It is the concept of 'energy' that is difficult to grasp since it is many faceted.

[The Greeks were, of course, aware that differences existed in 'earth' between named solids, but this was considered to mean that they contained more or less fire. Thus, for many centuries, following their thinking, attempts were made to convert different solids such as lead to gold, for example by heating. The fundamental problem was the inability to analyse such materials into a set of common units (now to be distinguished from states of matter), which we call 'chemical elements' today. In fact, we can see today that the physical states were confusingly muddled with chemical composition differences and the variables chosen were also a confusing set of properties related to 'energy' as well as composition, see Chapter 2].

In summary, the Greeks had no chemistry, but it would be quite wrong to push the idea of a purely philosophical (and physical) attitude in the Greek world too far—they were looking for cosmological explanations but also for logical explanations of the difference in behaviour of materials on Earth and their possible applications. Of course, they were not alone in this search—thoughtful man everywhere has always puzzled in this way about his environment. For example, a parallel set of ideas developed in China, but here there is greater difficulty in being sure of one's ground since the historical evidence is less securely based. The underlying concepts of the Chinese approach to the nature of the universe and the material world are contained in the *I Ching* (book of changes) ascribed to Wen Wang (*c.* 1200 BC) and in the *Shu-Ching* (book of stories) of the Chou dynasty (722–211 BC). The first of these introduces the *yin* and *yang* principles (combined in the *Ta-chi* symbol, Fig. 1.3), regarded as representations of opposed but complementary cosmic forces, which were the origin of all things that exist and cause change. (Today, attraction and repulsion would be one such pair). This again is an holistic concept. The second book refers to the 'five things' or 'five movers' (notice the idea of flow in 'movers') of which everything is composed—water, fire, wood, metal and earth—that might be changed into one another in a continuous and permanent cyclic manner (see Fig. 1.4). Here there is a parallel analytical reductionist approach to units and variables that is comparable with that of the Greeks, but a solid product from a living system, wood, and one from the non-living world, metal, are included. (The separation of wood from metal is known to us as a *chemical* composition distinction of units within 'earth'). These concepts were integrated into the Taoist philosophy, which originated in the writings of Lao-Tzu (sixth century BC), dealing mainly with ethics and social and political reform, and were later elaborated and extended by Chuang-Tzu (Fourth century BC), who was more concerned with the universe and the material world of observation or experience. While Lao-Tzu's emphasis is on 'permanence', 'spontaneity' and 'eternity', although the idea of constant 'flow' is present, Chuang-Tzu's emphasis is on cyclic 'change' within unity or oneness. (Note that flow and change introduce *time*, a variable with *matter* and *space* as of the essence of substance, which here is absolutely interlocked with 'energy'). The sense of constant cyclic change, frequently embedded in Chinese thought and expressed in the paradigmatic *Ta-chi* symbol (Fig. 1.3) is connected to the ideas of the flowing of 'material' from one shape to another and from one place to another, always in balance and always returning on itself.

With the insistence on flow, the Chinese in effect introduced a further variable in the classification of materials. The distinction rests in noticing that changes of physical state, for example melting and boiling, were of another kind from changes from a static to a moving (flowing) object. Thus a river was classified in their thinking as quite different from a lake though both are (liquid) water and both can freeze.

Within these classifications, by both Greek and Chinese, different minerals were recognised and named, i.e. speciated, and living organisms were separated as being quite distinct through physical appearance and mobility, e.g. plants, fungi and animals. In each class differences were observed, say between one kind of flower or animal and another. This led later to the classification of living organisms in the sixteenth to nineteenth centuries into families and then into species. If one wishes to say so, these are the units of biology. Today, much as we talk of *living species*, so we can describe 'speciation' of materials in chemistry. We, of course, like all earlier analysts, shall have to concern ourselves with the relationships of both these speciations to their separation into more fundamental units and the related variables, where this is possible. Today we believe that all living species have evolved, but until around AD 1800 this idea of change was not considered.

Implicit in both these early analyses, Greek and Chinese, is the idea that a reductionist approach must have an underlying holistic system, often associated with the notions of 'form' and 'motion' (Figs 1.1 and 1.4). This ancient set of ideas is not so different from the search by physicists today for a unifying theory of forces, despite the obvious separation of such a theory (of everything) from human experience based on our senses. One part of this description relates to the apparently fixed, static constructs around us, such as the minerals on Earth itself, affected by physical conditions, e.g. temperature, while another considerable part concerns the changing of 'systems' with time, evolution, or the idea behind the term 'flow', as observed in activities in stars in the universe (such as the sun) on Earth and in life, all of which are to be discussed in this book. As stated earlier, we shall not hesitate to look at physics and chemistry from an holistic as well as a reductionist point of view.

This brief and necessarily incomplete description of the Greek and Chinese approaches to the interpretation of the universe hopefully will have shown their similarities and differences—the first, which influenced occidental culture, is more rational, abstract, intellect based; the second, which impregnated oriental culture and the way of living, is more natural-istic, concrete, experience based. Curiously, the Greek approach put the emphasis on the static composition of matter and the Chinese on the dynamic character of its transformations, hence on flow, though clearly both features are present in some writings within either culture (see aside). Thus, it should be added that about the time (sixth century BC) the Chinese developed their approach to nature elaborating their concepts of perma-nence, spontaneity and flow, a Greek philosopher, Heraclitus of Ephesus, who proposed 'fire' as a primordial 'element', advocated a similar idea of a changing world, of eternal becoming, well expressed in his sentence 'every-thing flows'. For him, too, all changes in the world derived from the dynamic and cyclic conjugation of opposed pairs which, however, formed a unity containing and transcending them—the *Logos*. How curious it is then to observe that the most prominent Greek philosophers, e.g. Plato and Aristotle, rejected two major ideas (of today) that other less prominent colleagues—Democritus and Heraclitus—had advanced: atoms and flow. Perhaps it is even more curious to observe that, odd as they seem now to us, the concepts of Aristotle and his followers lasted for at least 2000 years without significant philosophical improvement. (We should be aware that we may have fallen into a similar paradigm trap, where our present-day models lead us to limit investigation of other possibilities).

The reason for the stagnation of development of ideas was that all through this historical period the role of man in the material world was regarded as a distinctly superior one in which he bore, through his self-conscious existence, a special relationship with the underlying holistic 'forces' (or form) in the universe. Some, in the tradition of Plato, believed in a rational world and in the power of abstract thinking which, without recourse to experiment (despised by many of the Greek philosophers, and their cultural inheritors to this day), could uncover the fundamental prin-ciples and patterns of nature; others sought life's meaning in religions, which allowed man contact with God, the omniscient and omnipotent prime mover, renouncing any search for other, necessarily 'blasphemous', explanations. Both of these attitudes inhibited development of an empirical scientific appreciation of nature, no matter what other 'truth' they may contain.

A radical change in attitude was caused by the slow development of the experimental, largely an improved reductionist, approach, which became of ever-increasing practical importance and put an end to simple faith in intellect-based or sense-based rational or mystical arguments for a descrip-tion of material around us. In the experimental approach no final system is supposed, *a priori*, to lie beneath the world observable through our *senses*, while models of parts of any system, which by their nature are reduction-ist, can be proposed provided they are consistent with observation. Observation now includes study with the help of powerful instruments external to man, often leading to exact mathematical descriptions but fre-quently with incomprehensible features. However, we must never forget that behind this approach there remains the ever-present wish and struggle to describe the objects and activities we observe within an holistic view which, we shall see, has to contain a description of the very factors ancient

Common classical features

Greek	Chinese
Water	Water
Earth	Earth
—	(Metal, Wood)
Fire	Fire
Air	—
(Flow)	Flow

thinkers uncovered. Above all we wish, like them, to discover *the minimum number of units and variables, and their inter-relationships, with which we can describe our observable environment.* These are then to be used to explain the properties of materials, which explanations we shall keep as simple as possible.

1.2 The development of modern views

It has become clear very recently, i.e. in the last 200 years, that early attempts, which are based on the human senses, especially touch, taste and sight, to give an impression of the whole universe in such simple systematic terms as those of the Greeks and the Chinese are inadequate. The difficulty of changing view was partly due to the size (scale) of the building units of the objects around us, (see aside) which have been revealed by experiment and are not open to man's senses, and partly to the misguided early wish to establish just a small number of useful units and variables (in the sense of useful for further development of understanding) to associate with observable phenomena (see Figs 1.1 and 1.4).

Thus, today, we need to be able to think of huge distances to comprehend the nature of the universe. The sun is approximately 100 million, i.e. 10^8, miles away and the moon is 2.5×10^5 miles from us. The next nearest stars are 10^{12} miles from Earth. Earth is about 10^4 miles in diameter. More important for this book are the sizes of small objects. The average human is 1.7×10^2 cm high and his finger nail is about $\frac{1}{30}$th cm thick. The thinnest hairs are 10^{-2} cm in diameter. The units of composition are atoms, the smallest of which, the hydrogen atom, has a diameter of 10^{-8} cm. Roughly speaking, 1 cm is to an atom as the distance to the sun is to a mile or kilometre! Mass is described in similar mathematical form. Thus we use a scale from 1 gram, a sugar lump, to 10 kilograms, 10^4 g, e.g. a young child. Going to small objects, a hair may be 10^{-3} g. On this scale an average atom is 10^{-24} g. (The number is one millionth of a millionth of a millionth of a millionth of a gram.) [Unfortunately, as we shall see, when we reduce scale below that of an atom to that of the electron (see below), mass, in the conventional wisdom of our senses, loses its meaning and so does the 'space' occupied by a very small mass. The observations on such small 'particles' cannot be understood from man's sense experience of larger ones. We shall try to explain and then to avoid this grave difficulty which lies at the heart of the 'nature' of matter]. Finally, the periods of time we have to handle are no longer thousands of years but billions of years, in which change has been continuous.

As we shall outline, the universe is now known to be exceedingly large and complicated, and yet explanations of it depend on the discussion of extremely small physical 'objects', much smaller than atoms, namely electrons, protons and neutrons, (see Table 1.1 and Section 3.2.1). (Note that there are other more fundamental particles which we shall not consider here, see Table 1.2). Our sight and touch do not really help us in this endeavour. Unfortunately, both the very, very large and the very, very small

our galaxy from the side

our solar
system....

our galaxy from the top

An impression or model of a galaxy taken using light, but many objects are missing. The size is unbelievable for the senses but found to be true by instrument measurement.

Table 1.1 Characteristics of the main, chemically important subatomic particles*

Particle	Symbol	Mass (kg)	Electrical charge (Coulomb)
Proton	p	1.67×10^{-27}	$+1.60 \times 10^{-19}$ $(+e)$
Neutron	n	1.67×10^{-27}	zero
Electron	e	0.91×10^{-30}	-1.60×10^{-19} $(-e)$

* Giving objects masses and charges is a way of expressing quantitatively their interactions that are observables. The fundamental character of mass and charge is extremely difficult to understand. Note that the mass of atoms is overwhelmingly that of protons plus neutrons (see Chapter 3). Mass and charge are the fundamental underlying units of all we describe in chemistry and biological sciences.

Table 1.2 Fundamental particles of protons and neutrons*

	Charge		Name	
Leptons	0	Electron–neutrino (?)	Muon–Neutrino (0.27)	Tau–Neutrino (<35?)
	$-e$	Electron (510)	Muon (106)	Tau (1777)
Quarks	$+\frac{2}{3}$	Up (5)	Charm (1500)	Top (180 000)
	$-\frac{1}{3}$	Down (8)	Strange (160)	Bottom (4250)

* Values in parentheses = mass at rest in megaelectron volts.

can only be dealt with using theoretical, physical and mathematical treatments, which are outside the reach of the comprehension of most of us, and can only be studied using instruments with sensitivities greatly outside the range of our senses. In large part, however, neither the very, very large, on the astronomer's scale, nor the very, very small, on the particle physicist's scale, need concern us greatly in this book, especially as we shall limit ourselves mainly to the context of Earth, and especially life on the Earth. We shall state unequivocally that, to the best of our knowledge, Earth itself, and then life, is where the natural selection of the atomic chemical elements has evolved and developed to its greatest degree of sophistication and must command our greatest interest. We live in a world that is dominated, therefore, by a somewhat less uncomfortable chemical atomic scale (where visual models are still extremely useful) than is required to describe the universe. In chemistry we can often assume that all matter is made of hard balls, atoms, of radius around 10^{-8} cm, which we scale up to about 1 cm to visualise in models, and which interact with one another (see Chapters 2 to 5). However, our sense-based concepts of space and time are not very useful in our understanding of interactions between such units and of changes of the units (see Chapter 3). In order to appreciate the modern ideas of units and variables (see Table 1.5 later) we shall start our description with the units of composition we use today.

Detailed examination of the materials available around us through chemical analysis and preparation over many centuries, and using more and more elaborate equipment, gradually showed that the composition of matter was far more complex than had been thought.* The search for understanding was (and still is) not just driven by scientific curiosity (of course) but by the use to which the knowledge might be put. But, whatever the reasons, the universality of this empirical analytical approach linked to chemistry was of major importance since it began to reveal a lengthy list of common substances, elementary species, the true units of composition to which all matter, animate or inanimate, could be reduced. Progress was slow because one weakness in the early chemical analyses, say until the end of the eighteenth century, was that they were at best qualitative, and thus did not allow clear-cut quantitative knowledge of the elementary composition of any of the materials. Most scientists were also hindered in their interpretation by mistaken viewpoints, for example, that matter was indefinitely continuous, i.e. *not* made up from small particles, atoms,[†] as proposed by some early Greek philosophers. Furthermore, it was thought that living organisms might contain a special extra feature, a 'vital force', which was treated as if it were open to discovery by analysis too. It was not easily realised either that an initial separation in thinking about matter and energy (compare Figs 1.1 and 1.2) would be helpful even if there would have to be a synthesis of ideas subsequently. [We shall show later that materials are built from units of mass while energy is a variable which adjusts their properties.]

The change in thinking about the substances around us, which can be traced back many centuries as it gathered acceptance, depended on increasing sophistication in the experimental study of materials. The main impetus came from more careful quantitative chemical analyses, i.e. breaking down all material, no matter which physical state it was in, into irreducible units, the 'chemical elements', which was developed over about 200 years, say from 1650 to 1850. Scientists used experiments such as the decomposition of solids by heating, or their reaction with acids, or dissolving them in water while following the weights and volumes of the reactants and the substances produced. It was finally realised that all of the materials around us (the 'species' of chemicals on earth, in air and water, including living organisms) were made from a limited number of 'chemical elements' which were represented by symbols, first circles with

---

*It often escapes notice that there is a fundamental change in the conception of matter (and charge) from the classical idea of continuous gradation to a formulation in terms of discontinuous units, specific atoms. We accept this atomic approach but no longer marvel that nature is like this. We find the discontinuous (quantum) approaches to energy conceptually more difficult, see later chapters.

[†]It was Boyle, who was born in 1627 and was later called 'the father of chemistry', who argued strongly against this idea and in favour of a corpuscular nature of matter, thus reviving the ideas of Democritus some 2000 years earlier. At about the same time Galileo insisted that understanding had to be based on quantitative knowledge which leads to mathematical formulation.

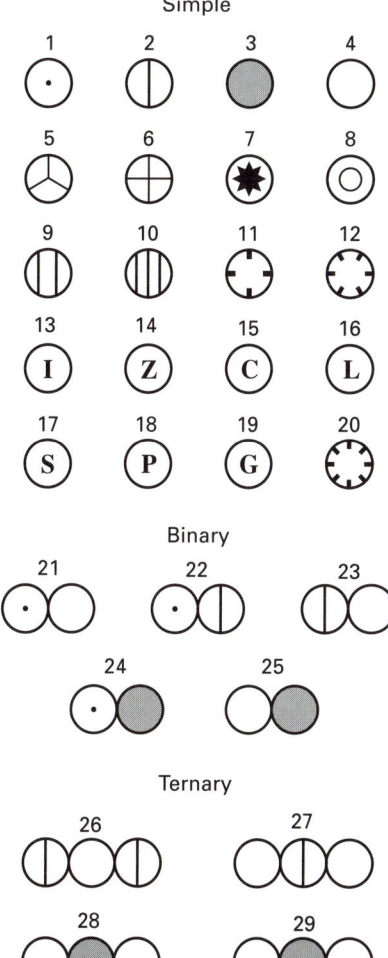

Fig. 1.5 Dalton's chemical symbols for atoms and molecules. (1) Hydrogen, (2) azote, (3) carbon, (4) oxygen, (5) phosphorus, (6) sulphur, (7) magnesia, (8) lime, (9) soda, (10) potash, (11) strontia, (12) baryta, (13) iron, (14) zinc, (15) copper, (16) lead, (17) silver, (18) platinum, (19) gold, (20) mercury, (21) water, (22) ammonia, (23) 'nitrous gas', (24) ethylene, (25) carbon monoxide, (26) nitrous oxide, (27) nitric acid, (28) carbon dioxide, (29) methane.

Table 1.3 The elements of major concern

H Hydrogen			
Li Lithium	Be Beryllium	B Boron	C Carbon
Na *Sodium*	Mg Magnesium	Al Aluminium	Si Silicon
K *Potassium*	Ca Calcium	Sc Scandium	Ge Germanium
Rb Rubidium	Sr Strontium	Y Yttrium	Sn *Tin*
Cs Caesium	Ba Barium	Ln (Lanthanides)	Pb *Lead*
			He Helium
N Nitrogen	O Oxygen	F Fluorine	Ne Neon
P Phosphorus	S Sulphur	Cl Chlorine	Ar Argon
As Arsenic	Se Selenium	Br Bromine	Kr Krypton
Sb *Antimony*	Te Tellurium	I Iodine	Xe Xenon
Bi Bismuth			
Ti Titanium	V Vanadium	Cr Chromium	Mn Manganese
–	–	Mo Molybdenum	–
–	–	W *Tungsten*	–
–	–	U Uranium	–
Fe *Iron*	Co Cobalt	Ni Nickel	Cu *Copper*	Zn Zinc
–	–	Pd Palladium	Ag *Silver*	Cd Cadmium
–	–	Pt Platinum	Au *Gold*	Hg *Mercury*

N.B. The elements not mentioned above will not concern us since they are very rare and little used. Note how elements known for a very long time have symbols unrelated to their names (in italic). The sources of these names are given in Table 1.4.

Table 1.4 The origin of some symbols for chemical elements

Antimony	Sb	(from stibium)
Copper	Cu	(from cuprum)
Gold	Au	(from aurum)
Iron	Fe	(from ferrum)
Lead	Pb	(from plumbum)
Mercury	Hg	(from hydrargyrium)
Potassium	K	(from kalium)
Silver	Ag	(from argentum)
Sodium	Na	(from natrium)
Tin	Sn	(from stannum)
Tungsten	W	(from wolfram)

dots, lines and letters (see Fig. 1.5) (Dalton, 1808) and then just by a letter code (Berzelius, 1813) (see Tables 1.3 and 1.4). These are the fundamental units of composition with which we must be concerned. The units could be combined to give compounds, e.g. $2H + O \rightarrow H_2O$ (water) or $3H + N \rightarrow NH_3$ (ammonia). The substances so analysed could pass through the states of matter, (solid, liquid and gas), by applications of variables such as temperature and pressure (see Chapters 2–5). As a result, we can now write down, on a chemical element basis and in a symbolic language, anything material that exists around us.

Using the techniques available in early days, around 1800–1850, the chemical elements could not be broken down further. It was concluded that they had to be made of some kind of indivisible pieces which, as stated above, were called atoms of the (chemical) elements by analogy with the ultimate material pieces of Democritus, i.e. they are atomic chemical elements, but different atomic elements had different sizes and atomic masses.

We stress that it has turned out that everything we see, smell, hear or feel on this Earth can be related to these atomic elements, without further reductive analysis. [This is due to the fact (see later) that even all biological systems, including ourselves, are made up in this atomic fashion.]

Thus, by the year 1860, it was known that all of one part, the material part [we come to fire (energy) in Chapter 2] of the observable world could be broken down into such chemically indivisible pieces (atoms), and at that time some 60 such chemical atomic elements had been discovered. All the material world became, therefore, open to us in a language of combinations of letter symbols for the units of composition, (see aside),

Typical combinations

H_2O, H_2O_2
NH_3, N_2H_4, NO
CH_4, C_2H_6, CO_2
HCl, Cl_2O, Cl_2O_7

$$mA + nB \longrightarrow A_mB_n$$
$$\text{atoms} \qquad \text{combination of atoms}$$

where the capital letters represent atomic elements and the italic subscripts are the numbers of atoms entering into combination. This 'alphabet' has subsequently been found to require about 110 symbols, equivalent to the known atomic elements (see Fig. 1.6) most of which must have two letters, e.g. Cu (copper) or Cl (chlorine), of course, but we do not need a more complex code for all of the chemical atomic description of living or dead matter. (Note that there are now known to be only 90 stable, 'natural' elements on Earth and no more, but the reason for this limitation is outside chemistry and is a subject into which we shall not enquire deeply.) Thus, *one major variable in nature is composition in terms of atomic element units.* The remarkable nature of this discovery must never be underestimated; it is of equal importance to the physical division of materials into gases,

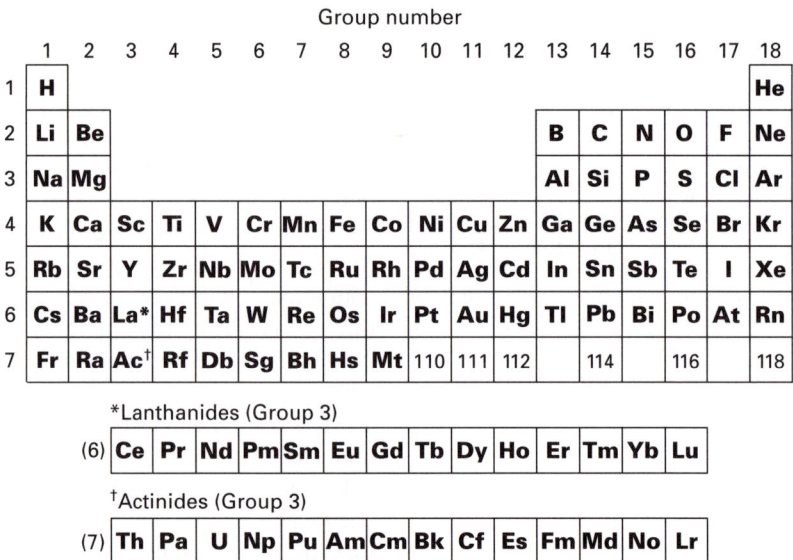

Fig. 1.6 The periodic table of chemical elements, displayed in the modern 'long' form. Each element is denoted by its symbol. From U (element atomic number, 92) the elements have been synthesised by man.

liquids and solids and to the understanding of energy, to which we turn in Chapters 2–5, but it was not discoverable from sense data.

Obvious questions arise immediately. How many different combinations of atoms, new chemical 'species', can be made? In a compound, A_mB_n, where A and B are elements and n and m are numbers, will any A join up with any B? Can there be an almost continuous range of n:m ratios or will n and m be simple? Will A_mB_n ... resist heat, water, etc., which is a question of the stability of a compound? The answers to many of these questions have been obtained, but the variety of ways of putting together these 90 stable elements, A + B + C etc., in any atomic ratio one wishes, is so immense that man is still searching for new properties and possible functional value of new chemical species. However, just as we now know that there are no more than approximately 110 chemical elements, 'natural' and man-made, so we know the major rules for the combining ratios of some (note, not all) of them. For example, Proust's law of constant combination (1797) and Dalton's law of multiple proportions (1830), apply to the familiar molecule H_2O (water), made from three atoms. [We shall see later that all such combinations are connected to loss or gain of energy (fire).] We have also discovered, quite probably, most if not all of the properties and major uses to which the combinations can be put!

A second type of question is why do atoms stick together in special combinations to form distinct chemical species, that is, why is composition not continuously variable? This question relates to the early discussion concerning 'form' and is answered today by proposing electrical forces between atoms. Before describing these forces in Chapters 2–4, we shall look first at the characteristic relationships between the naturally occurring 90 elements. In this way the problem of chemical forces reveals itself in a fascinating table of similarities and differences between elements and their combinations.

1.3   The periodic table

Let us see, in essence, how experimentalists discovered the systematic relationships between the elements shown in Fig. 1.6. Remember that they were looking for the number of fundamental 'elements' (units) by which *all* materials (species) can be described. Obviously, they could classify materials in many ways, such as by density (lead), by metal-like character (gold), by non-metal character (sulphur), etc., but they could also classify by the manner of combination.

Consider a simple material such as purified gold. No matter how chemists tried to melt and refine it or to cut it into smaller and smaller pieces, they found that there was no change in the properties. As an assumption, later shown to be correct, they therefore took gold to be one fundamental chemical element and considered that it could be reduced ultimately, and no further, to gold 'atoms'. Several other substances found naturally turned out to be similarly made of unchangeable atoms, e.g. silver and copper metals, although this is not the only state in which they are found in minerals. All such solid elements belonged in Fig. 1.1 under the category 'Earth', or in Fig. 1.4 under the title 'Metal'. The question arose as to how many atomic elements there were, i.e. which are the fundamental chemical 'elements' underlying the general Greek concepts of 'Earth', and similarly of 'water' and 'air', and in what way (and ratio) are they combined? Again, which are metals and which non-metals, and which atoms are most like which other atoms?

The next empirical findings concerned the several sources of copper, tin, lead and iron in mineral ores. The metals were obtained by taking these minerals and heating them in a wood or charcoal fire. Elementary metallic copper was formed from certain such minerals, like the other metals, so that it clearly had been in some combined form and firing had removed whatever was in combination with it. Firing of minerals gave access to a wider and wider range of pure metals which could not then be changed by further treatment. A natural but later assumption was that all these metals are immutable atomic elements which can occur in one or more combinations with other elements.

By collecting the gases given off in the firing while following weights of residues it was discovered that the metals in minerals were often combined with a very different type of element, later called the non-metal oxygen, and that the effect of heating with charcoal (carbon) could be written down, for example for iron oxide, as

$$2FeO \; + \; C \; \rightarrow \; CO_2 \uparrow \; + \; 2Fe$$

$$\text{iron oxide} \quad \text{carbon} \quad \underset{\text{dioxide}}{\text{carbon}} \quad \text{iron metal}$$

A further finding was that the heating of other substances in air, i.e. with oxygen present, could give such reactions as

$$CuS + O_2 \rightarrow Cu + SO_2 \uparrow$$

where a new gas was discovered which included a new element, sulphur (S). Minerals were therefore combinations, i.e. compounds, of atomic elements, often of metals with non-metals such as oxygen and sulphur. Similarly, some gases of the air were found to be simple combinations of

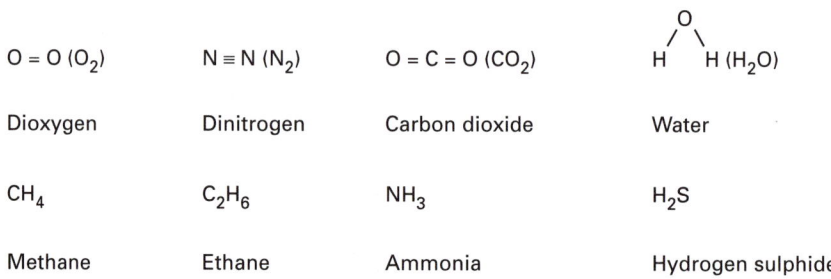

Fig. 1.7 Examples of some simple stoichiometric molecules. The top row have been given bond formulae (see Chapter 3).

Selection of stoichiometric combinations

Metal + Non-Metal
Non-Metal + Non-Metal
(i.e. common molecules)
but not
Metal + Metal

the same or different atoms, e.g. O_2, N_2, CO_2 and H_2O, which formed isolatable stoichiometric combinations of non-metals that were called *molecules* (Fig. 1.7), (see aside).

Without much elaboration it is possible to see that if by experiment you search for those forms of matter, now called atoms of the chemical elements, that cannot be split up by the procedures available before 1880, you will finish with a list of all the atomic-chemical elements on Earth. A final example is that of common salt, called sodium chloride. By experiment it was shown that this material could be broken down into two such elements—one was a continuous solid metal of sodium (Na) atoms, and the other a non-metal gas of chlorine (Cl) molecules, Cl_2. The analysis of the composition gave a constant proportion of the combining weights of these elements from which it was possible to derive the ratio of atoms of sodium to chlorine in sodium chloride as 1 : 1 and so the compound is NaCl, the formula of common salt.

The next step forward was the discovery that there were similarities and differences between the elements. For example, other metal elements were found that could be combined with chlorine in the same ratio as sodium and there were non-metals that could be combined with sodium, again in the same ratio as chlorine, to give yet other new minerals. It was found, for example, that from plants a slightly different mineral, potassium chloride, could be isolated. It was denser, but gave a metal very like sodium, namely potassium (K), together with chlorine (Cl). The formula was found to be KCl. Since the two compounds, NaCl and KCl, are very alike both in combining ratio and in physical and chemical properties but one is heavier than the other, it was logical to put the metal atoms in a so-called chemical group starting with Na since the atomic 'weight'* of this metal is less than that of K. (Groups were and are arranged vertically, with the lightest at the top.)

Na

K

Other minerals extracted from sea water were found to be salt-like (cf. sodium chloride), but as well as sodium they yielded on decomposition, respectively, a red non-metal liquid element, bromine (Br), and a black non-metal element, iodine (I). Since the combining ratio with sodium was 1 : 1 for NaBr and NaI and the density increased in the order Cl < Br < 1 another group of atoms could be arranged in a column of increasing atomic weight:

Cl

Br

I

*The atomic *weight* (more rigorously called relative atomic mass) is an empirical concept determined for each element obtained from natural sources. It is a number which expresses the atomic *mass* of an element in terms of the atomic *mass* of another element taken as reference and given a particular value. The reference is the hydrogen atom of atomic mass 1.00. Actually, the atomic weight is an average of the atomic weights of all the different isotopes (atoms of the same element) with different masses (whole numbers). To appreciate atomic weight, which is not a fundamental property of atoms, we have to analyse the isotopic composition (see Section 3.2.2). For this reason, atomic weights are presently calculated relative to the atomic mass of the ^{12}C isotope of the carbon atom, equal to 12 000.

It was observed that potassium gave KBr and KI as salts so the scheme was consistent with the previous grouping.

It did not take long to find other metals which combined with chlorine but in different ratios. For example calcium (Ca) gave $CaCl_2$, magnesium (Mg) gave $MgCl_2$ and strontium (Sr) gave $SrCl_2$. The density increased in the orders Mg < Ca < Sr, Mg > Na and Ca > K. If we call the Na, K group, group 1, combining ratio one, then a new group is group 2 of the metals, combining ratio 2. Arranging by chemical properties, metal character and density we have the atomic elements

Na   Mg
K   Ca
     Sr

The procedure, which the chemists of the nineteenth century followed, is now obvious. Thus, they found, by experiments on different materials, various non-metals which give atom ratios with Na or K different from the 1 : 1 of NaCl. The first two found were oxygen (O) and sulphur (S) which gave Na_2O, K_2O, MgO, CaO, SrO and Na_2S, K_2S, MgS, CaS, and SrS, respectively. Their chemical and physical properties led scientists to establish additional groups:

Na   Mg   O   Cl
K   Ca   S   Br
     Sr   I

Space has been left here between Mg and O and Ca and S to allow the possibility, which turned out to be reality, that there were more groups than we have indicated so far.

Later again there were found to be new combining ratios in yet other materials, for example Al(aluminium)Cl_3, C(carbon)Cl_4, Si(silicon)Cl_4, Na_3N (nitrogen), etc. In addition, it was discovered that the composition of water could be written as H_2O and there was a gas which analysed as HCl, both of which gave the very, very, light non-metal hydrogen. Slowly but surely it was seen that the combinations of the elements mentioned plus a few others (see Table 1.3) then allowed the simplest groups and *periods*, horizontal series, of atoms to be completed and established as

H
Li   Be   B   C   N   O   F
Na   Mg   Al   Si   P   S   Cl
K   Ca   Br
     Sr   I

Atomic weight increases sequentially along the rows up to calcium, see below, and down all groups. The pattern in the groups and periods was found to be the same for physical properties, e.g. densities and melting points, and for chemical properties. In fact, all investigations by around 1850 led to the same conclusion. Since then no elements have ever been found that interrupt these two first rows. However, around 1900 it was discovered that in the air there were other gaseous *elements*, not just oxygen and nitrogen, which would not combine with anything and were accordingly called *inert* gases (now called *noble* gases). On the basis of density they could be placed at the ends of the atomic series, i.e.

H							He
Li	Be	B	C	N	O	F	Ne
Na	Mg	Al	Si	P	S	Cl	Ar
K	Ca					Br	Kr
						I	Xe

Once such series were established elements could be given 'atomic numbers' starting from H = 1 and going to He = 2, Ne = 10 and Ar = 18. (The relationship between these numbers and the nature of atoms is described in Section 3.2.1.) Before going further we turn to the specific example of the variation of chemical combination in the groups. After hydrogen, which is the lightest and taken as the reference element, and helium, which is a noble (inert) gas, there is a period of eight elements, from lithium to neon. (In passing remember that hydrogen atoms combine to give H_2 but do not combine with helium.) Across this period it was found empirically that the combining numbers with hydrogen are in the order 1; 2; 3; 4; 3; 2; 1; 0, for the eight elements from number 3 to 10 (see Fig. 1.8). We represent the corresponding compounds as LiH, BeH_2, BH_3, CH_4, NH_3, H_2O, HF, (Ne). Thus we have first a sequence of two elements, H (giving H–H) and He, and then a sequence of eight elements. Both finish with gaseous monatomic elements, He and Ne, which, as mentioned above, are very reluctant to react and have been referred to as 'inert or noble gases'. For the moment we offer no explanation for these series of 2 and 8 but we observe that there is a rhythm of change along the period. Note especially that rules for chemical 'speciation' were being developed by these analyses.

A second way of seeing the rhythm of combining ratios is to consider binding to oxygen, O, rather than to hydrogen. Oxygen itself combines with two hydrogen atoms in the molecule H_2O. Thus, the combining ability of oxygen will always be twice that for hydrogen, that is, compared with hydrogen each oxygen atom needs to combine with twice the number of other atomic elements. So we find the formulae Li_2O, MgO, B_2O_3, CO_2, N_2O_3, O_2, F_2O, (Ne) (compare, for example, CO_2 with CH_4).

Now we can develop this table of weights and combining ratios for the next group of eight elements from sodium (Na) to argon (Ar) (see Fig. 1.8). Here we find that there is a second series of eight elements which fit in the groups of the series 3–10

Li	Be	B	C	N	O	F	Ne
Na	Mg	Al	Si	P	S	Cl	Ar

The fit is not perfect in all respects but it is extremely satisfying in several ways. First, the atomic weights increased systematically; secondly, the chemical combining ratios with oxygen and with hydrogen matched those of the series for Li to Ne; thirdly, the physical and chemical properties of not only the elements but also all their compounds, e.g. combinations with H, O, Cl and S, match in the groups; fourthly, *no* other elements were discovered to interrupt the series 2 (H to He) 8 (Li to Ne) 8 (Na to Ar). There

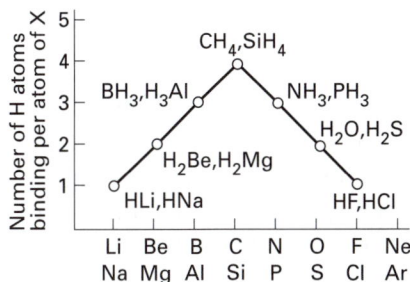

Fig. 1.8 Number of H atoms binding per atom of element X in the two eight element periods, atomic numbers 3–10 and 11–18.

had to be something fundamental about the series of numbers 2, 8, 8. What was it? Do not fail to notice that the series dominates the properties of all possible materials made from these 18 elements. We show, in Chapter 2, that there is a simple mathematical connection within these numbers in periods of 2, 8, 8 elements which has a powerful theoretical logic. The numbers are best seen immediately as (2×1), $(2 \times 1 + 2 \times 3)$ and $(2 \times 1 + 2 \times 3)$. This pattern does not allow there to be any other elements in these three series. There is, therefore, a very impressive mathematically essential logic to the series.

The remaining elements were discovered by series of experiments which included the determination of the atomic weights of new elements as they were found and also their combining ratios with, for example, H and O. However, the study revealed patterns that were not so simple as the above cases. For example, if we follow combination with oxygen as weight increases then we find that in the next series there are eighteen elements, i.e.

K　Ca　Sc　Ti　V　Cr　Mn　Fe　Co　Ni　Cu　Zn　Ga　Ge
As　Se　Br　Kr

and the compounds with oxygen are, at the beginning of the series,

$$K_2O \qquad CaO \qquad Sc_2O_3 \qquad TiO_2 \qquad V_2O_5$$

(and multiplying the observed combining ratio by two, as always for oxygen to relate its combining power to that of hydrogen) we have a series in the ratio 1, 2, 3, 4, 5, but then the following elements give oxygen compounds of (almost) constant composition, and combining ratio 2 or 3, before the 13th element, Ga, and then a trend parallel to that for Si, P, S, Cl, Ne follows:

Cr_2O_3　　MnO　　FeO　　CoO　　NiO　　CuO　　ZnO　　Ga_2O_3　GeO_2
As_2O_3　　SeO_2　　Br_2O　　Kr

In this series some obvious groupings and similarities of physical and chemical properties of elements were found which related back to the second series of eight given above. Thus, the first three conform to expectation:

Na　Mg　Al　　but then came　　Si　P　S　Cl　(Ar)
K　Ca　Sc　　　　　　　　　　　Ti　V　Cr　Mn

Highest oxides

TiO_2	V_2O_5	CrO_3	Mn_2O_7
Fe_2O_3	NiO	CuO	ZnO

Now, following scandium (Sc), titanium (Ti), vanadium (V), chromium (Cr) and manganese (Mn) are metals, while silicon (Si), phosphorus (P), sulphur (S) and chlorine (Cl) are non-metals, so that the matching of properties fails after scandium but see aside for matching higher oxides. Again Mn is not followed by an inert gas, krypton, Kr, until 10 elements later.

Working back from the inert gas Kr of the series of 18 elements and again comparing with the second series of eight, the elements based on combining ratios and physical and chemical properties which give rise to groups, are

(Al)　Si　P　S　Cl　Ar
(Ga)　Ge　As　Se　Br　Kr

but the elements previous to Al and Ga are Mg and Zn and then Na and Cu, respectively. The last two are obviously different even in combining ratio with oxygen since they give the oxide stoichiometries, Na_2O and CuO.

It was, therefore, no good pretending that two elements such as sodium and copper belonged in the same group. Accordingly, a table was written at first with the elements in groups divided now into two, A and B, subgroups with similarities, so that K, Ca, Sc before the elements difficult to classify up to manganese (Mn) were called A-subgroups and the elements from Cu on were called B-subgroups. The elements Fe, Co and Ni were placed in just one group (VIII) and the noble gases in a group of their own (0). This is a useful conventional procedure, keeping patterns of eight but without deep significance and was not universally adopted. To avoid confusion in recent times the elements have been placed in groups from 1 to 18 as shown below (see Fig. 1.6).

Na	Mg											Si	P	S	Cl	Ar
K	Ca	Sc	Ti	V	Cr	Mn	Fe	Co	Ni	Cu	Zn	Ga	Ge	As	Se	Br	Kr

A-groups			Transition elements							B-groups							
IA	IIA	IIIA	IVA	VA	VIA	VIIA		VIII		IB	IIB	IIIB	IVB	VB	VIB	VIIB	0
1	2	3	4	5	6	7	8	9	10	11	12	13	14	15	16	17	18

It took some time to fit in the remaining elements after Kr, which were discovered later, but in the end it was possible to see that the best matching in groups and periods was

K	Ca	Sc	Ti	V	Cr	Mn	Fe	Co	Ni	Cu	Zn	Ga	Ge	As	Se	Br	Kr
Rb	Sr	Y	Zr	Nb	Mo	Tc	Ru	Rh	Pd	Ag	Cd	In	Sn	Sb	Te	I	Xe
Cs	Ba	*Ln*	Hf	Ta	W	Re	Os	Ir	Pt	Au	Hg	Tl	Pb	Bi	Po	At	Rn

where the number in each row is 18 made up by $(2 \times 1) + (2 \times 3)$ and then $+ (2 \times 5)$. The groups of 10 elements, which were called *transition elements*, fitted between A- and B-subgroups. This is the series of *atomic weights* (with a few adjustments, see section 3.2.1) except at *Ln* (see below), and although chemical and physical properties do not follow such a simple pattern as at the beginning of this table there are obvious parallel patterns. The overall table thus built up is then 2 (H,He) 8 (Li period) 8 (Na period) 18 (K period) 18 (Rb period). The Cs period has 18 members, as shown above, but in it we have placed *Ln* in italics, since it was observed that there was a large jump in atomic weight from Ba to Hf. After much searching, 14 new elements, all very similar to the first, lanthanum, and for this reason called lanthanides (Ln), were discovered and placed properly in this *Ln* location.* Thus this Cs period really has $18 + 14 = 32$ members. In the lanthanides the weights (and atomic numbers, see later) increase without any real change in chemical properties, and accordingly they are all placed in group 3. Looking back at the K and Rb periods, it can be seen that in comparison with the Li and Na periods, where chemistry and physical properties change rapidly from element to element, the chemistry in

---

*These elements were also and are still called 'rare earths', but this description applies more properly to their oxides, Ln_2O_3. Note also that Tc, technecium, which is radioactive and unstable, was not fully authenticated as an element until 1938.

later periods, especially along the 'transition' elements, from Sc to Zn, from Y to Cd and from Ba to Hg (including Ln), all of which are metals, changes more slowly. This slow change is accentuated in the Cs period where the extra 14 lanthanide (Ln), elements, all very similar metals indeed, are inserted (Fig. 1.6.), now in a separate box. Thus, a group of 14 elements (i.e. 7×2) were even more exceptionally similar than those associated with the groups of 10 elements (i.e. 5×2) of the transition series. It was a triumph of chemical skill to put this table (Fig. 1.6), now called the periodic table of chemical elements,* together in these six periods which now are of 2, 8, 8, 18, 18 and *32* elements. No expectation was built into the pattern and at first no reason for it was forthcoming. No new elements have been discovered that break this pattern while new elements have been added to the end by synthesis and they conform to expectations (see Fig. 1.6).

We observe that this pattern is related to a very simple series of numbers, (2×1), $(2 \times 1 + 2 \times 3)$, $(2 \times 1 + 2 \times 3)$, $(2 \times 1 + 2 \times 3 + 2 \times 5)$, $(2 \times 1 + 2 \times 3 + 2 \times 5)$, $(2 \times 1 + 2 \times 3 + 2 \times 5 + 2 \times 7)$. It then became desirable to analyse these numbers, giving them a theoretical basis, and finally to use this theoretical basis to explain the physical and chemical properties of the elements throughout the table. This analysis shows that there are no further elements to be discovered below atomic number 86 $(2 + 8 + 8 + 18 + 18 + 32)$ (see Chapter 3).

As mentioned above, a few additional very heavy elements were in fact found which could be put in a last (seventh) period of 32 elements, beginning with francium (Fr), followed by radium (Ra) and actinium (Ac), and then by a further group of 14 elements similar to actinium (actinides) and including thorium (Th), protactinium (Pa) and uranium (U), as well as a series of man-made elements, but the atoms of all these elements are increasingly unstable, so that after U they do not occur naturally. This series of 32 elements is still not complete (see Fig. 1.6). The conclusion is simple—there are a limited number of 'natural' elements, 92, up to uranium (of which, as stated, only 90 occur on Earth).† That is final. The periodic table of the chemical elements contains the essential clue to all the material world on Earth (note, not in the stars) which man has longed to appreciate since he became self-conscious and started to observe. *From these (atomic) elements (the units of chemistry) materials of every kind, mineral and biological, in our surroundings, are made by adjusting the variable composition.* (The composition does, however, have an underlying complexity following from the basic nature of atoms which forces us to modify this statement in Chapters 2 and 3). This chemical knowledge, together

---

*The reader should note that the history of the periodic table covers a period of some 50 years, starting around 1800, before it was formulated in shortened form by Mendeleev in 1856(?) (see Section 3.2). The full periodic table took a further 50 years to complete and until 1999 new synthetic elements have been used to test that we do have a thorough understanding of it.

†Technetium (atomic number 43) and promethium (atomic number 61) are unstable and have disintegrated completely on Earth, but can be obtained in nuclear reactors and recovered from nuclear fuels. The first, Tc, can be obtained in ton quantities; the second, Pm, in milligram quantities.

with the understanding of the reasons for the physical division into gases, liquids and solids, which requires us to consider energy (see Chapters 2 and 5), provides the necessary instruments (principles) to cover completely the description of matter on Earth. *We shall show that this information leads us to a basic background logic for the full analytical description of all species of mineral or biological origin.* We do not really need further knowledge than this analytical composition, but we do need to probe further to uncover why the periodic table is in the form it is, why elements react with one another as they do (see Chapter 3–5) and why in certain combinations of associated units they form complex very dynamic systems such as living organisms. In all such quests the description will be as mathematically economical as possible. The discussion in subsequent chapters will, therefore, largely concern the possibilities of transformation, change, of combinations of atomic elements into the forms in which we see them and also the rates at which changes occur.

Now, the elements do not exist in the universe or on Earth in equal amounts, so that there is a practical limitation as to which elements can be readily used in combination with others, a question that we examine next.

1.4   The abundance of elements in the universe

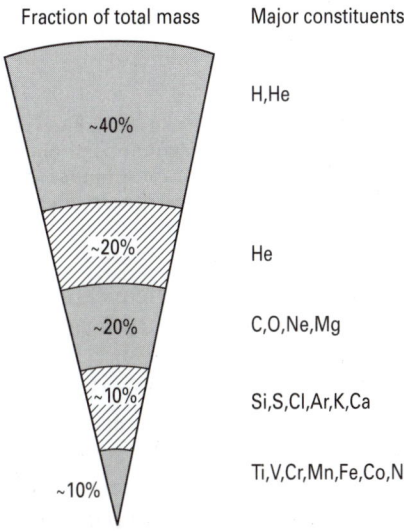

Fig. 1.9 Schematic diagram showing the 'shell' structure of a heavy star (around 25 solar masses) at the end of its evolution, just prior to a supernova explosion. The fraction of the total mass contained in each shell and the principal elements present are shown. (Of the 19 elements shown, 12 of them are essential for all life!)

It is usual today to postulate a beginning of the universe which scientists call the Big Bang. The Big Bang generated an unimaginably high temperature and after a relatively short time many primary particles, including those of the nucleus, protons and neutrons, as well as electrons. When the temperature decreased considerably, down to some 10^4 K, it then formed a considerable amount of H and some 10% of He atoms (plus a little deuterium and lithium). These were the first nuclei to be formed and describe the early composition of the universe. At first, this matter is considered to have been homogeneous, but it broke up (through expansion, turbulence and gravity) and gave rise to the stars in nebulae much as we see the universe today, but much less expanded. Heavier elements were formed some time later in giant stars.* The process involved successive nuclear fusion reactions, starting from the conversion of H into He, of He into C, of C into O, etc., and ending with the formation of Fe and related elements. The heaviest elements were probably formed in the explosion of these giant stars (then called supernovae). The elemental composition of some stars (high temperature gases) is known, and some segregation under gravity is seen in that the heavy elements such as iron have gone to the centre (Fig. 1.9). Owing to this process of synthesis, which has a logic of its own (see Section 8.4.8), the observed abundance of the elements in the universe is as shown in Fig. 1.10. Notice the very high abundance of H and He, the low abundance of Li, Be and B, the high abundance of C, N

---

*The stars up to 10 times the mass of the sun produce only He, C and N from H. It is only the really large, giant stars, which are some 0.05% of the total, that can produce the heavier elements.

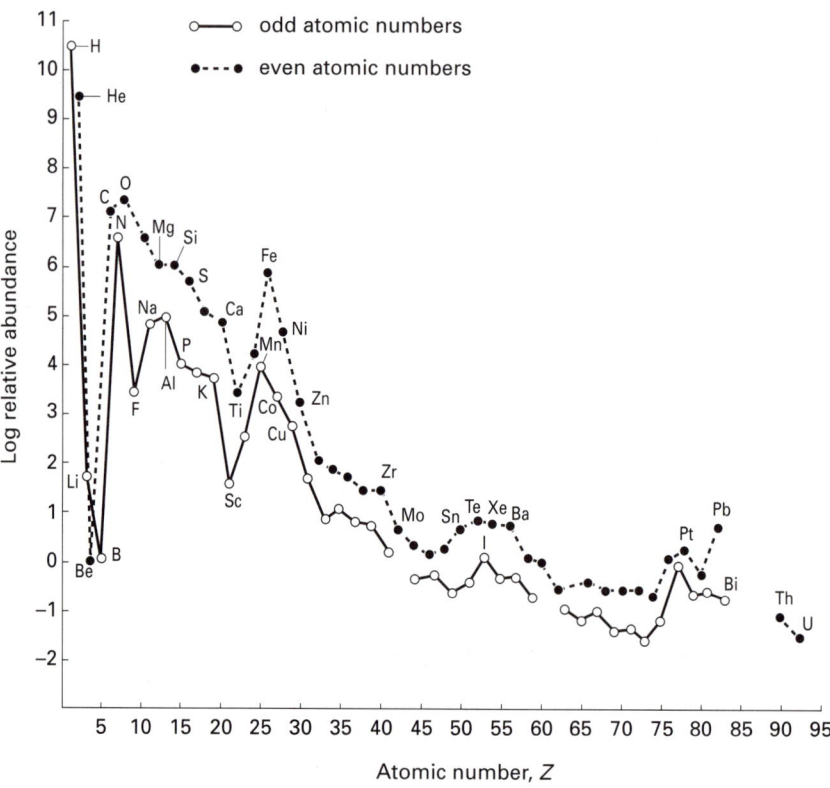

Fig. 1.10 Relative abundances of the 'unchangeable' elements in the universe [based on log (abundance of Si) = 6]. Filled circles (●), even atomic numbers; open circles (○), odd atomic numbers. We shall call this type of diagram a landscape diagram. ['Unchangeable' refers here to atoms of elements. Thus we ignore, for the purposes of this book, any transmutation of elements.]

and O, and later of Fe. These abundances dominated what chemical combinations could be formed when the atoms came together to form the Earth.

The Earth formed together with the solar system, including the sun, from a gaseous plume of very hot gas which erupted from an older giant star or originated from its explosion. Thus, the overall element content of the original Earth was achieved in this elementary particle (electrons, protons and neutrons) furnace and was dominated by the above pattern of conversion of hydrogen and helium to heavier elements. The reaction was incomplete, so that the abundance of the elements is in part a result of the starting material, in part of the rate of conversion to heavier elements and in part a consequence of the stability of the various atomic nuclei. For example, iron has the most stable atomic nucleus but it is not the most abundant chemical element. Thus, the abundances we observe are a result of rapid cooling before final conversion could take place. Since Earth is now a very cool place no further atomic fusion reactions can occur. It is for this reason that we do not consider any total content change of the types of atoms on Earth when describing the chemical composition of any material here. [In principle, we could have taken the very elementary particles

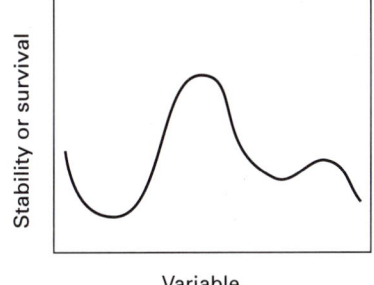

Speciation (chemical or biological) is based on the understanding of this type of diagram.

of Table 1.1 as the compositional variables of nature, but at temperatures below about 10^4 K these compositional variables become the chemical elements, our fundamental units in this book, which have gone on to react and give many 'speciated' materials.]

We shall plot the occurrence of species of all kinds, or the probability of the occurrence of these species, against one of their characteristic variables—such as composition—in several chapters of this book (see aside). Here we plot occurrence (abundance) of elements (chemical atomic species) against element atomic number (see also Chapter 9). In other chapters we shall plot, for example, stability (one criterion for occurrence) of chemicals against chemical composition (Chapter 5) or survival strength of a living system (evolutionary strength) against chemical composition (Chapters 10–16). These plots we shall call landscapes after the use of this word by Kauffman (see reference). Landscape plots then form a running theme throughout the book in our effort to describe the diversity of objects which we observe around us and the reasons for their existence. It is the higher levels of the landscapes that are most likely to be observed.

1.5   Summary

Table 1.5 Units* and fundamental variables described in Chapter 1

Units	Variables
Atomic elements	Relative amounts of each unit, i.e each element, giving % composition as variable

* In Chapters 2 and 3 we shall introduce charge, which also comes under units, so as to define composition fully. We shall also have occasion to refer to photons in Section 2.6.

In this chapter we have seen that, through analysis, the earlier schemes, both Greek and Chinese, for the classification of the observable material world and its variability (Table 1.5) were shown to be inadequate by scientists in the period 1600–1900. Analysts were able to show that, in fact, all materials at temperatures below about 10^4 K (roughly 10^4 °C) are composed of 90 *units*, chemical elements. *There are no other noteworthy elements occurring on Earth.* The *amounts* of the different materials around us are limited by the way in which these elements were formed in the universe. These amounts, abundances, mean that lighter elements (atomic number < 30) in combination, dominate the Earth. From the elements, mineral, biological and industrial combinations (compounds) are synthesised. The limitations of abundance affect the types of minerals in Earth but do not necessarily affect the local selective accumulation associated with man's activities or biological syntheses. An outstandingly important *variable* in the observed substances around us is therefore the *chemical composition* (Table 1.5) stated here in terms of the percentages of atomic elements in the material, e.g. H_2O is 66.7% H. (To this description we shall have to add an extra compositional variable, charge; see Chapters 2 and 3.) This description by itself does not allow us to understand why only certain combinations (compositions of fixed numeric ratio), e.g. H_2O but not H_3O or H_4O, are found to occur, why the observed materials are in different physical states of matter—gas, liquid or solid—or why they can be changed from one state to another. Therefore, extra variables (physical or chemical) must be uncovered and understood in fundamental terms, including potential energy, directed kinetic energy, temperature, pressure (Chapters 2–5) and, finally, time, which relates to both change and flow (Chapter 8). In Table 1.6 we summarise the changes in mankind's classification of the

Table 1.6 Description of the material world and its variables

	Greek (Chinese)	Modern man
States* Units† Variables‡	Air, water, earth, fire Air, water, earth, fire Moistness (fluidness) coldness, dryness, hotness Flow or motion (time)	Gas, liquid, solid Atoms, charges (and photons) Composition, Space and Time The derived variables are temperature, pressure, various energies, fields and momentum (see Chapters 2 and 5). (Size.)

* **States** are the observed physical condition of a material and are affected by the modern variables energy, fields, temperature and pressure.

† **Units** are the *ultimate* independent entities from which all material objects are made where materials vary only with unit composition, which we shall later relate to energy content. The deeper physical analysis of the units into particles such as electrons, protons and neutrons, and then to quarks, etc. (see Table 1.2 and Chapter 3) does not affect the descriptions in this book to any marked degree since prevailing temperatures are presumed to be lower than 10^4K. Spatial distribution will be analysed in Chapter 2.

‡ **Variables**. The two variables, size and fields, will be introduced in Chapters 5–7 and the variable time (motion or flow) will be described in Chapter 8. Because all units interact to some degree, the space (volume) occupied by units is not a variable that can be separated from size and energy. The description of energy (the capacity to do work) needs careful analysis (see Chapters 2 and 5).

units and variables of all materials, which provides the main theme of this book. We seek a quantitative logic for the description of everything we can observe which has to be based upon the chemical unchangeable units, together with the ways (the variables), and limitations on the ways, in which they can be put together to create both inanimate and animate material. In the course of this analysis we shall also be examining the properties resulting from changes in variables. A major concern will be the *mathematical relationship* of a given property to the units of composition and the variables. It is clear that the variable composition is a *linear* function of the amounts of each element present and so is open to reductive analysis. However, we shall find many circumstances in which such a simple relationship between a property and a variable does not apply and the connection between the observed property and the units plus the variables is mathematically complex. Thus, we have to be concerned with the way in which our reductive understanding is limited when we describe properties other than just composition. This will become increasingly important as we proceed towards organisation, an essential defining property of living systems. In Chapter 2 we begin the examination of the variables.

2

Forces and related energies

And now we might add something concerning a certain most subtle spirit which pervades and lies hid in all gross bodies, by the force and action of which spirit the particles of bodies attract one another at near distances and cohere, if contiguous ... and there may be others which reach to so small distances as hitherto escape observation ... and electric bodies operate to greater distances, as well repelling or attracting the neighboring corpuscles; and light is emitted, reflected, refracted, inflected, and heats bodies, and all sensation is excited and ... propagated along the solid filaments of the nerves.

Isaac Newton, *Philosophiae naturalis principia mathematica* (1687)

2.1   Introduction

In the previous chapter we described the change in knowledge over 2000 years with respect to the units of composition of all substances and materials around us. Whereas the Greek philosophers classified such substances and materials under the headings air (gas), water (liquid), earth (solid) and fire, and a similar classification was adopted by the ancient Chinese thinkers, we know today, leaving fire to one side, that the fundamental units of which everything is composed on Earth are the atoms of the 90 naturally occurring *chemical elements*. This is now irrefutable. There is a difference, however, in that the two classifications do not overlap since the classical description is one of *physical* variation (or physical state) while today's description is of *chemical variation*, i.e. composition. In fact, chemicals can be found which occur naturally as gases, for example oxygen, O_2, in the air, liquids, such as water, H_2O, in the sea, and solids, such as sodium chloride, NaCl, in rock salt. Yet we also know that on change of temperature, for example, all substances change physical state, becoming solids on cooling and gases on heating sufficiently. There must be a reason for this, so that we need to explore the variables which affect the physical state of substances and materials of known composition in our surroundings. This chapter, then, is focused on physical variables, as opposed to variation of the fundamental chemical compositional units. At the same time we need to notice that substances such as NaCl exist, but $NaCl_2$ and Na_2Cl do not. Hence, the fundamental atomic units do not always combine over a continuous range of composition. Why is this so?

The differences in physical states at any one temperature must be due, at least in part, to the way the constitutive units, atoms or combination of atoms, in substances such as rock salt, quartz or wood attract one another and are packed together in what we shall show is an *ordered* condition, while such units in gases do not interact to an appreciable extent and are, therefore, *disordered*. Liquids have their constitutive units constrained by limited volumes in containers, but are relatively disordered within these volumes. The obvious point can now be made that the ordering of constitutive units, atoms or molecules, must be due to 'forces' acting at a distance that pull them together, and that the forces between some atoms or combinations of atoms must be stronger than those between others (see aside). Thus, at room temperature and pressure strong forces give rise to solids, some of intermediate strength give rise to liquids, while, where forces are weak, gases only are observed. Later we shall see that there are physical variables that oppose ordering; this dichotomy is discussed in detail in Chapter 5. While we have referred here to forces giving rise to physical states, forces must also give rise to the chemical association of atoms. Clearly we must ask first what is the nature of the forces that act on or in materials, and, then, what do we understand by temperature and pressure,

One attractive force with which we are all familiar is that of gravity,* which pulls us and all other things around us onto the surface of the Earth, and in fact is easily detectable between any two large objects (Fig. 2.1). Experiments showed that this force can be related quantitatively to a property of all objects that is called their *mass* and is inversely proportional to, i.e. it varies with the inverse of, the square of the distance, r, between them. This may be expressed by the equation (for *two* interacting bodies)

$$Force = G.\frac{m_1 m_2}{r^2}$$

where G is a constant characterizing gravity and m_1, m_2 are the *masses*, amounts of matter, of the two interacting bodies. [Note that this equation also gives a definition of mass in terms of the attracting force between two objects, one of which can be the Earth itself, when we make $m_2 = M$, mass of the Earth; Fig. 2.1.] A very important feature of this experimental approach is that the observations are described by a *mathematically quantitative* equation, unlike the classical Greek approach to physical variables, which was purely qualitative. It is equally important to recognise that by using this equation we have introduced a quite new variable, *the spatial*

---

*Chemical forces**

Ionic	Na^+Cl^-
Covalent	$(H-)_4C$
van-der-Waals	$(He)_n$
Hydrogen bond	$N-H\cdots O$

* All forces are electrostatic in origin

---

*From the latin *gravis*, heavy body

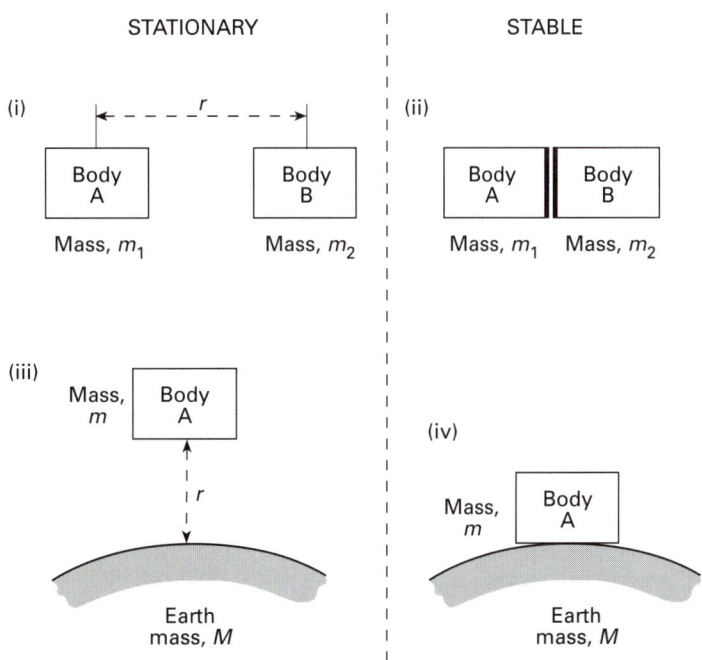

Fig. 2.1 Two bodies, A and B, interacting through their masses, (i) at a fixed distance, *r*, in a stationary state, (ii) in contact in a stable state, (iii) in a stationary situation where Earth is one body, where *r* is the distance from the centre of A to the centre of the Earth, (iv) in a stable situation with a body resting on Earth. (i) and (iii) should collapse to (ii) and (iv), respectively. The condition of a dynamic stationary state is shown in Fig. 2.2.

distribution of any two units of composition. [Note that by definition the force of attraction here is positive.]

While we experience such force, we do not really understand its origin or cause, that is, we do not know what is the essence of gravity, but it is clear that this is not the force that pulls atomic particles together, although it still acts at very short distances, since their masses are too small. We must, therefore, look for additional forces, strong enough even for very small masses, so that the formation of the materials around us can be explained.

2.2 Gravitational and electrical forces and fields*

At the time of Newton (seventeenth century) the existence of a natural 'gravitational' force between two (or more) large bodies (one of which could be the Earth or the sun) that caused them to be attracted by one another, was generally recognised. It is this force that keeps us standing on Earth, causes objects (such as apples) to fall to the ground and causes the Earth (and the other planets of the solar system) to orbit around the sun. It

---

*Until Section 2.5.1 we shall be concerned with pair-wise interactions of 'objects' and their sum and not with the behaviour of large numbers of particles ('objects' in the general sense of the word).

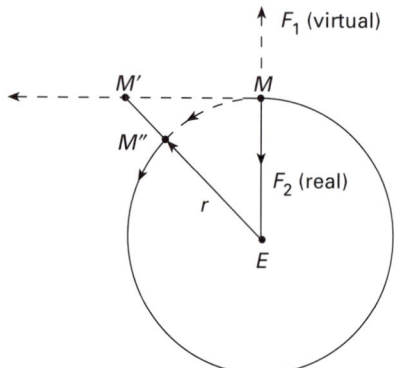

Fig. 2.2 In the absence of the gravitational pull of the Earth, F_2, the moon would have moved from M to M' along the dashed straight line. Because of the gravity, the moon actually moves along the circular arc M to M''. Relative to the Earth, E, points on the circular arc are closer than the points on the straight line. Viewed in this light, the moon, as it moves in the transverse direction, also falls towards the Earth (in this case from M' to M''). However, the Earth/moon (and Earth/sun) pair of bodies are in a dynamic stationary state since the radial distance, r, is fixed. [After Narlikov, J. V. (1996). *The lighter side of gravity*. Cambridge University Press, Cambridge.]

is not so obvious in the last case why the relationship is so persistent since we would expect the Earth to fall into the sun. In fact this is inevitable in the long term, but in the process of formation of the solar system there was an initial outward and rotating pulse of matter, which, together with the permanent gravitational force, generated swirls of particles and then, after progressive coalescence of these particles, fixed orbital motions of the smaller bodies around the largest. Since there is almost no drag (friction) in space, the initial rotatory impulse is sufficient to keep a planet in a fixed orbital motion for a very long time. We call such a condition a *dynamic stationary* state, to distinguish it from a truly *stable* static condition as when an object rests on the surface of the Earth,* which in this case of orbital motion depends on balances between attractive inward gravitational forces and what is effectively a repulsive 'force' resulting from tangential velocity. In Fig. 2.2 we illustrate this balance for the case of the moon and the Earth. *Static* stationary states correspond to the condition where *two* stationary masses are held apart at some fixed distance. We insist on dealing with two bodies at a time since we shall see that the description of dynamic states of more than two bodies is difficult.

Long before the time of Newton, however, in ancient Greece some 600 years BC, it had also been observed (see the quotation at the beginning of this chapter) that on rubbing a fragment of amber (called 'elektron' in Greek) with a piece of fur and then putting them apart there was a force which arose and caused the amber and the fur to move back together. The same happened on replacing the amber with an ebony rod. It was suggested that two bodies such as the ebony rod and the piece of fur had become able to attract one another because one (the ebony) had pulled out material of some kind from the other (the fur). But what was it? It was not known then, and the force of interaction was just described, even in the eighteenth century, in terms of the imagined creation of 'charge', which was not the same as mass. We say today that the ebony pulls out negative 'charge' (–) (electrons) from the atoms of the fur, which are then left positively (+) 'charged', and that oppositely 'charged' substances attract one another while similarly 'charged' materials repel. By observation we know that (+) and (–) charged particles attract one another more strongly the more the distance between them is reduced. This is called *electrostatic attraction* [or *repulsion* if the charges have the same sign, (+)(+) or (–)(–)], and applies equally to bulk bodies and to atoms (Fig. 2.3). There is, therefore, a force between two charged bodies, which has been demonstrated to be directly proportional to the charges and inversely proportional to the square of the distance between the charges, as expressed by the *quantitative* equation known as Coulomb's law:

$$Force = -A\frac{z_1 e z_2 e}{r^2}$$

---

*Theoretically, according to the equation for the gravitational force, the truly stable condition corresponds to the closest possible approach to the centre of the Earth, when $r = 0$, but the solid chemistry of the planet effectively hinders further approach through repulsive forces. If the crust melted all bodies of higher density would move towards the centre (see Chapter 9) until balance was reached.

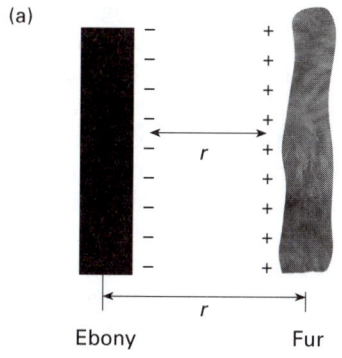

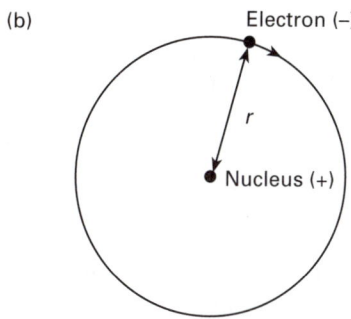

Fig. 2.3 (a) Charged materials such as ebony and fur interact at a distance, r. The example shown is of a stationary state. (b) The negatively charged electron moves in a fixed orbit, radius r, round the positive nucleus. This is a second example of a dynamic stationary, electrostatic, state. The two different values of r in (a) arise for two different charge distributions—on sufaces (above); internally (below).

where z_1e and z_2e are the charges (in terms of the charge of the electron, e) and A is a proportionality constant related to the so-called electrical permittivity of the medium. The force is positive for oppositely charged bodies. As stated in Chapter 1, the words negative and positive are conventional and indicative only of postulated uptake and removal of negative charge (electrons), respectively (curiously back to front in common language). As a direct consequence of the discovery of this force, the units of composition of a material, a fundamental variable, can no longer be described just by their atomic mass but we need also to describe their charge. (In passing observe that charge, like mass, comes in units of e and is not continuously adjustable.)

The above experiment in electrostatics can be carried out in many ways with different materials, and is basic to all chemical atomic combinations, where electrons (–) interact with nuclei (+) (see Chapter 3). Some atoms steal 'charge' from others and then are attracted by them in an effort to restore their original state.

Table 2.1 summarizes the characteristics of these two forces. While gravitational forces between masses are the basis of our description of the interaction between large objects and help us to describe the planetary system and other such interactions in the solar system and in the universe, electrostatic forces between charges, for example electrons and protons (in nuclei) (see Chapter 1 and 3) allow us to describe, amongst other things, the interaction of atoms, so making them the basis for 'chemical' forces and then the formation of mineral and biological species. (But note that a full description of electronic interactions between atoms will require us to discuss electron planetary-like *motion* in orbits, as well as *position*, so that we can describe both stable and different stationary dynamic situations under electrostatic as well as under gravitational forces.)

Table 2.1 The two major Forces*

Gravitational	The force of *attraction* between two bodies is proportional to the product of the masses, m_1 and m_2, and inversely proportional to the square of the distance between them, i.e. $F = Gm_1m_2/r^2$. [G is the proportionality constant]. F is large for bodies such as planets, is considerable (obviously) for man-sized objects on Earth (which 'fall'), but is of little consequence for atoms, where m is so small.
Electrical	The force of interaction between two bodies is proportional to the 'charge', z_1e and z_2e, they carry, and inversely proportional to the square of the distance between them, i.e. $F = -A, z_1z_2e^2/r^2$ (A is the proportionality constant.) Since z_1 and z_2 can be (+) or (–), there are combinations (+)(–) which are attractive and combinations (+)(+) or (–)(–) which are repulsive. The charge can be carried by atoms, which are very small, and is known to be due to local excess (–) or deficiency (+) of electrons. The force is very large between charged atoms but it becomes increasingly weaker as the ratio of charge to size increases. We hardly notice electrostatic interactions between bodies of man's size because they carry relatively small total charge.

* The major forces are here described in mathematical quantitative form just as was required of composition in Chapter 1. For atomic or molecular masses the electrical force vastly exceeds the gravitational force. This numerical approach is extremely powerful (compared with that of Fig. 1.1). We must also note the force related to change of velocity of a body (inertial force) (but see Section 2.3.2).

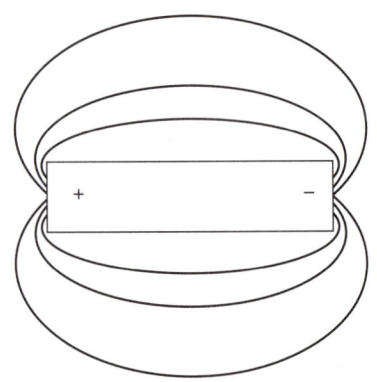

Lines of force around an electric or a magnetic dipole.

Another type of force which acts at a distance is the magnetic force exhibited, for example, by lodestone on iron fillings. Magnetism is a term which derives from the Greek coastal district of Magnesia where lodestone was found, and of which Thales said that 'it must have a soul because it moves iron'. Earlier, around 2700 BC, the Chinese referred to the iron oxide Fe_3O_4 (magnetite) as the 'stone that loves' and apparently they built floating 'compasses' of it, the precursors of magnetic needles which, as is well known, point towards the magnetic poles of the Earth. Later, at the beginning of the nineteenth century, it was found that electric and magnetic interactions could be integrated into just one *electromagnetic* interaction, a fact that led directly to the so-called *theory of fields* developed by Michael Faraday and James Clerk Maxwell in the second quarter of the nineteenth century.

What did Faraday and Maxwell mean by field or, more specifically, a field of a given force (see aside)? Quite simply, a field is a region of space in which the influence of some kind of 'force' is experienced. The force can be gravitational (originated by massive bodies), electrical (originated by charges), magnetic (originated by flowing charges) or any other that, as we have stated, is active at a distance. Why should masses attract one another, electric charges attract or repel one another and magnetic substances also attract or repel one another? In fact mass, electrical charge, magnetic properties and other even less familiar properties postulated in more recent days for subatomic particles, such as the 'colour' which accounts for the so-called *strong force* in nuclei (see Table 3.1) correspond to observable phenomena, but their real nature is unknown. The use of postulates in the development of science often confuses even scientists as to what is known. A word of explanation is appropriate here.

A particular stumbling block even for many learned people to an understanding of science is the use of the word 'force' which is a concept of action at a distance which pulls things together. It is thought that scientists *understand* 'gravitational force' or 'electrical (magnetic) force' and that this understanding is beyond the reach of others. The real situation is quite different. A scientist observes natural or contrived events and relates them to other observables that he can measure. For example, large dense objects attract one another more strongly than smaller, less dense ones. Thus, scientists say that the *gravitational interaction* of small objects is very small and not easily felt while planet Earth/man interaction is large and easily felt. However, the observations are found to be general to all very small objects (e.g. the molecules of our atmosphere are held by Earth) and all very large objects (planets held by the sun). Because this generality of attraction is observed and can be accurately mathematically formulated, scientists say that a gravitational force exists between material objects, i.e. there is action at a distance, and they convert the intensity of action into a quantitative descriptive word for each object, its 'mass'. This gravitational force is observed to vary inversely with the distance squared in all cases. Clearly this is in contradiction with the idea that objects only interact (as masses) by contact, which our senses feel (see Fig. 2.1). Although there is no deep understanding of the force of gravity for scientists any more than for anybody else, the concept of gravity is nevertheless very useful since the equation holds very accurately in many of our experiences and experiments. The same is, of course, true for the electromagnetic forces and for the other forces of nature (see Table 3.1). We also observe these forces and then relate their intensities to some property possessed by each object, such as its charge; we do not explain or describe them in terms of underlying causes well understood by our intellect. The scientific descriptions today are then related to symbols in mathematical formulae which reproduce the observed behaviour, but the symbols, for example mass and charge, although they can have quantitative values, are not open to explanation in depth. Scientist's statements are then just a way of correlating observables under 'natural laws' which are systematic quantitative *interpretations* of the phenomena studied. It is the business of science to try to reduce all observables to an understanding of simple units and their properties, and this may be a never-ending quest.

The bringing together of masses or charges of opposite sign leads clearly to a more *stable* situation just as does the separation of charges of the same sign, but again we have to ask a question: what do we mean by *stable*? In short, it means that 'the system' considered (we will come later to the definition of system in Section 5.2) is 'exhausted' and incapable of change and therefore of doing mechanical or any other form of *work*. One way of expressing this is by saying that *it has a minimum potential for work or that its work capability is as low as possible* owing to the position of the masses or charges. But this stable condition arises for masses, or charges of opposite sign, when they are closest together and the forces are greatest. The two forces, gravitational and electrostatic, are not therefore *directly* proportional to the potential for work of the system of masses and of charges in given positions. *Now, the capability for work is what we need to know since we wish to understand the capability of a system to drive change.* The combined *variables* of a system are obviously related to composition in terms of masses and charges, and to the *disposition in space* of such interactive units, which in this chapter we treat in pairs. It must be understood that this combination of variables, the potential for doing work, is equally fundamental to all chemistry and biology. It is very useful for us to refer to this *derived variable* under the term *energy*, as we outline in the next paragraph. It is the energy content, the capability of doing work, that reflects instability.

2.3   The capacity to do work: energy

The term 'energy' was first defined by D'Alembert (Encyclopedie Française, 1785) and then by Thomas Young (1787), and equated to the capacity to do *mechanical work** held within any stem. The term derives from the Greek word *energeia* (*en*, 'in'; *ergon*, 'work'). In order to analyse energy

---

*The mechanical work, W, performed by a force, F, is expressed mathematically by the product of the intensity of the force and the distance Δr through which an object of mass m is moved, i.e. $W = F \Delta r$. The unit of work is the joule (SI system) corresponding to the work done by a force of 1 newton over a distance of 1 metre.

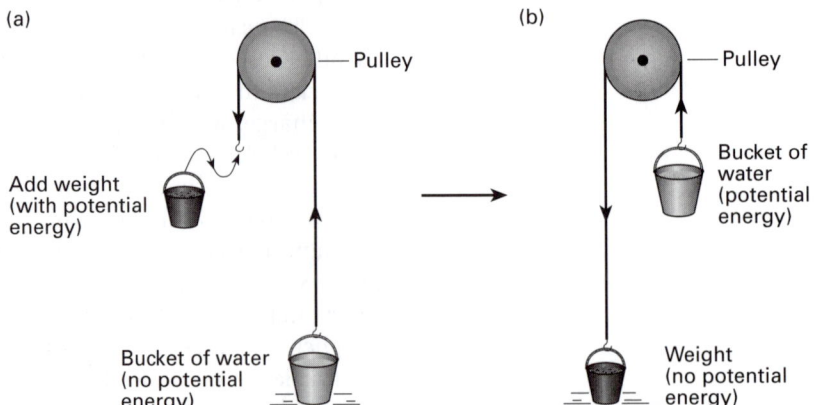

(a)

(b)

Pulley

Pulley

Add weight
(with potential
energy)

Bucket of
water
(potential
energy)

Bucket of water
(no potential
energy)

Weight
(no potential
energy)

Fig. 2.4 The transfer of potential energy to a stable object, a bucket of water at (a), by placing a heavier weight on a hook connected to a string over a pulley wheel so allowing the situation (b) to arise.

quantitatively we shall assume in this chapter, for mathematically simplicity, that in a system of many particles we may sum all interactions *pairwise*, but in Chapter 5 we shall correct this impression since we shall have to deal with very large numbers of particles—in one mole of a substance there are 6×10^{23} identical units.

Now, a body with a given mass suspended at some height can do work when allowed to fall back on to the ground under the action of gravity; for example, it can raise a bucket of water attached to a string passed over a pulley (Fig. 2.4). The latter capacity to do work is an exchange of *potential energy* in the Earth's gravitational field between the body, i.e. a given mass, and the bucket in two different positions. Notice that one or other of the objects is energized in the gravitational field of the Earth.

We wish next to relate the potential energy quantitatively to the forces we have described. If we plot force against (r) (Fig. 2.5), we see that the force is zero at infinite r and is infinity at $r = 0$, when the two bodies coincide in space. On the other hand, when we consider masses or oppositely charged bodies it is obvious that the greatest capability of doing work is when r is greatest. The force and the work capability, which we shall relate to potential energy, are therefore opposite in magnitude.

Now, the work done by a constant force acting on an object for a very small change in r, dr, is the product Fdr which, since F is inversely proportional to the distance squared, is proportional to $\frac{1}{r^2}$ dr. (For a change $\Delta r = r_2 - r_1$ this is the area under the upper, F-curve on going from r_2 to r_1 in Fig. 2.5.) It is then easy to show by integration that the capability of doing work on going from $r = r_1$ to $r = 0$ is proportional to $-\frac{1}{r}$ (not $\frac{1}{r^2}$) and it decreases (becomes more negative) as r decreases (see Fig. 2.5). We therefore say that the potential energy at a distance r is proportional to $\frac{1}{r}$. Thus, we can conclude that the variable, the capacity to do work due to the position of a pair of oppositely charged masses, is the algebraic sum of two terms described by changes in the gravitational and electrostatic potential energies, given by

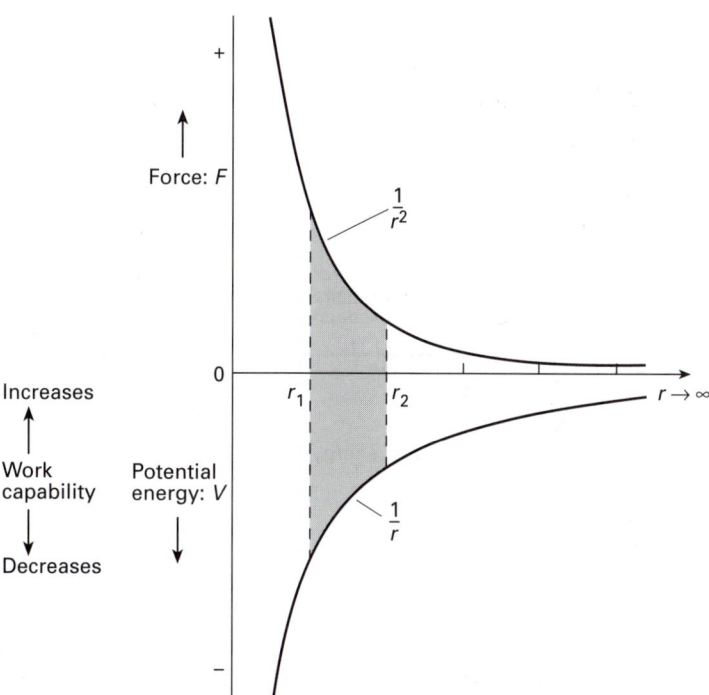

Fig. 2.5 A diagram showing the relationship between a force and the potential energy due to the interaction of masses or charges at a distance of separation r. The potential energy change is the area under the F curve or $V_1 - V_2$.

$$\text{Gravitational energy} = -G\frac{m_1 m_2}{r}$$

$$\text{Electrostatic potential energy} = A\frac{z_1 e.z_2 e}{r}$$

[The difference in sign arises here since opposed charges (+ –) attract.]

The smaller r, *the more negative the potential energy, the more the system becomes stable and the less it can do work.* Comparison of the expressions for force and for the potential energies shows that force is just the rate of change of the potential energy, V, with changing distance, i.e. $F = -dV/dr$. Now, we know, in fact, that when two bodies approach they do not coalesce, so that r never goes to zero. Why does this happen?

2.3.1 Repulsion at very short (contact) distance: sizes of objects

It is clear that while attraction exerts influence at all long distances of separation, when two bodies approach closely this attraction is met by a barrier (a so-called contact barrier), (see aside). This means, of course, that there are repulsive interactions which only come into play at very short distances. (Otherwise each one of us would sink into the Earth). The energy due to these repulsive forces is best described as being dependent on $1/r^n$, where n is now a large number such as 10, and ensures that as

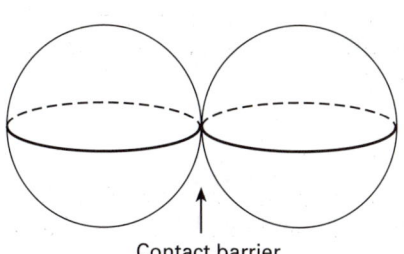

Contact barrier

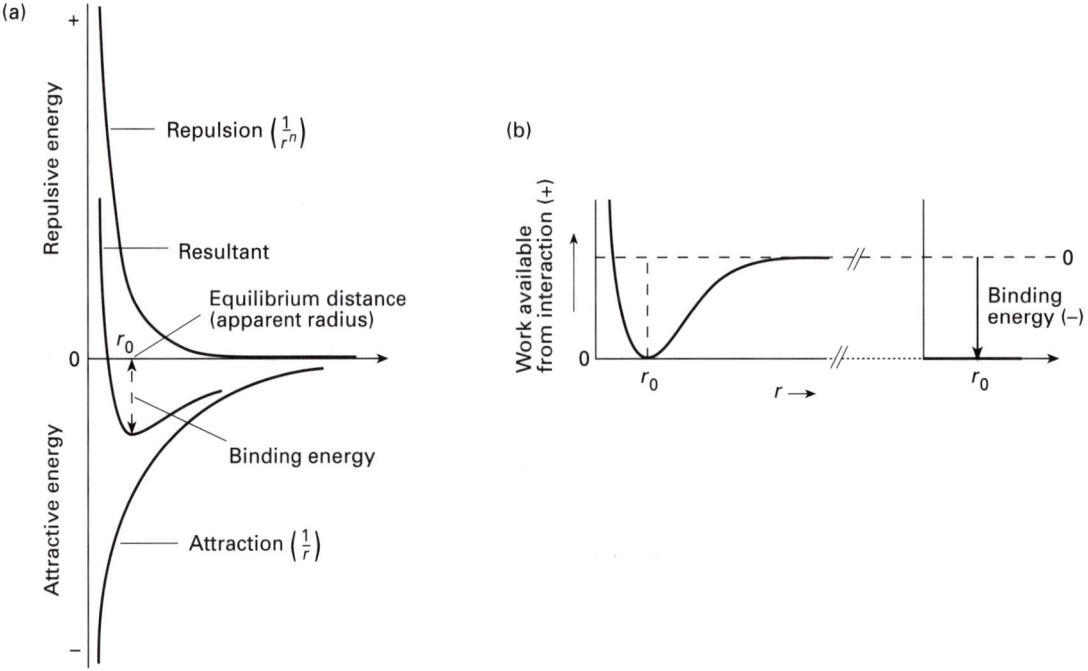

Fig. 2.6 (a) The introduction of short-range repulsive interactions to Fig. 2.5 results in a stable condition with objects separated at a distance r_0. (b) The work available from the interaction increases with the distance and reaches a minimum at r_0. The difference between the energy at infinite distance (r_a) and r_0 is the binding energy.

bodies come together there is a distance where balance between attraction and repulsion is established and no further potential energy can be lost. We can, therefore, write that for a stable system of two units:

$$Balanced\ potential\ energy = \sum(A\,/\,r - B\,/\,r^n)$$

so that the equilibrium (stable) distance and indirectly the 'size' of the objects are described by the values of A, B and n (Fig. 2.6). (Here A can be replaced by $-G$ for gravitational attraction energy.) The outer edge of all objects is really quite 'fuzzy', but our senses see edges due to the rapid changes of repulsive energy at close distances. We see that size is not a simple concept but a useful approximation to describe how space may be filled.

We shall give a more detailed description of repulsion, or the ability of one piece of matter to exclude another from the same volume, in Section 3.4.2 and Chapter 4, since the nature of repulsion is not a classical barrier in the case of atomic interactions.

Our discussion of potential energies in the previous sections focused largely on pair-wise interactions of *bulk* bodies, but can be extended to the 'internal' interactions of their constitutive units, atoms, described by 'chemical' potential energies (see Section 2.4).

2.3.2 The directional kinetic energy of masses

There is a second form of the capability of doing mechanical work, energy, now related to motion. Again we consider the interaction of two objects.

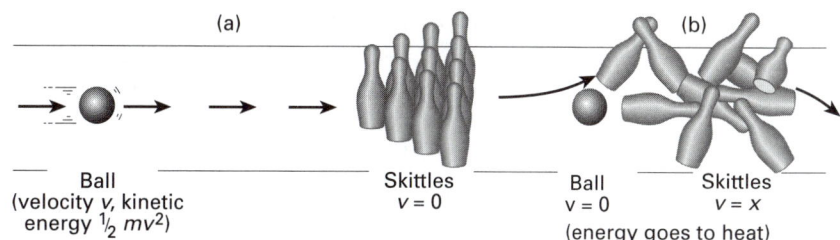

Fig. 2.7 The transfer of directed kinetic energy from a ball to a number of skittles. Gravitational forces carry the skittles to the ground eventually.

Constant frictionless movement or flow clearly does not require work to be done to maintain its motion since no force is opposing it, but a moving body can produce work since *stopping or deccelerating its movement* has mechanical effects: for example, consider the use of a battering-ram, the effect of a projectile or of a bowling ball (Fig. 2.7). Similarly, work has to be done to start movement or to accelerate an object, overcoming its so-called *inertia* (directly related to the gravitational attraction to which it is subjected), e.g. when shooting a rocket into the sky. It has been shown that in this case the capacity to do work is connected to the mass, *m*, and to the variable *directed* velocity, *v*, of the object; this is one form of what is generally called *kinetic* energy,* and found to be given by $^1/_2 \, mv^2$. (Note that a permanent force acting on an object changes its velocity and is related to the acceleration of the motion of the object; the mechanical effect it exerts is related to its decceleration).[†] *We have now come across a further fundamental variable, **time**, in that velocity is the rate of change of spatial co-ordinates of units with time.*

Therefore, there are here two very different *derived variables* associated with the capacity of a single mass or pairs of masses to do mechanical work, one related to the change of *position* in a 'field' (see Section 2.2) and the other to the relative loss or gain of *directed motion*. Both can do work, hence initially, by definition, they possess energy, which is called gravitational *potential energy* in the first case and *directed kinetic energy* in the second case. Of course, a pair of charged bodies can also possess electrostatic *potential energy*. Clearly, when no other forms of energy are involved, a body which has no potential energy and is not moving is *exhausted* of its ability to do mechanical work. The movement of charge with time gives

---

*Kinetic energy is here related to the velocity in a given direction, flow, but we can also consider a set of particles that do not flow but move randomly, so that we can also define a *random* kinetic energy (see Section 2.5).

†To change the state of rest or motion of an object we require the application of a force, *F*, for a given period of time, which causes a change in the 'quantity of movement' of the object (also called 'momentum') given by the product *mv*. Here *m* is the mass of the object [appropriately called *inertial* (rest) mass] and *v* is its velocity. Hence *F*. period of time = *m*. change in velocity, but since the ratio change in velocity/period of time represents the rate of change of velocity with time, which we call acceleration, *a*, we can also write force = *m*. acceleration or *F* = *ma*. When *m* is expressed in kilograms and *a* in metres per second squared (ms⁻²), the force is expressed in newtons, which is the unit of force in the adopted international system of units (SI). Notice that the units of force and energy are those of the fundamental variables: mass, charge, distance (length) and time.

rise additionally to magnetic interactions, magnetic force, and a corresponding energy, and so all forms of energy are related to the four fundamental variables: composition, defined by atomic matter and charge, spatial co-ordinates and their time dependences. The basic outlook described in these paragraphs, based on pair-wise interactions, was developed by Newton and is usually termed Newtonian mechanics. We shall need to change this approach somewhat in Chapter 3 when we introduce quantum mechanics in order to describe interactions between objects as small as atoms.

Now, we can apply these treatments to orbital as well as to linear motion. Consider a mass in a stationary state circling a central point of attraction with a radial velocity v_1 and at a distance r_1 (Fig. 2.2). We have shown already that there are two 'forces' operating in this particular state (see Section 2.2). Suppose there is another possible stationary state at some shorter distance r_2. The attractive force has now increased and so has the radial velocity, now v_2, since in order to keep the system in a stationary state the two forces must remain equal and opposite. Now, to increase the velocity, work has to be done on the system, while to reduce the distance the system does work. The total capability of doing work on going from r_1 to r_2, where r_2 is the more stable state, is, therefore, the algebraic sum of the changes in kinetic (KE) and potential energies (PE), i.e.

$$+ Change\ of\ KE - Change\ of\ PE$$

The equation applies equally to electrical interactions such as those in atoms (see Chapter 3).

To summarise, there are two different unchanging conditions of systems that we meet frequently and immediately in this book, which we call stable and stationary (Table 2.2). Later in this book we shall describe two other states of changing systems which we have just introduced and which we also list in the table for convenience.

Table 2.2 Definitions of conditions

Stable	A condition of a system* which cannot change spontaneously. It does not receive or donate material or energy. [The system can be dynamic in cycles (see Chapter 3), in non-classical mechanics.]
Stationary	A condition of a system which is unchanging but could change spontaneously. It does not receive or donate material or energy. The system has excess energy over a stable state. The system can be dynamic in cycles.
Steady[†]	A condition of a system through which energy and/or material flow continuously so that it is maintained constant. The system has excess energy over a stable state.
Developing[†]	A system which changes steady state due to the loss or accumulation of energy and material in it. The system has changing excess energy over a stable state.

* The word system will be discussed in Section 5.2 since it is used in a particular way in thermodynamics while we use it in a more general way for an ordered or an organised set of objects.
[†] We do not refer to these systems again until Chapters 8 and 10.

Before we go further we wish to relate the above observations on objects generally to chemicals in particular. Chemicals are specified by names such as sodium chloride and ethyl alcohol indicating that in these substances there are particular relationships between composition and structural features (interatomic distances) and that they are isolatable substances. Hence, before we turn to the analysis of temperature, pressure (and volume) (the three basic measurable *physical* variables of ensembles of units which we have not yet treated), we shall deal first with variation of electrostatic potential energy as we change distance between *atoms in stable and stationary states*, so as to appreciate a peculiarity of chemical combination not apparent from the above Newtonian treatment.

2.4 Chemical potential energy: components

The nature of objects with masses and charges in systems can be varied, in principle, by putting their fundamental constitutive units, atoms or molecules, (pair-wise with or without charges) into specified relative positions. Analogously to what we saw above, this positioning can be described by the potential energy of the system considered. Now, for large objects of a given mass, such as those we see around us, their distribution in space, in addition to their composition, is one variable that is *continuously adjustable*. While it is apparently true of atoms in molecules, or other types of compound, that in principle they too can be put in all different spatial arrangements, only a few arrangements are found.* It is useful, therefore, to separate these two cases, treating bulk distribution in space of objects under smoothly continuous physical variables as different from *chemical arrangement of atoms*, which is treated operationally under specific very local spatial variables of composition. Arrangements of atoms in space are given gross empirical molecular formulae, representing the element composition (one variable) but within any one formula there may be found to be a few (often very few) different substances, which have different *structural formulae* and consequently different properties, so that they are given separate names.†

For example, C_2H_6O is a gross empirical molecular formula, but it is known in two fixed spatial arrangements of atoms:

$$CH_3 \cdot CH_2 \cdot OH \quad \text{and} \quad CH_3 \cdot O \cdot CH_3$$
$$\text{ethyl alcohol} \qquad \text{dimethyl ether}$$

The two have different chemical potential energies since the electrons (−) and nuclei (+) interact at different distances and angles (see aside). In a

Examples of compounds of the same empirical formula

CH₃·CHO

H₂C———CH₂
\   /
  O

CHCl₂·CH₃

CH₂Cl·CH₂Cl

Co(NH₃)₄Cl₂,
cis

Co(NH₃)₄Cl₂,
trans

*The reason for this limitation is in part due to what is called the *valency* of atoms, which restricts their combining ability, and in part due to a directional dependence of attraction that gives atomic assemblies *shapes* (see Chapters 3 and 4). These properties arise in quantum mechanics and not in Newtonian mechanics.

†These named discontinuities in the variable *composition* and spatial formulation, which we relate to discontinuities in *energy of interaction*, parallel the discontinuities in *matter*, the appearance of atoms with mass numbers, and *charge*, which appears as multiples of the electronic charge. Perhaps we should not be so surprised by quantum phenomena.

mixture of the two, which can be varied in composition, the molecules *are in time-independent (stationary) states since they do not change, that is, they do not exchange atoms*. No other combinations of this elementary composition are found in a system of this composition. It is conventional to refer to such identifiably separate units of molecules as *components of an overall (mixed) system*. Relative to the basic units, the elements C, H and O, each of the above two combinations has the same composition, but because each has a different potential energy it is clear that *element composition* and *particular chemical potential energies* are two *separate variables of chemical substances*. Now, since the second variable is restricted in both ethyl alcohol and dimethyl ether, we may refer, for operational convenience, to just one variable—the composition of a mixture of them in terms of the two *components*—rather than stating one elementary empirical composition, C_2H_6O, and the total chemical potential energy (which is difficult to evaluate). In effect, by giving composition in terms of components rather than in terms of elements, what we have done is to acknowledge that there is, within the variable chemical composition in terms of *components*, a variable *electrostatic potential energy*, limited by certain combining rules for atoms in given spatial arrangements (see Chapter 4). (As we shall see in Chapter 8, it is essential that such components are totally independent and do not exchange atoms, since exchange would, in fact, remove their independence, and the total potential energy would no longer be a variable. It would then be sufficient to state the element composition since the system of elements C_2H_6O would collapse to a fixed scrambled mixture of alcohol and dimethyl ether.* We have *not* referred to any kinetic energies associated with either component, but we shall return to this topic in Chapter 5.

2.4.1 Components as variables of composition

By definition, a component then is a unit in a stable or a particular stationary potential energy trap or condition at a given elementary composition. To illustrate this statement again, consider the composition $H_4C_2O_2$. We observe this, for example, as acetic acid $H_3C \cdot COOH$, but there are other theoretically possible isolatable atomic arrangements with different atom arrangements in space, such as

$$H_2C = C(OH)_2, \quad H_2C = C \overset{O-OH}{\underset{H}{\diagup}}, \quad HOHC = CHOH \quad \text{and} \quad \overset{H}{\underset{H}{>}} C \overset{O}{\underset{O}{<}} C \overset{H}{\underset{H}{<}}$$

and mixtures of components such as ($H_2C = C = O + H_2O$), ($H_2C = CHOH + \frac{1}{2}O_2$), ($HC \equiv CH + H_2O_2$), ($H_2C = CH_2 + O_2$), ($CH_3OH + CO$), ($HCHO + HCHO$), ($CO_2 + CH_4$) and ($2C + 2 H_2O$). The most stable of these isolated

---

*Note immediately that any set of arrangements represented by letters, even linearly in space, corresponding to some empirical formula, is a description of different possible 'isomers' or components. Three-dimensional space gives rise to a rich variety of possibilities (Section 3.2.5). Variety can be extremely great in large molecules, and in living species this is one way of describing different DNAs. Thus, non-exchanging atomic arrangements (components) are a variable of speciation in living or dead matter.

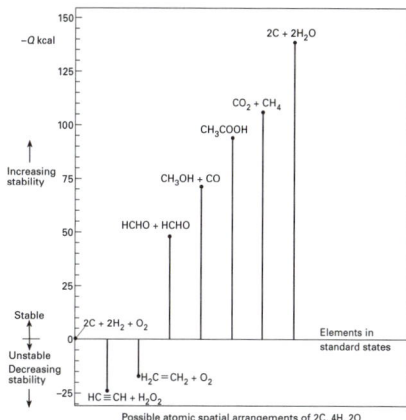

Fig. 2.8 A diagram of stability, in kilocalories, Q, due to binding interactions alone between atoms of chemicals of the same empirical C/H formula, 2C/4H/2O, (see also Fig. 5.21). This is a landscape diagram (see Section 1.4). Note that it is discontinuous on the x-axis due to quantisation of energy.

Components

Some sulphur compounds

S_8 H_2S SO_2 SO_3

Some nitrogen compounds

N_2 N_2H_4 NH_3 NO N_2O

Some oxygen compounds

O_2 O_3 H_2O_2 H_2O

systems (lowest potential energy), if we fix temperature at 300 K and pressure at 1 atm, is $2C + 2H_2O$. Figure 2.8, which is another landscape diagram (compare Fig. 1.10 and see Section 1.4), now of energies of isolated systems of one or more components, shows some of the other *possible* arrangements that obey combination (valency) rules, which in principle and often in practice can be observed if they are of sufficient persistence in the conditions chosen (see aside). They are all stationary states. In contrast, a peculiar arrangement such as $H_4C–CO_2$, which does not obey valency rules (see Chapter 3), is not observed. As stated, to be considered as an independent *component* of a given empirical composition, the possible atomic arrangement must be of sufficient stability and must not exchange atoms, so there are also kinetic conditions to be obeyed. Much of chemistry is, therefore, rather unpredicatable since the barriers to exchange, which will be discussed in Chapter 8, are difficult to assess.

A landscape diagram such as Fig. 2.8 can have any number of x-axes of composition corresponding to the maximum number of the given elements, up to 90. Since some peaks are of stable and some are of stationary substances for given compositions, chemistry is a highly empirical study of a vast multitude of species which can be made and kept or at least observed. We return to the further considerations of other variables that affect landscape diagrams of chemical stability in Chapter 5.

We have now defined two derived variables, *energy* and *component composition*, which are operationally useful for the description of chemicals around us and are combinations of the fundamental variables (Table 2.3). (Remember that the fundamental variables of a system are composition, in terms of mass and charge, the positions in space of the masses and charges and the time dependence of these positions. Note again that so far we have treated the sum of *pair-wise* interactions only.) In order to have a useful definition of chemical composition we shall not refer to the individual atomic masses and charges from now on but to chemical components, e.g.

Table 2.3 Fundamental and combined derived variables

Fundamental variable	Combined derived variable
Element composition (Atomic mass composition)	Gravitational and electrostatic *potential energies**
Charge composition	
Spatial distribution of elements and charges (Volume for bulk gases)	*Component composition* including local potential energies*
Time variation of spatial distribution	*Flow* of single objects (velocity: *directed kinetic energy* and *momentum*)* *Temperature (random kinetic energy)* and *pressure (random momentum)* in bulk gases

* These are most readily appreciated by considering single particles or pair-wise interactions.

ethyl alcohol, dimethyl ether, acetic acid or even compounds labelled by symbols such as PVC, DDT or DNA (note *no* atom or charge is necessarily shown). Again, to describe the ability of a system to change, or to do work, we refer not to the positions in space or to changes in space of the component units but to the energy associated with these variables (see Table 2.3), noting that the internal energy of the atoms and charges of each component has already been included in the definition of a component. Thus, the chosen variables used in chemical and biological science are matters of convenience for describing systems and operations performed on or by these systems. Stability, as yet just in terms of potential and directed kinetic energies, is also referenced to the components (Fig. 2.8, but see Fig. 5.21).

2.5 Additional energy forms and their interconversion

We recognize many different forms of energy (see Table 2.4), only some of which we have described so far, and we shall now relate them to the capability of producing useful mechanical work or of driving any chemical or physical process. Actually, any energy form can be converted into another, although there are inevitable losses in the conversion of energy into *useful* work. The efficiency of conversion varies with the form and the conditions. We shall examine this problem in Chapter 5. Here, by discussing how forms of energy that we have already described can be converted into other forms, we shall introduce forms of work capability or energy not yet analysed.

Obviously, potential energy can be transformed into directed kinetic energy by using it to change the rate of motion through acceleration. The intermediate kinetic energy is also transferable to potential energy, as in the classic case of raising a bucket of water attached to a string passed over a pulley or, to use a modern example, of putting a satellite into orbit. We can also transfer charge, and its flow, the electric current, is called *electricity*, while its storage is called *electrical potential*. Once again care needs to

Table 2.4 Types of energy

Gravitational potential energy (associated with the attraction between separated masses)
Electrostatic potential energy (associated with the interaction between separated charges)
Kinetic energy (associated with directed motion of masses)
Chemical energy (associated with transformations of chemicals, changes of internal potential electrostatic energy)
Radiant energy (associated with electromagnetic radiation)
Electrical/magnetic energy (associated with electrical currents)
Nuclear energy (associated with the nuclear binding forces)
Surface energy (associated with the surface of liquids and solids)
Mechanical energy (associated with stress in solids) and including pV energy
Thermal energy (associated with temperature, random kinetic energy of masses, i.e. random motion)

N.B. There is really a fundamental underlying unity in energy, which is always conserved although it can be interconverted amongst its different forms. All types of energy, except nuclear, are associated with biological as well as human machinery.

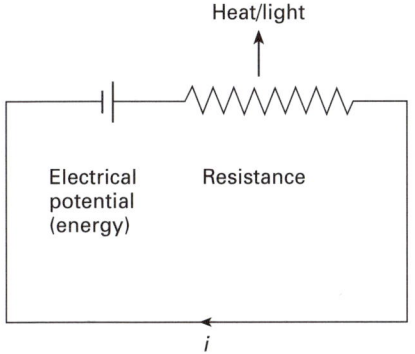

Heat/light

Electrical potential (energy)   Resistance

i

Transformation of energy in an electrical circuit.

be taken to notice that the flow of electricity, the current, is not itself energy but is related to energy only by deccelerating it through a wire called a resistance (see aside). (Therefore, once a current is achieved in an electric circuit of a superconducting material it flows with no friction so that the generation of heat and corresponding loss of energy are zero.) It requires energy to start or change directed flow of mass or charge, while deccelerating (or stopping) the flow gives out energy. As stated, the energy corresponding to the change of directed velocity from v_1 to v_2 is obviously proportional to the mass and it is the difference in $1/2mv^2$. Note that the *directed kinetic energy* is another *derived variable* (Table 2.3). But how are these potential and directed kinetic energies related to other forms of energy such as 'light', by which we see, and 'heat', which we sense as an increase in the hotness or warmth of our surroundings? Let us introduce these energies with a series of examples of interconversion of energy forms. In doing so we are, in some cases, describing properties which do not belong to single objects or pairs of objects, as in the above, but to large numbers of particles which should be treated statistically (see Chapter 5). We introduce them here in order to have an overall view of the different forms energy can assume.

2.5.1 Conversion of chemical and mechanical potential energy and directed motion to random motion (thermal energy)

Using energy, for example 'muscular' energy which comes from the chemical potential energy of food components such as sugar (see Section 2.4 and Chapters 10–14), we ourselves are capable of doing mechanical work, for example lifting an object from a lower to a higher place. Suppose we lift an object to a certain level above the ground and then let it fall. The more massive the object and the higher we lift it the more work (energy produced by our muscle force) is necessary for the purpose, but the more work (energy) we can get out of the fall if the corresponding energy is properly harnessed and utilized. Employing a more scientific terminology we would say that we have done work, using chemical energy, against the force of gravity to lift the object to a certain level whereby it will have acquired a given amount of physical potential energy. It will stay in its new position indefinitely (in a stationary, time-independent state) provided we keep holding it or leave it on a suitable platform. If, however, we remove the 'barrier' then it will fall to the ground under the gravitational force in what must be accelerated motion (since the gravitational force is acting permanently). It therefore acquires a directional (towards the ground) kinetic energy, given by $1/2mv^2$, where v is the velocity at each moment.* Here we see that the energy from our muscles has been transferred to the potential energy of the object, which in turn has been converted into directional kinetic energy so that we can write

Muscular energy (chemical) → Potential energy (physical) → Directional kinetic energy → ?

[There is, of course, an intermediate between muscular chemical energy and the raising of the position of the object which is the stretch or stress energy in the muscle fibres (see Sections 2.5.5 and 14.5).] But what is the energy represented by the question mark?

The directional kinetic energy can be used for some useful purpose, for example to lift another weight, e.g. a bucket of water, as described before in Fig. 2.4. (Another similar example would be that of water behind a dam, where the falling water could be used to turn an electrical turbine to produce electricity.) Staying with the simple case of the object falling on to the ground, what we would observe could be: (1) an eventual deformation of the ground and/or the object, which corresponds to mechanical work done (potential energy transferred); (2) a bounce due to the stress, mechanical energy, acquired by the object (imagine a rubber ball), followed by another fall (conversion of mechanical into kinetic, then into potential, then again into kinetic energy), a process which may be repeated again and again (but note that invariably the bounce is smaller and smaller until the object comes to rest, Fig. 2.9); (3) an inevitable, but not always easily detected, rise in warmth or hotness of the local environment, where the object, now still, has fallen, and of the object itself, that we shall relate to an increase of 'thermal' energy which we call heat content. (We shall ask later if all this heat can be reconverted into useful forms of energy for work.)

What we have illustrated here is a series of interconversions of energy, which is always conserved—it can neither be created nor destroyed, only transformed into other forms. In effect, this conservation of energy is a universal principle† called the first law of thermodynamics.

---

*Velocity is a directional, or vectorial concept, unlike speed, which is a so-called 'scalar' number, i.e. unrelated to the direction, although both are expressed by the same number.
†After the discovery that in nuclear reactions mass and energy can also be interconverted (according to Einstein's equation $E = mc^2$, where E is the energy, m is the mass and c is the speed of light), this effect can also be taken into account, but it is not significant in chemical changes.

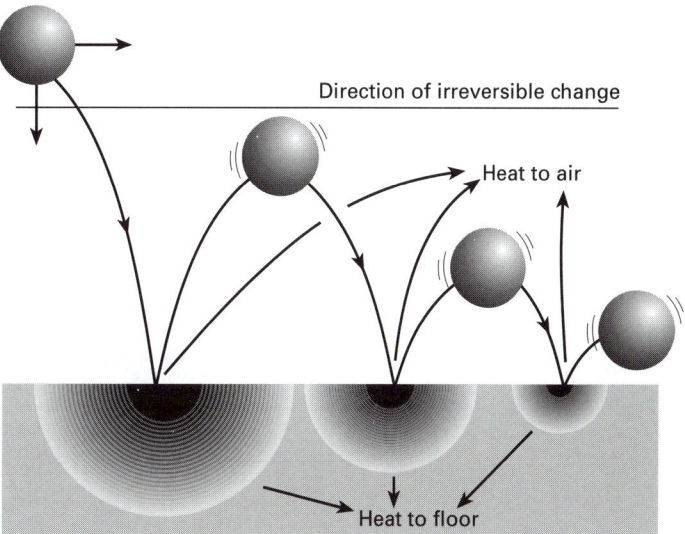

Fig. 2.9 On each bounce of the ball some potential energy and some directional kinetic energy is degraded to random kinetic energy of the earth and the air until the ball comes to rest. The reverse process has never been observed, see Section 5.4.2. [After Atkins, P. W. (1996). *Physical chemistry* (5th edn). Oxford University Press, Oxford.]

In this description we have introduced two new forms of energy—*stress*, a form of mechanical energy related to properties of condensed states, and *thermal energy* (or 'heat' content). Other, more common examples of mechanical stress are the compression of springs and the stretching of rubber bands. In the present text we will meet other cases of mechanical stress later (for example, in muscle contraction, see Section 2.5.5), but we must clarify and quantify now the concept of heat, which is central to our discussion.

Consider, as a second example, a water wheel that is turned by river water dropping down over it. During the turning the water loses gravitational potential energy and produces directed kinetic energy of the wheel (Fig. 2.10). The flow of water is not transmitted to any detectable extent since the river continues to flow at the same rate beyond the wheel. The water wheel can now be used to drive a paddle in a liquid held in a vessel. This will cause directed circular motion (flow) of the water in the vessel, of course, and stopping it will release energy. One way of stopping the circular motion is to change it to turbulent random motion in all directions. Fully turbulent motion, when summed, is not associated with any directional flow. Experiment shows that the measured (see below) degree of hotness (or coldness), which from now on we shall call the *temperature*, of the liquid then rises (slowly) and we conclude that some of the potential energy of that water in the river above the water wheel has been partly converted into what we call the 'heat' content of the liquid in the vessel by the fall of this river water (see Fig. 2.10). Similarly, as confirmed by experiment, the (turbulent) water at the bottom of a waterfall must be at a higher temperature than that at the top, but the water flows on at the

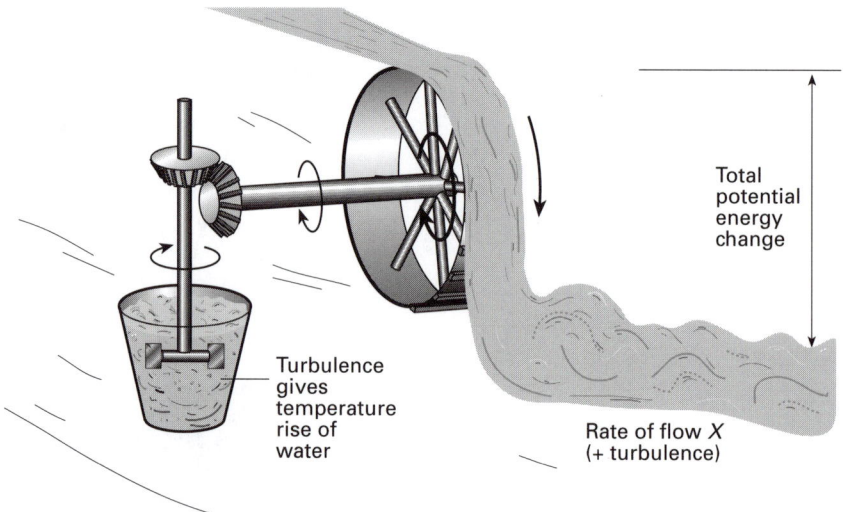

Fig. 2.10 An illustrative example of the transfer of gravitational energy to directed (circular) kinetic energy to random kinetic energy, using a water wheel (see text).

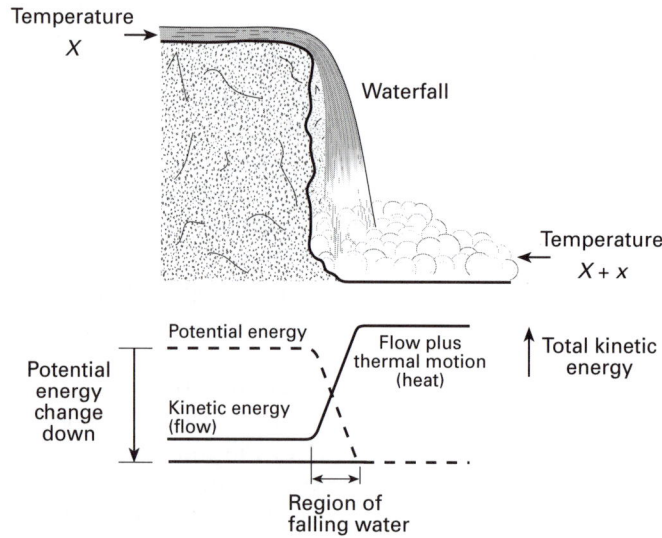

Fig. 2.11 A second illustration of the transfer of gravitational energy to random kinetic energy (change of temperature) of water at the bottom of a waterfall.

same rate (see Fig. 2.11). We write the conversion of energy as potential energy of river water → heat content of water. But what is *heat* in this sense? How does it differ from flow in a direction which clearly does not relate to the degree of hotness, i.e. temperature?

2.5.2 Heat and heat transfer by convection or conduction

Through a series of experiments carried out by various scientists, especially count Rumford and Sir Humphry Davy, at the end of the seventeenth century, and then J. R. Mayer and particularly J. P. Joule and Lord Kelvin in the middle of the nineteenth century, it was concluded that 'heat' (thermal energy) is related to the summed random motion, i.e. kinetic energy, of particles. The idea had already been considered by Francis Bacon.* In effect, it was possible to calculate a 'mechanical equivalent of heat' using experiments similar to that of the paddle turning in a liquid described above, but we had to wait until the end of the nineteenth century to find out that what actually moves randomly are the units that carry the random kinetic energy. These units are the constitutive atoms or molecules of the materials involved (in the above examples the turbulent motion of the molecules of the liquid in the vessel or in the river after the waterfall) and that this so-called convectional heat has no meaning in itself except in the sense of the change of *random kinetic energy*. Heat is not material like a solid or a fluid; it is just a variable of a substance related to the kinetic energy of random motions—vibration in solids (where heat transfer is called conduction), random translational motion of atoms or molecules, plus configurational, vibrational and rotational motion of atoms and molecules in liquids and gases (see Fig. 2.12).

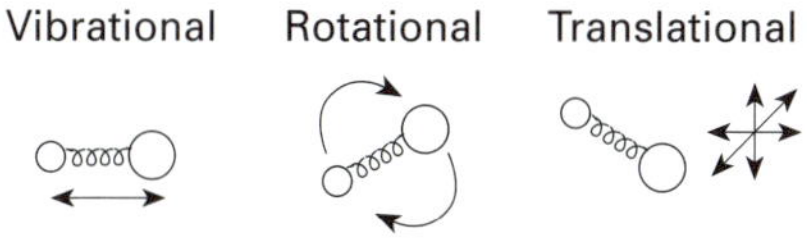

Thus 'to heat' (opposite of 'to cool') means to increase the temperature, that is, to increase the random kinetic energies of atoms and molecules without any change in their *average* positions, i.e. no change in potential energy, and independently from any bulk flow.

Following the approach to all observations from around 1600 onwards, it was clearly desirable to put temperature on a quantitative basis. This was done originally by devising different temperatures scales. The scale today (Celsius' scale or the centigrade scale) is based empirically on the freezing point of water, conventionally 0 °C in centigrade temperature units, and its boiling point, conventionally 100 °C. Clearly, molecules of water in ice have less motion in them than molecules in steam so the scale is a measure of kinetic energy. If the temperature is lowered so that gases of the air freeze, then it is observed that CO_2 freezes at −55 °C, N_2 freezes at −210 °C and helium freezes at −270 °C. At −273 °C it is not possible to lower temperature any further. There is, therefore, an alternative absolute scale called the kelvin (K) scale (see Section 2.5.3), in which 0 K = −273 °C, the absolute zero of temperature, when all classical motion ceases. On this scale ice melts at 273 K and water boils at 373 K.

These observations show that one of the variables causing change from solid to liquid to gaseous states is an increase in temperature which measures the random kinetic energy of constitutive particles, atoms or molecules, the units of composition. As stated, the motion can be vibrational, rotational or translational (see Fig. 2.12). We begin to see the confusion in Greek thinking between units of composition and variables of both

Vibrational   Rotational   Translational

Fig. 2.12 The three kinds of motion of a molecule.

---

*Heat itself, its essence and quiddity, is motion, nothing else (Francis Bacon, *Novum organum*, 1620).

composition and physical states, and why once it gripped peoples' minds it was hard to remove. In fact, the classical scientists had neither found the true units of composition nor the true variables of state which lie in separating out the contributions of energy (fire) to physical states of the materials from their chemical composition when described by *components*, (not now atomic elements). We must also keep in mind that heat *does not* include the kinetic energy of *directed* motion, flow, which we have described in Section 2.3.2. *There are, therefore, two separate useful derived variables, which relate to kinetic energy, one belongs to stationary or stable dynamic random systems, such as a gas, the other to flowing dynamic directed systems, such as a river* (see Table 2.3). The random kinetic energy, the temperature, does not involve any change of the system with time. However, we must also see that we have changed our description of systems from one based on pairs of interacting objects to one based on bulk properties of a many-particle system. The consequences are fully explored in Chapter 5, but we examine approximate ways of treating the new problem in Section 2.5.4.

It is clear that the same temperature change for different substances requires different energy to be put into them since the kinetic energies of a simple monatomic gas can only be in translational motion while that of a molecular gas such as water can be in such motion and in vibrations of bonds and molecular rotations (Fig. 2.12). A solid has its temperature raised by putting energy into vibrations only. Hence, any temperature change of one degree centigrade requires a different energy input for different substances dependent on structure and on physical state. This specific energy requirement is called the *specific heat* of the substance and is defined per gram (or per gram mole, when it is called *heat capacity*). The unit of energy is the quantity of 'heat' required to raise the temperature of 1 gram of water by 1 °C (1K) and is called the calorie. We have seen that mechanical energy is usually given in joules, but since mechanical energy can be converted directly to increase in temperature of water there is a simple relationship between the two units, namely 1 calorie ≡ 4.18 joules. In this book we shall use generally (but not exclusively) calories as the unit of energy since this remains the convention in many sciences. Furthermore, since all forms of internal energy can be transformed into one another, it is useful to refer all to these same units of heat, calories (see also Chapter 5).

2.5.3 Temperature, pressure and volume: *pV* energy

To complete the study of the physical derived variables that define the state of systems we recall the experiments of Boyle, published in 1602, when he demonstrated for the first time that, *at constant temperature*, the *volume* of a given sample of gas varies inversely with the *pressure*, that is $pV = C$, where C is a constant. This *quantitative* equation is known as Boyle's law, which introduces one new variable—the manner in which *bulk space* is occupied by particles in the absence of forces acting between them. Note that it, like temperature, is a property of a large number of particles. Later, in 1802–1808, Gay-Lussac developed a series of experiments demonstrating that several different gases, such as O_2, N_2 and H_2, showed the same dependence of volume on temperature, θ, and arrived at the equation $V = V_0 (1 + \alpha_0\theta)$ where V_0 is the volume of a sample of gas at 0 °C, α_0 is the so-called coefficient of thermal expansion, equal to $\frac{1}{273}$, and θ is the temperature, in °C. This relationship is called Gay-Lussac's law and states *that at constant pressure* a gas expands by $\frac{1}{273}$ of its volume at 0° for each degree rise in temperature. Both Gay-Lussac's and Boyle's equations are strictly valid for a few gases only, such as helium at and around room temperature, in which the interactions of their constitutive particles, atoms or molecules, are negligibly small. They are called *ideal gases*, meaning that they obey the above-mentioned laws (see aside). The majority of *real*, as opposed to ideal, gases, show deviations from this behaviour due to forces acting between particles (see Chapter 5).

Given the form of Gay-Lussac's equation it is convenient to define a new temperature scale, T, called *absolute*, such that $T = \theta + T_0$, (see Section 2.5.2). The unit of temperature in this scale is the degree kelvin, °K, or just kelvin, K. The value of T_0, the absolute temperature, corresponding to $\theta = 0$ °C was found to be 273.16 K. In terms of absolute temperature, Gay-Lussac's equation is $V = V_0 [1+(T - T_0)/T_0]$ or $V = V_0T/T_0$, that is, $V/T = V_0/T_0 =$ constant.

Since $pV =$ constant (at constant T) and $V/T =$ constant (at constant p) we can obviously write $pV/T =$ constant. The constant can be easily determined for a given set of p, V values, for example $p_0 = 1$ atm and $T_0 = 273.16$ K, when V_0, for 1 mole of gas, is 22 414 cm^3. (According to the so-called Avogadro's principle, this volume is the same for all ideal gases and contains the same number of particles, 6×10^{23}.) The calculated constant, which we call R, *the gas constant per mole*, is 8.314 J K^{-1} mol^{-1} or 1.98 calories K^{-1} mol^{-1} [but note that in this system of units (SI system) the pressure is measured in newtons and the volume in cubic metres; if p is measured in atmospheres and V in cubic centimetres then $R = 82.06$ cm^3 atm K^{-1}.] We can therefore write:

$$pV = RT \text{ (for 1 mole of gas) or } pV = nRT \text{ (for } n \text{ moles of gas)}$$

This is the well-known *quantitative equation of state of an ideal gas*, or *the gas equation*, which includes the results of three laws: those of Boyle, Gay-Lussac and Avogadro. For real gases some additional terms must be included, giving as one approximation the so-called van der Waals equation (see Section 5.8.2). (Note that the need to include these extra terms derives from the existence of significant interactions between their particles due to so called van der Waals forces, see Section 4.4). However, for most qualitative discussions, the simple gas equation, as given above, is sufficient. We stress that effort is made in all these analyses to derive *quantitative relationships, now between derived variables*.

While we have linked temperature to random kinetic energy we have not discussed the nature of pressure. The simplest way to visualise pressure is to see it as due to the effect of collisions of gas particles with the walls of a containing vessel. Newton showed that in a two-body elastic collision the so-called *momentum, mv*, was conserved and hence exchanged equally and oppositely. Therefore, the walls of a vessel suffer an impact as the momentum component for each particle perpendicular to the wall is reversed. This

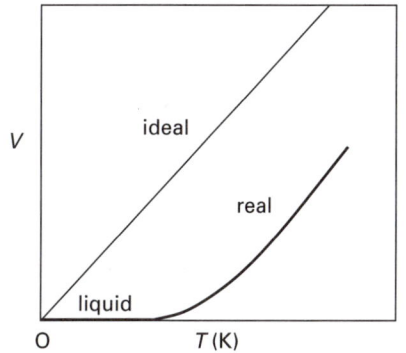

Ideal and real behaviour of gaseous substances at varying temperature.

acts as a force per unit area applied equally to all the walls of the container and, in the case where volume does not change, it can be said to be equivalent to an internal property of the gas, its pressure.* It is not associated with any movement (flow) and is independent of time (see below). Obviously pressure, like temperature, is a bulk property of a gas. Now we have to connect temperature and pressure to the fundamental variables *mass (m)* and *variation of position with time of individual particles* (velocity, *v*), but while we could describe variables such as composition, potential energy and flow in terms of properties of individual or pairs of atomic elements, ions or bulk objects, this is not so easily achieved for temperature and pressure. Obviously, the temperature and pressure belong to large numbers of particles and their interactions. However, not only is it impossible to describe multiple collision exchanges of energy or momentum using Newton's laws, but large numbers of particles have a wide (probabilistic) distribution of velocities. In the nineteenth century an approximate approach called the kinetic theory of gases was used to solve this problem.

2.5.4 The kinetic theory of gases

A useful theory of the properties of ideal gases, i.e. where molecules do not interact, was developed from the ideas that atoms and molecules in the gas phase occupied effectively no volume and could be treated as geometric point particles. They were then assumed to have *average velocity* (actually the root mean square speed[†]) which gave rise to a temperature, related to average randomised kinetic energy, and a pressure related to average randomised momentum. The theory allowed a deduction of the relationships between pressure, volume and temperature in terms of an averaged speed attributed to each and every particle.

While it is clear that in a system of a large number of particles individual particles cannot have a temperature or pressure, it is entirely legitimate to propose that all particles in a system are equivalent, and then their singular speed (energy and momentum) can be represented by the properties of a *hypothetical particle* which has no *directional* but only *random* kinetic energy or momentum. It does not flow and is time independent. This kinetic theory has been used to explain other properties of matter such as specific heats and viscosity. However, we note that it cannot give any explanation of properties such as melting and boiling points. We need to take into account the distribution of speeds and that of the occupation of space in a more fundamental, statistical way. Nevertheless, the kinetic theory of gases keeps a mathematical simplicity in the discussion of the derived variables temperature and pressure and their relationship, and is therefore a valuable introduction to them.

---

*Pressure is a force per unit area, hence its units are newton m^{-2}, which multiplied by units of volume, m^3, gives newton m, units of work. Hence the product pV has the dimensions of energy and is a form of mechanical energy (see Chapter 5). Commonly, pressure is measured in practical units such as atmospheres (atm), bars (~1.01 atm) or torr (abreviation of Torricelli, equal to 1 atm/760, i.e. 1 mm Hg).

[†]Speed is rate of movement without reference to direction: velocity is directional.

2.5.5 The derived variables of systems

Now that we have related the product pV to temperature it is obvious that there are only two equivalent ways of changing the energy content of a gaseous system using these derived variables: one, which we have described, is due to temperature change and depends on the specific heat of the substance, and the other is due to either pressure change at constant volume, or volume change at constant pressure, both of which involve change of concentration. (When volume is changed *anisotropically* the change pdV is seen as a mechanical change, as in the movement of a piston). Temperature and pressure (or volume) then define the physical condition of a gaseous system (but not the flow). Of course, these physical variables are independent of the chemical composition variable in terms of components (although to a certain extent they affect the potential energy of interactions). In Table 2.3 we have listed all the variables related to energy that we have discussed so far in this chapter. We stress that the first two (gravitational and electrostatic potential energy) are properties that we describe starting from pairs of objects, while the last two (temperature and pressure) belong to systems containing a large number of objects and will require reconsideration in Chapter 5. We shall leave aside variable flow until Chapter 8.

2.5.6 Chemical energy conversion into thermal energy (heat) and mechanical energy

As we have seen above, a chemical component of a given system can be characterized by elemental composition and total potential energy corresponding to a particular distribution of its constitutive units, atomic nuclei and electrons.

Thus, the gases methane (CH_4) and dioxygen (O_2) are made up of stationary state, time-independent, separate molecules, each with a given chemical potential energy. However, if we mix CH_4 with O_2 and make these gases react completely (that is, if we burn methane gas) then we get a rearranged alternative mixture of carbon dioxide (CO_2) and water (H_2O), again made up of separate molecules, each with its chemical potential energy, now in a stable state.

The sum of the chemical potential energies of $CO_2 + H_2O$* is lower than the sum of the chemical potential energies of $CH_4 + 2O_2$, hence on reaction there will be a release of energy (which cannot be destroyed), indicated by writing the reaction scheme as

$$CH_4 + 2O_2 \rightarrow CO_2 + 2H_2O + energy \text{ (for example, } Q \text{ calories)}$$

This energy will appear as '*heat*' (see aside), which we sense from an increase of *temperature*, which may be radiant heat (see Section 2.6) or

Heat from chemicals

conc. $H_2SO_4 + H_2O$
$CaO + H_2O$
Na (metal) $+ H_2O$
P (red) $+ H_2O$

---

*We can consider water to be in the gas or the liquid state. The total potential energies of the bulk materials are different, being lower in the liquid state since it takes energy to change the liquid (at 25 °C) to the gaseous state. In quantitative treatments we must fix a standard temperature (25 °C) and pressure (1 atm) to have a compatible energy book-keeping system.

thermal heat, as described above. Thus, component changes can be related to the temperature changes that these can produce. [Here, we again use components as a variable of composition instead of elements and so include the variable internal potential energy, but note that the number of variables remains the same: *three* elements (C, H, O) plus potential energy, or *four* components (CH_4, O_2, CO_2, H_2O) in any mixture of the gases.]

A further example, which leads us to say that *chemicals (food) give us (human beings) energy*, was illustrated by muscle energy, referred to in the example in Section 2.5.1. In fact, all the energy we use in our own metabolism and activities derives from the combustion of carbon compounds (see aside), for example, sugar from plants, with dioxygen, which releases energy and so enables us to do work. [As stated above, we use 'calories' as a measure of the energy we can derive from this reaction and this is why we also say (incorrectly) that food 'contains' calories.] The corresponding chemical equation for this combustion reaction is:

$$\text{Sugar} + O_2(\text{air}) \rightarrow CO_2 + H_2O + \text{energy (calories)}$$
$$(\text{e.g. } C_6H_{12}O_6 + 6O_2 \rightarrow 6CO_2 + 6H_2O + 672\,\text{kcal})$$

One of the things we have to know, therefore, is the energy released in the combustion of chemical compounds (Table 2.5), which can be determined experimentally, and how it is related to the (maximum) work that can be done on other systems, where the activity takes place in a specific environment. Relative to $6(CO_2 + H_2O)$, $C_6H_{12}O_6 + 6O_2$ has a higher content of internal chemical potential energy which is released in this reaction. Note that the formula of common sugar is identical to that of cellulose (wood) which is also a (poly)saccharide, but we are not able to digest wood and extract calories from it. Sugar and cellulose are components with the same empirical composition but different internal energies. [There is a different problem here, mentioned before but also needing to be stressed, concerned not with the formulation of chemicals or their energy states but with the *rate* of their change, i.e. with their *kinetic stability*. Neither wood nor sugar burns in the air spontaneously, but they do so in furnaces and in the human body (contrast metallic sodium, which reacts spontaneously on the surface of water). A much faster (higher) rate of burning, say of sugar,

Common fuels (+O_2)

Coal	C
Acetylene	C_2H_2
Bottle Gas	C_3H_8 or C_4H_{10}
Wood	$C_6H_{12}O_6$
Petrol	C_8H_{18}

Table 2.5 Some heats of combustion, *Q*, at 25 °C

Compound	−*Q* (kcal mol^{-1})	Compound	−*Q* (kcal mol^{-1})
Carbon (graphite)	94.1	Acetone (l)	428.2
Hydrogen	68.3	Formic acid (l)	61.0
Methane (g)	212.9	Acetic acid (l)	209.3
Ethane (g)	373.2	Oxalic acid (s)	60.8
n-Hexane (l)	995.9	Lactic acid (s)	321.5
Cyclohexane (l)	933.4	Ethyl acetate (l)	535.7
Benzene (l)	781.8	Tripalmitin (fat)	7510.0
Toluene (l)	945.7	Sucrose (s)	1350.4
Methanol (l)	173.6	D-glucose (s)	671.8
Ethanol (l)	327.7	Glycine (s)	231.8

depends now on assistance from other substances which help the change; these substances are called *catalysts* and we will consider them and their action in Chapter 8.]

We turn now to the transfer of the released energy from chemical reactions to the mechanical work (pV) which can be done. The easiest example is the work done in expanding a gas from a volume V_1 to a volume V_2, so let us connect this work to an operational engine using chemical fuels. In the petrol engine of a car, for example, we burn a gas (e.g. methane, CH_4, for the sake of simplicity) with dioxygen. The corresponding equation is that given above, i.e.

$$CH_4 + 2O_2 \rightarrow 2H_2O + CO_2 + 213\,kcal$$

when the thermal energy released is for a gram-mole of methane after complete combustion to CO_2, all at unit pressure and 298 K, so-called 'standard conditions'.

Due to the 'heat' released, forcing a change of temperature, we gain an increase of pressure in the car's cylinder (remember the gas equation $pV = RT$). This pressure is later partly converted into a volume increase and then in the mechanical motion of the machine. There are inevitable losses of energy since not all the 'heat' released can be converted into useful work—part of it is dissipated to the environment. However, the total amount of energy transformed plus that dissipated *remains constant*. We discuss the maximum useful work in Section 5.6.

We have demonstrated the following transformation: energy in chemicals in one stationary state ($CH_4 + 2O_2$) goes to another now stable state ($CO_2 + 2H_2O$) while giving thermal energy (largely change of random kinetic energy in the fuel), which goes in turn to changing configurational (potential) energy of the machine and then to *directional bulk motion*, i.e. directional kinetic energy.

If energy is applied from a source to any substance it is possible, at least in principle, to drive that substance uphill to a new chemical condition, necessarily a stationary state. Hence, just as we can consider driving physical changes, such as forcing an increase (or a decrease) of volume, so we can force chemical changes. For example, in a reaction where volume increases at fixed pressure in burning sugar, e.g.

$$C_6H_{12}O_6(s) + 6O_2(g) \rightarrow 6CO_2(g) + 6H_2O(g) + heat$$
sugar

the reaction goes spontaneously to the right, that is to a stable state from a stationary state. However, plant life takes the energy of the sun and reverses this reaction

$$CO_2 + H_2O \xrightarrow[h\nu]{sun} sugar\ (starch\ and\ cellulose)$$

Here, the sun's energy is trapped as chemical energy in the plant material. In order to understand this energy transfer we need to turn to a further variable, *radiant energy*, which finally brings us back to the non-convective part of the classical 'element' *fire*.

2.6   Fire and radiant energy transfer

As stated earlier, the attraction of amber for fur after being rubbed with it (involving electrical forces and the corresponding energy of interaction) was known at least by 600 BC, but amber was also known as a substance 'which sometimes emits sparks (light) after being rubbed with fur'. Again, dropping one flint stone on another had been seen to give rise to sparks, and knocking two flints together could be used to generate fire. Finally, the burning of fuels also gives rise to light and heat in radiant (fire) form. We must discover how the electrostatic energy of charges or the loss of kinetic energy of masses through friction have been converted to the energy of light and heat we can sense, much as we sense (see and feel) the energy (light and heat) coming from the sun's 'fire'. This transfer of energy does not require any transfer of massive particles or particle motion and is then unrelated to convection or conduction transfer of heat as discussed in Section 2.5.2. Accordingly, we call it *radiant* energy transfer.

Consider some further examples. Without any noticeable change in flow we saw that the potential energy of a mass of water in a stream could be transferred to the random kinetic energy of the particles of a liquid held in a vessel, that is to heat energy, by a water wheel (Section 2.5.1). Now, the same fall of potential energy of water, e.g. of a waterfall or over a dam, can be connected, in a hydroelectric power station, to generate electrical potential (see Section 2.5), i.e. converted constantly into charge separation, using a dynamo where two surfaces are constantly rubbed together. The transferred charge then flows equally constantly down connected wires. By allowing this electricity to flow through a resistance we can also create heat (this is the principle of electric heating), and create light (as in electric light bulbs). In effect, as stated in Section 2.5, this heat and light energy release is due to the *friction* which deccelerates the flow of electrons in a wire, called a resistance, but no mass is lost. Again, the light from the sun *heats* the (lower) atmosphere and the Earth's surface because it increases the random (turbulent) translational kinetic energy of the molecules of which the atmosphere is composed and the vibrational energy of atoms of the minerals that make up the surface of the Earth.* The sun does not transfer mass to Earth in this process and in the transmission there is no connection with the motion of any units of mass. So what is the nature of this radiant heat and light which does not involve in its transmission the movement of particles?

Now, we shall see in Section 2.6.2 that the radiant light and heat we observe from a bulk object such as the sun, a light bulb or a fire can be broken down into the emission of radiation from vast numbers of individual atoms or molecules. When we analyse it in this way we find that radiation is emitted as a flow (flux) of energy in a particular direction from each single unit and that it corresponds with a change in the potential energy

---

*In outer space there are very few particles, so there is no random kinetic energy and temperature must be very low of necessity. (This, of course, is not true for a body inserted in it.)

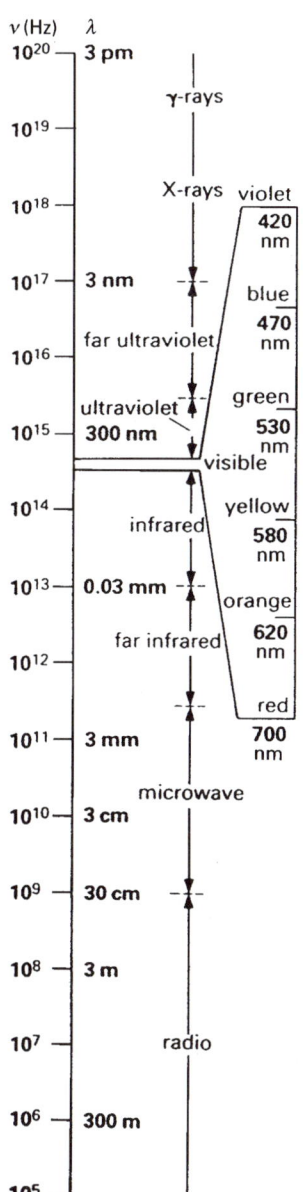

Fig. 2.13 The electromagnetic spectrum. Frequency (ν) is in hertz, cycles per second. Note the very small range of visible radiation.

states of single atoms or molecules. The fact that radiation appears isotropic, i.e. is equal in all directions, when emitted from a bulk object is due to the large number of different emission centres, each radiating in a given direction. This radiation is referred to as *black-body radiation* (Section 2.6.2), and it is not homogeneous but consists of a great variety of energy packets from very low to quite high energies. Examination of the spread of energies allows us to see that radiant heat and radiant light are both just part of a much greater variety of radiations. [Note that we refer to energy 'packets' which indicates a relationship between its discontinuous nature and that of matter and charge, see Chapter 3].

2.6.1 The variety of radiation

As stated above, light is the energy we *see* emitted as radiation from a source, e.g. the sun. Hence, knowledge of it is dependent on our senses. Study outside the *visible* region of this radiation of the sun using instruments shows that there is also radiation which we cannot *see* but which we feel as 'heat', compare an infrared lamp or a microwave oven. We do not see high energy radiation either, but it can burn us (UV light) or it can be used to see through some objects (X-rays). Finally, there is low energy radiation that we do not see or feel. In effect, there is a wealth of radiative energies (Fig. 2.13) which are again put on a quantitative scale. Apart from creating the possibility of vision over a narrow region, all radiation can also be converted to generate 'heat', as random kinetic energy of molecules in our body or in the air, noticed by a temperature rise. It is here obvious that our senses (Fig. 1.1) are poor instruments for studying what is around us in a scientific manner.

For some time, until this century, the treatment of 'light and heat' radiation has caused difficult problems for scientists in that its properties forced them to treat it as behaving both as if it were 'particulate' and 'wave-like',* i.e. as if it had a dual nature. The particles are called *photons* and unlike waves are in discontinuous units that can be counted as they are emitted from a source, but they are not to be confused with the material particles we have discussed so far since they have no *inertial (rest)* mass, although they are deflected by massive stars due to gravitational attraction. Perhaps we should call them energy 'packets'[†]. The particulate nature of light was the first idea to be advanced, but as early as the seventeenth century it was shown by Huygens that light also behaved as a wave form since it exhibited typical properties of waves of sound or water, for example, refraction and diffraction. The travelling waves of sound or water are due to the changes

---

*We create waves in a pond by dropping a stone into it. Such mechanical waves can be converted into electric potential energy by connection to a suitable dynamo, hence man can use sea waves as a source for electric power. The radiation waves are produced, for example, by energy loss of electrons in atoms, where the electron oscillation in a static atom is transferred to a moving wave, or by electric currents moving up and down vertical wires (long-wavelength 'radio' waves).

[†]The description of radiant energy as particles or energy packets requires us to introduce them as new fundamental variable units of composition, separate from chemical atoms and charges. We shall very largely ignore this complication since it does not affect considerations of systems at fixed temperature (see Section 2.6.2).

of *oscillation of mass perpendicular* to (not along) the direction of motion of the wave. (This is, in effect, a travelling gravity field generated by a loss of potential energy of, say, a stone falling in a pond.) The waves of light and heat radiation are due to the changes in *oscillation of charges* relative to one another, with loss of potential electrostatic energy, which give an electromagnetic travelling field called electromagnetic radiation. We discuss the problem in more detail in Section 3.3. It is possible to classify electromagnetic radiation generally by the wavelength of waves or by their frequency, as shown in Fig. 2.13. We now know all the possible forms of radiant energy from infinite to infinitesimal wavelengths, from below X-rays to beyond radio waves. In other words, just as in the case of chemicals where we know all we can know about atoms and composition, so we know all we can know of interest to chemistry about radiation energy. But just as in the case of atoms, some fundamental problems remain unsolved (see Chapter 3). There is always the question of the nature of the charges, their oscillations and of the dual wave/particle character of very small bodies as well as of radiation. Just as we found the composition could not be varied continuously on bringing atoms together so we find that radiation from isolated atoms is discontinuous (see Section 3.2.3). These observations have forced a revision of our views on the nature of atoms (Section 3.3).

2.6.2 Temperature and radiation

We measure temperature with a thermometer, often using the expansion of a liquid, e.g. mercury in a capillary tube. We have said that the temperature measures the random kinetic energy of the material and it is not unreasonable that a greater kinetic energy will cause a liquid to occupy more space (a gas, as stated above, occupies all the available volume). If we look at a molecular picture of the liquid (or of a gas) we find that the temperature represents the *average* random kinetic energy of the particles since some are moving more slowly and some more quickly. We can then draw a distribution curve of the random kinetic energy for all the molecules in the liquid (or gas) (see Fig. 5.4). Similarly, if at a given temperature we observe the colour or, better, the oscillation frequency of the radiation coming from the atoms or molecules in a solid incandescent bulk material, say a metal wire, we find not just one frequency but a distribution of oscillation which, for reasons we do not need to discuss here, is called a 'black-body spectrum' (see Fig. 2.14). (Notice again that just as the distribution of random kinetic energy derives from a statistical treatment of large numbers of atoms or molecules, the distribution of oscillation frequencies derives from the consideration of large numbers of photons, energy 'packets', according to the so-called Planck's quantum theory.* That is, just as matter is not continuous, radiant energy is not continuous—it is composed of *quanta* of radiation.)

---

*According to Planck's theory (1900), the energy of an oscillator is given by $E = h\nu$ where ν is the frequency of the oscillator and h is a constant (Planck's constant or quantum of action). In the SI system, $h = 6.626 \times 10^{-34}$ Js (Joule second).

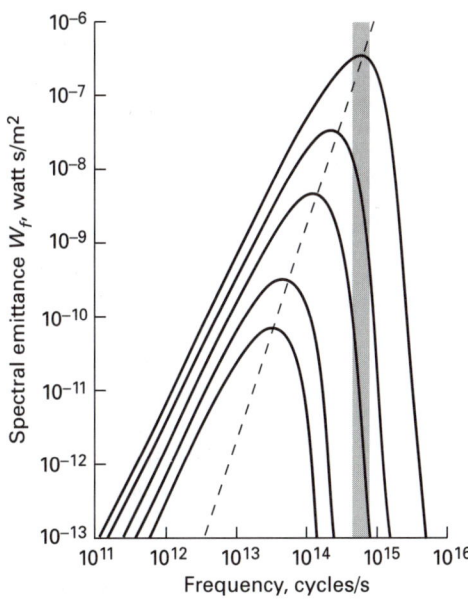

Fig. 2.14 The intensity of electromagnetic spectrum of radiant energy emitted at a given frequency plotted against frequency. Note how energy emitted increases with temperature from 500 K through 1,000 K, 2,000 K, 5,000 K to 10,000 K and more radiation occurs at longer wavelengths. The shaded band is the visible region, see aside. The total energy radiated, area under each curve, increases as T^4.

Animal detectors

Eyes: 'Visible' light
Snake: Infra-red
(Ears: 'Sound'
Bat: Ultrasonic)

The two distributions are just different ways of describing energies associated with a given temperature. Whereas we previously referred temperature to a mean random kinetic energy, clearly this is more accurately described by the distribution curves (see Section 5.3.1). As the temperature rises both distributions (and the averages) move to higher energies, either higher kinetic energy or higher radiation frequency, and broaden. Clearly, (averaged) temperature is characteristic of a given black-body radiation distribution.

However, a radiating body can only remain at a fixed temperature if, as it radiates, it receives energy to compensate for its radiation loss. All bodies at a fixed temperature, therefore, capture as much energy as they release through either convection or radiation. Since each system at a given temperature has a number of photons being released in any period of time it might be thought that they too constitute units of a system to be included in its composition. However, since the number of photons received is exactly the same as that lost, the net change in numbers of photons is zero in all states in temperature balance so that we can ignore them. We cannot see objects in the 'dark', no matter what the temperature. A very different situation arises when two bodies are at different temperatures (see Section 8.7.2).

As a further example, the 'temperature' of the space of the universe can be measured through its 'background' radiation. This temperature has been falling since the beginning of time owing to the expansion of space, and the wavelength of the initial radiation has increased. The temperature

Table 2.6 Temperature and visible appearance of a solid*

Temperature (K)	Visible appearance
1000	Dull red
1200	Cherry red
1500	Orange white
1800	White heat

* An experienced observer estimates correctly the temperature of a glowing body to within ± 100 K.

soon after the Big Bang was $>10^{10}$K, corresponding to radiation of very short wavelength, but today it is about 3 K, corresponding to the microwave noise in the universe measured by the Cosmic Background Explorer (COBE) satellite. This temperature is that of 'empty space', but the temperature of stars and even the Earth is much higher. The correlation between the age, the size and the temperature of the universe strengthens the theory of the origin of the universe as a Big Bang.

Radiated light and heat are, therefore, obviously connected through considerations of random kinetic and potential energies of different particles. We know this from experience since the hotter the body, the higher its temperature, and the shorter the wavelength of the radiation emitted, which we see as a change of colour (Table 2.6). Here we are stating implicitly that any body, no matter what its temperature, radiates energy. However, if the environment is at the same temperature it receives radiant energy equally. Of course, if two bodies are at different temperatures then radiation flows from one to the other and this radiation can be made to do all kinds of work at the lower temperature. We return to this problem in Chapter 8 since it concerns flow, for example from the sun to Earth. Before that, however, we must understand why the sun has such a high temperature.

2.6.3 Gravitational force, nuclear fusion and radiant energy

Everyone knows that the sun is our main 'source of energy'. From it we get light and heat, which are forms of energy different from any of the forms of potential energy, kinetic energy and mechanical energy described so far. As stated above, we call it radiant (or radiative) energy, which has been shown not to involve transfer of particles or particle motion, although it can be converted into other forms of energy (see Section 2.6.2). But, from where did the sun get the energy it radiates continuously since its formation, some 5 billion years ago?

By the middle of the nineteenth century, Lord Kelvin and Baron Hermann von Helmholtz proposed that it was the result of the continuous contraction of the sun's original gas cloud under the effect of its own gravity. As the volume decreased pressure increased, the gas particles moved faster and faster (randomly), so that the temperature increased and at high values the contracted cloud radiates 'heat' and even 'light' like any incandescent body. In other words, the sun shines because of its contraction under gravity (Fig 2.15). We can, therefore, imagine the energy conversion sequence:

Gravitational energy → Random kinetic energy (increase of temperature) → Radiant energy

This, however, is just one part of the story (and of the truth); calculations show that if contraction under gravity was the only reason for the radiation, the sun would have shone for just 30 million years, which is an absurd result given its age.

More recently it has been possible to estimate the temperature of the sun's surface—about 5500 °C—and calculations made by Eddington in the 1920s pointed to a temperature at the centre of the order of 10 million

Protosun >5 × 10⁹ years ago   4.5 × 10⁹ years ago

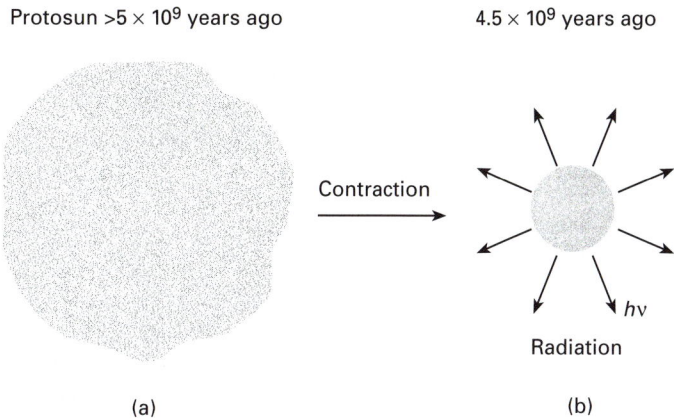

(a)   (b)

Fig. 2.15 Two stages in a stellar gas cloud that becomes the sun. Stage I, highly dispersed and starting to contract. Stage II, the cloud shrinks under gravity and begins to radiate as temperature rises. [After Narlikov, J. V. (1996). *The lighter side of gravity*. Cambridge University Press, Cambridge.]

degrees. As he proposed then, and as was confirmed later, this is a high enough temperature to overcome the electrostatic repulsion of hydrogen protons (+ charges) in collisions thus making them come close enough to feel the effect of the 'strong force' (se Table 3.1) and trigger the formation of helium, a nuclear fusion reaction which releases an enormous amount of energy (it has been estimated that about 0.7% of the mass of H has been converted into energy according to Einstein's equation $E = mc^2$). Gravity confines the gas undergoing explosive nuclear energy generation, allowing the process to be self-controlled.

Now the complete conversion sequence, which has lasted for almost 5 billion years and can last for another 5 billion, is:

Gravitational energy → Random kinetic energy (temperature increase) → Nuclear fusion (very high temperature) → Isotropic radiant energy (dissipated in all directions equally)

Note that for giant stars, much larger than the sun, the process of 'burning' hydrogen continues until all hydrogen has been converted into helium; then a similar 'burning' of helium follows to give carbon, burning of carbon gives oxygen, etc. This is in essence the process of synthesis of elements in giant stars which has led to the observed elemental abundances as described briefly in Section 1.4. We may be a little shocked to find that half the time, at least, for evolution on Earth has gone.

2.7   Summary

Whereas in Chapter 1 we showed by analysis that all materials were made from the naturally occuring 90 elements, in this chapter we have been concerned with looking at the reasons for observing certain combinations of them in chemicals and in certain physical states. After uncovering the two major potential energies, gravitational and electrostatic, which gave rise to pair-wise attraction now involving *charge* as well as elementary mass as fundamental units and so causing physical and chemical association, we found that these potential energies were derived variables, apparently continuous, of the *spatial relationships of masses and charges*, giving us three fundamental variables (see Table 2.3). Taken together with radial, time-independent cyclic motion and using Newtonian mechanics it was then possible to explain the bulk interactions of objects up to the level of planetary systems.

The treatment led us to describe isolated samples of materials of a given composition as being in one of three time-independent conditions: *stable states*, incapable of change; *static stationary states*, capable of doing work but trapped in a given energy condition; and *dynamic stationary states*, distinguished by their unchanging cyclic internal motion and potential energy traps (Table 2.2). Stationary states are capable of doing work, but are prevented from doing so by their trapped condition. It is fundamental to the study of all chemical systems that we understand this *capability to do work* (see Chapter 5) and later we need to turn to *the rate at which work can be done* (see Chapter 8). Only in this way can we understand why objects around us are as they are and why they might change in a continuous fashion. Historically this was assumed to be interpretable on the basis of Newtonian mechanics, until about the year 1900. However, these Newtonian treatments of interactions between two particles or two larger bodies fail to explain chemical interactions between atoms which, as we noted in Chapter 1, are of very limited kinds. For example, we find discontinuous combinations of H with O such as H_2O and H_2O_2 but virtually no others of formula H_nO_m. These discontinuous combining ratios which we shall analyse in Chapters 3 and 4, forced us to separate *local* interactions in atomic-dimension space from bulk interactions. It then became convenient to treat those chemicals such as H_2O that we can isolate and that mix with others unchanged as operational units of composition. We called those chemicals the *components* of the system. To explain them we need to use quantum mechanics, not Newtonian mechanics, which apply to bulk objects only. It will become apparent in the next chapter also that only quantum mechanics can explain the occurrence of the elements as seen in the periodic table. Component composition is then a useful *derived* variable (Table 2.3).

A further fundamental variable, *time*, was introduced to describe those systems in steady and developing states that have directed kinetic energy and momentum, but we have deliberately set aside the discussion of them until Chapter 8 onwards.

Throughout these descriptions of states and their energies we have had in mind that we must appreciate the quantitative connection of them all to the fundamental units and variables: mass and charge (units) and composition, space and time (variables). We found that Newtonian or quantum mechanics could deal with all states as long as the number of particles in the system, associated with compositional variables, was small, essentially one or two. Now, the derived and observable physical variables, temperature and pressure, are only describable in terms of very large numbers of particles in systems. To develop a quantitative theory of these variables we have used in this chapter the kinetic theory of gases, which makes all particles treatable artificially in terms of a hypothetical average particle which has a time-independent random kinetic energy (temperature) and random momentum (pressure). The success of the above quantitative treatment of variables allowed us to describe the conversion of energy from one form to another (see Table 2.4). This approach maintains mathematical simplicity, but fails to treat many properties of materials such as changes of physical state. The treatment required no new fundamental variables (see Table 2.3).

A further feature of our environment is that we observe what the Greeks called fire, which not only includes heat conduction by diffusion and random kinetic energy of particles, but also radiation. We noted that the description of radiation required us to introduce new units—'particles'—called photons. We could then describe radiation from a body in terms of summed atomic emission of photons and this emission, its black-body radiation, was shown to be equivalent to its temperature. Now, for most of our purposes the states we consider are of systems at constant temperature which gain and lose photons equally at all times. Thus we set aside consideration of photon reactions until Chapter 8 onwards where this condition of balance is not imperative.

The list of units and variables is not quite complete, as we shall see in Chapter 7 onwards. We have managed, however, to describe many given states in *quantitative terms* in equations ultimately related to the *fundamental variables* (Table 2.3). These equations are in simple linear form. If this situation was free from underlying and unreal assumptions, then, as was thought almost throughout the last century, the universe would be obviously deterministic and predictable. This is still thought to be so by some scientists today. In Chapters 3 and 4 we shall continue to analyse on this basis. However, in Chapter 5 we shall show that this is not a tenable position.

3

Electrons in atoms and their energetics

Nevertheless, for those who hope for something deeper than mere agreement between experiment and theory, quantum mechanics is sadly incomplete; it provides no intuitively satisfactory picture of the physical world—the great triumphs of modern physics have revealed a chasm between our everyday understanding and the uncomprehended reality that underlies it. After 100 years the electron is still a mystery.

Sir Brian Pippard FRS, Emeritus Professor of Physics, University of Cambridge, Lecture on One Hundred Years of the Electron, The Royal Institution, (1997)

3.1   Introduction

In Chapter 1 we showed that there were 90 stable elements on Earth, the fundamental units of matter at the chemical level which, in principle, could be put together to form compounds in whatever way we liked. Hence, the element composition was the *first fundamental variable* of all materials. In Chapter 2 we observed that, in fact, the atoms of the chemical elements could be charged, and so could bulk materials, and that charged bodies exerted an electrostatic force on one another. Thus, the units of composition of a material could no longer be described just by their atomic mass but we had to consider also their charge which then became the *second fundamental variable* of composition. Furthermore, it was found that many atoms of different elements tend to come together only in particular ratios, for example NaCl, CH_4, H_2O, etc. We deduced that this was due, at least in large part, to the preferred potential energy of these over other possible ratios and spatial arrangements of the same elements, where spatial position of atoms was the *third fundamental variable* (see Table 2.3).

Moreover, we found that on mixing certain different chemicals containing the same element but in different ratios (or even the same ratios) the atoms did not exchange at room temperature, 300 K, hence they are in a stationary (time-independent) state. We found it convenient, therefore, because of the restricted ways in which these chemicals appeared, to use an operational definition of the compositional units in a mixture which we called the *components*. In effect, this change of definition, corresponding to a very large increase in the units of composition, was to allow for the different uses of electrostatic potential energy, a *derived variable* in stationary atomic assemblies. As described in Chapter 2, the use of components in this way subsumes two variables into one. (There are other materials, combinations of elements, for example certain alloys, where no strong constraints apply to the ratio of different atoms in the product of mixtures of them and the elements remain the only units of composition. These materials will be described later in Chapter 4.) We also noted in Chapter 2 that certain atoms bind preferentially to certain other atoms. These properties were again stated to be related largely to the way potential energy is held between them. We remarked too, in that chapter, that there were other forms of energy, including those observed opposite physical properties of matter, namely random kinetic energy, and opposite flow, directed kinetic energy. We shall describe these energies further in Chapters 5 and 8, respectively. Before doing so, we shall explore in this chapter how energy affects atoms themselves, causing them to lose or gain electrons in preferred ways, a property which governs the formation of the components of particular atomic ratios and charges and their shapes. This will allow us to see why chemical trends, some of which were discussed in Chapter 1 and 2, and molecular shapes have appeared for the combinations of different atoms and molecules of the periodic table. What we need to show is where Newtonian mechanics fails us.

3.2   The basis of the structure of the periodic table

3.2.1 The breaking of the atom

In Chapter 1 we described the general way in which the present day form of the periodic table of elements was derived. However, the reason for the underlying pattern of the table, in periods of 2, 8, 8, 18, 18, 32, can only be illustrated after the nature of atoms is related to the more fundamental particles—electrons, protons and neutrons—which they contain, and their *constrained energies*.

Electrons, e, were discovered when it was shown that many substances emitted 'rays' when subjected to appropriate conditions such as high temperature, electric discharges in gases or strong electric fields. These rays were deflected on passing through electric fields, and thus behaved in electrostatic properties as very small *charged particles* for which the measured mass and the charge were always the same no matter the substance from which they originated. As mentioned earlier, the idea of such electrostatic charges to account for forces unrelated to gravity had been introduced before 1700 after the study of attraction (and repulsion) of amber, or ebony, rubbed with fur. The measured mass of the (discontinuous) charged particles was about $\frac{1}{1840}$ of the mass of one hydrogen atom and its measured charge was found to be and defined as the minimum unit of (negative) charge. (Note again that the assignment of negative or positive charge is just a convention; there are charged particles which behave in opposite ways in electric fields and they were distinguished by giving some a *negative* charge and the others a *positive* charge.) Shortly afterwards it was discovered that there were other particles that had unit mass, almost identical to that of the hydrogen atoms, and they were positively charged. They were called *protons*, p (see below). The obvious conclusion was that atoms, though fundamental to the classification of chemicals, had internal structure involving these smaller units. Since the number of electrons could be varied, in principle, relative to the number of protons, the units of composition are not atoms but, as so far defined, protons and electrons. However, the more useful units are atoms and excess or deficiency of electronic charge, which chemists and biologists use.

3.2.2 Classical models of the structure of atoms

Since the mass of the electron is so small, the idea was put forward that an atom, which is electrically neutral, consisted of a heavy, positively charged static sphere occupying all the volume of atoms in which electrons were embedded so as to make the whole neutral (J. J. Thomson, 1910). Subsequent experiments, performed by Rutherford, who bombarded thin metal foils with so-called α-particles [actually He^{2+} *ions* (helium nuclei), that is helium atoms which have lost their two electrons] gave, however, unexpected results. Most of the particles passed *unimpeded* through the metal foils, other particles were deflected at various angles and only a very few were sent back. These experiments provided evidence for the concentration of the mass of an atom in a *very, very, small* positively charged nucleus, much, much smaller than the atom it represented, surrounded by a number of electrons revolving around this nucleus (Rutherford, 1911) to keep the atom neutral (Fig. 3.1). All models of atoms from this time onwards are of *dynamic* stable and stationary states.* The radius of the nucleus was estimated to be of order of 10^{-12} cm while that of the atoms was of the order of 10^{-7}–10^{-8} cm. The nucleus was found to have a charge of +Ze, where Z is a whole number, the *atomic number*, and e is the charge of the proton (equal to the charge of the electron, that is, equal to the unitary charge). Each element has its own integral value of Z, giving a correlation with the sequence in the periodic table but no immediate explanation for the periods. Of course, this left an extremely difficult conceptual problem since the electron, with a mass $\frac{1}{1840}$ of the proton, filled 10^{12}–10^{15} times the volume of the nucleus!

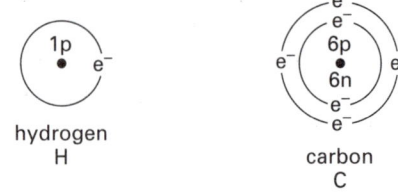

Fig. 3.1 An early representation of the outer and inner electrons (e⁻) and the nucleus [including protons (p) and neutrons (n)].

---

*The electron in this picture (Fig. 3.1) has a fixed orbit and hence a fixed attractive electrostatic interaction with the nucleus, but it also has a fixed tangential velocity round the nucleus, generating a virtual repulsive 'force' (Section 2.2). The two forces are in balance in any stationary state (compare planetary orbits). There could then be, apparently, an infinite number of possible orbits of different electrostatic and directional kinetic energies. There is, however, only one variable, the sum of the kinetic and potential energies, since these two must be in a fixed relationship in any one electronic state.

Obviously, the atomic number, charge of the nucleus or number of peripheral electrons was the relevant factor to take into account in the periodic build-up of chemistry in the table of elements. However, the correlation of numbers of protons with the relative weights of atoms was seen to be very poor. Further investigations showed also that many elementary atoms could exist in forms with the same atomic number but different (although closely related) *atomic weights*. These forms were called *isotopes* (same place) since they correspond to the same element, hence they had to be located in the same position in the periodic table. Uranium-235 (^{235}U) and uranium-237 (^{237}U), both corresponding to the same element uranium, atomic number 92, but with atomic weights 235 and 237, are examples of isotopes. There may be several isotopes of the same element, some stable and some unstable (radioactive) which undergo decay with time and may transform into other elements until some stable state is reached. Nuclear fission is based on this fact. Again, the ordering of atomic numbers and atomic weights was not always the same. As an example, the elements iron, cobalt and nickel, whose atomic numbers are 26, 27 and 28, have atomic weights 55.85, 58.94 and 58.69 which are not in increasing order. Thus, the charged protons, of mass one, could not explain all the mass of the atoms nor the presence of isotopes. This is why the first attempts to arrange elements according to the similarity of their properties in the order of their increasing atomic weights exhibited anomalies (as found by Mendeleev) which were corrected when atomic numbers were used instead.

Later experimental work concentrated on the structure of the nucleus which, as stated above, had the atomic number, Z, of positive charge. The existence of elements with atomic weights that were more than double their atomic numbers was puzzling, and it was originally thought that the nucleus should also contain internal as well as external electrons to minimize repulsion between the positive charge of the protons and to balance this charge so that it would be equal to that of the external electrons. It was only in 1932 that Chadwick discovered a new heavy particle in the nucleus, which has the same mass as the proton but no electric charge (today accordingly called a *neutron* and represented by n). The neutrons were later shown to interact with protons to form a stable structure through a new force that holds protons and neutrons together and is not 'electromagnetic' or 'gravitational' but of a different nature, called the 'strong force' (see Table 3.1). Thus, in very, very small 'particulate' materials (the nuclei of atoms) there is a force different from electrostatics or gravity which is not open to sense experience but is the basis of matter in this nuclear form.* [A fourth force is also operative in the nuclei of atoms (see Table 3.1) but is of no concern to us in this book.]

---

*The discovery of neutrons led to the increase in the number of underlying units of matter to three: electrons, protons and neutrons. It so happens that the influence of the mass of atoms, protons plus neutrons, is very small in chemistry and biology, so that we shall be able to ignore isotopic composition in this book and deal only with two underlying variables, electron deficiency or excess—defining charge—and the average atomic mass of the chemical elements with which we are concerned. It is the interaction of protons and neutrons which leads to the different stabilities of atomic nuclei and therefore controls, in part, their abundances. There is, however, another factor, which is the different rate of formation of the nuclei from protons and neutrons (see further reading).

Table 3.1 The four forces of nature (see aside)

Force	Relative strength	Particles acted upon	Range of action	Example
Strong	1	Protons and neutrons (quarks)	10^{-15} m	Holds nucleai together
Electromagnetic	1/137	Charged particles	Infinite	Holds atoms together
Weak	1/10 000	Protons and neutrons (quarks), electrons	10^{-16} m	Radioactive decay
Gravitational	6×10^{-39}	Everything	Infinite	Holds the solar system together

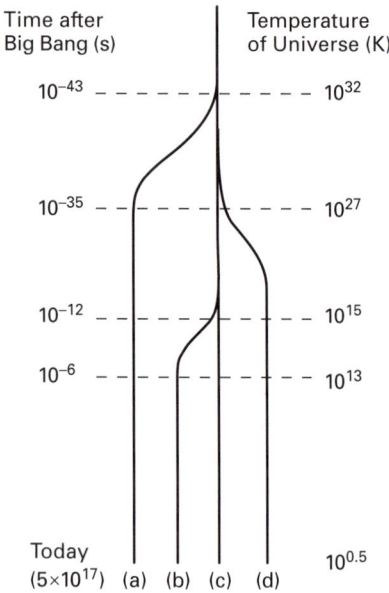

Time after Big Bang (s)

Temperature of Universe (K)

10^{-43} — — — — — — 10^{32}

10^{-35} — — — — — 10^{27}

10^{-12} — — — — — 10^{15}

10^{-6} — — — — — 10^{13}

Today (5×10^{17}) (a) (b) (c) (d) $10^{0.5}$

Separation of the four forces from the Big Bang: (a) gravity, (b) weak nuclear, (c) electromagnetic (d) strong nuclear

Protons and neutrons are therefore the main constitutents of the mass of the nucleus,* which together with the 'planetary' electrons form the known atoms (see Table 1.2 and Fig 1.6). The positive nuclear charge is now just the number of protons in it. This is the basic outline of Rutherford's atom, where the atomic weight of an isotope is just that of the sum of the number of protons and neutrons.

Although this model of the atom was not sustainable on physical grounds and in the light of further evidence such as line spectra (see below) so that a more complicated one had to be devised later, the division of atoms into a central nucleus with external electrons (in shells) allows us to appreciate that the forces that come into play when atoms are close to one another must be due to electrostatics. It is through these forces that compounds and extensive lattices are formed, which are the bases of mineral and living species. In essence, two or more positively charged nuclei interact co-operatively with electrons, while the whole is usually electrically neutral. All chemical bonding depends on these electrostatic interactions.

3.2.3 Quantum models of the structure of atoms

As stated, Rutherford's model of the atom was not sustainable on physical grounds since, in classical physics, electrons (electrical charges) moving in orbits should radiate energy, hence lose energy, and get closer and closer to the nucleus, finally falling into it and thus provoking a 'catastrophe'. It had, therefore, to be modified, and this task was undertaken by a Danish scientist, Niels Bohr (1913) who had worked in Rutherford's laboratory. Bohr started from the curious and until his time unexplained observation that the spectral light emitted or absorbed by atoms was in the form of lines of fixed wavelength (or frequency) and not in the form of a continuous range of energies. To 'explain' the observation he assumed that

---

*Today it is known that protons and neutrons are not elementary particles but are themselves constituted by 'quarks', which have fractional charge, positive or negative (see Table 1.2). These, however, cannot be obtained free and we need not take them into account in chemical phenomena. It is unfortunate, to say the least, that the more detailed the examination of materials below the level of atoms the more incomprehensible the explanations become.

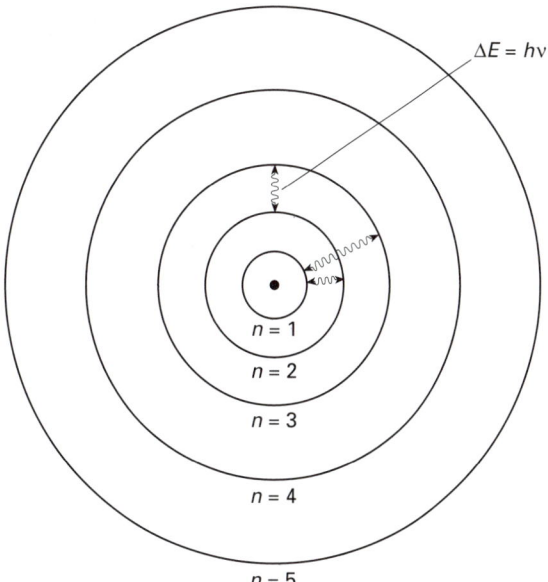

Fig. 3.2 Relative sizes of circular Bohr orbits for various quantum numbers. The arrows show electron jumps which may require (outward jump) or give out (inward jump) spectral energy (light) as photons.

atoms were composed of a small positively charged nucleus surrounded by point negatively charged electrons revolving now only in a *few permitted stationary* (non-radiating) circular orbits of different radii and different directional velocity, and consequently different energies. This assumption is non-classical, i.e. it is quantum mechanical, and in apparent disagreement with the properties of macroscopic objects. Excitation of atoms could, therefore, only promote electrons to *particular* permitted higher levels of energy (orbits of larger radii) which should then return to their original 'ground' level (orbits of smaller radii) by emitting the corresponding fixed difference as a quantum of energy, hν, a photon (see Fig. 3.2). The model, in essence, explained the *line* spectra of atoms in that the transitions of electrons between pairs of such orbits, say *i* and *j*, corresponded to discrete amounts of *radiant* energy $E_i - E_j$ (with frequencies ν, given by $E_i - E_j = h\nu$) and the oscillations of radiant waves, electromagnetic waves (Section 2.6) were therefore derived from changes in the oscillations of electrons around nuclei. The stationary states of electrons in atoms were called quantum states and characterised by their energy and their 'quantity of movement' (momentum, *mv*) around the nucleus at a distance *r*, which is called the *angular momentum, mvr*. According to the theory, to agree with the experimental data (line spectra), this angular momentum can only take certain values, which are multiples of the quantity $h/2\pi$, i.e.

$$mvr = nh/2\pi$$

where *n* is an integer that can only assume the values 1, 2, 3, 4 etc. It is said that the angular moment of the electron is *quantised* (in units of $h/2\pi$) and *n* is called the *principal quantum number*. Planck's constant, *h*, in this formula had already been associated with the quantisation of all radiation and hence its use gave additional standing to the theory.

Both the energy and the radii of the (circular) orbits of the electrons can be easily calculated in terms of n in simple cases such as that of hydrogen and hydrogen-like atoms (atoms with but one electron). For polyelectronic atoms the situation becomes more complicated (see below).

Bohr's model was subsequently extended by Sommerfeld (1915) to include elliptical orbits since this corresponds to the most general form for an orbital motion in a central field of force, as in the case of the planets of the solar system. (Note that in atoms we are dealing with an electro-magnetic, not a gravitational field, but this does not alter the argument.)

Now, for elliptical orbits two 'radii' are required, corresponding to the major and minor axes of the ellipse; hence, by analogy with the case of circular orbits, it was necessary to postulate a secondary (or azimuthal) quantum number, k, such that n/k = length of major axis of the ellipse/length of minor axis of the ellipse. Again, this quantum number, it was stated, could only take integral values: 1,2,......, n (not zero since this would correspond to a straight line passing through the nucleus). Of course, when $k = n$ the corresponding orbit is circular. In this way both the circular and the elliptical orbits are quantised.

For convenience here, the reason for which becomes clear later, instead of k we use the letter l such that $l = k - 1$. This becomes the secondary quantum number, but now it varies from 0 to $n-1$.

It is also convenient to replace the l quantum numbers of electrons in orbits by symbols (letters) as indicated below:

l	0	1	2	3
symbol	s	p	d	f

We can, therefore, describe any orbit stating its principal quantum number, n, followed by the letter corresponding to a given value of l, the secondary quantum number.

Since l can take values from 0 to $n-1$, for $n = 1$ we can only have $l = 0$, hence only one orbit, 1s, which is circular; for $n = 2$ there are two orbits, 2s and 2p, of which the first is elliptical and the second spherical; for $n = 3$ there are three orbits—3s, 3p and 3d—of which the first two are elliptical and the third is circular, etc. Figure 3.3 shows the types of orbits for $n = 4$.

Now, a moving electron (flowing charge) with a given n and l, can have $2l + 1$ spatial orientations in a weak magnetic field, which are charac-terised by a third quantum number, the magnetic quantum number, m_l, which can take the values 0, ± 1, ±, 2, ..., ± l. This means that s ($l = 0$) orbits can only have one orientation for $m_l = 0$; p orbits ($l = 1$) can have three orientations for $m_l = 0, -1, +1$; d orbits ($l = 2$) can have five orienta-tions for $m_l = 0, -1, +1, -2, +2$ and f orbits ($l = 3$) can have seven orienta-tions ($m_l = 0, -1, +1, -2, +2, -3, +3$). All the orientations are equivalent in energy terms and are said to be 'degenerate'.

If we now take into account the different types of orbit for each value of the principal quantum number, we have the following situation

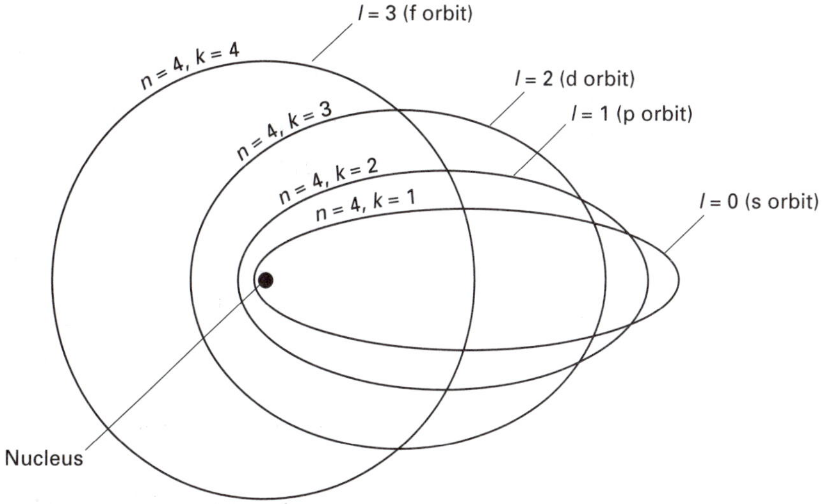

Fig. 3.3 Sommerfeld electron orbits for $n = 4$. Note how orbits penetrate for certain quantum numbers.

n	Types of orbit	Number of orbits
1	1s	1
2	2s, 2p(3)	4
3	3s, 3p(3), 3d(5)	9
4	4s, 4p(3), 4d(5), 4f(7)	16

This is a remarkable series of squares of simple numbers, and the periodic table (Fig. 1.6), has periods with 2, 8, 18, 32 elements, just twice these numbers! We see immediately that the number of elements in periods has been related to the energies of electrons classified by quantised motion. However, there were three remaining puzzles: (1) the number of states had to be doubled to fit the periods; (2) the electrons had to fill space; (3) the assumption of quantisation has no real theoretical basis. The first problem was solved by the discovery that electrons have another property: they behave as tiny magnets, a result that can be interpreted by admitting that electrons can apparently 'spin' in one of only two ways so that each n, l, m_l state of an electron has two new possibilities (recognised by magnetic interactions) and only two, which can be distinguished by assigning to each a fourth quantum number, m_s, the 'spin' quantum number, that can only take two values, one positive and one negative $(+\frac{1}{2}, -\frac{1}{2})$.

Thus, the periods of the periodic table were quantitatively related to the numbers of permissible energies of electrons in orbits, (see aside). Even then we see that the correlation only applies if each one electron alone could have one set of the four quantum numbers. A given orbit can be occupied by two electrons, but then they must have opposed 'spins'. This principle of allowed occupation of orbitals is called the Pauli exclusion principle.

Electronic configurations

	H	He	
	$1s^1$	$1s^2$	
Li	Be	B	C
$2s^1$	$2s^2$	$2s^22p^1$	$2s^22p^2$
N	O	F	Ne
$2s^22p^3$	$2s^22p^4$	$2s^22p^5$	$2s^22p^6$

If we, following Bohr and Sommerfeld, apply these principles to explain the long form of the periodic table, with the transition metal elements and

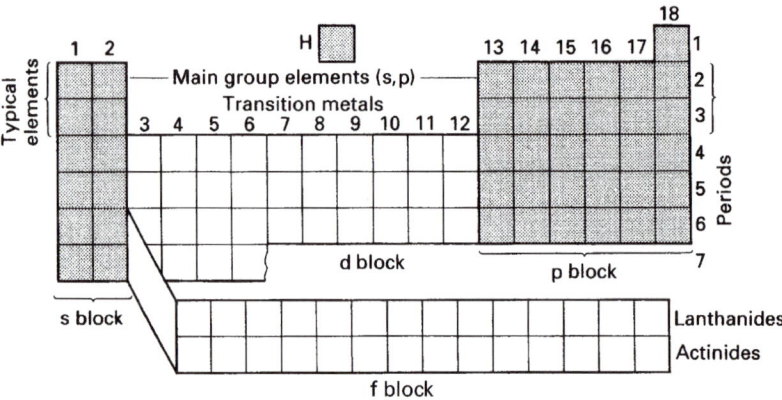

Fig. 3.4 The structure of the periodic table showing the subshell character.

Electronic configurations

Ti	V	Cr
$3d^2 4s^2$	$3d^3 4s^2$	$3d^5 4s^1$
Mn	Fe	Co
$3d^5 4s^2$	$3d^6 4s^2$	$3d^7 4s^2$
Ni	Cu	Zn
$3d^8 4s^2$	$3d^{10} 4s^1$	$3d^{10} 4s^2$

the lanthanides (and actinides) in separated 'boxes', we will observe that the peculiar format of this table is justified by the prior or preferential occupation of particular type of orbits by electrons when the *atomic number* (not the atomic weight) of the elements increases, that is, the table is made up of a particular arrangement of series of s, p, d and f blocks (see Fig. 3.4). Space is mysteriously occupied in these models by the restricted motions of electrons in orbits, not by 'matter' in the conventional sense of the word.*

It is now clear that the numbers 2, $3 \times 2 = 6$, $5 \times 2 = 10$ and $7 \times 2 = 14$ correspond to the complete filling of one s orbit, three p orbits, five d orbits, (see aside) or seven f orbits with pairs of electrons having opposed spins. The successive main 'energy levels' (or electron shells), defined by the values of n, will then have 2, 8, 18 and 32 electrons, the 'magic numbers' which were so intriguing in the initial experimental approach to the periodic table.

The ordering of the numbers 2, 8, 8, 18, 18, 32 result from placing the elements corresponding to the filling of 3d, 4d, 4f and 5d orbits within the periods of elements 4s 4p, 5s 5p and 6s 6p, respectively, due to inversions in the order of the energy of the orbits in atoms with many electrons (see Fig. 3.5). These inversions arise from the penetration of outer electrons through the shells of inner electrons.

---

*If we refer back to Chapter 2 we see that the energies of electrons in atoms are non-classical variables; they are quantised, i.e. they are no longer continuously variable. Again their directional velocities are non-classical, being restricted in both absolute value and direction also in quantised ways. This non-classical behaviour governs the nature of the periodic table, the formation of compounds between atoms and the shape of these associations. For this reason we treat potential and directed kinetic energies of electrons associated with protons in atoms quite separately from bulk fields and bulk flow. The latter are still treated as classical continuous variables. This distinction also allows a different statistical treatment of the properties of ensembles of atoms and molecules, as we shall see in Chapter 5. Furthermore, it lies behind the kinetic stability of combinations of elements of particular composition, e.g. CO_2 and CO, and not of continuously variable composition, $C_n O_m$. It then follows that these are usefully called components, as described in Section 2.3.

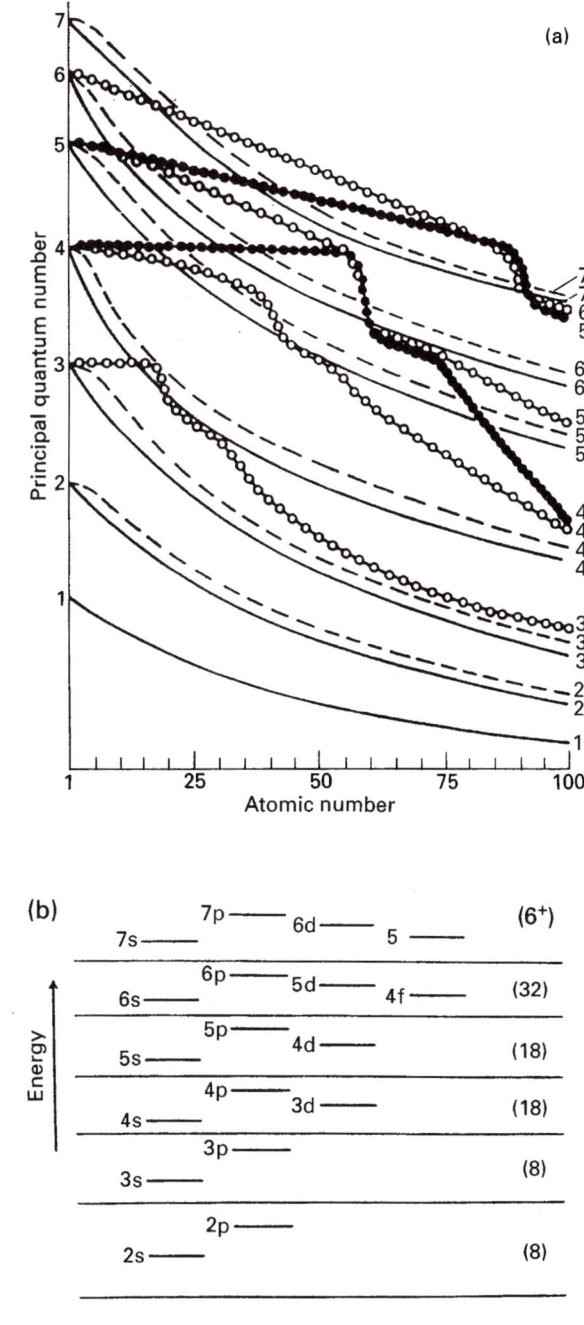

Fig. 3.5 (a) Energy dependence of atomic orbitals as a function of atomic number. (b) A simpler illustration of the predicted sequence of orbital energies for electrons in atoms; *s* levels can hold two electrons each; *p*, *d* and *f* levels, respectively, 6, 10 and 14. The diagram indicates the number of electrons that can be accommodated in each successive electron shell (row in the periodic table), ending with the filled-shell noble gas elements in group 18, and finishing at uranium (92). Note the way the energies cross-over due to penetration of wave functions.

It is also clear that there must exist a direct relationship between the outermost type of 'shells' filled with electrons and the chemical properties of the corresponding elements, since the table was built on this principle, i.e. elements with analogous chemical combinational properties were placed in the same column (group) (Fig. 1.6). There must, therefore, exist a relationship between the chemical combining properties of the elements and the energy of the electrons in those shells, specified by the (n, l) pairs, which may be obtained experimentally from spectroscopic measurements. Chemical properties are explicable, as we have seen in Chapter 2, by the description of element composition, including charges carried and appropriate energies of interactions between atoms to give a limited number of components. We aim today to explain the limitations in terms of *the quantised states of electrons in atoms* and the resultant electrostatic interactions of electrons shared between nuclei.

The Bohr–Sommerfeld model of the atom is sufficient for many qualitative and even quantitative purposes and has the additional advantage that it is experiment based (atomic spectra) and conceptually accessible to our senses (planetary motion is easily conceived). However, it is imperfect and inadequate; it starts from arbitrary postulates (for example, 'stationary' orbits in which the electron does not radiate energy, quantisation of energy states, etc.). Finally, it failed to explain further properties of electrons that were uncovered subsequently. As we shall see, a more incomprehensible model for the atom was to follow, which gave an answer to most of these problems, but introduces grave conceptual difficulties.

Most importantly, it is now clear why we had to separate the energy variables of atomic systems from those of bulk systems in Chapter 2. Variables of atomic systems move in quantum discontinuous steps; variables of bulk systems are treated as continuous. (This has a large effect on statistical treatments; see Chapter 5.)

3.3   The wave-mechanical model

We have seen in Chapter 2 that the properties of light could only be explained in terms of properties of particles (photons) and waves acting simultaneously. If wave-propagated light (electromagnetic radiation) behaves as if it is particulate, why cannot particles of matter (such as electrons) have some kind of wave behaviour? This was the starting point for new ideas on the nature of matter. In 1924, the French scientist Louis de Broglie asked himself this question and assumed it was a real possibility, thus associating the 'momentum' of an electron moving with a velocity, v, i.e. the product $p = mv$, to a wave with a given wavelength λ, where λ is now the wavelength of the wave *associated* with the electron which has a mass, m, moving with a velocity, v.

Louis de Broglie's intuition was confirmed a few years later by several experiments, amongst which that of Davisson and Germer (1927) showed that a beam of electrons could be diffracted by a nickel crystal lattice (that is similar to a grating in which the ions are separated by distances com-

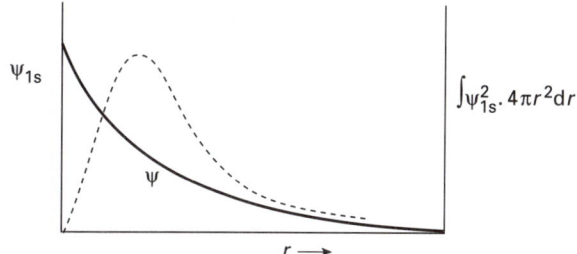

Fig. 3.6 Amplitude of 1s wave function, Ψ, as a function of distance, r, from the nucleus. The probability of finding an electron at a distance, r, is shown by the dashed curve and depends an integrating Ψ^2 over increasing spherical elements of radius, r. See aside for orbitals with planar nodes.

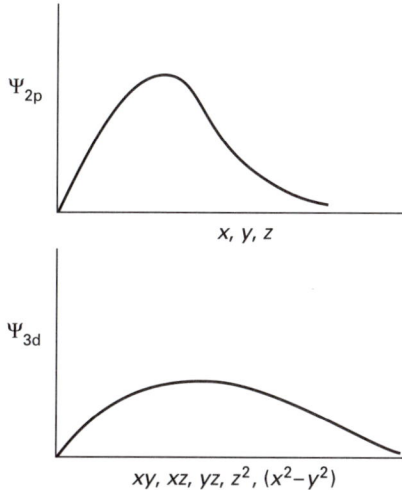

Orbital extension for
$l = 1$ and $l = 2$

parable to the expected wavelength of the wave associated with the electron motion) and exhibited interference patterns. (More recently, other experiments have shown that this behaviour is quite general and that even beams of atoms or molecules can be diffracted by appropriate gratings).

How can we look upon the quite extraordinary idea that a particle such as an electron of a determined inertial mass and charge can act as a wave? This is an impossible task for the human imagination based on sense experience of large objects. Bulk matter, as known to us, is such that a mass occupies a volume. That matter, when explored at the microscopic atomic level, is best described as occupation of volume by a waveform associated with fast motions of classical masses with charge (electrons), which themselves are far too small to fill the volume we ascribe to them, will always be puzzling. Moreover, all the masses and charges are in discontinuous units. The waves are not like waves on a string which are limited by the ends of the string where the amplitude of the waves is zero (so-called *nodes*). Since the wave of the electron is associated with a charge/charge (electron/ nucleus) interaction when it is trapped in an atom, the wave (the electron) is confined not by nodes at the ends of a string (the interaction is nowhere nil except at infinite distance so that an atom has no 'edge') but by the electrostatic force holding the electron and forcing it to be overwhelmingly near the nuclear positive charge (see Fig. 3.6). One node of all the waves is effectively at infinity. It is these waves that now occupy space. How can we represent such waves while we still maintain the classical idea of an associated charge and momentum (mv) of the electron? Schrödinger, an Austrian scientist, was able to derive an equation of a wave which includes its wavelength (or frequency) and its mass and velocity as a particle. Now, wave equations of this kind have solutions characterised by spherical nodes and/or nodal planes, which are automatically 'quantised' since only certain wavelengths can occupy space, i.e. are allowed, much as this is true of waves in musical instruments, e.g. on a drum surface (Fig. 3.7). The possible waves are solutions of wave equations. It can be shown (see further reading and textbooks on physical chemistry) that the characteristics of the waves can be related to the same set of 'quantum' numbers as we have described for electron motions in orbits, i.e. n, l, m_l and m_s. Due to this analogy, the solutions of the wave equation are called *orbitals*. Thus

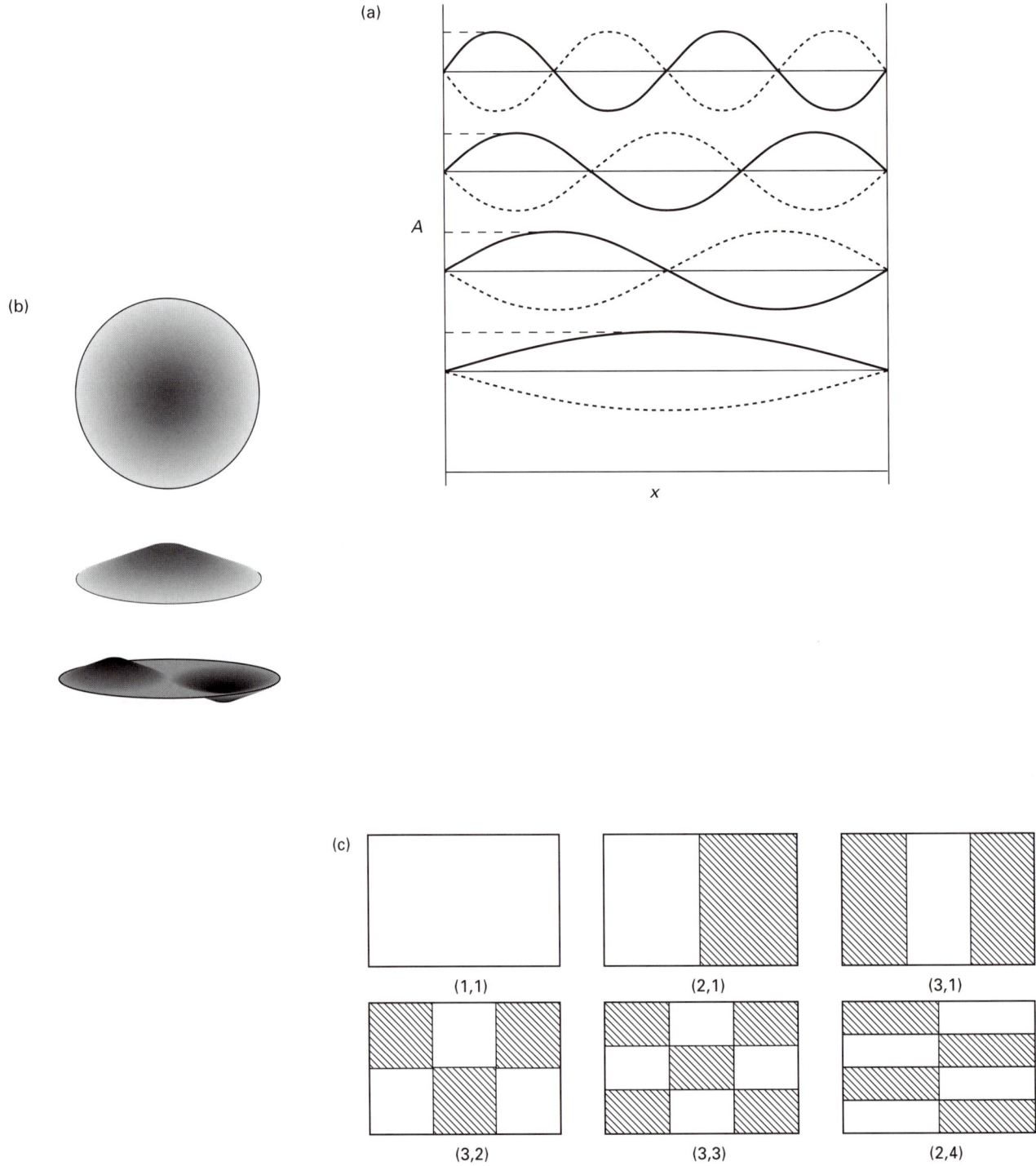

Fig. 3.7 Plots of amplitudes, *A*, and wavelengths: (a) on a simple string; (b) on a flat drum surface, all with terminal nodes; (c) more complex wave modes on a rectangular plate in which the unshaded parts are displaced oppositely, i.e. have opposite phases to the shaded and separated by nodal lines. The numbers give the number of half-waves in the x and y planes, compare the labels *s, p. d, f* for atomic orbitals in three dimensions. In (a) the two waves of opposite phase are shown.

s orbitals correspond to spherical waves with spherical nodes only (*n*) while p, d and f orbitals have increasingly complex radial patterns associated with planar nodes increasing in number from 1 to 3, respectively. Therefore, the periodic nature of the filling of magic number shells 2, 8, 18, 32 is found again. However, the fact that waves, by their very nature, *penetrate* one another radially, i.e. outer waves still approach the nucleus (see Fig. 3.9) also allowed an explanation of the order of the filling of the s, p, d and f sub-shells (Fig. 3.5) so that the periodic table could in effect be 'understood' in terms of a wave/mass/charge character for the electron. Note again that the observations are related *quantitatively* to the theory.

Despite being abstract and contrary to our senses,* this wave-mechanical model has many advantages over the classical or earlier quantum models since all physicochemical anomalies disappear and other problems can be tackled in more satisfactory terms, namely chemical bonding, stereo-chemistry and, generally, all aspects related to the motion of electrons in molecules. Generally speaking, *quantum mechanics* replaces *Newtonian mechanics* when we have to describe systems of very small units, but not for systems of bulk objects.

We shall not use extensively any of this wave-mechanical or quantum theory of the atom in this book and the reader may well ask why did we introduce it. One reason is simply to lead on to more advanced treatises on atomic properties (see further reading). Much more importantly we have now reached the conclusion that within the limitations of our com-prehension, mathematical modelling has established the periodic table at a *theoretical* level, while the last century established it on an empirical or practical level. Putting the two together we assert with complete confidence that the fundamental units of any material at a chemical or biological level are the atoms of the periodic table and their deficiency or excess of elec-trons, i.e. charge. Moreover, all properties of all materials are explicable in terms of the properties of atoms and charge, related to their electron distri-bution in shells, even though some may be co-operative properties. *There is no possibility of introducing new fundamental units or variables into the*

---

*Schrödinger's wave equation (in one dimension, *x*)

$$-\frac{h^2}{8\pi^2 m} \cdot \frac{d^2\psi}{dx^2} + V\psi = E\psi$$

where *h* is Planck's constant, states simply that

Directed kinetic energy + Potential energy = Total energy

(see Section 2.3.2).

The fact that it contains a wave function, Ψ, means that there are only certain allowed values of these energies and of *E*. It is an holistic atomic concept in that the uncertainty prin-ciple associated with it does not allow the simultaneous description of the exact position in space of an electron and its momentum. Also notice that the stable and stationary states of the electrons in atoms are not those given by consideration of bulk fields (Section 2.2); they are dynamic conditions, with the electron apparently 'moving' in a three-dimensional wave through and round the nucleus! Notice that the wave equation is a relationship between all four fundamental variables, mass, charge, position in space and change of position in space with time. Finally, this equation gives an explanation of the relationships of matter to electromagnetic radiation, (see Section 2.6).

discussion. By accepting that the theory* is correct, all the energetics and shapes of chemical combinations can be explained in principle. We are constrained, then, to understand all materials in chemical terms by three *variables; atomic* and *charge composition*, and *interaction energies internal to them* individually and when in combination (which we can now appreciate though do not comprehend). Obviously, the answer is related to the disposition of the units in space (averaged over time), the third fundamental variable of Table 2.3. We turn next, therefore, to a quantitative description of the properties of atoms so that the restricted variable energies of atoms in different states, including charged states, and in components can be fully appreciated. Notice that already we have a simple explanation of the maximum number of atoms that can combine with a central one in that it corresponds to the formation of complete shells of electrons in all the atoms, e.g. LiH, BeH_2, BH_3, CH_4, NH_3, OH_2, FH, where H, Li, Be and B have a 2-electron shell and C, N, O and F have an 8-electron shell (see Fig. 1.6).

3.4   Properties of atoms

Before we describe the ordered combination of atoms of different elements in any detail (Chapter 4) it is convenient to look at some physical properties of atoms themselves.

We ask first what is the spread of electrons away from nuclei, i.e. what is the effective size of atoms in combined states, and then what is the energy with which electrons are held by nuclei. Both will determine the way in which the atoms can form compounds.

3.4.1 Sizes of atoms

Using a variety of experimental data we now know that atoms in combination apparently have radii of around 10^{-8} cm [1 angstrom (Å) or 0.1 nanometer (nm)]. To illustrate just one (indirect) approach to the occupation of space by atoms and its variation in the periodic table, Lothar Meyer's atomic volume curve, of atoms largely in molecules or solids (Fig. 3.8), gives a good impression. The trends follow the periodic shell structure of atoms. (In passing, note that many other physical properties of *elementary substances* have been found to follow regular trends analogous to this: for example, density, melting points, boiling points, etc.) These atomic volumes, however, are obtained by dividing atomic weights by densities of the corresponding elementary substances, hence they do not refer to atoms in isolation. In effect, there is a fundamental problem concerning the meaning of the atomic radius since, as we have stated, the electrons are not strictly held close to nuclei but die away (expo-

---

*In fact, the theory does not allow one to distinguish the wave and particle character of electrons. They are jointly fundamental features of material at this level of study and our wish to relate their behaviour to that of macroscopic objects open to our perceptions is at fault. We are only able to make visual models of 'reality' based on our experience. Instrument measurements have altered the perception of what this 'reality' may be, but it cannot be given a truly representative image accessible to our senses.

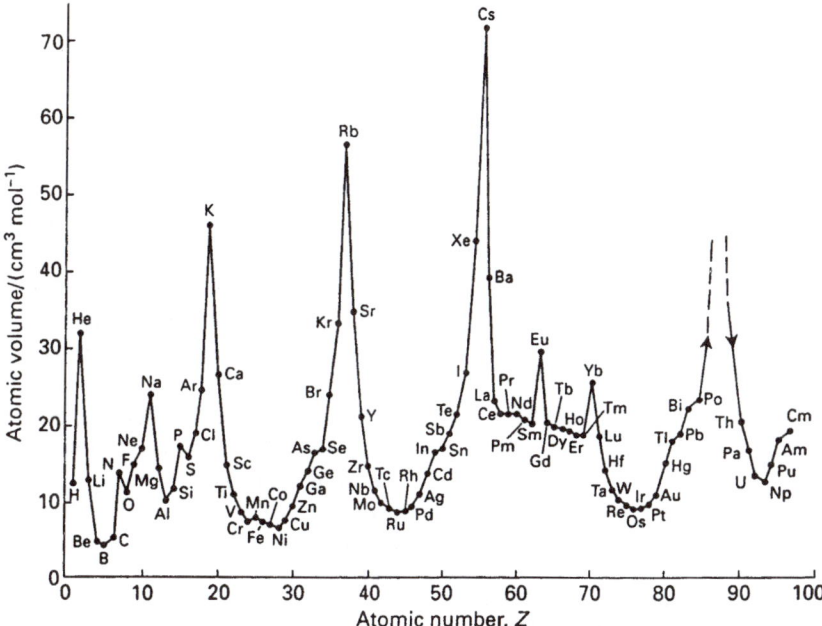

Fig. 3.8 The variation of atomic volume with atomic number.

nentially) with the distance from them in a probabilistic distribution cloud to infinity (Fig. 3.9);* consequently, isolated atoms have no defined boundaries. We could, of course, define theoretical radii corresponding to the distances to

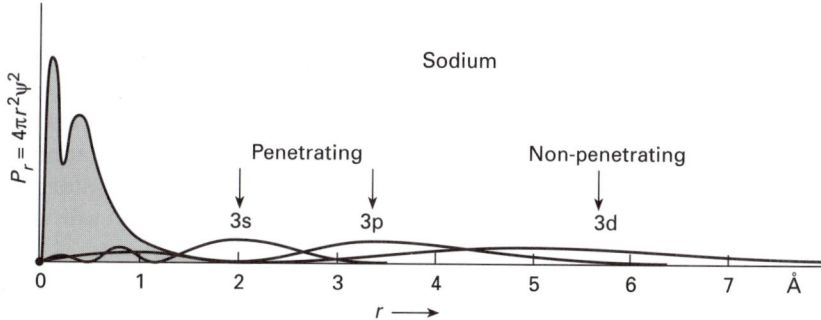

Fig. 3.9 Radial distribution functions for the 3s, 3p and 3d electrons of sodium. The shaded area is the electron density due to core electrons in the 1s, 2s and 2p orbitals. Note again the penetration.

---

*The reader may be worried by the observation that all electron waves in atoms extend to infinity. This problem was resolved by the realisation that it is the *square* of the amplitude of a wave that relates to the probability of the presence of an electron (see Fig. 3.13). An electron is then described as a particular charge and mass density ranging outwards from the nucleus but falling rapidly so that >90% of it is say within 2 Å of the nucleus, despite the fact that the amplitude is not zero until infinity (see Fig. 3.9). This density is used to define an atom in space for the vast majority of cases, but now and then, e.g. in electron transfer theories, a more extensive wave function is used. Today it is argued that it is incorrect to see 'particles' separately as masses or waves according to the experiment being performed; it is now said that the dual wave/particle character of matter is characteristic of it (see footnote to Section 3.3).

Table 3.2 Some metallic radii (Å)

Li (1.57)	Mg (1.60)	Fe (1.26)	Al (1.43)
Na (1.91)	Ca (1.97)	Cu (1.28)	Ga (1.53)
K (2.35)	Sr (2.15)	Zn (1.37)	Pb (1.75)

The values refer to co-ordination number 12.

Table 3.3 Some covalent radii (Å)

–H (0.3)	–O (0.66)
–C (0.77)	=O (0.62)
=C (0.67)	–S (1.04)
≡C (0.60)	=S (0.94)
–N (0.70)	Cl (0.90)
=N (0.63)	Br (1.11)
≡N (0.55)	I (1.28)

The values refer to common geometries.

Table 3.4 Some ionic radii (Å)

O^{2-} (1.45)	Na^+ (0.98)	Fe^{2+} (0.76)
S^{2-} (1.90)	K^+ (1.33)	Fe^{3+} (0.64)
F^- (1.33)	Mg^{2+} (0.65)	Cu^+ (0.95)
Cl^- (1.81)	Ca^{2+} (0.94)	Cu^{2+} (0.65)
Br^- (1.96)	Sr^{2+} (1.10)	Al^{3+} (0.45)

The values refer to coordination number 6.

Table 3.5 Some van der Waals radii (Å)

C (1.70)	H (1.20)
N (1.5)	F (1.35)
P (1.9)	Cl (1.80)
O (1.40)	Br (1.95)
S (1.85)	I (2.15)

maximum electron density, but what would be the practical interest of such a definition?

To make a long story short, rather than assigning definite theoretical radii to atoms we determine, from experimental data, several *operational* types of radii derived, in fact, from the ways in which atoms pack together and which are, therefore, dependent on packing (see Section 2.3.1 and Chapter 4). Thus, by treating atoms as hard spheres, we can obtain operational *metallic* radii for atoms in metals with 8 or 12 near-neighbours, or diatomic, tetrahedral or octahedral operational *covalent** radii, according to geometry, within molecular structures such as H_2, CH_4 or SF_6, or operational *ionic* radii for ions (charged atoms) in compounds such as $Na^+ Cl^-$, etc. Tables 3.2, 3.3 and 3.4 provide values of such radii for a set of selected elements. Clearly, when we put different elements together in combination we are now able to think in terms of fitting spheres of different sizes together, but while these sizes result from classical ideas of packing and forces for metallic and ionic interactions, in packed assemblies we must consider repulsion at short distances (Sections 2.3.1 and 3.4.2).

This idea can be extended to other situations, for example to the distance between equivalent atoms in near-neighbouring molecules, not the same molecule. Take, for example, solid iodine, I_2, where two different I–I distances are found, one in each individual molecule (corresponding to twice the covalent radius) and the other between pairs of neighbouring molecules (corresponding to twice the so-called *van der Waals* radius) (see Fig 3.10). Obviously, the van der Waals radius (Table 3.5) is larger than the covalent radius. The implication is that attraction between molecules is much weaker than that of atoms inside molecules so that optimal energy of binding is found at larger distances.

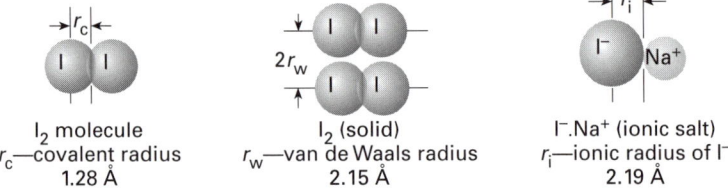

I_2 molecule
r_c—covalent radius
1.28 Å

I_2 (solid)
r_w—van de Waals radius
2.15 Å

$I^-.Na^+$ (ionic salt)
r_i—ionic radius of I^-
2.19 Å

Fig. 3.10 Comparison of the covalent and van der Waals radii of iodine, I, with the ionic radius of iodide, I^-.

*See Fig. 4.5 for the molecular structures in which bonding is due to sharing of pairs of electrons. Note that packing within molecules such H_2O is very different from packing in salts such as Na_2O.

One way of looking at these radii, therefore, is to say that they are the distances at which the longer range attraction of outer electrons for several nuclei (electrostatic forces) comes into balance with the repulsion of, especially, inner core electrons by the forces that generate the Pauli exclusion principle. Of course, the resultant filling of space by the electrons must simultaneously fit limitations due to their wave-like character. To understand the strength, not just the dimensions (structures) of these interactions, we must analyse the energies of electrons in different atoms and ions.

3.4.2 The exclusion principle: a new repulsion

We can offer the following explanation as to why this treatment of atoms, as if they were hard spheres, works in practice. We have stated that only two electrons can have the same first three quantum numbers, n, l, m_l, where the spin quantum number is different ($+\frac{1}{2}$ or $-\frac{1}{2}$). This means, of course, that parts of space (characterised not just by their co-ordinates, x,y,z but by the square of amplitude of a wave function associated with these co-ordinates) are occupied *maximally* by two electrons. They are held there by electrostatic attraction to nuclei, but what prevents other electrons from entering the same space? We do not know! We say that it is due to the so-called Pauli exclusion principle, a strange, only partly electrostatic repulsion which is found to fall off very rapidly with distance. For example, we know that two helium atoms, the simplest atoms with a filled outer shell (the first inert gas), can be brought together until they are some 3.5 Å apart. To bring them to 3.0 Å apart, which means that the outer electrons would be forced to occupy the same parts of space, requires enormous

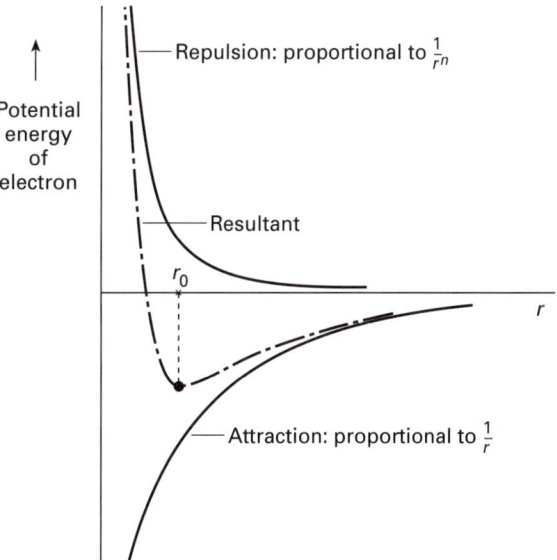

Fig. 3.11 The potential energy of an electron interacting with an atom with the electron affinity given at the equilibrium distance, r_0, between attractive and repulsive interactions (see Fig. 2.6). The interactions here are non-classical.

Anion Radii Å

SO_4^{2-}	1.51	HPO_4^{2-}	1.51
MoO_4^{2-}	1.76	WO_4^{2-}	1.77

N.B. Note ease of reduction

SO_4^{2-} > MoO_4^{2-} > WO_4^{2-}

compression energy and only happens by exciting electrons out of their lowest inert gas states! To the forces controlling a balanced distance with attraction we add repulsion forces and we give them a distance dependence of say $\dfrac{1}{r^{10}}$, (Fig. 3.11; see also Fig. 2.6). In other words there is almost a wall preventing approach at ~3.5 Å. This means that all atoms and ions (except H^+, which has no electrons) effectively have a radius between say 0.5 and 3.5 Å. It is this short-range repulsion which generates much selectivity in the size and packing of atoms. It decides in part the size of molecules and ions (see aside) and limits the binding energies, as much as the electrostatic attraction does, while the orbital character of waves gives the directional part of the whole, i.e. it gives shape (see Section 3.4.6). These principles apply throughout our considerations of both animate (living) and inanimate (mineral) substances.

3.4.3 Ionisation energies

We have shown how the electrons are built into atoms to give rise to the periodic table. We can now consider quantitatively the potential energy of electrons in atoms. For the last electron added this energy is equivalent to that necessary to remove that electron from the atom considered in its most stable (or 'ground' state). It is called the *ionisation energy*, usually expressed in electron volts (eV) but frequently replaced by the ionisation potential (IP_1), expressed in volts, which is numerically equivalent. (One electron volt/atom = 23.06 kilocalories/gram atom.) Naturally there will be successive ionisation potentials, but the second (IP_2), third (IP_3), etc. refer to the removal of electrons from progressively more positively charged *ions*, not from the neutral atom. As might be expected (and this is an example of the periodicity of properties), the experimental ionisation potentials vary regularly with the atomic number of elements (see Fig. 3.12) for IP_1, and their successive values will increase until some electron configuration of particular stability is reached, for example ns^2 or $ns^2 np^6$, cor-

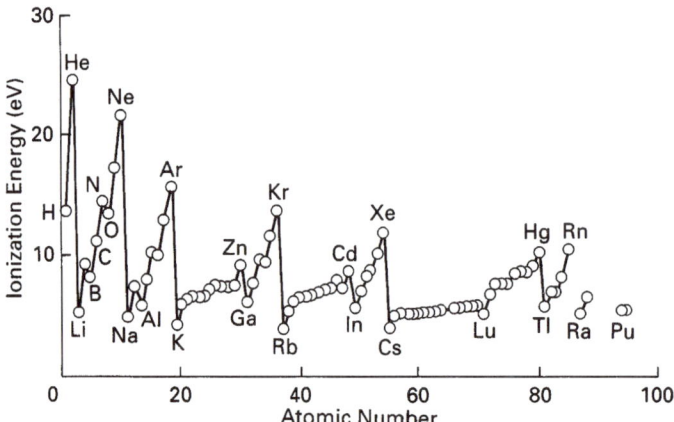

Fig. 3.12 First ionisation energies of the elements as a function of atomic number.

responding to those of the noble gases helium to radon, and then decrease sharply for the next electron added, the pattern being repeated for successive electron shells. Just as the number of electrons in an atom is important for the combining ratios of associated atoms in compounds, so the energies with which electrons are held is one important property when considering the strength of atom association of binding. Thus, when we look at binding strengths of hydrogen to all other elements these ionisation potentials become clearly important.

The energies of electrons in atoms given above are from experiment. It has proved possible to calculate the values very accurately, especially for lighter atoms, using a three-dimensional wave equation (see footnote to Section 3.3). In this way the electrostatic origin of the energies has been finally demonstrated. Moreover, it is possible to follow up this treatment with a calculation of the binding energy of electrons by two identical nuclei, say in H_2 or N_2, and so obtain proof of the origin of chemical bonds and the restrictions on their number and direction. In essence this treatment provides confidence that homonuclear chemical binding energies are understood to the extent that we understand the wave equation.

3.4.4 Electron affinity energies, electronegativity, atomic combinations and components

Now, as well as being able to lose electrons to give a *cation*, an atom can also gain electrons to become a negatively charged ion, which we call an *anion*. We have seen this in Na^+Cl^-. The energy released in the capture of an electron is called the *electron affinity* (Table 3.6). Just as with the ionisation potential, the electron affinity contributes to binding and it also follows in value the periodicity of the periodic table. The values can again be understood as deriving from electrostatic interactions calculable by the wave equation (Section 3.3). These dual properties of atoms can be combined into a single property. The average sum of the two is called the *electronegativity* (Table 3.7), i.e.

$$Electronegativity = \tfrac{1}{2} (Ionisation\ energy + Electron\ affinity)$$

which gives a way to assess the relative tendency to hold or give up an electron by an atom. This average can now be used in a semi-quantitative

Table 3.7 Pauling electronegativities (H = 2.1)

Li	1.0	B	2.0	O	3.5
Na	0.9	C	2.5	S	2.5
K	0.8	Si	1.8	F	4.0
Mg	1.2	N	3.0	Cl	3.0
Ca	1.0	P	2.1	Br	2.8
Al	1.5	As	2.0	I	2.2

Table 3.6 Electron affinities of some elements (in eV)

H	−0.75	C	−1.26	F	−3.40
Li	−0.62	N	+0.07	Cl	−3.62
Mg	+0.4	O	−1.46 (+8.75)	Br	−3.37
Ca	+0.3	S	−2.08 (+5.5)	I	−3.06

To convert to $kJ\ mol^{-1}$ multiply by 96.5. The first values for O and S corresponds to the formation of X^- from X and the second values to the formation of X^{2-} from X^-. Negative values correspond to energy released and positive values to energy absorbed.

H
|
$C^{3(\delta\oplus)}$
$^{\delta\ominus}Cl$ $Cl^{\delta\ominus}$
$Cl^{\delta\ominus}$

(a)

(b) $H^{\delta\oplus}$——$Cl^{\delta\ominus}$

$Cl^{\delta\ominus}$
|
$C^{4(\delta\oplus)}$
$^{\delta\ominus}Cl$ $Cl^{\delta\ominus}$
$Cl^{\delta\ominus}$

(c)

Examples of polar chloroform $CHCl_3$, (a) hydrochloric acid HCl, (b) and of non-polar carbon tetrachloride CCl_4. (c) Bond polarity is indicated by $\delta\oplus$ $\delta\ominus$.

way to understand combinations of elements. In effect, when two atoms combine, the difference in electronegativity gives an estimate of the asymmetry of charge distribution in each bond, called the *polarity* of the bonding. Thus, C and H have very similar electronegativity so that C–H bonds are not polar, in contrast with Na^+Cl^- which is highly polar since Na and Cl have very different electronegativities. Furthermore, it will also give a good estimate of the increase in binding energy on passing from bonding in elements X–X and Y–Y to bonding in 2(X–Y). This is why the difference in electronegativity is considered a 'criterion for the nature of chemical bonding' (see further reading). In the case of CH_4 and other 'molecular' compounds (or elementary molecular substances such as H_2, O_2, N_2, etc.) when the difference in electronegativity of two bonded atoms is small (or zero) there will be an approximately equal tendency of both to hold or give up their electrons, and the result is a share of pairs of electrons slightly shifted to the more electronegative of the two. This sharing of pairs of electrons is called *covalent* bonding (see Chapter 4). Where a difference in electronegativity exists there arises a more or less polar bond (see aside). If the difference in electronegativity is large, as for Na and Cl, then full transfer of electrons from one atom to the other can occur, in this case from Na to Cl, to give Na^+ and Cl^- ions, respectively, which tend to bind to one another by electrostatic attraction. This is called *ionic* bonding (see Chapter 4) and also gives a strong interaction.

The number of pairs of electrons shared, i.e. the number of covalent bonds formed, depends largely on the tendency of atoms to achieve stable electronic configurations, 2 or 8 electrons, corresponding to the filling of external electron shells of the inert gases. This is true also for the number of electrons transferred to form ions which, again, tend to achieve the stable outermost electronic configuration of the inert gases, compare Na^+, Cl^-, Ca^{2+}, O^{2-}, etc., all with 8 electrons in their outermost shells. Thus, we begin to understand why particular combinations of elements (components) are more stable than others when they show certain restricted valencies. Stability is due to energetics based on restrictions on ways of occupying space by electrons. It is a consequence of the properties of electrons, described by wave mechanics, which decrees that their potential energies, now in certain hetero- as well as homo-atom combinations, have a relatively high stability (high negative value) at a certain valency. Just because there are large variations in stability of atomic association, especially between different combinations of the same non-metal atoms, e.g. C_nH_m, it is not easy for atoms to undergo exchange or rearrangement. For example, C_2H_6 is persistent in the presence of C_2H_2 although two molecules of C_2H_4 are more stable. It is this inability to rearrange that allows chemical systems to develop using the derived variable of composition in terms of *chemical components*. Figure 2.8 illustrates the discontinuous variation in stability which arises through the nature of the wave functions of the electrons of non-metals. The same wave-mechanical treatment allows the description of their shapes. We then see why at a quantitative level only certain element combinations (components) are observed.

We shall give some examples of the success of this treatment in Section 3.4.6, but a full treatment must be sought in an advanced text.

3.4.5 Bonds

In this book we shall often use the word bond and we shall draw in connecting lines between atoms, such as in H–H, to represent a local binding interaction which we call a (covalent) bond. Weak bonds may be drawn as O...H, as in hydrogen bonds. Double bonds, in e.g. ethylene, are written as C=C, but in benzene, for example, are written C∴C round the ring (see Fig. 3.17 later). Now these drawings are convenient representations of interactions, but bonds located in this way do not exist. Interactions between atoms spread out to infinity in all directions if the wave theory of electrons is correct. Hence bonds represent the fact that the overwhelming percentage of interactions is mostly between the atoms concerned, say 99% within 8 $Å^3$ in H_2O gas at unit pressure (but less in H_2O as a liquid). The value of the description of a local bond is often doubtful when reactions are considered. However, with these caveats in mind, we shall follow the very useful tradition of drawing bonds as lines where and when we consider this appropriate.

3.4.6 Atoms and ions in fields (polarisability and shapes of molecules)

When an atom or ion is brought into a field of other electrostatic charges then, since its own electrons avoid external negative charge and its own

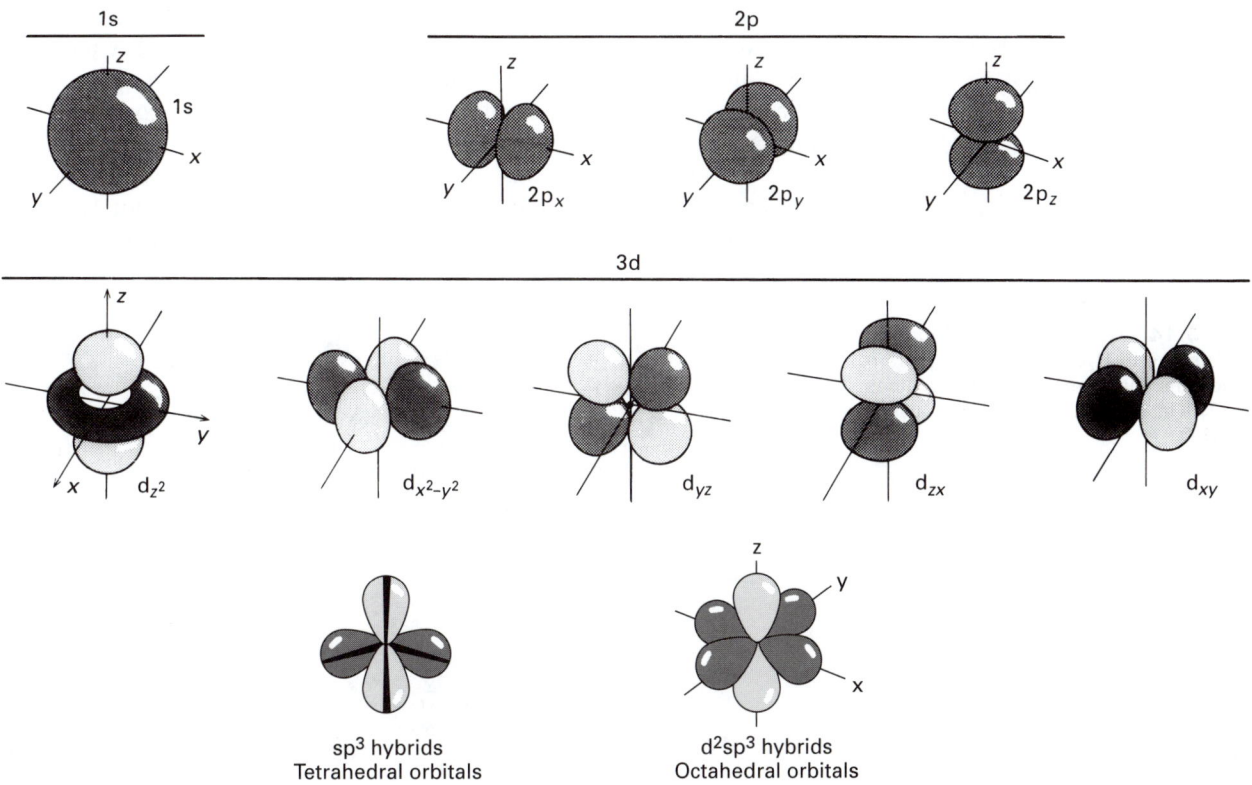

Fig. 3.13 Plots of electron density shapes of 1s, 2p and 3d orbitals referred to co-ordinates in space. Combinations of orbitals such as sp^3 and d^2sp^3 lead, respectively, to tetrahedral or octahedral electron density shapes.

nucleus, through empty orbitals, can attract such charges, the atom or ion becomes polarised. The electron cloud around the atom or ion takes on a shape, but there is a restriction on shape taken up due to the nature and occupation of s, p or d orbitals by the electrons (see Fig. 3.13). The field around one atom or ion can be due to other atoms or ions, of course. When several atoms or ions approach closely to a central one it is necessary also to take into account their mutual repulsion which is dependent on the size of the atoms or ions concerned as well as their own polarisation. Using electrostatic energies and the wave restrictions on electron motion, the shapes of molecules and lattices can be largely explained (see, for example, refs 1–5).

Consideration of two atoms bonded together leads to only one shape. Consideration of three or more atoms permits a much greater variety of shape. Now, if all the atomic balls interact equally then the 'best' shape for three atoms is an equilateral triangle. (This is the basic design on which many metal lattices are built.) If two atoms differ from the third then it might appear that due to polarity in the electron distribution the best arrangement is linear, $\bar{X}-\overset{+}{\overset{+}{Y}}-\bar{X}$; this shape is seen in CO_2. However, the water molecule, H_2O, which would have been expected to be linear, H–O–H, using classical electrostatics, is a bent molecule with an angle of approximately 105°. Here the shape is decided by the electron as well as the hydrogen nuclei distribution around oxygen. As we showed in Section 3.4.4, oxygen in water has a hold on 8 electrons, four of which, in two pairs, are bound to hydrogen, and four of which, in two other pairs, are not bonding (see Fig. 3.14). If we treat the distribution in space of the four pairs of electrons around a centre as equal in an sp^3 combination of four orbitals, then the best shape is clearly a tetrahedron, when repulsion between pairs of electrons is reduced to a minimum. This is very close to the shape of water. Deviations from the theoretical angle, about 109°, can be explained by repulsions of the unequally spread electron clouds (see Fig.

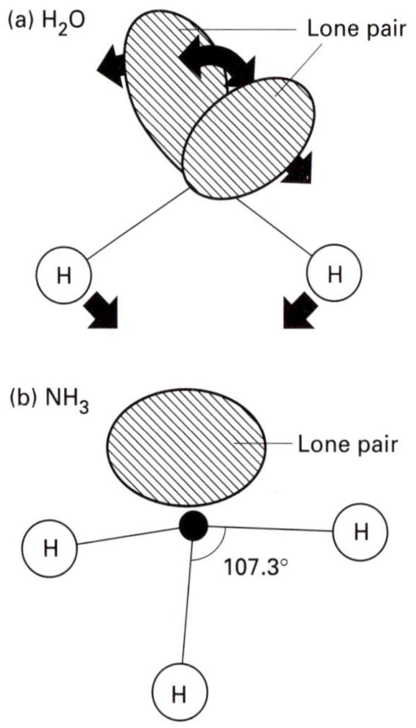

(a) H_2O

Lone pair

(b) NH_3

Lone pair

107.3°

Fig. 3.14 (a) The effects of lone pair repulsions in H_2O. Bonds with pairs of electrons are represented by lines. (b) The structure of ammonia, drawn as for H_2O.

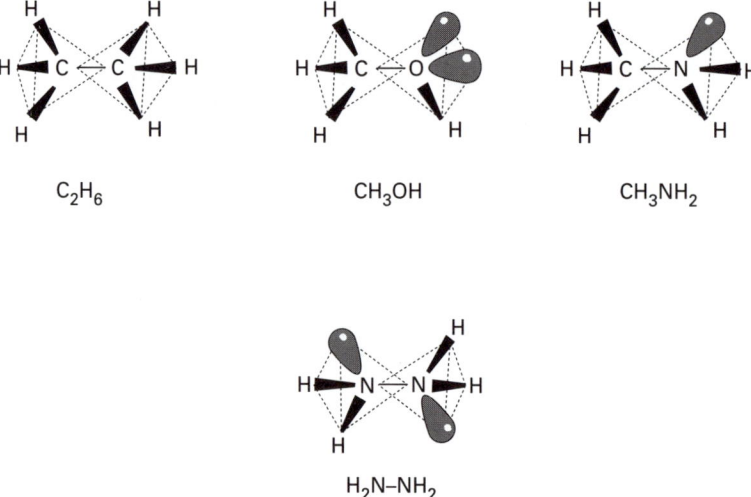

C_2H_6

CH_3OH

CH_3NH_2

H_2N-NH_2

Fig. 3.15 Stereodiagrams for some molecules, showing lone pairs.

3.14) and the analysis applies equally to the pyramidal shape of NH_3 and indirectly (since all the eight electrons in four pairs are bonded to hydrogen) to CH_4. The rule can be extended to two or more joined tetrahedra, as in C_2H_6, $CH_3 \cdot OH$, $\cdot CH_3 \cdot NH_2$, $NH_2 \cdot NH_2$ (Fig. 3.15) and so on. It also applies to other central atoms which build a full group of eight electrons, e.g. SiH_4. This type of singly-bonded structure is very important in the building of many polymers including those of living systems such as proteins.

Moving on to other types of molecules we have seen that oxygen gas is represented by O_2, which requires that it must be formed by having two bonds between O atoms so that each O has a share in four electrons to give it a total of eight. Following this description, the reason for the non-bent shape of CO_2 is that the four electrons originally from carbon are joined in two pairs to each of the two different oxygen atoms. The oxygens are slightly negatively charged, since they have a higher affinity for electrons than carbon, so that they repel and CO_2 is linear because this is the most favoured situation. Clearly the bonds are now double bonds.

Combining H, C and O in the form H_2CO, the shape is planar because this is the best way to reduce repulsion between the eight electrons on the central carbon, while two pairs are bound to one oxygen in a double bond and the other two pairs are bound separately to two H atoms. In a parallel way, shapes of larger and larger organic molecules are easily derived. The rules of occupation of space around central atoms using covalent bonds are now well known and can give rise to a variety of compounds with some very sophisticated properties.

Consider carbon joined by four single bonds to four different atoms, i.e. Cabcd. A tetrahedral shape is obtained, but it has two possibilities in space (Fig. 3.16). The two shapes are mirror images, called *laevo* and *dextro* (from left- and right-handed), which have some different optical properties and are called optical isomers. (Hands are mirror images of one another.)

Another example of further sophistication is found in the structure of benzene where the carbon atoms of C_6H_6 must form double bonds as well as single bonds for all of them to have a covalent share in eight electrons. Experiment shows that all six C–C bonds are equal, with a partial double bond character. Double bonded structures, such as ethylene and benzene, can give rise to geometric isomers. Figure 3.17 exemplifies the resulting structures.

We shall go on to consider the shapes of ions in complexes (see Section 6.11) before we discuss the way molecules or ions pack together in ordered systems. We draw attention here to the fact that the above considerations apply universally in all living and non-living substances.

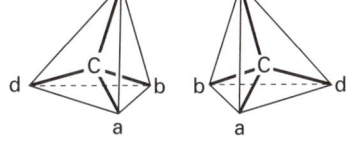

Fig. 3.16 In 1873, J. H. Van't Hoff explained optical activity for the two forms of lactic acid in terms of four groups attached tetrahedrally to a central carbon atom (a, OH; b, H; c, CH_3; d, HO_2C). These structures, based on asymmetric carbon, are mirror images of each other.

3.4.7 Oxidation states

As mentioned in Section 3.4.4, when an atom loses electrons it becomes a cation and when it gains electrons it becomes an anion. As stated, the number of electrons lost or gained depends on the distance to the nearest stable external shell of electrons, i.e. to the nearest inert gas electronic structure. (There are some exceptions, however, still based on the tendency to achieve stable internal electronic configurations. For example, NO and the anion $[SiF_6]^{2-}$ do not meet this criterion.)

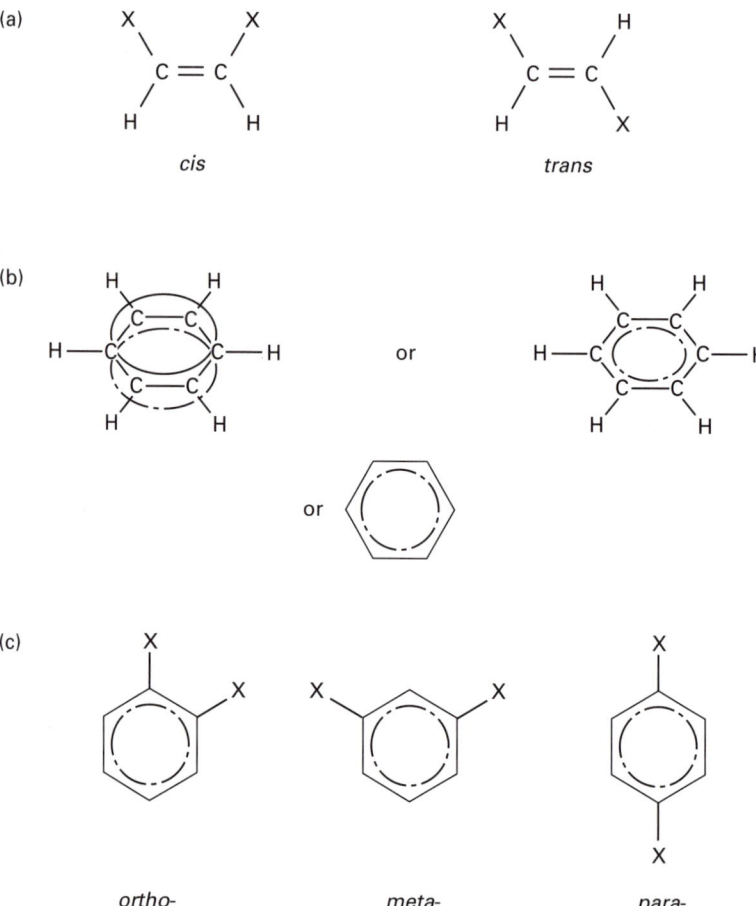

Fig. 3.17 (a) The *cis* and *trans* geometric isomers of ethylene. (b) Structure of benzene, showing that each carbon atom forms three single bonds (1 with hydrogen and 2 with neighbouring carbon atoms). The remaining electrons (6), one in each carbon atom, form delocalised orbitals over the entire molecule. (c) The geometric isomers of di-substituted benzene.

In many cases only a limited number of ions of each type of element are found, for example just Na^+ for sodium, just Ca^{2+} for calcium, both Cu^+ and Cu^{2+} for copper, both Fe^{2+} and Fe^{3+} for iron, just Cl^- for chlorine, usually O^{2-} for oxygen, etc. It is said that these ions have *real* oxidation states +I, +II, +III, −I −II, etc. (designated by roman numerals) which correspond to their charges. For the free elements, in say metallic Ca and Fe or gaseous Cl_2 and O_2, the oxidation state is zero by convention. This is a useful definition, not only to refer to the different ions but also to balance charges in formulae of ionic compounds which must exist as neutral species, e.g. NaCl (Na^+Cl^-), $CaCl_2$ (Ca^{2+} $2Cl^-$), etc.

But what about molecular compounds such as H_2O, or more complex ions such as SO_4^{2-}? Is the concept of oxidation state still useful? The answer is yes, although in this case it is not a *real* oxidation state but a *formal* oxidation state that we use. This is a convenient concept based on the fact that when two different atoms are bonded to each other the most electronegative takes a greater share of the electrons involved in the bond.

The definition uses a parallel with circumstances in which full transfer of the shared electrons occurs (which here, of course, is not true) so that the most electronegative atom is assigned a negative oxidation state (since it would gain one electron from each pair shared) and the less electronegative atom is assigned a positive oxidation state (since it would give up an electron from each pair shared). For example, in H_2O, H–O–H, the oxygen atom is said to have a formal oxidation state –II and each hydrogen atom is said to have a formal oxidation state +I.

Now, H_2O is, of course, neutral, and the oxidation states are balanced. If we take SO_4^{2-} the situation is a little more difficult to describe, but we know that oxygen is more electronegative than sulphur (see Table 3.7) and, generally, has the oxidation state –II. Since the overall charge of the species considered is –2, one easily concludes that the formal oxidation state of sulphur must be +VI in this complex ion (or +IV in SO_2, or –II in H_2S, since the oxidation state of hydrogen is generally +I).

Using this very simple set of rules, formal oxidation states are easily assigned in most situations for compounds of H, O, S or halogens (see Table 3.8) and this provides a useful conceptual instrument to predict formulae of compounds and complex species, to balance equations and to study a special group of reactions in which there is transfer of electrons or of atoms such as H, O and halogens between species (called oxidation–reduction reactions or *redox* reactions for short, see Chapter 6).

Table 3.8 Oxidation states

Nitrogen compounds				Sulphur compounds			Carbon compounds	
Substance	**Oxidation states**			**Substance**	**Oxidation states**		**Substance**	**Oxidation states**
NH_4^+	N = –III,	H = +I		H_2S	S = –II,	H = +I	HCO_3^-	C = +IV
N_2	N = 0			$S_8(s)$	S = 0		HCOOH	C = +II
NO_2^-	N = +III,	O = –II		SO_3^{2-}	S = +IV,	O = –II	$C_6H_{12}O_6$	C = 0
NO_3^-	N = +V,	O = –II		SO_4^{2-}	S = +VI,	O = –II	CH_3OH	C = –II
HCN	N = –III,	C = +II,	H = +I	$S_2O_3^{2-}$	S = +II,	O = –II	CH_4	C = –IV
SCN^-	S = –I,	C = +III,	N = –III					

Oxidation states (real or formal) are, obviously, properties of the atoms of the elements, depending primarily on their electronegativity, thus on the energetics of the electronic structure.

3.5 The energy states of atoms and atomic systems

We now see that atoms and sets of atoms can exist in more than one energy state or, in terms of energy, in more than one potential energy condition of its electrons. As discussed in Chapter 2, we may treat the atoms and charges as the fundamental units and the energy corresponding to their different arrangements as a variable. There is then a choice as to how to proceed. The way chosen in most textbooks today is to describe the combinational variation of atoms in terms of the wave functions of electrons, so called molecular orbitals or combinations of atomic orbitals. The analysis looks at individual molecules or structures. There are many advantages of such a treatment in that shape or structure (symmetry) is discussed readily, but one disadvantage is that mathematical insight is required. An alternative is to follow the purely empirical treatment outlined in Chapter 2, discussing energy content of atoms or atomic assemblies as a numerically discontinuous variable, in terms of pair-wise electrostatic interactions of atoms (nuclei and electrons). Within large numbers of units it becomes a *statistical* property of bulk chemical systems, now a continuous function, and is treated as a variable of a bulk system (see Chapter 5). We shall follow this procedure, noting that it generates the observed chemicals in nature or those which we synthesize as a consequence of permitted energy minima in plots of energy content versus percentage composition (Fig. 2.8). Thus, in Chapter 4 on structure, we shall appear to be using an almost classical approach to the binding energy of atoms because we shall have hidden the fact, for the most part, that we have allowed for the limitations of the wave nature of the electron upon acceptable energies in structures. Such a treatment fails to describe spectroscopic or magnetic properties, but fortunately these properties are of little consequence in this book. We are mainly concerned with the units and variables that lead to the bulk objects which we observe (see Chapter 5).

3.6   Summary

This chapter has given an historical introduction to our present-day, detailed view of the nature of the fundamental units of construction of all chemical and biological materials, the atoms of the periodic table and their charges, extended from the outline in Chapter I. It is not possible to see how this position can change. Logic demands that using these units and applying the *fundamental variables* composition and space (Table 3.9) which we have analysed in part in Chapter 2 but which we shall extend as we develop our discussion, we should be able to describe the combinational properties of atoms in everything static around us. In this chapter we have

Table 3.9 Fundamental units and variables in this chapter

Units*	Variables	Restrictions
Atomic nuclei and electrons	Composition based on atoms and deficiency or excess of electrons (charges)	Nuclear stability: quantum conditions
	Spatial arrangements of units	Electron/nuclear quantum conditions

* Later in this book we use atoms of elements and charge as the units of composition.

analysed the way potential and kinetic energies, derived variables, associated with the wave-like as well as the charged particle character of electrons, affect the properties of atoms themselves. A particular point which we have stressed is that the sum of the potential energy content and the constant directional kinetic energy content (dynamic stationary state) of the interaction of electrons with nuclei is a discontinuous variable which occurs only in allowed, discrete quantised values. It is this stepping of energy content which confines the elements to periods of the periodic table. The stepping applies to ground, stable, states and to excited, stationary, states of atoms and ions. The energy variable is here largely measured as ionisation and electron capture energy, and the effects of fields (due to atomic neighbours). Now, these sets of states also restrict allowed combined atom ratios to a few values (e.g. CO and CO_2), hence giving rise to landscape stability diagrams as in Fig. 2.8. Furthermore, they control the shapes of these associations, e.g. as molecules. Quantised states of electrons in atoms and atoms in molecules must, therefore, be very differently treated from the analysis of continuously variable bulk fields (see Chapter 6). This has led us to outline an explanation of why components, fixed associations of non-exchangeable atoms in combination and with a *particular* fixed potential energy and charge, if any, are the useful operational units of composition. Therefore, as stated in Chapter 2, the derived composition variable can be defined by the percentage of components in a mixture. In the next chapter we shall see how these energy variables lead to particular *structured (ordered) atomic and ionic associations* for the different atomic combinations. However, we must also be aware that some new properties arise in bulk assemblies of atomic, ionic or molecular structures, and again in large polymeric molecules, which do not follow from a treatment of single or pairs of atoms. We shall have to recognise co-operative phenomena, which are very different in solids, liquids and gases. Additionally, Chapter 5 will show that the description of the total energy of a system of atoms, and ions, besides requiring consideration of features relating to individual structures and pair-wise interactions (potential energies), must be analysed on a statistical probability basis, giving additional derived variables. Examples are the volume (pressure) and temperature constraints imposed on a material. After Chapter 6, in which we describe systems in solution, a subsequent chapter (7) introduces further derived variables such as compartments, external fields and the size of a system. We will then have much of the required knowledge to proceed to the study of the properties of all kinds of materials which are generated from combinations of atoms and ions (the components) of everything static we see around us on Earth. A final step involves us with the further variable *flow*, of both material and energy (Chapter 8), which is required to describe change with time, both in inanimate species and in particular biological organisms (Chapter 9 onwards). It also allows us to examine mankind's approach to chemicals (Chapter 15).

Finally, we observe that we have no reason to believe that there are any underlying variables required either from Newtonian or quantum mechanics other than *composition* (*atoms* and *charges*) and the distribution of units in *space* and through *time*, as outlined in this chapter. This, we assert, is true for all chemical and biological systems. What we observe directly or with the help of instruments is then a consequence of the inter-relationship of the variables, which can be very complex. To handle this situation we need to use operational derived variables such as component composition, energy, temperature, pressure and later flux, which will be developed as we proceed.

4

Order and stability of atom and component associations

The same letters variously selected and combined
Signify heaven, earth, sea, rivers, sun
Most having some letters in common.
But the different subjects are distinguished
By the arrangement of letters to form the words.

Lucretius, *De rerum natura* (c. 57 BC)

4.1 Introduction

When we look around us we observe different substances and objects, many of which are effectively static or are in dynamic stationary states in the sense that they do not translocate, i.e. flow. The analysis of all these substances and objects at a fundamental level shows that they are ultimately formed from atoms of the 90 naturally occurring chemical elements on Earth (see Chapters 1 and 3). Further vigorous reductive examination shows that, except for the rare gas elements, each atom is bound to at least one other, forming *ordered* atomic patterns which lie

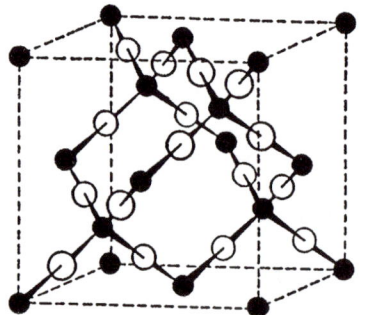

Fig. 4.1 The structure of SiO$_2$ (β-crysto-balite), a giant network.

behind the bulk appearance. When the atoms are bound to a relatively small number of others (but that 'small number' may well go up to thousands) we call the associated unit a *molecule* (see Fig. 3.15). When atom association is effectively of infinite extent we refer to the whole structure as a giant *network* (if covalent bonds are involved) (Fig. 4.1) or as a *lattice* in the case of ionic substances or metals (see Figs 4.2 and 4.6). The description of gases as molecular substances is obvious, but the liquid or solid state may be either molecular, and in the case of a solid structure is repetitive in space, or, alternatively, it may consist of a continuous atomic network or an ionic lattice* in which no identifiable molecules can be found. Thus, larger and larger constructs can be built from atoms, often brought together in molecules, or in networks or lattices, which are all *ordered* arrays. These structures extend to organic materials such as blocks of wood, minerals such as marble and so on. In Chapters 2 and 3 we have given the general theoretical and experimental background to the present ideas relative to the nature of the *internal* 'chemical' interactions, i.e. the binding forces and corresponding potential energies that give rise to these structures. In this chapter, we wish to give an outline description and an analysis of the causes and strengths of the *selective structural order* of these internal chemical interactions, which clearly lead to a kind of *classification* or *speciation* amongst chemicals additional to their composition, and, therefore, amongst the objects we see around us. When the nature of structures is well defined we shall describe briefly some consequential properties. This will lead us, in Chapter 5, to describe other variables that affect properties related to the internal dynamics of non-flowing systems. The discussion applies equally to chemicals of non-living and living systems, but the latter, which have non-repetitive ordering, also have additional features, which we discuss in Chapter 8 under the heading of organisation. In organisations there is movement (flow) in addition to static structures.

In previous chapters we have identified two useful variables which can be applied generally to a system of units, here the atomic elements (atomic masses and charges), to generate ordered matter of a given stability. The first is element composition and the second the potential energy arising from any combination of units in space. It is the fixing of these two variables which, from simple atoms (or ions), allows either stable or stationary state structures of atomic (ionic) combinations to arise and remain time independent, and therefore gives rise to shapes of molecules or, in lattices, to particular site geometries (site symmetries). As stated before, these two variables can be combined into a derived operational concept—that of components—corresponding to chemical species (now units) which on mixing do not exchange atoms or charge under the conditions existing there and then. Components, therefore, correspond to particular arrangements of atomic masses and charges, and our main interest in this chapter concerns the problem of what is the origin of the *particular* spatial patterns we observe, both in the components themselves and in their (co-operative) associations.

---

*We distinguish here atomic networks, such as those of diamond (a form of carbon) or silica (SiO$_2$) in which the atoms are bound by covalent bonds, from lattice structures and packed ions, such as Na$^+$·Cl$^-$ (common salt) or packed atoms in metals and alloys (see later).

4.2   Element combinations: qualitative analysis of binding energies and structures

There are two overriding factors which control the ways in which elements as atoms come together in a preferred ordered manner in the absence of motion: (1) the strength of bonding, i.e. the extent the potential energy becomes more negative as the units (atomic elements) combine in simple small molecules to give low molecular weight compounds; and (2) the co-operative modes of interaction between these molecules (or ions, or atoms themselves), related to the extra strength with which they attract one another, and sometimes rearrange, in a condensed, continuous solid state. [We assume that the repulsive barriers (see Sections 2.3.1 and 3.4.2) have been included, limiting the potential energy change at short distances.] The simplest of these two factors is the selective strength with which two atomic elements combine in a diatomic molecule, which is described by their binding (bond) energy. In more complicated polyatomic molecules an average bond energy between two atoms is calculated by dividing the total interaction energy by the number of bonds of the same kind, e.g. NH_3 has 3N–H bonds, CH_3OCH_3 has two C–O and 6 C–H bonds, etc. (The heat energy released simultaneously on formation of such molecules from their constitutive atomic elements is due, of course, to the selective change in electrostatic potential energy of the combined electron cloud around the nuclei.) We do not need to analyse this energy numerically just yet, but we must see its qualitative implications.

First, it is obvious enough that some atomic elements are strongly attached to themselves rather than to other elements, forming molecular elementary substances rather than compounds of more than one element. Two examples are provided by the air which contains both N_2 (dinitrogen) and O_2 (dioxygen) but very little NO (nitric oxide). A very different case is the widespread occurrence of the triatomic molecule H_2O (water) while there is virtually no H_2 (dihydrogen) on Earth or in the air. What decides these preferential *molecular* combinations? Clearly, it must be the relative strengths of N–N, O–O and H–H compared with N–O or H–O bindings. If we ask which partners do elements prefer *in simple molecules* then we find that the vast majority of metals and many non-metals (at least 80 elements) prefer to bind to oxygen rather than to any other element, including themselves, thus forming molecular oxide (M_n–O_m) species, although this is not true of nitrogen. All such combinations arise from this selective affinity of combining atoms and a corresponding decrease in electrostatic potential energy. (N.B. Any kind of binding leading to a more stable interaction *releases heat energy*.) In Chapter 3 we analysed the general way in which strong bonding could occur in terms of differences in atomic electronegativity, and oxygen is very highly electronegative. Thus, one important factor when we think of the universe as very young, and cooling from atoms at high temperatures (>3000 K) to form, at first, small molecules, was the formation of molecular oxides. Of course, it is necessary to know how much oxygen and other elements were available to so react at what were still very high temperatures. *Abundance* is clearly a restriction on the possible variation of composition, and temperature also limits the possibility of combination. Since, in fact, oxygen was less abundant than the sum of the abundances of the 89 other elements (Fig. 1.10) not all of them could form oxides. In effect, combination was controlled both by element abundance and affinity of the elements for one another, though later we shall see that some elements failed to find their most preferred combination (see below and Chapter 8). We need to go further than *molecular* species to understand the selection of partners when we consider normal Earth temperatures (which are *very low* when compared with the extremely high temperatures of the young universe) since most materials around us are in solid or liquid states and have condensed from the gaseous molecular state that prevails at higher temperatures. First, we have to consider why some molecules condense. Why is nitrogen (N_2) a gas, water (H_2O) an easily volatilised liquid, and yet iron (Fe), silica (SiO_2) and common salt (NaCl) are high-melting continuous solids? Clearly, if we start from molecules (obvious ones, such as N_2 or H_2O, or those of doubtful occurrence, such as NaCl or Fe_2) we must consider why some of them bind together more or less strongly to give liquids or solids while others do not and remain separate as gases, except at extremely low temperatures. Those combinations that condense as solids (or liquids) are preferred relatively to those that remain as gases owing to the extra potential energy stabilisation derived from the co-operativity of interactions between units on condensation. This returns us to a second consideration of chemical binding energy, as being between extended arrays of atoms or molecules as opposed to within single molecules. (All these bindings are, of course, electrostatic in origin, but the nature of electrostatic interactions at the atomic level is not simple; see Chapter 3.)

Consider the molecule NaCl, as opposed to N_2. We have stated in Chapter 3 that it formed because Na, as an atom of low electronegativity, was quite willing to become Na^+ (giving up an electron) to an element of high electronegativity such as Cl, which then became Cl^-. These charged species are called ions. Both Na^+ and Cl^- have an inert gas electronic structure with a full shell of eight electrons, thus NaCl should be written Na^+Cl^-. However, positive and negative centres attract one another in all directions so that 'molecules' or, better, ion pairs of Na^+Cl^-, will tend to come together first in a variety of linear co-operative structures $Na^+Cl^- \cdot Na^+Cl^- \cdot Na^+Cl^-$ or as sideways-bound monomers

$$Na^+Cl^-$$
$$Cl^-Na^+$$
$$Na^+Cl^-$$

If we extend these descriptions into three-dimensional structures it will be seen that the greatest stability is achieved by packing 6, or possibly 8, Cl^- around each Na^+ and, similarly, 6 or 8 Na^+ around each Cl^- (Fig. 4.2). These close-packed arrangements give very stable 'ionic' solids (salts) in which the average affinity of Na^+ for Cl^- is at least 1.5 times that in the 'molecule' Na^+Cl^-. In the resultant *co-operative* lattice the idea of a molecule is lost since no Na^+ pairs more with one Cl^- than it does with several others

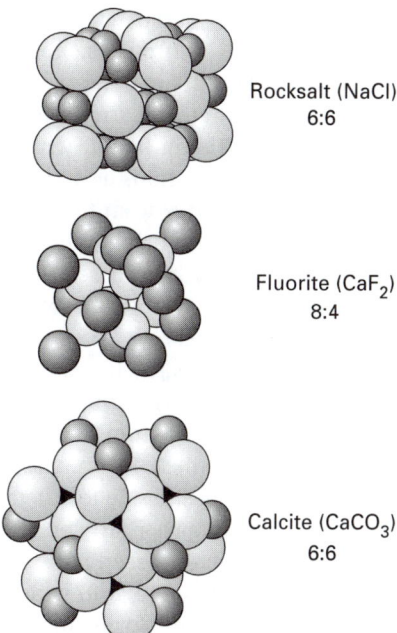

Rocksalt (NaCl)
6:6

Fluorite (CaF$_2$)
8:4

Calcite (CaCO$_3$)
6:6

Fig. 4.2 Some simple crystal structures showing space-filling. The metal ion is shaded.

(Fig. 4.2). Some other examples of salts are given in Fig. 4.2 and all are held together by extremely polar (charge) interactions, i.e. electrostatic interactions between ions. [In the case of ions we can treat them as charges spheres, of fixed ionic radii (see Table 3.4) and consider the electrostatic interactions of such spheres only.] When we talk of the bond energy of NaCl in crystalline NaCl we refer to the value of the energy 'per mole' of NaCl,* but note again that there are no local bonds.† This is an example where the co-operativity of large numbers of atoms in a material gives a resultant structure not simply related to the diatomic molecular form.

In contrast, N$_2$ has no asymmetric charge distribution and N$_2$ only attracts another N$_2$ molecule very weakly, for reasons we describe below. We must remember (Section 2.4) that all interactions between atoms, ions or molecules are ultimately due to the electrostatic binding of electrons to nuclei, and the intensity of the interaction depends on the degree to which an electron feels the field of more than one nucleus and the presence of other electrons. Thus, the formation of N$_2$ is due to the way electrons of each nitrogen atom see the nucleus of one other nitrogen only. This is called *covalence* and the corresponding bonds are called *covalent bonds*: single, double or triple according to the number of electrons involved (two, four or six) and their distribution (see Section 3.4.5).

The intermediate case of H$_2$O is of great importance since this is the dominant liquid on Earth. If we consider the electron distribution in H$_2$O then we can safely assume that the electrons in the two covalent bonds within the molecule are being somewhat pulled to the oxygen from the hydrogen atoms, since oxygen has the higher affinity for electrons (see Section 3.4.4). The schematic electronic distribution in the N$_2$ and H$_2$O molecules is

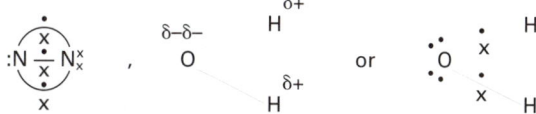

where δ signifies a fraction of the indicated charge, positive or negative. We observe that the structure of the water molecule is not linear, as expected for a simple electrostatic interaction H$^+$...O^{2-}...H$^+$, but is bent, with a H–O–H angle close to the tetrahedral angle of 109°, as stated in Section 3.4.5 (see Fig. 3.14). Although the electrostatic charge separation is now much weaker than in Na$^+$Cl$^-$ (which is formed from separated positive and negative ions), it again leads to a co-operative force between molecules

---

*The 'mole' or 'gram-molecule' corresponds to the molecular formula equivalent mass expressed in grams, which is the mass of the Avogadro's number of constitutive (repetitive) 'units', such as N$_2$, NaCl, SiO$_2$, etc.

†Bond energies can always be calculated from atoms in the molecule or lattice. However, it is conventional in tables to give the *heat of formation* of a compound from its elements in their natural state at 298K (or 25 °C) and at one atmosphere pressure (see Section 5.5, where the precise relationship of ΔH to bond energies is discussed).

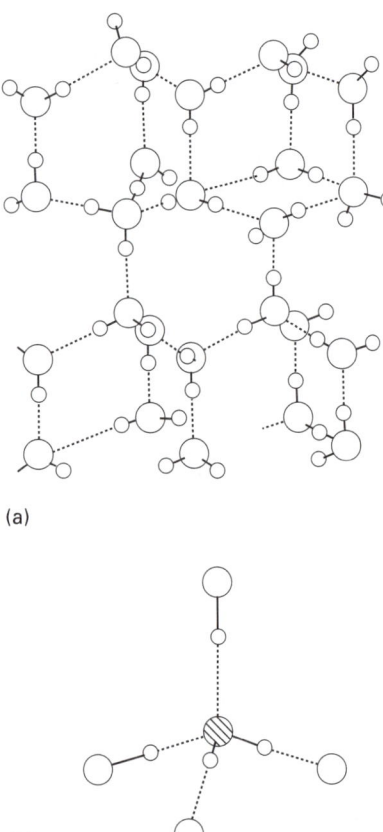

(a)

(b)

Fig. 4.3 (a) The crystal structure of ice in which (b) the H bonds are linear.

strong enough to cause condensation below 100 °C to liquid water, and then, at temperatures below 0 °C, to solid ice, whose structure is as shown in Fig. 4.3. Of course, at room temperature water is liquid, not solid. These conditions depend especially on considerations of the relative strength of ordering forces and the effect of temperature, an overriding factor, which favours disorder. We shall explain this order–disorder question, as well as the effect of temperature on 'affinity', in Chapter 5. Here we wish to stress the overwhelming importance of *co-operativity* of potential energies between the structural units which leads to the stability of ordered condensed states. In the two cases of N_2 and H_2O, as in all molecular solids, the simple molecular structure is still observed in the lattice.

Note that in H_2O we have two partially polar O–H *covalent* bonds and the H atoms cannot form further (covalent) bonds since H cannot bind more than 2 electrons (see Chapter 1). However, the H atoms will be 'unbalanced' because their electrons are somewhat pulled to one side, towards oxygen, and there will be a tendency on the opposite side for them to pull to themselves or accept pairs of electrons from appropriate donors, including the oxygen atoms of neighbouring H_2O molecules (see Fig. 4.3). This generates a considerable attraction called 'hydrogen bonding' to which we shall return later (Section 4.4). Oxygen in H_2O cannot form further bonds since it has the full complement of 8 electrons, but on binding covalently to 2H it is left with a negatively charged side of lone pairs of electrons away from the bonded H atoms (see Fig. 3.14), which it tends to donate to the unbalanced H atoms of neighbouring water molecules. Thus, the hydrogen bond has something of a covalent character.

If we consider next the solid carbon (diamond) structure we see that each atom can now form four covalent bonds (Fig. 4.4) to reach the neon (Ne) electronic shell structure, and the simplest arrangement is in a continuous 'covalent' series of tetrahedra sharing all electrons in covalent (non-polar) bonds, since this uses space (reduces repulsion) most effectively. It satisfies the affinity for electrons in all directions equally. Here there is no question of polarity in the bonds since all atoms are the same, but the result of forming such a continuous covalent bond (electrostatic) network is to make diamond (this form of carbon) a very stable *co-operative* solid. The network is much more stable at low temperature than a small C_2 molecule, since the electrons are less crowded, and is even more stable than in a larger molecule such as fullerene, C_{60} (see Fig. 4.4) again due to the greater satisfaction of electrostatic interactions in covalent bonds. In diamond all atoms then act *co-operatively* to hold one another in a three-dimensional network and each carbon has a complete 8-electron shell. Once again the sense of molecular units is lost in the diamond lattice. Note that as soon as there are (on average) more than four electrons per atom in the outer orbitals, e.g. at nitrogen, any co-operativity is virtually absent, e.g. in N_2. In a hypothetical continuous N atom diamond-like lattice, excess outer electrons (5 per atom) over bonding possibilities (4 per atom) would cause greater repulsion than in separated N_2 molecules and, therefore, nitrogen forms a 'triple-bonded' diatomic molecule (see above). The same diatomic structure appears in O_2, doubly bonded, while at F_2 one single bond gives the 8-electron surround for each atom. These stable homonuclear mole-

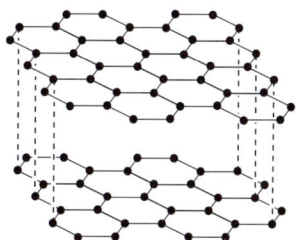

(a) Diamond (single bonds)

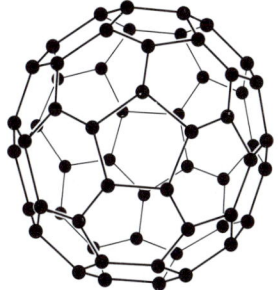

(b) Graphite (mixed single and double bonds)

(c) Buckminsterfullerene, C_{60} (mixed single and double bonds)

$-C{\equiv}C–C{\equiv}C–C{\equiv}C–$

(d) Polyethyne (triple and single bonds)

Fig. 4.4 Different forms of carbon.

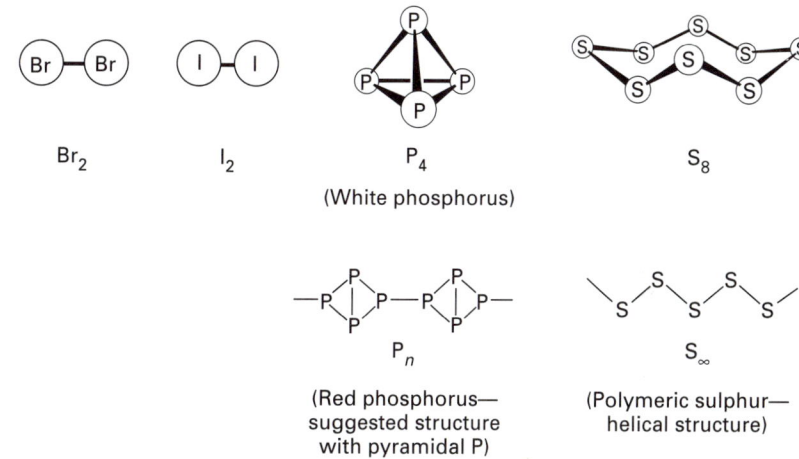

Br_2 I_2 P_4 (White phosphorus) S_8

P_n

(Red phosphorus— suggested structure with pyramidal P)

S_∞

(Polymeric sulphur— helical structure)

Fig. 4.5 Some examples of discrete molecules (Br_2, I_2, P_4 and S_8). Alternative forms for P and S are shown, but there are many other forms for sulphur, including linear polymers, and a third form for phosphorus (black phosphorus).

cules, gaseous at room temperature, are confined to the last three elements of the first row of light elements in the periodic table, together with hydrogen and chlorine.

In the structures of heavier elements, even where there are more than 4 outer electrons per atom, we often find alternative solid, ordered constructs, such as P_n and S_n, still using only 3 or 2 bonds per atom (Fig. 4.5). The reason why at room temperature N_2 is a triple-bond and O_2 a double-bond molecule, while P_n and S_n are lattices of single bonds, cannot be analysed here in any detail, but it can be said that it is due largely to the smaller size of the atoms N and O relative to the atoms P and S. Such factors become extremely important when we wish to explain the diversity of organic chemistry and why organic chemicals provide large numbers of components at room temperature and are the basis of living systems. In fact, the formation of triply bonded N_2, N≡N, and doubly bonded O_2, O=O, illustrates a general feature of bonding, most common in a few very light elements and their heteronuclear molecular combinations. They frequently form multiple covalent bonds, as in heteronuclear molecules, e.g. CO_2, O=C=O, and they are a strong feature of organic chemistry observable, for example, in benzene (C_6H_6) and ketones [$(CH_3)_2CO$, acetone]. Such multiple bonds restrict the geometry of molecules, often into a totally planar arrangement, as in benzene (see Fig. 3.17), or a partially planar arrangement, as in acetone or acetic acid. They are common features in polymeric biological molecules, including proteins, DNA, RNA and some lipids. Overall, it is clear that non-metals in combination form molecules, and their ability to give strong, continuous three-dimensional co-operative lattices is limited to carbon and the heavier elements of groups 14–16. Ionic metal/non-metal structures, such as NaCl, are very different from non-metal/non-metal structures.

From the examples above we see that at low temperatures (<2000 K) there are two ways of making continuous co-operative stable structures:

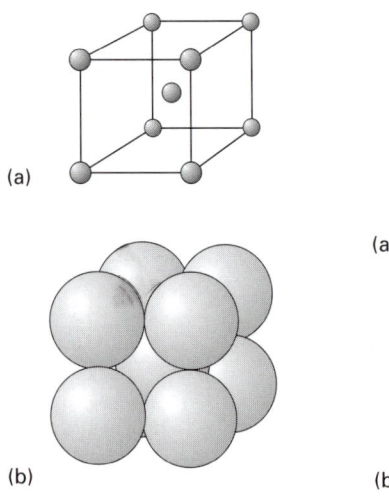

(a)

(b)

Fig. 4.6 (a) Packing of metal atoms in the body-centred cubic (b.c.c.) lattice. (b) Packing of equal spheres in a b.c.c. lattice but to illustrate space-filling properly all the balls should just touch.

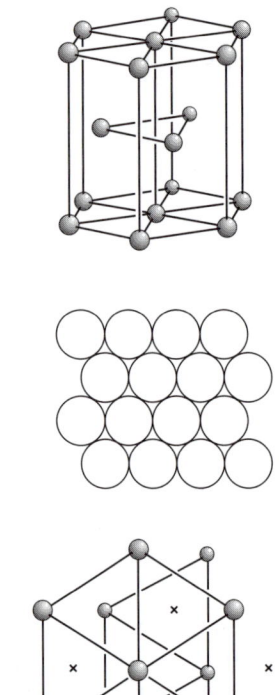

(a)

(b)

(c)

Fig. 4.7 Packing of metal atoms in a close-packed hexagonal (c.p.h.) lattice. (a) Stacking of close-packed planes in the c.p.h. structure. (b) A fragment of a plane of close-packed equal spheres showing space-filling. (c) Octahedral holes (x) in the c.p.h. structure.

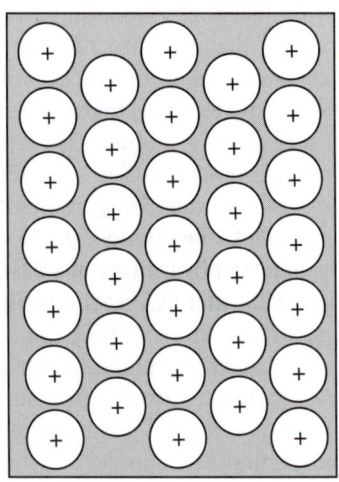

Fig. 4.8 A group 1 metal. Metal ions, M^+, form a lattice that is maintained co-operatively by a 'sea' of electrons (shaded part), i.e. electrons can have a range of energy values effectively in bands, not in well-separated discrete molecular energy levels.

(1) ionic interactions (Fig. 4.2), and (2) more than one covalent interaction per atom (Figs 4.4 and 4.5), to which we can add a third possibility, (3) metal–metal interactions (metals and alloys). Now, most metals, unlike the above non-metals, have less than four outer-shell electrons per atom, and by self-combination a filled outer shell of 8 electrons cannot be achieved. In elementary metals the atoms are seen to be as closely packed together as is possible for equal spheres, i.e. each atom usually has 8 (Fig. 4.6) or 12 (Fig. 4.7) near-neighbours. The number of external electrons available for bonding is now of little consequence for structure. Thus, having one, two or three electrons outside the nearest inert gas shell in metals such as sodium, calcium or aluminium does not affect structure. The external electrons are too few to form an atomic network such as diamond. Furthermore, we know that self-assembled metallic elements hold on to some electrons quite weakly since they conduct electricity, i.e. they allow electrons to flow very easily. The simplest way to describe metallic states then is to consider that their most stable condition is one in which the lattice of atoms is really a lattice of positive ions buried in a freely mobile, shared sea of some negative electrons (Fig. 4.8). The electrons are not associated with particular atoms or localised, as in covalent or ionic solids, but fill space around the positive ions, without structural implications. The

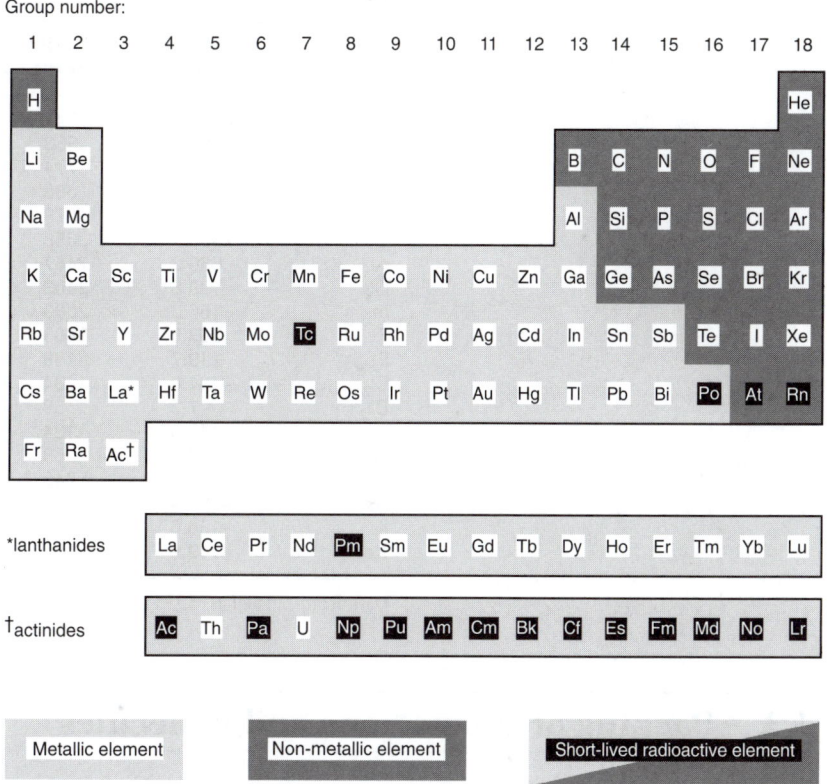

Fig. 4.9 The division between metallic and non-metallic elements in the periodic table. [From Cox, P. A. (1995). *The elements on Earth*. Oxford University Press, Oxford.]

electron 'atmosphere', which has a continuous range of energy values (sometimes called a 'band'), holds positively charged atoms again in a *co-operative* manner. The elements forming metals are shown in Fig. 4.9, and from the description given we can see why they are malleable and ductile, unlike diamond or silica, since there are no bonds broken if parts of the structure slide over one another. Various ways of classifying elementary substances and compounds arose first from these physical properties, in which selected species were grouped together as metals, salts and non-conductors before the atomic nature of their structures was known.

From the above, we see that bond strength in molecules and co-operativity between units at various compositions are dominant *selective* influences upon the nature of materials because they so influence the structure of an assembly. The nature of co-operativity is a special central feature of chemistry and biology since it is the major factor in deciding the ordering of atoms or components in a condensate. Notice how these bonding types match other physical properties in Table 4.1.

[In all the above cases we have noted how the variable temperature affected structure, but we stress again that we shall not examine the effect of this variable thoroughly until Chapter 5.]

Table 4.1 Trends in physical properties of the elements

Substance	Melting point (°C)	Boiling point (°C)	Density (g cm⁻³)	Heat of vaporisation (kcal mol⁻¹)
Li	180.5	1347	0.534	35.3
Na	97.8	883	0.971	23.7
K	63.2	774	0.862	18.9
Rb	39.0	688	1.532	18.1
Cs	28.6	679	1.873	15.9
B	2300	3658	2.34	120.7
Al	660.3	2467	2.70	69.6
Ga	29.8	2403	5.91	64.7
In	156.2	2080	7.31	55.5
Tl	303.5	1463	11.85	39.7
F_2	−219.7	−188	1.52 (liq.)	0.78
Cl_2	−101.0	−34	2.03 (liq.)	4.9
Br_2	−7.3	59	4.05 (liq.)	7.3
I_2	+113.5	184	4.93 (solid)	10.0
He	−272.7	−269	0.125 (liq.)	0.02
Ne	−248.7	−246	1.44 (liq.)	0.4
Ar	−189.4	−186	1.66 (liq.)	1.6
Kr	−156.6	−152	2.28 (liq.)	2.2
Xe	−119.9	−107	3.54 (liq.)	3.0

Data from Emsley, J. (1991). *The elements*. Clarendon Press, Oxford.

4.3   Packing of atoms, ions and molecules

As stated again and again, the major factor for the coming together of atoms is the attraction of electrons for nuclei, which determines their atomic affinity whether in molecules or lattices. In both molecules formed from different atoms and in the condensed state there is, however, a factor restricting combination which is the way in which the units—atoms, ions or molecules—fill space (see aside opposite). As stated, identical atoms, which are spherical, can be packed together in close-packed arrays and the array can give each atom 8 or 12 neighbours. However, it does not give the shortest distance between any pair of atoms; in other words, it does not allow good local bonds but only good general binding over the whole lattice. We have stated that the metals, which are close-packed lattices, are confined to elements with less than four electrons, but this is not the full picture. We find that elements of group 14 (C, Si, Ge, Sn, Pb) pack differently; at the top, diamond is a network of carbon atoms linked to four other carbon atoms, but, at the bottom, lead is an approximately close-packed metal with almost 12 close neighbours. Thus, there is a selective conflict between long-range *lattice* stability and short-range *bond* stability dependent on atom size. The smaller the atom and the greater its electron affinity and ionisation energy, or the electronegativity (see Section 3.4.4), the greater the tendency to form *bonds*, thus generating the lower number of near neighbours in the condensed state (see Fig. 4.5). At tin in group 14, both forms, metallic and continuous covalent solid states, are equally stable, and as a result tin money (metallic) often crumbled to dust (nonmetallic).

When atoms of rather different kinds and, therefore, of different sizes (see Section 3.4.1) are brought together, then close-packing of all in a random array is not always observed even for metallic elements—the larger atoms are close-packed, but the smaller fit into holes in the close-packing (see Fig. 4.7) so as to give optimal distances in as many directions as possible. Similarly, in salts, anions (X^-) are almost always larger than cations (M^+) so that M^+ ions fit into the holes of closed-packed X^- Lattices. The hole sizes then generate site symmetries. The reason certain atoms may prefer certain holes of a given symmetry other than on the basis of size concerns the importance of the types of atomic orbital and their occupancy which are relevant for the atoms under examination (see Section 3.3).

Now, as well as considering packing of atoms and ions we have to consider the packing of such components as *molecules*, either of the same kind or of different kinds. Turning to such molecules, the shape of the unit in the crystals is now idiosyncratic (see Section 7.2). Packing of individual molecular shapes in crystals is the result of a more complex balance between local interactions on molecular surfaces. We described the shapes of some molecules in Section 3.4.5 (and see aside) and some of their packing problems will be discussed again in Chapters 6 and 7. However, we make the point here that the stable arrangement still reflects an electrostatic potential energy, now small, so that both molecular shape and packing are a feature of potential energy and not separate variables (see Section 7.3). The energy now includes selective van der Waals interactions (see Section 3.4.1). If there is energy input then the shape can be changed, of course, to that of another stationary state. All aspects of structure are then characterisable by variation in composition and potential energy.

We turn to specific molecules such as methane to illustrate these points. One CH_4 molecule does not interact with other CH_4 molecules except very weakly and methane is, therefore, a gas at room temperature. When it condenses, the molecules, which are almost spherical, form a close-packed arrangement. We can consider next a larger C/H compound such as octane (petrol) whose formula is C_8H_{18}. The bond structure is

$$\text{H}-\overset{\displaystyle H}{\underset{\displaystyle H}{\text{C}}}-\overset{\displaystyle H}{\underset{\displaystyle H}{\text{C}}}-\overset{\displaystyle H}{\underset{\displaystyle H}{\text{C}}}-\overset{\displaystyle H}{\underset{\displaystyle H}{\text{C}}}-\overset{\displaystyle H}{\underset{\displaystyle H}{\text{C}}}-\overset{\displaystyle H}{\underset{\displaystyle H}{\text{C}}}-\overset{\displaystyle H}{\underset{\displaystyle H}{\text{C}}}-\overset{\displaystyle H}{\underset{\displaystyle H}{\text{C}}}-\text{H}$$

which again satisfies inert gas rules—all H and C atoms are satisfied as far as electron count is concerned. The molecular shape is, of course, three-dimensional, but here it can take on many configurations. The interaction between C_8H_{18} molecules is still quite weak, but stronger than that between CH_4 molecules. The molecules tend to lie parallel to one another. C_8H_{18} (petrol) is, therefore, not a gas but a liquid. We can follow the boiling points of these so-called saturated hydrocarbons (alkanes)—CH_4, C_2H_6, C_3H_6, C_3H_8 up to $C_{10}H_{22}$ (Table 4.2)—and note how the boiling point increases as the size (hence molecule–molecule interactions) increases. This implies that the ability to form condensed phases is due to increase in

Description of shape	Shape	Examples
Linear		HCN, CO_2
Angular		H_2O, O_3, NO_2^-
Trigonal planar		BF_3, SO_3, NO_3^-, CO_3^{2-}
Trigonal pyramidal		NH_3, SO_3^{2-}
Tetrahedral		CH_4, SO_4^{2-}, NSF_3
Square planar tetragonal		XeF_4
Square pyramidal		$Sb(Ph)_5$
Trigonal bipyramidal		$PCl_5(g)$, SOF_4
Octahedral		SF_6, PCl_6^-, $IO(OH)_5$

The description of some molecular shapes

Table 4.2 Some normal hydrocarbons (alkanes)

Molecular formula	Name	Boiling point (°C)	Melting point (°C)
Natural gas			
CH_4	Methane	−161	−184
C_2H_6	Ethane	−88	−183
Bottled gas			
C_3H_8	Propane	−42	−188
C_4H_{10}	n-Butane	−0.5	−138
Gasoline (petrol)			
C_6H_{14}	n-Hexane	69	−94
C_8H_{18}	n-Octane	126	−57
$C_{10}H_{22}$	n-Decane	174	−30
Kerosene			
$C_{11}H_{24}$	n-Undecane	194.5	−25.6
$C_{12}H_{26}$	n-Dodecane	214.5	−9.6
Gasoil (light)			
$C_{14}H_{30}$	n-Tetradecane	252.5	+5.5
$C_{16}H_{24}$	n-Hexadecane	287.5	18
Gasoil (heavy)			
$C_{18}H_{38}$	n-Octadecane	317	28
Biological lipids			
$C_{20}H_{42}$	n-Eicosane	334	36.7

N.B. Note how liquid ranges increase as molecular weight increases.

(weak) *co-operativity* with the size of molecules in a way additional but unrelated to charge distribution in bonds (polar interactions). The additional interactions between molecules or atoms, e.g. of the noble gases, are, as stated, the electrostatic van der Waals forces (see Section 3.4.1) which form a part of the so-called *hydrophobic forces*, interactions between *non-polar* molecules in water solution.* The fact that long-chain R–$(CH_2)n$–Y molecules are liquid or solid at room temperature owing to these forces creates structures with molecules packed side by side in close-packed arrays in a given plane. This structure gives the possibility for the formation of a living cell, since it can form closed membranes (Fig. 4.10).

As described in Section 3.4.1, the distances, r, between molecules interacting through van der Waals contact are large compared with the internal atom–atom bond distances and are correlated with the weaker interaction energy.

(Notice that these non-metal/non-metal compounds are often liquids and that this allows them to flow, e.g. H_2O and light oils. In turn they can form classes of material based on their motion, to which we turn to in Chapter 8.)

---

*Electrostatic potential energies have been related to charge–charge interactions in Chapter 3, but they also arise from interactions between charges and dipoles (see the water molecule in Fig. 3.14), between dipoles themselves and are even due to induced polarisation of both a static and a dynamic kind. However, in this series the interaction is progressively weaker and depends on a higher inverse power of distance r between the interacting species, atoms or molecules, i.e. the expression for the attraction energy is proportional to $\frac{1}{r^n}$.

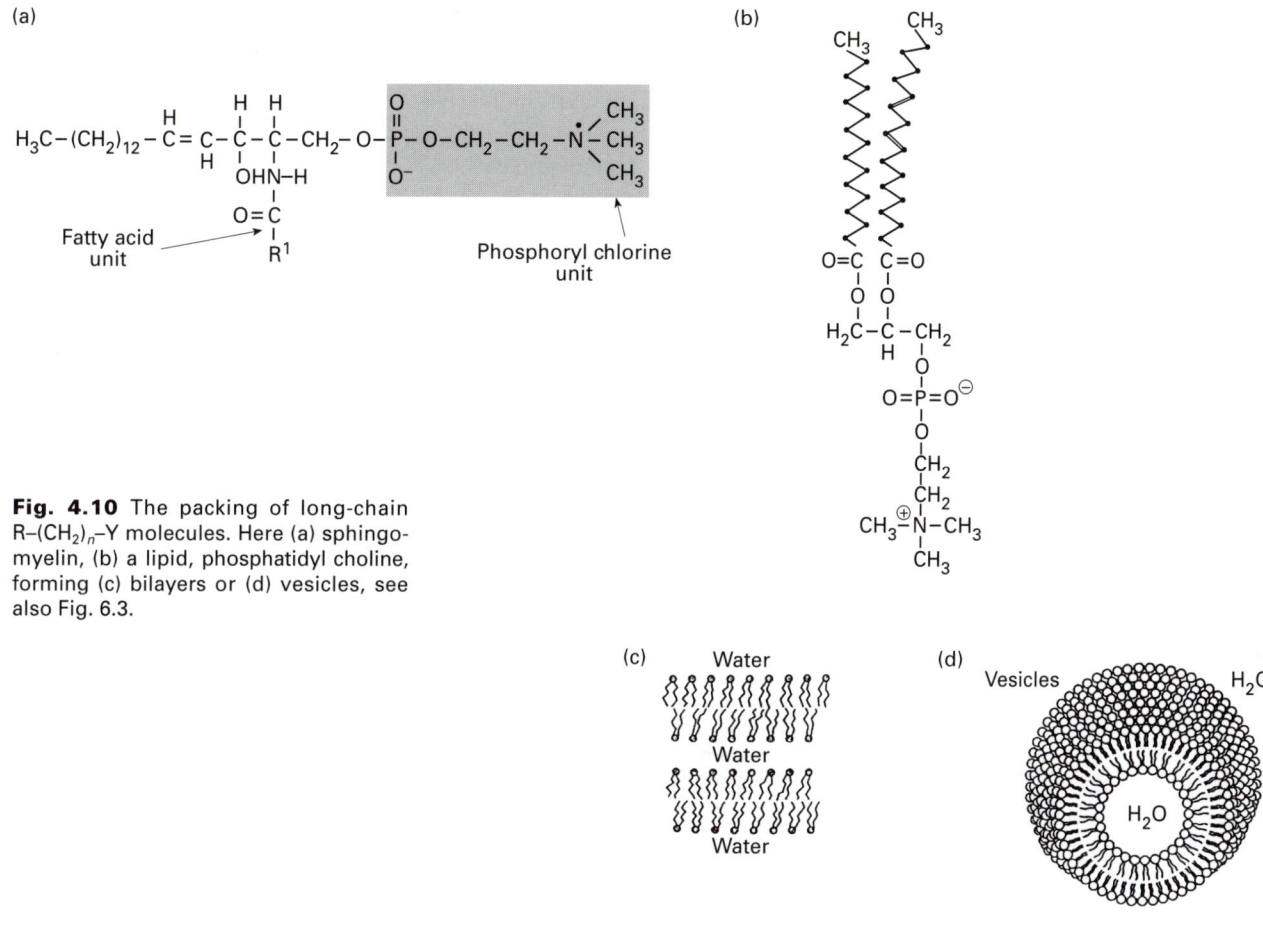

Fig. 4.10 The packing of long-chain R–(CH$_2$)$_n$–Y molecules. Here (a) sphingo-myelin, (b) a lipid, phosphatidyl choline, forming (c) bilayers or (d) vesicles, see also Fig. 6.3.

4.4   An aside: the value of hydrogen bonding

Above, we referred to a peculiar type of bonding, hydrogen bonding, that occurs when hydrogen atoms are covalently bonded to other atoms that exert a much stronger pull on the shared pair of electrons, i.e. strong electronegative atoms (see Fig. 4.3). Hydrogen bonding is then a special case of intermediate strength of co-operativity. As a stated result, water is a liquid at temperatures below 100 °C and solid at temperatures below 0 °C, not a gas like H$_2$S (hydrogen sulphide) which is a molecule larger than H$_2$O but with a smaller dipole and forming much weaker hydrogen bonds. Water is, of course, the dominant molecule of the sea and of all living organisms. However, this intermolecular association through hydrogen is not specific to the water molecule; the condition for its formation is, as stated above, that the hydrogen atom is bonded to a very electronegative element, which include not only oxygen (O) but also nitrogen (N), fluorine (F) and, to a lesser extent, chlorine (Cl) and sulphur (S). Thus, when we compare properties that depend on the degree of intermolecular association, such as

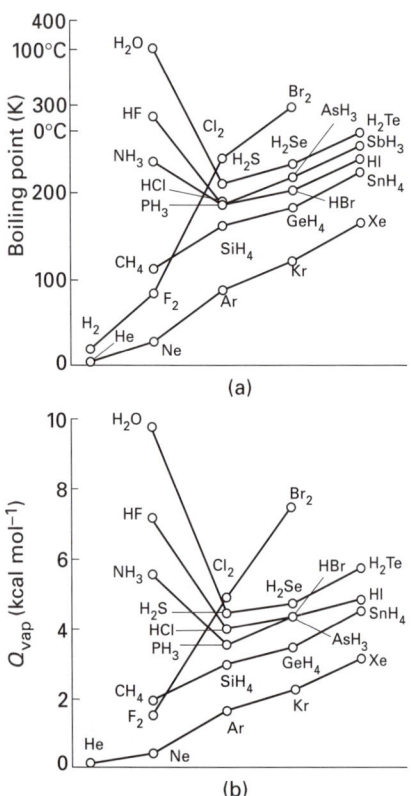

Fig. 4.11 (a) Boiling points and (b) heats of vaporisation, Q_{vap}, (see Section 5.8.1) of some substances that form molecular crystals. Note that the oxides of C, N, S and several halogens are gases at room temperature, unlike H_2O, while most other oxides are solids.

melting points or boiling points, we observe anomalous trends, as shown in Fig. 4.11, in contrast with the case of the noble gases or the hydrides of the carbon group of elements for which no hydrogen bonding is possible. Clearly, water is an extremely special liquid since here the H bond is especially strong.

Of course, this type of interaction is not restricted to simple small molecules. It is also observed in larger molecules and is present in all kinds of interactions of organic molecules, but is particularly relevant in biological polymers—DNA and proteins, for example—where it accounts for their structural folded characteristics: the double helix in DNA (see Fig. 10.7) and the so-called α helix and β sheet of proteins (see Fig. 7.6). Note that in these cases there are very many hydrogen bonds formed, so that although the energy of just a single bond is small (and the bond is weak), the *co-operative* effect of many such bonds leads to stable structures which pack together and form solids. Internally, each protein is a low-melting solid structure, but knowledge of its dynamics will also prove to be crucial for life. Importantly, protein melting points fall in the range of the liquid state of water i.e. between 0 and 100 °C. Table 4.3 gives an impression of how co-operativity changes in different compounds.

Table 4.3 Degrees of co-operativity in solids

Small (Weak van der Waals forces—polarisability)	Medium (Stronger van der Waals forces—dipole, polarisability and hydrogen bonds)	Strong (Ionic, covalent or metallic bonding)
Inert gases	H_2O	NaCl, MgO
Small hydrocarbons	Organic hydroxides	Diamond, silica
N_2, H_2, O_2	Carboxylic acids	Iron, copper
CO, CF_4	Ketones, etc.	
	Proteins, DNA	
M.p./b.p. < 0 °C	M.p. 0–200 °C	M.p. >> 200 °C
Molecular	Molecular	Continuous

4.5 Quantitative strengths of binding

The further analysis of the *energy* of atom–atom combinations requires us to examine, *quantitatively*, at a given temperature (298 K) and one atmosphere pressure why atoms of one element bind together. Now, we know that the affinity of atoms for electrons, i.e. the electron affinity energy, increases as we cross periods (see Section 3.4.4),

$$Li < Be < B < C < N < O < F >> Ne$$
$$Na < Mg < Al < Si < P < S < Cl >> Ar$$

Thus, this is the increasing demand to form bonds to other atoms. On the other hand, the tendency of atoms to retain their electrons, measured by the ionisation energy, follows the reverse order. This is also an order of willingness to form bonds. Up to the middle of each series, at C and Si, the elements tend to lose electrons, and after the middle they tend to capture electrons instead. Although it appears that none of the light atoms, up to atomic number 18, can form more than four sp^3 covalent bonds they are sometimes found with more than four co-ordinated groups, as in complex species, e.g. SiF_6^{2-}, and PCl_6^- (see Section 6.6 and references 1–5 for an explanation). This is due to the non-classical electron–electron repulsion of completed electron shells, here of 8 electrons, described by the exclusion principle (see Section 3.4.2). Since binding between two atoms must be related simultaneously to their tendency to give up their electrons and to accept electrons from others, i.e. their electronegativity (Table 3.7), and to this repulsion, it follows that the strength of binding of identical atoms to one another must reach its maximum when the two tendencies—to lose or to capture electrons—balance each other and the electron repulsion is not large, i.e. in the middle of each series:

$$Li < Be < B < C > N > O > F \gg Ne$$
$$Na < Mg < Al < Si > P > S > Cl \gg Ar$$

This is shown quantitatively in Fig. 4.12. Note that the co-operativity of binding leads to better bonding for Li, Be and B relative to F, O, and N, respectively. We also know that the elements after F and Cl, i.e. Ne and Ar, do not bind to themselves or to other elements. We can therefore rationalise Fig. 4.12 in terms of the combination of an energetic drive to reach a total of 8 shared electrons at C (carbon) against the increase in electron repulsion after this element, when there are too many electrons to share all of them. These bonds between identical atoms have no charge separation (polarity) and are called *homopolar* covalent bonds (see Section 3.4.4).

If elements of extremely different kinds are brought together, as in NaCl, then, as stated, the affinity for electrons of Cl is much greater than that of Na and a strong polar bond Na^+Cl^- is formed in the molecular state. It is

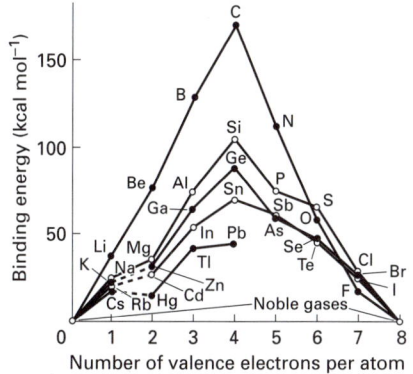

Fig. 4.12 Binding energy of solid non-transition elements relative to their free atoms. (See Chapter 3.5 of ref. 31 in Further reading.)

Table 4.4 A selection of lattice energies of ionic compounds

Salt	Lattice energy (kcal mol⁻¹)	Salt	Lattice energy (kcal mol⁻¹)
NaF	217	MgF_2	698
NaCl	185	CaF_2	631
NaBr	176	BaF_2	560
NaI	166	$MgCl_2$	592
KF	193	$CaCl_2$	542
KCl	168	$SrCl_2$	512
KBr	161	$BaCl_2$	489
KI	152	TiO	928
MgO	934	VO	936
CaO	845	MnO	911
SrO	789	FeO	938
BaO	751	CoO	954
MgS	807	NiO	974
CaS	740	CuO	970
SrS	696	ZnO	964
BaS	666	ZnS	864

Data from Ball, M. C. and Norbury, A. H. (1974). *Physical data for inorganic chemists*. Longman, London, and Waddington, T. C. (1959). *Advances in inorganic chemistry and radiochemistry*, Vol. 1. Academic Press, New York.

Table 4.5 Correlation of physical properties of ionic crystals and lattice energies

Salt	Lattice energy (kcal mol^{-1})	Melting point (°C)	Hardness (Mohs' scale)
NaF	217	992	
NaCl	185	801	
NaBr	176	755	
NaI	166	651	
BeO	1083	2530 ± 50	9.0
MgO	934	2825 ± 30	6.5
CaO	845	2615 ± 25	4.5
SrO	789	2420	3.5
BaO	751	1920	3.3

this polar 'molecule' that then gives rise to lattice structure in which there is co-operativity of electrostatic interactions but no simple representation in terms of bonds. As stated earlier, we can still describe the binding energy per mole, although there are no distinct molecules in the lattice. Some binding (lattice) energies are shown in Table 4.4, and a rough correlation with other physical properties in Table 4.5. We reserve further discussion of these trends until Chapters 5 and 9, since they are determinant in the formation of much of the Earth.

4.6   Stoichiometric and non-stoichiometric composition

In the above we have chosen to discuss combinations of atoms at given compositional ratios. We observe that while sodium is a metal and chlorine is a gas, bringing them together in any proportions yields only one compound plus one or other element in excess. Thus NaCl, with a 1:1 ratio, is known but no other compound, Na_xCl_y, is found. The quantitative reason for this fact is the balanced charge distribution in NaCl, i.e. $Na^+ + Cl^-$ lattices, as explained in Section 4.2, giving optimal co-operativity and the inability of this lattice to include extra Na or Cl atoms. Thus sodium plus chlorine gives only one compound which is a different chemical *species* from either Na metal or Cl_2 gas. The appearance of speciation and not a continuum is due to the energy of co-operativity of interactions, which is strongest within particular regions of composition that may be exceedingly narrow, as in NaCl, or broad (see Section 4.7).

Covalently bonded compounds give rise to a much greater variety of observed molecular speciation even for combinations of only two elements, X and Y, and the co-operativity between molecules is weaker. Thus, on combining C with H the compounds which have the strongest binding are given by series of hydrogen to carbon ratios such as

$$C_2H_2, \quad C_2H_4, \quad C_2H_6 \quad \text{and}$$
$$C_3H_4, \quad C_3H_6 \quad C_3H_8$$

Note that the numbers represent stoichiometric molecules, not lattice compounds. These are *not* all *stable* molecules since some can disproportionate,

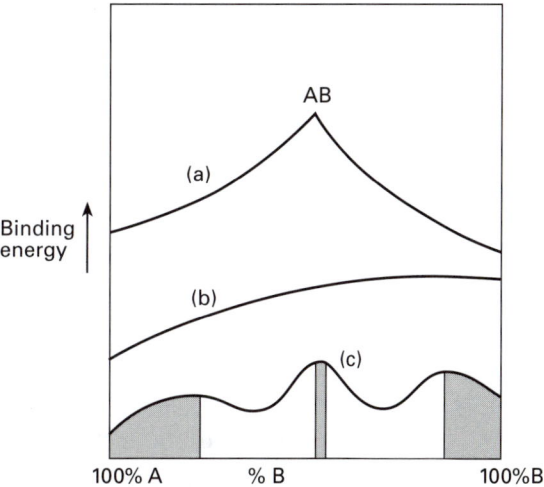

Fig. 4.13 The different cases of binding energy variation with % composition. (a) Formation of one stochiometric compound AB. The curve is convex. More than one peak can arise but all connections are convex. Examples: covalent molecules and simple AB salts. (b) Formation of a single continuous phase. The curve is concave. Examples—some alloys. (c) Formation of three phases of limited compositional ranges, A and B partially soluble in one another and in AB. The curve is alternatively concave and convex several times. Many phases may appear without relationship to stochiometry. These are land-scape diagrams of binding energy (see Section 1.4). Shaded areas are stable compositions, see also Fig. 5.19 (b). For examples see aside.

Examples of combinations

Stoichiometric

H_2O	CH_4	P_4O_{10}
NaCl	$CaCl_2$	AlF_3

Non-stoichiometric

Cu/Zn, Mg/Al, Fe/O

for example $C_2H_4 \rightarrow C + CH_4$. They do not do so readily, since the bonds do not break easily at 300K. Hence, very many covalent molecules co-exist as independent components. This *kinetic* resistance to exchange of atoms is the basis of organic chemistry (see Section 3.4.4). Thus, the difference between ionic and covalent 'compounds' is a crucial factor for the understanding of the variety of stable and stationary state speciation, which is not obvious when we examine plots of binding energy against composition. We can see this difference in the landscape of stability plots (Fig. 4.13). For combinations $A_mB_nC_p$, etc., in which there are covalent bonds only, the number of observed species of closely related composition, especially when m, n, p, etc. are large, is very great; consider, for example, the number of DNA genetic sequences obtained from combining four different molecular bases. The corresponding stability landscape diagram must be extremely rugged, compared with the case of simple ionic compounds, A^+B^- (see Fig. 4.13).

We can now see that the classical rules of Proust and Dalton of constant combination and multiple proportions, which all these components obey, arise from the nature of extreme ionic and covalent substances and are easily related to electron and empty orbital counting. These simple rules have no absolute validity, however, since it is only the variation in interaction energy within composition, together with its variation with temperature (see Chapter 5) that decides which materials appear. This interaction energy depends on factors other than the above for alloys and complicated minerals, or indeed for combinations of organic molecules.

4.7   Examples of phases* of continuously variable composition

We have described metal lattices, for example in the cases of elements such as sodium, magnesium, aluminium and iron, as being made of positive metal ions in an electron sea (see Fig. 4.8). It is then generally true that if we take, say, a mixture of two metal atoms, they will form associations where there may be no rules for compound formation discernible by counting electrons relative to atoms.[†] As stated, in metal elements there is always an excess capacity for electrons over the number of electronic orbitals occupied. In extreme cases, then, we expect and find that one metal may dissolve in another in a continuously variable way, e.g. K in Rb. Sometimes we speak of metal-metal 'compounds', but they are different from metal/non-metal or just non-metal compounds. Most metal-metal 'compounds', called alloys, are non-stoichiometric and exist over limited or even

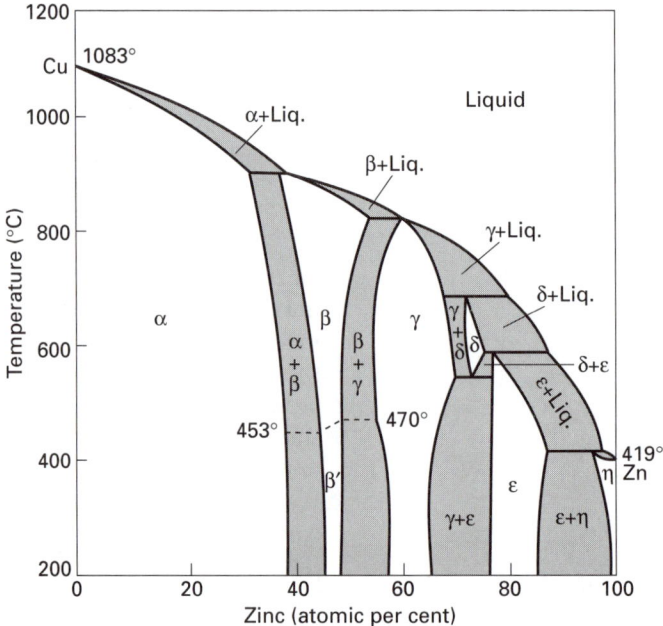

Fig. 4.14 The phase diagram for copper–zinc alloys. Note the temperature scale. The regions of composition α, β, γ, etc. are observed and are phases that are continuously variable in composition. Shaded regions do not exist but split up to phases to left and right. Thus two regions, e.g. β and α (so-called phases), can co-exist.

---

*A phase is a region of space which is homogeneous in all properties, including composition. Phases in contact such as ice, water and steam or crystals in water are treated here as being in a balanced exchange distribution of all constituents across their boundaries (see Section 5.8). In Chapter 7 we shall discuss homogeneous separate zones of space limited by boundaries which are not in balanced exchange and which we call *compartments*.
†There are cases of closely but not exactly fixed stoichiometry which obey roughly electron-counting rules (so-called Hume–Rothery rules). These are, more properly, called *intermetallic* compounds and current examples are $MgZn_2$, Cu_2Au, NaTl, etc.

considerable ranges of composition. Moreover, there may or may not be some ordering of atoms A relative to atoms B. The reasons why not all metal atoms of one kind mix with any amount of another are due to the fact that metal atoms are of different sizes (Table 3.2) and some metal atoms have a local charge higher than others through losing more electrons to the lattice, i.e. they differ in electronegativity. Clearly, balls of different size and charge pack less well than those of the same size and charge. Consequently, alloys have different limited ranges of composition and order over which they form. The very complicated picture for the possible landscape ranges of combination of copper and zinc is shown in Fig. 4.14, in which the different *phases* have a different atom packing, but there is no sense of stoichiometry. Notice that some phases, i.e. the continuous regions in which structure does not change with composition, are only found in limited temperature ranges. The treatment of binding in such lattices also considers *bands* of closely related energies for electrons, not *bonds* of any kind (see Section 4.2). A fuller discussion of such systems will be undertaken in Chapter 5 (see Section 5.9.1).

A little thought will show that in any continuous lattice of more than two elements we must not expect to have combinations of elements in strictly stoichiometric ratios or even strict order in structures, since two of the elements are likely to be similar and likely to be able to replace one another.

As an example, consider adding together MgO and Al_2O_3, magnesium and aluminium oxides. The simple substances are stoichiometric and formed when mixing the metals and dioxygen so as to balance charges strictly in their lattices $Mg^{2+}O^{2-}$ and $(Al^{3+})_2(O^{2-})_3$. However, if we put Mg^{2+} with Al^{3+}, ions of roughly the same size, and O^{2-} in one lattice, the charge balance can be managed in principle by increasing the amount of O^{2-} steadily from 1:1 for $Mg^{2+}O^{2-}$ to 1:1.5 for $(Al^{3+})_2(O^{2-})_3$ as we replace Mg^{2+} by Al^{3+}, and electrostatic balance is maintained but stoichiometry is not. Of course, we can also replace Mg^{2+} by Mn^{2+} and Al^{3+} by Fe^{3+}, so that an infinite variety of mixed oxides is possible. Equally, one O^{2-} can be replaced by, say, one S^{2-} or two F^- and so on. There is a limit to the stability of such combinations which is related to the *packing* of ions of different sizes and charges, so that stable combinations are in fact found in composition ranges, which are not necessarily narrow, much as for alloys (see Fig. 4.15).

Once again it is largely lattice co-operativity that decides the system of greatest stability. A very important case for us is the build-up of silicate lattices, which we can treat as derived from $Si^{4+}(O^{2-})_2$ in which we can substitute Si^{4+} by Al^{3+} and incorporate in the $Si^{4+}(O^{2-})_2$ lattice elements such as M^+ or M^{2+} (see Fig. 9.7). As long as the sizes of atoms give good packing, and hence large co-operative electrostatic interactions, these materials, which are the basis of rocks, soils and many of man's materials, are stable, although without stoichiometry, and have certain fixed and limited ranges of composition (Fig. 4.15). Much of the mineral crust of Earth, including all the clay minerals, is built up in this way because of the abundance of Al, Si, Mg and O, and the co-operative strength of their phases. In many of these examples there are ways of giving idealised structures,

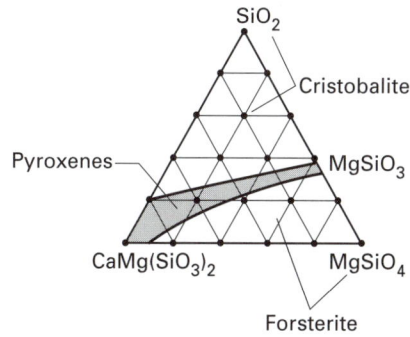

Fig. 4.15 If different ratios of SiO_2 and $MgSiO_4$, plotted on the right-hand of the triangle, are added to different amounts of $CaMg(SiO_3)_2$ (left-hand, bottom corner of triangle), then for some ratios of the three substances there is only a continuous range of calcium/magnesium silicates, shaded area, which are some of the minerals classed as pyroxenes. For all other ratios of composition of the three, either cristobalite (SiO_2) or fosterite ($MgSiO_4$) crystallise alone at first or later, together with $CaSi(SiO_3)_2$, a non-stoichiometric phase.

but on the whole such structures are limited to very local regions in a crystal. Note how very different these structures are from those of organic molecules.

A slightly different example of the way in which ionic lattices may produce compounds of unusual stoichiometry and even of non-stoichiometry is provided by the case of elements that can form more than one ionic oxidation state (see Section 3.4.7). An example is the case of iron oxides. Iron has two simple oxidation states, Fe^{2+} and Fe^{3+}, and forms two oxides, FeO and Fe_2O_3. However, there is a third oxide, magnetite, Fe_3O_4, which can be written $(Fe^{2+})(Fe_2^{3+})(O_4^{2-})$. Again, idealised structures can be drawn, but they are only valid locally.

Finally, we turn to the case of a continuous atomic network, i.e. continuous covalent bonding. Let us take the case of Ge, which forms a continuous diamond structure. Each atom has four covalent bonds, as for carbon. Now, this is the same structure as GaAs, thus substitution of a pair of Ge atoms by GaAs units in a mixture of the two is entirely possible; this substitution maintains covalent bond integrity in the lattice. Once again, size and the introduction of partial charge separation, none in pure Ge but a little in $Ga^{\delta+}As^{\delta-}$, will limit the amount of GaAs that can be put into Ge, but non-stoichiometric materials are readily made from Ge + Ga + As. These substances are of extreme importance in the preparation of, for example, chips of electronic semiconductors. Once again it is important to remember that the structures may contain locally ordered, but some disordered, regions.

The finding that solid substances can have ranges of composition but that the ranges have limits allows combinations to be made with useful properties, widely different from those of their pure components, that are put to functional use in very sophisticated ways. Examples are the use of copper in brasses, of carbon in steel, of Mg/Al combinations in light alloys and copper/lanthanide oxides in superconductors.

One may well wonder that if this is true of solids how much more so is it true of liquids and gases. The answer is that it is so, but there is a dis-

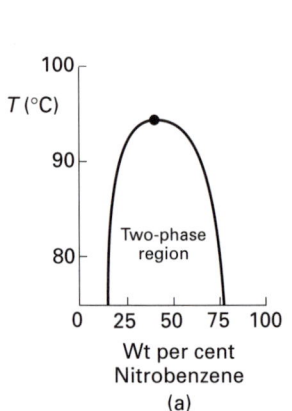
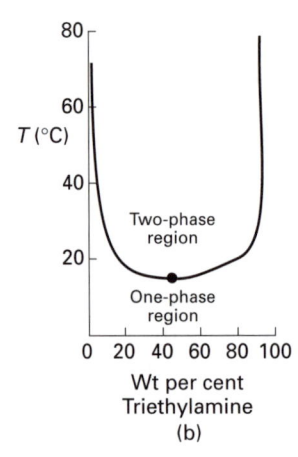
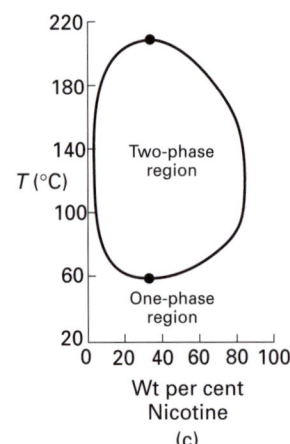

Fig. 4.16 Partial miscibility of two liquids. (a) *n*-Hexane + nitrobenzene; (b) trimethylamine + water; (c) nicotine + water. Critical solution temperature (CST) values are indicated (●).

tinction, however. Consider a gas. It consists of *separate* molecules, and although they can be mixed in all proportions the properties of the individual molecules are maintained. They can even be separated by suitably sized filters. Thus, *no co-operative property can develop* in mixtures of gases. A liquid is different again since, if it is composed of two kinds of discrete molecules or atoms, it can often behave as a sum of the two, as in a gas, yet in other cases it behaves in a somewhat co-operative manner. Liquids may, therefore, separate spontaneously from one another, e.g. oil and water, or can form limited compositional regions of one phase (Fig. 4.16) while gases cannot. (We describe the special case of dilute solutions in Chapter 6.) However, in all liquids any order is very local and all molecules are mobile. The description of biological membranes made from lipids is of a non-stoichiometric phase. (Note that the reason for the existence of the liquid state is not easily explained and requires considerable mathematical insight into the factors promoting order and disorder, as described in Chapter 5.)

Looking at the variety of examples given we may well consider that a major weakness in the teaching of inorganic chemistry, which has separated it inadvisably from geochemistry, is the stress in the former on valence theories based on simple stoichiometries, suitable for cases such as NaCl or MgO and for organic compounds, the logic of which fails for more complex materials. Many materials of practical interest are not stoichiometric and cannot be analysed using such theories since it is overall co-operative energies at given temperatures that decide stable systems and their properties. Which association will exist over what composition range is related to the discussion of the balance between energies aiding order and disorder (see Chapter 5). It is often hard to appreciate the properties that arise in these non-stoichiometric phases of variable composition. This is true also, in part, of the problem of what kind of phases will exist in living systems. We shall now illustrate this in the case of some biological compound assemblies that fall, apparently, into the same class of non-stoichiometric materials as alloy and mineral phases. It is extremely interesting to ask how well structured (ordered) are these biological assemblies. We consider some simple examples first.

4.8   Mixed bond types in larger molecular assemblies

Until now, we have treated ionic substances and continuous covalent lattices using only atoms, but we must also see that we can build substances from covalent *molecular ions* of opposite charge, or even covalent molecules, larger and larger in size if need be. A simple example is ammonium carbonate made of a positive ion, ammonium (NH_4^+) and a negative ion, carbonate, (CO_3^{2-}). Let us see why they exist.

The ammonia molecule is NH_3 and, like water, it has a polar character due to the two electrons not bound to H (see Fig. 3.14). If we add H^+ to it, then electron-pair donation from N to H^+ occurs and gives NH_4^+, the ammonium ion (Fig. 4.17). Now, consider CO_2 and note that it is a linear

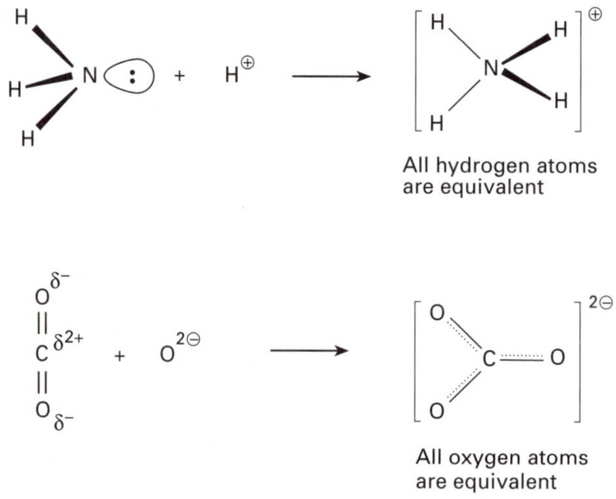

Fig. 4.17 Examples of formation of larger ions, NH_4^+ and CO_3^{2-}. Note the partial double bonds formed in the CO_3^{2-} anion.

molecule where the middle carbon has some positive charge, since the oxygen atoms have a stronger affinity for electrons (see Section 3.4.4). Adding O^{2-} gives us CO_3^{2-}, in which all the oxygens are now equivalent so we should not represent them differently (Fig. 4.17). Neither NH_4^+ nor CO_3^{2-} are stable in a gas phase, but if we consider a hypothetical experiment in which we mix them, then, just as we might mix Na^+ and Cl^- to give a lattice solid (Na^+Cl^-), so we get $[(NH_4^+)_2(CO_3^{2-})]$ which is a stable stoichiometric salt. Another example is provided by the different iron sulphides, such as FeS, comparable with NaCl and MgO, and FeS_2, pyrite, which contains the covalent S_2^{2-} unit, comparable with O_2^{2-} in sodium peroxide (Na_2O_2).

Given such starting examples we may extend them to other molecular ions and build many larger units, for example by inserting $-(CH_2)-$groups between $-NH_2$ or $-CO_2H$ units, making organic chemicals such as $(NH_3^+-CH_2-CH_2-NH_3^+)$ $(Cl^-)_2$ and $(^-O_2C-CH_2-CH_2-CO_2^-)$ $(Na^+)_2$. There is no reason for having a preference for a short chain in these molecules, so that we can generalise and write

$$H_3H^+(CH_2)_n \; NH_3^+ \; (A) \text{ and } {}^-O_2C(CH_2)_n CO_2^- \; (B)$$

Opposite charges will combine together and can form a hypothetical molecule $(A^{2+})(B^{2-})$. This then leads again to the formation of a stoichiometric ordered salt lattice like Na^+Cl^-. In some ways crystals of proteins are like this but with a big difference: they are formed from polymerisation of single bifunctional molecules called amino acids which have the general formula

$$H_3^+N(CHR)_n CO_2^-$$

where R can have any one of about 20 different chemical forms (see Table 10.3). Some hundreds of these units can be linked together to give polymers (polypeptides and proteins) that tend to fold into discrete molecular shapes. We describe them in detail in Chapter 10. There is a vast range of these and other polymeric molecules, leading to an extremely large series of chemical molecular species which have most irregular shapes in their folded forms. Amongst these polymeric molecular species are the nucleic acids RNA and DNA, polymers of nucleotides (see Section 10.4.1) which, like proteins, are all ions of salts.

Clearly, just as we consider stoichiometric solids based on $A^+ \cdot B^-$ we may extend our description considering the mixing of stoichiometric molecules of the kind just described above, when we could again obtain stoichiometric associations. As before, however, we may also consider the mixing of any number of units of, for example, different proteins, and see what speciation results due to co-operativity of packing involving many different such units. We expect that some fits in size and charge of these large irregularly shaped molecules will be better than others, so that they may form structures of no easily described stoichiometry $A_n B_m C_p$, etc., composed of many varieties of protein molecules, since proteins give rise, like metal ions, to a wide variety of charges and shapes (see Section 7.6). Such systems may well remain stoichiometric, in fact, but the stoichiometry can be extremely complex, or the combinations may exist over considerable ranges of composition. We can compare them to certain clay minerals.

Although it is difficult to make them, in principle it is possible to consider also covalent lattices comparable to GaAs, where Ga and As are replaced by molecules all covalently linked to one another in an assembly.

4.9   Structural isomerism

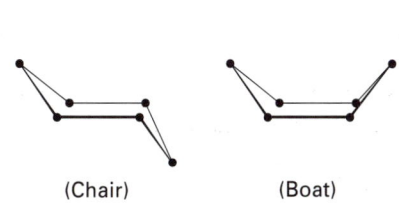

(Chair)   (Boat)

Two forms of cyclohexane, C_6H_{12}; carbon atoms only are shown.

The variety of combinations of elements is increased by the fact that, in principle, it can vary even for the same composition by altering the order of the bond arrangements, e.g. A–B–C is not the same as A–C–B or C–A–B; these are called structural isomers. If there is no exchange or rearrangement of atoms each one becomes a component, even though only one is stable. It is observed that this is a property of local (covalent) bonding, and in practice the occurrence of such structural isomers is not usually found in small ionic or metallic associations which form in the most stable state under given conditions (see aside). There are, then, structural isomers of limited kinds due to the geometric disposition in space of certain atoms and groups of atoms in (covalent) molecular structures, for example geometric and optical isomers (see aside and Figs 3.16 and 3.17). Notice that the angles between bonds are fixed, so that the number of such isomers is limited (see Section 3.4.5). In the case of optical isomers, the co-operative binding in a molecular lattice may be such that molecules of one optical handedness bind preferentially to their own kind. Such selectivity in solids is particularly strong, and is also extremely important in living organisms.

In bulk *solids* there may exist structural isomers that are called *allotropic* forms (different forms of occurrence of elements or compounds). Typical examples are the forms of phosphorus (black, red and white), $CaCO_3$ (aragonite, calcite and vaterite) and SiO_2 (quartz and cristobalite), and again relative stability is temperature (and pressure) dependent. For alloys and minerals, phase diagrams indicate the temperature range of stability of particular structures of the same formula. However, sudden cooling of a complex liquid system can result in an arbitrary element distribution in a lattice, a stationary state, of almost any variety.

4.10 Order and information: properties of linear assemblies

As we have stated, when atoms are joined together in a linear sequence or a more complex order by *covalent bonds* then they are not able to rearrange easily to form the most stable structure. Thus, if we construct a covalently ordered molecule (see aside)—no matter the complexity of A and B—such as

ABABABBA

or

ABBABAAB

Ordered sequences

DNA, RNA
Proteins
Saccharides
Some man-made polymers

we find that each form has its own one-dimensional pattern which it maintains in liquid and solid state, and even in the gaseous state, if they can be vaporized without decomposing. In this case, each molecule with a given sequence, and which may fold specifically, has a special 'information' content rather like that of the letters of a word. In this sense, a long string of units can carry information, i.e. a message, as long as the machinery sensing the order (sequence) behaves differently on reading different orders or sequences. The importance of such structural chemistry is that it forms the basis of the genetic code in the molecule DNA (see Chapters 10–16) which clearly has a parallel with a language in that it carries information that can be transcribed (into RNA) and translated (into proteins). It does require a reading machine.

4.11 Chemical speciation

At the beginning of this chapter we observed that the substances around us were not continuous in their elemental chemical variation so that not all possible numerical combinations of chemical elements were observed. We describe this observation by saying that chemicals are *speciated* at room temperature. We earlier used the word component to describe an independent combination of elements, so that *true* speciation, corresponding to the observation of independent components, results from an inability to exchange any atoms with other speciated associations. We have gone on to show that one factor leading to speciation, whether the material is gaseous, liquid or solid, is local preferential binding between atoms, which has both a strength and a direction, and gives a small number of stoichiometric links between neighbouring atoms. However, in the liquid, and more especially in the solid, state we found that speciation may also depend on long-range *co-operative* forces between molecules, atoms or ions. Speciation is not then restricted to a small number of near-neighbour atoms in stoichiometric array, but can be observed even over considerable parts of bulk space. The way in which co-operative forces act on a very large number of atoms often allows only *limited* variation of composition, e.g. in alloys or minerals, generating series of substances (Figs 4.14 and 4.15). These substances are usually non-stoichiometric and even disordered and we called them 'phases' (see also Sections 5.8 and 5.10). These phases then became the recognisable chemical species. The conclusion is that speciation can be dominant in solids as a result of the co-operative association of atoms in a way different from the simple cases of fixed combining ratios in small molecules. Clearly, this applies also to the minerals of Earth. As stated above, mixing of species of molecular gases leaves local molecular speciation but allows continuous variation in the bulk phase so that bulk speciation is not observed. Often this is also true of mixtures of molecules in the liquid state, i.e. solutions. We have also looked at the possibility that speciation, stoichiometric or non-stoichiometric, could occur through combinations of large molecules such as proteins and we shall return to this topic in Chapters 10–16 when we describe biological speciation. Chemical speciation is usually associated with physical form (shape).

Notice that all these descriptions of speciation, analytical or structural, are based on stable or stationary static states, but we shall have to consider later whether or not this approach covers all the possibilities of speciation open to a *dynamic* steady state. So far we have used the variations of composition, components and their distribution in space, but not the variation of this distribution in time, to define species.

4.12 Co-operativity and potential energy in crystal lattices: a question of mathematical linearity

We have described the co-operativity of interactions in liquids and solids as the increased stability resulting from additional negative potential energy relative to that internal to a pair of atoms or a molecule on forming an assembly. A question then arises: is this potential energy, variable from gas to liquid to solid, i.e. a *linear* property of the structure deducible from knowledge of simple parameters, where the parameters concerned must be those used in describing diatomic interactions such as charge and kind of atom, its radius, polarisability, electronegativity, etc.? If we take a simple ionic lattice as an example, the energy can be expressed in the form of a geometric expansion of a linear kind. A summation then gives the *linear* formula for the attractive electrostatic potential energy

$$\textit{Energy (attractive)} = N_A A \sum_i \frac{z_1 z_2 e^2}{d_i} = N_A A \frac{z_1 z_2 e^2 M}{d_i}$$

Table 4.6 Madelung constants for several crystal structural types

Structural type	Madelung constant
Sodium chloride (NaCl)	1.748
Caesium chloride (CsCl)	1.763
Fluorite (CaF$_2$)	2.519
Rutile (TiO$_2$)	2.408
Zinc blende (ZnS)	1.638
Wurtzite (ZnS)	1.641

where N_A is the Avogadro number, A is the constant for electrostatic (coulombic) interactions of pairs of charges (see Section 2.2), d_i are the distances between the pairs of ions considered and M is the so-called Madelung constant, a geometric factor that characterises the symmetry of the lattice (Table 4.6). In fact, this is the expression of the electrostatic potential energy derived in Section 2.2 but referred to a mole (hence multiplied by N_A) of a given crystal with a particular structure, i.e. a particular disposition of its ions characterised by the Madelung constant.

The same type of geometric sum, a measure of internal co-operativity, can be used for other (weak) interactions due to molecular dipoles, permanent or induced, which we have discussed under van der Waals forces (see Section 3.4.1). This treatment is, then, valid for crystals of neutral molecules, i.e. molecular crystals, but also for all lattices where transient mutually induced dipoles due to the polarisabilities of the ions have to be considered (giving rise to so-called dispersion, or London, forces, one of the components of van der Waals forces).

The expression for the corresponding potential energy is

$$Energy \text{ (attracting)} = -N_A \frac{CXX'}{d^6}$$

where X, X' are the intensities of the induced dipoles and C is a geometric factor which can be taken as equal to the co-ordination number since the energy decreases rapidly with the distance d so that only the immediate neighbours need to be considered. As expected, the energy of these interactions is comparatively low and is usually disregarded in the case of ionic crystals except when the ions are large and highly polarisable.

Now, as before (see Section 2.3.1), we must also take into account the short-range repulsion of the electron clouds. Here, again, since orbitals decay exponentially and rapidly with distance, we need only consider repulsion between neighbouring atoms and multiply by the Avogadro number to refer the value to one mole of the crystal.

Therefore, the corresponding repulsive potential energy is

$$Energy \text{ (repulsive)} = + N_A \frac{B}{d^n}$$

where n now depends on the type of electronic configuration of the ions or atoms (see Table 4.7) (and is used empirically in place of an exponential decay) and B is a geometric constant. Since atoms may have different

Table 4.7 Values for n for several electronic configuration types

Electronic configuration	Examples	Value of n
He	Li^+	6
Ne	F^-, Na^+, Mg^{2+}	7
Ar	Cl^-, K^+, Ca^{2+}	9
Kr	Br^-, Rb^+, Sr^{2+}	10
Xe	I^-, Cs^+, Ba^{2+}, La^{3+}	12

electronic configurations one usually takes an average value for n, for example $n = 8$ for NaCl.

The distance, d, to be considered in all these expressions is, of course, the equilibrium distance, r_0 (see Fig. 2.6) but note that the energy values are referred to one mole of the crystal, not to pairs of charges. As stated before, this distance corresponds to the minimum of potential energy and is easily calculated by making $dE_{total}/dr = 0$, which gives a value for B as a function of M and finally the expression for the 'lattice' energy:*

$$Energy_{tot} = - N_A A M \frac{|z_1 \cdot z_2| e^2}{d_0} (1 - \frac{1}{n})$$

As stated in Chapter 2, A is a constant, which in the international system of units has the value 9×10^9 N m²C⁻² (newton square metre coulomb⁻²).

Co-operativity of potential energy is, therefore, a linear function of the simple structure variables of charge distribution at 0 K. Given such linearity, the energy of binding, but not many other properties of co-operative structures, becomes reducible to the sum of the binding energies of pairs of atoms; however, on increase of temperature (which increases r_0) the ordered lattice loses stability and eventually melts to a disordered liquid. This order–disorder dichotomy requires consideration of new variables related to temperature (and pressure), and will be analysed further in Chapter 5.

4.13   Summary

We have now described the main types of chemical bonding and the stabilities of the resulting selectively *ordered* substances or compounds at given temperatures: these are summarised for convenience in Table 4.8. Depending on basic physical characteristics of the fundamental units (atoms, charged or uncharged) such as size, ionisation energy and electron affinity, (and orbital kind and occupancy), different types of substances are formed, ranging from close-packed metals to ionic crystals and covalent infinite networks, and from discrete molecular substances to infinite linear

---

*Without correction for van der Waals forces.

Table 4.8 Main types of chemical bonding and properties of substances

Nature of substance	Properties (at room temperature)	Examples: elements and binary compounds
Covalent bonding. Sharing of pairs of electrons between atoms (single and multiple bonds); directional bonding; usually low co-ordination number		
Discrete molecules	Gases, liquids or solids, depending on the intensity of intermolecular forces; relatively low boiling and melting points; non-conductors	H_2, CO_2, H_2O, I_2, C_6H_6, glycine, etc.
Atomic networks	Solid; usually high melting point; hard and insoluble; non-conductors	Graphite, diamond, quartz, SiC (carborundum), BN (borazon)
Ionic bonding. Non-directional electrostatic attraction between positive and negative ions; intermediate co-ordination numbers (usually 4 to 8)		
Ionic crystals	Solid; relatively hard and brittle; more soluble in polar than in non-polar liquids; high melting points. Non-conductors as solids but conductors as melts and, especially, in solution	NaCl, KBr, Li_2O, MgO, BaO, etc.
Metallic bonding. Non-directional attraction between positive ions and delocalised valence electrons; higher co-ordination numbers (8, 12)		
Metals and alloys	Solid (except mercury); high melting, malleable and ductile; high thermal and electrical conductance; insoluble unless reaction occurs with solvent	All metals, e.g. Fe, Cu, Ni, Ag, etc., and alloys
Mixed covalent/ ionic bonding	Solid, malleable, low melting, support ionic conduction	Many biological materials formed from, for example, fats and proteins giving membranes, flesh and skin.

or branched chains. There are still other intermediate types of association, for example, in the so-called liquid crystals, which we will not consider in any detail in this book. Remember that the only fundamental variables considered in the chapter are the composition based on units (atomic masses and charges) and the derived variable potential energy of the arrangements of electrons around nuclei of atoms in association that results from the spatial distribution of atoms in space, where space distribution is another fundamental variable. Effectively we are examining structure and properties at 0 K.

Amongst the compounds formed we observed that many had stoichiometric ratios and hence a series of discrete compound species A_xB_y could be obtained, for example $CH_3(CH_2)_nCH_3$. Rules for the combining ratios, *valence* rules, were easily found and explained in many cases. To a large degree, this formation of discrete compounds in series was restricted to molecules dominated by non-metals without regard to the phase they were in. Even very large stoichiometric molecules are readily observed for combinations of, especially, H, C, N, O, S and P in covalent compounds such as proteins and DNA. Discrete compounds are also formed when non-metals were bound to metals, for example in binary salts such as NaCl, but the identity of their molecules is lost in solids, melts or solutions. In these combinations, principles based on electrostatics, potential energy (attractions and repulsions), showed that the variability within chemical composition of the combination of *two elements* is again restricted, i.e. we observe particular ratios and ordered structures. However, this was not found to be a general rule where metal elements or where a larger variety of ions in metal/non-metal combinations are involved. When metallic elements are combined in alloys, A_xB_y, or in minerals, $A_xB_yC_z$, it is usual to find regions of *continuous* variation of $x:y$ or $x:y:z$ between certain ill-defined ratios which were called phases. Moreover, structure is now often disordered. As we increased the number of elements that we combine in this and other condensed systems, solids and especially in liquids, then continuous variation became more and more common. In particular, multi-component liquids are miscible over wide ranges of composition, many solids also dissolve in some liquids over wide ranges and 'solid solutions' are common in complex salts, alloys and related materials. Detailed examination of the nature of these different chemical assemblies led us to describe chemical speciation in different ways (see Table 4.8). We have seen too that the ranges of existence of phases is *temperature dependent*. This is the first physical variable we have introduced. The way in which temperature affects the coming together of atoms in molecules, lattices or networks will be examined further in the next chapter, but clearly it favours gases.

In this discussion we have described the properties of the substances as if they existed independently from any field external to them or from necessary contact with other chemicals or states of matter, and without any restriction as to the amounts present. Moreover, we have not described the effect of pressure, the second physical variable we have to introduce, upon them. For example, a given pure gas or liquid consists of identical atoms or molecules and we did not consider the volume of the container as either limiting or distorting its properties. The shapes of *molecules* were described as if they arise through internal interactions between atoms that are in one molecule alone, which is reproduced precisely by every other molecule in the system, and without regard to their environment. This is not our experience of the real world of objects, which have surfaces and which interact with each other. We shall return to the problem of these real systems in Chapter 7.

In such a discussion we could not handle properly other important features of the real environment and clearly we must now recognise additional features of everything that is around us, for example, the balanced co-existence of different states of matter (gas: liquid: solid) of any one substance. This is again part of the subject of the next chapter, which forces upon us a statistical approach to bulk materials rather than the one used in this chapter, which centres on single atoms, molecules or continuous crystals, describing them in terms of pair-wise interactions. Inadvertently, we see that we have been attempting to reduce everything we observe to simple, mathematically linear or additive relationships between the effects of the possible variables (here space) upon the fundamental units (the atomic elements of the periodic table and any charge carried). The next chapters will show the limitations of this approach.

5

The balance between order and disorder

He who wants to have right without wrong,
Order without disorder,
Does not understand the principles of Heaven and Earth
  Chuang-Tzu, Text. *Autumn Floods*, Section 17 (adapt. Thomas Merton)
                                                    (sixth century BC)

5.1   Introduction

In Chapter 1 we took, as an historical starting point for the discussion of all observations concerning the material objects we sense around us, the four 'elements' of certain Greek philosophers and scientists such as Aristotle (see Fig. 1.1). We also noted the relationship to similar thinking in Chinese philosophy (Fig. 1.4). We saw that both developed a logical connection between the observable material substances ('elements') and variables ('qualities') which adjusted them. In this book we wish to pursue

Table 5.1 History of units and variables

	Units	Variables
1000 BC to AD 1800	Air, water, earth, fire	Composition, moistness, hotness, dryness, coldness, (Derived from senses)
AD 1800–1900	Atomic elements and charges in chemistry and physics; photons	Composition, energy,* pressure and temperature derived from kinetic theory of gases. Functional thermodynamics; momentum
AD 1900–2000	Fundamental particles not useful in chemistry or biological sciences	Nuclear composition; energy*
	Atomic elements and charges. Molecular units in all sciences Radiation: photons and waves	Composition, energy;* statistical analysis of particle distributions. Fluxes related to space–time co-ordinates.

* The various forms energy can take are described in Chapters 2–5.

a similar logic while bringing it in line with modern knowledge, that is, we wish to describe the observable world around us in terms of 'chemical' composition and 'physical' variables, and to show why any substance or system can change. The development from the classical scheme is shown in Table 5.1. Later we can ask about the rate of change.

We have shown so far that, according to today's views, three of the classical Greek 'elements'—air, water and earth—are more or less directly related to the physical states of materials (gases, liquids and solids) (Fig. 1.2), which are dependent on the variables temperature (random kinetic energy), pressure (random momentum) and volume (see Chapter 2). We have also described directed kinetic energy (flow) but have delayed further discussion of it until Chapter 8. Moreover, as examined in Chapters 1 and 4, materials themselves of every kind are now known to be made from a limited number of units, the *atoms* of the *chemical elements* of the periodic table (Fig. 1.6), which *together with charge*, are the ultimate *units* to be considered in the fundamental variable *composition*. These elements came together in *ordered* constructs to make molecules, which appear as gases, liquids and solids, and other condensed systems, including giant networks and continuous metallic or ionic lattices, giving rise to all chemical material things on Earth (and elsewhere). As we have seen, the coming together of atoms of the same or different elements is due to what we call binding forces, which can be said to be related to the Greek idea of underlying 'form'. Note that this is a description of the observed distributions of atomic units in *space*, a further fundamental variable, and the 'reasons' for their appearance. The major forces of concern here are those we have labelled electrostatic (see Chapter 2) and the same is true of the major forces acting between small charged bulk objects (see below).

In Chapters 2 and 3 we found, however, that we had to describe internal electrostatic interactions of electrons with nuclei in a way different from bulk forces, since the first were variable in a discontinuous way (they are quantised, as we described in Chapter 3) while the bulk forces could be treated as smoothly continuously variable. As a consequence, we treated the sum of the internal energy variable together with the element composition and charge variables of ordered atomic assemblies in an *operationally separate way* by describing composition in terms of the *components* of the system. These are the common separatable chemicals we handle, together with any charge they carry, which do not exchange their elements or charge when mixed. They are discontinuously related to one another in composition and their concentrations can be varied independently. As pointed out in Chapter 2, this operational treatment of the composition has its own limitations since exchange of atoms or charge is not only temperature and pressure dependent but varies with the composition of the system under consideration, for example in the presence or absence of a catalyst. In a system which is not undergoing change, however, the definition is clear in that a component refers to chemicals in stable or stationary states, independent of all other chemicals present. Two circumstances will be left until later: (1) chemicals in equilibrium (see Section 5.9); and (2) chemicals undergoing change (see Chapters 8 and 10), where components are not independent. Under point (2) we shall be deeply involved with consideration of the third fundamental variable, *time*, the first two being *composition* (in terms of components) and *space*.

In Chapter 2 we also described, separately from atomic interactions, smoothly continuously variable *bulk* gravitational and electrostatic fields between objects of different masses and charges which may be at different temperatures and pressures. As well as describing stable states, such considerations deal with fixed large objects separated in space which have not been allowed to collapse to the greatest stability of the whole system. We included them again in stationary systems. This is the case for many objects we observe around us. For example the Earth lies in the gravitational field of the sun, the moon lies in the gravitational field of the Earth, and electrical condenser plates can be held apart while differently charged. In some of these examples, the bulk objects, e.g. the Earth and the moon, also had fixed *averaged* angular momentum. Such systems are cyclic dynamic stationary states, their properties are *time independent*, and are treated by classical Newtonian mechanics, using mass, charge, space and fixed cyclic velocity as the continuous variables.

The fourth Greek 'element'—fire (or heat and light)—can transform solids to liquids to gases; the Greeks connected it also to underlying 'form' and therefore made it similar in kind to the other three elements. As stated in Chapter 2, it was natural enough to consider fire in this way as it was given out as heat and light, felt and seen, so that it seemingly had to be an intrinsic part (of the composition!) of observable substances from whence 'fire' came. Today we say instead that 'fire' is related to another derived variable under the heading of *energy*, particularly *radiant energy*. We noted in Chapter 3 that radiant energy travelled as photons which should in fact be considered separately under units of the composition of the universe. In this book we shall largely avoid this complication.

The isotropic uptake of *energy* (e.g. heat or light) by materials, as we showed in Chapter 2, acts to increase their *internal energy* content, thus weakening bonds, and the random motion of their constitutive units, hence disorder, which opposes the forces that bring chemical (atomic) elements together generating order (see Chapters 2 and 4). Thus 'fire' is related in part to an increased intrinsic (positive) energy content of materials, which we called chemical potential energy (remember that stability is associated with negative potential energy), and to increased random kinetic energy, i.e. increased random motion of atoms and molecules (see Section 2.5). We had to stress, especially in Chapter 3, that all these properties of atoms and their combinations, unlike those of bulk objects, could only be analysed properly by quantum mechanics. In that chapter, while we described quantised stable and stationary states of electrons in atoms, we allowed such states to include both quantised angular momentum and potential energy of electrons in orbital motion, giving them a fixed directional kinetic energy and hence a fixed total energy. The changes in atomic states were then related to the uptake or loss of photons.

In Chapter 2 we stressed throughout that our discussion of both these cases, quantised or continuous, was centred on single-body properties or on pair-wise interactions (two-body problems). Now, for a large number of particles, particularly the case of atoms or molecules in even a tiny amount of matter, this treatment of dynamic properties was not solvable. In a gas we have to deal with an almost unimaginable number of multiple interactions of randomly moving particles (Fig. 5.1), and classical Newtonian mechanics is of little use in such circumstances, even though it is correct in principle. Despite the fact that the problem is apparently without solution, the behaviour of gases is regular and predictable in terms of simple gas laws involving the variables temperature, pressure and volume (see Section 2.5.3) so that there should be a way of relating the microscopic motion of the constitutive particles to the observable bulk properties. As mentioned in Chapter 2, this was achieved at one level by James Clerk Maxwell, a Scottish physicist, who developed the so-called 'kinetic theory of gases' in the years 1860–1865 based on the idea that the observable properties of gaseous bulk systems could be interpreted in terms of *statistical averages* of the properties, in particular of the root mean square speed of the constitutive hypothetical particles (assumed to have no volume) which give rise to a temperature and a pressure. In Section 2.5.4 we referred briefly to these average values together with Newtonian mechanics to describe the properties, and especially the energies, of gases

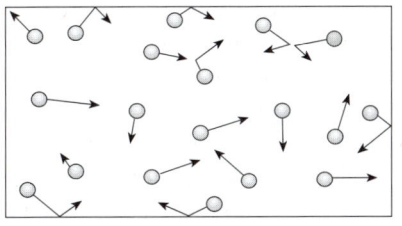

Fig. 5.1 Individual molecules in a gas have different space co-ordinates and different kinetic energies at one time. The assembly has a vast number of arrangements of spatial and energy distributions: each is a microstate within a macrostate condition. The macrostate is said to be increasingly disordered as the number of microstates occupied increases.

Table 5.2 Some quantitative expressions relating variables

Derived variable	Mathematical treatment
Potential energy	$\frac{Gm_1m_2}{r}$ (gravity); $\frac{Az_1ez_2e}{r}$ (electrostatics)
Kinetic energy	$\frac{1}{2}mv^2$
Momentum	mv
Volume	$pV = RT$ (perfect gas)
Composition	Linear sum of components or of atoms and charges
Pressure/temperature	Derived from kinetic theory of gases

with relation to their temperature and pressure; however, this theory did not lead to an understanding of changes of physical state or to the properties of liquids and solids. Always present in these developments was an effort to express properties of everything around us in quantitative mathematical equations (Table 5.2).

The introduction, around 1890–1900, of the full probabilistic/statistical approach to particle ensembles by Boltzmann led to the development of the new science of 'statistical mechanics', which has become the theoretical foundation of *thermodynamics*. Thermodynamics dealt initially with the transformation of heat into mechanical work and of mechanical work into heat, e.g. the Carnot cycle (see Section 5.4.2), but has been extended to a much larger range of phenomena (e.g. see Table 5.1). As we shall see, it leads to an understanding of, for example, changes of physical states of substances. The statistical treatment can be numerically evaluated using either quantum or continuous (Newtonian) mechanics.

In this chapter we use this statistical approach to uncover progressively a new treatment of the variables temperature and pressure (volume) which contribute to the properties of all *bulk materials* around us, such as melting and boiling points. *We shall see that this leads to an understanding of the ability of a system to do work, and its limitations to do so, which is so important since it tells us much of what generates the ability to cause change.* No new fundamental variables (see Table 2.3) are introduced in this equilibrium treatment, but we shall find it necessary to derive operational thermodynamic variables suitable for handling bulk physical chemical problems. We shall consider first the spatial distribution of large numbers of atoms or molecules in a single system (see Section 5.2), which is related to the pressure through their number in a given space. The total amount of material in, and the particular size and shape of, the system are not examined. (We shall return to these variables in Chapter 7.) We assume here that there is no influence of external fields (see Chapter 7).

5.2   Systems, change and variables (functions) of state

The problem which we must now face is that, since the description of large numbers of particles cannot be reduced to simple sums of properties of pairs of them, we need to formulate a new set of derived variables which will be operationally useful in handling the capacity for change of these large systems. This is the subject matter of *equilibrium thermodynamics* which starts by making distinctions between *systems in balance* examined under different conditions of the variables. Functional thermodynamics deals with derivations of relationships between these variables, while statistical treatment of the occupation of space and time co-ordinates gives these functions detailed meaning. In the latter case we have to handle quantised energies, but in this book we shall largely avoid this approach while using a classical treatment. *It must be understood that the variables we shall now define are variables (functions) of states defined per gram mole and not of individual units or pairs of units.* Before describing the different types of system which are analysed with particular variables, fixed at a constant value, e.g. pressure, we illustrate the change in attitude to variables by referring again to the energy of interaction between units in a system.

5.2.1 Thermodynamic variables of state of bulk multiparticle systems: introduction to work

In Chapters 2–4, we have described potential energies of interactions between pairs of particles and we have described temperature or pressure in terms of the averaged properties of single units, but in reality these are obviously variables that belong to multiparticle systems. For example, we need to have a variable *internal energy*, U, for a system which describes the potential energy of interaction *per mole at a given temperature and pressure*. This potential energy, which is a function of the state of the system, is not simply related to pair-wise interactions, since even they are temperature and pressure dependent. Change in U represents in part the *capability to do work*. However, the system also has a bulk molar volume, V, at a given pressure, p, and the *total* function of state, characterising these systems and due to these conditions is given by the sum

$$U + pV$$

Changes in this function represent a *total capacity* of the system for doing work, W, but we must not conclude without enquiry that it is all *available* to bring about change in other systems (see Section 5.6). In chemistry and biology we are usually concerned with processes at constant pressure, when a change in the sum, $\Delta U + p\Delta V$, is given the symbol ΔH, which represents now the *heat content* (or *enthalpy*) *change* per mole at constant pressure. (Note that we used measured ΔH, written as Q, in Fig. 4.11 as equivalent to pair-wise energies per mole, but this is only exactly true at 0 K. However, it is a good approximation.)

Now there are other contributions of systems to the availability of work which arise from considerations of the statistical probabilities of systems, which we have not yet examined. Before we deal with these statistical considerations (Section 5.3) we must define the different conditions under which we can examine the availability of work (or ability to impose change) of bulk systems, remembering that the two derived variables which involve transferable items within work are component compositional chemical and physical energy, where the energy concerned is not yet fully defined for bulk systems, see section 5.3.

5.2.2 Thermodynamic systems: definitions

We need first to define systems of large numbers of particles under different physical constraints. An immediate major difficulty in discussing 'systems' is that the word has a precise meaning in thermodynamics but a fuzzy one for the chemist or biologist! In books on thermodynamics a system is an object (or substance), or a set of objects (or substances), which is separated from all others by boundaries, real or imaginary (see Figs 5.1 and 5.2), and it is necessary to state which parts of it are in equi-

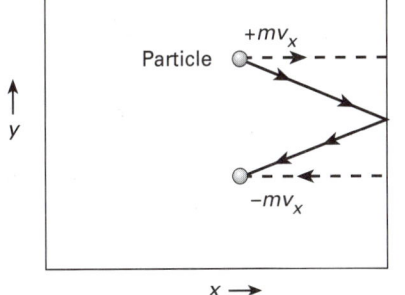

Fig. 5.2 The pressure of a gas (bulk property) arising from the collision of its constitutive particles with the walls of the container. In an elastic collision with the wall (perpendicular to the x-axis) the x component of the momentum is reversed but the y and z components are not altered. The change in momentum is, obviously, $2mv_x$.

Table 5.3 Situations used to define systems

System	Connection to environment	Internal constraint
(1) Static single body		
(a) Isolated	None	
(b) Closed	In energy exchange	Equilibrium
(c) Open	In material/energy exchange	
(2) Static multiple bodies		Each body in
(a) Isolated	None	internal equilibrium;
(b) Closed	In energy exchange	no overall equilibrium
(c) Open	In material/energy exchange	
(3) Dynamic, stationary multiple bodies		As for (2) but each body can have
(a), (b) (c)	As for (2)	fixed motion relative to others
(4) Steady state of single/multiple bodies		
(a) Sun/Earth	(a) Constant energy flow	Not in equilibrium
(b) Adult animal	(b) Constant material/energy flow	
(5) Developmental state of single/multiple bodies		
(a) All cells	(a) and (b) continuously changing	Not in equilibrium
(b) Young animal	flows	

librium or balance. A system can be *isolated*, when the boundaries do not allow transfer of energy or mass/charge to or from it, so that it is independent of even its environment. Alternatively, a system is said to be *closed* when only (the possibility of) energy transfer is considered, or *open* when (the possibility of) either mass/charge transfer or both energy and mass/charge transfer to the environment is allowed. The systems in all these cases are in balance, not changing, but the energies of particular possible changes can be calculated (Table 5.3). For example, in the simplest case of the expansion of a gas the analysis is of its energy change from initial to final conditions of V, p and T. However, we need a wider definition of systems, including those not in complete balance in space, that is, other than those that are describable by one simple equilibrium, for we must also include more than one set of objects interacting physically (each set separately in internal balance, but fixed in space or in constant relative motion). They too can be isolated, closed or open with respect to their environment (Table 5.3). Real processes have finite rates of change not described by this thermodynamic analysis. Consideration of rate of change introduces new kinetic (time-dependent) variables. Thus, systems undergoing finite rate of change cannot be defined in just the above terms (see Table 2.2).

Putting together the various aspects discussed in the previous chapters and the considerations above, we can define certain different general situations in which all materials can be considered to exist, related now to their capability to change and to their rate of change (Table 5.3). We deliberately consider the objects under study separately from their environment. This will link the description of the states shown in Table 2.2 to that of conditions describable by *equilibrium thermodynamic functions* of bulk systems as above, as well as others. Five different situations of systems are easily appreciated.

(1) The first, as mentioned above, is a single system which may be *isolated* (out of contact with its surroundings) so that its properties can be defined without reference to external conditions. The system has a particular physical state (defined by its physical variables) and a fixed chemical composition. (Imagine, for example, water isolated in an insulating Dewar vessel.) It is not subject to any force, for it is isolated from its environment, and it is also independent of time. This is a *stable* system fully describable by its physical and chemical composition variables (see below). Everything inside it is said to be in *equilibrium, an unchangeable, balanced condition.*

(2) A second situation is one of a system in which we consider two or more bodies or substances (which we shall generally call 'objects') fixed in space (static) but interacting with one another, so that they should move together or away from one another but, for some reason, e.g. the introduction of a barrier, this cannot happen. Internally, both bodies are in equilibrium. The objects in the system can be as large as Earth or as small as atoms. The condition of each object is relative, dependent on the presence of other objects, and is said to be frozen (energised), or *stationary*, independent of time. The ways in which it may be energised are described below. Note that overall the system has a pattern or 'shape'. It can be isolated from the environment or not, as for a single system. The interaction (or potential) energy can be altered if the barrier to relative movement is removed, when *the system, if isolated, can change and do work but only internally.* The different interactions in stationary conditions obviously introduce new possible physical variables, such as the distance apart of objects, which can affect internal energies, chemical atomic combinations, bulk interactions between bodies, temperature and pressure. Since, by definition, the separate systems are stationary, we do not allow change to occur, but we can *calculate* using *local balance in each system* the energy change which would arise if we allowed the relative differences in position of the objects to be altered or if we allowed both position and composition of the objects to change. In those circumstances, we could also consider transfers of energy to the environment bringing the whole, including the environment, into balance due to such changes. In this last situation the system would then be called *closed* (see above).

(3) A third situation is that of two (or more) objects in *constant frictionless and cyclic motion* with respect to the position of one to another, both (or all) of which are internally in balanced or stationary states. The two interact as in (2), but now there is also their relative directed velocity to be taken into account. We can consider that they form a condition of fixed cyclic relative motion which can also be said to have a constant pattern of motion, also sometimes called a 'shape' or 'structure', e.g. the planetary system (or, as we shall see, electrons around nuclei in atoms). They are in *stationary, now dynamic, states.* Since they are stationary we can again calculate the energy required to transform them or the energy they can give on changing to a new stationary state. They can again be looked upon as either closed or isolated systems depending on whether or not they exchange energy

with the environment. They are clearly capable of change. We can also treat them by equilibrium thermodynamics.

(4) A fourth situation is a so-called *steady-state system*, in which either or both energy (heat and/or light) and material constantly move through the system (from the environment). A river is one example of such a *flow* system and adult living organisms are also of this kind. These systems can be treated *as if they were closed*, when energy alone passes through them, or *open*, when energy and material moves through them; the discussion of these cases will be largely reserved to later chapters. [Steady states can be analysed approximately by equilibrium thermodynamic (activated state) considerations (see Section 8.2).] Such states are energised and can do several kinds of work since they operate in and on the environment, and the relationship with the environment can be changed. The variables now include the rates of the *in* and *out* flows of both material and energy, where time is included. The build-up and decay of such systems with time are also of interest [see situation (5)]. These conditions of material and energy can have 'shape', i.e. a pattern.

(5) The most difficult case to describe is a *developing* one in which any steady state can be visualised only momentarily before it changes into another. Around us we see many obvious illustrations, such as growth of geological, biological or man-made features. Today we are still not able to handle development with any great confidence (see Chapters 10–16).

To make for a simpler description at first in Chapters 5–7, we shall refer mainly to closed systems in balance with the environment, describable under situations (1)–(3). Our reason for classifying systems in this way is to allow us to find a suitable set of derived equilibrium thermodynamic variables, which are operationally useful for the discussion of changing of bulk systems, as opposed to the fundamental variables (see Table 2.3) which, despite the fact that they underly all our descriptions, are not so useful to us.

5.3   The physical states of bulk matter: a qualitative statistical approach to volume (pressure) and temperature

It is important to see that the revolution from the Greek view of matter as continuous to a view based on atoms has consequences far beyond the considerations we have so far introduced under the general themes of the potential energy of ordered structures and of *averaged* random kinetic energy of the material units in liquids and gases, and while treating single hypothetical averaged particles or interacting pairs of them. The consequences include a re-evaluation of the usefulness of energies associated with volume (pressure) and temperature in bringing about change. Curiously, it concerns the *relative probability* (a statistical concept) of existence of gases as opposed to liquids and solids. We must look again at the nature of bulk matter.

Taking an atomic gas as an example there are, obviously, empty spaces between occupied spaces in a given volume, but the occupation changes continuously and extremely rapidly. Thus, there are many instantaneous ways of representing the gas, which are all equivalent in potential energy (see Fig. 5.1). However, as stated above, a bulk property of a gas, e.g. temperature or pressure, is not one of these instantaneous snapshots, called 'microstates', which are unobservable (not measurable). It is the averaged sum of the properties of all the possible microstates that gives the observable (measurable) property of the so-called bulk 'macrostate'. We, therefore, imagine all the instantaneous fluctuating microstates of a gas contributing to the overall (observable) condition due to very rapid fluctuations of the distribution and velocities of gas particles. For example, at any one instant there could be many or a few atoms of the gas colliding with the walls of the containing vessel, but the property which is measured, the pressure of the gas, is a statistical description of its macrostate, i.e. the average of all possible microstates describing collisions with the walls. (N.B. As stated in Section 2.5.3, pressure is not an energy, it is a *force per unit area*, proportional to the mean change in the momentum of the molecules or atoms of a gas at the walls of a vessel per unit area and unit of time; see Fig. 5.2.)

Now, a solid, ordered material, does not exert a pressure, does not expand to fill a volume and clearly does not have the same number of microstates since its atoms are fixed and can only vibrate in the lattice to some extent. We can then ask a question: if we examine a particular system of atoms or molecules at any fixed temperature or pressure do we expect it to be a gas, a liquid or a solid? To analyse this situation we postulated the existence of attractive *forces* in Chapter 2, whose corresponding energies are here included in U, but now we have to consider the volume occupied at a given temperature since experience shows that the more we expand the volume (or reduce the pressure) or increase temperature, the more we favour the gas state. In the opposite sense, if a gas is strongly compressed or strongly cooled it forms a liquid or, eventually, a solid.

How can we explain the observation that increase in volume (decrease of pressure) favours the gaseous state, while we clearly saw in Chapter 4 that the electrostatic energy of interactions between atoms or molecules favours the condensed condition? Consider first the situation for an atomic substance, e.g. an inert gas such as helium. Clearly, if we divide space up into equal volume cubes each of which can be occupied by one atom (or molecule), then when we take a fixed number of atoms (or molecules) there are different, *equally probable* instantaneous distributions (or microstates) open to the system. Only one such distribution has all the atoms close-packed in a small region in what is the ordered condensed solid (Fig. 5.3). Several other distributions have atoms close-packed and some not, and most of them will have all atoms distributed randomly. Since every cube in a given volume can be occupied by one atom (or molecule) it is immediately clear that the same number of atoms (or molecules) has many possible ways of occupying a larger number of cubes in space, i.e. the number of ways will increase if the volume increases. If the atoms (or molecules) do not inter-

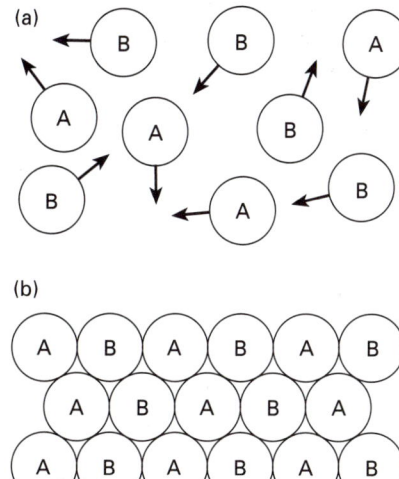

Fig. 5.3 (a) Disorder, spatial and kinetic. Note there are many possible arrangements. (b) Order. Note there is but one arrangement.

act, i.e. are not held together, they will occupy, *on average*, all the cubes equally. We refer to this as a disordered arrangement of atoms or molecules in the gas relative to an ordered state in the solid, which is very improbable in the absence of interactions between the constitutive particles. (There are only a few 'ordered' arrangements compared with far more 'disordered' arrangements.) On the other hand, if they interact strongly, then it is probable that most (not all of them) will be held together in the condensed state, that is, most of the possible arrangements are predominantly close-packed. We see that, by doing so, the number of unoccupied volume cubes will increase (and the pressure decreases). Thus, the fewer the atoms that are not condensed the more arrangements they have open to them. So there will always be a pressure of gas atoms (a so-called vapour pressure), often exceedingly small, above any condensed region, which is less the greater the interaction in the condensed state. Hence, either in the complete absence or in the presence of interactive forces between units, the more ways in which a system can exist the more statistically probable is its existence.* Random disorder is expected in the distribution of atoms and molecules in space, in a gas phase, in the absence of interaction between them, while the opposite, an ordered state, is expected if there is strong interaction between particles.

In fact, we see everywhere around us materials that co-exist apparently in balance between ordered states (due to favourable potential energy) and disordered states (due to statistical weighting) and where the balance responds to two variables, temperature and the volume open to it, as constrained by external pressure. An obvious example is water, which is simultaneously present at a given temperature and external pressure as water vapour, as liquid water and/or as ice (Fig. 5.4). Here, the water vapour is a disordered gas (except for the 'structure' in the H_2O molecules), the liquid water is in a more limited disordered state and the ice is an ordered solid. Clearly, the higher the pressure the more favourable are the condensed states. Of course, the water molecules themselves are assumed to have the same molecular structure in solid, liquid or gaseous states. Other substances, such as dioxygen, are found as disordered gases almost exclusively, and yet others, for example minerals, are in essence ordered solids. In contrast, living systems are clearly only partly ordered since they have within them much liquid-disordered water solutions. In order to understand the change from gas to liquid to solid we have to calculate the work which has to be done to reduce a gas to a condensed state. This will allow us to assess the statistical probability of the gas in the same energy units as the potential energy of the condensed phase since we have seen that this potential energy is related to a work capacity. We then have a completely fresh insight into the properties of substances, both physical and chemical.

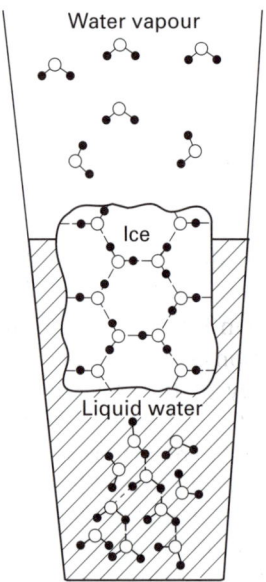

Fig. 5.4 Increasing disorder : ice < liquid water < water vapour.

---

*To see the probability argument more easily, consider coins labelled head and tail, on opposite sides, thrown on to a table. It is easily shown by repeated throwing that, for any large number of coins, the most probable condition of them is an equal number of heads and tails, not all tails or all heads. If all heads happened each time we would say that order is favoured by some special interaction of, say, tails with the table. If we call all throws that do not yield all heads or all tails disordered, then it is obvious that the ordered condition is highly improbable in the absence of some special interaction.

We shall next approach this problem of ordered–disordered conditions in a quantitative way, to complement the approach to binding energy of atoms in molecules and lattices given in Chapter 4. We turn then to one main purpose of this book, which is to understand how preferential affinity of the chemical elements for one another has generated the static physical *order* $\rightleftarrows$ *disorder* balance, i.e. a balance of variables, now collective variables of the bulk substances we observe around us, treating the question of whether particular chemical elements (in compounds) favour a particular state so that some are overwhelmingly gases, some liquids and some solids under defined conditions of the variables temperature and pressure. In other words, we shall subdivide the vast range of chemically specified substances into their different physical states using derived thermodynamic variables. This division has dominated both inanimate and animate evolution. When the question of the association of particular *physical* states with particular groups of elements and their compounds has been analysed, we go on to ask (in Section 5.6), in a parallel statistical or probabilistic analysis, about the *chemical* composition balances in materials. For example, why is water (a liquid) more stable than a mixture of two gases (H_2 and O_2) at room temperature, and why is this relative stability of water dependent on temperature. We shall be concerned only with systems (1), (2) and (3) in Table 5.3, using equilibrium thermodynamics.

5.4 Order–Disorder balance (equilibria)

We shall next consider *quantitatively* the balanced equation

$$\text{ORDERED} \rightleftarrows \text{DISORDERED}$$

for a given system, where we are concerned with a large number of particles in a system.

Whenever two conditions such as these are in balance we refer to them as *in equilibrium* and we represent this condition by the double arrow $\rightleftarrows$. We have already described the binding energy of the ordered condition quantitatively in Chapter 4, and evaluated its contribution to the bulk properties ΔU and ΔH per mole in Section 5.2.1. The factors contributing to the disordered condition [related to the variables available volume (or pressure) and temperature] have just been described qualitatively in Section 5.3. To see the influence of these factors in quantitative terms we will consider the disordering of a given mass made of particles taking into account separately: (1) the different possible *spatial* arrangements of the particles of the system, related to the pressure or volume of the system (see above); and (2) the various possible distributions of the *energy*, related to temperature (mainly the random translational kinetic energy) of the system amongst its constitutive particles. To proceed with this statistical or probabilistic analysis we wish to show that while the pure ordered state, for example a crystal, has a certain amount of potential energy per mole associated, in principle, with a perfect single arrangement due to forces of interaction between its units (atoms, ions or molecules in the cases dealt with in

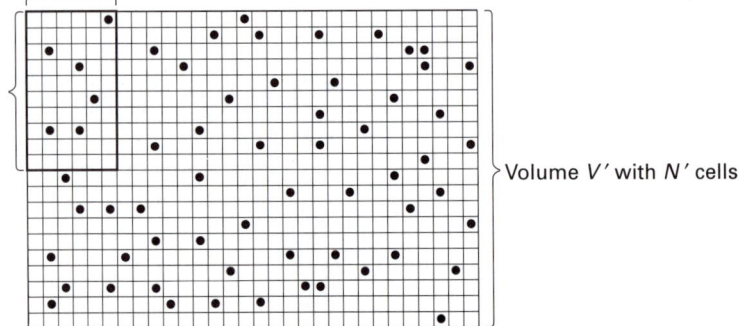

Volume V with N cells

Volume V' with N' cells

Fig. 5.5 The figure shows the relationship between the number of cells, N and N', and the corresponding volumes, V and V'. If all molecules (black circles) in V' were put into V there would be a single possible ordered state.

Chapter 4), a typical disordered state, for example that of an ideal gas, has a very different 'energy' (to be determined) related to the statistical weighting of its many microstates. As stated above, by relating both order and disorder to the same variable (energy) we can discover which state of a material—gas, liquid or solid—is the most stable under particular conditions of the measurable variables, temperature and pressure.

5.4.1 Quantitative considerations of disorder

As stated above, as the volume occupied by, say, one mole of gas increases, i.e. as the number of boxes in a figure such as Fig. 5.5 increases, then the number of arrangements of particles, i.e. the number of possible microstates, increases. The increased probability of existence of a gaseous macrostate must be related to this volume increase. We shall not do any calculations in this book (see references 41 and 42) but give the answer directly, corresponding to the statement that the probability of existence of that gaseous macrostate, at a fixed temperature, is related to the molar volume by the expression

$$S_{conf} = \text{constant} \times \log V$$

where this probability, S_{conf}, is called the 'configurational' *entropy* of the system, referring to the spatial distribution of its constituent particles.

Change of volume from V_1 to V_2 then gives the change of probability as

$$\Delta S_{conf} = \text{constant} \times \log V_2/V_1$$

Now, *work* needs to be done by the system to change the volume against the fixed external pressure. This work corresponds to a change of energy, pV, and is just the function

$$Work\ done = -\int_{v_1}^{v_2} p\mathrm{d}V = -\int_{v_1}^{v_2} \frac{RT\mathrm{d}V}{V}$$

since, from the gas law, $pV = RT$ (for one mole of gas) (see Section 2.5.3).

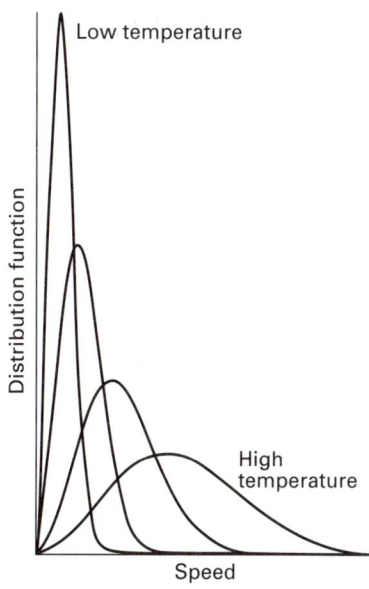

Fig. 5.6 The distribution of molecular speeds with temperature. Note that the most probable speed (corresponding to the peak of the distribution) increases with temperature and, simultaneously, the distribution becomes broader.

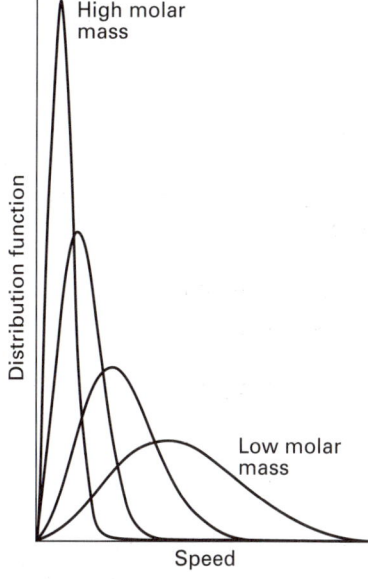

Fig. 5.7 The distribution of molecular speeds with particles of different mass at a fixed temperature.

Hence

$$Work\ done = -RT\ 2.303\ \log V_2/V_1 = -T\Delta S_{conf}\ (*)$$

(compare the treatment of the Carnot cycle in Fig. 5.9). Thus, the work necessary to increase the volume (favouring the gas state) is equal to the product of a change of the probability factor, ΔS_{conf} (the configurational entropy change), times the temperature. The work, $T\Delta S$, is expressed in units of energy.

This work can be performed at the cost of an equivalent amount of energy (heat), Q, *transferred from the surrounding environment*, which will be equal to the increase of 'disorder' energy, $T\Delta S_{conf}$, i.e.

$$Q = T\Delta S_{conf}$$

[N.B. Q is the heat transferred assuming that the temperature of the surroundings is the same as that of the system, i.e. in idealised closed equilibrium conditions (Table 5.3) also referred to as conditions of reversibility.]

In the above paragraphs we have been concerned with the examination of the way component units, where components are the variables of composition, are distributed statistically in space. Now, just as we examine the changed probability of the existence of a gas with change in molar volume, we can examine its change with temperature. In effect this is a change of the number of random kinetic energy states with temperature. Following the procedure of the treatment of volume, we can consider the temperature as an averaged sum of a set of microstates. The microstates are now formed from particles placed in boxes with random translational kinetic energy labels, $\frac{1}{2}mv^2$, and with directions along all axes x, y, z of space (see Fig. 5.1). The particles in the gas phase occupy a larger number of microstates the higher the temperature; see the distribution curves of Fig. 5.6 (and observe that speed varies with mass; Fig. 5.7), hence a gas is favoured over a solid, which changes only vibrational states of the particles in the lattice, the higher the temperature (Fig. 5.8). Since the occupation of an energy state, E (translational, vibrational or rotational), is dependent on temperature, each state of different energy has a probability of being occupied, which was deduced by Boltzmann to be dependent upon $e^{-E/RT}$. The weighted probability of occupying the huge variety of states open to a gas is, therefore, a more difficult summation than the probability for the occupation of volume elements. We can show that this weighted probability of existence of the gas at different temperature is given by a probability function, S_{th}, called the 'thermal' entropy, i.e.

$$S_{th} = constant \times \log T$$

Of course, S_{th} is higher the higher the temperature, and it can be shown (see below) that, at two different temperatures,

$$\Delta S_{th} = constant \times \log T_2/T_1$$

---

*The constant in the expression S_{conf} = constant. $\log V$ is thus found to be R 2.303, where R is the so-called perfect gas constant (see Section 2.5.3).

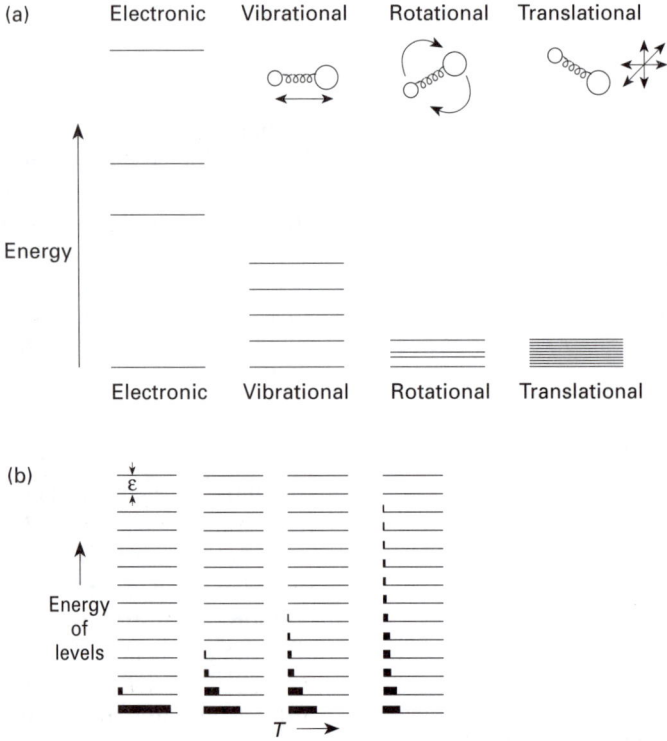

Fig. 5.8 (a) The quantised energy levels for different forms of molecular motion (compare with Fig. 2.12). (b) The distribution of particles over equally spaced energy levels becomes broader as the temperature, *T*, increases (see Fig. 5.6).

In effect, to change the temperature of the gas a given amount of energy (heat), Q', must be supplied *from the environment*. The amount of energy necessary to increase the temperature of one mole of gas by one degree is called the molar heat capacity, C^*, so that the energy required to increase the temperature by dT will be CdT.

This energy will increase the 'thermal' entropy of the system by a differential amount dS_{th}, and the corresponding energy change will be TdS_{th}, just as for the configurational entropy change.

Since $CdT = TdS_{th}$ we can easily derive, for a temperature change from T_1 to T_2, the expression

$$\Delta S_{th} = C \int_{T_1}^{T_2} dT/T = C \times 2.303 \times \log T_2/T_1$$

and we may write for the energy $Q' = T\Delta S_{th}$, where Q' is the heat taken in from the environment again in idealised equilibrium closed conditions (so-called conditions of reversibility) and T is the average temperature (a satisfactory approximation provided the range $T_2 - T_1$ is not very large). Again we pause to notice that what we have examined here is the way com-

---

*The molar heat capacity can be at constant volume, C_v, or at constant pressure, C_p.

ponent units are statistically distributed over individual variation of their *space co-ordinates* with *time* (physical variables), which, being treated statistically at fixed volume, do not give rise to *net* movement (flow). Although each particle has a directed kinetic energy and momentum, the collective system does not; we say that it has *random* motion of which the resultant directional character is zero (see Section 2.5.3).

Now, when both configurational and thermal effects are considered, the total change in entropy of the closed system will be

$$\Delta S_{sys} = \Delta S_{conf} + \Delta S_{th}$$

and the total amount of energy transferred from the environment to bring about the change in equilibrium conditions, $Q_{tot} = Q + Q'$, will be

$$Q_{tot} = T \Delta S_{sys}$$

We have therefore, $\Delta S_{sys} = Q_{tot}/T$, which allows the total entropy change of the system to be calculated from the knowledge of the total amount of energy (heat) transferred from the environment in so-called conditions of equilibrium (or of reversibility). Alternatively, ΔS_{sys} can be calculated from the change in the total number of possible microstates of the system obtained from statistical considerations, but this is a difficult task. For this purpose it is necessary to know all the quantised states of molecular assemblies (as in Fig. 5.8), so that so-called 'partition' (distribution) functions can be calculated.

5.4.2 Irreversibility of change: the hidden (dispersed) energy

All the analysis so far may have led to the impression that all forms of energy could also be reversibly converted into work, i.e. a capability of changing systems, apparently in keeping with the facts that energy is conserved and the internal energy of a system can vary by doing work or having work done on it, or by exchanging heat across its boundaries, i.e. $\Delta U = Q + W$, which expressed the so-called *first law of thermodynamics*. A revolution in thinking was brought about when Carnot, a French engineer, showed, however, that when a gas cycled through a particular set of T, p changes "at equilibrium" (see aside), while doing $p \Delta V$ work in which heat is given to and taken *from the environment*, this is not so. The cycle is shown in Fig. 5.9(a), but since it is a theoretical process it is not as easily understood as some real parallel examples (see text to figure). One such simple and obvious illustration is the fact that we cannot take heat generated at the bottom of a waterfall, (see Section 2.5.1), and use it alone to generate work (increase potential energy) to lift the water back to the top.

Denbigh (see ref. 45) lists (Table 5.4) several more examples of changes leading a system from state A to state B but which cannot be reversed. The general implication is that when in a change we distribute heat (energy) or particles in space or over energy levels more widely we cannot simply reverse the distribution. *Since energy is conserved in every process there is a property of systems which is not conserved and in which the energy becomes 'hidden', so that it cannot be used to do work. This is a remarkable and funda-*

"Changes at equilibrium"

Here change occurs by infinitesimal steps so that the change takes infinite time. This is a hypothetical description to allow the use of equilibrium thermodynamic analysis of change.

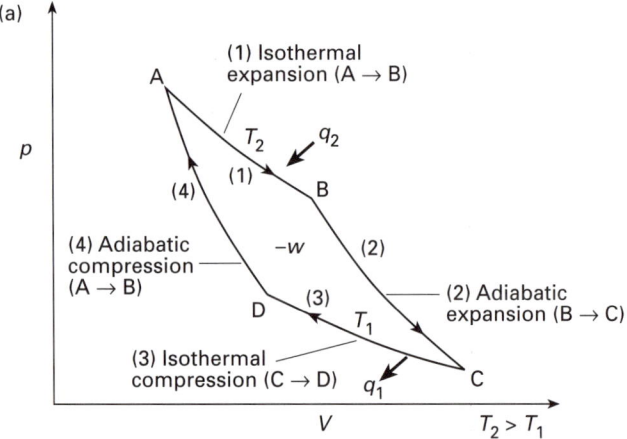

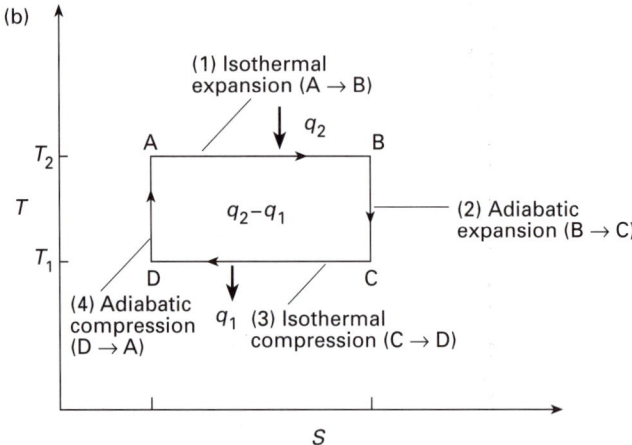

Fig. 5.9 (a) The Carnot cycle shows that, in a cycle of the pV kind, the heat required for the isothermal step* (1) differs from that given out in the isothermal step (3). The difference in work done† in the four steps is just that for the isothermal changes since the work involved in the two adiabatic changes is equal and opposite, so the total work done is

$$-W = RT_2 \ln (V_2 - V_1) - RT_1 \ln (V_4 - V_3) = q_2 - q_1$$

Since $(V_2 - V_1) = (V_4 - V_3) = \Delta V$ we can deduce that

$$-W = q_2 - q_1 = R(T_2 - T_1) \ln \Delta V$$

This means that not all the amount of heat taken by the system (q_2) at a higher temperature (T_2) is used to do work; part of it (q_1) is given out to the environment at a lower temperature (T_1). If $T_1 = T_2$ no work is done, i.e. work requires a source of heat and a sink of heat at different temperatures. In (b) the cycle is described in terms of entropy (S) changes and temperature. ΔS for the cycle is $\Delta q/T$.

---

*N.B. An *isothermal change* is a change in p and/or V without a change in temperature, i.e. $dT = 0$. It involves heat transfer from or to the outside if the system is closed (see Section 5.4.1). An *adiabatic* change is a change of p and/or V with a compensating change in temperature, T (see Fig. 5.2); no heat from outside the system is involved, i.e. $Q = 0$.
†By convention, work performed by the system is given a *negative* sign. Generally, any transfer of energy into the system is given a positive sign, and any transfer from the system to the environment is given a negative sign.

Irreversibility

State A

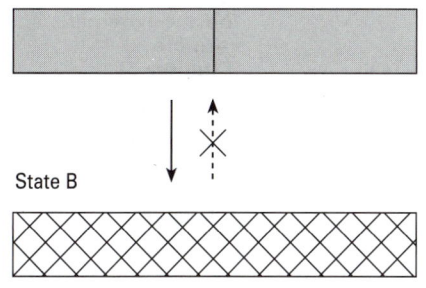

State B

Note. The two compartments of A and that of B are each separately at internal equilibrium.

Table 5.4 Irreversible changes, see Aside.

State A	State B
Two equal blocks of copper are connected by a wire. One block is at 20 °C and the other at 30 °C	The blocks are each at 25 °C
A dilute gas at a temperature T occupies one-half of an adiabatically enclosed vessel* and the other half is a vacuum.	The gas at the same temperature, T, occupies the whole of the vessel, and p falls.
A dilute gas, X, occupies one-half of an adiabatically enclosed vessel and a dilute gas, Y, occupies the other half. The temperature is T and the pressure p is fixed.	The gases are uniformly mixed throughout the vessel and the temperature has the same value, T. The pressure p is unchanged.
An adiabatically enclosed vessel contains hydrogen and oxygen and a catalyst. The volume is V and the temperature is T.	The vessel contains the same amount of hydrogen and oxygen, partially or totally combined as water, together with the catalyst. The volume is V and the temperature exceeds T by an amount corresponding to the heat of reaction. The pressure p changes.

* An adiabatically enclosed vessel does not allow heat transfer from or to the environment.

mental discovery concerning *systems*. This property is in fact the *entropy, S* [see Fig. 5.9(b)]. The inability to convert all forms of energy *reversibly* into work is seen to be a consequence of the obvious inability to reverse an increase in the probability, i.e. an increased statistical weight, of a system. This property belongs to a system, is analysed per mole and is not directly attributable to single or pairs of particles. This and all other equilibrium thermodynamic functions (see next section) are collective *functions of state* and are given per mole of the system.

5.5 The complete list of thermodynamic variables (functions) of states of bulk systems

In Section 5.2.1 we defined, for bulk systems, the internal energy, U, and the enthalpy, H, where $H = U + pV$, so that these equilibrium thermodynamic variables, also called *functions of state*,* could be appreciated in their relationship to the descriptions of internal energies, pressure and volume (now per mole) as defined by theories based on idealised particles in Chapter 2, Section 2.5.4. In Section 5.4.1 we have re-examined the occupation of volume, V, and the temperature, T dependent dynamic states, and have thereby given pressure, p, and temperature, T, a statistical

---

*Functions of state, as the name indicates, depend only on the *state* of the system considered, so that in any changes the values of such functions depend only on the *final* and *initial* states and not on the path followed for the change. Note that Q and W are *not* state functions.

Table 5.5 Thermodynamic collective variables of state

Fundamental variables	Derived variables (Chapter 2)	Collective derived thermodynamic variables (Chapter 5)
Composition by mass (atomic) and charge spatial distribution	Composition defined by components including local atomic potential energies	Composition by components, internal energy, U,* enthalpy, H,* entropy, S*
	Random kinetic energy (temperature)	
	Random momentum (pressure)	Free energy, G* (and A*), temperature, T, pressure, p (or volume, V)
Mass (atoms), charge	Directional kinetic energy	
		Flux, $J^\dagger$
Space (distance) and time	Directional momentum	

* Defined experimentally per mole, not directly related to theoretical quantities.
† Not a thermodynamic variable but a collective *rate* variable (see Chapter 8)

meaning. This has led us to introduce a new contribution to the total collective energy, given by $T\Delta S_{sys}$, where ΔS_{sys} is the so-called *total entropy change* of the system considered, a state function related to the measurable variables volume (or pressure) and the temperature of the system considered. We list the set of collective variables in Table 5.5.

Now, while the first property, U, could perhaps be treated at 0 K and only at that temperature (see Section 5.2.1) by simple linear addition of pair-wise interactions (only in principle since it becomes an impossible task for large atomic or molecular systems), this is not so at higher temperatures. The second property, H, involves terms dependent on T, p and V, and the third, S, involves terms depending on log V (or log p) and log T and cannot be treated in such a simple manner under any conditions. Such *complex* mathematical relationships of bulk systems allow an analysis of properties of substances. However, the analysis shows that, for example, transition points, phase changes, etc. are not simple linear functions of variables such as pressure or temperature. These properties are not predictable from the examination of single atoms or molecules and cannot be estimated by addition of pair-wise interactions; they are properties of large ensembles or whole systems. We cannot ask what is the temperature or pressure of a volume occupied by one or two units even though they have given masses, charges and velocities! Analytically and reductively only compositions of systems can be described in terms of sums of single units; other properties of materials follow from the nature of ensembles, not of units. We see why the kinetic theory of gases fails to reproduce many bulk properties such as melting or boiling points. We shall now explore the value of this equilibrium thermodynamic treatment of the energy available for work in the description of change, both physical and chemical.

5.6 Available (free) energy

Rate of change

No process which goes at a finite rate of change can be examined by equilibrium thermodynamics since connected parts of the system cannot be at equilibrium. Such systems are described by irreversible thermodynamics and change then has a larger ΔS, larger lack of reversibility.

Quite generally we wish now to evaluate the total energy available to do *work in a closed system* as we go from one equilibrium condition of a system to another, as this will tell us the capacity for change in this operation. Examples of this were discussed in Section 5.4.2 (see Fig. 5.9 and Table 5.4). In that section it was shown that certain changes of bulk systems could not be reversed, so that not all capacity for change is available to do work. We now see that they involve the redistribution of particles into more probable situations as judged by distribution of pressure or temperature (Table 5.4). Now, these are just the processes we have evaluated under the heading of entropy, S, which for a given change has an equivalent energy, $T\Delta S$. Clearly, this increase of probability favours disorder while the decrease of potential energy (now in ΔU and ΔH), in part only described in Chapters 2–4, acts in favour of order. We wish, therefore, to analyse together the values of $T\Delta S$ and ΔH since, commonly in any process, both order and disorder changes are involved, e.g. in the melting or boiling of substances and in chemical reactions, to see how much work is available. However, it is obvious that they are opposed to one another, so that the total available energy for work (excluding changes in volume, which are included in ΔH) is the difference in energy between them. Of these two terms only ΔH is, in principle, fully reversible, while $T\Delta S$ is 'hidden' in the increased statistical probability of the system and cannot be reversed.

A consequence of the above description of the distinction between the energy which can be used to do work and the energy which cannot be used, associated with ΔS, is that, if possible, all closed systems collapse to a state where no work can be done by them, i.e. they release the maximum amount of energy as heat or work into the environment. Thus, the system is said to go to a state of maximum entropy so as to be fully at equilibrium both internally and with the environment. We return to this point in Chapter 16 and see aside.

We may, therefore, define a total *available* (or *free*) energy change, ΔG, of a system at *constant pressure** as the algebraic difference of ΔH and $T\Delta S$, i.e.

$$\Delta G = \Delta H - T\Delta S$$

This is effectively a difference, in terms of energy, between the degree of change of order and disorder energies of a system. If the change leads to a system of lower internal potential energy then ΔH is negative (energy released) while it is positive in the opposite case (energy absorbed). On the

---

*N.B. We could also define the total available energy change at constant volume per mole, ΔA, but this is not so useful since we are usually interested in systems at constant atmospheric pressure, $p = 1$.

other hand, if the final system is more disordered $\Delta S > 0$, while if it is less disordered $\Delta S < 0$.

The balance, ΔG, can therefore be negative and there will then be a given amount of energy available to do work on a second system or to be transferred to the environment, or it can be positive and energy needs to be introduced (from a second system or from the environment) to effect the required change. (Change in a closed or in an open system always involves the 'environment' and a change of disorder in it).

If $\Delta H = T\Delta S$ for a change then $\Delta G = 0$ and the system will be at equilibrium. No internal change can occur and no energy will be available to do work. This, by definition, is a stable system. It can be seen that this is a condition of maximum entropy. (Notice that in all the above considerations we are concerned with *change* but not with the *rate* of this change. This further aspect will be considered in Chapter 8.)

From this discussion we have arrived at a very simple conclusion which allows us to describe any stable system (and indeed any stationary system) and its possible changes in terms of its compositional units (analysed in Chapter 1 under fundamental elements and charges and in Chapter 2, operationally, as components) and the other variables within the free energy ΔG. ΔG is an algebraic sum of the effects of changes of composition, potential energy, temperature and pressure (volume) from one equilibrium state to another.

It is frequently convenient to refer to the new set of derived molar variables, ΔH and ΔS, while also defining p and T, so that we can analyse the changes in ΔG under given conditions. Tables of these quantities at one atmosphere and 298 K (25 °C), so-called *standard conditions*, written $\Delta G°$ and $\Delta H°$, are given in compilations from which we have selected a few values in Tables 5.6–5.8. The values of $\Delta G°$ and $\Delta H°$ are for the formation per mole of the listed compounds from their elements not as free atoms but in their standard conditions, e.g. for water

$$H_2(g) + \frac{1}{2}O_2(g) \rightarrow H_2O(\ell)$$

where g = gas and ℓ = liquid. The values in Table 5.8 are *absolute* entropies ($S°$, not $\Delta S°$) at 298 K for the substances listed.

We now wish to summarize and stress the new numerical understanding we have gained from the analysis of systems of many particles, which was not discoverable from examination of single units or from the interactions between pairs of units. As stated above, the part of the energy (heat) which goes into changes of pressure or temperature, i.e. increasing the probability of a system, and which is a part of the total conserved energy, is not reversibly transformable into work. [Note that the total energy of everything within an isolated system (e.g. the universe) is conserved.] This heat is the 'hidden' energy, $T\Delta S$ referred to in Section 5.4.1. Stability is now represented by $-\Delta G$ (not $-\Delta H$) and all movement towards stability, increase in $-\Delta H$, i.e. higher negative values of ΔH, will generate heat. This heat may be either retained in the system (if it is isolated) or transferred to the environment in part (if the system is closed), but it cannot be used to

Table 5.6 Enthalpies of formation, ΔH_f^o per mole, at 298K

Compound	ΔH_f^o (kcal mol^{-1})	Compound	ΔH_f^o (kcal mol^{-1})
Inorganic compounds*			
H_2O (g)	−57.79	CaO (s)	−151.8
H_2O (l)	−68.32	MgO (s)	−143.8
H_2O_2 (g)	−32.53	MnO (s)	−92.2
HCl (g)	−22.06	CoO (s)	−57.2
SO_2 (g)	−70.96	NiO (s)	−56.5
SO_3 (g)	−94.45	CuO (s)	−37.6
H_2S (g)	−4.81	ZnO (s)	−83.2
NO (g)	21.60	ZnS (s)	−48.5
NH_3 (g)	−11.04	Fe_2O_3 (s)	−196.5
CO (g)	−26.41	Al_2O_3 (s)	−399.1
CO_2 (g)	−94.05	SiO_2 (s)	−209.9
Organic compounds†			
Gases			
Methane, CH_4	−17.89	Ethylene, C_2H_4	12.50
Ethane, C_2H_6	−20.24	Acetylene, C_2H_2	54.19
n-Pentane, C_5H_{12}	−35.00	cis-2-Butene, C_4H_8	−1.36
Isopentane, C_5H_{12}	−36.92	trans-2-Butene, C_4H_8	−2.40
Liquids			
Methanol, CH_3OH	−57.02	Benzene, C_6H_8	11.72
Ethanol, C_2H_5OH	−66.35	Chloroform, $CHCl_3$	−31.5
Acetic acid, CH_3COOH	−116.4	Carbon tetrachloride, CCl_4	−33.3

* g, gas; l, liquid; s, solid.

Table 5.7 Free energy of formation per mole, ΔG_f^o, at 298K

Compound	ΔG_f^o (kcal mol^{-1})	Compound	ΔG_f^o (kcal mol^{-1})
Inorganic compounds			
Gases		**Solids**	
H_2O	−54.64	CaO	−144.4
H_2O_2	−24.7	MgO	−136.2
HCl	−22.77	MnO	−86.8
SO_2	−71.79	CoO	−51.2
H_2S	−7.89	NiO	−50.6
N_2O	24.9	CuO	−30.4
NO	20.72	ZnO	−76.05
NH_3	−3.97	Fe_2O_3	−177.1
CO	−32.81	Al_2O_3	−376.8
CO_2	−94.26	SiO_2	−192.4
Organic compounds			
Gases			
Methane, CH_4	−12.14	Ethylene, C_2H_4	16.28
Ethane, C_2H_6	−7.86	Acetylene, C_2H_2	50.00
n-Pentane, C_5H_{12}	−2.0	cis-2-Butene, C_4H_8	15.74
Isopentane, C_5H_{12}	−3.5	trans-2-Butene, C_4H_8	15.05
Liquids			
Methanol, CH_3OH	−39.73	Benzene, C_6H_6	29.76
Ethanol, C_2H_5OH	−41.77	Chloroform, $CHCl_3$	−17.1
Acetic acid, CH_3COOH	−93.8	Carbon tetrachloride, CCl_4	−16.4

Table 5.8 Absolute entropies per mole, $S°$, at 298K

Substance	$S°$ (cal mol^{-1} K^{-1})	Substance	$S°$ (cal mol^{-1} K^{-1})	Substance	$S°$ (cal mol^{-1} K^{-1})
Solid elements		**Solid compounds**		**Liquids**	
Al	6.77	CaO	9.5	Br_2	36.4
C (diamond)	0.6	$Ca(OH)_2$	17.4	H_2O	16.73
Ca	9.95	$CaCO_3$	22.2	Hg	18.17
Cu	7.97	CuO	10.4		
Fe	6.49	Cu_2O	24.1		
Na	12.2	Fe_2O_3	21.5		
S (rhombic)	7.62	SiO_2	10.0		
Si	4.51	ZnO	10.5		
Zn	9.95	ZnS	13.8		
Monatomic gases		**Gaseous diatomic molecules**		**Gaseous polyatomic molecules**	
He	30.13	H_2	31.21	H_2O	45.1
Ne	34.95	D_2	34.6	CO_2	51.1
Ar	36.98	F_2	48.6	SO_2	59.4
Kr	39.19	CO	47.3	H_2S	49.1
Xe	40.53	NO	50.3	NO_2	57.5
H	27.39	N_2	45.7	NH_3	46.0
F	37.92	O_2	49.0	O_3	56.8
		HF	41.5		
Organic compounds (gases)					
Methane, CH_4	44.5	Isopentane, C_5H_{12}	82.1	1-Butene, C_4H_8	73.48
Ethane, C_2H_6	54.8	Neopentane, C_5H_{12}	73.2	cis-2-Butene, C_4H_8	71.9
Propane, C_3H_8	64.5	Ethylene, C_2H_4	52.45		
n-Pentane, C_5H_{12}	83.4	Acetylene, C_2H_2	49.99	trans-2-Butene, C_4H_8	70.9

reverse change completely as we can see from the alternative description of the Carnot cycle in Fig. 5.9(b). The so-called *second law of thermodynamics* is often stated as follows: spontaneous change always leads to an increase in entropy, i.e. it is irreversible. [It is tempting to couple this observation with the peculiar nature of the fundamental variable *time* of an expanding universe, which is apparently irreversible (this is why some authors consider entropy change to be the 'arrow of time'). Is it inevitably related to the fall of temperature and pressure in the universe? These are deep philosophical problems which we shall not consider further.] We are now in a position to consider the drive towards change in both the physical and chemical conditions of any system of materials, but we shall not be able to reduce this description, except qualitatively, to the fundamental factors underlying equilibrium thermodynamic variables much though we know that these derived operational variables are dependent on the fundamental variables. This is the dilemma of complexity.

We shall, from Section 5.7 onwards, turn to examples of the treatment of ΔG and the two composite variables, ΔH and ΔS, for some important specific cases. Before proceeding with closed systems, in which the environment is always involved and which are common, we take the easier case of an *isolated* system, where there is assumed to be no involvement of the environment.

5.7   Expansion and cooling in an isolated system

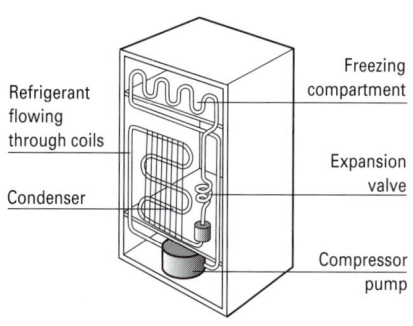

Fig. 5.10 In a refrigerator, a liquid called a refrigerant is first evaporated (turned to gas), then condensed by being pumped and turned back to liquid. As it evaporates inside coils in the refrigerator, it takes in heat. As it does this, it cools the contents of the refrigerator.

From what has been said we can also understand a very important consideration from an evolutionary (and a technical) point of view: the fact that on expanding an *isolated* gas (no environmental influence, see Table 5.3) it cools. Let us see why this is so by considering a gas in an isolated vessel, which expands very slowly in such a way that no heat is exchanged with the surroundings (so-called *adiabatic reversible expansion*). In this case energy does not enter or leave, $Q = 0$, so that $\Delta S_{sys} = Q/T = 0$, i.e. the total entropy of the system remains constant. However, since the volume increases, the spatial distribution broadens and the configurational entropy increases, hence the thermal entropy must decrease by the same quantity to keep the total entropy constant. This corresponds to a decrease of random kinetic energy, hence to a decrease of the average temperature. That is, the work of adiabatic expansion of a gas is done at the expense of the decrease of random kinetic energy of its molecules, as in a refrigerator (see Fig. 5.10). This is in part analogous to what happens in an isolated expanding system, such as the universe may be imagined to be, which has cooled from an exceedingly high temperature to the present 3K. *This cooling of the universe due to expansion was and still is today one major cause of the (natural) selection of the chemical elements into their compounds and into their observed physical conditions.* They condensed and became more ordered *locally* on *cooling*, but as a whole the universe, on expanding, has become more disordered so that its configurational entropy increased, i.e. $\Delta S_{conf} > 0$. Of necessity, the increase in configurational entropy overcomes the decrease in thermal entropy, ΔS_{th}, so that $\Delta S_{sys} > 0$.

5.8   Physical states in balance: phases

5.8.1 Latent heats and physical states of substances

The major result of the above analysis is readily stated. We have a way of appreciating the balance order $\rightleftarrows$ disorder for any system of chemicals in any physical states since both are now related to energy. In Chapters 2 and 4 we have shown how to think about ordering interactions and energy and we now appreciate macroscopic disorder and its associated energy. We can, therefore, treat separately the balance between physical states (the Greek 'elements') and that between chemical components that react. Thus, we shall discover the drives to (but not the *rates* of) change in the universe, which have helped to bring about observable materials and conditions as they are today. We turn next to two simple examples concerning physical states in balance in *closed* systems, i.e. systems that exchange energy but not material with the environment, at atmospheric pressure.

From what has been stated above, it is expected that if we measure the energy at equilibrium of a standard amount, say a gram-mole, of a liquid (or a solid) separately from that of a gram-mole of its vapour, there will always be found, at a given temperature and *external* pressure, a difference in enthalpy energy, ΔH, between the two states, which is related to the strength of the interactions between particles, atoms or molecules, in the condensate and in the vapour. There will however be quite another difference in *free* energy, ΔG, since part of this energy is in the form of the entropic energy difference, $T\Delta S$, which for a gas, relative to a liquid (or a solid) includes terms dependent on volume (partial pressure) and temperature related to $\log V$ and $\log T$. When the system is in balance $\Delta G = 0$, hence $\Delta H = T\Delta S$ on moving from condensed to gaseous states.

Now, assuming ΔH is independent of temperature, ΔS changes until at some temperature, T (the boiling point) the value of p (the vapour pressure) is that of the atmosphere. Above this temperature only a gas can exist, quite obviously, and hence the temperature and pressure of it are independent variables. Below this temperature there is always a balance, a *phase equilibrium*, between the vapour and the liquid, with the vapour pressure below one atmosphere. This balance is then a restriction on the total number of variables of the whole system, the so-called number of degrees of freedom or variance (F), since because the phases, the liquid and the vapour, cannot be varied independently (by definition they are in equilibrium) the vapour pressure and the temperature are linked. The more phases, P, where the components C are in balance across the phase boundaries the more restricted is the system (see Section 5.8.4). When three phases, e.g. ice, water and vapour are present it is at a unique combination of T and p.* This explains why substances have several fixed points which characterise their changes of physical state.

Latent heat of vaporisation

Trouton's Rule states that this heat is 90–100 kJ/mole independent of the substance involved

The amount of energy, $Q_p{}^\dagger$, necessary to change 1 gram-mole of liquid into vapour *at constant* p *and* T is called the *latent heat* of vaporisation of the liquid (see aside). (It is usually defined at external pressure = 1 atm, but can be determined at any external pressure.) There is a parallel latent heat of fusion for the transition of a solid to a liquid. In each case the required heat is supplied from the environment since we are dealing with a closed system.

The liquid range is not easily calculated, but is of obvious importance; in particular note that *co-operative* dependent interactions stabilise solid states (Chapter 4) and, not so strongly, liquid states. Given the nature of one major problem we are discussing in this book—the selection of the elements into components in a given physical state—a question which we then can ask is: are particular physical states of matter favoured by particular elements in chemicals? Table 5.9 and Fig. 5.11 give some data for liquid ranges.

In Chapter 4 we showed that atoms have preferences for one another, represented by extreme possibilities. In one extreme case covalently bonded small *molecules* are formed with no polarity, e.g. H_2, N_2, O_2, CCl_4, C_6H_6, etc., and with very little interaction of one molecule with another. (The

---

*According to the Phase Rule, section 5.8.4, F = C + 2–P. This is why for C = 1 and P = 3 one gets F = 0, i.e. the system is invariant and there is just one possibility for the coexistence of the three physical states.

$\dagger$At constant T and p this amount of energy, Q_p, equals ΔH, see above. It is a definite amount that depends only on the initial and final states (remember that H is a state function).

Table 5.9 Range of existence of liquid form (at atmospheric pressure)

Substance	Melting point (°C)	Boiling point (°C)	Range of existence of liquid form (°C)
Oxygen, O_2	−218	−183	35
Nitrogen, N_2	−210	−196	14
Water, H_2O	0	100	100
Ammonia, NH_3	−78	−44	33
n-Hexane, C_6H_{14}	−95	69	164
n-Octane, C_8H_{18}	−57	126	183
Benzene, C_6H_6	5.5	80	74.5
Ethanol, C_2H_6O	−114	78	192

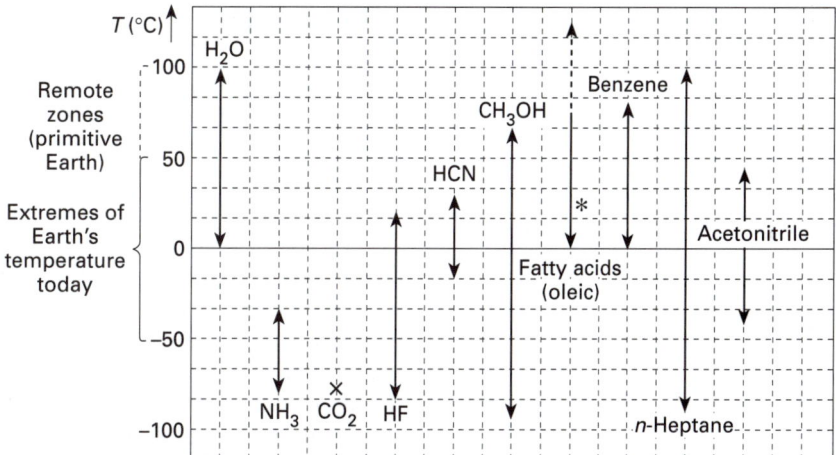

Fig. 5.11 The liquid ranges of a variety of compounds. (*) Includes liquid crystal range.

best examples are the noble gases whose atoms have no affinity for one another). A different case is that of the covalently bound atomic networks, such as diamond or silica, which are, of course, solids with extremely high co-operative interaction. In the third case, always involving hetero-nuclear compounds, the species are strongly polar (ionic) and form solid *co-operative* units, for example, $[Na^+ \cdot Cl^-]_m$, rock salt. In between them there are cases, for example of somewhat polar covalent molecules, such as HF, H_2O and NH_3, where a main consideration is the *moderate co-operative* association through the formation of hydrogen bonds. The fourth and last case is that of metals, which also form solid, *strongly co-operative* lattices, involving the metal ions and their 'free' electrons $[M^+, e^-]_m$, which could be looked upon as comparable to electron–cation 'salts'.

Naturally, these differences in strength of co-operative binding are reflected in the values of the latent heats, i.e. by the energy necessary to take one gram-mole of the solid, or a liquid, to the state of gaseous molecular or atomic vapour, that is, from the ordered to the disordered state (Table 5.10). The largest ΔH values are for the strongly associated solids or liquids, particularly atomic networks, ionic and metallic substances, which

Table 5.10 Data on melting and vaporisation*

Substance	Enthalpy (kJ mol⁻¹) of		Melting point (K)	Entropy (J K⁻¹ mol⁻¹) of	
	Fusion	Vaporisation		Fusion	Vaporisation
Metals					
Na	2.64	103	371	7.11	88.3
Al	10.7	283	932	11.4	121
Fe	14.9	404	1802	8.24	123
Ag	11.3	290	1234	9.16	116
Hg	2.43	64.9	234	10.4	103
Ionic crystals					
NaCl	30.2	766	1073	28.1	456
AgCl	13.2		728	18.1	
BaCl₂	24.1		1232	19.5	
Molecular crystals					
H₂	0.12	0.92	14	8.4	66.1
H₂O	5.98	47.3	273	22.0	126
Ar	1.17	7.87	83	14.1	90.4
NH₃	7.70	29.9	198	38.9	124

* 1 J = 4.18 cal.

do not become gaseous until very high temperatures. The smallest values are for inert gases that remain in the (atomic) gaseous state even at low temperatures. A few molecular substances, mainly those forming hydrogen bonds, which have intermediate co-operative energies, corresponding to moderate values of ΔH, condense at close to room temperature. As we have seen, a particular and almost unique case is water, which has a relatively high boiling point considering the number of atoms per molecule. Many larger organic molecules, e.g. petrol or octane (C_8H_{18}), also fall in this class due to the larger number of interactions (London forces, not H bonding) between atoms in larger molecules, increasing ΔH per mole. [Special mention should be made of individual biological polymers, such as proteins and nucleic acids, where extensive H bonding leads to 'solid' folded molecular products (see Chapter 4, and Section 5.8.5). Folding is equivalent to a condensation in that there is a loss of configurational entropy of the polymer chain.] It is not difficult to see now that these observations are a consequence of the fact that the conversion of a condensed state to a gas gives approximately a fixed change of entropy at the boiling point, when the partial pressure is equal to one atmosphere, when it follows that T of boiling is related to ΔH of conversion since $\Delta H = T\Delta S$. Since the entropy change on going from a solid to a liquid is a small fraction of the entropy of vaporisation, the range of temperature over which a liquid exists is usually small. (Note that the gaseous state is favoured by increase of T and ΔS.)

This leads us to Fig. 5.12, which shows that at relatively low temperatures of the Earth, say −50 to 100 °C, we have the following situation.

1. Permanent gases are largely formed by elements in the top, right-hand corner of the periodic table, whether or not they are in mutual combination in small molecules, i.e. H, C, N, O, F, Cl and the noble gas group.

Gases
H_2, O_2, N_2, F_2, Cl_2
NH_3, CO_2, CH_4
Liquids
H_2O, HF, HCN, C_6H_{12}
CH_3OH, many oils, Hg
Solids
Metals, metal oxides, silicates
(most minerals and higher molecular
weight organic compounds)

Fig. 5.12 Some substances found as gases, liquids or solids in the temperature range 273–323 K.

2. Liquids are formed *in this temperature range* by a very small group of elements in compounds, largely from combination of a few elements in the same part of the periodic table as in (1). Examples include a very few *small* molecules, e.g. H_2O, and a considerable number of larger molecules, e.g. C_6H_{12}, and molecules of heavier elements, e.g. CCl_4. They differ from the molecules in (1) due to greater polarity or larger size, which favour intermolecular interactions. Note that H_2O immediately appears to be exceptional.

3. Solids are formed from the vast majority of elements and simple combinations of elements, forming especially metals, alloys, salts and a small number of continuous covalent lattices, for example silica, diamond, etc. (see Figs 4.1 and 4.4) and organic polymers of high molecular weight. Covalent lattices only occur for elements able to form three or more covalent bonds, i.e. especially group 13, 14 and 15 elements.

4. H bonded polymers, proteins and DNA, 'melt' at a similar temperature to that at which water changes physical state, 273–373 K.

We can now state that the electronic structures of different elements discussed in Chapters 1 and 3 have led them to appear in different physical states of structured substances or compounds, as described in Chapter 4, i.e. some as gases, molecules of low polarity and low co-operativity, some as solid atomic networks, some as co-operative metallic solids, some as polar co-operative ionic solids and a few as liquids of lower co-operativity. However, we also see that the characterisation of all objects around us into the units of physical states, gas (air), liquid (water) and solid (earth), is not as fundamental as is the characterisation into atomic elements since the physical states are temperature and pressure dependent. It is the understanding of both the nature of the *units* (the chemical elements and the charges) and of the fundamental *variables* that has brought us to this appreciation.

It must always be remembered, however, that elemental composition as a variable on Earth was and is always limited by the *abundance* of each element present (Fig. 1.10). The consequences of these differences in chemical co-operativity and abundances are profound and dominate the physical basis of the constitution of Earth and the chemical nature of life. Life is, of course, totally dependent on the elements which can give rise to mobile phases, either gases of the atmosphere and/or liquids such as water and oils. Gases, and especially liquids, are only a very small percentage of what is around us but they can *flow*, a necessity for life. Thus, it is not the boiling point of water that matters so much, but its liquid range. Later we shall see that it was most importantly the matching of the liquid ranges of water and those of certain fatty materials (oils) (Fig. 4.10), together with the 'melting' ranges of biological, molecular weight polymers, which allowed life to appear. (N.B. Life is also dependent upon the appearance of special compounds which form these polymers, e.g. amino acids, but these are relatively unstable and could not have helped in the initial formation of components on early Earth, see Chapter 9.)

The selective affinities of the chemical elements based on free energy changes and abundances, which forced some non-metals into combination with one another, e.g. CO_2 and H_2O (but note the homonuclear species such as N_2 and the uncombined noble gases), decided that the only liquid on Earth's surface some $4-5 \times 10^9$ years ago was water and the only relevant gases were CO_2, H_2O and N_2, with some CO, H_2S, CH_4 and NH_3. If the temperature had fallen lower, say to around $-90\,°C$, and CO_2 had been the only gas available, then carbon dioxide would have formed as a solid, a very real possibility observed in other planets, e.g. on Venus and Mars at the polar caps. Again, the vapour pressure of water largely controls the ability of the Earth's surface to maintain the temperature, still $25\,°C$ today, through its greenhouse effect (see Section 15.13).

5.8.2 Real systems: van der Waals equation, changes of state and complexity

In order to see how our discussion of the relationship of properties of systems to fundamental units and their variables can become quite complex mathematically when *real* as opposed to the *ideal* simple cases are considered, we take again the example of a gaseous system which we have treated in a simple classical (non-statistical) fashion (see Section 2.5.4) using the kinetic theory of gases. We there derived the so-called perfect (ideal) gas equation $pV = RT$ (for 1 mole of gas) which we have used subsequently. This equation shows a *linear dependence* of p or V on the temperature T, when V or p, respectively, are constant (Fig. 5.13). Now, if such a law held generally, there would be no condensation, since by definition of

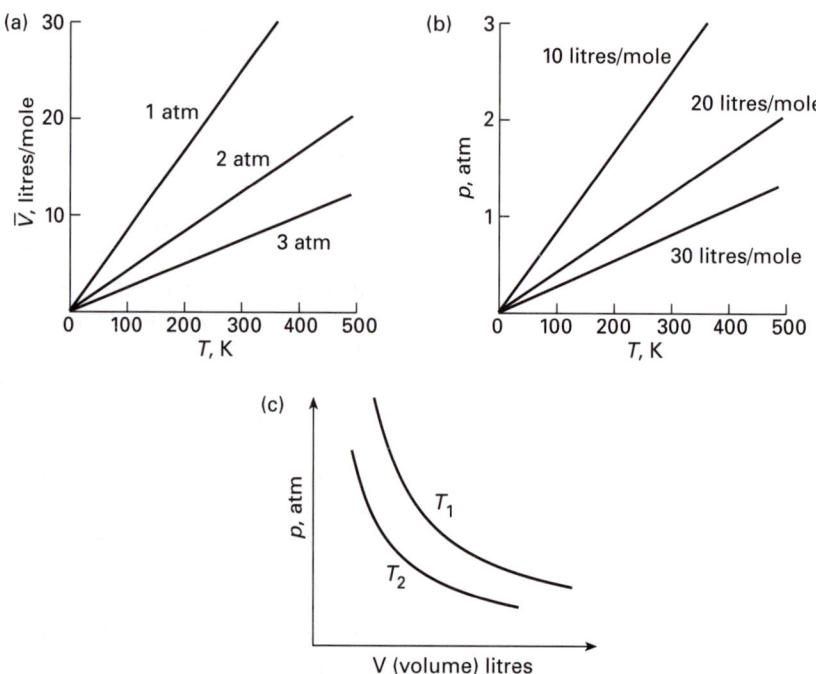

Fig. 5.13 (a) Isobars of the ideal gas. (b) Isometrics of the ideal gas. (c) Pressure/volume relationship at different temperatures, $T_1 > T_2$.

an ideal gas its constitutive particles do not interact. For example, the equation predicts that the volume of a gas approaches zero as T approaches 0 K or the pressure becomes infinitely large, which obviously is not true since in both cases the gas liquefies at some T or p values and thereafter not much decrease of volume is observed. Clearly, the equation is an approximation, valid only for a limited number of cases within a limited range of values of the variables, and must be modified if we want a closer approach to the behaviour of common non-ideal gases.

Several modifications of the equation (all approximate and non-statistical) have been proposed to take into account the existence of interactions between the particles of the gas (van der Waals forces, see Chapter 4) and the fact that these particles (atoms or molecules) have a volume. We use the van der Waals approach here since it immediately reveals the difficulties in real systems.

The existence of forces between the particles of the gas will decrease the pressure exerted by the gas on the walls of a container and this effect will be proportional to the intensity of the forces, which can be shown to vary with the inverse of the square of the volume, i.e. proportionately to a/V^2. As to the volume, we can use a finite volume, b, for the gas at 0 K (or infinite p), approximately equal to the volume occupied by one mole of the liquid or solid (see aside), which must be subtracted from the total volume, V.

The modified ideal gas equation will therefore be

$$p = RT/(V - b) - a/V^2$$

or

$$(p + a/V^2)(V - b) = RT$$

which is called the van der Waals equation. The constants a and b can be obtained experimentally for each concrete case—they are different for every gas.

The form of the van der Waals equation can be rearranged to give

$$V^3 - (b + RT/p)V^2 + (a/p)V - ab/p = 0$$

Since this is a cubic relationship between V and p it will have three roots for V at given values of p and T. That is, for a given temperature there are three possible values of the volume for the same value of p. Actually, only the two extreme values are significant—the smallest, corresponding to the condensed state and the largest corresponding to the gaseous state; the intermediate solution has no physical meaning (the volume falls as the pressure falls). This can be seen by comparing the graphical representation of the van der Waals equation (Fig. 5.14(a)] with the experimental isotherms (p, V curves at constant T) of real gases [Fig. 5.14(b)] where the condensation regions (plateaux at constant p for variable V) can be observed, i.e. any value of V between the extremes is possible.

However, the van der Waals equation is but an approximation to reality and this equation gives no detailed explanation of pressure or temperature dependences of change of state since a and b are adjustable parameters, different for each gas. Clearly, the behaviour of bulk materials can only be

Volume approximation

The volume occupied by a molecule , b, becomes a serious problem in living cells, see Chap. 10.

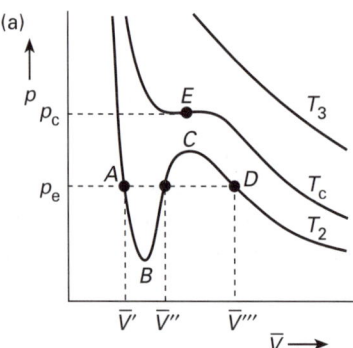 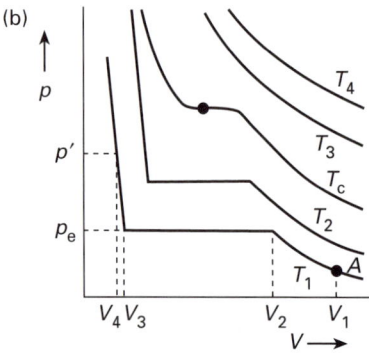

Fig. 5.14 (a) Isotherms of the van der Waals gas. (b) Isotherms of a real gas. T_c, the so-called *critical* temperature, is the maximum temperature at which liquid and vapour can co-exist at a *critical* pressure, p_c. The corresponding volume, V_c, is the critical volume.

approached by more complex dependencies of variables in the context of the collective properties of large ensembles. Many such relationships have been proposed which give closer agreements with the real behaviour, e.g. the so-called general virial equation and the Dieterich equation which involves a Boltzmann-type function (see references 36, 37). A full treatment of condensation phase changes has been given by Wilson (Ref. 50).

A possible way of looking at the changes of state, and why they are so difficult to handle, is to see the variables as describing different and competing ways of occupying collective microstates due on the one side to attractive forces, i.e. tendency to order (enthalpy decrease), and on the other to statistics describing the tendency to disorder, i.e. entropy increase. The stable condition employs both together to maximum advantage, but the variation in the two with energy put into the system, a third variable, is different. The overall equation of free energy then goes through critical points which cannot be calculated in any simple way, since we are dealing with non-linear systems. In a sense we have not been able yet to give a full quantitative treatment of such systems. We trust, however, that the reader is now impressed by the value of a statistical thermodynamic approach to the collective properties of bulk materials, even though there is loss of insight into the contributions of pairs of atoms or molecules.

5.8.3 Mixing of substances in one phase

A second example of physical change in a closed system is that of mixing or dissolving one substance in another. As we have seen, entropy is clearly going to a maximum in a system of a single substance the larger the number of possibilities there are open to it, that is, the larger the number of microstates compatible with a given macrostate characterised by its pressure (volume), temperature and composition.

If more than one substance is considered, e.g. two ideal gases, we must consider the additional variable composition. The number of microstates (possibilities of different distributions in space) increases on mixing (since the particles are non-identical) and the maximum increase is reached when

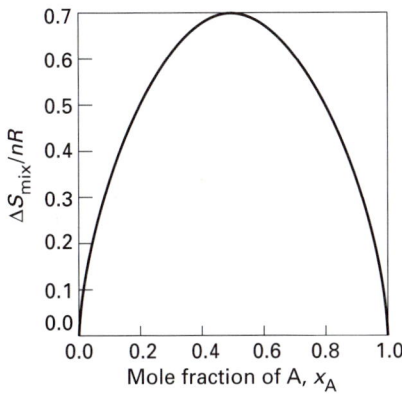

Fig. 5.15 The entropy of mixing of two perfect gases and of two dilute solutions that form an ideal solution. The entropy increases for all compositions and temperatures, so perfect gases and solutions mix spontaneously in all proportions at all temperatures.

the two are present in equal amounts (Fig. 5.15). Once again we do not consider any energies of interactions. For example, the most probable situation for a mixture of nitrogen and oxygen gases is that they will mix thoroughly, as in the air, and will not form separate layers. [The fact that there is an ozone layer high in the atmosphere, i.e. not mixed, means that the upper atmosphere is not obeying balance (equilibrium) rules, but is in a steady state (see Sections 5.2.2 and 8.6.4).] The logic applies to all states of matter, so that in the absence of preferred binding there is a general tendency to mix in liquid (solution) or solid (solid solution) states, as well as in gaseous states, all of which are, therefore, stabilised, i.e. their existence becomes more probable (see Section 5.4 for a more quantitative treatment). Of course non-ideal (real) systems deviate from the expectations based on the behaviour of ideal systems.

5.8.4 Limitations on variables (phase rules)

The minimum number of (intensive) variables, independent of the amount of matter, which define the state of a given system in equilibrium is called its *number of degrees of freedom* or its *variance* (number of variables), F. Now, if fields are absent, a system of C components, distributed in P phases is characterised by its composition in each phase (expressed in percentage or molar fractions of the components) and by its temperature and pressure. That is, the number of variables of *any phase* is $C - 1$ (only $C - 1$ components need be specified since the % composition of one is the difference from 100%), plus 2 (temperature and pressure). The total number of variables is then $P(C - 1) + 2$.

Now, if the system is in equilibrium, the number of variables of distribution is reduced by $(P - 1)C$ restrictions since there are $(P - 1)$ equilibrium conditions for each component across phase boundaries [i.e. $(P - 1)$ equality conditions between P phases].

The number of degrees of freedom will therefore be

$$F = P(C - 1) + 2 - (P - 1)C = C + 2 - P$$

which is the expression of the so-called *phase rule*. It defines the variables in an equilibrium system. The difficulty in its use lies in determining the number of components. However, one point needs stressing: *all phase equilibria reduce the number of degrees of freedom (variance) of the system.* We have already seen how this applied at the fixed melting or boiling points of all substances.

Throughout Section 5.7 and this section we have considered the separation in space of phases with boundaries but in balanced exchange of units across the boundaries. As stated, the balance restricts the variance of the system. When balance is removed the divisions of space become compartments and variables increase (see Chapter 7). The evolution of the universe, of Earth and of life depended critically on this loss of equilibria (balance) across phase boundaries.

5.8.5 Thermodynamics of polymeric molecules

Individual polymer molecules, such as proteins and DNA, have an additional entropy term associated with the possible configurations of the polymer chain. Each molecule could, in principle, form a folded unique structure when it has low probability and, of course, it will only do so since there is a highly advantageous and unique co-operative interaction between units of the polymer chain in a structured fold. This is the enthalpy change, ΔH, of the fold at a given temperature—compare a perfect single crystal. A much more probable condition is given by the sum of the completely disordered microstates of random configurations of the chain weighted by the $T\Delta S$ term (configurational energy corresponding to an entropy change on forming a liquid from a solid). There is then a temperature at unit pressure where the polymer loses its structure and melts. Of course, a wide range of different types of polymers may exist, with different melting points, and there is also the possibility of allotropy—interconversion from one folded state to another. This is called an allosteric change. However, a polymer may have, associated with one macroscopic state, a limited number of microscopic states so that it changes its structure somewhat with conditions, for example increasing internal motions with temperature, like a rubber. Note that most proteins melt over a small range of temperature in the region 40–100 °C. There is value in all these possible different behaviours of polymers and it is this balance of order and disorder, with its variation with temperature and pressure, that has made the construction of biological machinery, and therefore life, possible. We see more clearly in Chapters 10–16 that the biological polymers DNA, RNA and proteins are special in that they are very well tuned to function in water. This analysis of the properties of *single molecules of polymers* shows that they have complex properties, not linearly related to their sequences. A theory of their folding has to be based on the parallel with that of the theory of phase changes (see Wilson, ref. 50).

We have now described all we shall need concerning balances of physical states and must turn to balances of chemical species.

5.9   Chemical systems in balance (equilibria)

Let us consider a chemical reaction, e.g. the combustion of hydrogen with oxygen, which gives water

$$2H_2(g) + O_2(g) \rightarrow 2H_2O(liq)$$

together with the reverse reaction

$$2H_2O(liq) \rightarrow O_2(g) + 2H_2(g)$$

Let us assume that the temperature and the pressure are kept constant, say at 25 °C and 1 atmosphere. In balancing the reactions so that $\Delta G = 0$ we have to consider different chemical bond energies, largely dominant in ΔH per mole, and different physical states, largely dominant in $T\Delta S$ per mole. In the first reaction the entropy change is negative at room temperature and atmospheric pressure because the total number of particles of the system (especially those in the gaseous state) is reduced, so the entropy change is against the reaction. This is counterbalanced by the large favourable heat evolved, ΔH, due to the higher bond strength of 4H–O against one O–O and two H–H and the greater order in H_2O (liquid). Starting from H_2 and O_2 the reaction proceeds as written at room temperature and atmospheric pressure until balance is reached with virtually 100% water, but with very, very low pressures of H_2 and O_2 present. At very high temperatures, however, the reaction is balanced differently since $2H_2 + O_2$ have higher disorder than $2H_2O$ and multiplication by a higher T and change in ΔS itself causes $T\Delta S$ finally to overcome ΔH, so that it is the reversed balance that it is favoured. Of course, there is always a position of balance between the chemicals at all temperatures, hence we can write the system as an equilibrium between two tendencies denoted by $\rightleftarrows$, that is

$$2H_2(g) + O_2(g) \rightleftarrows 2H_2O(liq)$$

Anaerobic/aerobic planets

It is important to see that the accidental atmosphere of early Earth was essential for the beginning of Life. As an exercise consider atmospheres dominated by O_2, N_2, H_2, or H_2S. Could life have begun?

The balance applies to the stable condition of the three chemicals involved at the Earth's temperature, assumed to be 25 °C, and atmospheric pressure, $p = 1$ atm (standard conditions). We can now extend this observation and find out which combinations of elements in which physical forms are the most stable in the standard conditions, i.e. at 25 °C (298K) and atmospheric pressure; this will be shown to be a major factor in the selective combination of the chemical elements. We choose to concentrate first on the dominance of the oxides of the elements since, as shown in Chapter 4, oxides are amongst the most stable forms for all elements, and dioxygen, O_2, was of quite high abundance in the gases that formed Earth (see aside).

Table 5.6 allows us to establish $\Delta H°$ orders for oxides *per mole* of dioxygen incorporated on reaction of an element with dioxygen (gas) ($\Delta H° = \Delta H$ under standard conditions, p = 1 atm and $T = 298K$). To see how these values can be used to predict which state is favoured we can consider that a certain material can be in one of two situations in a closed system.* For example, it could be as Mg metal plus dioxygen gas or as magnesium oxide:

$$Mg + \frac{1}{2}O_2 \text{ or } MgO$$

We wish to know the difference in free energy (capability of change) of the two situations at standard conditions. We can do this by allowing the reaction to take place while removing all heat to the environment where it can do work or be dissipated.

$$Mg + \frac{1}{2}O_2 \rightarrow MgO + \Delta H \text{ (environment)}$$

(note that, by definition, ΔH is *negative* since it is given out by the system). Now, in any reaction system at a given temperature we must also consider

---

*As stated in Section 5.2, a closed system is one in which only energy, not material, can be lost to (or received from) the environment.

the weight of the term $T\Delta S$ from one side of the reaction compared with the other. For example, in the reaction $Mg + \frac{1}{2}O_2 \rightarrow 2\,MgO$ it favours $Mg + \frac{1}{2}O_2$ while in the reaction $C + \frac{1}{2}O_2 \rightarrow CO$ it favours CO. We then find the following orders for $\Delta G°$ $(= \Delta H° - T\Delta S°)$

$$Mg > Al > Si > P > S,$$
$$Mg > Mn > Fe > Co,\ Ni\ > Cu\ < Zn$$
$$\text{and}\quad H > C \gg N > Cl$$

These orders define the sequence of decreasing affinity* of the elements for oxygen, at equilibrium in the standard conditions. If we raise the temperature (Fig. 5.16) those oxides that are most favoured are those that are gases due to the weight of their larger $T\Delta S$ term (i.e. disorder). Note how the lines of Fig. 5.16 cross one another.

Once again remember, too, that the relative abundance of oxygen limited the ability of lower elements in the series to become oxides when Earth was forming. We must not forget, therefore, when thinking of natural processes, that there can be, simultaneously, competition of some of these elements for another element rather than for oxygen, for example hydrogen, when the order of affinity is

$$O > Cl \gg C > N > S \gg \text{all metals}$$

Such competing sequences of *affinities* take on particular importance when discussing, for example, the formation of the Earth from a gaseous mixture of elements (see Chapter 8). Thus Cl prefers H to O while all metals prefer O to Cl. *These affinities govern a major step in the natural selection of the chemical elements based on selective tendencies of a multiplicity of reactions taken together at equilibrium. The drive of heat loss, ΔH at constant pressure (corresponding to entropy gain of the environment), is the essential directing principle against the background of preferential combination, i.e. binding energy, and internal change in entropy, as well as the fall of temperature. Thus ΔG values show us the capacity for change.* In a qualitative way, Table 5.11 summarises our knowledge of the stability of some relevant types of compounds.

Above we stressed that the orders of stability corresponded to balanced ΔH or $T\Delta S$ energies per mole in *standard conditions*. As stated, the free energy per mole available for change is the difference between the reactants and products

$$\Delta G° = \Delta H° - T\Delta S° \text{ (in standard conditions)}$$

If the conditions are different, e.g. if the temperature becomes much higher, the orders will vary; this was and is of deep significance in natural processes as it is in the laboratory or industry. To enable the understanding

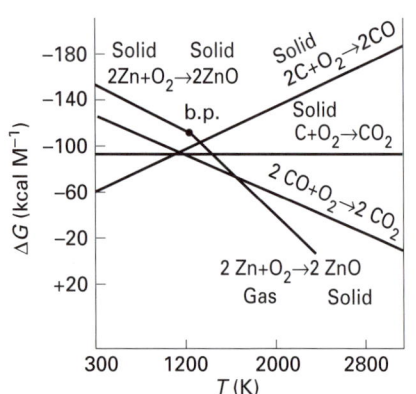

Fig. 5.16 The change in free energy, ΔG, with temperature, T, for the oxidation of Zn, C and CO; for C there are two different products. Above the boiling point (b.p.) of Zn the oxide loses stability more rapidly. Note that the more gas molecules in the chemical changes the more $T\Delta S$ (disorder) favours that change. (Stability increases upwards.) The figure is part of the Ellingham diagram (see refs).

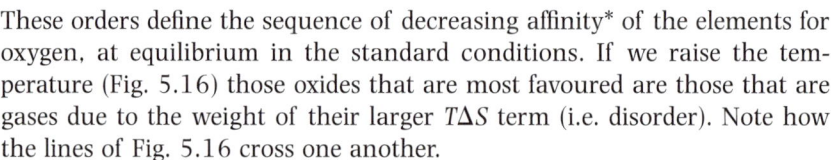

*This concept of internal affinity is, of course, related to the electrostatic potential energy of the interaction between the elements considered. The higher the affinity the larger (more negative) will be the potential energy of the interaction, the more stable will be the combination and the higher the tendency to achieve it.

Table 5.11 Compounds of special thermodynamical stability

	Stable	Very unstable
Hydrides	O, N, C, (S), halogens	Heavy elements, metals
Oxides	H, C, Si, P, (S), metals	Halogens, N, O
Sulphides	Metals, (H)	Non-metals

Brackets refer to compounds of somewhat lower stability.

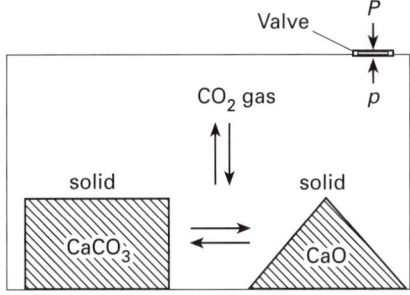

There are three phases and three compounds but only two components

Fig. 5.17 There are three phases and three compounds but only two components in calcium carbonate decomposition at equilibrium. (Calcium carbonate, calcium oxide and carbon dioxide are in equilibrium.)

of these processes we consider two examples where the stable form changes with temperature.

The formation of lime. The reaction in a factory for making cement (Fig. 5.17) is

$$CaCO_3(s) \rightarrow CaO(s) + CO_2(g) \uparrow$$

The disorder increases to the right as CO_2 is a gas. The co-operative energy of binding increases to the left. High temperature favours the disordered situation due to the contribution of the $T\Delta S$ term. If the external pressure is kept at one atmosphere then at every temperature there will be some CO_2 in balance with $CaCO_3$ and CaO. That amount of CO_2 exerts a given pressure. At a certain high temperature this pressure becomes equal to one atmosphere. At any higher pressure all $CaCO_3$ must go to CaO since the excessive pressure allows CO_2 to blow off through the valve in Fig. 5.17 and the reaction proceeds as written. At one atmosphere, calcium carbonate has a decomposition temperature of 750 °C. Thus calcium carbonate could not have formed on Earth until after the Earth reached a temperature of below 800 °C.

The preparation of zinc metal. Zinc can be obtained naturally as an oxide. The zinc oxide reacts with carbon (coal) to give zinc

$$ZnO(solid) + C(solid) \rightarrow Zn(solid) + CO(gas)$$

Following the argument above, the metal zinc will be preferentially obtained at high temperatures since the $T\Delta S$ term increasingly favours the CO(gas) the higher the temperature (see Fig. 5.16). A furnace at 1000 °C easily brings about the reaction, but at equilibrium at room temperature no zinc as metal is expected or found on Earth. However, gold, with a low affinity for oxygen, is found as a metal.

5.9.1 The relationship between non-stoichiometric substances and temperature

What we have described in the previous sections is the combined balance between states of matter—gas, liquid and solid—and between compounds of different chemical composition. So as to have just one word to include all the ways in which a chemical can exist we refer to its *phases*, the different chemical and physical states open to it, which may only exist (*in balance*) at particular temperatures and over a limited range of composition. This term includes the three physical states of matter of any one substance, of course, but is more general, since it includes allotropes and substances of different composition. We can draw *phase diagrams* where we show the change of physical and chemical state with temperature, or more complicated ones in which changes of composition are also conceived. One such diagram is given in Fig. 4.14. These are illustrated in many chemistry books. Whereas in Chapter 4 we considered only the way the binding energy, now incorporated in ΔH (at constant p), affected such diagrams, it becomes progressively obvious that the most important energy variable is ΔG, the free energy, which also takes into account the entropy, or statistical weighting, of each state, $T\Delta S$, as well as ΔH. We turn then to an analysis of ΔG in different circumstances, asking, as we observed in Chapters 2–4, why only some substances are stoichiometric.

5.10 Mixing and stoichiometry: chemical speciation and free energy considerations

As stated in Section 5.8.3, if we add together two sets of ideal gas atoms, A and B, which do not show A-A or B-B or A-B interactions separately, then the outcome, when the mixture has reached its best possible condition is a complete mixing of A and B. Complete mixing gives no specific ordering of A in B (or *vice versa*) and, as we have seen, this condition is driven by probability considerations. Alternatively, some degree of coming together of A and B in what we have generally called a 'compound' may occur. Only when A and B have an affinity for one another, A–B, greater than any affinity within A–A or B–B, will there be a preferred formation of A–B in a 'compound'. We wish to give these conclusions a pictorial (graphical) representation in terms of ΔG; in effect, these are landscape diagrams, since ΔG gives the thermodynamic probability of observing an association.

As stated, there are two competing tendencies; to disorder and to order. Disorder favours mixing and it will increase per summed unit of A + B as the ratio A to B approaches one, since we are adding to any disorder in A alone or B alone. Thus we draw the picture of the additional disorder, $T\Delta S$, for unit amount of A + B as in Fig. 5.18. If ΔH is positive but small, or if ΔH is negative, Fig. 5.18(a), or if ΔH is zero, Fig. 5.18(b), then the mixture is completely miscible. It is seen here that ΔG, the determining *free* energy, is of continuously diminishing gradient.

Consider next that there is a strong interaction between A and B; now ΔH is negative and large. We have seen this in the case of NaCl, common salt, which we described as a lattice of Na^+ and Cl^-. Remember that in the lattice each unit of Na^+Cl^- is bound to another unit in a co-operative manner (Fig. 4.2). Clearly, the greatest energy of interaction arises when in the system of A [Na atoms (ions)] and B [Cl atoms (ions)], always with the same total amount of A + B, the A and B ions are in the ratio of 1:1. The crystal so formed is completely ordered so that its energy opposes the dis-

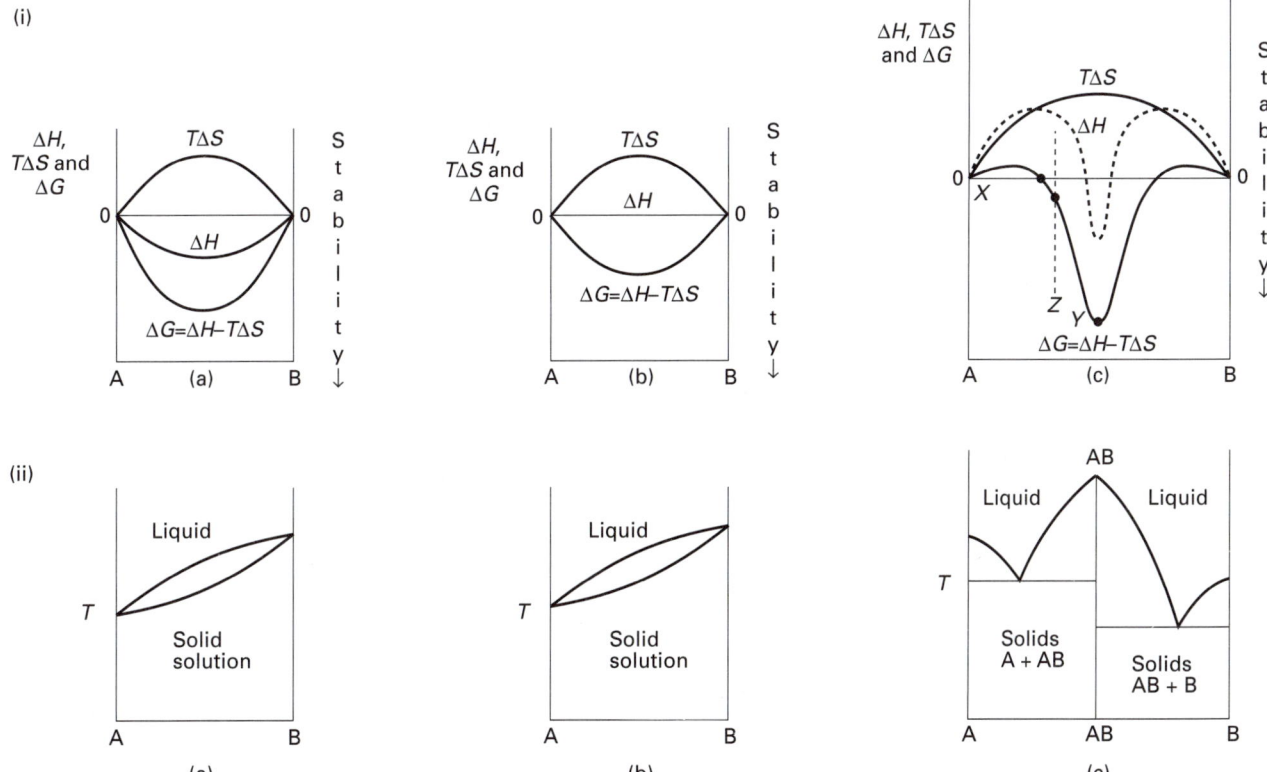

Fig. 5.18 (i) The sum of entropy and enthalpy changes to give the free energy of combination of mixtures of A + B in solid phase. (a) ΔH and $T\Delta S$ favourable—complete miscibility; (b) ΔH is zero (ideal behaviour)—complete miscibility; (c) ΔH is only favourable when A and B are nearly equal due to co-operativity. Here, a compound AB of narrow composition range forms at Y. (N.B. $T\Delta S$ is always favourable to mixing but is shown here with the + sign for clarity.) (ii) The expected phase diagrams for cases (a) and (b)—complete miscibility in both fluid (liquid) and solid phases—and for case (c) formation of a compound AB that is completely immiscible with either A or B in the solid state.

order represented above. In this case ΔH undergoes a rapid change of gradient, Fig. 5.18(c) (and see Fig. 4.13). Combining the energies, $\Delta G = \Delta H - T\Delta S$, we have the possible results shown in Fig. 5.18 where we show also the three corresponding phase diagrams for equilibrium between solid and liquid phases, plotting composition against temperature of melting.

Let us now consider what will happen if there is only a much weaker interaction in A–B (as compared with that in M$^+$–Cl$^-$) and that in the solid it is also possible for A and B to mix as in a gas. We have described this possibility in alloys of two metals in Section 4.7. The two tendencies (shown in Fig. 5.19) oppose one another and we can see that for small amounts of A in B, or B in A, disorder may win, while for equal amounts (as in the case of NaCl) some ordering gives the more stable situation. There are now *composition regions* of complete mixing and of 'compound' formation. However, the 'compound' is not so well defined any more; it is to some degree internally disordered, A can replace B to a small degree on the B lattice points, and A_nB_m has a range of compositions.

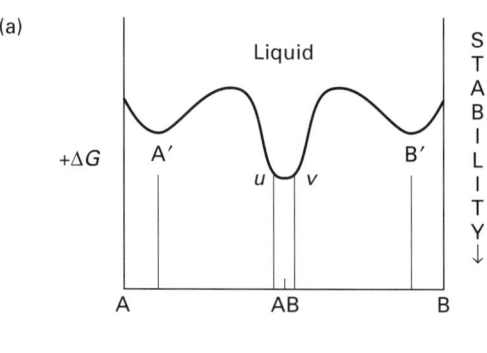

(a)

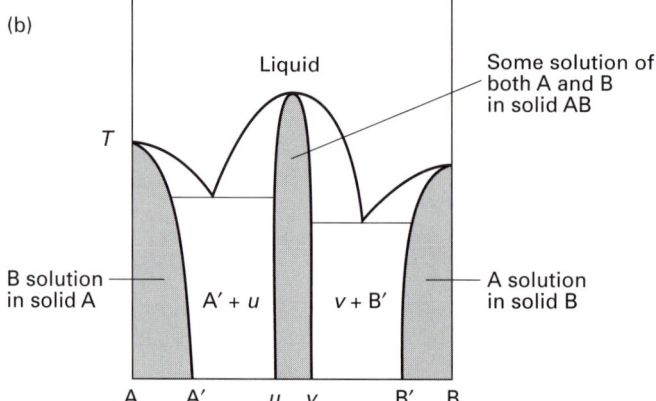

(b)

Fig. 5.19 (a) Free energy composition diagram for partially miscible phases of B in A, A and B in AB and A in B, giving rise to non-stoichiometric compounds. (b) Typical phase diagram arising from such a stability–composition diagram. The ΔG-composition curve corresponds to the formation of a non-stoichiometric compound in the range u to v.

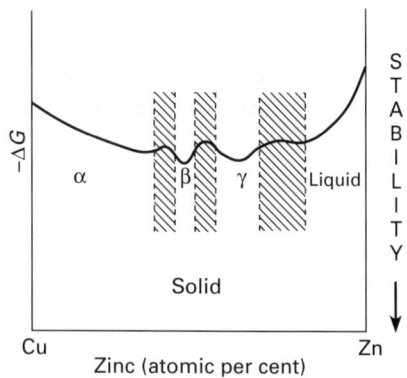

Fig. 5.20 A free energy versus composition diagram for the Cu–Zn alloys at 780°C (see Fig. 4.14). The shaded zones split into phases of composition at their two extremes due to lattice co-operativities. This is a landscape diagram.

From these considerations we can see that whenever interactions are strongly selective, giving a preference for one type of order or another, then the simple stoichiometric compounds formed will be of the kind AB (NaCl) or AB_2 (SiO_2), etc. with virtually no disorder. If, however, the binding in A–B is weak, then internally in the solid there will be the possibility of some disorder and a solid can exist over a range of compositions. The typical case is that of two metals, say zinc and copper. Thus the breakdown of disorder to give 'alloy phases' does not have to follow simple principles related to any stoichiometry. It has been found that all kinds of 'alloy phases' of ill-defined composition can form due to selective interactions within A + B. The zinc/copper alloys, for example, give a remarkable series of phases of different composition. We illustrated this in Fig. 4.14, where we see that, as we vary composition from 100% Zn to 100% Cu, there are regions of composition in which a certain structure gives stability followed by regions of compositions that are not stable. Figure 5.20 gives a stability, $-\Delta G$, against composition diagram at a particular temperature (780°C). It can be seen that the composition–stability diagram is broken up into zones.

It will be useful to have the image of *stability composition landscapes* in our minds in later chapters since they give the thermodynamic probability for existence of particular compositions.

So far we have only considered A + B. What happens if we add a third, C, fourth, D, and so on, kind of atom? Clearly, the more kinds of atom we add the more likely pairs of atoms will be similar. It is unlikely, then, that a substance ABCD will exist as a simple compound (Sections 4.6 and 4.7). In fact, if A and B are alike and C and D are alike the most probable result is a composition based on some ratio of A + B to C + D in which the relative amounts of A and B on one hand and C and D on the other can vary. When we come to look at real complex chemical systems such as rocks, soils and living things, all of which may contain as many as 20 different elements, we shall find that such materials are open to variation, sometimes over narrow and sometimes over wide ranges of composition. We see, therefore, that speciation of a phase can be in terms of a narrowly limited composition, e.g. NaCl, or may be in terms of a wide range of composition in minerals or organisms. Thus, speciation does not exclude variety of composition. Now, this description of the balance of $T\Delta S$ and ΔH in composition does not apply just to inorganic substances and we next look at organic chemicals.

5.11   Free energies and stationary states (components)

Until now we have only described *stable states* of various mixtures of elements in free exchange (equilibrium) at all given compositions, at different values of the other variables, temperature and pressure. Earlier we stated that many complex substances around us are in *stationary states*, where for a given elementary composition there had been introduced an extra free energy (largely internal chemical energy) generating chemicals showing no exchange of atoms, which we have called components (see Section 2.4). By definition, they are not equilibrated. We can still determine values of the free energy of formation, $\Delta G°$, for these *components* relative to the elements in their composition (for which, by definition, $G° = 0$), where $\Delta G°$ is defined for a fixed temperature (298K) and pressure (one atmosphere). These values give us the relative capability for change of components. Such stationary states dominate especially the chemistry of non-metals and a very important case for us is that of organic chemicals. Note that the removal of equilibrium greatly increases the number of variables in a system.

5.11.1 The nature of organic chemicals

We can now plot $\Delta G°$ for carbon compounds (components), in different oxidation states. This diagram (Fig. 5.21) indicates the relative stability, $-\Delta G°$, of a large number of isolatable *stationary states* of combinations of C, H and O. They are all possible components in mixtures and there are no element exchange equilibria between them under ambient conditions. This is a very different situation from a diagram for alloys, see Fig. 5.20 (and see

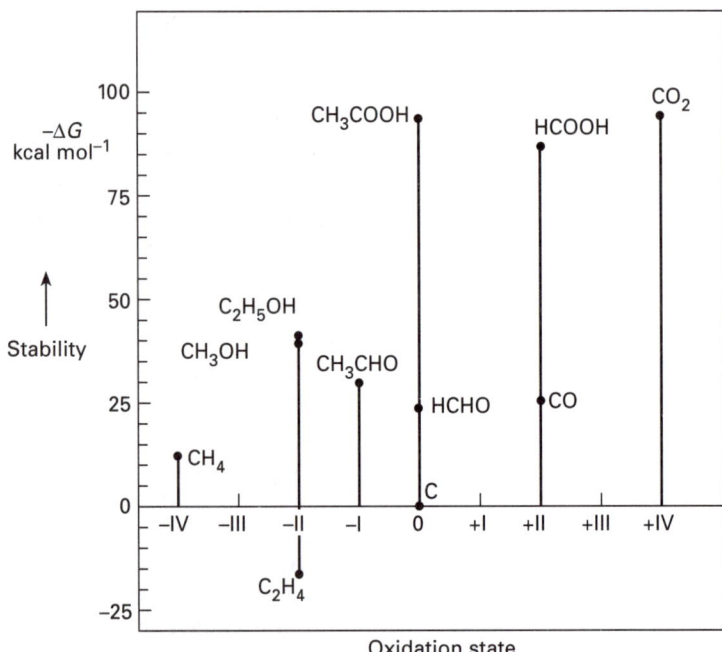

Fig. 5.21 Landscape diagram of free energy for some carbon compounds with hydrogen and oxygen. Here $-\Delta G$ corresponds to the thermodynamic probability of formation of different C/H/O 'species'. Note that this type of diagram can be related to the conventional 'oxidation–state diagram' in which we plot nE (volts) against oxidation state (see Fig. 6.13). The plots for covalent *molecules* show discontinuities in composition not observed for alloy phases (see Fig. 5.20).

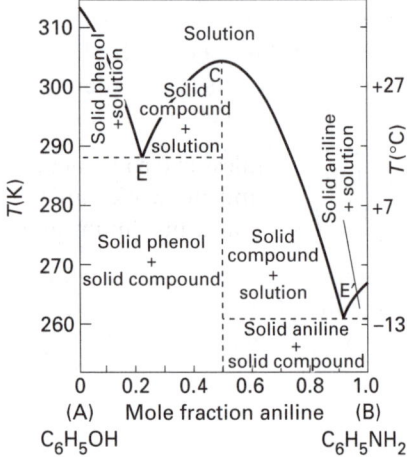

Fig. 5.22 The system phenol + aniline, illustrating the formation of an intermolecular compound. Note the temperature scale due to low co-operativity in condensed organic phases. The composition of the compound formed here is slightly variable.

Fig. 4.14 where we look at Zn mixed with Cu at equilibrium). There, the different atomic combinations, *phases*, are formed in exchange at equilibrium, and then the system is defined by the % composition of the elements alone: the only 'components' there are the atoms Cu and Zn. Thus, organic chemistry readily gives rise to complicated components, which may also mix, and form different phases (Fig. 5.22) ([Of course the same approach applies to many groups of non-organic compounds.])

The oxidation state diagram of Fig. 5.21 is a disrupted contour of free energy content ($\Delta G°$) against composition at a given temperature (25 °C) and pressure (1 atom) showing only a few of the thousands of possible stationary states of C/H/O combination, since we can have extremely high molecular weight compounds. The value of $\Delta G°$ includes all the relevant variables (see Section 5.6, and for large molecules see Section 5.8.5). Appreciating the relative $\Delta G°$ values in this diagram is essential for an understanding of the chemistry of living organisms. Of course, many inorganic compounds are also in stationary states, as are all organometallic compounds. In such diagrams the variation of $\Delta G°$ with element composition is very sharp, as in NaCl, since valence considerations have to be met. Of course, we can also draw $\Delta G°$-composition diagrams for two (or more) organic components; then the variation of $\Delta G°$ with composition may be very small (no phases separate) or large (when phases may separate), and may even lead to stoichiometric association (see Fig. 5.22).

5.11.2 Components and equilibria

Consider a fixed empirical composition of atoms, say $C_2H_4O_2$. Separable chemicals exist in what we call *stationary states* at unit pressure and at 298K. Two such states are CH_3COOH and $(CH_4 + CO_2)$. The most stable of these standard conditions is $(CH_4 + CO_2)$. It follows that we can mix them in any ratio we wish, keeping composition constant. There are then two variables, in that we can vary $(CH_4 + CO_2)$ independently from CH_3COOH, keeping the empirical composition fixed.

Now let us allow exchange at equilibrium of one mole of total composition $C_2H_4O_2$

$$CH_4 + CO_2 \rightleftarrows CH_3COOH$$

Although the amount of CH_3COOH present will be minuscule, this equilibrium system is the most stable, i.e. ΔG is slightly lower than for $(CH_4 + CO_2)$ alone and much lower than for CH_3COOH alone (Fig. 5.23). There is now only one variable since we cannot vary $CH_4 + CO_2$ independently from CH_3COOH, and the total amount of substance is fixed. Since the system at equilibrium is fixed we could define all other free energies of the same composition relative to its free energy. Here, the difference between the three conditions is the different *fixed* amount of free energy dissipated in going from equilibrium to each of the two stationary states, both of which could be defined in this way or by direct reference to formulae, i.e. to the amounts of components.

Since it would be very difficult to define the energy of an equilibrium mixture of all possible arrangements of a given composition, say, $C_xH_yO_z$ where there could be hundreds of combinations of different formulae, it is convenient instead to refer all energies to *standard states of elements at one atmosphere pressure and a temperature of 298K (25 °C)*, i.e. relative to the free energies of C(graphite), H_2(gas) and O_2 gas (see Table 5.7). Differences in ΔG, ΔH or $T\Delta S$ for $CH_4 + CO_2$ and CH_3COOH are then differences between stationary states. Such differences in energies of components decide where the equilibrium lies between them. The equilibrium (stable) condition is the one to which all systems tend to go unless external energy is applied to the system. At equilibrium, $\Delta G = 0$ and no work can be obtained from the system, when the entropy must be at a maximum and cannot change further.

We can apply this logic to any system of elements and charges, when *at full equilibrium* the elements and the total charge could be the only variables of composition since all possible combinations of them in all possible potential energy conditions would be in balance.

Now, all chemical compounds on Earth, in life or anywhere else are formed from the 90 naturally available chemical elements. If all reactions between them and all physical transfers between their phases were in balance, then, if the *relative amount* of each was also fixed, there would be no way in which the system as a whole could change at fixed temperature and pressure. The variance would be zero. This is called a fully balanced or equilibrated state. Had it occurred, and given that the Earth has kept a fixed temperature and pressure, roughly as it has done for 4×10^9 Years,

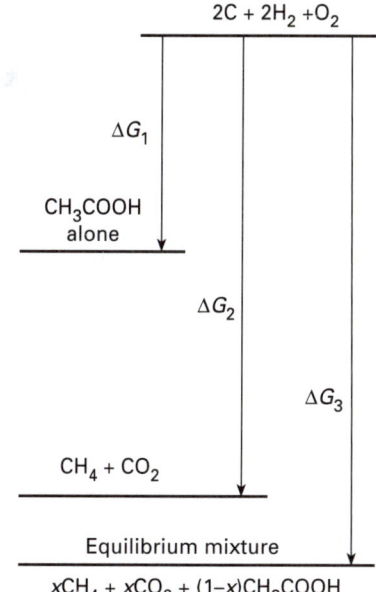

Fig. 5.23 The free energy at 25 °C and one atmosphere of separate gram-moles of $2C + 2H_2 + 2O_2$ (conventionally $G° = 0$) and one gram-mole of CH_3COOH, or gram-moles of $CO_2 + CH_4$. The equilibrium $\Delta G°$ for $CO_2 + CH_4 \overset{\rightarrow}{\rightleftarrows} CH_3 COOH$ is also shown.

nothing could have changed. Our experience is quite different. We know that the atmosphere, the sea, the composition of the Earth's crust and the nature of living systems have all evolved. The evolution corresponds to and derives from a developing complex chemical system.

What we are saying above is that if all the elements came together in a balanced physical and chemical condition, say at the temperature and pressure we experience on Earth, then physical and *chemical speciation* would have been decided once and for all time *assuming that no external energy is applied to the Earth and that Earth did not lose energy*. This did not happen, but the analysis of this assumption gives us a very safe basis from which to consider the circumstances that have arisen.

(1) Although much of the solid material of Earth conforms closely to a balanced physical and chemical condition, which allows much geochemistry to be rationalised, the immediate solid surface, e.g. soils, the waters and the atmosphere, are not in this condition.

(2) Since Earth has had a fixed temperature gradient, not a fixed temperature, from the time of its formation, we can still analyse, in terms of equilibria, zones at different temperatures (see Chapter 9) as long as mixing exchange is complete. This is not true for the crust.

(3) While the whole of the Earth's crust has remained in an approximate temperature balance for 4–5×10^9 years (there have been periods of minor cooling and heating) whence it can be described as in thermal physical balance, it actually is in a steady state (see Chapter 8). Locally, the sun's radiation and energy from the core have affected chemical conditions at the surface of the Earth and in particular have greatly affected organic chemicals, generating life. The result has been continuous evolution of the crust and life.

Taking points (1)–(3) together, the principles described in Chapters 1–4 and now in this chapter allow us to see how *physical* and *chemical speciation* could have arisen initially throughout much of the Earth, especially in its gases and solids. Thus we have a thorough appreciation of systems (1) (2) and (3) in Table 5.3. [We need an additional chapter (Chapter 6) to show how these principles of speciation apply to equilibrated solutions such as the sea.] However, looking more closely at the top layers of Earth and then at living organisms in it, we see that out-of-balance systems become more and more dominant and quite different considerations will have to be appreciated in order to understand the chemistry of some materials and especially those involved in biological speciation (see Chapters 9–15).

5.12 Predictability of properties: chaotic systems and the interdependence of variables

There is a final condition to consider in the discussion of order–disorder, which is that found in a chaotic system. In effect, it might be thought that since we have been able to describe forces and potential energies in terms of two-body interactions in Chapters 2–4, and we have found that pressure and temperature are related to random motion, i.e. random velocities of particles, we should be able to describe a system completely using either Newtonian or quantum mechanics, or both, for different bulk and small systems, respectively. As we have seen, this is not possible *in practice* and we had to define operational bulk thermodynamic functions which could be measured but could only be explained in terms of statistical mechanics and their calculation requires the use of very powerful computers. This does not remove the apparent theoretical possibility of predicting the future from the present. In 1776 Laplace stated already that

The present state of nature is evidently a consequence of what it was in the preceding moment, so that if we comprehend all the relations (forces) of the entities in this universe, then we could state the respective positions, motions and general effects of all these entities at any time in the future or in the past.

This implies that we could be able to understand *in principle* all observable states of systems and how they will behave in the future.

This logical position relied apparently on a confidence in Newton's laws (as we have described them) and now relies at the microscopic level on quantum mechanics. It is a view still held by many scientists. In fact, the view is incorrect for two reasons. The first was stated by Poincaré in 1903

Even if the natural laws were no longer hidden from us, it may happen that very small differences (*undetected by us and perhaps not detectable*) in initial conditions can produce very great ones in the future.

The smallest error in position or velocity of individual particles has dire consequences for the behaviour of a system. This then affects our ability to describe dynamic stationary states such as planetary systems with absolute accuracy. The problem increases as the number of particles increases and they are allowed to collide. The effect of small differences in states will become extremely important from Chapter 8 onwards when we look at *the survival of kinetic patterns*. In passing, we note that a conclusion from quantum theory is that position and momentum are always associated with uncertainty (Heisenberg's principle), and we are forced to use a probablistic approach even at the level of single particles.

A second problem, which belongs really to movement along pathways (see Chapter 8), concerns not the inability to know any initial circumstance sufficiently well, but the fact that in certain situations a system can change arbitrarily in, say, one of two ways. Such a situation arises when there is no difference in the energies or probabilities that allow change in two alternative ways. Consider, for example, an oscillating frictionless pendulum: its stationary state is fixed, for small angles of oscillation, θ, and easily described mathematically (Fig. 5.24). However, let the pendulum start in the upside down position. How will it behave? Will it oscillate, or travel in circles, or will it do both in arbitrary ways, or finally will it give rise to oscillations and circles in patterned succession? The problem here is that the pendulum starts from and has to pass through a point where *two* different motions can result and where an infinitesimal energy input is critical (see also Fig. 5.25). This is called a point of

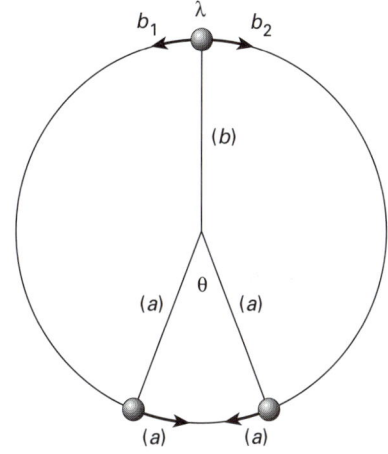

Fig. 5.24 The swinging of a pendulum through a small angle θ is entirely predictable (and can be expressed by a linear equation relating force to displacement). The ability to predict the behaviour of a pendulum starting from an inverted position can only be expressed in terms of the probability of its behaviour. (The motion is a non-linear relationship between force and displacement.)

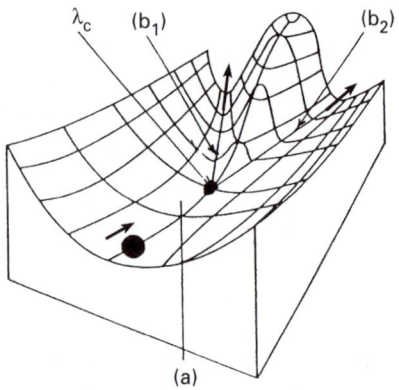

Fig. 5.25 Mechanical illustration of bifurcation. At point λ_c the ball can follow two equal alternative paths — b_1 and b_2. This is also true of the pendulum in the inverted position (Fig. 5.24).

bifurcation, λ, and there is no certain answer to the problem. The link to the argument of Poincaré is that an infinitesimal (unknowable) difference between velocities or positions at the top of the pendulum swing changes the pattern for the future. In fact, the system could behave irregularly if there were even the smallest fluctuation in the environment. A second, perhaps clearer, example is that of a ball rolling over a surface, when at a given point if can follow two alternative paths (Fig. 5.25). Which one will be preferred? Obviously neither, and we are unable to predict which one will be selected in any given experiment. We have to be aware that the problems described here are frequently met, so that prediction of how a given system will behave is not possible, but a picture can be given for the *probability* of behaviour of a system, although this is not, of course, the picture scientists like to present of the reductive power of their analyses. This behaviour is typical of so-called chaotic systems. However, it has been observed that such chaotic systems can and do develop persistent dynamic patterns, and thus some kind of 'order', organisation, (Chapter 8). As expected, systems composed of many particles are more susceptible to chaotic behaviour. This has caused great difficulty in understanding the relationship between biological and physical systems (see Chapter 16).

5.13   Summary

In this chapter we have considered the major variables of many-particle *bulk* material systems of kinds (1) (2) and (3) in Table 5.3, especially so as to appreciate the significance of the commonly used variables temperature and pressure (volume) in terms of statistical occupancy of spatial positions and of thermal states. The assumption is that all such occupancies are in rapid exchange so that we need a collective treatment of particles. The collective property is defined per mole of substance. When we considered the energy required to alter the statistical probability of the occupancies we found that we could relate them to log p and log T. This led us to define a new concept, that of entropy, as a collective derived variable summing the effects of p and T affecting the energy available from a system. Knowing the energy equivalent to the changes in statistics of space occupation and random motions of a system expressed in this collective disorder term, $T\Delta S$, we could add it algebraically to the term ΔH, containing the sum of all the changes of ordering energy of the pair-wise interactions and others, incorporated in ΔU, described in Chapters 2–4, plus the work of expansion $p\Delta V$. We have shown that the disorder collective term $T\Delta S$ is not reversibly convertible into work, so that the available *free* energy change, capability of doing work, i.e. ability to bring about change *at constant pressure*, is then defined by the equation

$$\Delta G = \Delta H - T\Delta S$$

Notice that ΔG includes the variables of composition, with their interactive energy in terms of components largely in ΔH. In this equation involving just *equilibrium thermodynamic variables (state functions)*, we have avoided

Table 5.12 Units, variables and restrictions introduced up to Chapter 5

Units	Variables	Restrictions
Atomic elements and charge	Relative amounts of each unit giving % composition as a variable	
Components	Potential energy of interaction between units in stable and stationary states (spatial relationships)	Equilibrium exchange between units reducing the number of components
	Angular momentum* in stable and stationary states	
	Random momentum* associated with pressure	
	Random kinetic energy* associated with temperature	Equilibrium exchange between phases
	Directed kinetic energy* Directed momentum*	See Chapters 8 and 10–16

This form of presentation must not give the idea that the restrictions correspond to particular variables opposite to them in the table, e.g. angular momentum and equilibrium exchange between phases.
* All these derived variables contain the fundamental variables, time and spatial position, although they refer to unchanging stationary states.

direct reference to the generally used underlying variables—atomic element and charge composition, pair-wise internal binding energies, mechanical energy (pdV), random kinetic energy (temperature) and also random momentum (pressure)—and to a large degree we no longer refer to the derived variables of Table 2.3. In fact ΔH and ΔS contain appropriate though complicated sums of all these variables (Table 5.12) so that ΔG provides an extremely valuable quantitative way of evaluating comparatively the stability of systems and especially their capability for change. This gain is countered by the loss of ability to reduce the system to the even more simple fundamental variables of composition and space–time distribution.

Immediately we found an explanation of the physical states of matter and their changes for different materials in terms of the relative values of ΔH and $T\Delta S$ at different temperatures. We assumed of course that all physical states were in equilibrium, i.e. in fully exchanging phases. Thus the explanation of the Greek and Chinese 'elements'—earth (solid), water (liquid) and air (gas)—is complete. These 'elements', states of matter, and fire (energy), are due to a complex association of several variables with known components and energy. Now, just because we have separated out the units and variables of all materials we must not assume that analysis of all we observe is open to simple explanation. We cannot understand the formation of different states of matter even in balance (equilibrium), i.e. the balanced formation of gas, liquid or solid phases, without analysing the interactions between the underlying variables. It would have been very rewarding if we could then have shown that the relationships were ideal and linear. In fact, the relationships are non-ideal and non-linear.

Non-linear or complex dependencies give rise to collective properties of assemblies where there is no obvious way of appreciating the system in terms of pairs of interacting single particles.

Extending the analysis we looked at the relationship of the change of ΔG for chemical reactions of components, including changes with temperature, where ΔG is measured relative to standard states of elements per mole, but excluding the rates of any such changes. We could then examine, quantitatively, equilibria between stable compounds and/or stationary state compounds over wide temperature and pressure ranges. We observed that both phase transfer and chemical exchange equilibria restricted variability, that is the number of degrees of freedom (also called variance). In conclusion, *we are now able to appreciate speciation and both changes of chemical and physical systems as we vary composition, temperature and pressure, in terms of available energy, i.e. work capability, ΔG. We drew suitable landscape diagrams illustrating this point. We can, therefore, claim to have now a great insight into why non-living materials at local equilibrium* are found with given compositions, structures and physical properties, even when only some of them can be traced to interactions between small numbers of units. However, since chemical change is often accompanied by statistically complicated physical change, we must again expect complex, not simple behaviour.

Two problems remain to be analysed before we look at rates of change, i.e. the involvement of *time*.

1. The application of the analysis of ΔG to conditions in dilute solution (Chapter 6) since this will help us to understand, in part, variation of speciation in biological systems.

2. The limitation of any system to a 'small' volume when it can interact with other such systems through space. This will re-introduce bulk interactions, i.e. fields (Chapter 7) and shapes of systems.

These two additional chapters will conclude our analysis of static systems. We have come a long way in the understanding of the factors underlying the diagram in Fig. 1.2, while removing the original classical ideas of 'elements' and 'qualities'. Unfortunately, this still leaves us without explanation of another feature of the world around us—flow in inanimate, or more especially in animate (living) systems (see Chapters 8–16). Flow involves consideration of rate of change as well as of change itself. The equilibrium thermodynamics of this chapter only concerns change and not the rate of change. However, the reader should see that we are beginning to move from an appreciation of the *nature of primitive matter* in the universe to an uncovering of *the nature of man*.

6

Dilute solutions and
order–disorder balance

---

*Water is the blood of the Earth, and flows through its muscles and veins.
Therefore, it is said that water is something that has complete faculties ... It is
accumulated in Heaven and Earth and stored up in the various things (of the
world). It comes forth in metal and stone and is concentrated in the living
creatures. Therefore, it is said that water is something spiritual.*

Kuan-Tzu (Chapter 39) (*c*. Third century BC)

6.1   Introduction

In Chapters 4 and 5 we have been concerned with systems in which there
was either only one component or in which there were several com-

ponents, each present in considerable amounts. In Chapter 4 our major concern was with ordered structures while in Chapter 5 it was with the balance of order with disorder within such systems. In both chapters we considered the effect of variables on the material present. The units of these materials were ultimately chemical atomic elements (mass and charge) but it was found to be operationally easier to use as units the so-called 'components', which were chemical combinations that did not exchange any of their elementary atoms with one another. The variables were then composition, in terms of components (all present in large amounts), temperature and pressure. In essence, in Chapter 5 we described these variables through the available free energy function, ΔG, which gave quantitative expression, *at constant pressure*, to the effect of the variables composition and temperature. (It is normal for us to work close to constant pressure.) In this chapter we wish to use the same approach in the consideration of dilute solutions since much of the chemistry of the sea and the laboratory is carried out under such conditions. It will be seen in Chapters 10 onwards that this is also the case for living systems. The value an element or a component has in a living system, although restricted ultimately by abundance, is often relative to its presence in air or its presence in dilute solution in water, i.e. its availability.

6.2 Ideal solutions

The simplest picture of a dilute solution is that obtained by considering mixing a liquid, the major component, called the solvent, with a second substance, the minor component, called the solute, which may be a gas, a liquid or a solid, and which leads to a perfectly random arrangement of both solvent and solute. The tendency to go into this *ideal solution* is due to the increase of entropy on mixing (randomness giving increased probability) (Section 5.6.2) while the resistance to dissolving is due to the (chemical) bonding between the molecules of the solvent, or between those of the solute molecules separately. The solubility of the solute in a litre of the solvent is then a measure of the relative stability of the solution compared with that of the pure solute and solvent, which can again be expressed by a free energy difference, $-\Delta G$ ($= RT\ln c$), between the free energies of the solution and the sum of those of the unmixed solute and solvent (see Section 6.4). Deviations from ideality are due to solvent and solute interactions in the solution phase. If there is more than one solute present in the solvent, then the two, or more, solutes may form 'complex' species (usually of low molecular weight) due to favourable chemical interaction between the solutes or their constitutive parts, a further deviation from ideality. This is, of course, again a problem of chemical bonding. We now consider the influence of these factors in selected systems. We wish to show that the properties of chemicals in solution, including their solubilities, are describable by the same variables as were covered in Chapter 5 (see Section 6.11).

6.3   Water as a solvent

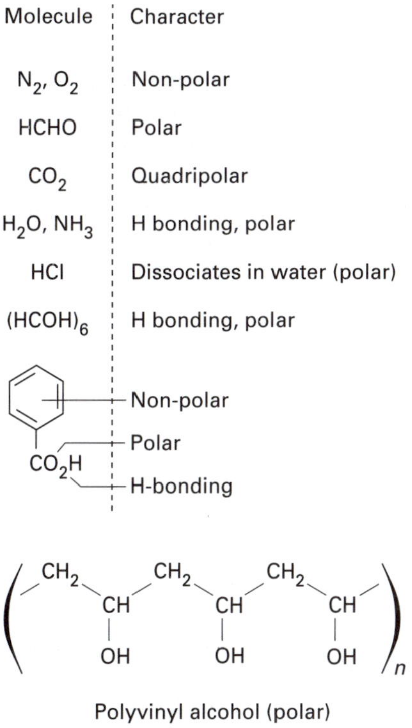

Molecule	Character
N_2, O_2	Non-polar
HCHO	Polar
CO_2	Quadripolar
H_2O, NH_3	H bonding, polar
HCl	Dissociates in water (polar)
$(HCOH)_6$	H bonding, polar

Non-polar
Polar
CO_2H
H-bonding

Polyvinyl alcohol (polar)

Fig. 6.1 The character of some typical molecules.

Table 6.1 Solubility of some gases in water at $T = 0\ °C$

Gas	Solubility in water (mol l^{-1})
Oxygen	2×10^{-3}
Ozone	2×10^{-2}
Nitrogen	1×10^{-3}
Nitrogen monoxide	3×10^{-3}
Carbon monoxide	1.6×10^{-3}
Carbon dioxide*	7.5×10^{-2}
Methane	1.5×10^{-3}
Sulphur dioxide*	1.0×10^{-2}
Hydrogen sulphide*	2.0×10^{-1}
Ammonia*	>50
Hydrogen chloride*	>20

* These gases react with water.

As stated, a solution consists of a homogeneous bulk solvent with minor components dissolved in it. In this book we shall stress the nature of water as the bulk solvent since it is the major liquid on Earth and is the solvent of life. It is clear that it can dissolve some compounds in an almost limitless way, e.g. low molecular weight alcohols and sugars, while other compounds have a limited solubility, e.g. salts, and others hardly dissolve in it at all, e.g. silica, diamond, most metals and oils. Again there are a few substances which on mixing with water form peculiar film-like structures, e.g. soaps (see below). We must understand the reasons for these different behaviours. A rough principle of solution is that 'like dissolves like' material. Water has the formula H-O-H and forms associated units in the liquid state due to interaction of partial charges in the hydrogen and oxygen atoms (see Section 4.2). It dissolves very well, therefore, other substances carrying similar *partial charge* separation and/or that can also form hydrogen bonds, such as methanol, CH_3OH. The formulae of some such substances are shown in Fig. 6.1. Here, favourable solvent-solute interactions are often an additional factor in the free energy change of solution, on top of the entropy of mixing change (Section 5.6.2), so as to increase solubility.

As the partial charge separation and the tendency to form hydrogen bonds is reduced in series such as CH_3OH, H_2S, CO_2, CH_4 so the solubility in water decreases very rapidly. Simple molecules such as O_2, N_2 and H_2 have no charge separation and like CH_4 have a very low solubility in water (Table 6.1). If we combine, in one molecule, a head group which has partial charge separation with a tail which has no such charge separation, as in molecules such as

$$CH_3 - CH_2 - CH_2 - CH_2 - CH_2OH$$

then the longer the oily tail the less the solubility in water of the substance, until we reach almost insoluble fats which may have over 20 $-(CH_2)-$ units in their tails. Such fats as butter and margarine, which have no polar head group, do not dissolve appreciably in water, while fats with polar head groups *disperse* on agitation as tiny colloidal drops in water (Fig. 6.2) and are called soaps or detergents. Soaps usually have a carboxylate, while detergents have a sulphate head group. In these cases solvent-solvent (water-water) interactions prevent solution overcoming the favourable entropy of mixing even though the solute-solute interaction is usually weak. This exclusion from water is sometimes called a hydrophobic interaction between solute molecules!

Some salts, such as sodium chloride, are quite soluble in water, but they have a complete and greater charge separation than is present in H_2O. As we have seen, we write the formula of sodium chloride as Na^+Cl^-. Here, and in similar salts, where the separated charges are not larger than units of one, we find fairly soluble compounds, which dissociate into separated ions in water. However, as the charges increase the salts become less soluble. We can describe magnesium, aluminium and silicon oxides as

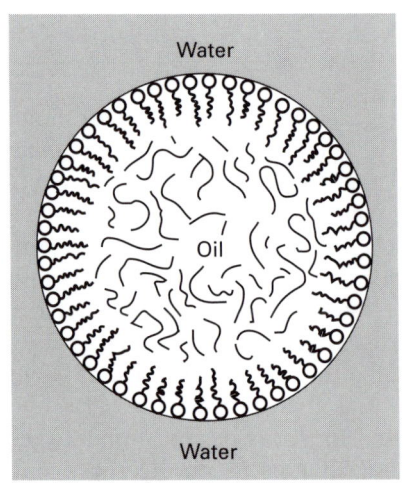

Fig. 6.2 Example of a colloidal drop of oil in water.

$Mg^{2+}(O^{2-})$, $Al_2^{3+}(O^{2-})_3$ and $Si^{4+}(O^{2-})_2$ lattices, but we remind the reader that the lattice of a salt is strongly co-operative relative to the molecular state (Section 4.2). Thus, oxides of highly charged ions such as Mg^{2+}, Al^{3+} and Si^{4+} are virtually insoluble and, in fact, they form the soils, clays and rocks around us, remaining in very dilute solution only, while the sea is a fairly concentrated solution of dissolved Na^+Cl^-, common salt. Here we see that solute-solute interactions reduce solubility, overcoming the entropy term of mixing.

Returning to organic molecules, there are some which carry balanced charge on a framework. Of particular importance in this book are the amino acids described in Sections 4.8 and 10.4.1, which have the formulae

$$^-O_2C - CH - NH_3^+$$
$$|$$
$$R$$

where R is a small organic side-chain. These substances carry relatively low charge, like Na^+Cl^-, and are usually soluble in water, unless the R group is large. The same is true for charged derivatives of fatty acids such as phosphatidyl serine or choline (Fig. 6.3) when the molecules form micelles, vesicles or bilayer sheets which are often found in living organisms. Another soluble group of organic chemicals are the nucleotides of RNA and DNA. These are only some of the organic molecules that are common in biological systems (Table 6.2).

The fact that ions (charged atoms or molecules) are soluble in water reminds us that it is essential when describing composition as a variable to give the charge character of each chemical species as well as its chemical atom content. The fact that charges may be separated allows electrical fields to be built up in aqueous media and these charges can then carry current (see Sections 6.13 and 6.14).

Table 6.2 The major organic molecules of biological systems

Small molecules	Large molecules
Charged amino acids	Charged proteins
Charged sugar phosphates (DNA and RNA bases)	Charged DNA and RNA
Saccharides, i.e. sugars (charged and uncharged)	Polysaccharides (charged and uncharged)
Charged dicarboxylic acids	–
Charged monocarboxylic acids	Charged fatty acids (insoluble)
Phosphate esters	Charged lipids (insoluble)
Long-chain ethers (archaebacteria)	Uncharged lipids (insoluble)

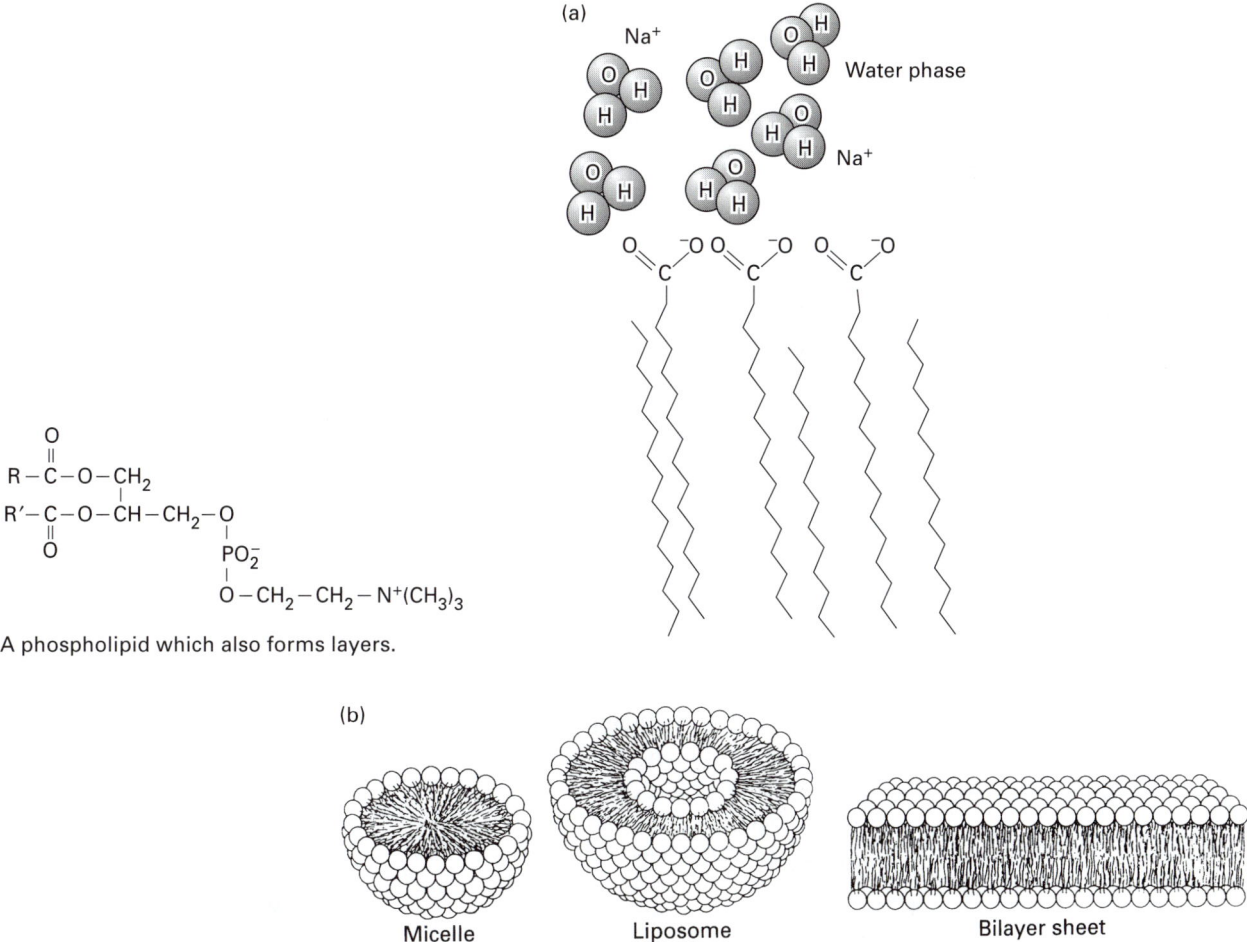

A phospholipid which also forms layers.

Fig. 6.3 (a) A picture of water and sodium ions near the surface of a fatty acid anion lipid layer. (b) Cross-sectional views of three structures that can be formed by phospholipids in aqueous solution: spherical micelles with hydrophobic interiors; sheets of phospholipids in a bilayer; and spherical liposomes comprising one phospholipid bilayer (see Fig. 4.10 and Section 10.4.1). Biological membranes are formed from bilayer sheets.

6.4   General considerations of solubility in other solvents

When we consider different solvents of low or no polarity, i.e. without partial charge separation, then the rules concerning solubility are quite the opposite to those above. Thus, few of the small molecules that contain polar groups are soluble in oils and neither are salts—solute-solute inter-actions are too strong. Considerable solubility is now confined to non-polar solutes in non-polar solvents. This is why we use 'oily' soap, not just water, to remove oil from our hands and clothes. Note that since biological chem-istry is largely organic chemistry in water it could only exist using polar organic molecules, unlike most of man's organic chemistry which uses molecules of lower polarity in non-polar solvents. Even the isolation of bio-

Surface ions

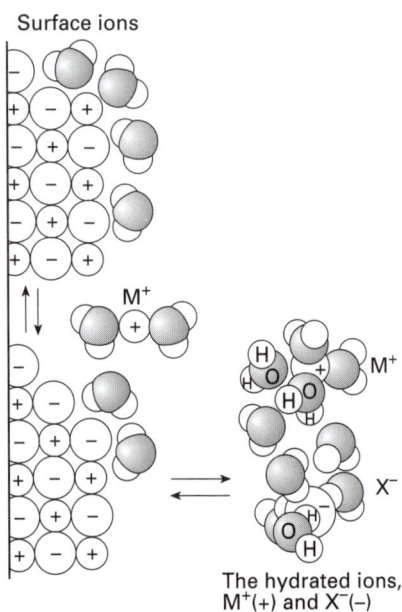

The hydrated ions,
$M^+(+)$ and $X^-(-)$

Fig. 6.4 Equilibrium between a pre-cipitate MX and ions M^+ (+) and X^- (–) in water, illustrating solvation.

logical cells is due to an organic solvent, a so-called lipid, which surrounds their internal aqueous solutions, as in Fig. 6.3

The solubility of a substance is, therefore, clearly decided by the two opposing considerations of the energetics of order and disorder discussed in the previous chapter. The ordering interactions, which oppose solution, are due in the first place to the relative co-operative binding of atoms or components together in individual and separated phases, whether these are moderately strong as in a solid such as sodium chloride, Na^+Cl^-, or very strong as in $Mg^{2+}O^{2-}$, or not so very strong as in water itself, or weaker as in methyl alcohol, or very weak as in petrol, C_8H_{18}, or higher molecular weight, detergent-like molecules. Besides these ordering interactions in the separated pure substances, there is also some local ordering of solvent, e.g. of water molecules around ions (solvation; Fig. 6.4), which now favours the solution. Thus we find that: (1) water molecule-water molecule interactions are stronger than, say, methane-water interactions, hence methane and other very non-polar molecules are largely forced out of water, leaving even ordered water around the few molecules remaining in solution; but (2) water molecule-polar molecule or water molecule-ion interactions are somewhat comparable in strength to water-water, polar molecule-polar molecule and ion-ion interactions, whence polar molecules and salts are often soluble in water.

Other materials which are virtually insoluble in all solvents are those that form strong continuous networks of low or no polarity. We described such materials, like diamond, sulphur, silica and many metal elements, such as gold, in Chapter 4. Many liquid metals do dissolve to a considerable extent in one another to form alloys, but, for example, diamond does not dissolve in anything.

Whether or not a substance dissolves depends, therefore, on a balance between these electrostatic interactions (enthalpy terms) and the increase of disorder on solution (entropy terms). In an ideal solution (no interaction between particles) we just consider the disorder of mixing the solvent and solute as represented by a (statistical) energy term, $RT(2.303)\log c$, where T is the temperature and c the concentration of solute* (Sections 5.3 and 6.2). If this were the case, then disorder would be simply calculated as for a mixture of gases (Section 5.9). However, in real solution cases there is little possibility to examine the algebraic sum of these heat and entropy terms since in such non-ideal cases the influence of disorder is very difficult to assess. For example, the solvation of solute by the water molecules mentioned above may create ordering of the water at close range but disordering at long range. Moreover, the solute may be associated to some degree, even in solution, so that the amount in solution, the solubility, is not simply related to a concentration of one particular species. There is, in effect, an overall energy change helping solution, but it cannot be described straightforwardly in terms of order and disorder, $\Delta H^\dagger$ and $T\Delta S$.

---

*Note the analogy of the concentration of the solute, c, to the partial pressure of a gas, p, when the statistical term is $RT(2.303) \log p$.

†It happens that frequently the value of ΔH is temperature dependent as shown by changes in heat capacities, C_p, which causes severe problems with any simple explanation of ΔH in terms of chemical bonds.

Table 6.3 Solubility of classes of compounds in water and in organic solvents

Considerably soluble in	
Water	**Organic solvents**
Salts, organic and inorganic	Alkanes and alkyl halides
Complex ions (salts)	Ethers
Short-chain acids, amines, alcohols, thiols, aldehydes and ketones	Long-chain acids, amines, alcohols, thiols, aldehydes and ketones
Amino acids (and many proteins)	Alicyclic compounds
Nitrogen bases, e.g. pyridine, nicotine, adenine, thymine, etc.	Aromatic compounds
Sugar (polysaccharides)	Lipids
DNA^{n-} and RNA^{n-}	Organometallic compounds

For these reasons we shall not try to give a detailed theoretical description in thermodynamic terms of the quantitative solubility of substances in different solvents and shall be content with a numerical tabulation of solubility trends amongst different molecular species (Table 6.3). For an understanding of why biological systems of organic and inorganic species in water are as observed, this knowledge of the limitations on solubility of a vast range of chemicals is essential. It is also essential for an understanding of the evolution of these systems. It is solubility that determines the availability (not the abundance) of elements. *Availability is then a heavy environmental constraint on the composition variable for all of life, just as there is a decided limitation on the temperature variable due to the use of liquid water with its narrow liquid range.*

6.5 Quantitative treatment of solubility: precipitation and solubility products

So far we have described the sea as a solution of common salt, Na^+Cl^-, in water, but of course it contains many other elements in smaller quantities. They can associate (see below) with the Na^+ and Cl^- ions, increasing the apparent solubility of NaCl. The solubility, the solution limit opposite precipitation, is better expressed now as grams or gram-molecules (moles) per litre in water, so as to give a quantitative description of the limitations to relative amounts of elements in the sea.* In Table 6.4 we list solubilities of common salts. It is immediately obvious that there is a huge range of values, from very soluble to virtually insoluble salts. If we turn now to living systems which, as stated, are based on water solutions, the (unfortunate) fact is that many essential chemical elements are locked in the

---

*In many books on the environment the concentrations are expressed in parts (grams) per million grams of water (ppm) which corresponds to milligrams per litre.

Table 6.4 Solubilities in water at 25 °C

Anion*	Solubility in water at 25 °C (mol l⁻¹)													
	Li^+	Na^+	K^+	Rb^+	Cs^+	Be^{2+}	Mg^{2+}	Ca^{2+}	Sr^{2+}	Ba^{2+}	Al^{3+}	Sc^{3+}	Y^{3+}	La^{3+}
OH^-	5.6	27.0	20.0	19	25	1.3×10^{-5}	3.2×10^{-4}	1.8×10^{-2}	8.6×10^{-2}	0.27	$<10^{-4}$	$<10^{-4}$	$<10^{-4}$	$<10^{-4}$
F^-	0.05	1.0	17	12	24	5.5	1.9×10^{-3}	3.1×10^{-4}	9.3×10^{-4}	1.2×10^{-2}	$<10^{-3}$	$<10^{-3}$	$<10^{-3}$	$<10^{-3}$
CO_3^{2-}	0.2	2.0	8.1	9.6	8.0		1.3×10^{-4}	6.2×10^{-5}	3.9×10^{-5}	4.4×10^{-5}	†	†	†	†
Oxalate	0.7	0.3	2.0	1.5	1.6	3.4	2.7×10^{-3}	6.7×10^{-6}	2.4×10^{-4}	5.0×10^{-4}	$<10^{-2}$	$<10^{-2}$	$<10^{-2}$	$<10^{-2}$
Formate	8	15	40	‡	‡		0.71	1.3	0.53	1.4	§	§	§	§
Acetate	5	6	29	‡	‡		4.5	2.2	2.0	2.9	§	§	§	§
Cl^-	19.0	6.0	4.6	7.5	11.0	Large	5.7	7.3	3.5	1.8	2.0		3.0	
Br^-	18.8	9.0	5.5	6.5	6.0	Large	5.5	7.0	4.0	3.5	>5.0		2.5	
I^-	12.4	12.3	9.0	7.0	3.0	Large	5.0	7.0	5.3	5.3	>5.0			
SO_4^{2-}	3.2	4.3	0.6	1.8	5.0	4.0	3.0	7.8×10^{-3}	6.2×10^{-4}	9.5×10^{-6}	1.0	0.6	0.2	0.04
NO_3^-	10.5	11.0	3.0	3.6	1.0	8.2	5.1	7.4	3.3	0.35	2.0		5.0	0.2
ClO_4^-	5.7	16.0	0.1	0.07	0.09	7.1	4.5	7.9	10.8	5.9	2.0			
ClO_3^-	48.0	10.0	0.6	0.3	0.3		7.7	9.7	6.9	1.1				

* NB. With these anions transition metal ions in salts follow a similar pattern of solubility to that of M^{2+} and M^{3+} salts in the table.
† Not stable.
‡ Very high solubility.
§ Fairly soluble.

rocks in rather insoluble forms, e.g. Mg^{2+}, Fe^{3+} and Zn^{2+} in oxide or sulphide lattices, thus they are of low availability.

In order to appreciate the problem of the availability of individual ions from these solids, we consider that at equilibrium there is a quantitative connection between the contents of a solution and the solids with which they are in contact, for example the sea and the nature of the minerals on the sea bed. This is easily seen on the tidal coasts of the ocean, especially in hot climates where common salt (NaCl) is deposited as water evaporates. We have, then, the equilibrium

$$(\text{sea}) \ Na^+ + Cl^- \rightleftarrows NaCl \ (\text{solid deposit}).$$

The two sides of the equation are in balance, a fact that we express by stating that the product of the concentrations of the ions in solution when solid appears is constant, that is

$$[Na^+][Cl^-] = K_{sp}$$

where K_{sp} is the so-called solubility product (constant).* (Here we have loss of degrees of freedom; $[Cl^-]$ and $[Na^+]$ are not independent variables in the presence of NaCl crystals due to formation of a new phase; see Section 5.10).

---

*The concept of solubility product is strictly applicable to sparingly soluble salts only. Note too that the solubility, or the solubility product, are equilibrium restrictions on variability since they represent limits to concentration in solution. In fact, this is just another case of a new condensed phase acting as a restriction on variability (as we saw in Section 5.10). The solubility product is directly related to a free energy difference, $\Delta G = RT \ 2.303 \ \log [\text{solubility product}]$ and can be analysed in terms of ΔH and ΔS, as in Section 6.4.

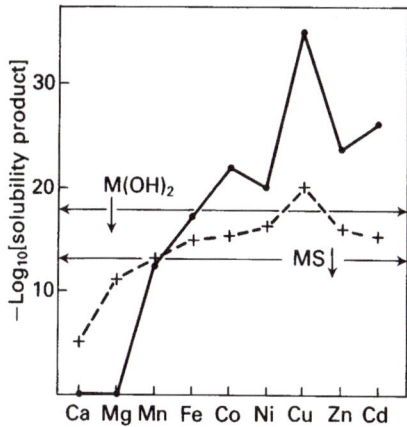

Fig. 6.5 The solubility products of hydroxides (broken line with crosses) and sulphides (solid line with circles), which have been of extreme importance in the evolution of life. The horizontal lines M(OH)$_2$ and MS indicate the solubility product which gives a precipitate at pH = 7.0 when [M] = 10^{-4}M and, for the sulphides, when [H$_2$S] = 10^{-3}M.

As stated, sodium chloride is quite a soluble salt, so the sea is very salty. Calcium carbonate is much less soluble, so the sea contains much less Ca^{2+}, CO_3^{2-} and HCO_3^-. For M_iX_j precipitates, generally, the solubility product K_{sp} is defined as $K_{sp} = [M]^i[X]^j$ where we express concentrations in moles per litre. In some cases it is very small; for example, that of iron(III) hydroxide, is

$$K_{sp} = [Fe^{3+}][OH^-]^3 = 10^{-39}(M^4l^{-4})$$

and consequently there is very little free ferric iron in the sea since its pH is about 8, when $[OH^-] \sim 10^{-6}$M. Table 6.5 gives many solubility products.

In Fig. 6.5 we plot the solubility products of some hydrated oxides (hydroxides) and some sulphides. It is clear that the amounts of divalent ions in the sea depend on whether sulphide (HS^- and H_2S) is present locally or not. Hydroxide is always present since it comes from the dissociation of water itself which gives a small amount of this ion

$$H_2O \rightleftarrows H^+ + OH^-$$

We return to the problem of the availability of elements in the sea under different, including primitive (H_2S), conditions in Section 7.14 and then in

Table 6.5 Solubility product constants, K_{sp}

Substance	Formula	Solubility product*
Aluminium hydroxide	Al(OH)$_3$	2×10^{-32}
Barium carbonate	BaCO$_3$	5.1×10^{-9}
Barium oxalate	BaC$_2$O$_4$	2.3×10^{-8}
Barium sulphate	BaSO$_4$	1.3×10^{-10}
Cadmium hydroxide	Cd(OH)$_2$	5.9×10^{-15}
Cadmium oxalate	CdC$_2$O$_4$	9×10^{-8}
Cadmium sulphide	CdS	2×10^{-28}
Calcium carbonate	CaCO$_3$	4.8×10^{-9}
Calcium fluoride	CaF$_2$	4.9×10^{-11}
Calcium oxalate	CaC$_2$O$_4$	2.3×10^{-9}
Calcium sulphate	CaSO$_4$	2.6×10^{-5}
Cobalt sulphide	CoS	8.0×10^{-23}
Copper(II) hydroxide	Cu(OH)$_2$	1.6×10^{-19}
Copper(II) sulphide	CuS	6×10^{-36}
Iron(II) hydroxide	Fe(OH)$_2$	3×10^{-15}
Iron(II) sulphide	FeS	6×10^{-16}
Iron(III) hydroxide	Fe(OH)$_3$	4×10^{-38}
Magnesium ammonium phosphate	MgNH$_4$PO$_4$	3×10^{-15}
Magnesium carbonate	MgCO$_3$	1×10^{-5}
Magnesium hydroxide	Mg(OH)$_2$	1.8×10^{-11}
Magnesium oxalate	MgC$_2$O$_4$	8.6×10^{-5}
Manganese(II) hydroxide	Mn(OH)$_2$	2.5×10^{-14}
Manganese(II) sulphide	MnS	3×10^{-15}
Nickel sulphide	NiS	2×10^{-21}
Strontium oxalate	SrC$_2$O$_4$	5.6×10^{-8}
Strontium sulphate	SrSO$_4$	3.2×10^{-7}
Zinc hydroxide	Zn(OH)$_2$	1.2×10^{-17}
Zinc oxalate	ZnC$_2$O$_4$	7.5×10^{-9}
Zinc sulphide	ZnS	4.5×10^{-24}

Data from *Stability constants* (1964). Spec. Pub. No. 7. The Chemical Society, London. Note the preferred use of Roman labels of oxidation states.
* The units of concentration are (moles)^1 (litre)$^{-1}$.

Chapter 9. Later we shall ask how these conditions and therefore availability have changed with time and how such changes have affected the Earth's surface, the sea and, hence, living organisms. It is not necessarily the case that all substances in the sea form saturated solutions, of course.

Before proceeding, notice that when discussing solubility it is assumed that if the solvent is removed the material will be recovered unchanged—reactions are excluded. This is not so in many cases; for example, if metallic iron is exposed to water it is slowly converted (not dissolved) to $Fe(OH)_2$ and then to $Fe(OH)_3$, which is insoluble

$$Fe + 2H_2O \rightarrow Fe(OH)_2 + H_2$$
$$2H_2O + 4Fe(OH)_2 + O_2 \rightarrow 4Fe(OH)_3$$

Therefore, metal iron cannot be recovered from solutions of either kind of iron salt just by removing water. Note that another factor controlling solubility is the oxidation state of the element, here iron, due to the presence or absence of O_2. [We stress again that the amount of an element in solution (the observed solubility) can be very different from the free ion concentration of it, as allowed by the solubility product, due to association of the ion with other groups present in solution.]

Another series of salts of interest, since they are present in the sea, involves larger charged anions such as in the carbonates (e.g. Na_2CO_3), sulphates (e.g. Na_2SO_4) and nitrates (e.g. $NaNO_3$). They all dissolve in water to give Na^+ ions and complex negatively charged non-metal species: CO_3^{2-}, SO_4^{2-} and NO_3^-. Salts such as $Ca^{2+}CO_3^{2-}$ have stronger electrostatic interaction than $(Na^+)_2(CO_3^{2-})$ which makes them of lower solubility, that is they have smaller solubility products in water. Of course, the less soluble salts precipitate first; this is why $CaCO_3$ is the basis of huge mountain ranges such as the Dolomites and the white cliffs of Dover, as well as being the material of shells. In fact, as stated, most salts of metal ions and anions, both of which have a charge of greater than one, are rather insoluble, hence strong solutions of non-metal anions are based on Na^+, K^+ or 'complex' positively charged ions such as the ammonium ion, NH_4^+. An exception is $Mg^{2+}SO_4^{2-}$ which is quite soluble, so that there is a considerable amount of Mg^{2+} and SO_4^{2-} in the sea. These solubility considerations have again greatly affected the way in which living systems have been able to evolve. They affect the ionic media of life, and the availability, hence the utility, of the elements, and they restrict the kind of salts which can be made into insoluble protective matrices—$CaCO_3$ and SiO_2 to give but two examples. Abundance and solubility are two major limitations on variables and therefore on composition.

6.6   Combination of species in solution: complex speciation and stability constants

In aqueous solutions containing both negatively and positively charged ions, i.e. anions and cations, they may bind to one another to some degree, thus forming ionic or (partially) covalent bonds and leaving only some ions free. There are then the two possibilities

Metal ion + anion → 'precipitate', e.g. $Ca^{2+} + CO_3^{2-} \rightarrow CaCO_3 \downarrow$

Metal ion + anion → 'complex', e.g. $Cu^{2+} + Cl^- \rightarrow CuCl^+$

A soluble complex such as $CuCl^+$ is an example of apparently new speciation in solution. A more complicated example, which is important in the sea, is the reaction of Cl^- from dissolved Na^+Cl^- with other metal ions. We take the case of lead, Pb^{2+}, where the reactions with chloride can be written as

$$Pb^{2+} + 2Cl^- \rightleftharpoons PbCl_2 \downarrow \text{ insoluble precipitate}$$

$$PbCl_2 + 2Cl^- \rightleftharpoons PbCl_4^{2-} \downarrow \text{ soluble 'complex'}$$

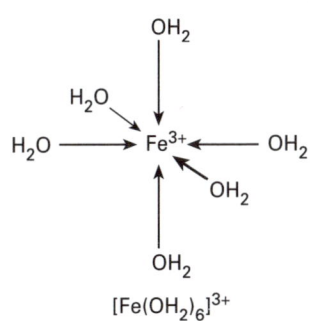

$[Fe(OH_2)_6]^{3+}$

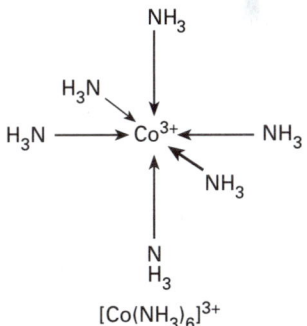

$[Co(NH_3)_6]^{3+}$

Fig. 6.6 Octrahedral co-ordination in complex ions.

We shall refer to all *association reactions* between ions (or between ions and molecules) as acid-base reactions, where the acid is the group accepting pairs of electrons, partially, and the base (usually called the 'ligand' in complex species) is donating them (Fig. 6.6). This type of bonding is often called a *co-ordinate bond* and the complex is referred to as a *co-ordination complex*. Complete transfer of electrons or atoms is different and will be described under oxidation–reduction in Section 6.13.

The formation of complex species becomes particularly important in biological organisms, where especially the anions are complicated molecules, e.g. proteins, which can bind small cations and anions. It is to such molecules that we turn in the next section since this will allow us to develop an aspect of the natural selection of the chemical elements in living systems (see ref. 52 in Further reading).

Now, just as we expressed solubility as an equilibrium between species in solution and in the precipitate, so we can express complex ion associations

$$X + Y \rightleftharpoons XY$$

by the *stability constant*

$$K = \frac{[XY]}{[X][Y]}$$

Or, more generally,

$$K = \frac{[X_m Y_m]}{[X]^m [Y]^m}$$

for a reaction

$$mX + mY \rightleftharpoons X_m Y_m$$

It is not material to this discussion if X and/or Y carry charge.

Notice that for the proton this association is called the acid constant and is conventionally written as a dissociation not an association constant, i.e.

$$K_a = \frac{[H][Y]}{[HY]}$$

Table 6.6 Stability constants (log K) of some complexes of biologically relevant ions in aqueous solution ($T = 25\,°C$)

Ion	Log K						
	Acetate (ML)	Ammonia (ML$_4$)	Glycine (ML)	Cysteine (ML)	Histidine (ML)	ATP (ML)	EDTA (ML)
Ca^{2+}	0.57		1.31			4.25	10.65
Mg^{2+}	0.55		2.22	< 4		4.55	8.85
Co^{2+}	0.7	5.53	4.64	8.00	6.90	5.10	16.45
Ni^{2+}	0.84	8.12	5.78	9.8	8.67	5.22	18.4
Cu^{2+}	1.82	13.00	8.15		10.20	6.42	18.8
Zn^{2+}	1.20	9.65	4.96	9.17	6.55	5.16	16.5
Cd^{2+}	1.56	7.38	4.22		5.39	5.31	16.5

Data from Martell, A. E. and Smith, R. M. (1989). *Critical stability constants*. Plenum Press, New York.

However, it is also usual to refer to the pK_a of an acid, which is $-\log K_a$. In this way, $\log K$, for metal ions and other associations, and pK_a, for protons, are both association constants. The larger these values, the stronger the binding free energy, given by $\Delta G = -2.303\, RT \log K$. Some examples of stability constants of complex ions are given in Table 6.6. The values of ΔG are a complicated combination of order (ΔH), and disorder ($T\Delta S$) terms, as before (see Sections 6.4 and 6.5).

In passing, notice that any ion that associates in a complex or forms a precipitate has its free ion concentration relatively controlled or buffered by these reservoirs.

6.7 Competitive affinity between metal ions and non-metal compounds in solution

We have seen in Section 4.8 that complicated sequences of atoms, charged and uncharged, can be joined together in molecules such as

$$^-O_2C - CH_2 - CH_2 - CO_2^- \quad (1)$$
$$\text{or} \quad H_2N - CH_2 - CO_2^- \quad (2)$$
$$\text{or} \quad HS - CH_2 - CO_2^- \quad (3)$$

These (small) molecules are all soluble in water. Just as we asked in Section 5.8 which metal elements preferred to be bound to oxygen rather than to sulphur in solids, so we can now ask: if we put a salt in water, say NaCl or CuCl, with a mixture of organic molecules such as (1)–(3) above, does Na$^+$ (from NaCl) selectively bind the same organic structure as Cu$^+$ (from CuCl)? The answer is no. Atom for atom the selectivity is much the same as that which we observed in solids, except that the organic molecules supply a greater diversity of possibilities and here we deal very frequently with charged species as products as well as reactants. If we examine molecules (1)–(3) and so forth separately, and ask with which relative strengths they

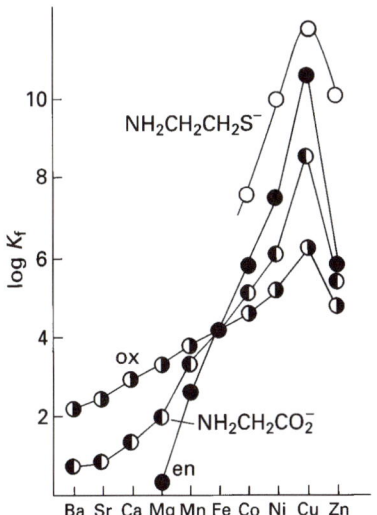

Fig. 6.7 The variation of stability constants, K, for the complexes of M^{2+} ions of the Irving–Williams series: ox, oxalate; en, ethylenediamine.

bind metal ions, that is, the value of their stability constants (Section 6.6) then the results are as shown in Fig. 6.7. The sequences are always the same

$$Ag^+ > Cu^+ > Na^+ > K^+$$

and

$$Cu^{2+} > Zn^{2+} > Ni^{2+} > Co^{2+} > Fe^{2+} > Mn^{2+} > Mg^{2+} > Ca^{2+}$$

The metal ions which bind best to these *small* organic molecules of flexible structure are the smallest, and for a given size the ones that form the strongest covalent bonds. (Observe that the formation of covalent bonds is related to the electron affinity of the ions, see Section 3.4.4.)

Interest now centres on the differences in strength of co-ordinate covalent bonds along the series of donors S, N and O, see examples in Table 6.7. The order is that of the slope in Fig. 6.7 which is

$$S \geqslant N > O$$

It follows that if we take an excess of all three chemicals (1)–(3) listed above and add a little of, say, Ag^+, Cu^+, Na^+ and K^+ salts (see the first series listed above) then the final balanced state will have Ag^+–S and Cu^+–S, but Na^+–O and K^+–O links. Compare the fact that Ag and Cu may be found in sulphide ores, but Na and K are in oxide clays of the Earth (Section 9.8). If

Table 6.7 Selective ligands based on different donor atoms

O-donor	$^-O_2C-CH_2-CO-CH_2-CO_2^-$
N,O-donor	$^-O_2C-CH_2-NH-CH_2-CO_2^-$
N-donor	$H_2N-CH_2-CH_2-NH-CH_2-CH_2-NH_2$
O-donor (phenolate)	
N-donor (aromatic)	
O,S-donor	$^-O_2C-CH_2-S^-$
N,S-donor	$H_2N-CH_2-CH_2-S^-$
S,S-donor	$^-S-CH_2-CH_2-S^-$
O-donor (hydroxamate)	
EDTA N,O-donor	

Table 6.8 Ligands preferred by different metal ions in simple co-ordination compounds

Metal ion	Ligands
Na^+, K^+	Oxygen-donor ligands, neutral or of low charge (–1)
Mg^{2+}	Carboxylate, phosphate and polyphosphate (total charge >–2); N-donation (special)
Ca^{2+}	Carboxylate, phosphate (less than Mg^{2+}), some neutral O-donors
Mn^{2+}, Fe^{2+}	Carboxylate, phosphate and nitrogen donors combined, (thiolate)
Fe^{2+}, (special)	Unsaturated amines (particularly porphyrins)
Cr^{3+}, Mn^{3+}	Phenolate (e.g. tyrosine), hydroxamate, hydroxide
Fe^{3+}, Co^{3+}	Carboxylate, N-donors, polypyrroles, (thiolate)
Ni^{2+}	Thiolate (e.g. cysteine), unsaturated amines, polypyrroles
Cu^{2+}, Cu^+	Amines, ionised peptide >N^-, thiolate
Zn^{2+}	Amines, thiolate, carboxylate

The case of Fe^{2+} is especially complicated by spin-state changes and polymerisation of mixed oxidation states.

Other possible ligands

Metal ions	*Ligands*
Fe, Co, Ni	CN^-, CO Carbanions
Mg, Ca	F^-, BO_3^{3-}

we take the second series listed above we will find the preferred binding sequences for divalent cations, M^{2+} (see aside also):

M – S	M – N	M – O
Cu, Zn	Ni, Co, Fe	Mn, Mg, Ca

The more sulphur donor ligands are present relative to the metal ions the more the central metal ions are bound to sulphur, but the metal ions Na^+, K^+, Mg^{2+} and Ca^{2+} are always in M–O complexes. Since M–O bonds can be provided by water by forming $M-OH_2$, it is often the case that in such mixtures, and even in the presence of the organic molecules, Na^+, K^+ and to some degree Mg^{2+} and Ca^{2+}, are found free, i.e. bound to water in quite high concentration, much as they are found free in the sea even in the presence of CO_3^{2-}.

The separation of metal elements (as ions) in biological systems, the most refined example of natural selection, is based partly on these abilities to bind different organic molecules (see Table 6.8) and, of course, will depend not just upon the nature but also on the amounts of these organic molecules that are synthesised. Thus, competitive binding in a biological medium is a problem related to competitive binding much like that in analytical chemistry. When comparison is made with the distribution of the elements in minerals the association is Cu, Zn, Ni, Co sulphides; Fe, sulphides and oxides; Mn, Mg, Ca oxides, in close agreement with the above. We have described these problems in our book *Biological chemistry of the elements* (see ref. 52 in Further reading).

6.8   Competition at equilibrium in solutions

The discussion of competition in solution for different combinations, once the concentrations are known, is therefore that of effective binding of say A to B in the presence of other competing chemicals but also under particular conditions. This is a discussion of equilibrium not unlike that of the evaluation of how the elements came together with partners during the formation of the Earth, but in that case we started the analysis from the association in the gas phase under conditions of constrained amounts (abundances) of the elements and under circumstances of different temperatures during cooling (see Section 5.8). Now, water is the totally dominant liquid on Earth. Whenever we describe water itself and chemicals in contact with it (precipitates or complexes in equilibrium) we invariably have to consider chemical speciation as part of a balance between the available chemical species in more than one condition, i.e. as a solid (e.g. $CaCO_3$), as a gas (e.g. CO_2) and in solution and in a variety of soluble combined forms also in balance (e.g. HCO_3^- and H_2CO_3). As stated, for some elements the equilibrium constants of their complexes are so small that the elements can almost invariably be treated as free hydrated ions in water, e.g. Na^+, K^+ and Cl^-. Other elements are overwhelmingly combined, especially if there are suitable organic molecules present also, e.g. copper and zinc, but some free ions always remain in solution. Intermediate cases are those of elements such as Mg^{2+} and Ca^{2+}. Speciation is then a complicated balance between many chemical forms (see Section 6.15).

We now see that the combination of elements with one another or with larger molecular components in solution obviously affects their availability and may well protect them from forming insoluble salts. Within a chemical system or living organism such combinations can be utilised to increase the functional value of the element. Thus complex formation is at the very heart of living matter. Before continuing, notice that every equilibrium is a constraint on the variables of the system (Section 6.11 and see Section 5.9).

6.9   The binding together of organic molecules: self-assembly in solution

Now, any study of speciation in aqueous solution can be dealt with in terms of the stability constants of the various species that can be formed, taking into account the prevailing conditions, especially the relative concentrations of the reagents A, B, C, etc., which can be metal ions, non-metal ions, molecules or charged molecules, etc. We have analysed above the association of metal ions and organic molecules in water. Let us now consider the association of organic molecules only, which is of relevant interest in biology.

We have seen that organic molecules can be built from a variety of units. We shall come across the following, particularly in proteins and nucleic acids

1. uncharged hydrocarbon groups $-CH_2-$, $-CH = CH-$, CH_3-, and C_6H_5- (phenyl).

2. charged groups $-NH_3^+$, $\left[-NH-C\begin{smallmatrix}NH_2\\NH_2\end{smallmatrix}\right]^+$, $-CO_2^-$, and $-PO_2^- -$

3. neutral groups with a large dipole: $-OH$ and $-NH_2$.

Molecules built from units under (1) are called hydrocarbons, or fats if the number of units is large. They self-associate in water to form membranes (Fig. 6.3) or oil-like solvents. Such solvents dissolve other oil-like molecules.

Examples of molecules built from $-CH_2-$ and $-OH$ groups under (3) are the polysaccharides. Even as polymers they tend to be random or form simple open linear networks soluble in water, such as sucrose, or insoluble, such as starch or celluloses.

Molecules containing all three classes of group form a diverse set of compounds, including the amino acids of proteins and the pyrimidine and purine bases of DNA and RNA. The more charges there are in the polymeric chains the less they tend to fold (see Section 7.6) but they may form linear assemblies, as in the DNA double helix. Where there is a higher percentage of uncharged hydrocarbon units the molecules (proteins) fold into particular and selected shapes.

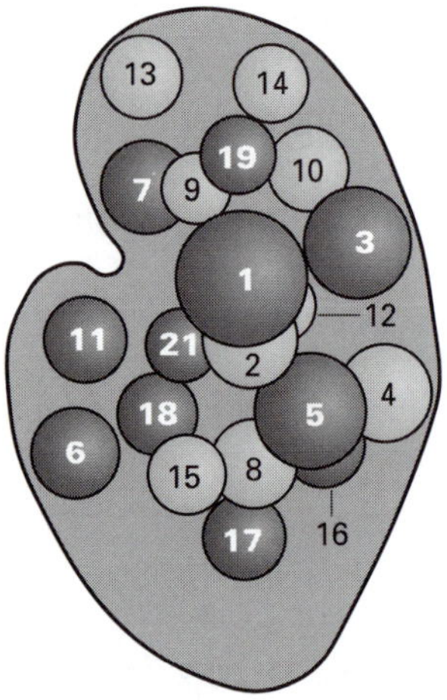

In water, or in membranes, the polymers can bind to one another or to themselves so as to give self-assembled, stable, soluble 'particles'. Thus, DNA in cells of higher organisms is associated to and protected by proteins called histones (see Fig. 7.11), while RNA forms, with a variety of proteins, the machinery for protein manufacture, the so-called ribosomes (Fig. 6.8). Still other proteins associate to form the machinery for energy capture, often within membranes (see Section 10.8). Polysaccharides bind to proteins and help to hold cells together while the 'poly-saccharide' framework of DNA and RNA (pentose phosphate polymers) binds to many types of protein. Thus, associated structures of an ordered kind can be formed at equilibrium in solution and they can be of any size one wishes to imagine. (Note that there is another form of non-equilibrium self-assembly called programmed self-assembly which we describe in Chapter 10.) It is important to take into account that in all these large molecule associations the *surface* energies are as important as the interior. A degree of freedom is the amount of each component (protein, RNA, etc.) produced, which at equilibrium could also decide the shape of the associated unit (see Chapter 7).

Note that biological polymers differ from man-made polymers in that they are not built from repeating monomers but from ordered sequences. This raises a problem to be discussed in Section 10.13 and in Chapter 16.

Fig. 6.8 Twenty-one proteins in a bacterial ribosomal particle. Unnumbered shaded space is RNA.

6.10 Partition of species between immiscible solvents

In Sections 6.1 and 6.2 we pointed out that some bulk solvents were virtually immiscible, for example, petrol and water. Placed together they form separate layers or, in the cases of long chain lipids, vesicles. Some solutes overwhelmingly prefer one solvent when the second is a barrier to their diffusion. Biological membranes formed from fats prevent loss or entry of ions across themselves because the partition coefficient, the ratio of solubilities, is small

$$\frac{\text{solubility in membranes}}{\text{solubility in water}} \ll 1.0$$

Other uncharged substances, such as alcohols, pass through membranes easily. The differences in partition coefficient are used in chemistry and living organisms to extract some molecules from water. Complex ion formation may assist or prevent such extraction. When small molecules or ions bind to proteins they may be effectively extracted into a hydrophobic 'solvent' or site.

In these cases, where two or more phases are present, there may or may not be equilibrium across the phase boundary. Molecules can associate with one another in either phase and this must be taken into account when considering partition.

6.11   Species, equilibria and variables

In this chapter we have discussed chemicals in equilibrium in solution or that between solutions and precipitates (or gases). It is important to notice that the variables of chemical composition in solutions in which there are equilibria (balanced reactions) forming soluble complexes such as

$$A + B \rightleftarrows AB$$

are just the concentrations of A and B, and the concentration of AB is then not variable since it is related to those of A and B through the stability constant K. Here, increase in 'species' in solution does not affect the variance. When there is also a precipitate of AB then, at equilibrium, the concentration, corresponding to the solubility, of AB is fixed and hence the concentrations of A and B can only be varied relative to one another, not independently. The criterion of equilibrium exchange between units across phase boundaries is again a direct restriction on the variable composition. The chemical variable composition is then restricted by every equilibrium.* We can, therefore, write

No equilibrium, AB is a component.	A, B, and AB	:three variables
Complex formation in equilibrium	$A + B \rightleftarrows AB$	:two variables
Precipitation (with or without complex formation) equilibrium	$A + B \rightleftarrows AB \rightleftarrows$ precipitate	:one variable only
Partition between phases	$A_{S_1} \rightleftarrows A_{S_2}$ (s = solvent)	:one variable

In the final analysis, if every chemical element in the periodic table naturally occurring on Earth (90) was present in one solution and *all were in equilibrium* with the compounds, complexes and precipitates they can form, then there would still be only $90 - 1 = 89$ variables in the percentage composition. (It is $90 - 1$ since we define a composition by percentages.) On the other hand, if the associated units such as AB above were not in equilib-

---

*In this chapter remember that we are considering only *equilibria* between *all components* present in solution. If equilibration fails then speciation has to be described in a quite different way (see Chapter 8).

rium, then the variables would correspond to the number of elements plus the number of non-equilibrating associated units present, that is, the number of components of the system. The variables in solution are then the same as those discussed in Chapter 5: free energy in terms of composition, the percentages of components, temperature and concentration (pressure and volume), as restricted by chemical and physical phase equilibria. It is essential that such considerations are understood if living systems are to be appreciated. Self-assembly, as described above at equilibrium, generates an ordered system which does not increase the variables since it is a consequence of the composition in terms of the free components alone. When the equilibrium condition is removed, the variance increases.

6.12   Shapes of molecular species in solution

In Section 6.6 we drew attention to the formation of 'complex' species in solution and gave some examples, such as $CuCl^+$, $PbCl_4^{2-}$ and NH_4^+. More generally, we can refer to species such as $ML^{n\pm}$ and $ML_m^{n\pm}$, where M is a metal ion and L stands for a 'monodentate ligand' (forming just one co-ordinate bond) or a 'polydentate' ligand (forming more than one

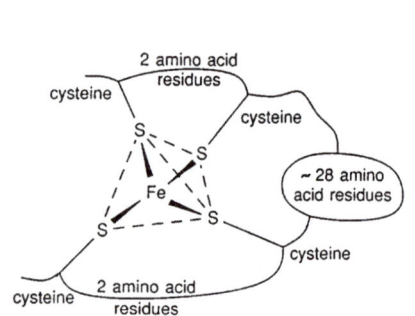

The co-ordination of iron in a thiolate centre of a protein, rubredoxin.

Fig. 6.9 (a) The haem group and (b) iron co-ordination to a protein, cytochrome *c*, (below).

co-ordinate bond). The ligand L may be neutral or charged and may have just one type of donor atom or more than one type. Complexes formed with polydentate ligands are called 'chelates'. The shapes of such species, see asides, are, like the shapes of neutral molecules, decided by bonding considerations, both electrostatics and covalency, and by the need for electrons, including non-bonded electrons, to occupy space to decrease repulsion (see Section 3.4.5). In the simplest cases, such as of $Zn(NH_3)_4^{2+}$, which is symmetrical and has no unpaired electrons, the complex ion is of tetrahedral geometry, as in the case of CH_4. However, complications often arise due to the competing demands of electrostatics, covalence and the presence of unpaired electrons in 'd' and 'f' subshells of atoms, so that structures are often different even for periodic table neighbours, as can be observed in a series of ML_m complexes of nitrogen-donor ligands formed with Fe^{2+}, Ni^{2+}, Cu^{2+}, Zn^{2+} (see Figs 6.9 and 6.10). Elsewhere we have described the reasons for the differences which become important in the way atoms or ions can be selected in combination with non-metals (see reference 52 in Further reading).

Now, from elements such as H, C, N and O there can be made large molecules that themselves fold into shapes (see Sections 6.9 and 7.6). Let us suppose that we can select a molecular framework made from these non-metals so as to create a cavity. The cavity can be large or small and can be shaped. Thus it can be dove-tailed to generate preferential binding of one atom, molecule or ion over another. Some examples are shown in Fig. 6.9 and its aside. This basic principle of building frameworks to fit a group or to create a property in a group is often called molecular recognition and is one of the major arts of chemistry—an art which has been perfected in minerals, man-made binding systems and especially in biological systems, as we shall see.

In this context there is now a need to consider the building and shapes of considerable, or even very large, frameworks formed from non-metals, e.g. proteins, a topic which we turn to in Section 7.6.

Tetragonal Cu²⁺ bis-ethylenediamine

Tetrahedral Zn²⁺ bis-ethylenediamine

Octahedral Ni²⁺ bis-ethylenediamine dihydrate (high-spin)

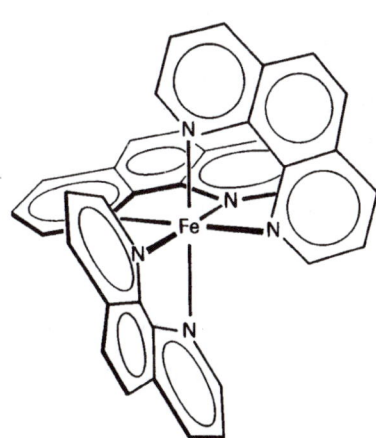

Octahedral Fe²⁺ tris-1,10-phenanthroline (low-spin)

8-coordinate, naturally, occurring, vanadium chelate

Fig. 6.10 Examples of chelate formation where a metal binds to several ligands displacing (many) protons from the protonated ligands.

6.13 Oxidation–Reduction equilibria in solution and electrochemistry

In the above, the considerations of solution–precipitate equilibria (Section 6.5) have concerned the association reactions to form salts and complexes. A quite different consideration is the *reaction* of pure elements with water (H_2O), in which water is split up into separate hydrogen- and oxygen-containing compounds. The reaction is a competition for oxygen between hydrogen and another element, e.g. copper

$$Cu(s) + H_2O \rightleftarrows H_2(g) + CuO$$

Now we look upon the solution of copper in water by considering an acid solution $2HCl + Cu(s) \rightarrow CuCl_2 + H_2$. Since HCl in water dissociates into $H^+ + Cl^-$ and $CuCl_2$ into $Cu^{2+} + 2Cl^-$, the Cl^- cancels out and is not relevant. We can then write

$$Cu(s) + 2H^+ \rightleftarrows H_2(g) + Cu^{2+}$$

Clearly, the charges on the atoms have changed in the reaction, i.e. their oxidation states have changed. In such cases we call the reactions oxidation–reduction or redox reactions (see Section 3.4.7). The question here is which element, H^+ or Cu^{2+}, holds on to electrons more strongly to become H_2 (g) or $Cu(s)$, respectively, relative to the hydration of the respective ions, since we can divide the reaction up into

$$Cu(s) \rightleftarrows Cu^{2+} + 2e \text{ plus } Cu^{2+} + n(H_2O) \rightleftarrows Cu^{2+}(H_2O)_n$$

and

$$H_2(g) \rightleftarrows 2H^+ + 2e \text{ plus } 2H^+ + m(H_2O) \rightleftarrows 2H^+(H_2O)_m$$

where e is the electron held in the metal by metal lattice energies and in the H_2 by covalent bonds. We can measure their relative electric potential energy, i.e. the potential energy of, say, a copper metal atom to become a copper ion in water, relative to that of H_2 gas to become H^+ in water, in an electrochemical cell (see Fig. 6.11).* Using this process we can establish a comparative series which shows which element prefers to hold electrons and remain as the free element (H_2 or Cu metal) rather than enter solu-

---

*In any equilibrium the free energy change from standard states of components on one side of an equation to standard states on the other can be given in units of kcals, kjoules (1j = 4.18 cals) or in units of volts, E, where 1 volt = 13 800 cals. The relationship to the equilibrium constant is

$$\Delta G = RT \times 2.303 \log K, \text{in cals}$$

$$\text{or} \quad E° = \frac{RT}{n\mathscr{F}} \times 2.303 \log K, \text{in volts}$$

where $\mathscr{F}$ is the Faraday.

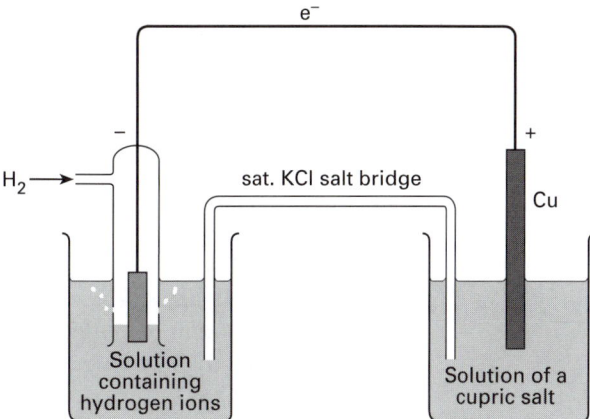

Fig. 6.11 A simple electrochemical cell. H_2 bubbles over a spongy Pt electrode, the hydrogen electrode, which is connected by a salt bridge to a copper electrode using 1 M HCl–1 M $CuCl_2$ as electrolytes.

Table 6.9 Redox potentials of some ion systems at pH = 0, T = 25 °C

Element	$E°$* (v)	n	Main mineral source	Order in which metal was used by man
(a) Metal ion systems				
K	−2.92	+1	KCl, KCl.MgCl$_2$	
Na	−2.71	+1	NaCl	
Ca	−2.87	+2	CaSO$_4$, CaCO$_3$, Ca$_3$(PO$_4$)$_2$ CaCl$_2$ (Solvay process)	
Mg	−2.36	+2	Mg salts, MgCO$_3$	5/6
Al	−1.66	+3	Al$_2$O$_3$	5/6
Mn	−1.18	+2	MnO$_2$	
Cr	−0.74	+3	FeO·Cr$_2$O$_3$	5
Zn	−0.76	+2	ZnS	3
Fe	−0.44	+2	Fe$_2$O$_3$, Fe$_3$O$_4$	4
Co	−0.28	+2	CoAsS, Co$_3$S$_4$	5
Ni	−0.25	+2	Sulphides	5
Sn	−0.14	+2	SnO$_2$	3/2
Pb	−0.13	+2	PbS	3/2
Cu	+0.34	+2	Metal, sulphide	2/3
Hg	+0.85	+2	HgS	3
Ag	+0.80	+1	Metal, Ag$_2$S, AgCl	2
Au	+1.7	+1	Metal, tellurides	1
(b) Some non-metal ion systems‡				
O	+1.23	+2	Water (H$_2$O)	
S	+0.14	+2	FeS	
Se	−0.40	+2	–	
Te	−0.60	+2	–	
F	+2.87	+1	Na$_3$AlF$_6$	
Cl	+1.36	+1	NaCl	
Br	+1.07	+1	Seawater	
I	+0.54	+1	Seawater	

* $E°$ for the reaction M^{n+} (aq) + $ne \rightleftarrows$ M where n is given in the next column.
‡ For non-metals, $E°$ for the reaction, X_2 (aq) + nH^+ + $ne \rightleftarrows 2H_nX$ (aq).

tion as a hydrated ion. The observed 'electrochemical series' in order of preference to be a free element (Table 6.9) is

$$Ag^+ > Cu^+ > H^+ > Na^+ > K^+$$
$$Cu^{2+} > Ni^{2+} > Co^{2+} > Fe^{2+} > Zn^{2+} > Mn^{2+} > Mg^{2+} > Ca^{.}$$

[Elements that tend to form ions (negative redox potentials) are called electropositive, relative to H^+.]

This is a series very similar to that which we saw before for the insolubility of salts of the corresponding ions, e.g. hydroxide and sulphides, and the formation of complex ions in solution. There is a simple reason. The atoms that wish to be out of water to the largest degree, i.e. bound not to O of H_2O but to themselves, give up their electrons reluctantly as judged by ionisation energies, i.e. they prefer to be in a situation where they retain their electrons completely as in the metallic state ($M^{2+} \xrightarrow{2e} M$). The parallel with complex ion or precipitation reactions arises since the metal element that prefers the state of association with itself as a metal rather than becoming an ion also seeks association as an ion with the non-metal which gives electrons back to it most readily in a bond ($M^{2+} \leftarrow X$). The donor power of X is in the inverse order of the non-metal atom electron affinity, that is

$$S \geqslant N > O(H_2O)$$

The least electropositive elements such as copper and mercury are, therefore, usually found bound to elements that give back electrons readily, e.g. sulphur, while the most electropositive elements, sodium and potassium, are held, albeit weakly, by oxygen, e.g. from water. The degree of covalence in different circumstances decides all orders to a large degree.

For non-metals (electronegative elements) the question of which state they prefer, as elements or ions in water, is similar since in water they have too many electrons and the ones which hold electrons best

$$F > Cl > Br > I,$$

are the very ones which prefer to be in the state $2X^-$ rather than X_2. Thus, according to the electrochemical series, oxygen (O) becomes $2O^{2-}$ not O_2, more strongly than S becomes S^{2-}, not S_2. In other words, it is easier to reduce O_2 than to reduce Sn, or, as we say, O_2 is a better oxidising agent than sulphur. All this information can be put on a quantitative basis, (see Table 6.9).

6.13.1 Redox potentials of complex species

There are some metals which have several ionic oxidation states open to them, (Section 3.4.6). For example, we have to consider Cu^+ and Cu^{2+} for copper. Others are listed in Table 6.10. There are now two equilibria of interest. Which electrochemical process is less difficult?

$$Cu \rightleftarrows Cu^+ + e \quad \text{or} \quad Cu \rightleftarrows Cu^{2+} + 2e$$

Table 6.10 Some electrode potentials in aqueous solution at 25°

Electrode	$E°$ (Volts)*
$Cr^{3+} + e \rightarrow Cr^{2+}$	−0.41
$V^{3+} + e \rightarrow V^{2+}$	−0.26
$H^+ + e \rightarrow 1/2H_2{}^1$	0
$Sn^{4+} + 2e \rightarrow Sn^{2+}$	0.15
$Cu^{2+} + e \rightarrow Cu^+$	0.17
$MnO_2 + 4H^+ + 2e \rightarrow Mn^{2+} + 2H_2O$	1.23
$Cr_2O_7^{2-} + 14H^+ + 6e \rightarrow 2Cr^{3+} + 7H_2O$	1.33
$ClO_4^- + 8H^+ + 8e \rightarrow Cl^- + 4H_2O$	1.39
$MnO_4^- + 8H^+ + 5e \rightarrow Mn^{2+} + 4H_2O$	1.51
$BrO_3^- + 6H^+ + 5e \rightarrow 1/2Br_2 + 3H_2O$	1.52
$Co^{3+} + e \rightarrow Co^{2+}$	1.95
$Ag^{2+} + e \rightarrow Ag^+$	1.98
$S_2O_8^{2-} + 2e \rightarrow 2SO_4^{2-}$	2.01

Table 6.11 Redox potentials for the Fe^{III}/Fe^{II} couple (volts)

Ligand	Potential
(o-phenanthroline)$_3$	+1.10
(Dipyridyl)$_3$	+0.96
$(H_2O)_6$ ($a_H = 1.0$)	+0.77
$(CN^-)_6$	+0.36
(Oxalate^{2-})$_n$ $\begin{cases} n = 3\ Fe^{III} \\ n = 2\ Fe^{II} \end{cases}$	+0.02
EDTA^{4-}	−0.12
(8-hydroxyquinolinate$^-$)$_3$	−0.20
Cytochrome c	+0.25
Myoglobin	+0.10
Rubredoxin	−0.30

Table 6.12 Some standard redox potentials of copper proteins at pH = 7.0

Protein	Redox potential (mV)
Lactase	
Type I (blue)	+785
Type II	+500
Type III	+400 (?)
Azurin	
Type I (blue)	+330
Plastocyanin	
Type I (blue)	+370
Tyrosinase	
Type III(?)	+370
Haemocyanin	> +800 (?)
Dopamine	
monooxygenase	+310

Note that the redox potentials are higher than that of the copper aqueous ions (+ 170 mV) which means that Cu^I is bound with greater strength than Cu^{II}.

The problem can be restated in terms of the reaction

$$2Cu^+ + 2H^+ \rightleftarrows 2Cu^{2+} + H_2$$

so that we can place the Cu^+/Cu^{2+} redox couple in the same series as the electrode potentials. The reactions relate to the strength with which two different ion species of the same element in water have a desire for electrons, (Table 6.10). Non-metal species in water can be treated in a similar way.

Since both M^+ and M^{2+} (or any other M species) in water can also bind organic molecules, then the series in Table 6.10 is adjusted by this binding

$$2H^+ + 2M^+ \cdot (\text{organic 'ligand'}) \rightleftarrows 2M^{2+} \cdot (\text{organic 'ligand'}) + H_2$$

By choosing the organic 'ligand' carefully the energy of this reaction can be changed in many ways. Tables 6.11 and 6.12 give examples of iron and copper complexes including some metalloproteins. In living systems the redox potentials of elements are set at particular values by complex formation.

At equilibrium the presence of one set of components including the electron decides the composition of the other set of substances which can be formed in the redox equation, and the component composition is then the variable of the system. Since the two oxidation states in these equations may be differently combined with other molecules in solution we must be very careful in defining the conditions under which oxidation–reduction potentials are measured.

6.13.2 Oxidation state diagrams

As stated, for many elements a whole range of oxidation states exist. In the case of manganese they are

$$\text{Mn(metal), } Mn^{2+}, Mn^{3+}, MnO_2, MnO_4^{2-}, MnO_4^-$$

with oxidation states 0, II, III, IV, VI and VII, respectively. In the case of nitrogen they are

$$NO_3^-, \ NO_2^-, \ NO, \ N_2(gas), \ N_2H_2, \ N_2H_4, \ NH_3$$

with oxidation states V, III, II, 0, –I, –II and –III, respectively. Now, if we wish to show the relative stability of all these states, which is a quantitative measure of $\Delta G°$ (per mole) of each in water at 298 K and 1 atmosphere, we can plot $\Delta G°$ against the oxidation state as an 'oxidation state diagram'. The $\Delta G°$ values can be given in kilocalories (kilojoules), or the volt equivalent scale where 1 volt equivalent = 13 600 kilocals (times 4.18 for kilojoules). We plot graphs of volt equivalent, nE, at pH = 7.0 and 25 °C

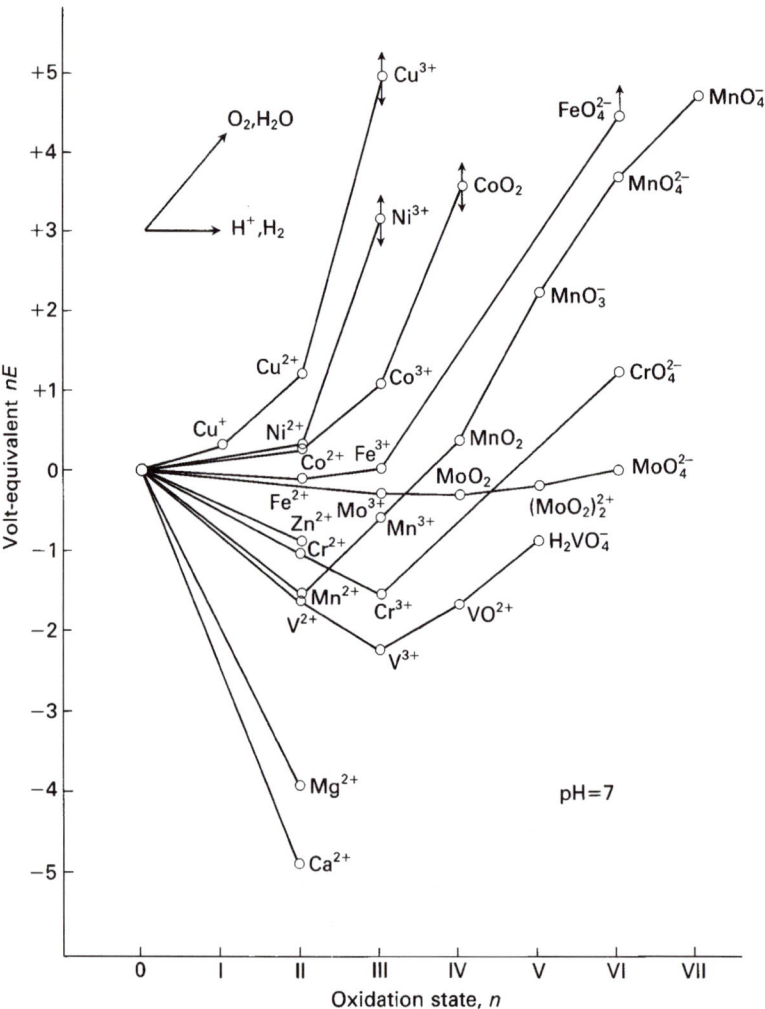

Fig. 6.12 Oxidation state diagram for some metals at pH = 7 (nE versus n) taking into account the formation of hydroxides, where n is the oxidation state and E the electrode potential from the element to that state. Note that the reference potential is Pt·H_2/10^{-7} M H^+, at pH = 7.0, shown top left, and not that of the standard H_2/H^+ one molar electrode.

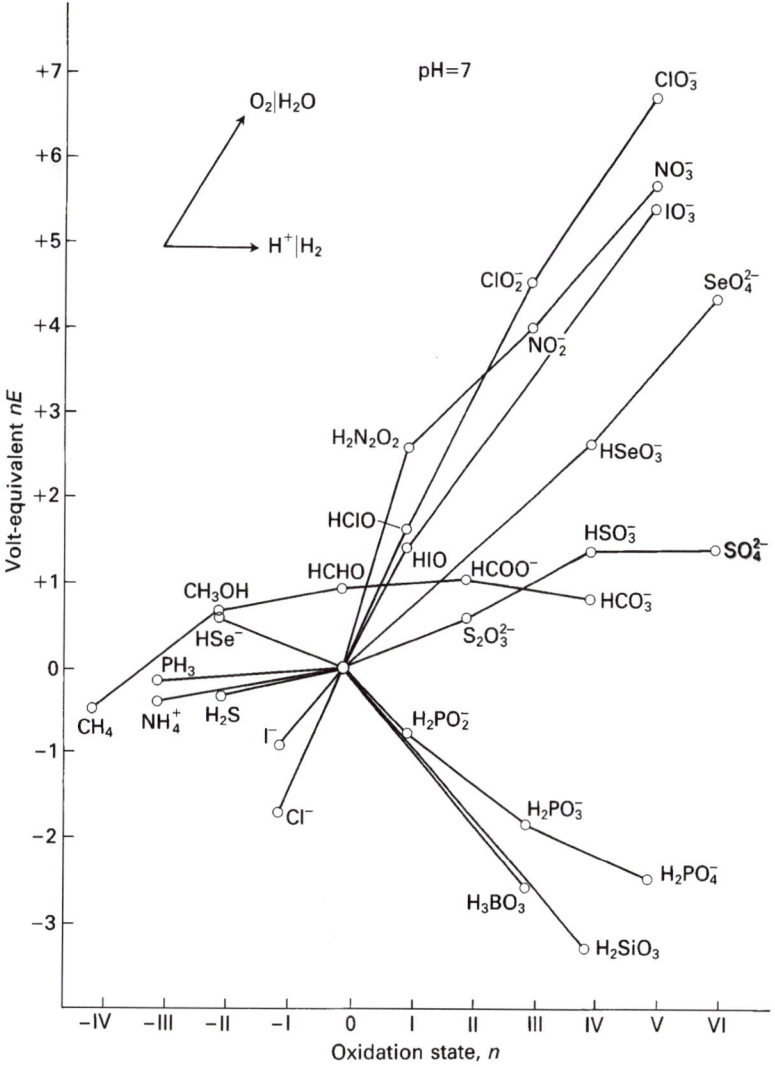

Fig. 6.13 Oxidation state diagram for non-metals at pH = 7 (nE versus n) taking into account protonation reactions. Note that the reference electrode potential is Pt•H_2/10^{-7} M H^+, at pH = 7.0, shown top left, and not that of the standard H_2/H^+ one molar electrode.

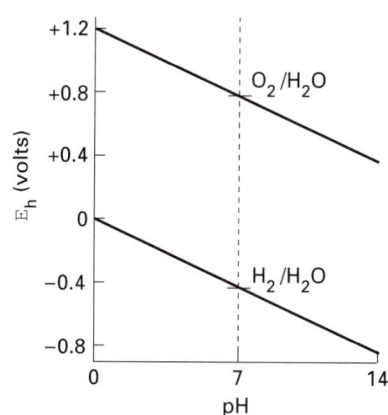

Fig. 6.14 The potential of the water redox system at 25 °C, H_2 pressure = 1 atm and O_2 pressure = 1 atm, respectively. E_h stands for the redox potential relative to the standard hydrogen couple $H^+ + e \rightleftarrows \frac{1}{2} H_2$ at pH = 0 and H_2 pressure = 1 atm.

against oxidation state for metals and for non-metals in Figs 6.12 and 6.13, respectively. The volt equivalent for $2H^+ \rightleftarrows H_2$ is plotted as 0.0 volts and that for $2O^{2-} \rightleftarrows O_2$ as 1.23 volts at pH = 0, but both systems vary with pH (see Fig. 6.14). The redox potential between any two couples is the slope of a line connecting them, i.e. volt equivalent per oxidation state.

Outside man's activities, these selectivities, together with those between all acids and bases in solutions where there are organic molecules, are very largely found in living organisms, while selectivity amongst inorganic substances occurs mainly in soils. Water is the connecting medium. Thus, we are interested in the way the environment of living systems affects the possibility of the organisms to collect elements (Chapters 10–12). Man has evolved quite different strategies for getting elements for his own use which avoid water but the new materials he produces are then in contact with both water and air. We shall examine the different problems that arise in all such cases in Chapters 15 and 16.

The different oxidation states of all elements or components in a mixture may be treated in these *equilibrium* terms, when individual complexes of different oxidation state are not independent variables. For example, there can be set equilibrium anaerobic or aerobic conditions given by dioxygen content when the H_2O/H_2 or H_2O/O_2 equilibria set all other redox equilibria (Fig. 6.14). If electron (or atom) exchange is not permitted, however, they become separate components of a system, thus increasing the variance.

6.14 The conditions controlling redox potentials in solution

Just as conditions control the binding constants of elements or molecules to one another so the oxidation potential of the environment controls whether or not one oxidation state or another of ions and molecules is stabilised, and therefore available to living organisms. A major concern in this book will be the oxidising or reducing conditions in aqueous solutions but

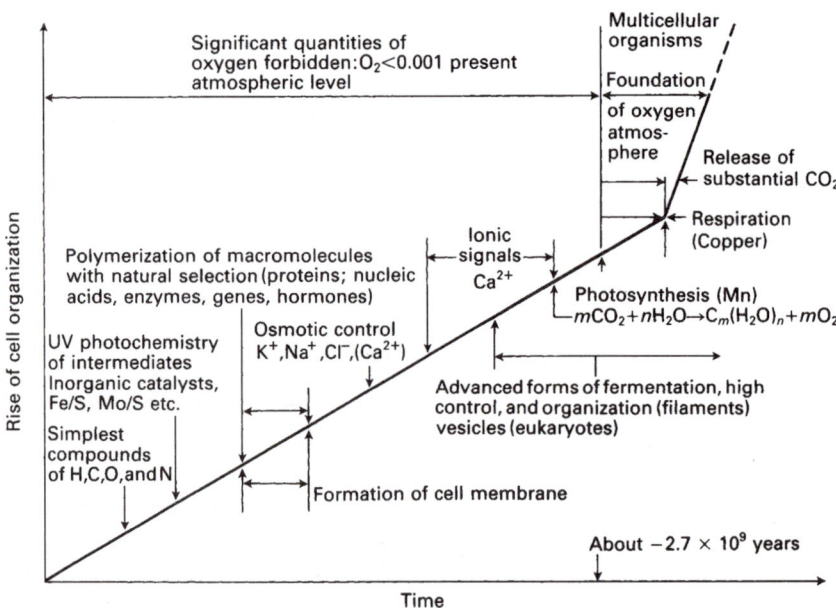

Fig. 6.15 A picture of evolutionary development where a decisive role is filled by the generation of dioxygen. The break in evolution which may well have been caused by the changed levels of the *available* chemical elements dependent on dioxygen is reflected in the development of eukaryotic compartments and multicellular organisation. The very earliest cell–cell organisation may have involved inorganic surfaces (not shown) as has been frequently discussed (see text). The figure is obviously related to Fig. 13.2. Note that the time-scale is not linear.

we need to observe once again that they are not independent of pH; for example, there is the equilibrium

$$\text{(reduced) RSH + RSH} \rightleftarrows \text{RS} - \text{SR} + 2H^+ + 2e \text{ (oxidised)}$$

which involves the concentration of H^+ and is therefore, pH dependent. The sea, for example, is at a fixed pH of around 8.0 today and is exposed to air containing dioxygen, but it was not so in the one or two billion early years of Earth. Another consideration arises from the fact that when Earth formed hydrogen dominated all abundances and hydrogen is a reducing gas, hence oxidised dithiolates would not have existed

$$H_2 + \text{RS–SR} \rightarrow 2\text{RSH}$$

Thus the *environmental equilibria* impose strong limitations upon the possible chemistry and biochemistry on the Earth's surface.

The beginnings of life were still in a reducing atmosphere which changed very slowly, over billions of years, to today's oxygen-rich atmosphere (see Fig. 6.15). This change affected elements differentially and in order of their ease of conversion to oxidised forms (Fig. 6.16). But notice too that the removal of sulphide, for example, allows some elements to be freely dissolved in water since the precipitation of their sulphides is prevented, e.g. Zn, Cu and Cd. While such equilibrium conditions may be approached in theory, but not so easily in practice, it is of the essence of many chemical, and all living, systems to avoid equilibration while utilising current circumstances.

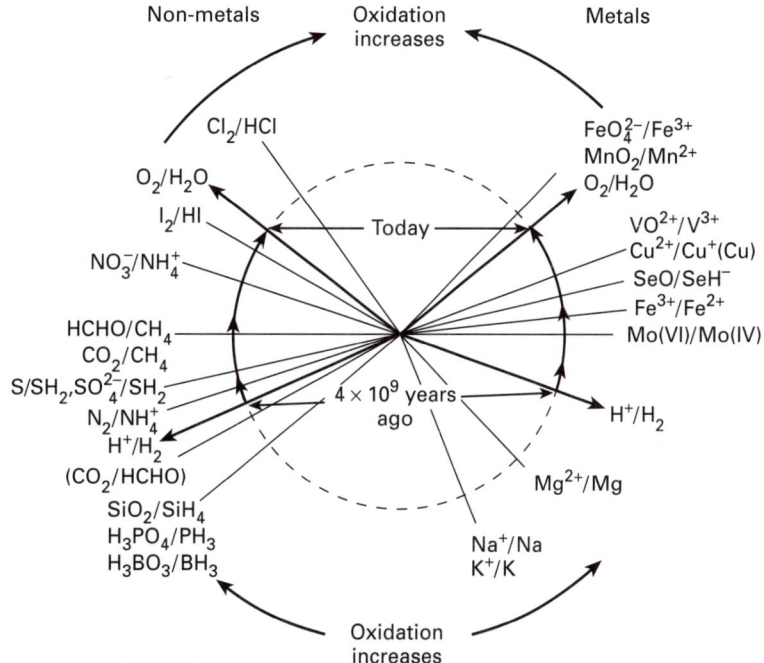

Fig. 6.16 The sweep of rising redox potential at pH = 7.0 with the evolution of Earth's surface showing changes imposed on metals from lower to higher positive oxidation states and on non-metals from higher to lower negative oxidation states.

6.15   The speciation of elements in the sea

The sea is a solution of many elements, often in ionic forms. Everyone knows the sea is a salt solution, but it is not a fully saturated solution of common salt (NaCl) (see Table 6.13). As stated above, the sea also has a high content of Mg^{2+} and Ca^{2+} ions and the amounts of CO_3^{2-} (carbonate) or SO_4^{2-} (sulphate) do lead to precipitation of the carbonates of Mg^{2+} or Ca^{2+}. The sea is quite close to a saturated solution of CO_2 and, at the surface, of O_2 and N_2. It is also not far from saturated with SiO_2. Of the very insoluble substances present, the hydroxide of iron is also close to its solubility limits, both with respect to Fe^{3+} and Fe^{2+}, with a redox potential of around +0.8 volts (pH = 8) at the surface.

Table 6.13 Inorganic speciation (calculated) of some trace elements in sea water (25 °C, 1 atm, 3.5% salinity, pH = 8)

Metal	Main species
Al^{3+}	$Al(OH)_3$ 100%
Cr^{3+}	$Cr(OH)_3$ 100%
Mn^{2+}	Mn^{2+} 58%, $MnCl^+$ 37%, $MnSO_4$ 4%, $MnCO_3$ 1%
Fe^{2+}	Fe^{2+} 69%, $FeCl^+$ 20%, $FeCO_3$ 5%, $FeSO_4$ 4%, $Fe(OH)^+$ 2%
Fe^{3+}	$Fe(OH)_3$ 100%
Co^{2+}	Co^{2+} 58%, $CoCl^+$ 30%, $CoCO_3$ 6%, $CoSO_4$ 5%, $Co(OH)^+$ 1%
Ni^{2+}	Ni^{2+} 47%, $NiCl^+$ 34%, $NiCO_3$ 14%, $NiSO_4$ 4%, $Ni(OH)^+$ 1%
Cu^{2+}	Cu^{2+} 9%, $CuCO_3$ 79%, $Cu(OH)^+$ 8%, $CuCl^+$ 3%, $CuSO_4$ 1%
Zn^{2+}	Zn^{2+} 46%, $ZnCl^+$ 35%, $Zn(OH)^+$ 12%, $ZnSO_4$ 4%, $ZnCO_3$ 3%
Mo^{VI}	MoO_4^{2-} (100%)
Cd^{2+}	Cd^{2+} 3%, $CdCl^+$ 97%
Pb^{2+}	Pb^{2+} 3%, $PbCl^+$ 47%, $PbCO_3$ 41%, $Pb(OH)^+$ 9%, $Pb(SO_4)$ 1%
Si^{IV}	H_4SiO_4 92%, $H_3SiO_4^-$ 5%

Data from: Turner, D., Whitfield, M., and Dickson, A. G. (1981). *Geochimica et Cosmochimica Acta*, **45**, 855.

Because the sea varies from place to place (especially its temperature and its redox potential) it is not saturated with many of the other chemicals present in it and so variation of living forms is increased. We shall discuss this in more detail in Chapter 9.

6.16   Restrictions on availability

Just as we plotted abundance against atomic numbers, (Fig. 1.10) so we can now plot availability in waters on Earth, e.g. in the sea, against atomic number (Fig. 6.17). This is another landscape plot (see Section 1.4). There are conditions that affected availability but did not affect universal abundance. One major consideration is the escape from Earth of those elements, mainly non-metals, that are to be found at room temperature as gases, H_2, N_2, CO_2. The major gases to escape then are H_2, He, Ne, N_2, CO_2 (CO) and to some degree H_2O and O_2. The reserves of H and O in H_2O and of C (in

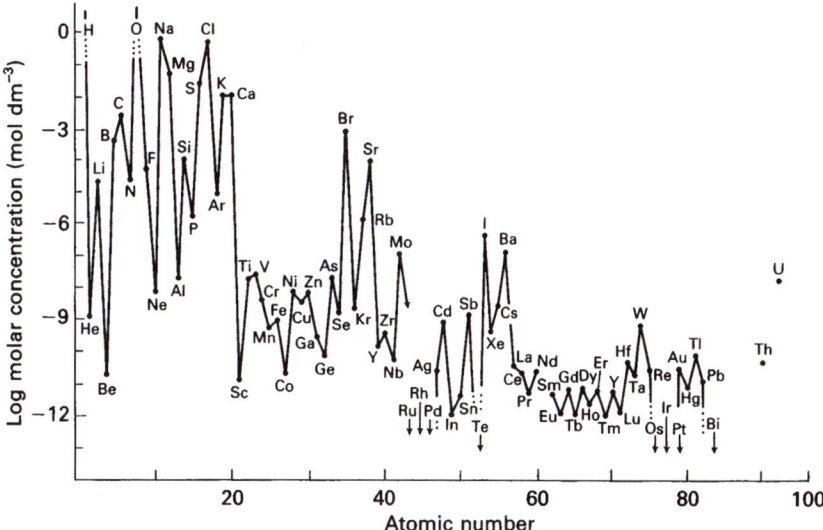

Fig. 6.17 The concentrations of elements in the *sea*. (After Cox, P. A., ref. 2 in Further reading.) Note this availability plot is a landscape diagram.

living organisms and their residues coal, oil and gas) make the loss of H, O and C much less significant than the loss of N, which is only retained in living systems and in a few deposits, e.g. guano salts.

A second consideration is the solubility of materials in water, which is made difficult to appreciate in detail by the intrinsic insolubility of many simple salts (Table 6.14) and the increase in solubility due to complex formation. Each oxidation state of an element must be considered separately.

The third condition limiting availability is the environmental reducing or oxidising atmosphere. The reducing atmosphere limits heavier elements, including Cu, Zn, and Cd. The oxidising atmosphere limits the element Fe, but in particular it changes the available states of N, C, S and Se. When we consider evolution, availability will be a major constraint. A detailed account is given in Chapters 10–14.

Table 6.14 Elementary species remaining after mixing cations of the first 30 elements of the periodic table at 10^{-1} M, pH = 7.3 (25 °C) with a small excess of anions as X^- or weak acids (HX)*

Soluble (10^{-1} M)	Li^+, Na^+, K^+ Cl^-, NO_3^-, SO_4^{2-}
Relatively soluble (10^{-4} to 10^{-1} M)	Mg^{2+}, Ca^{2+}, Mn^{2+} F^-, HCO_3^-, HPO_4^{2-}, $B(OH)_4^-$, VO_3^{3-}
Sparingly soluble (10^{-8} to 10^{-4} M)	H^+, Ba^{2+}, Zn^{2+}, (Fe^{2+}), Co^{2+}, Ni^{2+} OH^-, (SH^-), (H_2S), $Si(OH)_4$
Almost insoluble (< 10^{-8} M)	Al^{3+}, Sc^{3+}, Ti^{4+}, Cr^{3+}, Fe^{3+}, Co^{2+}, Ni^{2+}, Cu^{2+}, Zn^{2+} OH^-, SH^-, H_2S, $Si(OH)_4$

* It is assumed that the anions are in slight excess in each category of solubility as X^- or HX. (X) indicates a borderline case.

While we were discussing the elements in the Earth and the formation of compounds we stated that there were 90, less one, compositional variables—all the naturally occurring stable elements of the periodic table. When we consider chemical systems in water, such as organisms, it is unrealistic to consider all 90 elements as variables due to both abundance and chemical equilibrium considerations. However, it is not obvious which elements are made available in sufficient amount by the activities of organisms themselves. Hence the variables of composition in organisms will be treated empirically in later chapters. Once again remember that the variables of the environment are reduced by equilibria between chemicals in a homogeneous system and remain limited by equilibration between different physical states, phases.

6.17 Conclusions

This chapter has to be seen in the context of Chapters 4 and 5. In those chapters we showed how elements compete for stable or stationary combinations, mainly in the solid state but also in simple liquid and gaseous conditions, leading to speciation. In this chapter we have shown how they compete in all dilute (aqueous) solutions. Both analyses are studies in part of structures and properties but also of 'speciation' at equilibrium (or in stationary conditions) where there is an order$\rightleftarrows$disorder balance, now in a homogeneous system. A 'species' here is quite different from a component in that equilibria remove independence from some 'species', leaving, as before, only the true components as the variables of composition. In living organisms the competing reactions between elements are found both in solids (shell and bone formation, for example) and in solution (in blood, for example), just as they are observed in the formation of rocks, but in solution availability, not abundance, is dominant. Moreover, the competition in all these reactions has to be seen against the prevailing limitations on the variables temperature and pressure, as well as on the relative amounts of substances present. These are the same variables as those described in all our considerations of balanced (equilibrium) states. It has to be observed, however, that such equilibria in solution may involve many combinations of individual molecular or elemental substances giving rise to a large number of larger particles, since many organic components may be present, as in a living organism. The situation differs from that which we outlined for the distribution of elements in minerals, where one element was overwhelmingly placed in a small number of phases and zones, e.g. the oxide or sulphide layers of Earth, often separated by gravity. All the same, if we could have fixed temperature, pressure and relative amounts (% composition) of all elements and assumed equilibrium between all possible combinations, then the nature of both the Earth and the sea would have been fixed. Earth could not then have evolved. Observations on evolution imply that this did not happen and while there is the need for understanding of variables at equilibrium we need also the understanding of variables of change both in solid and solution states.

Now and then in this chapter we have noted correlations between a bulk property, such as the formation of a precipitate or of complexes, and characteristics of single atoms or ions, such as their ionisation potentials. Now atoms and ions are the fundamental units of composition and we must therefore stress that these correlations are not likely to be exact since the bulk properties examined are collective thermodynamic properties, e.g. ΔG, and bear a complex, not a simple, mathematical relationship to the fundamental units, as we have shown in Chapter 5. Hence, while this reductive analysis is helpful as a guide to the behaviour of bulk systems, it cannot be used with precision. It is often the case in the discussion of chemical systems and more so in the discussion of many geochemical/environmental systems and also of biological organisms, that we are able to see only qualitatively the important contribution to *observed features* of systems around us by reference to the simple fundamental units and variables (Table 2.3). For quantitative descriptions we shall have to refer to the derived complex variables of thermodynamics (see Table 5.2) for inanimate systems. (We shall need other derived variables for the treatment of living organisms.) As complexity increases we have to accept loss of detailed insight. *Importantly, however, we have a very deep appreciation of the chemical speciation of solids, gases and solutions comprising everything around us.*

Before we proceed to discuss the chemistry of change, especially in living forms where the equilibrium condition does not apply generally (see Table 5.3) we must introduce two extra sets of variables. In the next chapter we shall examine first the novel possibility of dividing space locally so that equilibrium (balance) exists in each local zone, see Aside but not between adjacent spatial zones. This is a discussion of compartments in stationary states (see Chapter 7) and involves limited amounts of material. Thus, the total amount of material becomes an extra variable on top of the percentages of materials. It is then necessary to analyse the shape of compartments, since limited zones have shapes. Secondly, a zone adjacent to another one affects it through its fields, e.g. gravitational or electrical, so that further new variables are generated, i.e. novel distributions of mass and charge of compartments in space. Clearly, the increase in variables arises from a removal of space equilibration. Greater variability, still, will be introduced by introducing exchange at limited rates, when we will have to analyse change and rates of change, allowing the separate compartments to exchange either energy and/or material in a controlled way with time. Time is then a further variable (see Chapter 8) which can also be treated in different compartments. Remember always that, on the surface of Earth, everything changes! This leaves again the deepest problem, which is that, despite the apparent large increase in variables, living organisms (non-equilibrium systems) develop and reproduce almost exactly, implying that they are highly constrained. We observe that variables are very limited within each biological species just as they are limited by equilibria in inanimate systems. What then are the constraints?

Sizes of zones

Zones	Radius (μ)
Test Tube	10^4
Cell (Bacteria)	10^1
(Plant)	10^3
Nucleus (Cell)	10^{-2}
Protein	10^{-3}–10^{-2}

7

Systems with boundaries: compartments

He was in logic a great critic
Profoundly skilled in analytic
He could distinguish and divide
A hair 'twixt south and south west side

S. Butler, *Hudibras* (1612–1680)

7.1   Introduction

In the previous chapters of this book we have analysed the variables of unlimited chemical systems: percentage composition (in terms of components), temperature and pressure, but we have not considered the *size* or *shape* of any system we observe. In effect we have often drawn hypothetical boundaries around liquids and solids or considered gases in containers to separate them from a passive environment with which only energy (heat) could be exchanged (see Fig. 5.2), but we did not ask if any consequences stemmed from the existence of real boundaries or from the limitation of their volumes. That is, we discussed essentially internal potential energies and other variables in isolated or closed systems (see Chapters 2 and 5) without reference to other possible independent neighbouring systems. It is true that in particular cases we defined *mass* and *charge*, but these were assumed to be concentrated at fixed points in space, or points in space in random motion in a given system, for example atomic and molecular constructs or the solar planetary system. Fields—gravitational and electrostatic—were then discussed as *internal* variables of the system considered,* and conclusions were derived without taking into account the actual size of the objects or substances involved and independently from any relationship with other systems which would require us to introduce further *external* variables, especially external fields or differences of temperature between systems.

Now, examination of the world around us shows immediately that individual substances, bodies or 'objects', both living and inanimate, are normally constrained in size or volume by barriers or boundaries, and have a shape. A second consequence of the limitation of amounts is that we must consider several co-existing systems when space is split into *compartments*† which will then interact with one another. The interaction will depend on the extent and nature of their boundaries. This interaction will act as a limitation on the *sizes* and shapes of the whole system of compartments. The fields that are generated will be just the external bulk electrostatic and gravitational fields we described in Chapter 2. They are new variables.

The constraints of size or volume apply not only to large bodies or large masses of substances, e.g. the Earth and its liquid water and solid rock, and to all of the bodies in the universe—planets, stars, galaxies, etc.—but also to the Earth's atmosphere, a gas.

Similar considerations apply to small objects, but while large bodies such as the sun, the Earth and the moon interact through gravitational fields, the forces acting between small objects are dominated by electrostatic fields, particularly if they are charged, as is the case for surfaces and for many molecules. In the context of this chapter we are not dealing with those interactions that form bonds (see Chapter 2) but with long-range, continuously variable forces which we associated in Chapters 2–5 with bulk materials. Of course, this distinction is blurred at intermediate distances, but we shall use distinctly different theory in the two cases— quantum mechanics for bond formation, Newtonian mechanics for long-range interactions. Note that in all cases the variables are the disposition of atoms and charges in space, but now we have divided space up into compartments.

We shall begin this discussion by giving examples of shaped objects before we consider compartments generally and how they interact with the environment, including with one another. In this chapter we shall still be looking only at ordered stationary or stable systems, not at steady-state

---

*As stated in Chapter 2, the definition of system is very much a matter of convenience in the interests of analysis, so that we can choose to isolate, say, a chemical reaction, or the Earth, or the planetary system as a separate 'system'.

†In this book we shall use the term *compartment* rather loosely to refer to a given *isolated* system limited by boundaries. By definition, compartments do not exchange material or energy with one another *in a balanced way*; they must not be confused with *phases*, defined as material in different physical states in a space with boundaries *at equilibrium*, i.e. which exchange both energy and material in a balanced way (see Section 5.8.4).

patterns of flow.* Therefore, we shall not be concerned with time-dependent processes, such as *patterns* of rivers or living organisms, which are left to one side until later chapters since they are not stable or stationary—they are 'open' and often evolving steady-state systems through which there is flow of matter and energy. The consideration of these aspects requires the introduction of further variables (see Chapter 8).

7.2   Shaped objects

The assumption we shall make at first is that static, stable shapes, such as those formed after crystallisation of salts from solution, as shown in Figs 7.1 and 7.2, are in balance and have no tendency to change to

Fig. 7.1 Examples of some observed crystal shapes.

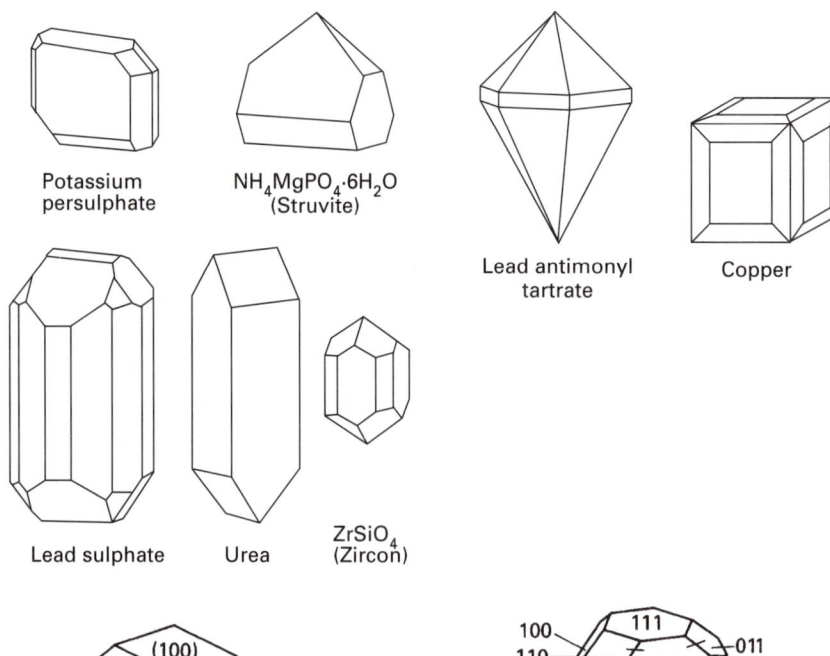

Potassium persulphate

$NH_4MgPO_4 \cdot 6H_2O$ (Struvite)

Lead antimonyl tartrate

Copper

Lead sulphate

Urea

$ZrSiO_4$ (Zircon)

Fig. 7.2 The crystallography of: (a) a conventional cubic crystal, Fe_3O_4; and (b) a crystal of Fe_3O_4 from a magneto-tactic bacterium, the shape of which is decided by surface interactions with biological structures. The numbers in the faces are Miller indices references 36, 37.

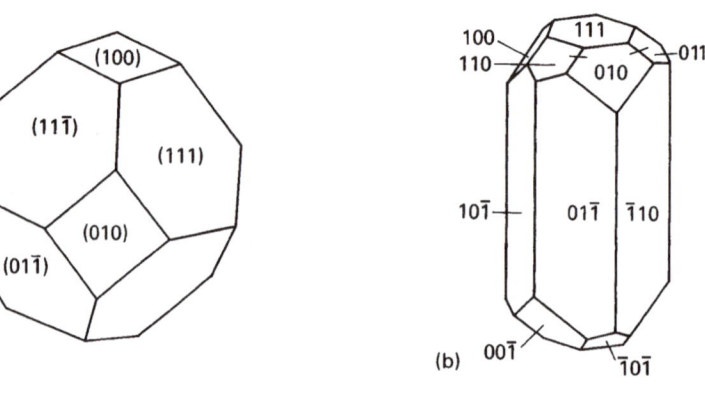

(a)

(100)
(11$\bar{1}$)
(111)
(010)
(01$\bar{1}$)

(b)

100
110
111
011
010
10$\bar{1}$
01$\bar{1}$
$\bar{1}$10
00$\bar{1}$
$\bar{1}$0$\bar{1}$

---

*As stated before, a planetary system or a system of electrons in atoms is considered to be in a stationary state since no matter or energy flows through it. They have fixed (angular) momentum (see Chapter 3).

another shape if the external conditions are held constant. The reason for this stability of a shape is easily seen.

Consider a crystal. If all its properties are in equilibrium then its shape is decided by the balance of energy between the atoms in the interior and those on the surface. Obviously, the best place for an atom to be from the energy point of view is in the interior of the crystal since it will be interacting fully with its neighbours and consequently has a lower potential energy, but just because there is a limited amount of material some atoms must be on the surface and then they will have a higher potential energy, particularly at the edges and corners. It can be shown that for atoms of equal size (and where there is a considerable amount of material) planar surfaces give the greatest stability so that the best arrangement is a close-packed cube when the faces are at right angles. if close-packing is not possible because the internal units have a non-spherical (molecular) shape, then surfaces other than those at right angles may be preferred (Fig. 7.1). Here there will be a variety of different surfaces with different areas and energies per atom, which sum to give the most stable (lowest potential energy) condition of the crystal. Obviously, the size of the crystal depends primarily on the amount of material which comes out of solution, but does its shape also depend on the amount of material, i.e. are size and shape related to one another?

7.3   Size and shape

Now, it is a common observation that the initiation of a crystal, i.e. nucleation, from a small number of units does not give the same shape as that finally achieved after growth to a considerable size.* Let us assume that the crystals are always in internal equilibrium. Obviously, the importance of the surface energy diminishes relative to the internal energy of a crystal as the crystal grows, since a larger number of units become internal. After a certain size there is a certain shape which has the particular ratio of areas of surfaces that gives maximal stability. Thus, after an initial growth period, further growth is of constant shape. It is also the case, for the same reasons, that at each earlier stage of growth the developing crystal has a fixed shape though it is different from the final one. (N.B. This is also a common observation of living objects but for a different reason.) In Chapters 5 and 6 we could describe a stable, unlimited single system, i.e. one at internal equilibria, in terms of a complete set of variables: temperature and pressure (physical variables) and percentage composition, i.e. chemical variables. Once we limit the system by the amount of material then this amount is an extra variable. However, once amount is defined and the system is in a balanced state, a new characteristic of the system, its shape, is also defined, but as we have seen it undergoes transformation

*For very small objects the most favoured shape may well be that of a sphere since edges and corners are most unstable; atoms in these positions have lower interaction with their neighbours and higher interaction with the environment.

during a change of size. We ask next: is the balanced shape open to variation according to its environment? [N.B. Apparently shape is not defined for a system which is not at equilibrium, (Fig 7.3). We shall be forced to enquire later why biological organisms have shape (Chapter 10).]

7.4   The environment of a system and its shape

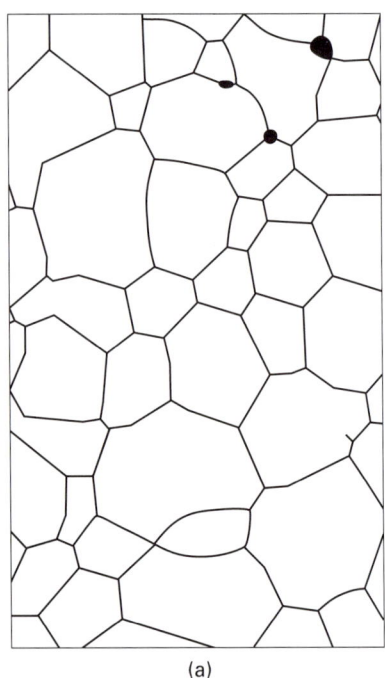

(a)

The surface energy is clearly dependent upon the environment, for example the solvent in which a crystal sits, for the obvious reason that the solvent (the environment) interacts with the surface. The shape of a crystal in a solvent is, then, solvent dependent. The reverse situation is that of a drop of liquid in contact with a given environment, for example a solid surface (effectively a field) (Fig. 7.4). The simplest case is that of an isolated drop in the absence of a field, when the drop is spherical. When the drop lies on a surface, and according to the strength of the interaction between the liquid and the solid surface, the drop will spread to a greater or lesser extent. Drops of oil spread on water, while drops of rain water do not spread on glass (Fig. 7.4). Stable shapes of liquid drops are therefore dependent on the fields they experience. External and remote fields may also affect surface energies and therefore shape, but once the environment and the fields are fixed, the shape at equilibrium is again fixed by the amount present. The problem is general, so that in the absence of fields the Earth, which passed through a melted state, should have been a sphere, but fields (and motion) have distorted it. Variation of amount also changes the shape of a liquid drop since the percentage of material in contact with the surface changes. Of course, all these remarks again concern equilibrium situations only. A dynamic spinning system also has a time-averaged shape.

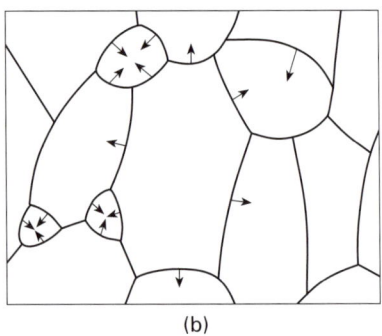

(b)

Fig. 7.3 (a) Grains in typical mixed mineral. (b) Drop or grain growth mechanism via contact. The boundaries move towards the centre of curvature (arrows). As a result, the small units eventually disappear. Contact is not necessary since growth can occur through the vapour or liquid phases.

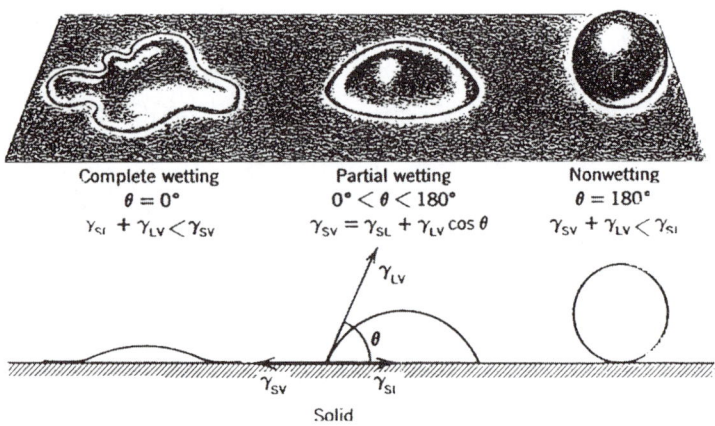

Complete wetting	Partial wetting	Nonwetting
$\theta = 0°$	$0° < \theta < 180°$	$\theta = 180°$
$\gamma_{Sl} + \gamma_{LV} < \gamma_{SV}$	$\gamma_{SV} = \gamma_{SL} + \gamma_{LV} \cos \theta$	$\gamma_{SV} + \gamma_{LV} < \gamma_{Sl}$

Fig. 7.4 Wetting of a substance by a liquid, showing balance of different forces, γ, and the wetting angle, θ: S, Solid; L, liquid; V, vapour.

7.5 The shapes of crystals of particular substances

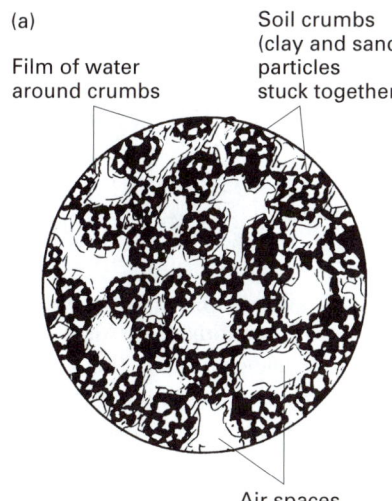

(a)

Film of water around crumbs

Soil crumbs (clay and sand particles stuck together)

Air spaces

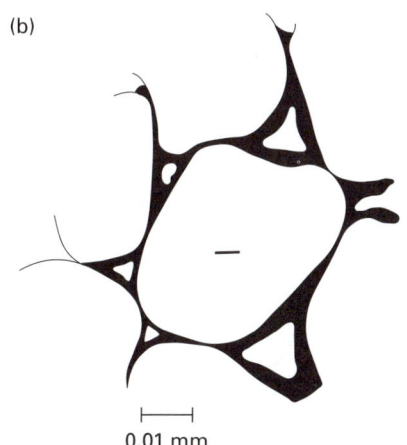

(b)

0.01 mm

Fig. 7.5 (a) Magnified (diagrammatic) view of soil crumbs. (b) Clay particles. Shaded areas represent water held by capillary action.

The earliest shapes to describe, which we referred to above, are those formed from equally sized spheres, e.g. atoms or ions. Two classes of compounds, as well as many metal elements, may fall into this group—alloys and salts (see Chapter 4). Equal-sized atoms or ions in these lattices (with metallic or ionic bonding, respectively) give rise to cubic shapes of crystals. However, atoms in alloys and salts are, in general, not all of the same size, so that the packing is often dominated by the arrays of the larger atoms (ions) when the smaller atoms (ions) fit into the holes generated by the larger ones (see Section 4.3). Once again it is easily seen that cubic structures can result, but the occupation locally of individual holes allows the formation of octahedra or tetrahedra around individual small atoms (ions). We note, however, that the relationship between local (or point) symmetry and crystal shape is not an obvious one, since now the atoms on the surface may be of different ratio from those in the interior, even at equilibrium. By affecting surface energy the distribution could affect shape.

As discussed in chapter 4, a third type of bonding is covalent bonding. We showed in Section 4.9 that non-metal atoms do not pack together in the manner of alloys and salts, but because of the similar and strong affinity for electrons of all these elements they share localised electrons in bonds which are constrained in their numbers by the filling of electron shells. The shells are mainly of eight electrons (see section 3.4.4). The covalent molecules formed have shapes that are not spherical; packing considerations are now quite different and many different shapes of crystals can be found. Small molecule shapes are described in Chapters 3 and 4. Now we must consider separately the shapes of larger molecules, even polymer molecules, which fold and form their own shapes.

We may also have to consider shapes which are not the most stable, that is, are not formed at equilibrium but arise through the way in which the crystals have been prepared, e.g. by varying the temperature and pressure of their preparation (see Chapter 8). A particular natural example is the formation of the minute particles of the soil (Fig. 7.5), and another industrial example is the preparation of icing sugar or common 'crystalline' sugar, though all such systems tend to change to more stable shapes.

7.6 The shapes of folded polymers

We take the example of proteins as particular, well-studied cases of a shaped polymer. When we examined the shapes of crystals of such salts as sodium chloride the numbers of units of Na^+ and Cl^- which came together were very large. On the other hand, the folding of a protein is of a molecule of the order of 100–1000 continuously linked amino acids (Fig. 7.6). While in both cases the shape that evolves is a stable co-operative property

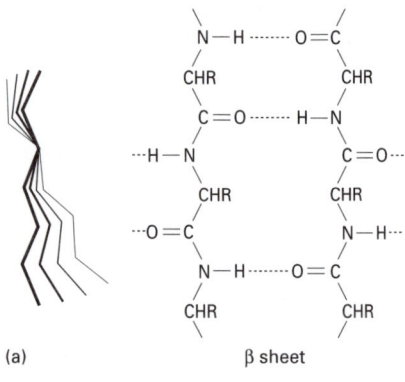

(a)   β sheet

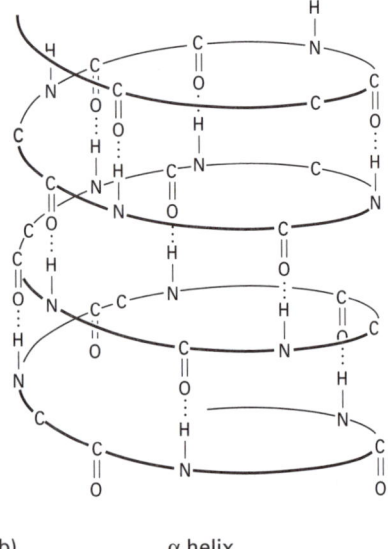

(b)   α helix

Fig. 7.6 Representations of the protein secondary structures of: (a) a β sheet, as in silk; and (b) an α helix, as in wool. The dashed lines represent hydrogen bonds between CO and NH and the C symbols represent the α-carbon atoms (the H and R groups attached to these are omitted, for clarity). The packaging of β strands in the curved sheet is shown in (a).

of the whole, there are striking differences. The protein fold has three dominant features.

(1) The H bond networks link the peptide –CO–NH– bonds into, for example, sheets and/or helices (see Fig. 7.6). This is a co-operative H bond contribution to stability, as in ice.

(2) The so-called hydrophobic effect, which is a gain in stability by hiding in the interior those side-chains such as $-CH_3$ that would otherwise destabilise water.

(3) The hydrophilic surface of side-chains such as $-NH_3^+$, $-CO_2^-$, $-OH$, which interact strongly with water and are placed on the external surface.

Of course, because it is a polymer and not 100–1000 separate amino acids, the linear covalent string helps folding enormously but does not affect shape directly. There is no symmetry consideration in the shape, which is idiosyncratic. It is the particular *sequence of monomers* that controls the shape (Figs 7.7 and 7.8), although the relationship is not thoroughly understood.

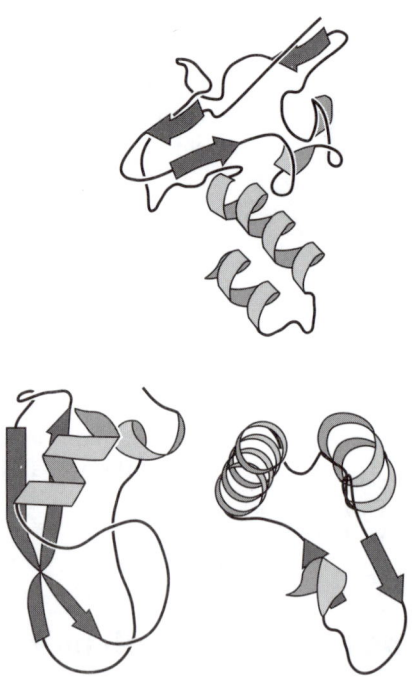

Fig. 7.7 Diagrams for the folds of a miscellaneous set of small proteins. The backbones of these polymers fold into helices, which are clearly seen, and almost straight strands, which come together to give β sheets, shown as arrows (see Chapter 9). The number of different folds known is considerable—perhaps 50–100—but the surfaces generated are of almost infinite variety since each helix or strand may have more than 20 amino acids, which can be varied in sequence amongst 24 amino acid types. Clearly, the surfaces of the fold can bind other molecules uniquely and may help to stabilise this fold or change it to another. (After work of J. Thornton.)

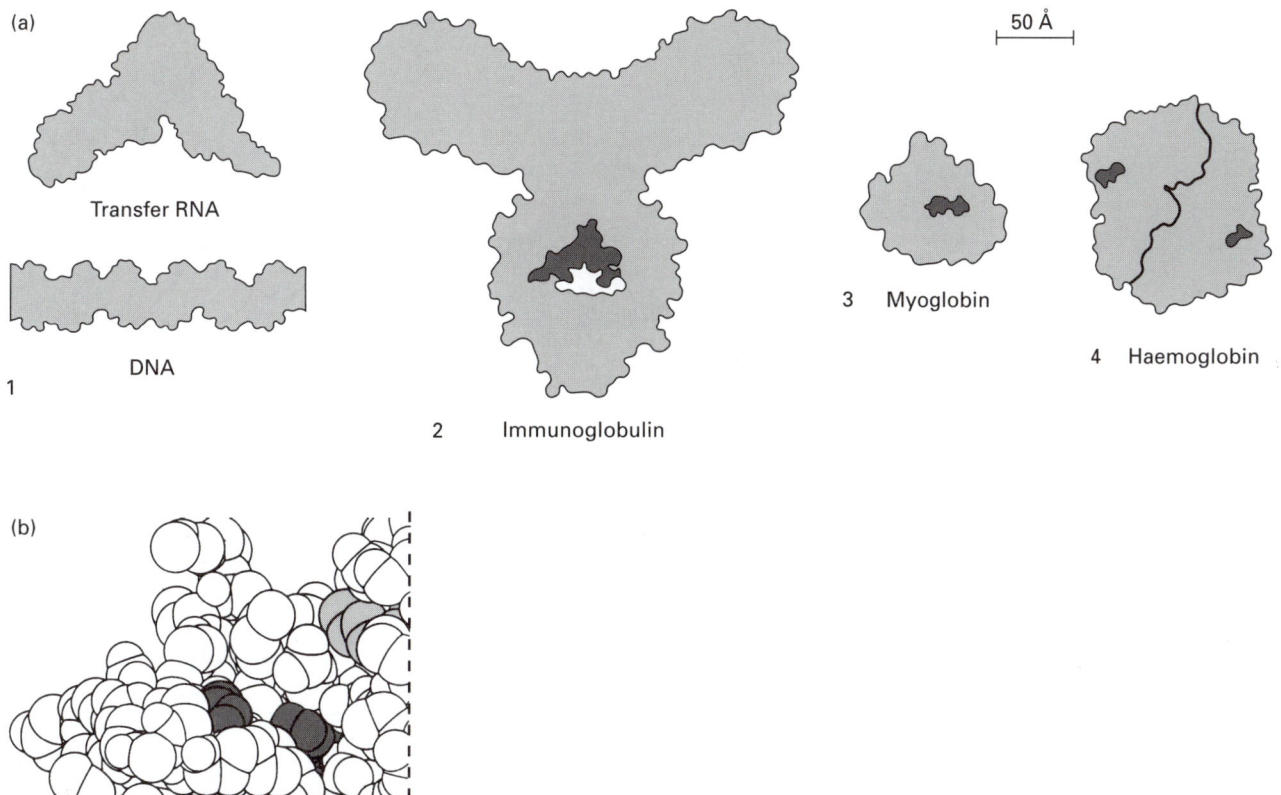

(a)

Transfer RNA

DNA

1

2 Immunoglobulin

50 Å

3 Myoglobin

4 Haemoglobin

(b)

Fig. 7.8 (a) Shapes and sizes of some biological polymers. The shapes are functional much as is the shape of an organism. (Scale bar, 50 Å.) Haemoglobin has a molecular weight of 64 000 dalton. (b) A more detailed view of the way in which amino acids in proteins are close-packed such that surfaces are unique to sequences and individual hot spots (shaded) can arise which also have functional value. To some degree these shapes are averages over fluctuations.

Now, in itself, a protein in water can still be treated both as a molecule in solution and as a separate phase which has a melting point. We have to consider both the statistics of the configurational properties of the folding $\rightleftarrows$ unfolding equilibrium of a protein molecule (order $\rightleftarrows$ disorder balance, see Section 5.8.5) and those of the solution properties, with all the complications of association and assembly into larger bodies. At equilibrium the shape of an isolated protein is decided by the sequence, but this shape need not be the equilibrium shape in an assembly (Fig. 7.7). Therefore, there is a need to describe not only shapes of proteins but also shapes of assemblies. When proteins fold they leave cavities that are often sites of activity in enzymes, while when proteins assemble they can form channels or spaces to be filled selectively by other large molecules or even minerals. In effect they form microporous solids much as are observed in silicate minerals. We see this most easily in ferritin, a combination of 12 protein subunits which forms a protein vesicle (Fig. 7.9). The vesicle is a container of precipitated $Fe(OH)_3$. In fact, this is the iron storage system of very many living cells.

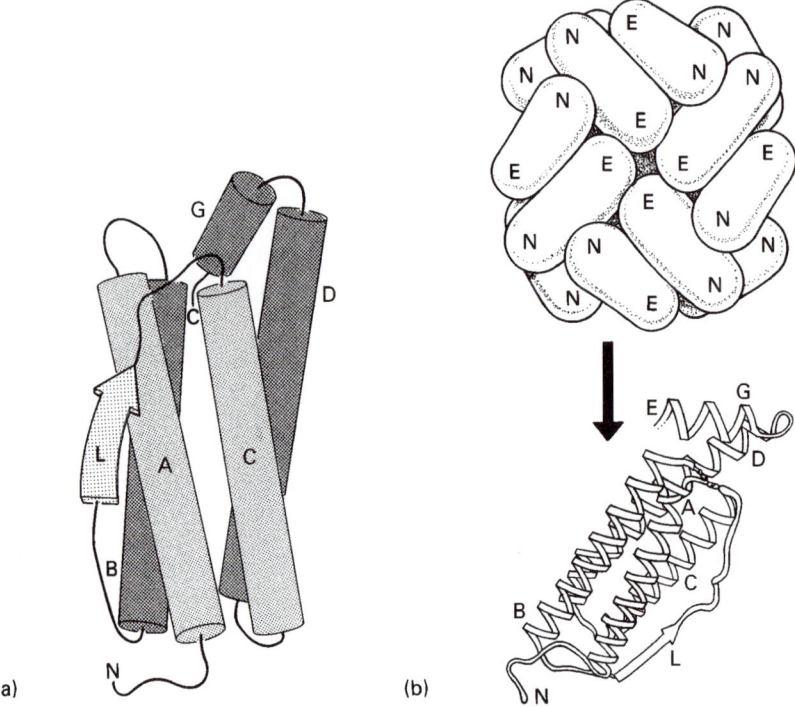

(a)　　　　　　　　　　　　　　　　　　　(b)

Fig. 7.9 (a) Schematic representation of the ferritin subunit. Four long helices, A, B, C and D, of one protein form a bundle, with the fifth helix, G, lying at an acute angle to the bundle. There is only one region of intermolecular anti-parallel β pleat in the loop L. The way in which the subunits self-assemble is shown in (b) where the subunit [see (a)] are illustrated differently. (After P. Harrison.)

Protein cages of this kind are the basis of many viruses (Figs 7.10 and 7.11), which in the central space contain genetic information in a folded polymer of RNA or DNA. The shapes of the viruses and of ferritin are based on quite simple symmetry considerations since they are made from a

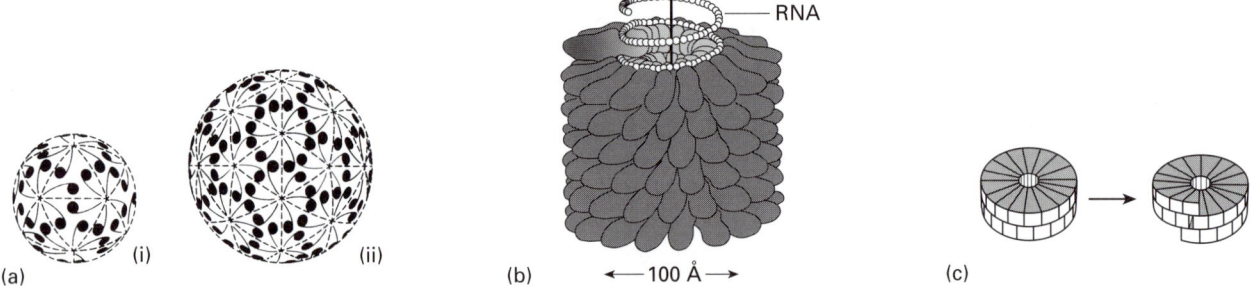

Fig. 7.10 (a) The tomato bushy stunt virus. Icosahedral surface lattices showing the packing of (i) 60 strictly equivalent subunits and (ii) 180 quasi-equivalent subunits. Note that all of the tail-to-tail contacts in part (i) are in rings of five, whereas some of these contacts in part (ii) are in rings of five and others are in rings of six. [After Harrison, S. C. (1978). *Trends in Biochemical Sciences*, **3**, 4.)], (b) Model of a part of tobacco mosaic virus (TMV), showing the helical array of protein subunits around a single-stranded RNA molecule. [After Klug, A. and Caspar, D. L. D. (1960). *Advances in Virus Research*, **7**, 274.] (c) Schematic diagram showing the conversion of a TMV protein disc into the helical lock-washer form. [After Klug, A. (1972). *Fed. Proc.* **31**, 40.]

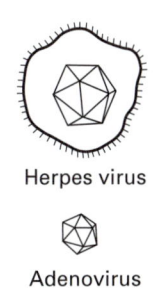

Herpes virus

Tipula iridescent virus

Adenovirus

Polyoma virus

|← 10 micron →|

fixed number of identical protein units. The shape is then that of classical polyhedra. Note how the shape is decisively controlled by the amount, i.e. the number, of protein molecules in the cage. The surfaces of proteins are built so that they bind and assemble together in these patterns.

Other ways of building particles from proteins are found in the nucleosomes and the ribosomes. Here, larger lengths of DNA and RNA are built around protein-made structures (see Fig. 6.8), and the final shape of the proteins is decided by the whole assembly. Notice how DNA wraps itself around the surfaces of a few histone proteins (Fig. 7.12) while RNA interacts in a very different way with a much larger number of proteins (see

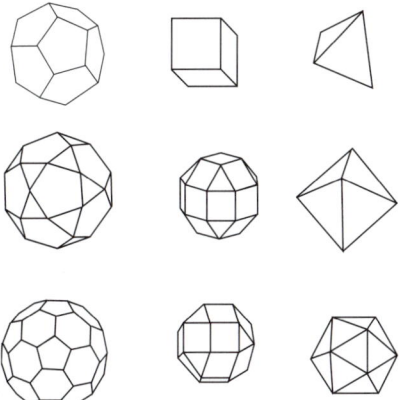

Fig. 7.11 Some shapes built from single units, and which can be made from, for example, simple triangles or squares on a closed surface.

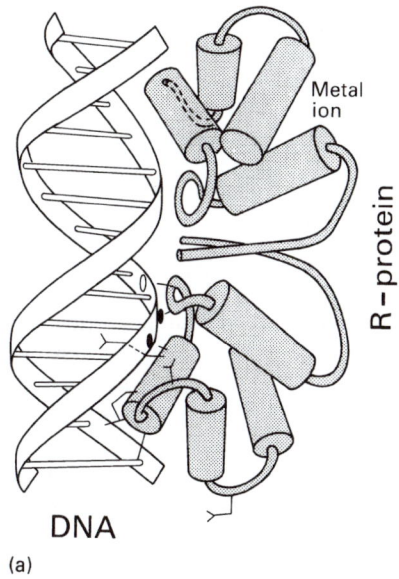

(a)

(b)

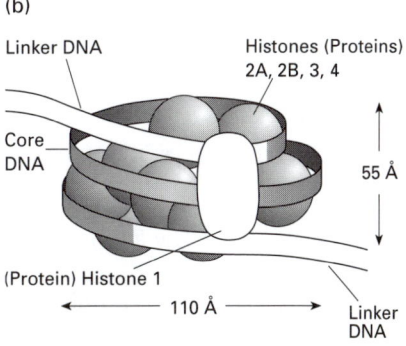

Fig. 7.12 (a) The way in which a helical set of two protein segments could fit into the grooves of DNA to stabilise both. Note the value of mobile helices. Several metal-binding regulatory proteins may be of this kind (see Section 10.13). The structure is from the works of B. W. Matthews and T. A. Steitz. (b) Schematic diagram of a nucleosome. The DNA double helix (shown as a band) is wound around an octamer of histone proteins (two molecules each of 2A, 2B, 3 and 4, shown as spheres). The unit is quite stable. Histone 1 binds nucleosomes together from outside.

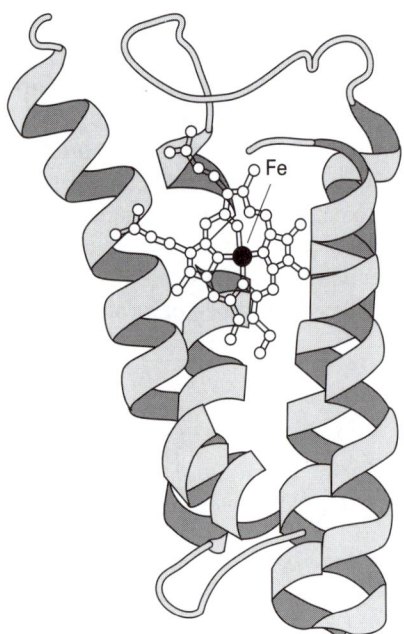

In proteins like cytochrome c', which are mainly helical, the helices may well be relatively mobile with respect to one another.

Fig. 6.8). One question that must be asked is 'How is it possible to make these multi-unit particles and leave so little excess of the building blocks?' This is a question of controlled synthesis and is described in Chapter 9. In passing, note that smaller particles remain dispersed in aqueous solution but really large particles sediment under gravity so they must be held in place by filaments. Thus gravitational forces come into play in biological constructions much as they do in the constraints on man's buildings.

7.6.1. The curious nature of biological structures

At first sight the structure of molecules of proteins, RNA and DNA seem to be no different from the ordered structures we have described in Chapter 4. Closer inspection shows, however, that the individual units, i.e. each molecule, have three peculiarities.

(1) There are no internal repeats in the structure formed from the folding of a linear polymer.
(2) The structure is based upon a sequence which has a beginning and an end.
(3) The polymer has an exact molecular weight.

The synthesis of these polymers is brought about on a preformed mould or matrix and each synthesis has a sequential development in time. Again, the activities of the DNA, RNA and proteins (especially enzymes) are each linked to syntheses or degradations along pathways in time. The whole of the cell is linked of course to a timed activity and cycle, and we note that the coupling of time to structure is inherent in both the synthesis and activity of every polymer. This is a very distinctive and curious feature and arises since all the units in a cell are part of organised flow not static structure. We shall find it important to distinguish between a biological polymer, e.g. a protein whose structure can be given as for any other molecule, and such a polymer in a functioning cell (see aside, Section 10.11 and Chapter 16). After all, an isolated component of a machine, e.g. a bicycle chain, has a structure that does not describe its cyclic activity when part of an organised machine, a bicycle. See note on page 213.

7.7   Composites and their shapes

The surfaces of small crystals of minerals or proteins allow them to interact with other exposed surfaces. If the second substance is a polymer then large numbers of crystallites can interact with the polymer to form so-called composites. The shells and bones of animals are made in this way, but so too are some structures in plants, such as those incorporating opaline silica and calcium oxalates. The two surfaces interact largely through electrostatic attraction of a relatively non-specific kind. Man-made materials of this kind include various plasters and cements, and even fillings for teeth. Matching of the polymer and crystal surfaces is a major although very empirical industry, where in place of biological polymers synthetic fibres can be used which may be mineral or organic. Now, as stated, these constructions are based on relatively non-specific interactions and frequently, therefore, shape is not defined except by the use of a mould before the mixed composite sets. In fact, here we deal with amorphous materials where shape is controlled entirely by external factors that introduce limitations on size. In effect, these moulds are repulsive fields.

7.8 Fatty acids and lipids: shapes of liquid crystals and their co-acervates

A quite different way of building shaped material from that of making a salt crystal or from the folding of a protein, is to make two-dimensional mono- or bi-layers and then for these to form enclosed volumes. The molecules that build such structures are long-chain fatty acids with terminal polar groups (see Section 6.3). Now, just as polymorphic forms occur in salts, so polymorphic structures can be built from these two-dimensional layers. The structures observed are planar sheets and spherical vesicles of all sizes (compare soap bubbles, Fig. 6.3). The final form of these structures depends upon the mode of preparation. These spherical vesicles act as containers for many chemicals in living cells and at their largest are the containing structures of cells themselves, Fig. 7.13. Note, however, that they are often stationary state structures rather than stable structures. Within cells these vesicles must be retained in position since otherwise they could fall under gravitational pull.

Of special interest here are the surfaces that are negatively charged so as to keep the vesicle structures in extended forms. Now, the head groups of the lipids can interact more or less specifically with certain proteins and

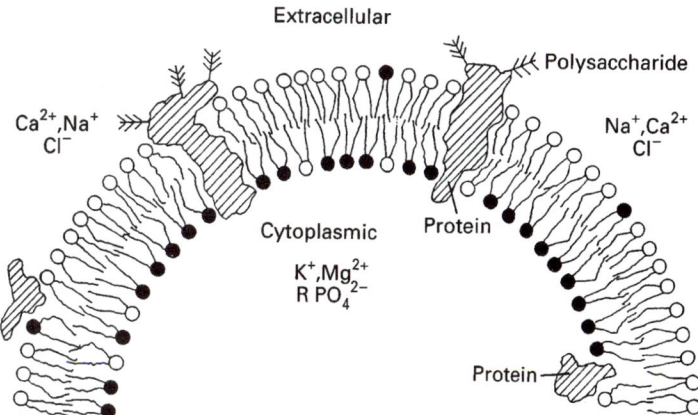

Fig. 7.13 A schematic representation of the assembly and distributional asymmetries of some biomembrane components. Lipid asymmetries are shown as a difference in the distribution of the polar head groups, with the negatively charged lipids (filled circles) localised predominently in the cytoplasmic leaflet. (After D. Chapman.) The lipids of archaebacteria differ from those of other bacteria, which may also be an adaptation to environmental conditions.

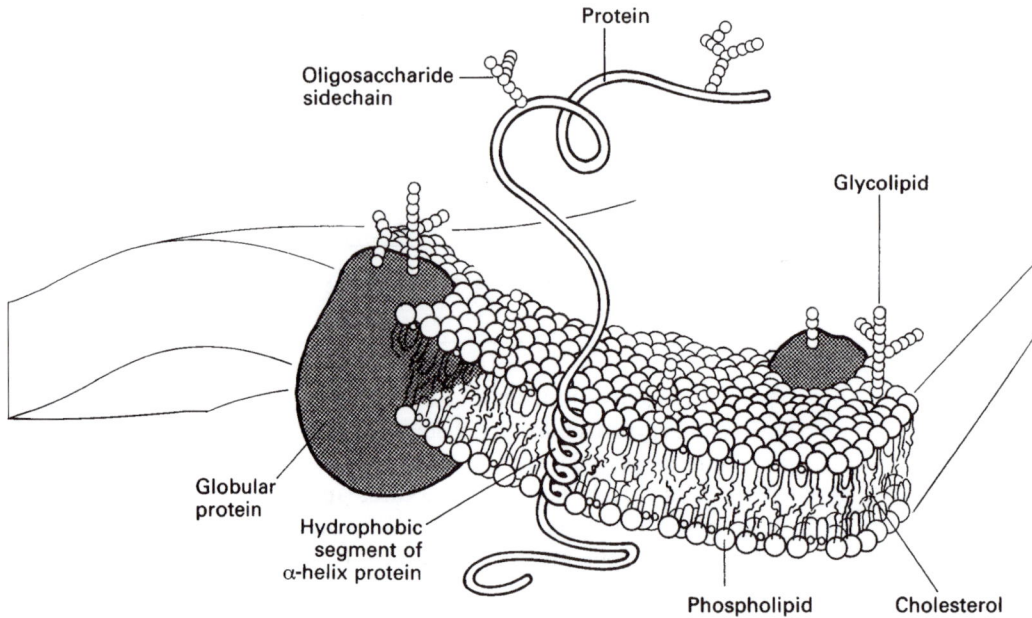

Fig. 7.14 The organization of proteins in membranes, now seen in three dimensions.

their constructs, e.g. ribosomes, so that instead of protein particles becoming fixed to polymer fibres they can be fixed to the extended, two-dimensional, curved surfaces of lipid vesicles. This is one way in which local gross structure is built in a living cell. Alternatively, the protein constructs may enter or even penetrate across the lipid layers so as to form quite different structures, pores, pumps or other contacts between aqueous phases inside and outside the vesicles. The design of the protein particle is, of course, critical. To get a protein to cross a hydrophobic membrane so as to link two aqueous layers it must have a sectionally specific structure, hydrophilic–hydrophobic -hydrophilic, to favour the interactions with the aqueous layers and with the membrane between them (Fig. 7.14).

Now the very fact that these lipid layers close off to form a vesicle means that the aqueous zones inside and outside are no longer identical and can differentiate, forming separate compartments. This division of space into separate compartments introduces quite novel ways of developing the distribution of material and energy in space and generates fields. We turn now to this question.

7.9 Local balance and physical fields

When describing systems held next to one another but which cannot mix, like the inner and outer aqueous zones of lipid vesicles, we have to consider the degree to which the contents of a given isolated system, which we have called a compartment, may restrict, through the effect of gradients of the

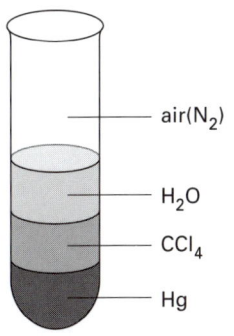

Fig. 7.15 The layered separation of some substances in a gravitational field.

fields it generates, the spatial distribution open to the contents of a second isolated compartment. As stated in Chapter 2, a field is a long-range effect now operating across a barrier. For example, everyone knows that gravitational fields cause a settling into ordered layers of materials of different density that do not mix (Fig. 7.15). Even large denser molecules tend to 'sink' in water rather than being evenly mixed in it.

We have also pointed out that electrostatic fields are observed between charged bodies, thus, as we mentioned in Section 2.2, ebony rods attract fur after they have been rubbed together. Similarly, vesicles that are negatively charged repel one another; they pack together only when charge is compensated by positive ions. The balance finally achieved is between dispersion (disorder) and association through improved interaction with the field (order). Note that the field in each case is generated by a distinct and separated compartment. In fact, under these conditions, final balance is prevented by barriers that do not allow the field to be nullified, but we have to be able to describe the balance locally. We still call the whole systems *stationary* since their condition is independent of time, but this would not be the most stable situation if thorough 'mixing' had been allowed.

7.10 Osmotic pressure and electric potential differences between compartments

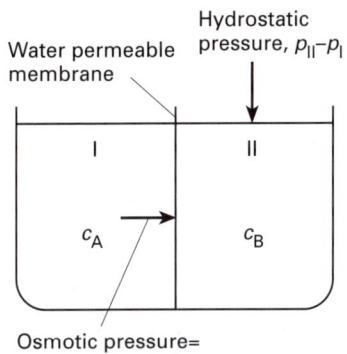

Fig. 7.16 Osmotic pressure. Here, solutions A and B are separated by a membrane that is permeable to water but impermeable to all solutes. If c_B (the total concentration of solutes in solution B) is greater than c_A, water will tend to flow across the membrane from solution A to solution B. The osmotic pressure between the solutions is the hydrostatic pressure that would have to be applied to solution B to prevent this water flow and is given by the van't Hoff equation as $\pi = RT(c_B - c_A)$.

Consider two ideal solutions separated by a so-called semi-permeable membrane, i.e. some boundary that allows the passage of the molecules of one compartment, the solvent, but not the molecules of the others, the solute. Let the two solutions be at different concentrations, c_1 and c_2. Since transport of the molecules of the solutes is not possible, the system will be out-of-balance with respect to the distribution of the solute between the two compartments and the solvent will tend to flow from one compartment to the other until equilibrium is re-established. A stationary state can, however, be achieved if a hydrostatic pressure is applied to the solution of higher concentration to prevent the in-flow of the solvent. The hydrostatic pressure is called *osmotic pressure* and is usually denoted by π which can be shown to be equal to $RT\Delta c$, where Δc is the difference in solute concentrations (see Fig. 7.16). Such a compartment system stores energy related to pressure (a concentration difference). In living systems the two compartments are the inside and the outside of a cell with a lipid membrane boundary. [Notice the relationship between $p = RT/V$ (Section 2.5.3) and $\pi = RT\Delta c$]

Now, the osmotic pressure can be replaced by adding to one or other of the solutions, inside or outside the cell membrane, a second chemical, for example NaCl outside and KCl inside, at a suitable concentration so as to give osmotic balance (Fig. 7.17). The system now stores energy relative to a mixed solution of both salts and the free energy difference is given by $\Delta G = RT \ln (c_1/c_2)$. The separation of inorganic ions, Na^+, K^+, and Cl^-, largely overcomes the destructive effects of osmotic swelling due to the organic content of the cell.

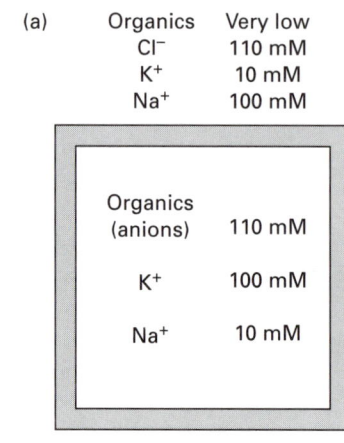

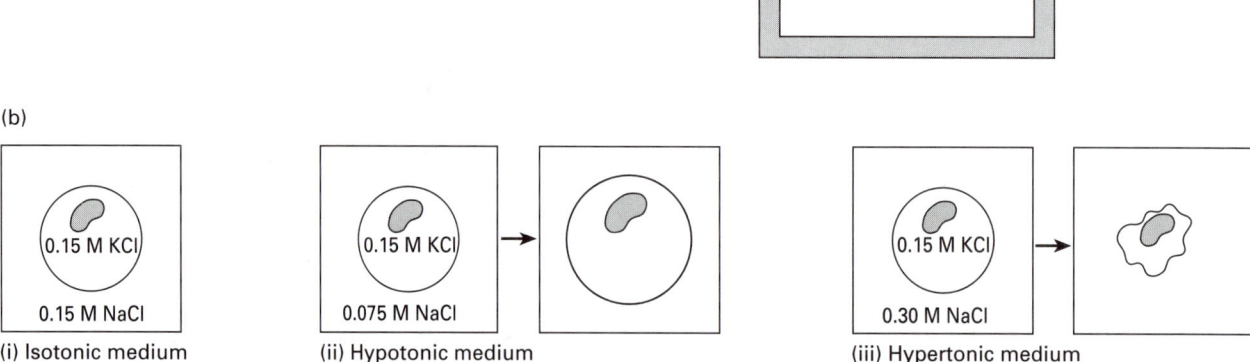

Fig. 7.17 (a) An idealised description of osmotic balance in a cell. Note the charge balance too. (b) Animal cells respond to the osmotic strength of the surrounding medium. Sodium, potassium and chloride ions do not move freely across the cell membrane, but water does (arrows). (i) When the medium is isotonic, there is no net flux of water into or out of the cell. (ii) When the medium is hypotonic, water flows into the cell until the ion concentration inside and outside the cell are the same. Here, the initial cytosolic ion concentration is twice the extracellular ion concentration, so the cell tends to swell to twice its original volume, at which point the internal and external ion concentrations are the same. (iii) When the medium is hypertonic, water flows out of the cell until the ion concentration is the same. Here, the initial cytosolic ion concentration is one-half the extracellular ion concentration, so the cell is reduced to one-half its original volume. [After Darnell J. Lodish, H. and Baltimore, D. (1990). *Molecular cell biology*. Scientific American Books, New York.]

As well as a difference in concentration, there can be an applied or internally generated electrical field difference, ψ, acting on two compartments across their boundary. This is due to the compartmental separation of charge as opposed to mass.

Any charged species, any ion, even in equal concentration on both sides of the boundary, feels the difference in potential, ψ, so that the free energy difference between compartments derived from the electrostatic field is

$$\Delta G = z\mathcal{F}\psi$$

where z is the charge of the ion and $\mathcal{F}$ is the Faraday constant (96/500 coulombs per mole); thus, when ψ is expressed in volts, ΔG is obtained in joules.

Such an electric field is generated by all known biological cells since the distribution of oppositely charged ions is not equal—there is a forced separation of ions such as H^+, Na^+, K^+, Ca^{2+}, Mg^{2+}, and Cl^-. This electrical field

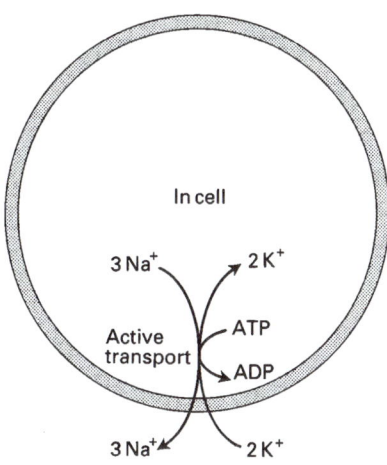

Fig. 7.18 The pumped movements of sodium ions out of and potassium ions into the cell. ATP hydrolysis (chemical energy) is required for the movements of the ions against their gradients. The actual coupling is of two K^+ for every three Na^+ so that the pump is electrogenic, that is, it develops charge and chemical gradients. We describe a resting state in terms of the free energy status of the inside and outside of the cell considered separately.

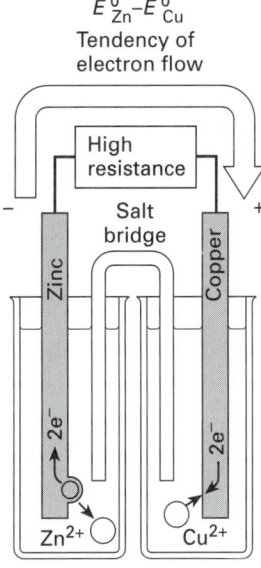

Fig. 7.19 The relative electric potentials of electrodes and the tendency to electron flow. The bridge is a KCl solution in an agar gel. Here we do not discuss rate of flow (see Chapter 7).

is the basis of stationary energy storage: to be used later in, for example, nerve messages (Fig. 7.18), and energy transduction from primary sources in living cells (see Chapter 10 onwards).

Summing up concentration and electrical field differences, we have a free energy difference for each compartment distributed between two compartments given by

$$\Delta G = RT \ln(c_1/c_2) + z\mathcal{F}\psi$$

This energy difference can be used in biological systems to do work, for example to pump required materials selectively in cells.

A further example of electric potential differences is useful. Let us consider the case of two metals, copper and zinc, dipped in water solutions of their own metal ions each at 1 molar concentration but isolated in separate compartments. In each case that metal tends to dissolve in the solution leaving a negative charge in the solid (the *electrode*); this is opposed by the tendency to deposit metal ions from the solution which would give the electrode a positive charge. The two opposed processes eventually reach a balance at the metal–solution phase boundary, but there will be a potential difference between the electrode and the bulk solution which is measured by the so-called *electrode potential* (see Section 6.13). As described, the electrode potential depends both in sign and value on the nature of the metal and the concentration of its solution. Although there is internal (local) balance in each compartment, between the two compartments (*half-cells*) there is no equilibrium and we can evaluate the out-of-balance potential (the difference of potential) by measuring the electromagnetic force (EMF) between them (see Fig. 6.11). For this purpose we insert a wire connection with a high resistance galvanometer between the two electrodes and establish a contact barrier between the solutions, a sintered glass or a salt-bridge, to allow electrical continuity (Fig. 7.19). Here, we see ways of creating zones of different redox potential. The difference in redox potential is a measure of the charge distribution in equilibrium. It is commonly found across cell membranes in organisms.

The high resistance prevents effectively any current flow so that we measure a property of a stationary condition. Thus, we will find that the two local electric fields differ in this case by 1.1 volts. This design is the basis of an electrolytic (car) battery. (Note that in the present example zinc is the negative electrode and copper is the positive electrode so that the current tends to flow from zinc to copper.) Many different cell types exist, as is shown in Table 7.1.

Table 7.1 Types of electric current generators based on oxidation–reduction reactions

Electrical current generators	Elements involved in redox reactions
Danniell cell	Copper/zinc
Fuel cell (Bacon cell)	Oxygen/hydrogen (carbon electrodes)
Car battery	Lead (0, II, IV) (sulphuric acid)
Dry cell (Leclanché)	Zinc/manganese (MnO_2)
Sodium/sulphur battery	Sodium/sulphur (solid electrolyte)
Nickel/cadmium battery	Nickel/cadmium (hydroxides as electrodes)
Mercury battery	Mercury/zinc
Lithium battery	Lithium

Next, let us make a similar cell with the same metal, e.g. zinc, but in this case the aqueous compartmental solutions, both of zinc ions, have different concentrations and are not allowed to mix. The two liquids give different electron densities at the phase boundary on the same metal, but both are at local equilibrium with solutions. Thus, the energy difference can be measured as above, as an EMF, or by considering the difference in order (entropy) of the zinc ions. In Chapter 5 we have given the answer as

$$\Delta S = R\ln\frac{c_1}{c_2}$$

and the ordering energy is $Q = T\Delta S = RT\ln\frac{c_1}{c_2}$. Since the energy (in joules or kcal mol^{-1}) is related to the electrode potential E (in volts) by $Q = n_i \approx E$, where n_i is the number of electrons involved (2 in the case of Zn) and $\mathcal{F}$ is the Faraday, the charge of a mole of electrons, i.e. 96 500 coulombs, we can write, generally

$$EMF = \Delta E = \frac{RT}{n\mathcal{F}}\ln\frac{c_1}{c_2}$$

but it is obvious that the energy difference $RT\ln\frac{c_1}{c_2}$ exists without the electrodes, that is, wherever there is a gradient of concentration. All we must have is a barrier between salt solutions as in Fig. 7.19, see Table 7.2.

Energy differences can always be converted to electrical potential differences. Some examples are given between different types of electrical current generators in use by man in Table 7.1.

Now, as we can convert chemical (potential) energy differences into electrical potential difference, we can also use electrical potential differences to cause chemical changes. This is the basis of electrolysis, widely used today in industry.

Table 7.2 Materials used to make compartmental barriers

Type of compartment	Materials
Geological compartments	Minerals*
Man-made compartments	Metals* Minerals* (clays, glasses) Plastics (wood, paper, etc.) Composites
Biological compartments	Organic polymers (plastics) Composites Fats (membranes)

* High temperature preparation.

7.11   The nature of physical traps

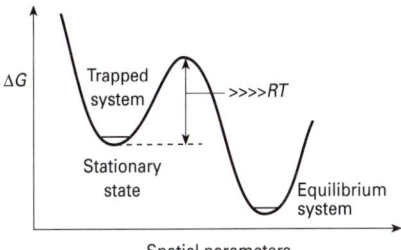

Fig. 7.20 A stationary state, which is not an equilibrium state, is defined by a system that is not at equilibrium with its surrounding material and is not gaining or losing material or energy.

We are now in a position to generalise our considerations to systems of multiple compartments that are co-existing in space, each internally at local equilibrium or balance but out of equilibrium with one another due to the existence of barriers (Table 7.3). We shall call them stationary trapped states (Fig. 7.20). Such traps are well-known in our experience. Thus, a balloon has a pressured gas inside which is not in balance with the gas outside it. Similarly, water held by a natural or artificial barrier, a dam, is in internal balance in a lake or reservoir but not in balance with the sea, although the sea itself is a system in local balance. The out-of-balance potential may be in the field applied (gravitational or electrical), in pressure or concentration differences (i.e. solute concentrations) or in temperature. To the various local equilibria in separated compartments there must be added differences in bulk electrostatic charge carried by compartments and, in fact, any other differences. The concept of local balance in limited volumes is very useful since it allows us to calculate the free energy (relative to a defined standard state) of each compartment and so establish the maximum energy available for work, ΔG, from the difference in energy of separate components. It is clear that any (partially) open connection we may make between the compartments will allow flow towards final, complete equilibrium mixing. From the flow, energy can be extracted and work done until the whole of the combined system is in balance (equilibrium). Real machines where there is flow cannot operate at 100% efficiency for reasons discussed in Section 5.4, but the maximum energy available is always calculable (see Further reading for Chapter 5).

Table 7.3 Some modes of energy storage in compartments

Compartments	Storage	Barrier
Reservoir versus ocean	Pressure under gravity	Dam
CH_4/O_2 versus CO_2/H_2O	Chemical energy	Chemical bond
Centre of Earth core Fe/air, O_2	Chemical energy (and heat, temperature)	Rocks
Biological liquids/external chemicals	Pressure, chemical energy	Lipid membranes
Metal plates in electrical cell	Electrical potential	Air Plastics
Aqueous solutions	Electrolytic gradients	Plastics Lipids

7.12   Summary of variables of compartments and fields

In order to generate a compartment, the size of a system has to be defined. This results in an increase in the number of variables by one, since we now include the *total content* of the system whereas in unlimited systems we describe only the percentage composition of components. By defining the total content (and consequently the mass and charge of each component) and its free energy, whether it is in a stable or stationary state, we define implicitly the size of the system (compartment) as well as its shape, as in the case of crystalline systems. Thus, shape is a useful variable for classification purposes and has been generally used in mineral science since it incorporates all the variables within composition and free energy. (Biological shapes of organisms, although frequently used for classification purposes, cannot be included here since living systems are steady state, evolving systems, not stable or stationary.)

Now, if there is more than one compartment, then the variables are given by the sum of all the variables in *each* compartment, which now includes total content besides percentage composition, and the effect of fields originated by other compartments. Since each compartment can be changed independently, the number of variables increases dramatically once equilibrium between them is lost (remember that the equilibrium conditions restrict the number of degrees of freedom) so that only a very descriptive account of the nature of the whole can be given. This is the case, for example, of the huge diversity of minerals that constitute the surface of the Earth, for reasons which will become apparent in Chapter 9. Many of the minerals are in internally equilibrated states although they are in stationary conditions with respect to their environment. In effect, they have originated in processes that prevented further change from occurring, either because of compartmentalisation or because the rate of internal change has become too slow. Table 7.4 summarises the situation for the variables of systems at the end of this chapter. In effect, we have now described all the static variations of the distribution of atoms and charge (the two variables of composition) and their possible distribution in *space* (a third variable). If we set compartments at different temperatures

Table 7.4 Units, variables and restrictions in Chapter 7

Units	Variables	Restrictions
Atomic elements and carried charges	Absolute amount of each unit giving composition of limited system	Balanced relationship between attractive and tangential forces
Components	Potential energy of interaction between units in stable and stationary states	Equilibrium exchange between atomic elements in components
	Angular momentum in stable and stationary states and associated directional kinetic energy	
	Temperature and pressure Isotropic motion and kinetic energy	Equilibrium exchange of matter and energy (radiant) between compartments, generating phases
	Division into compartments and consequential fields between compartments	

and pressures we have also defined their time-averaged randomised condition. Once again we remind the reader of the limits of our understanding. Throughout this chapter we have used components as our units, and our variables are the local, compartmental, thermodynamic variables within ΔG. This available free energy is a complex mathematical function of the fundamental variables, atomic and charge composition and temperature (pressure is fixed). As stated in Chapter 5, our difficulty is that, although in principle we can reduce the collective thermodynamic variables to the fundamental ones, the analysis is too complicated to give useful operational value. Thus, it is extremely difficult to explain why a given shape of an object is observed. This does not detract from the usefulness of shape as a descriptively valuable way of classifying objects. Here, another word of caution is necessary. Where shape describes an object which has a function we must be sure that we know if the object is in a stationary state, or is transitory (see Chapters 10–16).

Our next task, therefore, is to show how rate processes themselves generate diversity of stationary states, but also give rise to quite new states (steady states of flow) and hence we need to consider a new set of variables dependent on *time*. We must remember that everything changes with time so that our simplification, the omission of time dependence, up to this point in the book, has been a deliberate one to increase our ability to analyse the transformations of systems as we proceed from matter to man. In fact, we can claim a great appreciation if not full understanding of all stationary conditions around us in physical–chemical terms.

Notes on the Nature of Biological Polymers
A biological polymer such as a protein, DNA or RNA is conventionally represented by a structure and an explanation of its properties including activity is then made in terms of local features. This is often extremely useful, however, it has its dangers. These polymers have considerable internal mobility which varies greatly from one to another. Many physical responses of the molecules are then due to cooperative global reorganization from one set of mobile states to another. The processes may well resemble (second order) phase transitions. One such possibility could involve small adjustments of many H-bonds, as well as of side-chains, as the energy of a transition distributes itself cooperatively. Here a protein would behave as a very small phase or system. The energetics of processes based on rigid structures can be misleading particularly when treating for example changes such as electron transfer coupled to proton movements within a protein. The structural changes could be almost insignificant, but the energy changes could be critical. Larger adjustments of conformation are common in other proteins acting as pumps, gates, receptors and so on. (At this level of description a molecule as large as DNA may never visit the same conformatiion twice during a cell's life cycle). See Chapters 10–14.

8

Change and its control

Tao generates the One.
The One generates the Two.
The Two generates the Three.
The Three generates all things.
All things have darkness (chaos) at their bulk
But strive towards the light
And the flowing power gives them harmony

Lao Tzu, *Tao Te Ching* (sixth century BC).
Transl. Richard Wilheim/H. G. Ostwald[†]
(English version, adapted)

[†]The One is unity (oneness); the Two is duality with its partition into yin and yang; the Three, the 'flowing power', is the unifying medium of the two dualist powers.

8.1   Introduction to time and flow

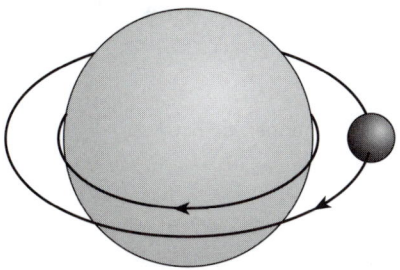

Fig. 8.1 Two methods of counting time: (1) the rotation of Earth and (2) the movement of one body round another, e.g. the Earth round the sun.

In Chapters 1–6 we have described systems in equilibrium (stable) or in stationary states, where changes are not possible or so slow that they can be disregarded. We have also described systems in which there is fixed cyclic motion within stationary states. In Chapter 7 the discussion was extended to small amounts of materials in separated compartments at *local* equilibrium and included the effects of fields. In those discussions *time* was not a variable; yet it is clear that, in fact, everything around us changes with time. In this chapter we enter the new realm associated with one of the great philosophical problems of which the Chinese, rather than the Greeks, seem to have been most aware, that of systems in 'flow'. Such analysis will lead to a discussion of the *evolution* of species, both animate and inanimate, on Earth.

As with many major underpinning parts of our experience, such as the real nature of matter, energy and fields of force, we cannot expect a thorough, deep understanding of time. In effect, one way to approach time is to consider it to be measured by rate of change of position of mass (*motion*) in a set direction, i.e. directed, not random, motion. For example, local time periods we measure by the movement of the Earth round the sun (the yearly period), by the moon's movement round the Earth (the old month) and by the rotation of the Earth on its own axis (the day) (see Fig. 8.1). These are examples of *repeated* cyclic motions, which we have described in Chapter 2 under stationary states. If, however, we consider that the universe started from the 'Big Bang' at time zero, then there has been *continuous outward flow* of matter which is a measure of continuous time to the present expanded state of the universe. [There is, in effect, an equivalent measure of time that considers the expansion (stretching)* of the wavelength of the original photons in the universe, that is, of radiation energy released some one million years after the big bang and correlated with the expansion of the universe, which can be equated with a fall in the temperature to about 3K. In this way the cosmic background noise, i.e. radiation detected in the microwave (sound) region at about 1 mm wavelength (about 1% of which can be heard as a hiss in a common TV set) reflects the age of the universe.] Time is then a fundamental concept associated with vectorial flow of matter and radiation, and was created with them.

A dominant feature then is *flow* due to the initial explosion and its consequences, which we shall assume cannot be reversed. Within the initial overall outward flow turbulence developed, and with it local flow in almost isolated systems, which was and is not outward in a universal sense. This

*The concept of 'stretching' the wavelength comes from the analogy with the so-called Doppler effect of sound waves. Photons with larger wavelength have lower energy and correspond to the emissions from a cooler body. As explained in Chapter 5, we can simply say that the distribution of the same (approximately) number of photons from the initial state to a much larger volume (of the universe) increases their *configurational* entropy but lowers their *thermal* entropy. Thus, generally, the universal temperature of space gets cooler, although *locally* (e.g. in stars) it may still be very high. The whole universe is out of balance, equilibrium.

Table 8.1 The evolution of flow systems

Year ago	Flow system*
15×10^9 ↓ onwards ↓	Big Bang: expansion of universe—creation of 'space' and 'time' Formation of 'mass'/'energy' Turbulent flow Galaxies, stars Basic atoms in stars Heavier atoms in stars (nuclear reactions; see Fig. 8.2)
5×10^9	Solar system Earth Cycling of planets (stationary states) Geochemical reactions
4×10^9	'Living' steady states Evolving developmental states on Earth
200	Man's industry

* Flow involves bulk transport or local rearrangement of particles to form atoms and then larger associations, see text.

turbulence created local patterns or structure in the universe, such as the nebulae, stars and planets. The separated systems have then undergone different rates of internal change (see Table 8.1) and this includes the continuous change of Earth's surface and the evolution of life in one chemical system. The question is which variables control these rates of change, the central problem of *kinetics*, an understanding of which would permit us to go deeper into the analysis of the (natural) selection of the chemical elements from the Big Bang to activities on Earth and even to those within ourselves.

This chapter is concerned with the nature of such kinetics in the first place, and of its consequences subsequently. We shall find that, although the overall flow in the universe is increasingly disordered in a general sense due to turbulence, local fields and flows came about that are quite different and became *organised* into patterns, at first physical, e.g. solar nebulae and rivers on Earth, and then biological, apparently 'purposeful' and 'reproductive', and hence controlled. This evolution of systematic flow, together with selection of chemical elements, reaches its most sophisticated state in higher living organisms, but it is just the general nature of such states that we attempt to uncover in this chapter.

We start this analysis by noticing that there are four kinds of changes to be considered, all of them involving flow: (1) downhill change towards stable states; (2) uphill change, which requires energy to generate stationary states; (3) steady conditions in a system with unchanging flow, where the condition does not move of necessity towards a stable state over a long period of time although it can move to another steady state; and (4) developing systems such as living organisms. It is the general nature of downhill, time-dependent systems that we shall examine first, before we turn to uphill flow and then to the flow patterns that can only be maintained in fixed forms by the constant input (and dissipation) of energy, the dynamic maintenance of organisation being compensated for by the overall increase of entropy (disorder). Developing systems such as living organisms only appeared late in time. In all cases interest will be centred on the systematic

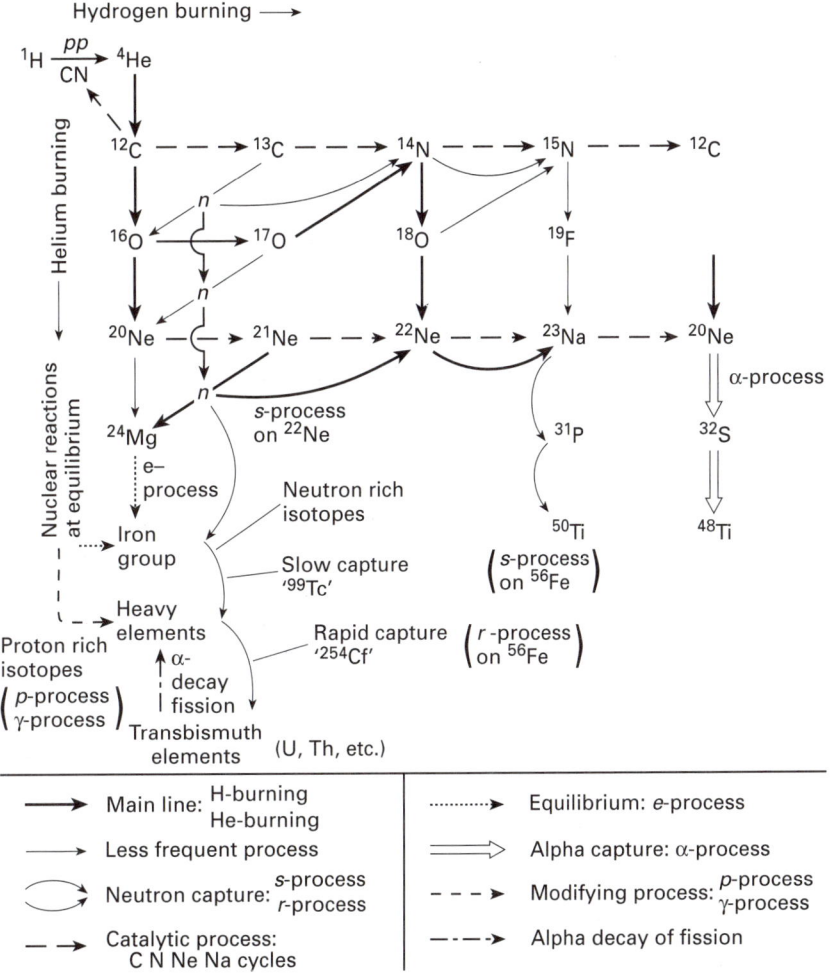

Fig. 8.2 The formation of the elements. [After Burbidge, E. M., Burbidge, G. R., Fowler, W. A. and Hoyle, F. (1957). *Reviews in Modern Physics*, **29**, 547.]

change of space co-ordinates either of bulk movement, i.e. *physical transfer*, or of *chemical* exchanges within 'components', which involves spatial atomic or group rearrangement in *local chemical space*. Immediately, new variables must be introduced (those of flow) associated with *rates of change of position of matter and energy in given directions*. We enquire first as to what prevents systems from spontaneous collapse to their most stable condition, which implies that certain values of variables have time restrictions upon them and do not easily change. While we uncover the new variables we must also find how they became constrained in steady or developmental states much as we found that equilibria restricted degrees of freedom of systems of chemicals reducing the number of components, i.e. compositional variables. If we apply simple kinetic theory to perfect gases then flow is a linear property of composition, space and time. However, for real systems flow can only be analysed in terms of derived variables, much as we have seen for time-independent bulk properties.

8.2   Barriers to change

Before we consider the rate of directed change, flow, in bulk or molecular space we need to examine the rate of change from one randomised condition to another. This is the problem usually examined under the heading of chemical kinetics. Biological *pathways* may include bulk directed changes, flow through physically constrained routes, and/or localised chemical reactions or a succession of chemical reactions in a particular sequence with random diffusion between steps.

If we leave iron objects exposed to air and water they rust, a chemical change expressed by the equation

$$4Fe + 6H_2O + 3O_2 \rightarrow 4Fe(OH)_3$$

iron + water + oxygen → rust

There are here two ways of arranging electrons and positively charged nuclei in local space, one on the left and the other on the right. The more stable situation is that on the right, but iron in the presence of air and water does not spontaneously and rapidly go to the hydroxide on the right. We say there is a *chemical barrier* to the reaction. The 'size' of the barrier is a control on the probability of change at any time and, therefore, of the rate of change with time. The explanation is clear. On the left are molecules of H_2O and O_2, in which the arrangement of electrons of atoms in

(a) Physical change (transfer)

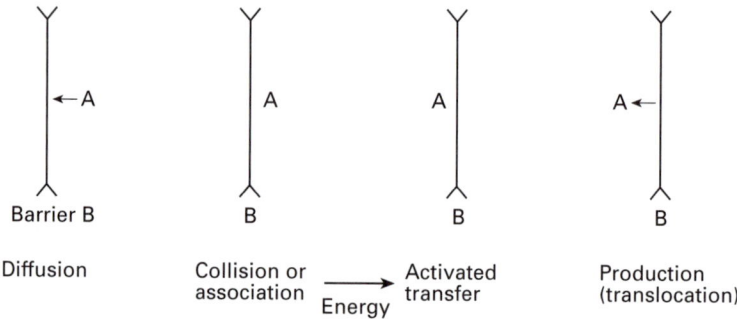

(b) Chemical change

Fig. 8.3 Schematic examples of (a) bulk transfer of a unit, A goes through a barrier B (bulk spatial change); and of (b) a chemical reaction (local spatial change), A_2 and B_2 undergo rearrangement of atoms.

molecules is quite stable in isolation, and likewise there is solid iron, Fe. To re-order the atoms to give $Fe(OH)_3$ in solid rust there is required to be a breaking of one O – H of H – O – H (water), a breaking of O = O in the molecules of oxygen and of Fe–Fe in the iron lattice. The obvious way to do this is to weaken the electrostatic binding in all three so that they have stretched bonds. Energy input to make the change rapid is necessary (Fig. 8.3). That is, before the most stable condition is reached the electrons and nuclei must be put into a much less stable condition than that of Fe (metal) plus O_2 + H_2O (molecules) at 300K. As explained in Chapter 5, the energy distribution in a system of atoms has some atoms (and molecules) with a higher energy than others. With or without energy input it is only the ones with higher energy that can react and change partners. The rate of change is proportional to the number of these 'activated' atoms. Since the number of energised atoms depends on temperature, (Fig. 5.8), the reaction goes faster the higher the temperature. Notice that although energy is needed to maintain reaction there is an overall downhill change in energy from reactants to products, i.e. there is release of heat. Observe too that on the surface of the iron, as rust forms, the atoms move in particular ways to new positions *locally*.

Now, if you want to avoid rusting, the iron can be protected from water and air by enclosing it in a material not permeable to these two chemicals. This is a physical barrier, one way of preventing physical flow. There will then be two barriers to change—a *physical barrier* and a *chemical barrier*—both involving inhibition of movement (Fig. 8.3), but here direction of movement overall is not controlled.

To understand the rate of such changes we need to give a general treatment of the energy required to cross both physical and chemical barriers and to consider the effect of other factors (variables) (Table 8.2) affecting physical and, within components, chemical change. Both changes require physical movement of atoms relative to one another. This is seen most clearly for chemical change in stereochemical changes of chemicals such as *cis–trans* isomerism or inversion of optically active isomers (see Fig. 8.4 and Figs 3.16 and 3.17).

Table 8.2 Variables of flow systems

Variable common to all systems	Additional variables of flow* only
Composition in terms of components, temperature (random kinetic energy), pressure (random momentum), total size, compartments, fields (bulk)	*Directed momentum* or *directed kinetic energy*, which can become an applied force generating *flux** in other bodies
Restrictions Phase equilibria Chemical equilibria	*Energy barriers* (a) Between compartments (b) Within components (c) Organisation (feedback)

*Flow is quantitatively described by the flux (see Section 8.6.5). Generally, flow of a *liquid* within a constrained space is a non-linear property of the fundamental variables, composition, space and time.

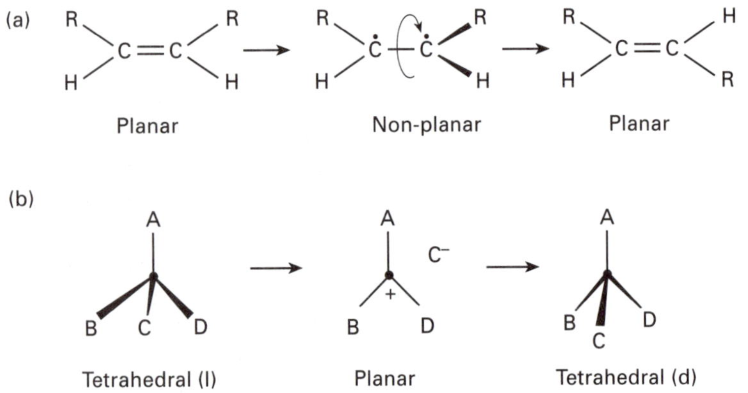

(a)

Planar   Non-planar   Planar

(b)

Tetrahedral (l)   Planar   Tetrahedral (d)

Fig. 8.4 Examples of pathways of internal local change: (a) geometrical, *cis–trans*, and (b) optical isomerisation.

8.3   Variables affecting rates of physical and chemical material change

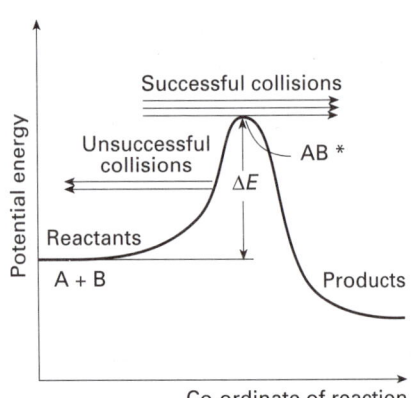

Fig. 8.5 The activation energy barrier. Successful collisions have activation energies, ΔE^*, at least equal to the energy barrier, ΔE.

The limitations, kinetic traps, on changes of chemical elements on Earth can be treated in two theoretical ways. The first, so-called collision theory, states that, generally, in order to change, particles (atoms or molecules) must come together and collide with one another, i.e. reactions are of the kind (see Fig. 8.3),

$$A + B \rightarrow AB \rightarrow \text{products}$$

A and B may both be atoms or molecules, or A can be an atom or molecule while B can be a barrier to diffusion. The collision can result in activation of A alone, for a unimolecular reaction, or of AB, for a bimolecular reaction, or of A with the barrier for physical transfer, depending upon the nature of the reaction. Change can only occur if the collision is of sufficient energy to overcome the barrier.

In Fig. 8.5 we draw a diagram for the progress of a reaction over an energy barrier, ΔE, in terms of the approach of A and B to one another, and then their further change of atomic co-ordinates (Fig. 8.3). As represented, only those collision energies that overcome the energy barrier, that is, that have an *activation energy*, ΔE^*, at least equal to ΔE, can result in products. One aspect which we need to know, therefore, is how to cross the activation energy barrier so that change to products occurs. (Note again that 'products' here can represent the result of a chemical transformation or a physical transfer over a barrier.)

In all such cases, the *chemical* reaction rate in a single phase, gaseous or liquid (solid–solid reactions are extremely rare), depends directly on the partial pressures (or concentrations in solution) of the reactants A and B, written [A] and [B], and on temperature, which controls their rate of diffusion and the energy of collision. We express this is by writing for the rate of reaction as $v = k[A][B]$, for simple bimolecular processes, where k is a temperature-dependent constant, the so-called *rate constant*, which contains the collision frequency, Z, related to the velocity of the molecules and to the

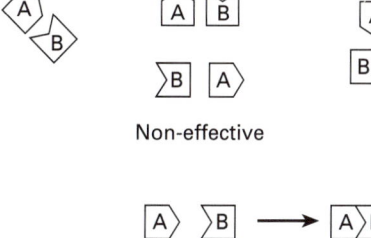

Non-effective

Effective

Fig. 8.6 Non-effective AB associations (above) and effective association leading to reaction (below). Collision has occurred in all cases; see Fig. 8.3.

effective cross-section for collision (see step (1) in Fig. 8.3‡). Generally, the probability of having the required activation energy, ΔE^*, at a given temperature, is given by the exponential $e^{-\Delta E^*/RT}$ which is a typical Boltzmann factor (see Section 5.3). Thus, k varies with $e^{-\Delta E^*/RT}$ and with Z.

There are two further factors affecting the rate constant, namely the relative probability of occurrence of the required state of association AB, in which A and B are correctly oriented for reaction, i.e. there may be only one *effective* particular orientation or collisional association of A and B when they come in contact (Fig. 8.6), and there may be only one particular directed movement of the atoms using the energy ΔE^* that allows the rearrangement of bonds in A and B to lead to the products of the reaction. Both can be taken into account by introducing a combined *steric* probability factor, P, a term related to the effective collisions, which is independent of temperature and pressure. We may therefore write for the rate constant the expression

$$k = ZPe^{-\Delta E^*RT} \tag{1}$$

where P includes all terms associated with atomic positions independent of temperature. It is readily seen that this treatment is based on components as units and apparently on molecule–molecule collisions. It is related to the classical theory of gases in that it does not deal with statistical distributions in this simple form. (It is clear that 'physical' movement through a barrier can be treated in a similar way and, as yet direction is not involved.

A second treatment of reaction rates is the 'transition state theory' (or 'activated complex theory') of chemical kinetics, based on concepts of statistical thermodynamics. In this theory we may characterise the probability of the equilibrium concentration of a required activated complex, AB* (Fig. 8.5) relative to the probability of absence of association, by the equilibrium constant of the activated complex, K^*. The rate of reaction can be related to this constant by considering the equation:

$$A + B \rightleftarrows AB^* \xrightarrow{k} \text{products}$$

where the rate is given by $v = -d[A]/dt = k[A][B]$, which must equal the product $[AB]^* \times$ frequency, Z', of passage of this activated complex over the activation barrier, i.e. the frequency with which the complex breaks apart into products. Thus

$$k = Z'[AB^*]/[A][B]$$

Since $[AB^*]/[A][B] = K^*$, where K^* is the equilibrium constant of the activated complex (see Section 6.6) and $-RT\ln K^* = \Delta G^{\circ*} = \Delta H^{\circ*} - T\Delta S^{\circ*}$, then

$$k = Z'e^{-\Delta H^{\circ*}/RT} e^{-\Delta S^{\circ*}/R} \tag{2}$$

---

‡Generally, the expressions of the rate of reactions are more complex and must be determined experimentally. For example, for the reaction $H_2 + I_2 \rightarrow 2HI$, the expression $v = k[I_2][H_2]$ is observed, but for the apparently similar bimolecular reactions $H_2 + Cl_2 \rightarrow 2HCl$ or $H_2 + Br_2 \rightarrow 2HBr$, the rate expressions are quite different, meaning that the pathways (mechanisms) for these three reactions differ considerably.

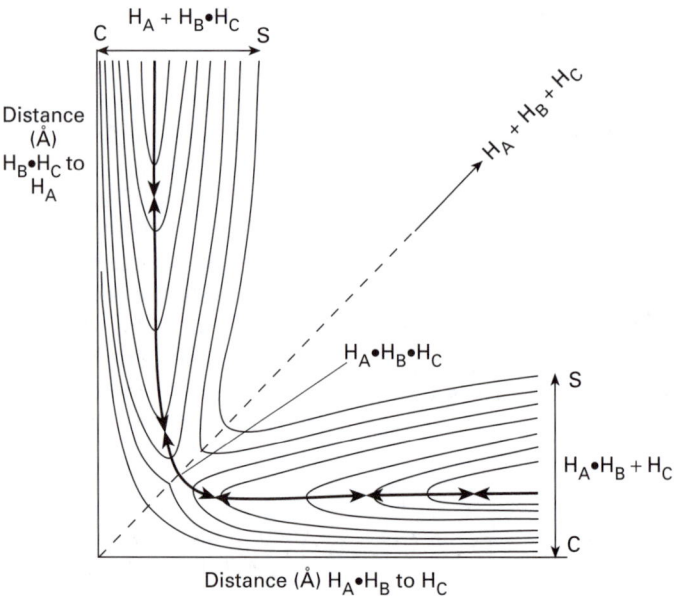

Fig. 8.7 The potential energy surface for the exchange reaction between H atoms and H_2 molecules. Continuous pathway (arrows) shows energy increases to three associated H atoms, by compression of $H + H_2$. The reaction path goes over the saddle between contours of a landscape in energy. (S) and (C) represent directions of splitting or compression, respectively, of H_2.

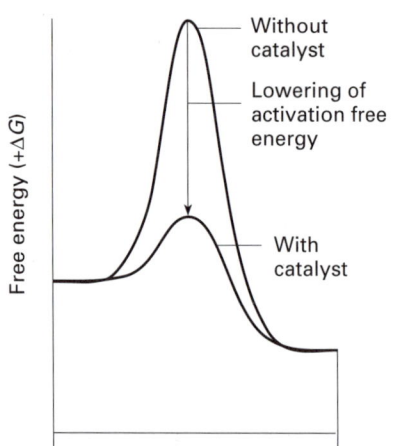

Fig. 8.8 The effect of a catalyst or carrier on the activation free energy, ΔG^*, for chemical reaction or transfer.

(The frequency term in the activated state may be looked upon as a vibrational frequency which turns into a directed local translational motion along the direction of one of the bonds holding the complex together, causing it to split into products that fly apart (Fig. 8.7‡.) (This applies to a chemical change or a simple directed diffusional rate over a barrier.) Of course, for many reactions, and particularly for reactions in solution, when pressure and volume are constant, $\Delta H^* = \Delta E^*$, so that this result has the same form as that obtained using the collision theory, where the steric factor P is now seen as an entropy factor. In both theories the crossing of physical barriers may also be described in a similar fashion. Note, however, that the variables in transition state theory are those of bulk systems (thermodynamic variables) not of individual units.

We see that both theoretical approaches will apply (with adequate adaptations) to either physical transfer of material or chemical change, since both involve motion along a reaction co-ordinate over an energy barrier (Fig. 8.7). Thus, our problems are to understand the requirement for energy, for orientation in an appropriate structure for reaction and for restriction to particular movement in space. Clearly, control over change is exerted by two factors: (1) by the physical barriers to motion in bulk space

---

‡This approach using reaction co-ordinate diagrams allows us to describe chemical transformation as a *directed* motion over a local barrier in space, such as that in Fig. 8.7, when the reaction rate is seen as a flow of particles along a path in a valley, with a given velocity reduced by the energy barrier.

to be overcome by energy, generating diffusion in a direction; and (2) by chemical bond barriers in a complex to be overcome by energy to give directed *local* motion of atoms or groups. Catalysts (enzymes in biology), which we shall discuss later, are substances, metals or compounds, that help to overcome one or all of the barriers to chemical change in local bond space (Fig. 8.8), while pumps (transfer catalysts), which transport units without chemical change, create modes of movement, if necessary with applied energy, across physical barriers in bulk space, e.g. membranes. There is then a very large variety of processes that we shall have to consider in order to cover the control factors within kinetics (Table 8.2). We can see, however, that some of the new variables of systems with which we are concerned are (directed) *vectorial motions* and *energy input*, both included under flux. Energy can be put in by modes other than by bulk temperature and pressure change, since directed motion can be used. We shall take the physical barriers first.

8.4 Rate control by physical barriers

A simple example of a physical barrier is the separation of two liquids by a membrane, when at one extreme no mixing can occur (Fig. 7.16). On the other hand, the membrane can be semi-permeable to some or all of the chemicals present. In this case we are interested in the controlled *rate of transfer*, the flow, of chemicals through the membrane. Such flow occurs downhill towards an equilibrium state unless energy is applied so as to pump units in the opposite direction. We consider energy input in Section 8.7.

8.4.1 Selective pathways through physical barriers

If mechanical *flow* is to be controlled then the barriers to the mobile chemicals, gases or liquids, must have controlled entrances and/or exits, perhaps

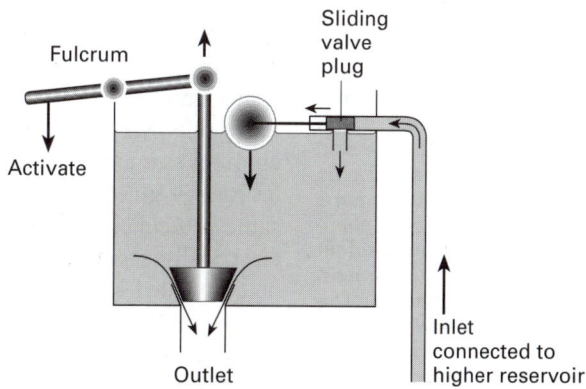

Fig. 8.9 A ball valve-operated pump used in many storage systems for water, e.g. in lavatory or sewage plants. The inflow acts to feedback and close the valve.

aided by pumps through the barriers. Man-made macro-devices are obvious in the sluices and pumps used to manage water levels for canals or water supplies to houses and factories. The devices inserted include taps, also prominent in chemical equipment, which are controlled by the external agency—the human operator. A sophistication in such supply lines is a feedback valve such as is used in lavatory cisterns, when the filling of a tank is shut off by its own rising water level using a stop-cock ball valve (see Fig. 8.9). There is no selection of the liquid which can flow in such a system. The obvious question is how can this vectorial material flow be managed on a molecular scale and made selective.

The first concern is to examine the properties of the material, the flow of which is to be controlled, since this may well decide the type of control barrier to be used. Commonly, man-made equipment allows restricted flow of bulk fluids using metal valves as described above, but equipment for controlled flow of such different entities as ions and molecules uses, for example, mineral molecular sieves as barriers (see Fig. 9.8), in addition to valve control of bulk flow. However, they are not very selective, though selectivity is increased in man-made microporous solid silicates. In living organisms, gross pumping, as in the arteries and veins, is carried out by muscles such as the heart in which flow is controlled by large valves, while flow of individual ions and molecules in water solution is tightly controlled by channels with gates in membranes made of organic lipid and protein molecules (see Fig. 8.10). The scale of the control machinery can be down to micrometres in much of man's equipment, but must be down to nanometres in microporous material and in living organisms. Let us consider ion or molecule flow through membranes first, and then electron flow and the flow of photons (light) in membranes later.

8.4.2 Ion or molecule selective pathways through barriers

Whether ion flow is generated uphill by a pump or is downhill through a channel in a membrane, the 'hole' through which the ion passes has to be of similar size to the ion that flows if it is to be selective. Moreover, we shall see that an *essential* feature of organisms is that their pumps and channels must discriminate between sizes of ions differing by much less than 0.5 Å ($1 \text{ Å} = 10^{-8}$ cm) in diameter (see Table 3.4). In outline we know how this can be achieved. As mentioned in Section 7.6, proteins can be made into cylindrical rods, and packing such cylinders together in a regular way forms small channels (compare the packing of spheres) leaving continuous holes (Fig. 8.10). When the angle of packing of cylinders is changed the channel size can be adjusted, so that not only are such channels selective for ions of given sizes but they can also be gated in open or shut states. The same principles apply to the flow of molecules. If the channel structure is electrically asymmetric across the membrane the open or shut states can be controlled by imposing favourable or unfavourable electrical potentials across the length of the channel. Again, added chemicals which bind to one end of the channel may open or close it and then allow their flow, selectively. Details of molecular design are not appropriate in this book, but a little imagination will show that the shaping of the surfaces of the protein cylinders may allow flow of different molecular shapes whether

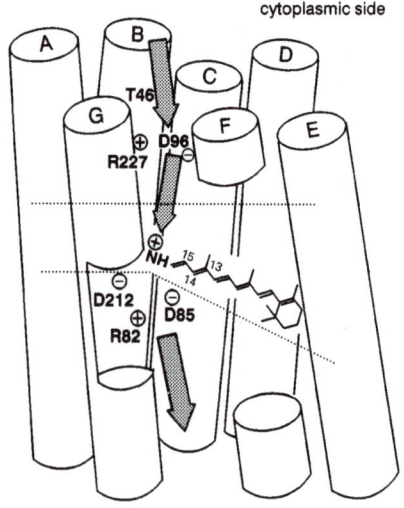

cytoplasmic side

Fig. 8.10 The proton channel in the bacteriorhodopsin protein where D = aspartate, R = arginine and T = threonine, which are amino acids in the protein chain capable of proton transfer. A similar channel in a membrane can be built for other ions leaving a larger pore. There is a suggestion that some channels are built from multiple protein domains rather than from the protein helices shown here as cylinders. The rhodopsin molecule is shown. Energy (light) pumps the protons across the membrane. The organic molecule blocks flow when de-energised and causes flow when energised. [See *Nature* 1997, **389**, 206.]

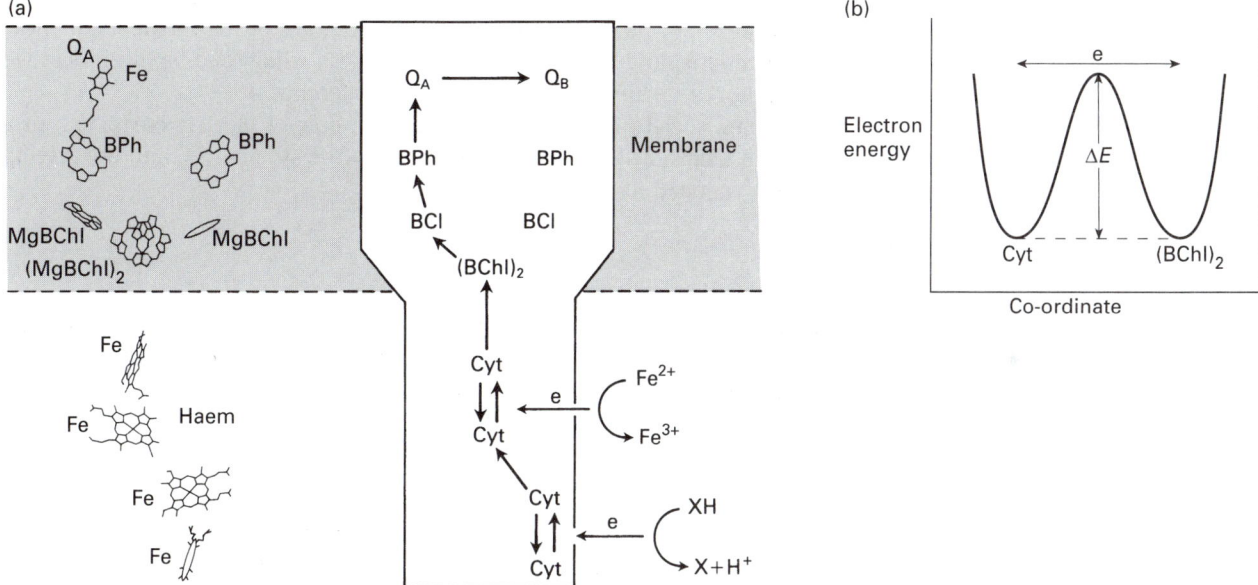

Fig. 8.11 (a) The electron transfer scheme of the bacterial reaction centre equivalent to photosystem I. Cyt, cytochrome; BChl, chlorophyll; Q, menaquinone; BPh, phaeophytin. (b) The barrier between electron transfer centres, ΔE, is penetrated since the electron can tunnel.

they are ions or not (Fig. 8.10) and the flow can be controlled both in rate and in chosen direction between compartments. Deliberate application of energy to such systems will make pumps.

Rates of flow of ions (molecules) from water into a channel depend in part on how quickly the water bound to an ion or molecule is lost so as to allow the entity to enter the channel (see Table 8.5 later). The rate of flow through a channel itself depends not only on the size of the channel but also on the strength of binding of such species by the channel walls, and therefore the channel acts as a filter. Notice how certain ions, e.g. Na^+, K^+ Cl^-, readily lose water while others do not, e.g. Al^{3+}, Ni^{2+}. The former are used as ion current carriers in organisms; the latter are not since they bind too strongly. Intermediate cases are H^+ and Ca^{2+}.

8.4.3 Electron pathways: conduction in a matrix

Again we refer to human equipment first. A wire is a continuous solid metal cylinder. As described in Chapter 4, a metal contains a continuous cloud of electrons. If an electrical field is applied along the wire, i.e. a charge distribution difference (difference of potential) from one end to the other, then electrons flow in the wire to neutralise the field. Another device requires electrons to jump from one metal atom to the next in a wire made from a metal compound such as magnetite, Fe_3O_4. In biology very short 'wires' are made somewhat differently by placing metal atoms at short distances from one another in protein matrices, when electrons can also pass from one metal atom or organic molecule to another (Fig. 8.11 and see Fig. 10.12). It is quite common for these electron-hop centres to be placed in a series that crosses a physical barrier such as a membrane. Notice that the direction of flow is decided by the redox potentials of the members of these constructions, and the barriers are resistances which can be given detailed description in terms of the material* (see reference 48).

We can refer to the variables of directed flow of electric current, i, in a metal in terms of the number of carriers in the system, N_0, and the driving force, V, applied to the system.

$$i = \frac{VN_0}{R}$$

(3)

where R, the resistance, accounts for scattering of the flowing electrons.[†] In a semiconductor there is a temperature-dependent jump or hop from site to site involving, again, a Boltzman factor:

$$i = \frac{VN_0}{R} e^{-\Delta E^*/RT}$$

(4)

The relationship of this equation to eqn (1) in Section 8.3, is clear.

All electronic circuits can be controlled by condensers and switches as well as by driving force and resistances. They can also be managed by feedback control (see Section 8.4.5). In a hop conductor the distance apart of centres is very important and conduction in protein matrices is also limited by separation of centres.

8.4.4 Pathways for photons (lights) in a matrix

In man's modern equipment there are bulk designed materials manufactured into narrow transparent tubes which act as light pipes. In biology the light is 'piped' by a series of chemical molecules; chlorophyll, the green pigment of leaves, is an example (see Fig. 10.12). In photosynthesis membranes the light is caused to 'jump' from one molecule to the next until it reaches a place, a reaction centre (see Fig. 10.13) where it causes a chemical change, i.e. a change of chemical bonds. We see again the need to have special molecules for special functions, here to guide light on a path, and this will be a common theme in biological devices. Again, direction can be managed by the energetics of the molecules in the device. The process does not have an energy barrier ΔE but only a probability barrier which increases the more molecules are separated.

8.4.5 Feedback control of physical pathways

In the above discussion of flow the variables have been the different energy gradients, positive or negative, and the number of particles involved, while

---

[†]Electron transfer in many materials is not a simple, activated hop process, since the electron can tunnel through barriers of the kind shown in Fig. 8.11. This quantum-mechanical process is described in the Further reading (Marcus theory). It is important in matrices made from proteins in organised arrays in living organisms, as well as in conventional semiconductors.

[†]The usual form of this expression, known as Ohm's law, is $i = V/R$, where the units of i, V and R are, respectively, amperes, volts and ohms. In this case N_0 represents one mole of electrons, equivalent to 96 500 coulomb.

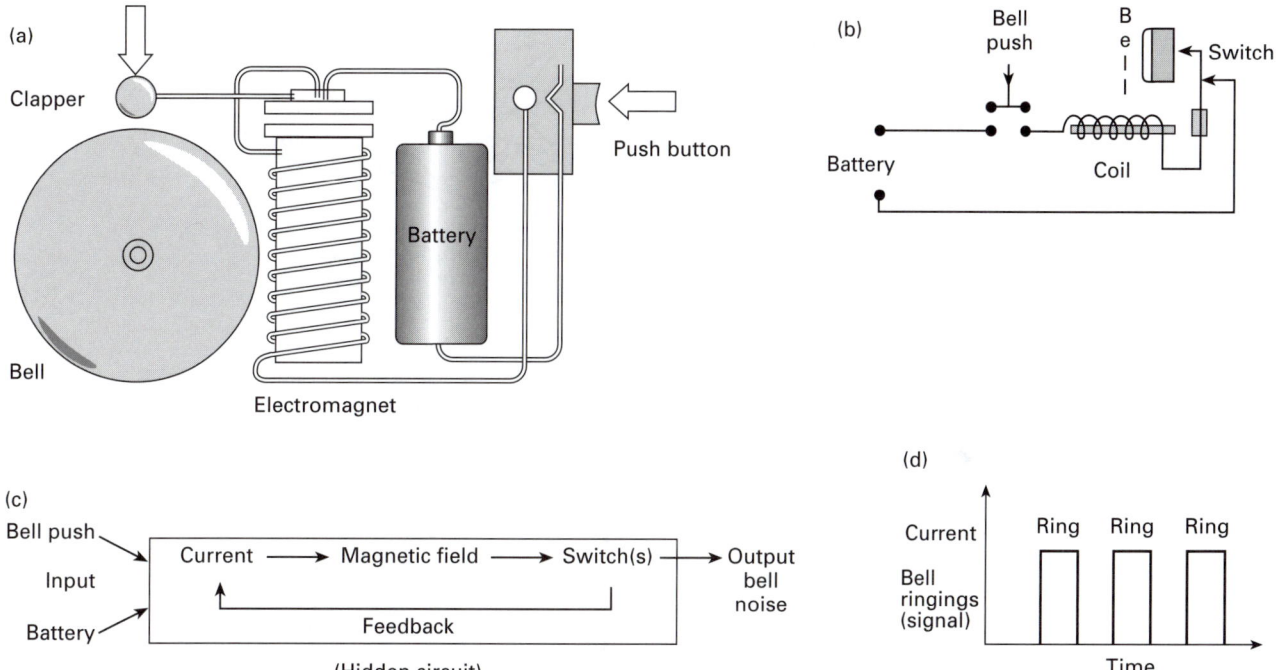

Fig. 8.12 (a) A door bell system. (b) A line wiring diagram for (a). (c) A feedback diagram of the non-observable parts of the system. (d) The output as a square-wave pattern.

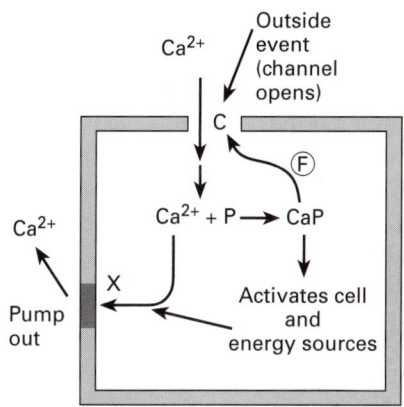

Fig. 8.13 An outside event causes the channel C to open allowing Ca^{2+} ions to enter. Inside the cell they bind to a protein P which, as CaP, activates the cell but at the same time it causes the closing of the channel by feedback F and removal of Ca^{2+} at X, a pump.

we have introduced various impediments to motion. The question arises as to how they can be used to create steady, regulated flow. We have given the example of a stop-cock in a water supply which closes off the supply when the water in the reservoir reaches a given level (see Section 8.4.1). Now, the same effect can be produced on the flow of electric charge. A simple device is that of a door bell (Fig. 8.12). The pressing of the bell button allows electron current and causes the bell to ring, but simultaneously the movement of the bell hammer breaks the circuit and stops the flow, only for the return of the hammer to reactivate it. Thus, an intermittent or square wave-like signal is produced (Fig. 8.12). It is also possible to use the intermittent flow of chemicals through a channel to act on that channel so that at a given concentration they block flow in the channel. The resultant ion current is also in the form of waves, e.g. calcium waves. It requires, of course, a special chemical binding to a specific surface to give such a mechanical switch action (Fig. 8.13). Here, again, where there are barriers to flow, then the application of energy, e.g. increase of temperature, allows atoms, ions or molecules to cross the barrier more readily. The specificity of action and the placing in space of activity for different elements and chemicals are seen now to be critical factors in physical flow.

The full description of physical barriers will need us to consider the positions of pathways, channels, barriers to pathways, gates and ways of opening and closing gates (reducing barriers), which in turn will need controlled (energised) operation. Inspection of any *organised* flow in man's operations shows that exactly these principles are applied, for example on a

(a)

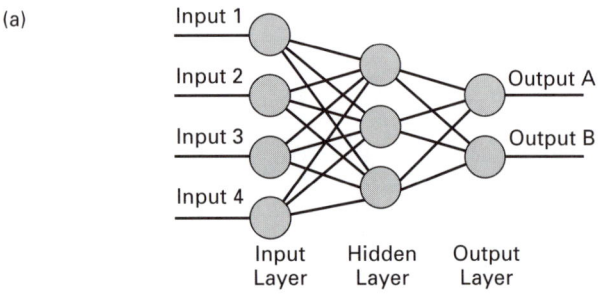

(b)

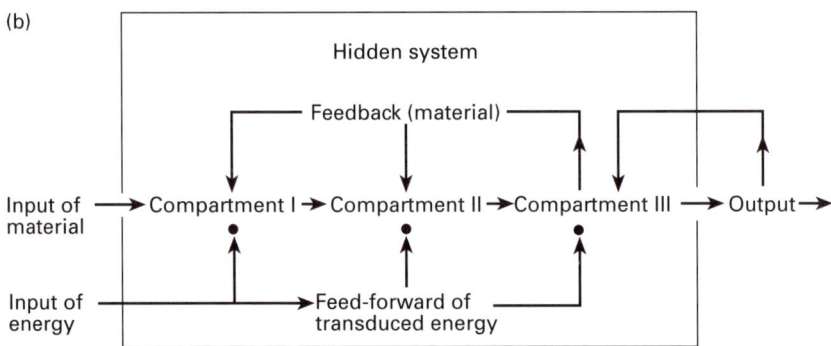

Fig. 8.14 (a) A representation of a small part of a computer circuit. Note the network of connections. (b) A parallel representation of a cell indicating a network construction.

macroscale in a modern chemical factory or on a microscale in a computer network (Fig. 8.14). In effect, a multiplicity of feedback and feedforward controls link flow in a 'hidden' circuit from the input to the output (Fig. 8.14). This is the essence of a computer system, where patterns of electron flow act in a unified way so that a particular input of information can only yield, through the application of energy and the peculiarities of the hidden circuitry, a particular output. To make an operationally effective system we need to appreciate not only the intrinsic variables of flow, amounts and energy input, but also the nature of constraints so as to generate a collectively successful activity of the units in the system.

We turn next to barriers to uphill movement and then to chemical change.

8.4.6 Uphill physical transfer

Uphill movement requires energy and, in effect, a machinery to prevent the immediate return of the material that is moved. An example is a pump for moving air through a non-return valve, as used when pumping up a bicycle tyre, or a water pump that lifts water against gravity to spill it into a channel at a level somewhat lower than the highest point of the lift (see Fig. 8.9). Huge irrigation schemes use such devices. We describe more complicated machinery such as a steam engine later (see Section 8.13), since it is not a simple physical pump. A molecular pump in a membrane also has

Table 8.3 Chemical materials for physical control of the direction of flow

Material		Flow
Man-made		
Metals	⎫	Electrons, ions
Plastics	Wires	Fluids (water
Clays and glass	and pipes	and other liquids)
Concrete	⎭	Gases
Biological		
Metal atoms in proteins		Electrons
Lipid/proteins (membranes,		Fluids (water), ions,
veins, air tubes)		molecules (air)

to operate through a non-return valve, and does so, we believe at present, by the motion of α helices of proteins (see Fig. 8.10 and Section 8.4.2). Notice that direction is secured in part by the construction and in part by the sidedness of the application of energy. The interest is now in the possible sources of energy for work of this kind, see Section 2.5.4. In many systems the energy comes from the change of chemicals, e.g. ATP, from a stationary to a stable or different lower energy stationary state, e.g. ADP + P_i. The application of energy is guided in space by the positioning of an initiating mechanism (e.g. a spark) or a catalytic site (enzyme) for converting a stationary chemical to a stable (or lower energy stationary) one. Obviously we need to describe rates of chemical change, to which we turn next.

Before doing so we remind the reader to observe how the selected constructions of the chemical components are being used in these flow systems, together with applied energy to constrain the variables of flow rate and direction. The selection involves materials such as those given in Table 8.3, which are used to 'pipe' chosen flowing chemicals, or, as we see below, to act as containers for chemical reaction purposes.

8.4.7 Conclusion to physical flow in a chemical system

At this stage of the discussion we can see that the *flow rate* of material can be continuously varied as we apply energy or remove barriers. At the same time flow can be constrained to any particular *direction* we choose, but these variables are not as yet set to any particular values since we have given no reason for any selection. Note that the situation differs from that in Chapters 5–7, where selection arose for speciation of static materials by the changes of free energy, ΔG, with variables that were not of a simple gradient. We take up this point again in Section 8.9.2. The most important consideration is that a system of flows can be organised (by feedback) so that selected activity is developed in particular constructions, machines.

8.4.8 Nuclear barriers

A case of barriers to flow different from all the above is the disintegration of nuclei, i.e. radioactivity. Here, nuclei give out electrons (β-particle), helium nuclei (α-particle) or they split into two pieces. These processes are

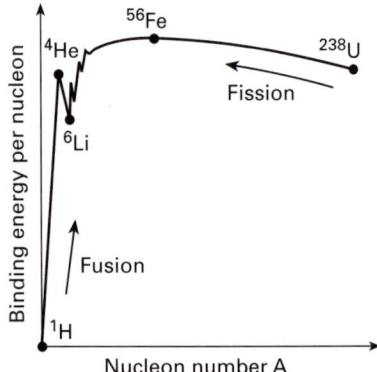

Fig. 8.15 The binding energy per nucleon of atomic nuclei, a landscape diagram reflecting the ultimate likelihood of the abundance of the elements (iron would dominate).

not thermally activated; they are temperature independent and are due to 'tunnelling' of particles out of nuclei. Thus the radioactive disintegration rate is typical of a given nucleus and has been used to characterise elements and time.

Nuclear synthesis (fusion) does depend upon collisions, but is opposed by an entropy loss. Fusion requires a high temperature but not an excessively high temperature. The degree to which different nuclei formed in giant stars therefore depends on cooling to a sufficiently low temperature of the initial proton/neutron gas of the early universe, and then on the stability of the nuclei formed to further collisions (synthesis or fusion) or to disintegration (fission). The intrinsic stability of nuclei is shown in Fig. 8.15. The diagram shows that at low temperature Fe should be the dominant nucleus in the universe, but this has been prevented by cooling and expansion, which have greatly limited approach to equilibrium. The pathway of synthesis of the elements is shown in Fig. 8.2. The elements Li, Be and B have intrinsically unstable nuclei and are rare in the universe, as are the unstable very heavy elements around uranium. The abundance of the elements, a decisive factor in the evolution of everything chemical, animate or inanimate, is given in Fig. 1.10, a frozen landscape diagram.

8.5 Chemical bond barriers: equilibrium, stationary states and flow

In Chapters 2–7 we have considered two possible situations in the discussion of chemical compounds. In the simplest case, a stable system, the compounds arose through the optimal interaction in local atomic space of every atom with every other atom at some temperature and pressure. This is a full equilibrium (balance), including all reactions, and no change is then possible (unless the physical conditions of the system change, e.g. temperature). The second case was the formation of compounds (also called components) by atom combinations entering permanent chemically energised traps. Here the optimal energy condition of the particular chemical composition was not reached, but a stationary state was obtained due to the inability to escape from non-optimal partnerships.

In this section we wish to consider how these energised, non-optimalised chemical arrangements can be relieved at controlled rates and in particular ways, and, conversely, how optimalised stable chemical systems can be rearranged using energy input so as to generate non-optimal (energised) chemical compounds in stationary state traps. Furthermore, we wish to consider how a steady state of flow of material and energy can produce a system that is in a physical and chemical *steady state*, e.g. an adult living organism, and finally we shall look briefly at developing systems as seen in the growth of living organisms. (Note, in passing, that the local atomic space occupied by atoms is limited by quantum restrictions not applicable to bulk space operations. *Chemical (biochemical) pathways are not just bulk space movements but are successions of controlled steps of the relative positions of atoms flowing through selected series of molecules i.e. local bond changes*).

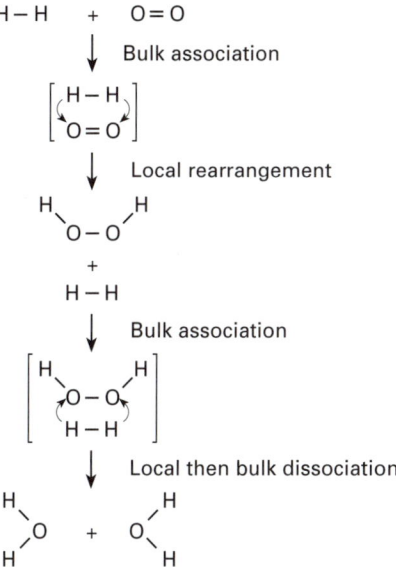

Fig. 8.16 Example of a vectorial, *local* spatial rearrangement of atoms to a new condition and *bulk* movement. Here the hydrogen atoms of the H_2, and the oxygen atoms of O_2, move as shown to give two water molecules eventually (see Fig. 8.3).

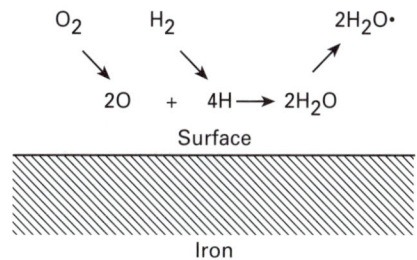

8.5.1 Selection of chemical pathways by reactivity controls: catalysts

A chemical reaction involves the vectorial, *local* spatial rearrangement of atoms to a new condition in which no further rearrangement takes place. At first we wish to consider control over a simple change from molecules in a stationary trapped state, such as a mixture of two gases, to a more stable product. As an example we can consider the reaction of hydrogen and oxygen to give water (Fig. 8.16)

$$2H_2 + O_2 \rightarrow 2H_2O$$

The reaction is not noticeable on mixing the gases alone at temperatures below 500 °C, but occurs in the presence of certain additional substance, usually a metallic surface, which is called a *catalyst* (see Section 8.3). A catalyst lowers the energy required to go over the barrier (Fig. 8.8) but although it may help to break chemical bonds, here in H_2 and O_2, it is unchanged at the end of the reaction. We can imagine a process as follows:

$$H_2 + \text{metal Fe} \rightarrow \text{H atoms on the surface of Fe}$$
$$O_2 + \text{metal Fe} \rightarrow \text{O atoms on the surface of Fe}$$
$$\text{Surface (2H atoms + O atoms)} \rightarrow H_2O$$

The H_2O molecules leave the Fe surface. The Fe surface is then clean again once the water molecules leave, (see Aside). The reaction has been catalysed.[*]

An obvious starting point for the discussion of the selectivity of *rates* of chemical change is therefore the size of the barriers to reaction. Now, in Chapter 4 we saw that some elements, notably the lighter non-metals, when bound to one another, hold electrons in localised (somewhat inaccessible), covalent bonds. It is not surprising, then, that even when a particular combination of two such elements is not the most stable, e.g. $H_2 + O_2$, the combination persists. To give another example, methane gas (CH_4), a combination of carbon and hydrogen, does not react with the

Table 8.4 Some persistent but unstable compounds in air (O_2)

Compound	Stable product
CH_4 gas (methane)	$CO_2 + H_2O$
C_8H_{18} liquid (petrol)	$CO_2 + H_2O$
Candle grease ($C_{20}H_{42}$)	$CO_2 + H_2O$
Solid sulphur (S_8)	SO_3 (H_2SO_4)
Alcohol (C_2H_5OH)	$CO_2 + H_2O$
Proteins ($C_nH_mO_pN_q$)	$CO_2 + H_2O + NO_2$
Sugar ($C_6H_{12}O_6$)	$CO_2 + H_2O$
Silicon (Si solid)	SiO_2 (silica)
Iron (Fe metal)	Fe_2O_3 ($Fe(OH)_3$ (rust))

---

[*]Alternatively, the reaction mixture can be sparked to produce radicals which continue to produce more radicals until reaction is complete (see internal combustion engines).

oxygen of the air (O_2) although the reaction is favourable to formation of CO_2 and H_2O

$$CH_4 + 2O_2 \rightarrow CO_2 + 2H_2O + \text{heat}$$

In fact, gas cookers and gas-fired generators of electricity work on the basis that the reaction needs an initiator since C–H and O–O bonds are persistent. The same is true in C_8H_{18} (octane or petrol) where C–C as well as C–H bonds persist in air, which contains oxygen. Petrol and gas need an initial spark (a high temperature kick) to get the reaction going, after which the reaction is kept going by chains of radicals, or by self-induced heat release, or by repeated sparking which raises the temperature locally of the incoming gases. Table 8.4 lists some inert (unreactive) or persistent (in stationary states) compounds in air and water. The basis of biological structures lies in these covalently linked, light, non-metal compounds, but in this case large molecules are needed (compare plastics). Both the large and small non-metal compounds allow control over the reaction—the basis of organic chemistry—especially through the resistance to change of covalent bonds. Living systems, as well as man-made apparatus, need to convert these covalent substances both for synthesis, of proteins for example, and to obtain energy. As they cannot be sparked, they need catalysts in solutions. The choice of solvent is now a particularly important consideration, before we turn to catalysis.

8.5.2 Solvent constraints on reactions: reactivity in water

To advance the discussion towards the topic of selection of elements in organic chemistry, and especially in living systems, the next consideration must be the examination of chemical species in solvents, especially water, and then the description of catalysts for their transformation in this solvent. It is obviously possible to consider mutual exposure of any mixture of substances in chosen solvents and to ask what is the final stable chemical composition and which are the barriers that limit the rate of change to that composition. Clearly, limitations include solubility and solvation of reactants (see Chapter 6). We concentrate attention on solutions in water.

In Chapter 6 we discussed the solubilities of various atomic and molecular species in water, noting that they were either charged or polar. One example of a substance made of charged species was sodium chloride, composed of sodium ions (Na^+) and chloride ions (Cl^-). Given suitable partners, a great variety of simple ions can be obtained in water solution. As explained in Chapter 6, these ions are surrounded by water molecules which usually exchange rapidly, but, as mentioned above, in some cases they are not released easily. Thus, many hydrated ions cannot react immediately on collision just as they could not enter channels quickly (see Aside). The orders of speed of reaction are approximately

$$M^+ > M^{2+} > M^{3+}$$
$$\text{e.g.} \quad Na^+ > Mg^{2+} > Al^{3+}$$

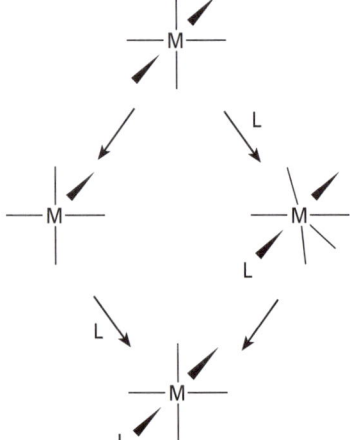

Two different ways of reaction of metal ions with ligands.

Table 8.5 Grouping of ions according to water exchange

Fast (~10^9 s^{-1})	Medium (~10^5 s^{-1})	Slow (~1 s^{-1})
Na$^+$, K$^+$, Ca^{2+}	Mg^{2+}, Ni^{2+}	Be^{2+}
Cl$^-$, Br$^-$, I$^-$	Ga^{3+}	Al^{3+}
NO$_3^-$, SO$_4^{2-}$ HPO$_4^{2-}$		Fe^{3+}, Sc^{3+}
Zn^{2+}, Cu^{2+}, (La^{3+})		M^{4+}
Organic molecules		

Ion size (Å)

Na$^+$	Mg^{2+}	Al^{3+}
0.98	0.65	0.45

K$^+$	Ca^{2+}	Sc^{3+}
1.33	0.94	0.81

Rb$^+$	Sr^{2+}	Y^{3+}
1.48	1.13	0.93

and the larger the ion of a fixed charge (see aside) the faster the rate at which water leaves, e.g.

$$Ca^{2+} > Mg^{2+} \text{ and } La^{3+} > Al^{3+}$$

If, therefore, functional value is to be found in rate of water loss, corresponding to the rate of binding to an anion or other molecule in solution (Table 8.5), M$^+$ ions (Na$^+$ and K$^+$) and larger M^{2+} (Ca^{2+}) ions will be particularly useful, but ions such as M^{3+}(Al^{3+}) will not.

Inorganic anions are all large and all exchange water from around themselves quickly. Likewise, all polar organic molecules such as sugars and simple alcohols, acids and bases (whether they carry positive or negative charge or not) exchange water rapidly. These are the major substrates of living cells. The value of the fast removal of water is fast binding, which will be seen in Section 12.7, to give rise to the development of messengers in water that must diffuse rapidly and bind quite well and quickly. In what follows we stress organic and inorganic reactions in water and their catalyses since they dominate living organisms. In this way we show how rate and direction, the variables of chemical change, are managed. Very similar principles apply to reactions in all other solvents.

8.5.3 Barriers to hydrolytic reactions

A different type of reaction of water exchange involves the attack of water (here in water) on an organic molecule; this is called *hydrolysis*, and its reversal is called *condensation*. Here, there occurs the rearrangement of covalently bonded atoms in space, which is normally slow

$$X–Y \quad + H_2O \rightarrow X–OH \quad + HY \text{ (hydrolysis)} \tag{1}$$

$$X–OH \quad + HY \quad \rightarrow XY \quad + H_2O \text{ (condensation)} \tag{2}$$

Many organic molecules undergo such transformations, but while reaction (1) is going to a more stable condition, reaction (2) goes to a more unstable one and requires a source of energy. Nevertheless, reaction (2) is the main route to the formation of biological polymers such as DNA, polysaccharides and proteins. Before we examine the coupling of energy to such reactions as (2), we shall consider the rate of hydrolysis, (1), in some detail in order to show the type of catalyst needed. Note that if reaction (2) is to be

Ester hydrolysis

$$\overset{\delta-}{C_2H_5O} \cdot \overset{\delta+}{COCH_3}$$

$$\uparrow \quad \downarrow$$

$$H \cdot OH$$

$$\overset{\delta+}{} \quad \overset{\delta-}{}$$

$$\downarrow$$

$$C_2H_5OH + HO \cdot COCH_3$$

allowed and reaction (1) prevented there must also be control over these two pathways.

As stated above, hydrolysis is the attack by one somewhat inert molecule, H_2O, on another, say X–Y (a real example is ethyl acetate, $C_2H_5O \cdot CO \cdot CH_3$). The two are in a stationary condition in the presence of one another. To *catalyse* the reaction we observe that X–Y contains a polar bond which can be made more reactive by increasing its polarity. Thus, selective introduction of a metal cation (M^+) and a non-metal anion (Z^-) can help the reaction by associating in the following way

$$X^{\delta-} - Y^{\delta+} \quad \rightarrow \quad X^{\delta\delta=} \ \text{----} \ Y^{\delta\delta+}$$
$$\vdots \qquad\qquad \vdots$$
$$M^+ \quad Z^- \quad \rightarrow \quad M^+ \qquad Z^-$$

Now, dipolar water can attack as follows

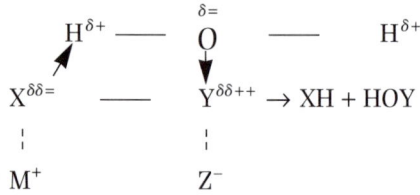

Obviously, to be functional M^+ and Z^- must be available in water, but must lose hydration water easily (quickly) and bind (but also exchange rapidly from bound conditions) with X and Y. As we saw in Chapter 6, free M^{2+} ions are available, as are anions such as Z^-, but M^{3+} ions and Z^{2-} anions usually are not. Therefore, favourite catalysts in water are ions such as Zn^{2+} (or Mg^{2+}) and OH^- or SH^-. Just these groups are widely available in biological protein catalysts, i.e. enzymes. Why do we mention Zn^{2+} (Mg^{2+}) and OH^- specifically? Because they have all the requisite properties of easy loss of water, quite strong binding to X or Y and rapid exchange from bound conditions. We say that they have *functional value* relative to other possible choices of cations and anions.

It is then possible to list ease of hydrolysis in terms of polarity of bonds. For example, $H_3C–NH_2$ and $H_3C–O–CH_3$ are not easily hydrolysed while $CH_3–CO–O–CH_3$ is. The introduction of heavy atoms often increases reactivity, so that *neutral organic esters of sulphates and phosphates are hydrolysed readily*, though not when negatively charged. We will discuss later the importance of this fact in living processes. Notice that the reactions are downhill, but they do not occur *rapidly* without catalysts. However, selective weak catalysts can be used to hydrolyse the weaker bonds, while stronger catalysts are required to break the stronger bonds, e.g. $Zn^{2+}>Mg^{2+}$. Selectivity of catalysis is also necessary and is achieved by the particular surface of the catalyst: a protein* in biological systems or, especially today,

---

*In a metalloprotein which is also a metalloenzyme the active metal ion is often in a peculiar geometric state, an entatic state, 'designed' to assist reaction, i.e. to be a particularly effective catalyst.

a microporous solid in industrial equipment. These surfaces recognise the shapes of molecules to be catalysed in specific ways, even at the level of D and L optical isomers (see Section 8.2). Here we see control by a catalytic surface of the direction of chemical change, i.e. the rearrangement in selected ways of atoms in local space (chemicals).

Since catalysts only increase speed, the reaction of hydrolysis can be reversed (which we now call a condensation reaction) by introducing an energy bias in favour of XY (see Section 8.6).

8.5.4 Oxidation and reduction

We explained in Section 8.5.1 that many stationary chemicals, unstable but incapable of reaction, are vulnerable to oxygen as well as to water, and some are vulnerable to hydrogen. Although hydrogen is not now present in our atmosphere, some was available when Earth formed, while at that time no oxygen was in the atmosphere. There are then, further possible reactions in addition to hydrolysis: attack by oxygen, one form of oxidation, and attack by hydrogen, one form of reduction (see Sections 3.4.7 and 6.13.2 for a description of oxidation states). Straightforward examples are

$$\text{Oxidation} \quad 2O_2 + CH_4 \rightarrow CO_2 + 2H_2O$$
$$\text{Reduction} \quad CO_2 + 4H_2 \rightarrow CH_4 + 2H_2O$$
$$\text{compare (Hydrolysis)} \quad CO_2 + H_2O \rightarrow H_2CO_3$$

In the last reaction, in contrast with the other two, the oxidation state of carbon does not change (see Section 3.4.7), see also aside.

8.5.5 Barriers to oxidation and reduction

The barriers to reactions such as

$$CH_4 + 3O_2 \rightarrow CO_2 + 2H_2O$$

are again the strength of the covalent bonds C–H and O=O. To assist reaction the gases can be heated or sparked, as explained above. They can also be exposed to a catalyst which, clearly, cannot be of the same kind as that used to cause hydrolysis reactions,* such as that of ethyl acetate, i.e.

$$C_2H_5OCOCH_3 + H_2O \rightarrow HOOCCH_3 + C_2H_5OH$$

which we have just discussed. In the reaction above, involving oxidation of carbon and reduction of oxygen, we must induce a complete change of partners by reducing the numbers of hydrogen atoms or by increasing the numbers of oxygen atoms associated with carbon. In the intermediate condition the catalyst has to form catalyst–O and catalyst–H bonds, and transfer O or H atoms in specific directions of local space. The elements, as ions, which assisted the group of hydrolytic reactions described in the above

Addition and elimination

$C_2H_4 + Cl_2 \rightarrow CH_2Cl.CH_2Cl$

$CH_3 + CH_2Cl \rightarrow CH_2:CH_2 + HCl$

---

*A catalyst for hydrolysis is not required to change oxidation state while an oxidation/reduction catalyst is required to do just that.

Activation of dioxygen

$$Fe^{2+} + O_2$$
$$\downarrow$$
$$Fe^{3+}.O_2^-$$
$$\downarrow 2H$$
$$FeO^{2+} + H_2O$$

$$FeO^{2+} + C_6H_6$$
$$\downarrow$$
$$Fe^{2+} + C_6H_5OH$$

paragraph, i.e. Mg^{2+} and Zn^{2+}, do not form these bonds. Instead, the catalysts must now contain transition metals, such as Mn, Fe, Co, Ni, Cu, either in molecular compounds or on surfaces, which give metal–oxygen and metal–hydrogen bonds. Notice that in forming bonds to O or H atoms the metal atom changes its oxidation state (see aside). As explained in Chapter 3, it is the transition metals and their ions that can undergo such changes easily, not Mg^{2+} or Zn^{2+}. Then we can write a scheme of reaction such as

$$2M + O_2 \rightarrow 2MO$$
$$4M + 2CH_4 \rightarrow 2M\text{--}H + 2M\text{--}CH_3$$
$$2MO + 2MCH_3 \rightarrow 2M\text{--}O\text{--}CH_3 + 2MH \rightarrow 4M + 2CH_3OH$$
$$(+2M)$$

[Notice that writing the reactions of C–H (in CH_3OH) with O_2 three more times we get to $C(OH)_4$ which is not stable and decomposes to $CO_2 + 2H_2O$, the thermodynamically favoured condition.] These considerations tell us that certain elements, or in particular cases certain kinds of organic molecular centres, are required in catalysis for certain kinds of reactions, which are examples of functional value. *We shall find that as both hydrolytic and oxido-reductive reactions are essential in biological transformations, several different catalytic elements are, therefore, required.* In fact, this same statement applies to man's organic chemistry and indeed to parts of inorganic chemistry. As stated above, to control the direction of change the surface associated with the metal ion or organic centre is of extreme importance, e.g. in enzymes or zeolites. A suitable choice of catalyst can even allow optically active isomers to be differentially synthesised or degraded.

8.6 Organic reactions and energy requirements

To appreciate the energy available from organic reactions, or conversely the energy required to drive these reactions uphill, we refer to Fig. 8.17

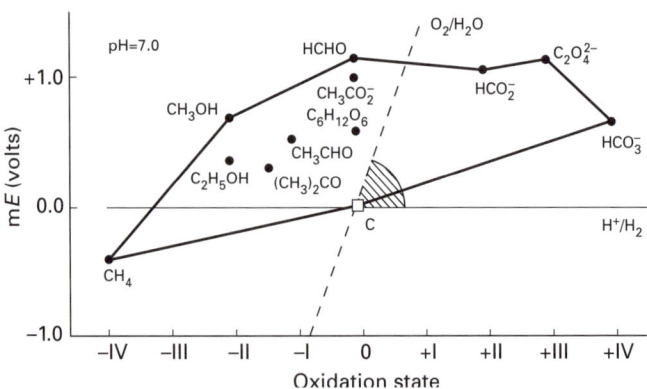

Fig. 8.17 The oxidation state diagram for some carbon compounds at pH = 7 (see Chapter 5).

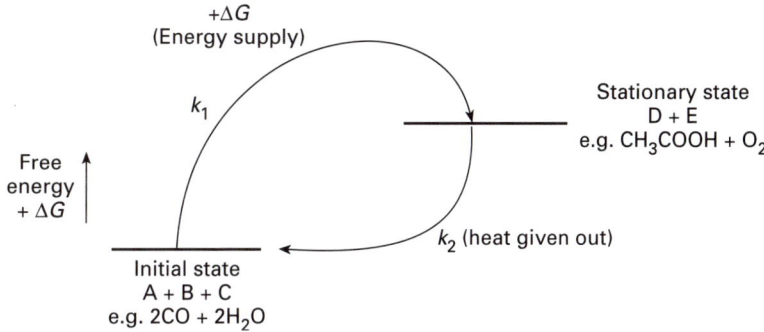

Fig. 8.18 A simple diagram showing the free energy (ΔG) relationship of two combinations of 2C + 4O + 4H. To pass from the more stable to the less stable requires a second reaction system to give +ΔG to the more stable combination. The less stable can return to the more stable by a different catalysed route with heat release.

which is an oxidation–reduction state diagram for carbon/hydrogen/oxygen compounds (compare Fig. 5.22). The stable states of carbon are, in the presence of hydrogen, CH_4, and in the presence of dioxygen, CO_2 (HCO_3^-). In any mixture of oxidation states, then, the most stable condition is $CH_4 + CO_2$. [N.B. under pressure, carbon (coal) is the most stable product, all H and O leaving as gases.] All the other compounds in Fig. 8.17 are (thermodynamically) unstable to disproportionation to CH_4 and CO_2. On the other hand, at room temperature, they are stationary state chemical *components* once formed, since they do not exchange C or O, and frequently not even H. They are kinetically persistent compounds. Thus, organic and biological chemistry is a practice whereby change within organic molecules is brought about using deliberate control over conditions to guide reactions along *pathways* to desirable products only. The pathway control for synthesis rests in the use of energy input, while heat or light by themselves cause degradation. Selective control is achieved using solvents and functionally effective catalysts as well as energy-containing stationary states of molecules, the degradation of which is coupled to the conversion of other stable molecules to stationary states (Fig. 8.18). (Note that the activity of organic chemistry is a very empirical one since catalysts can be very selective but not always specific, for obvious reasons, given the variety of possible products.) Thus, biological chemistry demands extreme control over pathways and we turn to some examples in Chapter 10. It will be seen that living organisms in particular have developed the most fantastic set of catalysts (enzymes), and carriers to ensure that fragments of organic molecules move along productive pathways without interfering with one another. (We give an example of uphill chemical change in the next section.) Although we have used carbon as an example, all other elements in combination with hydrogen and oxygen atoms can be analysed by similar oxidation state diagrams (see Figs 6.12 and 6.13). Hydrolytic changes involve changes at fixed oxidation state, that is, up and down in Fig. 8.17, e.g. from CH_3COOH to HCHO. We turn first to reactions in which water is removed—condensation reactions, which are so essential in organisms.

8.6.1 Removal of water in synthesis: condensation

A major reaction in organic chemistry is the removal of water (or alcohol) from between two reactants. As stated above, it is called condensation, and in the case of removal of water it is the reversal of hydrolysis (see Section 8.5.3). Clearly, the reaction requires a drying agent to remove the water. A drying agent is obviously one that has a greater stability after combination with water. Examples are

$$CaCl_2 + nH_2O \rightarrow CaCl_2(H_2O)n$$
$$CaO(lime) + H_2O \rightarrow Ca(OH)_2$$
$$P_2O_5 + 3H_2O \rightarrow 2H_2PO_4$$

but notice that the preparation of the dehydrating agent costs energy to reverse the above reactions. The reaction can be carried out in the solid state using separate compartments in a laboratory desiccator. However, such reactions can also be carried out in water solution given the appropriate conditions. For example, many anhydrides are kinetically stable in water despite the fact that in terms of thermodynamic stability, ΔG, they are unstable to hydrolysis, i.e. in the absence of catalysts their stationary states persist. In chemistry we treat them as components. Important examples of these materials and their hydrolyses are

$$\text{Acetic anhydride: } (CH_3CO)_2O + H_2O \rightarrow 2(CH_3COOH)$$
$$\text{Pyrophosphate: } [(PO_3)_2O]^{4-} + H_2O \rightarrow 2HPO_4^{2-}$$

Thioester-driven condensation

CH₃S.COCH₃ + 2CH₃COOH

↓

CH₃SH + HOCOCH₃ + (CH₃CO)₂O

Both of these reactions are slow in the absence of an adequate catalyst, but can be made to proceed if a suitable catalyst is added to the reaction medium. A less common reaction is shown in an aside.

It is then a matter of which dehydrating agent or anhydride is most useful for the organic chemist in a given reaction pathway when provided with a proper catalyst in a given solvent. The reagents then need to be activated to react with water, but only in the presence of catalysts and the compounds to be dehydrated. Once again note that the precise direction of change to products is under control of the catalyst surface and now the supply of substrate.

8.6.2 Chemical speciation in aqueous reactions

In the above sections we have shown that just as the description of a stable or stationary chemical system was dependent on the composition variable, so composition is a critical variable of change. Certain elements combined in particular situations form unchanging (stable or stationary) patterns. Some, when put together, change spontaneously, but the vast majority of chemicals require specific compositional additives (catalysts) to generate change. It is through these chemicals (catalysts) that the possibility of *controlled steady states* can be generated by the use of feedback or feed-forward interactions upon the catalytic power. In other words, the local flow of atoms in the kinetic scheme of Fig. 8.18 is manipulated through the use of catalysts and substrates acting on them. The chemical speciation of the system is *not* describable by the major components, the *substrates*, alone, but requires the analytical content of the catalytic chemicals as well. When considering free energy content, major components weigh by composition only (Chapter 5), but when flow is under consideration, as in survival of a biological system, minor components are equally vital and must be weighed by functional value. We shall need to be very careful in the description of constraints on the variables of flow since 'value' may refer to use or to survival of a system.

In our previous books (refs 68 and 69) we have shown why certain elements can perform particular roles in chemical systems due to the particular nature of their thermodynamic and kinetic properties, (see Table 8.7 in Section 8.9). Outstanding are the use of metal ions in catalysis, or light non-metals in the construction of large linear molecules and of heavier non-metals such as phosphorus and sulphur in the transport of organic fragments and energy. All are needed to give the complement of reactions in a living cell.

8.6.3 Control of organic reactions: a summary of single pathways

We cannot describe the methods of either man's or life's organic chemistry further here. Instead, we advise the reader to consider the way in which the organic chemist, or a cell, goes about controlling different organic reactions, noting especially the following.

1. The manner in which a route to a desired product is mapped out so as to avoid side-reactions. Here the choice is of *the sequence of intermediates*. This is an organisation of flow of chemical atoms, i.e. rearrangement of atoms in directions in local space. This route, often called a pathway, has to be directed; energy is applied (Fig. 8.18).

2. The route to and from the *intermediates* is directed partly by selecting reagents for carrying fragments of molecules, the use of protecting groups and the choice of the atmosphere and/or solvent in which the reaction is carried out.

3. The selection of chemical route is also aided by the choice of special catalysts for given steps (see Table 8.7).

4. During the course of chemical transformation physical separation or transfer of intermediate is often designed. This is, of course, organised physical flow (see later, Fig. 8.21).

We shall find parallels with this laboratory chemistry in biological and industrial systems, but here the degree of *internal* control is greater. In effect it is not possible or desirable to keep on isolating intermediates by outside intervention as in laboratory chemistry. Instead, reactions are made to follow prescribed flow sequences with or without physical isolation and transference across boundaries. The ways of manipulating the sequences are fully described in organic, biochemical and chemical engineering textbooks (see refs 68 and 69, in Further reading). Two sequences of biological reactions, the glycolytic pathway and the citrate cycle are given as examples in Figs. 10.17 and 10.18 (later). (Note that the catalysts for forward and back reactions in some steps are different, whence it is possible to control the direction along the pathway by feedback or feed-forward signalling as we explain in the next section.) We return to the nature of such 'cycles' in Chapter 10 [They are not the same as cycles described in Chapters 2–5.]. We stress again that every step is a manipulation of either atoms or groups in local molecular space, or of bulk transfer between compartments across a spatial divide. The application of energy and of directed motion are common to both (see the variables of flow in Section 8.2).

8.6.4 Steady states of chemical change: feedback control in single pathways

We showed in Section 8.4.5 that rates of *physical* transfer in spatially divided systems, for example, in computers or between cellular compartments, could be managed by feedback and feed-forward connections so that variability of the flow could be extremely constrained in rate and direction by the control system within structured bulk space. Of course, the field or driving force was applied also under feedback control. Thus the system as a whole became almost invariant once initial inputs and power were stated (Fig. 8.14). It could also fluctuate regularly, as in a doorbell circuit. However, as yet there is no reason for the selection of one rate or one direction of flow of chemicals over all others.

Within these multi-compartmental structures we now need feedback and feed-forward control over changes of their *chemicals* (components), not limited by equilibria. Given that the system in any particular steady-state flow must have a fixed input and output, the internal condition of chemical change must be fixed, i.e. we need controls over rates of chemical reactions in series. In effect, the schemes needed are those in Figs 8.19 and 8.20, where the hidden processes are chemical catalysed steps in a single sequence or, at most, a divided pathway (see Section 8.6.3). The feedback from product or feed-forward from reactant acts now on the catalysts (cat, Fig. 8.19) thus adjusting not just the diffusion rate but chemical reaction rate constants, k_f and k_b, to particular values by changing the activation energies ΔG^*. In living systems, the feedback internal to the reaction

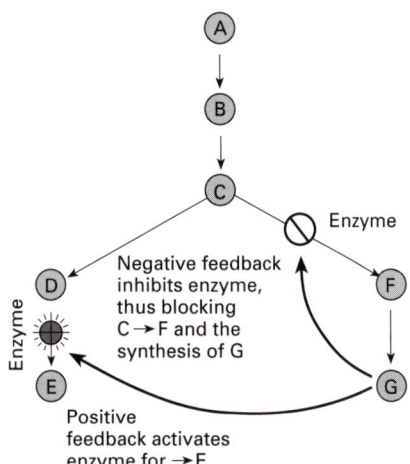

Fig. 8.20 Allosteric feedback regulation. Compound G inhibits the enzyme for the conversion to F, blocking that reaction and ultimately its own synthesis, demonstrating negative feedback by allosteric modulation. Compound G also provides positive feedback to the enzyme catalysing the step from D to E, changing that enzyme to a form that will catalyse the reaction. (After Purves *et al.*, see ref. 74.)

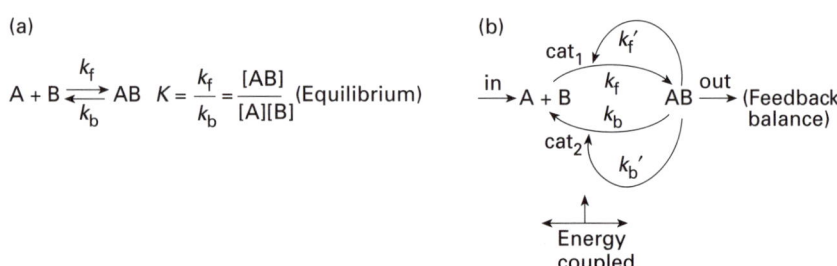

Fig. 8.19 Schematic illustration of how a constant state is maintained: (a) at thermodynamic equilibrium, and (b) by feedback from product to catalysts; the constant levels of A and B are based on controlled supply of material (and frequently of energy) and the rate of use of AB.

sequence may be due to different in-cell concentrations of reactants, that is, of small molecules (substrates and co-enzymes). In essence, feedback of this kind, utilising biological polymers as catalysts, is brought about by those polymers (enzymes or proteins) that exist in two or more states, one state driving, the other stopping reaction. We then have a switch called an allosteric change of the enzyme (Fig. 8.20).

State A $\rightleftarrows$ State B

Catalytic   Not catalytic

Positive message   No message or inhibitor message

8.6.5 Rates of chemical reactions in physical flow systems

In Fig. 8.21 we show a system through which material flows. In this system it can be that there is a reaction so that material leaving the tube differs from that entering. After an initial period the flow system will settle down and will be such that the whole has a steady-state gradient along its length. Thus, change in composition takes place along a space co-ordinate. As stated before, the steady state does not correspond to an equilibrium condition. Of course, we could consider the tube as being split up into a series of compartments, each different in composition, in the steady-state flow situation. If the rate constants in all the processes involved are first order, then the whole system is readily defined. However, where there are second-order rates, the problem of solving the analytical composition of the compartments is difficult. [Note that energy is required to maintain flow and while the free

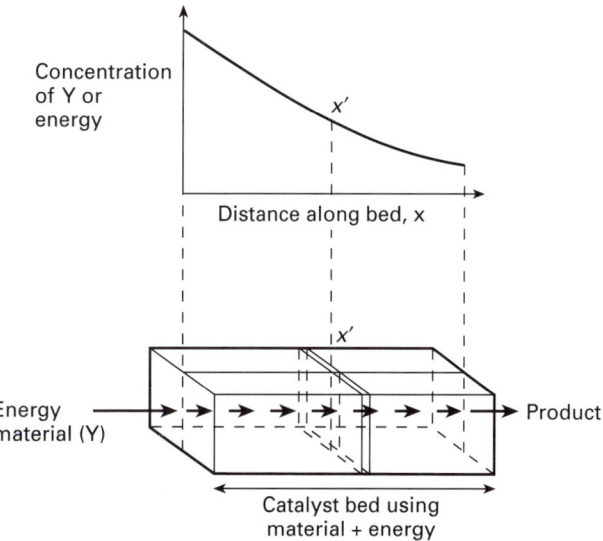

Fig. 8.21 The flux of particles down a concentration gradient. Fick's law states that the flux of matter (the number of particles passing per unit area per unit time) is proportional to the concentration gradient at that point, see x'. Equally, the energy flux will be proportional to the temperature gradient.

gas diffusion rate is linearly related to the force applied (also note Ohm's Law), liquid diffusion in a pipe is more complex.]

Turing and Prigogine have both separately considered general series of single-pathway reactions (not of first order) where A converts to B in the compartments and through steps

$$A \rightarrow X_1 \rightarrow X_2 \cdots \cdots \rightarrow B$$

This type of system stores some of the free energy of conversion of $A \rightarrow B$ in X_1 and X_2, etc. and can be well controlled when products of later steps feedback on rates of earlier steps. There can appear patterns in these compartments rather than smooth gradients of change (see also Section 8.10). Such patterned systems have been called *dissipative*: they lose heat to the surroundings while patterns in $X_1 \rightarrow X_2$ ... etc. appear. It is sometimes considered that such dissipative systems are similar to living cellular activities. This may well be true for steady states, which many cell hold over fairly long periods, but all cells actually develop from initial conditions and many proceed to reproduction and death. The nature of such living cellular processes will be looked at again in Chapter 10 under the heading of developmental states.

Our major interest in such schemes of chemical reactions in this book is their value, primarily to the survival of the system, whether it is just in a steady state, with perhaps self-repair, or whether it is of a living organism which is repetitively developmental. We stress that the constraints on the variables of flow must be collective features of systems not readily reducible to descriptions of rates of individual unit reactions. Notice also that we are presently only concerned with single pathways, while later we must observe intricate multi-pathway reactions.

8.6.6 Effects of temperature on feedback

We mentioned in passing (in Section 8.3) the variables temperature and pressure, while describing rate equations. In effect, eqn (2) shows that through the expression

$$k = Z'e^{-\Delta G^*/RT}$$

each rate constant (like each equilibrium constant) is temperature dependent. Thus, every single pathway, with or without feedback, is dependent on temperature, and as for equilibrium balance, if the kinetic system is to be fully defined (i.e. it is to be rigidly maintained), then the whole system must have good temperature control. Feedback thermostats are well known in man's equipment, e.g. in the central heating system of a house, in computers and in all modern chemical factories. Equally, the most advanced animals have good temperature control. Humans, for example, are controlled at $310\pm2K$, and while this fixed temperatures ensures, for example, a constant value of the solubility product of bone and hence the Ca^{2+}, HPO_4^{2-} and H^+ concentrations of circulating fluids (Section 14.5), it also keeps all rate constants fixed, giving feedback homeostasis.

8.7   Energy and its distribution

Initial energy for life

(1) Earth's Unstable Chemicals
(2) Light from the Sun
(3) Local Gradients of Chemicals

In the description of chemical change we saw that atoms of groups could be given direction locally by supplying energy (Section 8.6) just as the bulk flow of material in space could be directed by gravitational or electric fields or, in fact, by chemical change (Section 8.4.6). Both could be driven uphill by the provision of appropriate energy when aided by catalysts. For any *downhill* chemical pathway there will be an energy release which in biological systems can be the difference in free energy say between $C_6H_{12}O_6$ (sugar) and $3CH_3COOH$ (acetate), as in glycolysis, (Fig. 10.14) or between $CH_3COOH + O_2$ and $CO_2 + H_2O$, as in the citrate cycle (Fig. 10.13). An uphill process in biological systems, such as that towards ATP synthesis, can only run if it is coupled to such an energy-producing change (Fig. 8.18 and see aside). A living cell generates, stores and uses energy, but is a net energy consumer. The energy supply must be connected to a control mechanism and coupled to chemical pathways if the whole system, energy input and material synthesis, and energy output and waste disposal, is to be held at a controlled rate. The principles of control can be applied to any source of energy, e.g. a fuel cell, an electric battery or radiant energy, and any number of energy-requiring processes can be connected by energy transfer between them. The control utilises a switch-on or switch-off of energy when the given pathway is underemployed or overemployed, respectively. The feedback control is applied by altering the structure or inhibiting the action of proteins, 'enzymes', involved in energy transduction (see Fig. 8.19), or energy transfer as in allosteric system (see Fig. 8.20). Thus, energy supply needs to have both feedback control and coupling to bring about desired rates of synthesis.

8.7.1 Feedback control over energy in chemical pathways

The major sources of energy for industry or biological cells depend on the supply of fuels or radiation. In a steady-state system this supply to each pathway has to be managed (increased or decreased), to match the input and output requirement of materials and to control the physical state variables (temperature and pressure) throughout a pathway. Thus, a message system has to connect the internal (chemical) production with the fuel supply. There must be, therefore, communication networks based on sensors that relay the condition of the system to the energy distributing modes (see Table 8.6).

Table 8.6 Circuit components in message systems

Transmission	Carrier	Power	Sensor
Electronic	Electrons	Charge storage (fields)	Electrode or thermo-couple
Electrolytic	Ions	Ion gradient (fields)	Allosteric protein (electrode)
Chemical	Molecules	Molecule gradient	Allosteric protein
Mechanical	Stress	Strain (compression)	Touch-sensitive cell
Radiative	Photons	Emitter	Photocell (eye)

In industry the sensors can be thermocouples and heat can be supplied electrically or by heated steam, and this system can be independent of the material throughput and its controls. The thermocouple switches energy off or on in a certain temperature range and the energy is not usually 'self-generated' by the reaction system. In self-sustaining biological systems there are two major modes of energy distribution which are linked to one another: chemical bond energy (e.g. in pyrophosphate bonds of ATP) and ion gradients (H^+, Na^+, K^+ and Ca^{2+} mainly). We shall see how each and every biochemical pathway is related to these energised chemicals, which are also used as the messengers of the condition of the system. However, the general principle is that the excess free energy, in the form of the amounts of pyrophosphate (ATP) and ion gradients, is held constant at any one time. (The equivalent situation in industry is to hold the potential energy constant, e.g. the electrical mains potential.) The ATP and the gradients are produced by the metabolic material pathways themselves, so that *energy and material are linked*. Coupling is achieved by feedback using ATP or ion gradients simultaneously as modulators of cellular pathway catalysts, e.g. the enzymes of Fig. 10.13. Thus *energy flow* is an essential feature of cellular activity and industrial plants and a new restrictable variable affecting the *material flow* inside a cell or a chemical factory. We see that the controls of chemical single pathways described in Section 8.6, or of physical transfer described in Section 8.4, all have to be coupled *in selected ways* to energy creation and distribution.

8.7.2 Radiation flow (fire) and the overall free energy of life

It is necessary now to make an insertion to see the relationship of the nature of flow of radiation (fire) and the overall energy drive for present-day life (see aside). A striking point is that we do not 'see' objects around us at temperature equilibrium with us. We are then in darkness. Fire is the descriptive word for the propagating radiation from a body hotter than ourselves. Thus fire (as an observable) is seen when a body has a relatively high temperature; it belongs to two compartment systems which are both at least closed, not isolated, i.e. energy passes between them. The difference in temperature is observed as a flux of photons from a high to a low temperature. Three features of the radiation flow are important: (1) it allows energy to be transferred without transfer of matter; (2) like all flow it is directional, outwards from the high temperature source; and (3) any absorbing body, at the lower temperature, may well be in an environment of still lower temperature, e.g. in the relationship of the sun to Earth and Earth to outer space. The consequences are: (a) the intermediate body, here Earth, gains energy and is heated somewhat; and (b) the radiation lost by the intermediate is of longer wavelength than the absorbed radiation. In a steady state of flux, therefore, the number of low intensity photons leaving the intermediate system, i.e. the Earth, may exceed, many-fold, those leaving the hottest system, the sun, (in the sun/Earth case 30-fold). This flux (which increases the entropy of the environment) then produces a huge capacity for work between the time energy is received and the time it is lost, despite the fact that only about 1% of it is retained. Life is a large part of the work achieved, but a second essen-

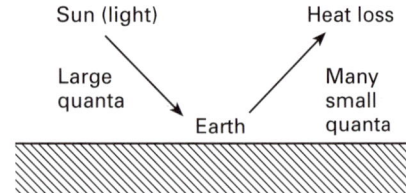

Sun (light)   Heat loss

Large quanta   Many small quanta

Earth

tial effect for life is the maintenance of the physical steady-state temperature, around 300K for the inorganic Earth. Radiation transfer, then, gives the possibility for doing work in living organisms (see Section 5.11).

8.8   Conclusion to physical, chemical and energy flows

It is now seen that linked material and energy flow can be controlled both as to rate and direction in a multiplicity of pathways. The direction here represents both selection of product by rearrangement of bonds and the appropriate choice of physical transfer of material in bulk space. However, it is clear that while the rate or the direction could be set at any value and maintained in a steady state, we have given no reason for a preferred rate or direction of change. As with physical transfer there is no purpose associated with the rate or type of change. We shall, therefore, adjust our approach to those variables requiring that the uses of space, chemicals and energy are dedicated to the particular products that we see around us as part of living organisms or of industrial chemical activity.

8.9   Functional significance

Prominent amongst the objects in our changing surroundings are parts of the Earth (apparently inert) and living organisms (active). The general collapse of chemicals towards equilibrium states, as happened in the formation of the Earth and, in part, in its subsequent changes, appear to be accidental and not co-operative. It is difficult to assign a 'purpose' or functional significance to such unco-ordinated activities. Much more interesting are the systems such as living organisms and man's industry where the processes are energy demanding and where the product of any and every given pathway is of direct value in sustaining a whole cell or a society. The question is then how is it possible now to direct changes of all the different pathways of material and energy to the development of a self-sustained system in an organism or a society. We shall refer to this purpose as generating *survival value.** The need is clearly for all the individual pathways to be linked to form a coherent organisation* (Fig. 8.22).

This leads us to quite a new question which is: 'Is it a feature of these organisations that only certain values of the rates of the different transformations and transport paths are mutually supportive?' This would imply that the variables of change (of flow) have optimal values for survival in a particular system. We can then consider plots of survival value against the variables, such as those in Fig. 8.23, much as we plotted stability, $-\Delta G$, against several variables for non-living systems in Chapters 5–8. Where

---

*The fact that much of man's activity could be said to be directed towards pleasure today must not be allowed to hide the basic drive, which is towards survival.

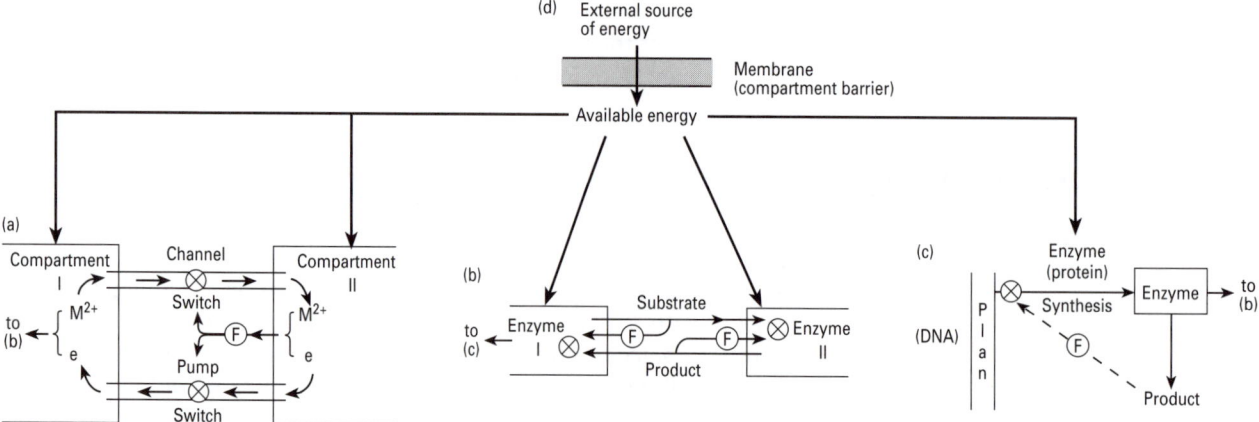

Fig. 8.22 Illustrative parallel examples of cellular feedback in: (a) a physical circuit; (b) a chemical circuit; (c) an information circuit for expression of catalysts (enzymes); F, feedback to control X. In each cell there is a network of reaction paths linked not only internally as in (a), (b) and (c) but additionally (b) and (c) themselves are also linked. The compartments of a cell or cells themselves are also linked together, as in (a). One or all of the compartments should contain a source of energy (d) which supplies all systems and is under feedback control (not shown) from (a), (b) and (c).

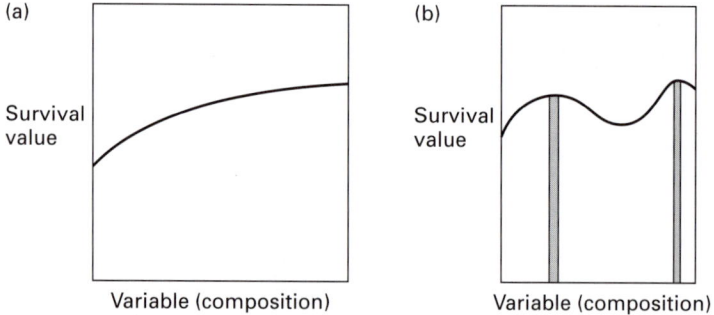

Fig. 8.23 Plots of survival value against a variable such as composition, energy supply or temperature showing (a) continuous variation, (b) the development of distinct species, shaded regions. Note that the peaks in this graph correspond to higher survival values while in the case of stability ($-\Delta G$) composition plots (Figs 4.13, 5.18 and 5.19) we have plotted in just the opposite way; there is no conflict whatsoever between the two representations—it is all a matter of definition of the reason for existence, thermodynamic stability or kinetic strength.

such plots are not smoothly continuous there should be found separate regions of the variable where survival value is high and others where it is low. This is observed. Unfortunately, to date, study of such systems has not generated an assessment of survival value to parallel ΔG, where the survival value would be related to effective collective terms (compare with maximum entropy) to ensure quantitative success. This is not only a problem for living organisms but also one of the economic viability of societies.

In this chapter we look next at the general way of building an inter-communicating and selected system of paths. In Chapters 10–15 we will consider whether we can discern the way variables such as composition, temperature, energy supply and spatial separations are limited in real living and man-made systems so that we can appreciate increase of survival

Table 8.7 Selection of chemicals for functional value in living organisms

Chemical	Function	Amount required
C/H/N/O components	Structures	Large
Na$^+$K$^+$Cl$^-$	Osmotic and electrolyte balance	Moderate
C/H/N/O/P/S	(Co-enzymes) Transfer of fragments and energy	Moderate
Ca/P, etc.	Signalling	Moderate
Fe	Electron transfer	Small
Fe, Co, Ni, Cu, Zn, Mn, Se, Mo, W	Catalysis	Traces
Zn, Fe, P, etc.	Regulation	Traces or moderate

value. We can also suggest that organisms are split into species because each one has a special dependence on variables so as to generate high survival value. We can then see how such logic applies to man's industry and later how we might learn to create a sustainable society. As we shall see, collective co-operativity becomes a key factor.

An important point now is that when we describe the involvement of an element or component in the composition of a system it is not just its percentage contribution to the whole that matters, but this percentage multiplied by a weighing factor representing its functional significance for the whole system. This affects our perspective of the compositional variable. We shall see that while some elements are required in large amounts, others, equally essential, are needed in traces only (Table 8.7). It is also the case that excess of many elements can be deleterious. There is, in effect, a balanced homeostasis of the elements within any biological system to optimise the whole. This is also true of an industrial system.

8.9.1 Feedback in concerted chemical reaction pathways

If a series of pathways are to be timed to occur together in a reaction vessel, then there has to be some exchange of *information* of the rate of one pathway to the rate process of the other. An obvious way to adjust rates is for products of one pathway (A) to slow down its own rate while speeding up that of another (B)

$$(A) \ X + Y \xrightarrow{\ F_b\ } XY$$
$$(B) \ U + V \xrightarrow{\ F_f\ } UV$$

where a continuous arrow shows slowing (here feedback) and a dashed line shows acceleration (here feed-forward). These interactions are called crossed feedback controls, which knit together separate systems into an organisation. For example, it is found in biological systems that many intercommunications between pathways are made by: (i) fragments of substrates carried by co-enzymes; (ii) energy carried by co-enzymes (ATP); and (iii) ions that modulate paths differently, e.g. Ca^{2+} (see Fig. 8.13). We have described these messenger activities extensively in our previous books (see refs 68 and 69) and will turn to features of their evolution in Chapters 10–15.

8.9.2 Biological chemical pathways and coupling to bulk spatial constraints

Biological chemical pathways are frequently shown as linear lists of substances connected by enforced changes due to enzymes (see Section 8.6.3 and Fig. 10.14). They are pathways in the sense that barriers to particular *local* directional re-arrangement of atoms within molecules are lowered, so that atoms in space are moved *locally* along a given route. They are not pathways in *bulk* space unless: (1) enzymes are physically coupled in bulk space and substrates are handed on physically from one catalyst centre to another; (2) the enzyme is a device that acts while also moving a substrate through bulk space, e.g. through a membrane in a direction. Such an enzyme is simultaneously a pump. Gross patterns of chemicals in space are set up by the controlled movements in bulk space [see Fig. 8.22(a)]. The more usual situation of many enzymes is that they do not lend any directed *bulk motion* to substrates since the enzymes rotate and diffuse relatively freely in solution. The two senses of a pathway must be clearly separated. Since energy is distributed in living systems *within chemicals*, transfer of energy can also be directed by changes in local bondings, e.g. in pathways or in bulk space. Networking is frequent and implies branching and coupling of chemical and physical paths. Feedback or feed-forward involves the adjustment of both barriers by products or reactants acted on and acting upon catalysts so as to generate coupled local chemical and bulk physical molecular patterns. Movement in bulk space is obviously more simple and we observe it everywhere around us. At the present time we are not fully aware of the way chemical reaction pathways and bulk movements are integrated in cells, but we know that they are. There are many kinds of bulk trafficking, for example from the Golgi, along tubules and by endo- and exo-cytosis (see Section 14.10). These are examples of joint physical transfers, chemical transformations and energy couplings to form a network much like that in a computer designed to fulfil a function (see Fig. 8.14).

8.10 Cooperative kinetics: patterns and dissipative systems

It has been shown by experiment and by computer simulations that in a number of situations, when energy is fed into a complicated man-made physical or chemical system which is under feedback constraints, then the rates of change within the spatial parts of the system can form a *pattern* of chemicals in space (see Fig. 8.24). We have already mentioned this possibility for a single pathway in Section 8.6.5. This is apparently contrary to the expectation based on the thinking of Chapter 5, where we showed that the introduction of energy should lead to more random or chaotic behaviour. The systems are usually called *dissipative*, for obvious reasons. There is not a real contradiction since we are describing here a condition of activity intermediate between initial and final states. We can then write

Initial inert stationary state of high persistence

Energy →

Possible co-operative active state (plus chaotic environment) →

Final move to chaotic state (equilibrium with environment)

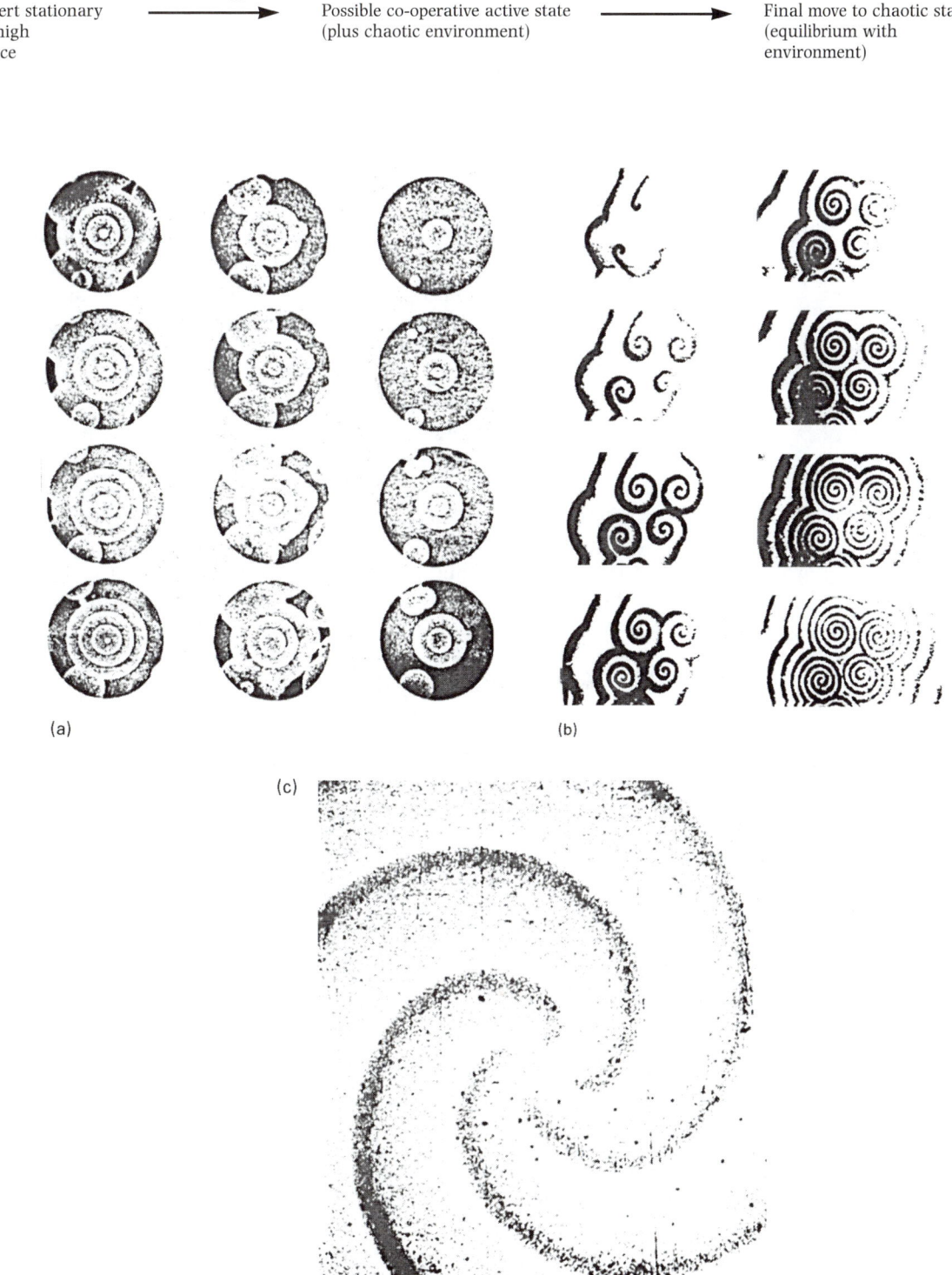

(a)

(b)

(c)

Fig. 8.24 Wave propagation in a two-dimensional layer of Belousov–Zhabotinskii reagent. (a) Target patterns. (b) Spiral waves. (c) Multi-armed spirals.

A pattern can be in microspace, i.e. of chemical bond changes in sequence (a reaction pathway), as well as in bulk space (transfer of material).

We need to ask why the intermediate is not chaotic. The patterned state is, in fact, dependent upon the rate constants within the activity being correlated in particular ways (co-operatively) by feedback. Such active states have been observed in localised heating of liquids (e.g. the so-called Bénard convective cells), in the mixing of certain chemicals (clock reactions) and in some isolated systems removed from biological organisms, as well as in life itself. The very nature of these patterned states is such that, as can be seen in the above scheme, 'they exist on the edge of chaos' and can easily collapse. However, they can be stable for long periods as long as the required energy (and materials) are supplied without perturbation. These systems are also adjustable through perturbations which may be small in themselves, but the resultant switch from one steady-state kinetic pattern or active condition to another may well appear as a jump. We are extremely tempted to compare the situation to that in a living system. An example is the succession from a grub, or caterpillar, to an adult insect, or butterfly. A small perturbation of the active pattern rate constants can also result in complete collapse to a chaotic final state. We may wonder at the small number of changes in a cell chemical (DNA) that can generate a disastrous mutant or cancer, and believe that this is a result of lack of control of cell growth leading to failure—in this case death—but in the context of 'living on the edge of chaos' these features are unavoidable.

8.11   Chemicals and self-assembly of equipment

The final step in the description of an organised system is the development of a complete and contained structure. We see this in both laboratory and industrial equipment, which are put together in an *ordered* way by

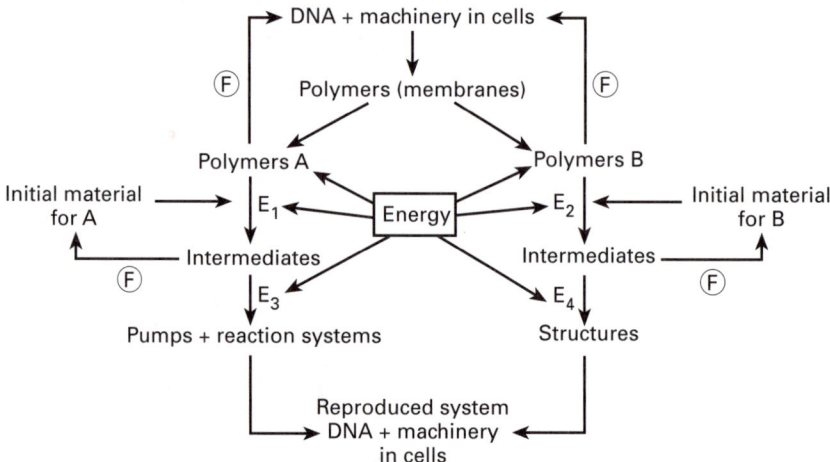

Fig. 8.25 The cell reproduces itself in the above way, which requires both chemically active enzymes and more passive structures. F = Feedback. See also Fig. 8.22.

mankind. In living organisms this structure requires a self-management of the synthesis and positioning of polymers of many kinds (proteins, lipids and polysaccharides) using transfer of small groups to construct large units. We return to such transfer in Chapter 10 and to the energy input, where needed in biology, in Chapters 10–12. It is often their combination under feedback control that allows self-assembly of biological objects, though some assembly is downhill. The internal activities of self-assembling systems must be linked to the structures they build, which requires consideration of the style of *feedback*, F, acting on the enzymes E_1 to E_4, and cross-talk between the system (Fig. 8.25).

For the pieces of structure to be compatible with the unification of the whole there has to be limited production of each piece. The basic proposition is that synthesis of materials, even for structure in a complex organisation, must be based on management of each reaction pathway so that it is in communication with, say, at least two other pathways (forming a network) (see Fig. 8.14). The pathways of organic, and much of biological, chemistry are all based in H, C, N, O, P and S elements, so that it is the management of the small fragments produced from these elements, e.g.–H,–CH_3,–NH_2,–OH, SO_4^{2-} and HPO_4^{2-} by (1) carriers, (2) catalysts and (3) mobile components of the catalysts, such as other elements or cofactors, which is relevant. Where structure involves other elements such as calcium (e.g. in biominerals) and zinc, their management is equally important. Here again we see how controlled local synthesis and bulk movement in space are needed to create the objects we see that have shape.

8.12 Dynamic shape

A brief word is now required about the shape that is useful to an organised system, before we examine the problem in more detail in later chapters. The shape of a system, e.g. that of any living organism, is not controlled by a repeating order, as is the case for the atomic and molecular structures described in Chapter 4, and therefore must be due to fields, e.g. mechanical tension or electrical constraint. Thus, a liquid shape is controlled by surface tension (see Section 7.4). Similarly, the shape that a living system has in freely available space is a reflection of tension in polymers within it. This tension in a flexible body must be in feedback relation with energy processes linking it to the metabolism of the organism if the shape is to be appropriately designed and maintained, and yet be manipulable. More important still, is control of shape during growth. This is not to deny that certain shapes found in organisms do not arise directly from simpler considerations of equilibria as described in Chapter 6 (see D'Arcy Thompson, reference 58, in Further reading). However, the major feature is feedback control over production of polymer units which then decides dynamic shape by the manner of binding them together and by the energy input to them. A typical example is the filamentous structure internal to cells, e.g. in muscles of vertebrates. Thus, living shape is a consequence of cooperative organised reactions, now including energy input, just as the shape of inanimate objects was due to co-operative binding of limited amounts of material (see Chapter 7). Now, these dynamic shapes arise because they too have survival value, see note at the end of Chapter 7.

8.13   Connected shapes and functional value for change: machines

A container obviously has value in that it protects its contents in a compartment. Such protection is a major consideration for organisms and for man's activities. However, other values are connected to objects of useful shapes within the container, especially where continuous flow and/or change is desirable in a directed way. For example, a pipe or tube is a container that allows flow along a specified path and this construction of shape is paralleled in an electric wire. More complex shapes are needed to control flow in devices such as switches and pumps, and these also involve dynamic features. For example, in man's mechanical machines, solid constructs such as cylindrical bars are used to connect moving parts and these are often articulated at ball and socket joints or as hinges (see Figs 8.9 and 8.26). Alternatively, there are energy-absorbing structures such as springs which also control moving parts, as in clocks and watches. Inspection of any machine reveals that shape has to be designed, i.e. selected, opposite dynamic function, and most frequently the design has both structurally static features (platforms or frameworks) as well as moving parts. In this way machines are built from parts as we have described for separate

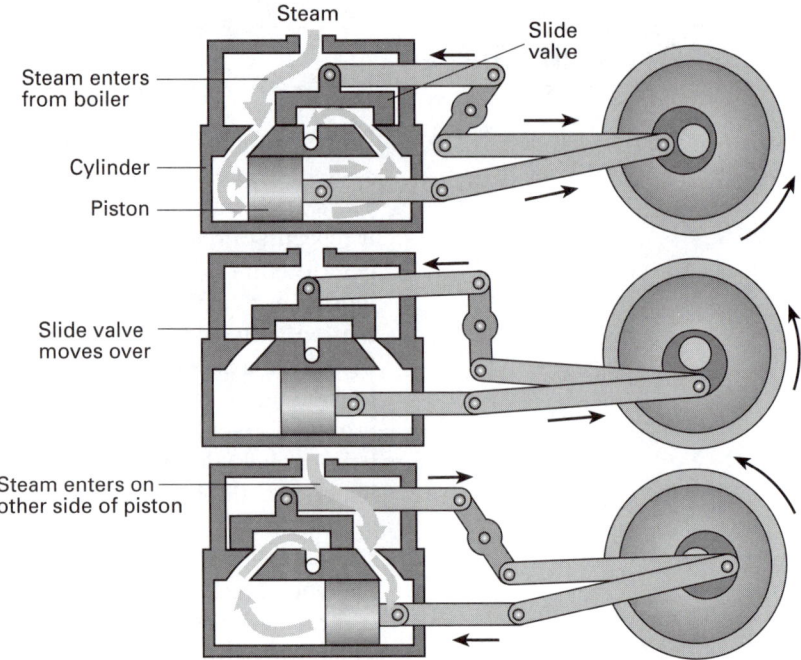

Fig. 8.26 The simple operations of a steam engine showing the controls over flow in each cycle of the engine. It is a good exercise to relate this engine to a biological pump.

structural units in Chapters 2 and 7, *but in a machine they are organised in a sequence in space to act in timed relationship.*

Figure 8.26 represents a steam engine. One feature of this machine is the coupling of momentum (flow) in particular (designed) ways. Gas flows along tubular constraints and through valves to push pistons, the connecting rods move on the basis of the piston's guided movement in cylinders and finally a cam or wheel is rotated giving a specific directed movement (momentum) to the whole body of the machine. Hence, the gas flow drives the wheel of the engine in a chosen direction. The momentum coupling, control over direction, amount of energy transfer and the material movement are fashioned in a desired, organised way. The couplings constrain the variable possible motions of the isolated parts so that a machine is a unity, identifiable with a purpose. Note that both fixed and moving parts form the whole and that energy is required in any machine.

The parallel with biological activities is not quite exact since a purpose is now described by a function. A biological machine is chemical in nature and transfers material and energy in coupled systems of two kinds. First, there is the allosteric (conformation change) activity of proteins due to the guided relative motion of their α helical constructs which parallels the coupled rod-like activity of a mechanical machine (see Fig. 8.10). Note also the presence of structural units, β sheets, with or within which helical rods drive motion (Table 8.8). Then there is the seemingly random motion of small molecules such as ATP (carrying energy) and co-enzymes (carrying material). However, these activities are all coupled since the reactions of decisive importance occur only at specific sites on special proteins. The overall coupling results in co-ordinated motions, as occurs in the cylinder of a motor car. It follows that transport by either pumps or mobile enzymes has a constrained momentum in local and bulk space, respectively. Thus, as in any machine, the whole is 'purposeful'. When we move up-scale from proteins to organs, such as muscles, the parallel with a machine is obvious.

It is of no consequence that the original biological machinery was established by random searching while man has designed his machine. In effect, *though the probability is very very low,* any machine could have arisen by the accidental coming together of parts in a particular way. This is the suggested way biological cells evolved, but much empirical study was also needed in the development of man's machinery. The history of any of man's devices is usually a long one of trial and error, e.g. that of the aeroplane over two centuries at least and dreamt of even in classical times. The probability of such events by accident is very very low, of course, but once a system has been created, no matter how simple, it can be improved upon provided it has modes of survival and/or reproduction with modification. We shall see that there are two modes of development—one based on biological coding, the other on man's rational construction—both are forms of planning. One is apparently random 'planning' based on survival value through modifying a DNA code, the other is 'rational' planning based on interactions of another coded store of information in the brain. The immediate feature is that the machinery has selected parts allowing selected paths of material and energy. What is required is instruction towards an overall purpose or function.

Table 8.8 Machine parts

Man-made	Proteins
Platforms	β-sheets
Rods	α-helices
Hinges	'Random' stretches
Rising hinges	Helical pairs
Valves	Multiple helices
Wheels	Flagellum base
Pumps i.e.	Molecular pump
platform + rising	β sheets + helical
hinges + valves	pairs + multiple
	helices

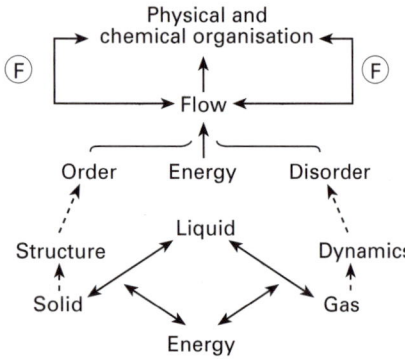

Fig. 8.27 A reminder of the central problems known in classical times (see Fig. 1.1 and Section 1.1), but now expressed in modern language, and with the connection to flow, leaving inherent unresolved problems of the evolution of the materials around us. F = Feedback. Chemical organisation includes DNA or planned activity by man.

8.13.1 Control over machines: information and instructions

A machine is very restricted in its momentum variables by the instructions that have been applied to its construction. Instructions decide the informed state of the machine (see Fig. 8.26) but are not part of the machine. In living systems, DNA is the basic information that restricts the building of the machinery; DNA is then a constraint on variables, generating a given species (see Fig. 8.27). Variety of machinery is obviously possible using different sets of instructions; similarly, variety of species is related to variety of information from DNA. There are, therefore, varieties of constraints on living machines that produce species, but the instructions in DNA are but a coded form of the way variables are constrained to make a given living machine. The information is not part of the machinery, much though it may tell how to make or repair it. We return to these problems in Chapter 16.

The enforcement of a purpose, a design or a coded instruction, is a one-off restriction on variables which is holistic by its very nature. This is in no way to be confused with the introduction of a vital force, or to be attributed to an extra agency other than those we know of in physics and chemistry. Always remember that any functional building can be described and labelled by a plan (DNA in living systems), but the operational dynamics represent the reality. Of course, the functional building (organism) can be defined (described) also by its shape, hence classification (speciation) of constructs on different bases have arisen. It can also be analysed by its physical or chemical parts, but these only give analytical labelling and do not, even cannot, describe the way they are put together. The whole is more than the sum of parts.

8.14   Developing systems

All cellular development is continuous towards cell division, though many cells are stopped at some intermediate stage. The progression is not simply cyclic in that the cell doubles its content in a complete 'cycle' and *reproduces*. Thus, there is a series of steps of synthesis and degradation that requires signals. It is believed that these signals in all cells involve phospho-

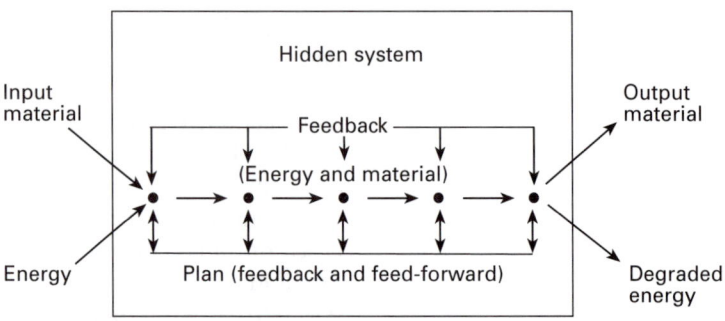

Fig. 8.28 The outline plan of all flow systems to be described in Chapters 10–15.

rylation and dephosphorylation control processes and novel protein degradation and synthesis. In simple bacterial cells this mode of control may be totally dominant. In cell multiplication in higher, multicellular organisms, differentiation occurs simultaneously, so that a growth pattern is observed that involves signalling between cells and is dependent, as cells accumulate, on the spatial relationships between them. The signalling is now complicated and involves hormones going between cells, for example thyroxine and various peptides. We are not yet in a position to give a detailed account of such development (see Chapters 10–15).

We have now concluded our description of the kinetics of change (flow). We see that from the initial outward-directed, but turbulent (chaotic) flow (Table 8.1) there has developed organisation (Fig. 8.28). It is organisation that man must learn to control (see Chapter 15). (Notice particularly that we have distinguished order in static systems from organisation in dynamic and living systems.)

8.15   Summary: the units and variables of flow systems and their restrictions

We know that the fundamental units of all chemical systems are elementary atoms and charges. The *local* change of their position during chemical reaction is equivalent to flow along a reaction co-ordinate and this we refer to as microscopic, or local, flow. The fundamental variables of changing composition of flowing chemical (molecular) systems are, therefore, as before, the elements of the periodic table and charge. As we have seen, bulk flow of material can be more easily described operationally in terms of the conventional spatial co-ordinates of the movements of *components*. The material in any flow system is, therefore, expressed in the *same units* as those described in Chapters 1–7. The further common variables, other than composition, are temperature, pressure, overall amount and physical bulk fields. The potential of any field is now a driving force of flow as well as a thermodynamic variable (Chapter 7). In microscopic change, chemical bond energy or chemical potential (free energy difference) provide the drive for flow.

The variables of local or bulk flow relate directly to *motion in a direction*, which for a given mass, m, is included in the momentum $\overrightarrow{mv}$ (the arrow signifies a vector) and for the flow of electricity, electrons or other charges, is given by the intensity of the current, i. Two new variables associated with the fundamental variable time are then the *rate* of flow and *its direction*. This applies to atomic elements, charges, components, (even radiation, photons) or bulk flow. The direction of flow and the rate of flow are limited by high barriers along that direction, parallel to the line of flow, which prevent escape, while across the line of flow there may be smaller energy barriers (see Fig. 8.9), which restrict rates. Successive ordering of barriers gives *organised* flow in a sequence, much as a river flows through a series of lakes. (Note that we have shown that chemical change can be considered in the same way as bulk flow by describing local reaction

Table 8.9 Types of gradient causing flow

Field gradient	System open to flow
Gravitational	Layering of Earth, sea and atmosphere; separations in chemistry; biological polarity of masses in cells
Chemical concentration	Storage across boundaries, batteries
Electrical	Storage of charge (many electrochemical cells and all biological membranes)
Mechanical tension or pressure	Deformation of structures
Radiation causing a temperature gradient	Excited absorber; convection currents in all materials

co-ordinated barriers and their control by catalysts.) To overcome barriers in continuous flow there is a *continuous energy input* (an energy flow) which is then an additional observable variable that must come from a field, even of radiation. The rate of flow, or flux, is a combination of material injected and the energy applied to it. The unimpeded flows are bulk linear properties of the variables.

Now the chemical local or bulk flow can occur in two ways along a path, either spontaneously (down an energy or field gradient, Table 8.9) or by pumping (using energy, up a gradient). The rates of microscopic or local flow are expressed by chemical reaction rate constants, k, in the following way, where energy can be injected by several means (Table 8.9).

$$A + B \underset{k_b \text{ (energy input)}}{\overset{k_f \text{ (energy release)}}{\rightleftarrows}} AB$$

(When k_f equals k_b this simple reaction is in balance, or equilibrium, which we have written as AB $\rightleftarrows$ A + B in Section 5.7, when there is no energy input or output.) The constraints on the variables of flow using both forward and back reactions will include controls over both k_f and k_b. In a net bulk flow system we also have to include the physical movements *in* and *out* of the system. We then have

$$A + B \xrightarrow{k_{in}} A + B \underset{k_b \text{ (Energy input)}}{\overset{k_f \text{ (Energy release)}}{\rightleftarrows}} AB \xrightarrow{k_{out}} AB$$

where we see that k_f and k_b refer to the microscopic chemical flow of A and B units but k_{in} and k_{out} refer to the flow of units through a bulk system. If the movement in the bulk is random then these quantities are diffusional rate constants; however, in living systems we more often meet circumstances in which k_{in} and k_{out} correspond to channelled or pumped paths in specific directions. All of these flows are described as pathways by biochemists. The concentrations of AB, A and B in the system are at this stage all variables dependent on the rate constants, unlike the situation at equilibrium. Nevertheless, under constant, *controlled* equal supply and removal of energy and material, a given *steady state* of flow is reached. We may call this an organised pattern rather than an ordered structure. Thus, control systems have removed the variables, so that a fixed steady-state system emerges even in a long succession of steps on a pathway.

To ensure a steady state of the system we then use feedback control, which cannot operate directly on the rate constants k_f or k_b, or on k_{in} and k_{out}, but only on the catalytic or pump centres that are associated with these transformations.

Here F represents a feedback (or feed-forward) control process in which either A or B or both act on the catalysts and on channels or pumps, but F can act on energy input, which is also controlled by proteins. The controlled passage (conversion) of A + B to AB, which may be closely directed in space as well as controlled in rate, can also be written as in Fig. 8.28 where now there are multiple inputs of materials and energy all linked in a collective network to make a product which we draw as

$$A(C) + B(D) \dashrightarrow \underset{\text{Hidden systems}}{\underline{A(C) + B(D) \rightarrow AB(CD) \text{ product}}} \dashrightarrow \text{waste}$$

We see that if the internal products are to be related to some specific function, even merely survival of a self-sustaining system, then there have to be instructions incorporated into the hidden machinery. The instructions must be read and followed if the machine is to be repaired or reproduced. In living machinery the cell reproduces the instructions (DNA) but once made the instructions are not part of the machinery.

The question arises: is it possible to give linear analytical mathematical expression to the relationships of the derived variables (the flows or rate constants in these controlled systems) or are they intrinsically non-linear and only open to computer analysis (see also Section 5.12)? It is found that all feedback systems are non-linear. Only the definition of the whole system is possible, much though we can appreciate features of it in reductive terms of derived or fundamental units and variables. The properties of these systems are emergent, e.g. living. At present we do not have ways of

Table 8.10 Units, variables and restrictions in Chapter 8

Units	Variables	Restrictions
Fundamental (atomic elements and charges) or *Derived* (components)	Absolute amount of each unit giving total composition of limited system	Equilibrium exchange between atomic elements in components
	Potential energy of interaction between units in stable and stationary states	Equilibrium exchange between phases
	Angular momentum in stable and stationary states Associated kinetic energy	
	Division into compartments by barriers	
	Isotropic motion and kinetic energy (pressure and temperature)	
	Fields between compartments (time independent)	
	Flow in components (chemical change)	Organisation based on instructions (information)
	Supply of energy (continuous) to a compartment	
	Supply of material (continuous) to a compartment	Feedback time-dependent activities (steady and developmental states)

relating survival value in Fig. 8.23 even qualitatively to those features we observe in living objects (see Chapter 16). It is very probable that understanding the relationship will force us to devise new derived variables, much as in thermodynamics, but which are not very usefully broken down to the fundamental units and variables (see Table 8.10 and compare Table 8.2). We turn in the following chapters to the description of particular inanimate and living systems and how they have evolved, so that we can see how the variables of flow have been adjusted over billions of years. In the final chapter we shall review the whole of our effort in order to put as logical a construction as possible upon the connections in flow systems. The ultimate aim is to find a link between matter and man which is the story of the evolution of physical–chemical systems to date.

9

The evolution of Earth

Come and I shall tell thee first of all the beginning of the sun and the sources from which have sprung all the things we now behold—the earth and the billowing sea, the damp vapour and the titan air that binds his circle fast round all things.

Empedocles of Agrigentum (*c.* 450 BC)

9.1 Introduction

The sun and its planetary system formed some 5×10^9 years ago from the association under gravity of an elliptical disc of nuclei and electrons. The central zone of the disc, at a temperature around 10^7K, became the sun and has remained at this temperature while the outside edges cooled progressively to give the planets. The intermediate stages of cooling led to the formation of small molecules of many elements, mainly oxides of hydrogen and carbon, of several non-metals such as Si and P and of most metals such as Na, Mg, Al, K. There was an excess of elements over oxygen so that, much though most elements prefer binding to oxygen over any other combination (Section 5.8), those with weaker binding entered other associations. As a consequence, non-metals such as nitrogen and chlorine combined with the excess of hydrogen or remained free (N_2), and excess metals bound either to sulphur or remained free as metal associates. In a rough and ready way the whole system of elements could be said to have been equilibrated in a gas phase at a temperature above 3000K.

Now, as the temperature dropped further the equilibrium could have adjusted itself to a final condition at the present-day temperature of Earth, as described in Section 5.11.2. However, and fortunately, the state of many of the elements became trapped in their higher temperature combinations due to condensation as liquid or solid particles. Of course, many of the very stable combinations at 3000K remained so at 300K and are present today. In particular, most of the solids now in the mantle are silicates, i.e. mainly complex silicon oxides with aluminium and magnesium and many other incorporated metal elements from groups 2–10 of the first long row of the periodic table. Nitrogen and chlorine remained at first as gases: N_2, NH_3 and HCl. By way of contrast, frozen stationary states became established for metals such as iron and alloys of it (now in the central core of Earth, which were deprived of oxygen) and as a mixture of mainly carbon dioxides and water as gases (now in the atmosphere). The water eventually formed liquid drops or ice particles. The force of gravity first gave rise to larger conglomerations of oxides, of iron and its alloys, and some sulphides in meteorites. The further effect of gravity was to give rise to accretion and then the planets, including Earth, as a conglomerate of meteorites. The planets themselves are very different in composition due to gravitational interactions with the sun and their own fields.

It is thought, then, that the Earth that formed was mainly a mixture of large, meteor-size lumps of solid, of iron and of some iron oxides and silicates, some sulphides, ice (or water) and a large atmosphere of nitrogen, some hydrogen, methane, ammonia and hydrogen chloride, and large quantities of carbon oxides. At this stage there was sufficient energy, largely from radioactivity in the core, to remelt the solids, with the consequences that the heaviest fractions, iron and its alloys settled as liquids in the centre with a mainly oxide, partly liquid, mantle, but with a colder, solid, surface crust. The moderate temperature was not high enough, or did

Table 9.1 Development of the inorganic chemicals of the Earth from the solar system space

Element	Compounds
H	H_2 and H_2O, mostly lost (10^{-8})*
He	Lost
C	CO, (CO_2), mostly lost (10^{-4})*
N	N_2, (NH_3 and HCN?), mostly lost (10^{-4})*
O	MgO, SiO_2, Al_2O_3, CO, (CO_2) and now O_2
Ca, Mg, Al, P, Si, Na, K, Ti, Cr, Mn	Oxides, silicates, $Ca_3(PO_4)_2$, in crust and mantle (and in meteorites)
S	SO_2, much lost, FeS
Cl	NaCl, HCl
Fe/Ni	Alloys based on Fe, since all oxygen was removed by C, Mg, Al, Si in the core (also in certain meteorites)

* The approximate remaining fraction of elements is given in brackets.

not last long enough, to allow equilibrium to be established. During this process the atmosphere thinned out losing much of the gaseous compounds, especially of H, C, and N, to outer space. Residual water evaporated only to form clouds in the Earth's cold outer atmosphere and finally condensed to give the oceans as the least dense (liquid) layer. Table 9.1 summarises this development over the first one billion years.

The most remarkable feature of this situation is the presence of water on the surface. As stated in Section 5.11.2, if all phases of Earth had equilibrated then at low temperature the excess of iron in the core would have reacted with all the water according to

$$Fe + H_2O \rightarrow FeO + H_2$$

The gravity field of Earth cannot retain hydrogen, which would have escaped continuously, and hence we should have expected much more iron oxide in the mantle and somewhat less iron in the interior of the Earth. There would have been no water. Notice too that H_2O as a gas is poorly retained if the atmosphere is hot, $> 100\ °C$, so that water on the surface is unlikely to have been present in large amounts unless combined, e.g. in hydrated salts. One guess as to the origin of the sea, which at first covered virtually the whole of the Earth, is that it formed from water released as vapour when hydrated or hydroxylated silicates were remelted (see above), which then condensed during the final stages of cooling. In any event, the sea (H_2O) is held away from the hot iron core by the mantle in a non-equilibrium system of compartments.

The appearance of land has been gradual and represents another non-equilibrium situation in that some dense solids rose high above the low density sea, see asides. The upward thrust of the land was due to the turbulence of the hot liquid inner zones of the Earth. The upthrust with cooling then generated compartments out of physical balance (balance in the gravity field) and out of chemical balance, since they generated small compartments in the outer crust systems that belong to previously well-equilibrated, largely interior core phases. *The surface of the Earth* then became a multitude of small and large non-equilibrated compartments and chemical components which represent a considerable source of energy in storage.

The above sequence left Earth with zones of different composition shown in Fig. 9.1 (compare with Fig. 7.15). Of course, this is a very generalised description and eruptions through the boundaries of the zones were constant so that through volcanic action many solid particles of all kinds have been continuously generated, thrown on to the land, into the sea and into the atmosphere. Before going forward towards the present day we wish to look in more detail at Earth when it was no more than 1×10^9 years old, that is at about the time life appeared some 4×10^9 years ago. We treat each zone as a separate *stationary* chemical system. In doing so we are entitled to put to one side the deep 'equilibrated' regions of the Earth, which have remained as described in Table 9.1 and Fig. 9.1 and which we believe we understand in essence, and to concentrate upon the surface. Given the many uncertainties that remain concerning the process of formation of this surface we can only outline an account of its probable early character and then describe how it evolved to its present state.

Composition of volcanoes

Sulphurous Gases
Mantle Minerals
(Pyrolite, Alkali
Basalt etc)

Unstable volcanic minerals

$FeS + H_2S \rightarrow FeS_2 + H_2$

$Fe(Ni) + H_2O \rightarrow FeO(NiO) + H_2$

$Heat + MCO_3 \rightarrow MO + CO_2$

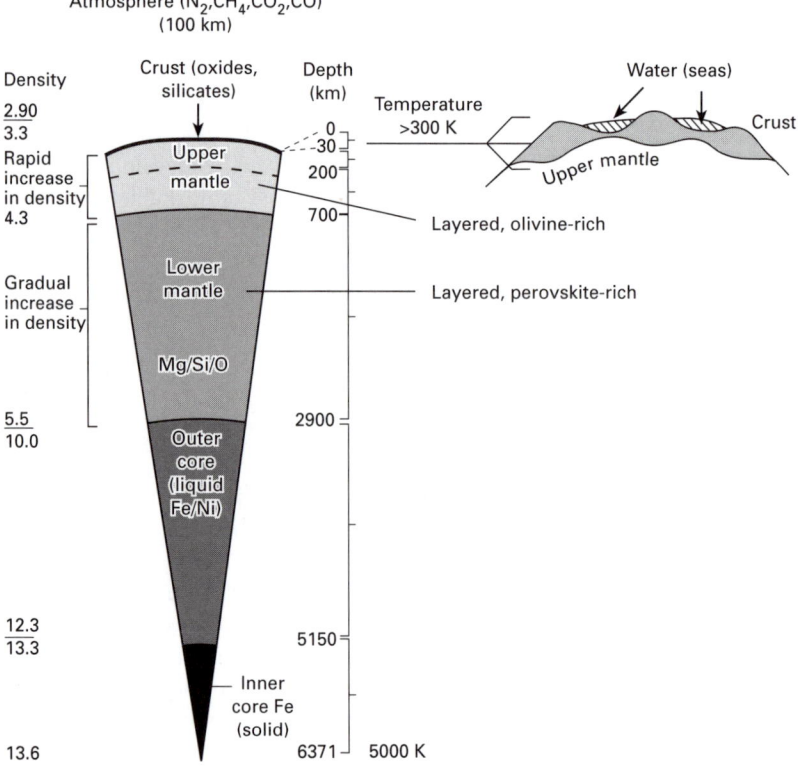

Fig. 9.1 A schematic section through the Earth. Notice that there is a huge source of energy in unbalanced chemical and physical (high temperature) gradients of the Earth. This out-of-balance is due to the route of formation of the planet, that is, it is a kinetically controlled situation due to gravity. Some dramatic consequences of this physiochemical out-of-balance are earthquakes and volcanic eruptions and (possibly) life itself (see Fig. 5.10). The out-of-balance shown in this figure has been increased later by living systems via the production of dioxygen (O_2) and carbon (coal and oil) deposits. In this last series of steps sunlight energy is trapped in the crust.

We do this by looking at its zones—the atmosphere, the hydrosphere (sea) and the (crust) surface—in turn, taking primitive conditions of all three first. Throughout the description we shall be aware that as well as the absence of chemical element and phase equilibration (Table 9.2), there are considerable gradients of temperature and pressure even in the surface zones. The vertical temperature gradient arises from two sources—the very hot internal iron core and the sun. The total effect of the sun is to raise the Earth's temperature some 50–60 °C higher than the core can maintain. The combined effect of changes in the sun, some cooling, volcanic activity on Earth and the change of the atmosphere, e.g. especially loss of CO_2, has meant that over 4×10^9 years the *average* temperature of the surface of Earth has remained approximately constant, although there have been periods (for example, ice ages) when it has changed locally within relatively small limits. Moreover, the sun heats the atmosphere and the surface unequally, so that Earth has horizontal polar and tropical regions. Pressure

Table 9.2 Non-equilibrium inorganic phases on Earth

Phase	Nature of non-equilibrium
Many aqueous solutions	Fresh water brought in contact with land
Clay minerals	Eroded by weathering or activity of plants (see asides)
Unoxidised rock	Product of volcanoes, ocean, vents, etc.
Oceans	Circulation with life forms dominant in upper regions (Temperature switching)
Air	Lack of catalysts for N_2/O_2 reaction; O_2 generally present
Frozen rock	Many examples, e.g. zeolites

rises steeply at depth in the sea and considerably affects the chemical mixture as sediments settle. Edges of continental plates drift and are driven under and over one another. This submerging and emerging of chemicals has influenced the chemistry of the oceans, and the minerals that appear on the surface from original sediments, which have been melted and equilibrated in the mantle surface. Clearly, the surface of the Earth can only be examined in a rather empirical manner as it has evolved.

9.2 The primitive atmosphere

Given the complexity of the stages in the formation of the Earth as it cooled down to temperatures below 100 °C, we have to more or less guess the nature of the early atmosphere. It seems probable that in a first stage, at temperatures <100 °C, there was a high CO_2 pressure (10 bars),* some N_2 (1 bar), CH_4, H_2O and CO, and somewhat less H_2S, HCN and NH_3. It contained virtually no O_2 (perhaps 10^{-13} bar) (see Table 9.3 and Fig. 9.2). While H_2 was being lost at a fair rate, there was always a reserve of reactive combined hydrogen in H_2S and in H_2O. The chemical activity at the surface could have been intense when life began around 4×10^9 years ago. The compounds in the atmosphere were subjected to the effects of storms too, probably of unimaginable strength. Small amounts of many C/H/N/O compounds were undoubtedly formed (see Section 10.4.1) which built up through their kinetic stability at these low temperatures, and these later dissolved in the sea and may well have assisted life's beginnings. These compounds would have been energised so that they could change with time. Of particular interest is the amount of NO (nitric oxide) which may have been generated at that stage (see Chapter 11) and possibly HCHO.

We have to observe next that the formation of the atmosphere at <100 °C corresponded to the creation of a planetary compartment far from temperature equilibrium with the neighbouring star (the sun). Like the formation of stars, the formation of planets is another result of the action of gravitational forces, which, in the case of the solar system, leave these bodies relatively cold but bathed in the very high radiation field of sun, which itself is trapped in its galaxy's gravity field. The radiation

*1 bar is approximately 1 atmosphere.

Table 9.3 Summary of data on the probable chemical composition of the atmosphere during stages I, II and III**

Stage I (early Earth)	Stage II (~2 × 10⁹ years ago)	Stage III (Today)
Major components ($p > 10^{-2}$ atm)		
CO_2 (10 bar)		
N_2 (1 bar)	N_2	N_2
CH_4		O_2
CO		
Minor components ($10^{-2} > p > 10^{-6}$ atm)		
H_2 (?)	O_2 (?)	Argon
H_2O	H_2O	H_2O
H_2S	CO_2	CO_2 (10^{-3} bar)
NH_3	Argon	
Argon	(CO?)	
NO(?)	(NO?)	
Trace components ($p < 10^{-6}$ atm)		
He	Ne	Ne
Ne	He	He
	CH_4	CH_4
	NH_3 (?)	CO
	SO_2	NO
O_2 (10^{-13} bar)	H_2S (?)	

* We are able to give a good account of stage III (Section 9.6) and a good estimate of stage I, but the evolutionary period, stage II, is hard to describe with any accuracy.

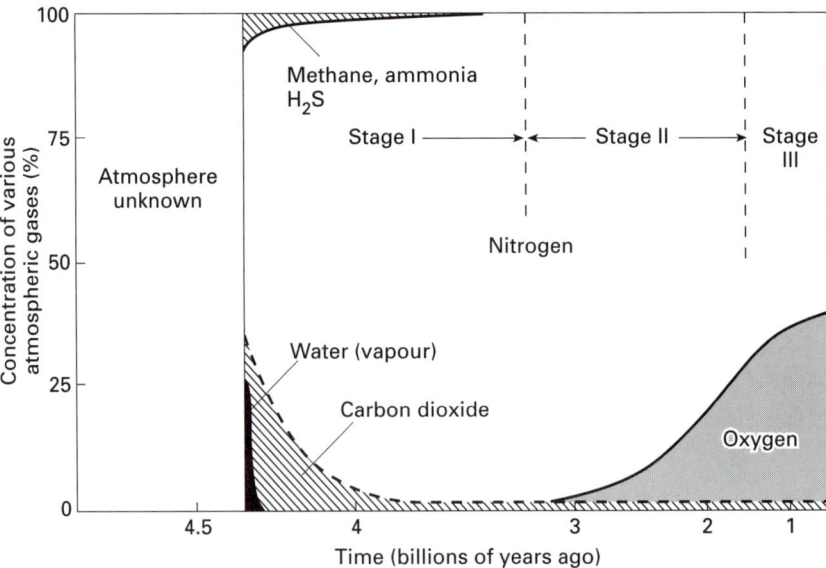

Fig. 9.2 Atmospheric composition, shown by the relative concentration of various gases, has been greatly influenced by life on Earth. Note that the total pressure has fallen greatly, and light gases have been lost in part, but nitrogen always dominates.

initiated chemical reactions of the atmospheric gases, for example of CO_2, N_2 and H_2O, to give some organic compounds and perhaps some dioxygen.

The atmosphere, therefore, constantly interacted with the surface of the Earth (including primitive living systems) and with the sun, and suffered

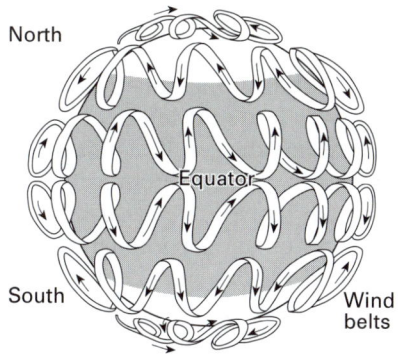

North

Equator

South

Wind belts

Fig. 9.3 Temperature differences around the world give rise to the winds. If the Earth did not turn, winds would tend to blow in a north–south direction, but the spinning of the Earth deflects them so that they blow east or west.

outpourings from inner zones. On the other hand, as stated, the sun's energy fell unequally on Earth, so that the atmosphere had both a horizontal and vertical temperature gradient and a composition gradient. The fact that the atmosphere as a whole was (and is) far from equilibrium in various dimensions and directions, means that it was (and is) an unstable mixture, always changing with time. It was and remains a dynamic system, and in flow, due to these large energy-generating inhomogeneities (see Fig. 9.3). The flows produced within the Earth's atmospheric systems were constrained by the Earth's surface and their impact on its solids caused erosion and so affected the Earth's surface physics and chemistry (see Sections 9.3 and 9.4).

9.3   The nature of the early sea

Table 9.4 The characteristics of the early ocean and of the ocean today

Proto-ocean (?)
pH = 2.0 (initial); T = 80 °C
CO_2 and SO_2 not very soluble
HCl gives the acidity
Initially weak content of cations, but increasing to Ca^{2+}, 115 mM; Mg^{2+}, 95 mM; Na^+, 120 mM; K^+, 60 mM
Redox potential around −0.5 to 0.0 volts

Early ocean
pH = 8.0; t = 55 °C
HCO_3^- (CO_2) high; SO_4^{2-} low; H_2S high
$Ca^{2+} \geqslant$ 10 mM
Fe^{2+}, 1 mM; $Zn^{2+} < 10^{-10}$ M
Redox potential > 0.0 rising to < 0.4 volts

Late ocean (today)
pH = 8.0; T = 25 °C
HCO_3^- (CO_2) high, and SO_4^{2-} (not H_2S) present
Average concentrations of cations are Ca^{2+}, 10 mM; Mg^{2+}, 105 mM; Na^+, 470 mM; K^+, 10 mM
Redox potential up to 0.80 volts at surface (O_2)
Fe^{3+}, 10^{-17}M; Cu^{2+}, etc., see Fig. 6.17.

The very earliest sea, as far as we can judge, must have been very different from that of today owing to the peculiar initial conditions when water condensed. As stated above, it is thought that, initially, this sea covered virtually all of the mineral crust. Note that the temperature would have been close to 100 °C. It is quite possible that the rain that fell then was acid, mostly from HCl. Indeed, the original source of the chloride of the sea may well have been gaseous HCl since, as stated, chlorine has such a low relative affinity for most elements other than H. In addition, there could have been some H_2S and perhaps SO_2, and much CO_2 in the atmosphere, which would also have contributed to acidity. The acid rain would have eroded the surface and submerged minerals, some of which are basic, giving progressively a less acidic solution.

The best guess at the proto-ocean composition, estimated to be initially at pH 2–4, is that of Table 9.4. Very quickly, further erosion moved the pH up to around 5 and then, more slowly, to the neutral and then to a slightly alkaline region (pH ~ 8) where it has remained. During the rise in pH, land masses separated from the sea and one another because of fluctuations of the surface and there was also precipitation of carbonates as well as sulphides (later to become sulphates). (The solubility products of such salts and their dependence on pH are described in Section 6.5.) The sea itself was most likely black with metal sulphide particles and clusters from volcanoes. The major precipitate was $CaCO_3$, eventually removing most of the calcium from the sea and most of the CO_2 from the atmosphere. This removal was continuous owing to the constant transport of $CaCO_3$ from the dissociated condition in the sea to huge deposits on land. The sea quickly became, effectively, a saturated calcium salt solution, and maybe it had not changed dramatically before life was engaged in the carbon cycle. The redox potential of the early sea, shown in Fig. 9.4, was

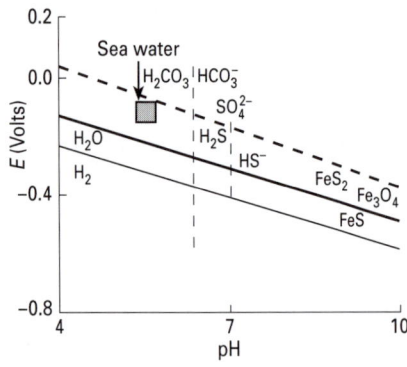

Fig. 9.4 A pH versus oxidation potential, E, diagram for some components of (early) sea water with special reference to sulphur/iron chemistry. (After M. Russell).

Table 9.5 Some trace elements in the early sea*

Elements present
Fe^{2+}, Mn^{2+}, (Mo^{6+}), V^{4+}, (Ni^{2+}), W^{6+}, (Co^{2+}), Se as H_2Se

Elements largely absent
Cu^{2+}, Cd^{2+}, Zn^{2+}, Cr^{3+}, Ti^{3+}

* The assumption is that the pH $\geqslant$ 5 and the amount of H_2S kept the sea as a reducing medium (see Fig. 9.4). The concentration of Mo^{6+} may have been lower than that of W^{6+} as Mo is precipitated as MoS_2 at low pH

very reducing, between 0.0 and –0.5 volts at pH = 5.5, owing to the presence of sulphides and bound hydrogen. The resulting trace elements in the sea are given in Table 9.5 (see Fig. 6.17). We have to assume that there were available sources of N and C (NO and HCHO) to start life.

Note, however, that the sea has never been just the result of this acidic rain and/or runoff from rivers of fresh water. Rivers are far from being saturated salt solutions and contain small particles as well as salts. The small particles settle at the river outlets to the oceans and develop into large delta lands, which, together with wind-borne particles, form new land masses. All of these 'compartments' affect the sea as do its own tidal and streaming flows, the movement of continental masses, and of tectonic plates, as well as internal volcanic submarine activity. It is believed that the entire volume of the oceans is 'recycled' through the flanks of ocean ridges every 10 million years or so.

As a consequence of the upsurging of molten rock from the fissures of the solid crust there were, and to this day there continues to be, deep local zones of the sea, which are under quite considerable pressure, at a temperature in the region of 100 °C (Fig. 9.5). At such temperatures the solubilities of many substances in water are enhanced, while for others they are reduced. At the same time, reactivity both for reactions leading to complex molecules and for their decomposition was also increased. It is sometimes thought that primitive life may have begun under these conditions (see Chapters 10 and 11) but the conditions are equally biased both for it, from synthetic reactivity, and against it, from degradativity activity. In this book we shall not stress these peculiar regions of the sea since we are not concerned so much with the origin of life but with its evolution under more normal conditions.

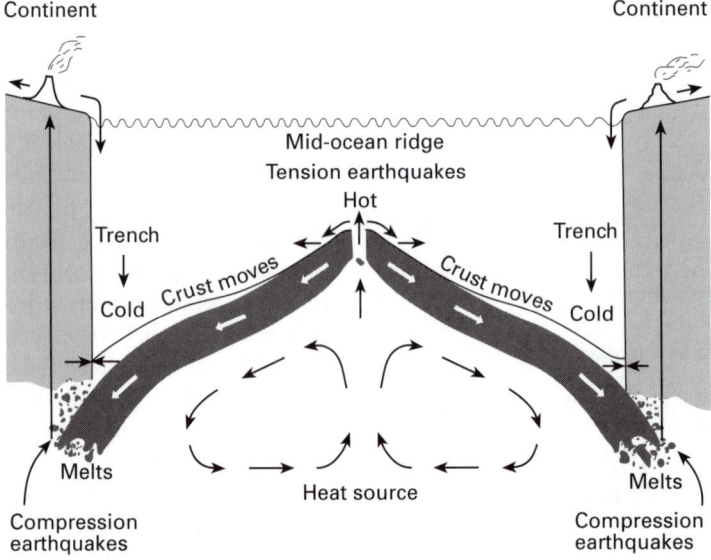

Fig. 9.5 The flow of heat and molten rocks to and from the mid-ocean ridges.

9.4 The surface of the early Earth

While we have been able to give a good impression of the whole Earth and of the early atmosphere, and a somewhat poorer one of the early seas and of the layering of the deep inaccessible ocean regions of the Earth near ocean trenches, we will not have such an easy task with the rocks, soils and clays of the Earth's surface (see Table 9.6), which have often emerged from the crust. There are three problems that complicate the situation. First, sudden eruptions with violent local increases of temperature force solids to dissolve into molten phases which, on cooling rapidly, do not come to equilibrium—all kinds of frozen solid solutions and mixtures are then expected on the surface, containing many components. Secondly, wind and rain generate deposits of small particles. These soils, formed by erosion and deposits of solids due to gravity, can be very fertile since they can contain the necessary diversity of inorganic elements for life, and maybe it was the development of fertile sands and soils that allowed plant life to move from the sea. However, sands can also be extremely infertile, as in the case of spreading deserts. The deserts are often very short of trace inorganic elements, apart from their shortage of water and organic compounds, so that they may not sustain life easily. We return to a brief description of soils in Section 9.9. Thirdly, the subjection to high pressure on folding and unfolding under the surface at the edges of continental plates, and re-emergence later, led to the formation of the peculiar sedimentary and metamorphic rocks.

Scattered on the surface of the Earth there is, therefore, a huge variety of mixed mineral deposits (Table 9.6 and see Section 9.10). In the early periods sulphur would have had a considerable involvement in these

Table 9.6 The geological history of surface ore formation

Era (billion years before present)	Geological and chemical features	Major ores formed
Early Archaean (3.8–3.0)	Submarine trench formation; basic magma flows give *primary greenstones**	Fe, Ni, Cu sulphides, Au
Late Archaean (3.0–2.5)	Recycling of primary greenstones, hydrothermal processes	Cu, Zn hydrothermal sulphides
Early Proterozoic (2.5–1.7)	Uplifted crust erodes O_2 produced by photosynthesis, oxidation of Fe^{2+}	Au deposits Banded Fe formations
Mid–late Proterozoic (1.7–0.7)	Thick continental crust forms Atmospheric O_2 increases; active sulphur redox chemistry	Ti, Cr oxides, Fe sulphide, Pt metals Co, Cu, U deposits
Phanerozoic (0.7–present)	Extensive crustal recycling	Hydrothermal Cu, Zn, Mo, Sn, Pb
	Tropical weathering conditions Secondary enrichment	Al, Fe resistates[†] Co, Ni, Cu minerals

* Greenstones—named after deposits in South Africa.
[†] Resistates—rocks resisting weathering.

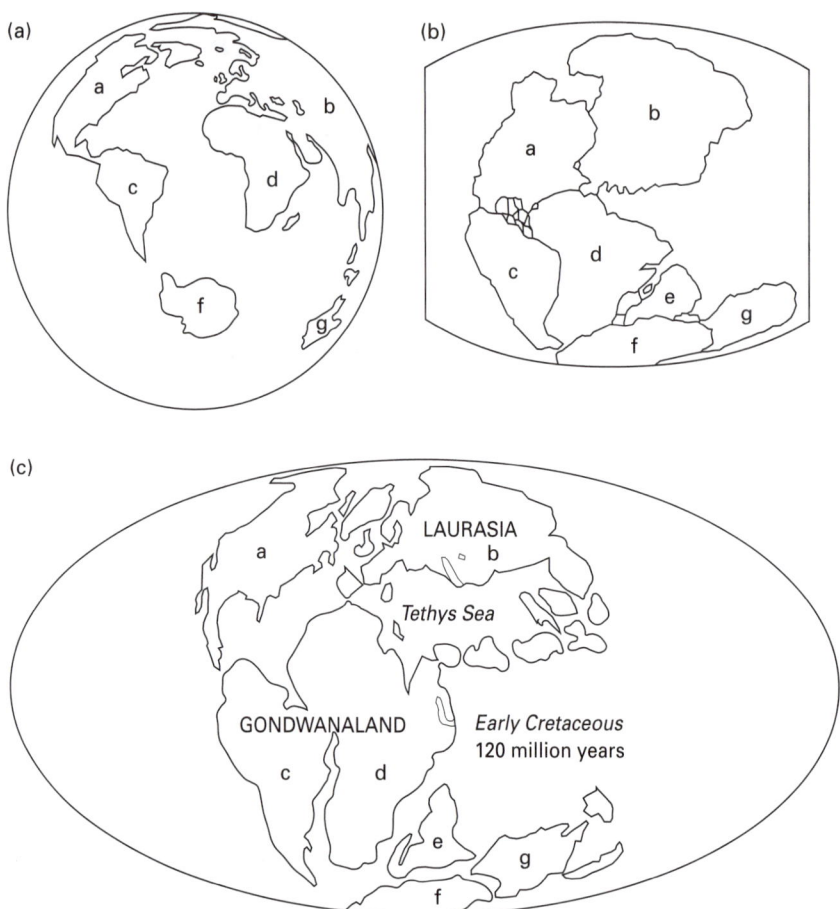

Fig. 9.6 The development of the large, major continents: (a) today; (b) > 200 x 10^6 years ago; (c) 120 x 10^6 years ago. Fragments of the original super-continent are the continents we know today: a, North America; b, Asia; c, South America; d, Africa; e, India; f, Antartica; and g, Australia.

surface reactions although at all times oxides dominated. Owing to all this relative movement it is even uncertain how the continents of today developed in relatively recent times, though they seem to have started as one land mass, then as separate lands called Gondwana and Laurasia, which split up in many ways over the geological periods (Fig 9.6). The land masses drifted into different temperature zones and have thus suffered erosion and volcanic activity in diverse ways. This separation has greatly affected the evolution of living organisms.

Unfortunately, therefore, when dealing with the natural selection of the chemicals of the Earth's early surface, the convenient and simple approach based on chemical equilibria that we have used in Chapters 5–7 and for bulk considerations of Earth in this chapter, is largely lost; it becomes only useful background knowledge. In part, the problems stem from the nature of the conditions in complex phases such as solutions and melts (see Table 9.1). We must see the Earth's early surface zones as a vast dynamic system in both a chemical and physical sense. Flow (movement) is seemingly random for the tectonic plates, causing continental drift, but flow in the atmosphere and sea (see Chapter 8) is an effect of the gravitational and radiation field gradients in which Earth was generated and now sits. Fortunately, for the nature of our story, the bringing of chemistry to life, a much more dramatic and systematic development has occurred in the sea—the beginnings of life. The sea remains a more easily appreciated environment than the sands and soils of Earth.

9.5 Some early non-equilibrated inorganic compartments of interest: an aside

The most obvious of the non-equilibrium inorganic solid structures are the clays and zeolites (see Table 9.2). These structures are based on Al/Si/O/M lattices with open frames (Figs 9.7–9.9). The zeolites (see Fig. 9.9) can be

Geometry of linkage of SiO_4 tetrahedra	Si/O ratio		Example mineral	Formula
Isolated tetrahedra: linked by bonds sharing oxygens only through cations	1:4		Olivine	$(Mg,Fe)_2SiO_4$
Rings of tetrahedra: joined by shared oxygens in three-, four-, or six-membered rings	1:3	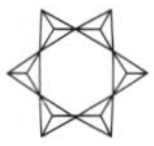	Beryl	$BeAl_2(Si_6O_{18})$
Single chains: each tetrahedron linked to two others by shared oxygens. Chains bonded by cations	1:3		Pyroxene	$(Mg,Fe)SiO_3$
Double chains: two chains joined by shared oxygens as well as cations	4:11	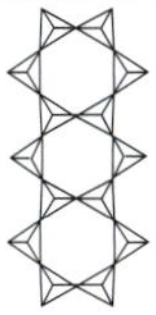	Amphibole $(Ca_2Mg_5)Si_8O_{22}(OH)_2$	
Sheets: each tetrahedron linked to three others by shared oxygens. Sheets bonded by cations or alumina sheets	2:5		Kaolinite	$Al_2Si_2O_5(OH)_4$
Frameworks: each tetrahedron shares all its oxygens with other SiO_4 tetrahedra (in quartz) or AlO_4 tetrahedra	3:8		Feldspar (albite)	$NaAlSi_3O_8$
	1:2		Quartz	SiO_2

Fig. 9.7 Some major silicate structures. [After Press, F. and Sievers, R. (1986). *Earth*, 4th edn. W. H. Freeman and Company, New York. With permission.]

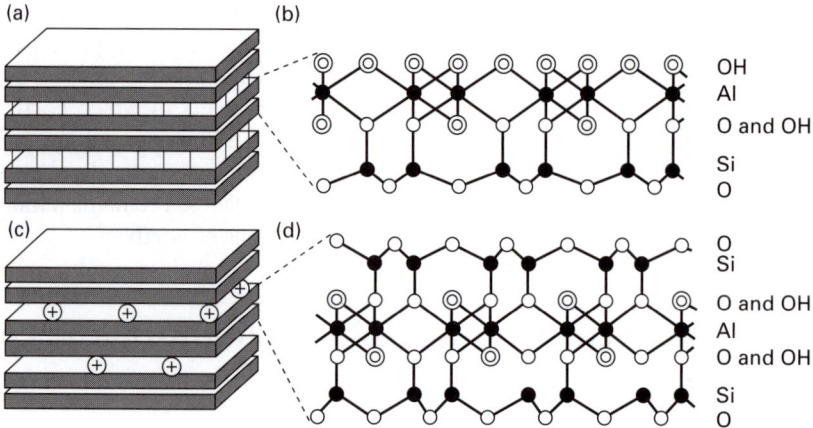

○ Oxygen
◎ Hydroxyl group
● Silicon
● Aluminium
⊕ Positive ion

Fig. 9.8 Most clays are made up of stacks of silicate layers. In kaolinite (a) asymmetric layers are linked by hydrogen bonds. Each layer consists of a net of aluminium atoms and hydroxyl (OH) groups fused to a net of silicon and oxygen atoms (b). Other clays have symmetric layers, in which a silicon–oxygen net is fused on both sides to a metal–hydroxyl net; these layers are negatively charged and linked by positive ions (c). In illites (d) much of the negative charge arises from the substitution of aluminium for silicon atoms.

made in a variety of forms and, curiously, they have some features in common with proteins and polysaccharide, such as:

(1) preparation by energised condensation polymerisation;

(2) relatively ready hydrolysis;

(3) side-chains that give acid-base catalysis [see Fig. 9.9(b)]; and

(4) easy incorporation of metal ions for catalysis, now including reductive–oxidative reactions.

In one sense they are less complex than proteins in that they have fewer building blocks, just $Si(OH)_4$, but the blocks can be assembled in many non-linear ways and they are truly three-dimensional, not linear covalent systems [contrast with proteins but compare also with polysaccharides (Chapter 10)]. Interestingly, there are theories of the origin of life that start from these silicate clays (see Section 9.12 and refs 87–89 and 118 in Further reading). A feature that they lack, and that is of great importance in proteins, is flexibility. The treatment of their thermodynamic properties is very different from that of organic polymers (see Section 5.8.5).

There are also may other components locked in compartments, such as the sulphides. Many are quite reactive but remain available for reasonable lengths of time. Of course, the components and compartments are energy stores. Here, the most important is FeS which is unstable in the presence of

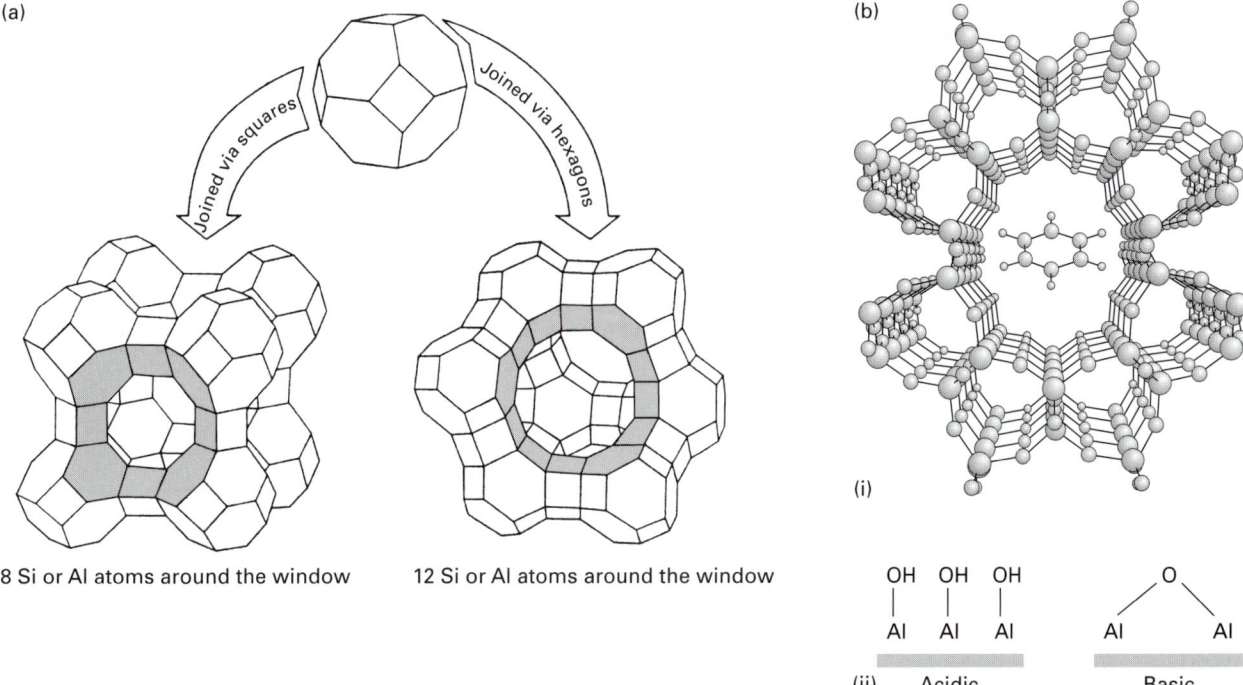

Fig. 9.9 (a) An example of a zeolite showing the distribution of silicon and aluminium atoms around the window. (b) (i) The view through the channels in Theta-1 zeolite with an absorbed benzene molecule in one of the channels. [Reproduced by permission from Dyer, A. (1988). *An introduction to molecular sieves*. John Wiley, Chichester.] (ii) Catalytic sites on the surface of a zeolite. Compare with protein channels in Chapter 8.

H_2S giving pyrite, FeS_2 and H_2, plus energy, a reaction which could have been part of the initial steps towards life; the hydrogen being trapped in a bound form.

Much more dramatic than the development of interesting inorganic chemicals *per se*, has been the development of inorganic chemical reactions leading to life processes (see Chapters 10–13). Now, this development requires energy. As stated above, there are many available sources, including direct effects of lightning and the sun on molecules such as CO_2, N_2, H_2O. The formation of NO is particularly interesting since it binds to Fe^{2+} and to some sulphide centres so that it could have given both the nitrogen for synthesis and the energy required for life through inorganic reactions. We have mentioned above other energy sources such as unstable silicates and iron sulphides. All such considerations are, of course, speculative and only given some credibility by present-day experiments.

We stress here the value of seeing all these zones in the light of the treatment of variables in Chapters 1–8. Much though the diversity of these surface zones of Earth leads to difficulties in description, there are but two variables, composition and energy, which change locally. For the most part, then, our analysis of the primitive Earth's surface reveals an initial system of inorganic stationary states, based on all 90 available elements, very differently distributed in two large compartments, the atmosphere and the sea, and in a multitude of small compartments from rocks down to small particles in soils. Putting together all the material evidence available, we have an appreciation of what could have been, but little certain knowledge.

Now that we have described the major partitioning of the elements at the early period of the Earth's formation, we shall turn to a more detailed analysis of the atmosphere, the water and the crust on the Earth's surface today. In large part the change was generated by the large-scale production of dioxygen by living organisms that converted the Earth's surface into an oxidised rather than a reduced layer (Section 9.10). The oxidation may have come close to completion some 700 million years ago, but this is not certain and can only be considered in terms of a dynamic steady state, not a thermodynamic balance.

9.6   The atmosphere today

Although the analysis of the present-day air was not easily accomplished, it was largely achieved before 1800. The air has two 'element' gases (N_2 and O_2), two 'compound' gases (CO_2 and H_2O) and some noble gases, particularly argon which is even more abundant than CO_2 (Table 9.3 and Fig. 9.2). Thus, it has kept all the inorganic compounds necessary for the production of those organic, life-giving, compounds made from only H, C, N and O. We stress again that the air we know today is not the 'air' present when life started; the original 'air' contained virtually no O_2 (Table 9.3 and see Fig. 6.15). Again, today's air does not have a balanced composition in a gravity field; for example, in the stratosphere it contains much ozone, O_3, produced by the action of the sun on O_2. Here is another example of vertical energisation, compare the crust of the Earth and the sea. Clearly, the composition of the air today does not relate to the composition of gases that should exist in equilibrium over the surface of the Earth (see Table 9.7). It is true that the water vapour pressure is not far from saturated, that is, humidity varies from 50–100 per cent, and this itself is the

Table 9.7 Some chemically reactive gases of the air today*

Gas	Abundance (%)	Flux (10^6 tons)	Extent disequilibrium
Nitrogen	79	300	10^{10}
Oxygen	21	100 000	Taken as reference
Carbon dioxide	0.03	140 000	10^3
Methane	10^{-4}	500	'Infinite'
Nitrous oxide	10^{-5}	30	10^{13}
Ammonia	10^{-6}	300	'Infinite'
Dimethyl sulphide	10^{-8}	70	'Infinite'
Methyl chloride	10^{-7}	10	'Infinite'
Methyl iodide	10^{-10}	1	'Infinite'

* Abundance and flux (rate of annual flow through the atmosphere) of some of the gases of the air, and the extent to which this composition seems to violate the ordinary rules of equilibrium chemistry. (The word 'infinite' in this context means beyond the limits of computation.) [Lovelock, J. (1979). *Gaia—a new look at life on Earth*. Oxford University Press, Oxford.]

cause of water cycles. However, the nitrogen in the air should form some nitrogen oxides and the dioxygen of the air should, as well as contributing some nitrogen oxides, more obviously consume all the carbon compounds on the surface, including all life. Even carbon dioxide is only close to equilibrium with its solution in the sea or the formation of carbonates. Table 9.7 gives examples of all these out-of-balances in the atmosphere. We shall ask, in Chapter 13, how this atmosphere arose from the original atmosphere and we shall find that it is due to life and to the fact that light gases that do not condense can escape the gravitational field of Earth.

The situation is that the gases, especially H_2O, N_2 and CO_2, in the air are not very reactive at normal atmospheric temperatures, $\sim 25\,°C$. Even O_2, although a quite reactive substance, does not start to 'burn' organic matter easily in the temperature range $25–100\,°C$. In other words, air (O_2) requires catalysts in order to react. Again, H_2O, N_2 and CO_2 are stable molecules and to force them to become part of organic compounds, such as sugars, DNA, RNA and proteins, it is necessary to supply energy from the sun (as in biology), lightning or from appropriate devices for chemical synthesis in a laboratory or factory. There is no longer easy access to energised minerals, either sulphides or silicates, as in the primitive environment.

9.7   The nature of the sea today

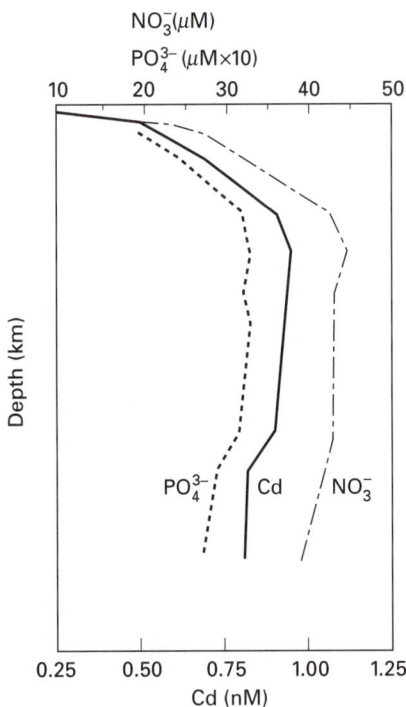

Depth profiles for NO_3^-, PO_4^{3-} and Cd in the Pacific Ocean

The inorganic elements of the sea today (Fig. 9.10), and of the lakes and rivers, accumulated after the development of sedimentary land, of course, and after much interaction with the changing gases of the air and of eruptions of the solid crust. (Notice how thin is the layer of water at the surface of the Earth; sea water accounts for about 90 per cent of all water on the Earth, but oceans occupy only a tiny part of the volume of the planet. If the Earth is represented by a sphere with a diameter of 30 cm the oceans form only a surface layer as thin as this sheet of paper and the largest abyss, ~ 11 km, would be no more than 3 mm deep! It is then easy to see that it can change rapidly with changes in other zones.) Certain general chemical features are worthy of note since, as we shall see, they affect element accumulation and combination in life. We treat the sea separately from fresh water.

The ease of extraction of elements into the surface of the sea from the land, i.e. in the presence of dioxygen, is shown in Figs 9.10 and 9.11. The composition differs greatly from that resulting from the activity of the early sea on the early minerals (see Fig. 6.17). The removal of oxidised iron (Fe^{3+}) and other ions (e.g. Mn^{3+}) from sea water by precipitation as oxides/hydroxides contrasts with the relatively unchanged availability of less abundant elements, such as vanadium, molybdenum and selenium, which remain soluble in oxidised forms (the absolute values are given in Fig. 6.17). However, over 4×10^9 years, the changes have been very considerable (see Table 9.4 and 9.8). The situation at the bottom of the sea or of a lake can be very different (see aside) since sulphide removes elements such as copper, and to some extent zinc, but does not remove reduced iron,

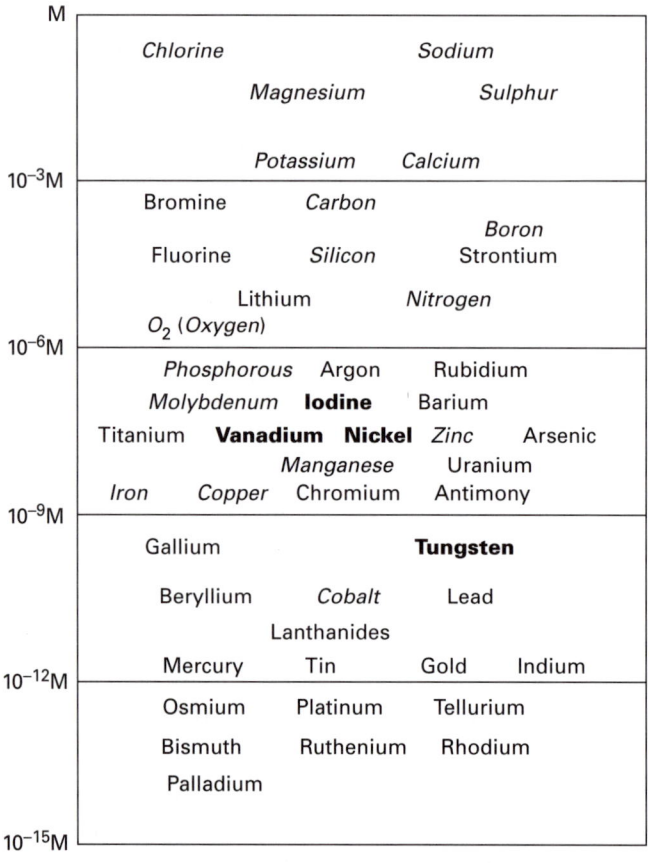

Fig. 9.10 Sea water is more than just salty. It is a selection of many elements, although some of them are present in minute quantities. See also Fig. 6.17. Usual biological elements, *italics*; rare biological elements, heavy type; non-biological elements, normal type.

Fe^{2+}, or manganese, Mn^{2+}. It is believed that the conditions of deep *static* water are close to those of much earlier times when life began (Fig. 6.17). (However, much of the deep ocean is not anoxic since the ocean waters circulate.) Under these conditions, a column of water in the natural world has a steep gradient of elements (and other nutrients). The consequence is that different life forms appear in the different zones (see Fig. 13.14).

The seas (and lakes) are thus structures varying vertically and horizontally, based on energy flow from the sun and the mantle. However, the seas also flow round the Earth in huge currents, such as the Gulf Stream, and the seas are tidal too, whence there is a constant circulation of contents from zone to zone. All these features affect the chemistry in given localities in different ways, but the flows ensure that the different regions are of relatively constant analytical composition. Notwithstanding, the sea is never at equilibrium with the Earth, although some reactions are quite close to equilibrium. The composition given in Table 9.8 is then very much a general average. The organic chemical content of the sea is given in Table 9.9.

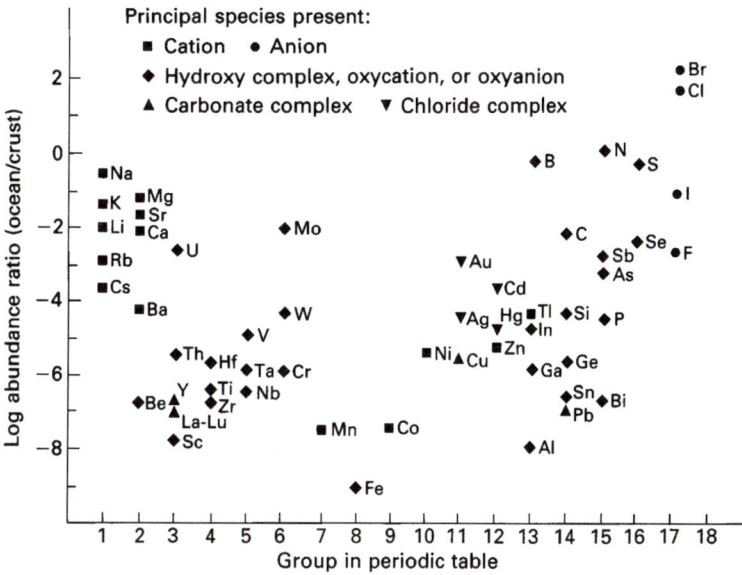

Fig. 9.11 This figure shows the ease of extraction of elements from the Earth by the sea *today* and is a guide to the ease with which biological systems will have access to elements, no matter what the absolute abundance. The top of the figure indicates the speciation. [From Cox, P. A. (1989). *The elements*. Oxford University Press, Oxford. With permission.]

Table 9.8 Available free concentrations in the sea as they changed with time

Metal ion	Original conditions (molar)*	Aerobic conditions (molar)*
Na^+	$>10^{-1}$	$>10^{-1}$
K^+	$\sim 10^{-2}$	$\sim 10^{-2}$
Mg^{2+}	$\sim 10^{-2}$	$\sim 10^{-2}$
Ca^{2+}	$\sim 10^{-3}$	$\sim 10^{-3}$
Mn^{2+}	$\sim 10^{-6}$	$\sim 10^{-8}$
Fe	$\sim 10^{-7}$ (Fe^{II})	$\sim 10^{-19}$ (Fe^{III})
Co^{2+}	$\sim 10^{-9}$	$\sim (10^{-9})$
Ni^{2+}	$<10^{-9}$	$<10^{-9}$
Cu	$<10^{-20}$ (very low), Cu^I	$<10^{-10}$, Cu^{II}
Zn^{2+}	$<10^{-12}$ (low)	$\sim 10^{-8}$
Mo	$\sim 10^{-9}$ (MoS_4^{2-}, $Mo(OH)_6$)	10^{-8} (MoO_4^{2-})
W	WS_4^{2-}	10^{-9} (WO_4^{2-})
H^+	pH low (5.5?)	pH 8.5 (sea)
H_2S	10^{-2}	Low [SO_4^{2-} (10^{-2} molar)]
HPO_4^{2-}	$<10^{-3}$	$<10^{-3}$ molar
$O_2^\dagger$	$<10^{-6}$ atm	$\sim 10^{-1}$ atm (21%)
CO_2	>10 atm	10^{-2} atm
N_2	$>?10$ atm (NH_3?)	~ 1 atm (78%)

* Except where other units are specified.
† N.B. Changes in the air.

9.7.1 The nature of fresh water today

The problem of fresh water solutions, especially those running over the land in rivers and lakes, is that there is a variation from the extremes of

Table 9.9 Average concentrations of organics in sea water

Components	Concentration in sea water (μg carbon litre^{-1})
Free amino acids	20
Combined amino acids	50 (to 100?)
Free sugars	20
Fatty acids	10
Phenols	2
Sterols	0.2
Vitamins	0.006
Ketones	10
Aldehydes	5
Hydrocarbons	5
Urea	20
Uronic acids	18
Approximate total	340

'pure' water, coming from rain or ice to soil water and then to water that is more closely related to sea water or even exceeds it in saltiness, for example in the Dead Sea (which is a lake). Although rain water is not free from dissolved compounds, it is nearly so. Once the water starts to flow over the land its composition changes due to its effects on rocks (erosion) and to biological contamination. There is, then, a local condition of fresh water, just as there is a local condition of the life in it, dependent on temperature, rates of flow and evaporation, and so on. We have to treat each system separately, but we can discern the principle that, to a good approximation, each system is constant under given conditions. This is attested to by the fact that year in and year out the flora and fauna over millions of years are seasonal and local—the cyclic flow of inorganic chemicals and life is never-

Table 9.10 Forms in which the main biological elements occur in aerated soil, water, rivers, lakes and sea (or in blood plasma)—simple species

Cations	Anions	Neutral species
NH_4^+, H_3O^+	HCO_3^-, CO_3^{2-}, NO_3^-	H_2O, $B(OH)_3$
Na^+, K^+	$H_2PO_4^-$, HPO_4^{2-}	CO_2, $SiO_2.nH_2O$
Mg^{2+}, Ca^{2+}	OH^-, F^-, Cl^-, Br^-, I^-, SO_4^{2-}	N_2, NH_3, O_2

Table 9.11 Mean chemical contents (mg/Kg) of world river water

Cations		Anions		Neutral species	
Na^+	6.3	HCO_3^-	58.4	SiO_2	13.1
K^+	2.3	SO_4^{2-}	11.2	$B(OH)_3$	0.013
Mg^{2+}	4.1	Cl^-	7.8		
Ca^{2+}	15	NO_3^-	1		
Fe^{2+}	0.7	Br^-	0.02		
Ba^{2+}	0.05	I^-	0.002		
V, Ni	0.01	F^-	<1		
Cu, Zn		$H_2PO_4^-$	(1)	Total 120	

Data from Henderson, P. (1982). *Inorganic biochemistry*. Pergamon Press, Oxford.

Table 9.12 Abundances of the elements in the Earth's crust by weight (after Gibson). See Fig. 1.10

Relative abundance by weight

Element	%	Element	%
O	50	K	2.4
Si	26	Mg	1.9
Al	7.5	H	0.9
Fe	4.7	Ti	0.6
Ca	3.4	Cl	0.2
Na	2.6	P	0.1

Abundance expressed in g/(metric) ton*

<1 kg/ton	Mn, C, S, Ba, Cr, N, F, Zr, Zn, Ni, Sr, V
<100 g/ton	Cu, Y, W, Li, Rb, Hf, Ce, Pb, Th, Nd, Co, B
<10 g/ton	Mo, Br, Sn, Sc, Be, La, As, Ar, Ge
<1 g/ton	Se, Nb, Sb, U, Ta, Ga, In, Tl, Cd
~mg/ton	I, Pt metals, Ag, Bi, Hg, Te, Au, noble gases
~mg/1000 tons	Ra, Ac, Po

* The abundance of the remaining elements are conveniently expressed in g/(metric) ton 1 kg/ton = 0.1%) of crustal rock.

ending. (The Chinese, see Chapter 1, were very aware of this and perhaps their resulting belief of an inability to change things caused their society to stagnate for a long time. It seems likely that the Western world today will reach the same conclusion by trial and error!) How should we look upon such chemical systems? Perhaps the simplest way forward is to note that, no matter what the source, the water, in contact with the land, though a dilute solution, becomes much richer in certain elements than others. These are the common abundant elements that have reasonable solubility at pH = 6–8. They are listed in Table 9.10. (Also see Table 9.11 for averaged element content of rivers.)

The difference between fresh water and the sea may appear to be of little concern to the overall purpose of this book, i.e. 'bringing chemistry to life'. However, it is thought that, although life started in a calcium carbonate-saturated sea, it was possibly its migration to the rivers that allowed the development of bony phosphate animals (modern fish) and thus invasion of the land by animals, following that by plants. It may well be that controlled use of Ca^{2+}, CO_3^{2-} and HPO_4^{2-} in different waters has allowed animals to emerge from life in water.

Finally, we must refer again to soil water which remains for long periods locked into clays (Section 9.5). This water is the source of nutrients for land plants. Before we can discuss it thoroughly, however, we need to know the composition of soils (discussed in Section 9.9) after analysing the composition of the Earth's crust today and the trace element fractionation in rocks and soils of the crust (Section 9.10), which greatly influences availability in any given era.

9.8 The nature of the crust today

Since, at low temperature (less than 100 °C) most of the more abundant elements have a higher affinity for oxygen* than for any other element (see Section 5.9) and since there is plenty of O_2 today, this element, held far from equilibrium, now totally dominates the composition of much of the top layers of Earth's solid surface (or crust) in metal silicates and carbonates, where the metal is mainly Al, Mg, Ca and Fe or even some Na in granite, soils and clays (see Table 9.12 and Figs 9.7–9.9) and in Earth's only liquid, H_2O. Oxygen represents 46.6 per cent of the calculated weight of the crust, 62.6 per cent of its atoms and 91.7 per cent of its volume (see Table 9.12). With oxygen in excess, locally, at the surface, new oxides have formed, such as those of iron and tin, and deposits of sulphates and nitrates have appeared. Much carbon, originally as CO_2 in the atmosphere or dissolved in water, partly as H_2CO_3, is associated now in carbonates such as $CaCO_3$ and $MgCO_3$. Some of it is also found in coal and in the hydrocarbons of oil and natural gas. These are, like oxygen, products of life's chemistry (see Section 9.11.4), but they have formed new isolated com-

---

*At room temperature the decreasing affinity for oxygen follows the order Ca, Mg, Al, Ti, Si, Na, K and then Mn, Zn, C, Fe, H, Cu.

partments, as already stated, and their gases are not in equilibrium with O_2. (This is a clear case of compartmental kinetic isolation.) Finally, because surface oxygen is largely mopped up by light elements such as H, Al, Si, C, P, and now Fe, there remains an excess of S, particularly in deeper zones of water, and there several heavy metals that have lower affinity for O remain as partners of that element, for example as Fe, Co, Ni, Cu and Zn sulphides, or occur as free metals or alloys.

9.9   The nature of soils and soil water

While we have often attempted to describe the composition of the Earth's surface in terms of the phase rule (Section 4.7) and aqueous solution equilibria (section 6.15), the possibility of doing so was lost once we came to examine the uppermost layer of the crust—soils. Soils are formed from sediments, but they are today grossly mixed with organic compounds. They are extremely heterogeneous and even one sample from any particular field contains many different phases. The appearance of the soil and the water held in it is shown in Fig. 7.5. The size distribution of particles varies from gravel (with stones of around 1 cm and greater) through sand to clays (with particle sizes around 0.001 cm or less). From place to place soil differs in grain size and composition, so that different life forms can benefit in different ways (compare deserts and forests). In many areas soils are deficient in particular elements valuable for certain plants and then animals. Other soils contain high levels of toxic metals such as nickel and chromium and few plants can grow on them.

The difficulty in describing soil is compounded by the presence of varying amounts of water. Thus soil water is different from that of the sea and from the fast runoff in lakes and rivers. In Table 9.13 we give the availability of elements in water from soils at very different pH values. The problems of describing speciation in such solutions are immense since there are highly active surfaces of clay minerals and organic compounds present, such as humic and fulvic acids, derived from lignins, (see Section 13.10), as well as a variety of anions and cations, all of which can form 'complexes'. (See Sections 6.6–6.8 for a discussion of the applicable conditional binding constants). The following points may be usefully kept in mind.

1. Lower pH values mean higher concentrations of free cations and anions; the problem of waters arising from acid rain, with its higher aluminium content, should be mentioned here.

2. Lower pH values allow faster equilibration between soil and soil water.

3. Low pH will hydrolyse esters easily.

4. Oxidation is greater at higher pH.

The only realistic way to approach the quality of a given soil is through empirical study. A cycle of elements in soils is shown in Fig. 9.12. A dis-

Table 9.13 The availability of elements from soil waters of different pH*

Element	Availability of element (μg l^{-1})	
	pH, 3	pH, 7.5
B	20	500
Cd	100	–
Mn	6000	500
Fe	2000	100
Co	1–5	1–5
Ni	1–5	1–5
Cu	750	50
Zn	7000	100
Mo		5
Se		1

* From ref. 20 in Further reading. Bulk elements such as Na, K, Mg, Ca are present in excess of 500 μg l^{-1} at pH = 7 and about equally so.

cussion of such cycles has been given in our previous book, *The natural selection of the chemical elements* (see Further reading).

9.10   The evolution of Earth's Land: summary

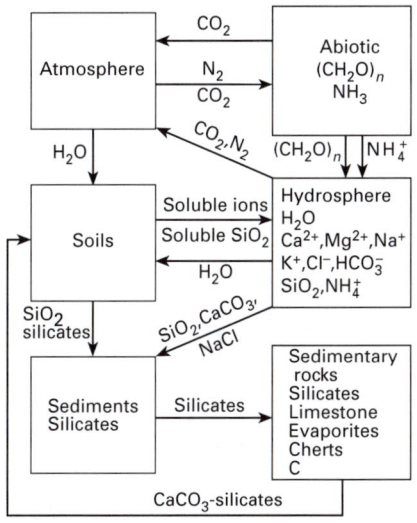

Fig. 9.12 Main abiotic exogenic or surface geochemical cycle. Little of this cycle has changed since Earth formed; O_2 is omitted. Note that sulphur and heavy elements are also excluded.

Earth has changed in two very distinct ways, but only seriously in the crust, crustal waters and the atmosphere. The first change was largely physical and began with the appearance of land at least 4.2×10^9 years ago. Subsequently the land broke into continents and in turn they drifted and collided with one another (Fig. 9.6). Thus, large new compartments developed.

The second change is chemical and it has taken place continuously over the whole period of Earth's existence (see Fig. 9.13). There were and are major changes due to geochemical and geophysical changes. These include the enforced mixing of the upper layers of the mantle, due to continental drift and volcanic action, and the losses of gases from the atmosphere, e.g. of H_2, CH_4, and some N_2 and CO. By far the greatest changes, however, have been the production of organic matter and dioxygen arising from life's chemical activity and reactions due to these chemicals. The variation in oxygen content of the atmosphere is shown in Figs 6.15, 9.2 and 13.2. It forced great changes on the surface rocks, so that few sulphides are present. The minerals formed a huge diversity of types making for local conditions in soils and waters. These have greatly affected life's local chemistry and subsequently man's industry.

We see, therefore, that there have been great changes in the atmosphere, the sea and the crust, but the changes are different locally. *It is these changes that, in our opinion, have brought about the possibility of evolution in living forms, especially via the natural selection of the chemical elements in both inanimate and animate forms.* We shall develop this theme throughout Chapters 10–15.

9.11   The rate of change of element distribution

Once dioxygen began to accumulate in the atmosphere we have to consider many thermodynamic and kinetic factors in the changes of the Earth's surface. In order to probe the rate of these changes we can examine the types of deposit on Earth in rocks laid down at various times (see Table 9.6) as well as the isotope ratios of the elements in these deposits. These ratios allow estimates to be made of the ages of the minerals. The isotope ratio in any deposit depends, of course, upon the initial ratio when Earth formed (say, of $^{13}C/^{12}C$ for carbon) and also on the chemical processes through which the element has been. A kinetic pathway always discriminates to some degree between isotopes. [(In chemical kinetic studies great use has been made of the differences in rates of reaction of ^1H, ^2H and ^3H (hydrogen, deuterium and tritium).] The deposited form of the element is also a useful indication. Thus, the formation, locally, of banded

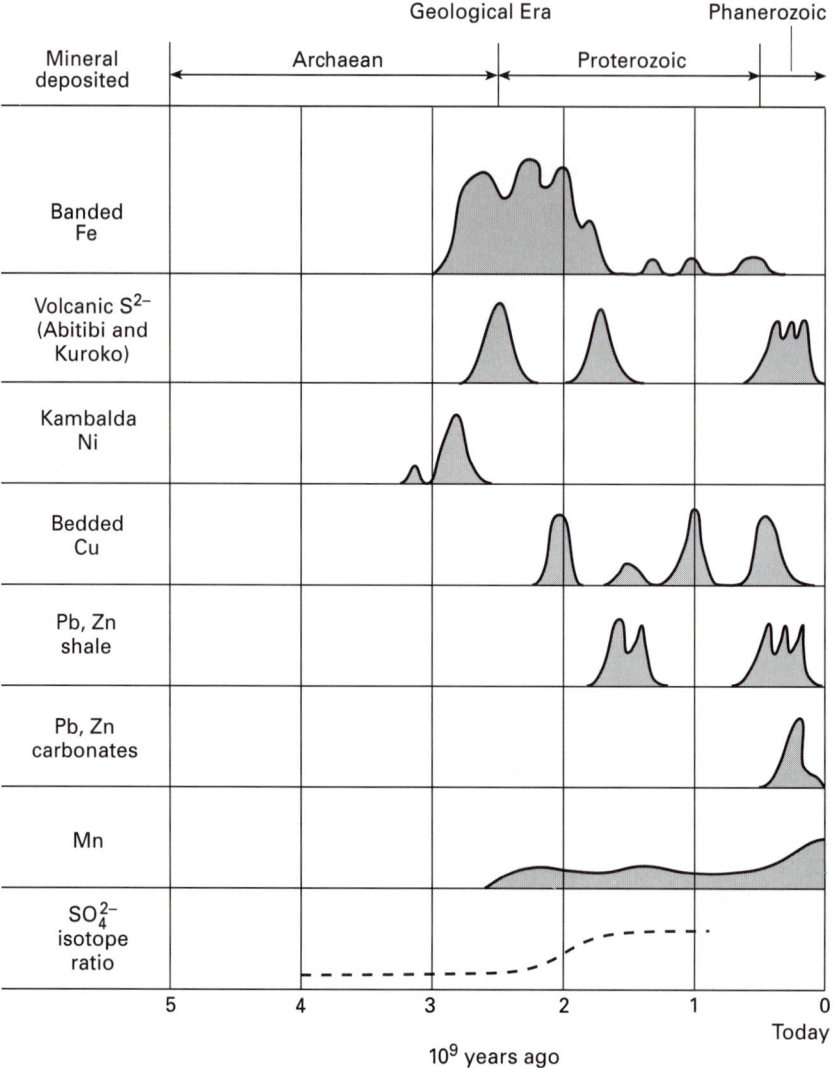

Fig. 9.13 The formation of mineral deposits on Earth since its beginning. [After Lambert, I. B., Beukes, N. J., Klein, C. and Veizer, J. (1992). In *The Proterozoic Biosphere*, (ed. J. W. Schopf and C. Klein) pp. 59–62. Cambridge University Press, Cambridge.]

red iron oxides would indicate that in that locality ferric oxides had formed in preference to ferrous sulphides. The discovery of the nature of the process through which an element has passed is aided by knowledge of its association with other elements. For example, it is expected that barium would be found with the earliest sulphate deposits since barium is the most insoluble sulphate. In fact baryte is the commonly observed barium mineral in early stages of the Earth's change from strict anaerobic conditions when sulphide, not sulphate, was present. We show the estimated times of appearance of several types of mineral in Fig. 9.13.

Before summarizing aspects of the changes of the Earth's composition we indicate briefly the time course of the major changes in element availability which parallel the changes in the nature of these deposits. (Isotope fractionation studies, it should be noted, may well lead to much more detailed understanding of this evolution of the Earth than we have now.)

9.11.1 Dioxygen

There is little doubt about the *slow rise* in dioxygen partial pressure in the atmosphere, as was described in Sections 6.14 and 9.2 (see also Fig. 13.2). There may have been some O_2 released earlier due to photolysis of water, but the major source is and was due to photosynthetic activity in organisms. Dioxygen is thought to have reached its present levels less than 700 million years ago.

9.11.2 Sulphur and selenium

Initially, almost all sulphide would have been found in sulphides, S^{2-} or S_2^{2-} salts. Some oxidation would have occurred as soon as dioxygen exceeded some 10^{-10} bar. It is not surprising, then, that baryte, $BaSO_4$, dating from around 3.5×10^9 years ago has been found. However, the isotope ratio of the sulphur in the sulphate (Fig. 9.13) would indicate that sulphate-using bacteria did not appear until some time later, say 3.0–2.5×10^9 years ago (see Section 11.2). These bacteria use the energy of reduction of sulphate back to sulphide (see aside and Chapter 11), while cycling carbon through oxidation and reduction.

Selenide would have been oxidised at a date later than 2.5×10^9 years ago (see Fig. 6.16).

Sulphur isotope fractionation (F)

$$H_2S + 2O_2 \longrightarrow H_2SO_4$$
$$MO + H_2SO_4 \longrightarrow MSO_4 + H_2O$$

Geochemical: Low (F)

- - - - - - - - - - - - - -

$$O_2 \left(\begin{array}{c} \nearrow SO_4^{2-} \searrow \\ \\ \nwarrow S^{2-} \swarrow \end{array} \right)$$

Bacterial cycle: High (F)

9.11.3 The oxides of nitrogen

There is considerable disagreement concerning the evolution of these oxides. Once earth had formed there were undoubtedly enormous atmospheric storms while the mantle constantly broke through the crust. In the air, lightning could have formed a considerable amount of nitric oxide from nitrogen gas and water vapour according to the reaction

$$N_2 + 2H_2O \rightarrow 2NO + 2H_2$$

Even in reducing conditions, nitric oxide generated by lightning could have accumulated in solution, either in combination with sulphide.

$$RS^- + NO \rightleftarrows [RS \cdot NO]^-$$

or in combination with ferrous iron

$$Fe^{2+} + NO \rightleftarrows [Fe \cdot NO]^{2+}$$

No comparable species are available from dioxygen chemistry in the presence of reducing agents. The later loss of H_2S and sulphides and the rise of O_2 in the atmosphere meant that NO from storms could react preferentially with O_2 rather than reduced sulphur, giving some NO_2

$$2NO + O_2 \rightarrow 2NO_2$$

and then in water

$$H_2O + 2NO_2 \rightarrow HNO_2 + HNO_3$$

The result was a soluble form of nitrogen oxides as the anions NO_2^- and NO_3^-. The anion NO_3^- is quite stable and so, like sulphate, would accumulate until there was a biological activity to suppress it. Notice that, just like sulphide, nitric oxide would help to keep the free dioxygen low, while both sulphate and nitrate increased in water. Nitrate, however, could not have been present in noticeable concentrations until the dioxygen pressure rose to say 10^{-3} bar, i.e. relatively recently (see the oxidation state diagrams in Figs 6.13 and 6.16).

9.11.4 Abiotic and biotic organic compounds

We have already described the general oxidation of atmospheric carbon compounds from some CH_4 and CO to CO_2. There has always been a variety of other carbon compounds present in the atmosphere, as we can surmise from those present in outer space. Most of the lighter gaseous compounds were slowly reduced in quantity (lost in space or oxidised), so that few C/H compounds remain today in the atmosphere. A contrary development has been the gradual increase in trapped organic compounds in minerals, e.g. gases (CH_4), oils C_nH_{2n+s} and coals. Life itself is a huge store of a multitude of other half reduced carbon compounds.

9.11.5 Mineral elements

The slow transformation of minerals with time, and as dioxygen accumulated, is shown in Fig. 9.13 (see also Table 9.6). It was not only minerals themselves that became oxidised, but the newly formed salts also changed the composition of the sea owing to their solubility. The loss of iron and the gradual rise of zinc and copper are but two features of the general changes that took place over billions of years. We shall return again to the effect of changes in trace elements upon evolution of organisms in Chapters 10–15.

9.12   Clays and sulphides: origins of early life?

Some of the most difficult inorganic systems to explain are the clay minerals, yet they are amongst the most important for life. Even before living organisms left the water, they, e.g. algae, began to bind strongly to sediments in the shallows. At that time the sediments (mud) would be and have remained the major source of very many essential elements (Fig. 9.14). Today, the soils are still the source of water as well as of the trace elements for land plants, and man's agriculture is totally dependent on this thin mineral layer. The soils are mainly composed of aluminium, magnesium, ferric iron and other metal silicates, with many other elements present in small amounts. They also retain organic and inorganic compounds by strong adsorption or even located in aqueous pockets or channels.

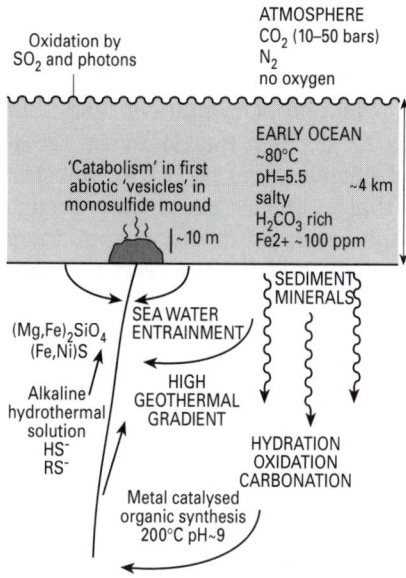

Fig. 9.14 A geologist's view of the interrelationship of the atmosphere, the sea and the sea bed when life began, 4 x 10^9 years ago. (After M. Russell.)

Some facets of soil chemistry do require further stress since, as mentioned above, it is believed by some that the surfaces of clay particles may have assisted very early life forms. Notable features are the small particle size, the heterogeneity of phase structures in a given volume and the fact that these clay structures are not only far from physical equilibrium but are also far from chemical equilibrium. Clearly, the difficulty of reaching equilibrium in soils now arises from the prevailing low temperature and the very slow rearrangement of silicate lattices, even down to the loss of water. [The loss of water from them at high temperatures formed the basis of man's first industry—making pots and bricks (Chapter 15).] Some of the forms in which the clay structures appear are truly striking and life-like (Fig. 9.15). The long worm-like unit, shown in Fig. 9.15, which is called imogolite, is found in many clays. (The central hole is of ~ 10 Å diameter.) When we look at the earliest forms of life based on cylinders (see Fig. 11.1) we cannot help but wonder if these minerals assisted stabilisation of early organic forms. The tube-like structures grow and on fracture they multiply—their surfaces are also catalytic. The zeolites are related minerals with open structures (see Section 9.5). [Many of man's industrial processes use these frameworks (see Chapter 14).] Some sulphides seem to be able to

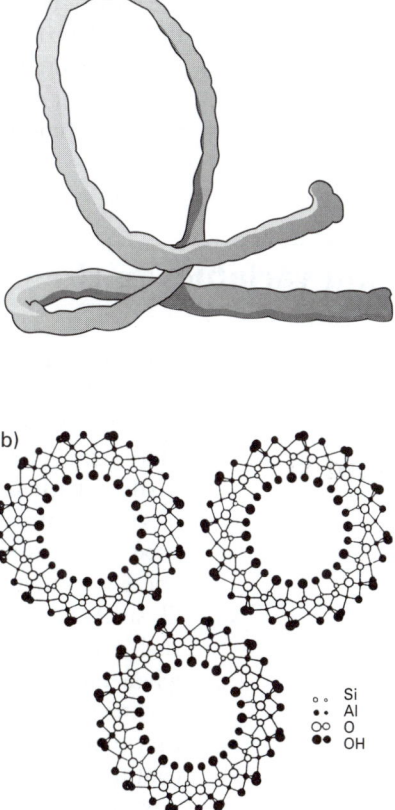

Fig. 9.15 Imogolite. An unusual clay structure. (a) Observed as a worm-like mineral in soil; (b) the atomic structure in cross-section.

form similar structures (see reference 81) and these have also been associated with life's beginnings.

Other authors (refs. 87–90 in Further reading) have stressed the potential roles of sulphides, not only in the formation of cylindrical tubes, but also in giving rise to membranes suitable for dividing space into the necessary compartments for accumulation of chemicals, as occurs for silicates. In particular, there is the suggestion that sulphides could have been a primary chemical energy source and that, coupled to this energy, there would have been a natural catalytic surface for C/H/O/N reactions. The basic energy comes from the reaction.

$$Fe^{2+} + HS^- \rightarrow FeS + H^+ \text{ (proton gradient)}$$
$$H^+ + FeS + HS^- \rightarrow FeS_2 + H_2 \text{ (bound in organic molecules}$$

and

$$H_2S + CO_2 \rightarrow [S] + C/H/O \text{ compounds (C capture)}$$
$$[S] + FeS \rightarrow FeS_2, \text{ pyrite (energy source)}$$

(Pyrite is a peculiarly stable product, containing *low-spin* Fe^{2+}, see ref. 68.)

This subject remains speculative and has only slight relevance to the present book which concerns *the logic of the observed* development of selection of the chemical elements in all systems, biotic and abiotic, so as to generate speciation. We shall see that there are other possibilities relating to organic compounds directly. We can never do more than guess at the origin of life (see Schmeller, G. (1998) New Scientist, 12th Sept., pp. 30–35).

9.13 Units and variables in the evolution of Earth's chemistry

The logical analysis which we have used in this book shows that by considering restricted equilibrium between chemicals, i.e. limited atom exchange, and restricted equilibrium across physical boundaries, we could use operational units (components) and physical divisions (compartments) to describe the Earth at particular periods in time. Operating in this way, the *local* variables of composition, temperature, pressure, size and fields allow us to give good general explanations of all the substances and objects (solids, liquids and gases) that we observe. Thus, we find speciated different minerals, as is illustrated in many a geological museum, which can be understood in such terms. However, it is doubtful whether this approach is equally useful when handling such materials as soils and clays. Here, the sizes of compartments are so small and even local chemical equilibrium is so often absent that characterising a soil can only be a gross approximation. Soils grade continuously into one another, much as do aqueous solutions and gaseous mixtures. Now, maybe it was out of these very zones, perhaps at the bottom of the sea, that living organisms appeared and they are very clearly speciated.

The description of Earth is also time dependent in several ways due to the lack of equilibrium, and these time dependences are only systematic to some degree. They are as follows.

1. The upsurge of material from the deep molten zones due to volcanoes, which are unpredictable in magnitude and effect. On the whole the activity has reduced over 4.6 billion years. This is to be taken together with burial of material due to folding of the mantle.

2. The generation of dioxygen has slowly changed the surface layer. The process is continuous while reduced materials are continuously provided by upsurge of materials (1) and biological activity (5).

3. Gases such as H_2, CH_4, CO_2, N_2 and O_2 escape to interplanetary space. The process is continuous. The reserves of N_2 are only in the organic material of life.

4. The sun's radiation fluctuates, but as a star it dies slowly. Temperature fluctuates but eventual cooling is inevitable.

5. Biological activity, see also (2), is a continuous reducing activity of carbon-oxygen to carbon-hydrogen compounds which have been deposited in the crust, as oil, coal and gas. (Note also carbonates).

In terms of our four fundamental variables of Chapter 2, mass, charge, space and time, much of Earth can be treated without reference to time, but the smaller part, the crust, is time dependent and locked into life.

There is no doubt that the surfaces of the Earth have evolved very slowly and, very fortunately, with (relatively) small changes in temperature. The gradual change in its atmospheric, aqueous and mineral surface has allowed, in our opinion, the gradual evolution of life from matter to microbe to man, which we shall describe in the next chapters.

However, the changes of the Earth are inevitably towards equilibrium, in contrast to life's activity, which is seemingly away from it (see subsequent chapters).

Clearly, in order to proceed with the analysis of the changes of the surface we need to refer to the chemistry of living organisms. This approach is strengthened by the general hypothesis that perhaps the mutual development of the surface of Earth and the chemistry of organisms has by now come into some steady state. The idea, developed by Lovelock (see refs 93 and 94) has been named Gaia (the name of the Greek earth goddess).

9.14   The resulting steady state today: Gaia?

As the dioxygen pressure rose through life's pressure, i.e.

$$nCO_2 + nH_2O \xrightarrow{\text{life}} (HCHO)_n + nO_2 \uparrow$$

it was possible, as we described above, for it to react first with the sulphides and reduced metals of the Earth's crust so that, for example, much of the surface sulphur became sulphate, SO_4^{2-}, and much of the iron became Fe_2O_3.

Let us assume the reactions are complete. Dioxygen could also react with reduced carbon compounds, which we may represent as HCHO (or, for example, CH_4) produced by living organisms, which also reverse the reaction.

$$HCHO + O_2 \rightarrow CO_2 + H_2O$$

Life

where carbon, hydrogen and oxygen are now cycling.

This activity, it is proposed, generates *a steady state* catalysed by life and energised by the sun at present-day levels of CO_2 and a dioxygen pressure of around one-fifth (21%) of one atmosphere

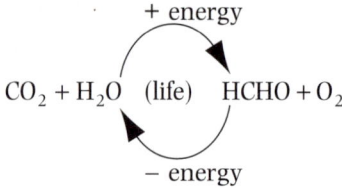

$$CO_2 + H_2O \quad (\text{life}) \quad HCHO + O_2$$

+ energy

− energy

This steady state could be self-regulated, see aside, and corresponds to one manifestation of the so-called Gaia hypothesis. This hypothesis, or concept, is hard to explain in just a few paragraphs but, as originally proposed, Gaia was a self-regulating Earth (including the biosphere) *as if* it was itself a living organism. This proposal has a loose connection with the themes of the present book. What one realises now is that both living organisms and the Earth are complex interactive realities which, through multiple crossed feedback and feed-forward control mechanisms (the essence of 'life'), might reach an overall self-regulated *steady state* in aspects that go far beyond the simple cycle of carbon as described above. We must always be aware, however, that we are dealing here with a far-from-equilibrium, complex, dissipative system, so that any apparent steady state can evolve to a different steady state since immeasurable energies and material resources, for example in the mantle and core of the Earth, are not considered. These and one or two other considerations could undermine the hypothesis, namely:

Earth's temperature regulation

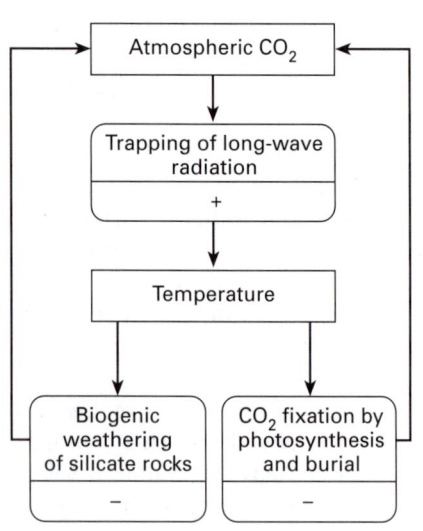

1. The molten core of the Earth still has a huge reservoir of reduced metals as sulphides which could break through the surface and remove much of the dioxygen, forming sulphates. In fact, there is still much sulphide and reduced iron near the surface.

2. The escape of CO_2, CH_4 and O_2 from Earth's atmosphere is continuous and faster than that of H_2O; there is also the loss of N_2 to be taken into account.

3. Impacts of meteorites can cause changes in the Earth's energy capture and could distribute unusual elements in dust.

4. Global variation in temperature can affect the catalysts of the steady state. Here the activities of man must be included.

5. Man increasingly interferes with the element availability just as dioxygen-producing organisms did before him.

Finally, we must also remember that the nature of a complex system is always vulnerable to a 'chaotic' switch to a novel condition about which we may be unaware. Thus, although Gaia is an interesting idea that can be used as synonymous of the capacity of the Earth (as an open system including the crust, the hydrosphere, the atmosphere and the biosphere) to keep an overall steady state, particularly on the surface, it cannot be taken as a kind of natural law or proven rule of general application.

Clearly, our next step in the analysis of the complicated interaction between the crust of Earth and its changing environment is to examine living organisms so as to appreciate their continuous and evolving involvement. Here we are concerned not with a steady state but with developmental states, or evolution. Only then can we see if a steady state is likely at the present time.

10

The principles of the chemistry of living systems

We may furthermore conclude that the evolution of life, if it is based on a desirable physical principle, must be considered as an inevitable process despite its indeterminate course ... It is not only inevitable in principle but also sufficiently probable in a realistic space of time. It requires appropriate environmental conditions (which are not fulfilled everywhere) and their maintenance. These conditions have existed on Earth and must still exist in many planets in the Universe.

Manfred Eigen (1971). *Naturwissenschaften*, **58**, 465–523.

10.1   Introduction

We shall assume that living systems are chemical in their material nature. Much as there is a puzzle as to how inanimate objects come to form separable speciated chemical compounds, as we have discussed in Chapters 2–9, so there is a parallel puzzle concerning the species of living forms. The diversity of life, as illustrated in Fig. 10.1, is very, very great indeed. We shall find it convenient to divide the species into anaerobic prokaryotes, anaerobic eukaryotes, aerobic prokaryotes and eukaryotes (single cells), multicellular organisms and self-conscious man. It is generally believed, on the basis of convincing but not overwhelmingly rich evidence, that life developed in that order (Fig. 10.1). Before we deal with specific organisms under any of these headings it is worth stressing the common chemical features of cells, since all organisms are cellular. In this chapter, therefore, we shall give the very complicated cellular systems of living organisms a common, simplified logical description in chemical terms. That is to say, we wish to help our reader to appreciate how life's chemistry is managed whether or not chemicals are all that there is to life.

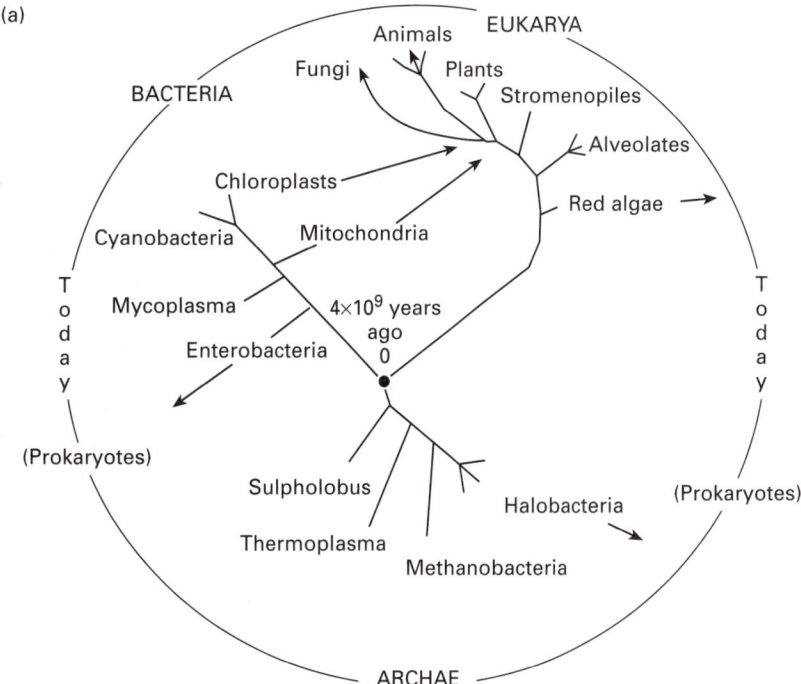

(a)

Fig. 10.1 (a) The three kingdoms of life in a radial time sequence based on genetic (RNA) analysis. The distinction between anaerobes and aerobes is not made here. All the kingdoms advanced with time but in very different ways. [See Woese, C. (1990). *New Scientist*, 11 August; Sogin, M. (1993). *Science*, **260**, 340.]

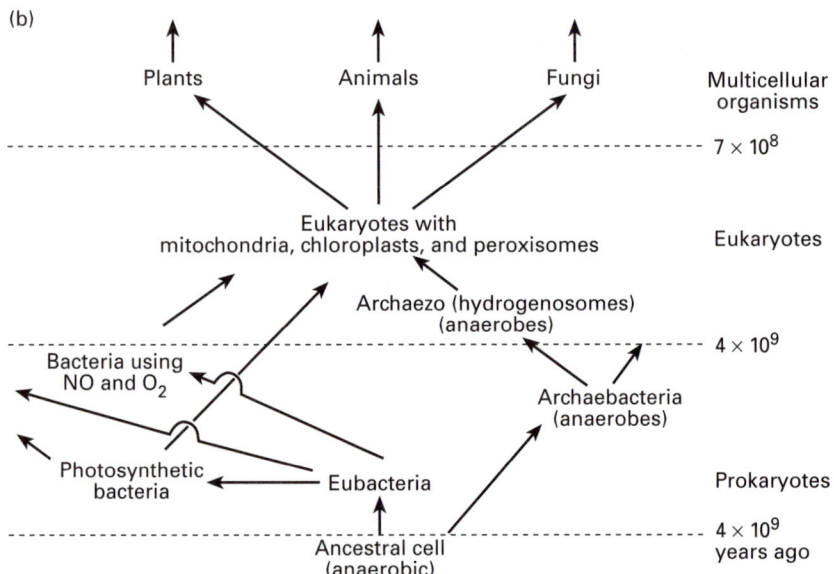

Fig. 10.1 (b) An alternative way of looking at evolution, due to T. Cavalier-Smith (see ref. 142.) See also ref. 145 and W. Martin, BioEssays (1999) *21* 99.

We know for certain that living organisms are, in essence, flowing, managed, chemical organisations, in that they take in simple chemicals which they then process and finally decompose back to simple chemicals (Fig. 10.2). Within the organism, the simple chemicals are built up into a complex machinery of large molecules which develops and reproduces itself before it dies. Thus, there is a *steady state* on Earth of 'living' chemical material at any one time, but every cell or organism is itself in a *developmental state*. These states have evolved in complexity and diversity, which implies that there are three different problems of this chemistry of organisms that need to be appreciated: (1) the origin of life; (2) the essential organization of living organisms, including development, reproduction and death; and (3) the evolution of life. Given the general nature of the problems, illustrated in evolutionary diagrams such as that of Fig. 10.1, we shall start our discussion in a reductive analytical chemical fashion much as was necessary for the understanding of the mineral material world, i.e. in terms of chemical elements, before we turn to an appreciation of the functional value of their compounds (see Chapter 1). Therefore, we ask first which atomic elements are known to be required by *all* living systems and are, therefore, essential for life. Once we have established which are the *essential chemical elements* (*units*) of living organisms, then we need to ask whether it is operationally advisable to describe many of their combinations in compounds as independent components, as we have done in inanimate chemical systems (see Chapters 2–5). After that, we enquire which are the *variables* and restrictions on these variables that have allowed the units to come together to give a living, flowing system and allowed it to evolve.

Living is an emergent property of such systems. It has no new fundamental variables, and no new separate units, but is a resultant of a combi-

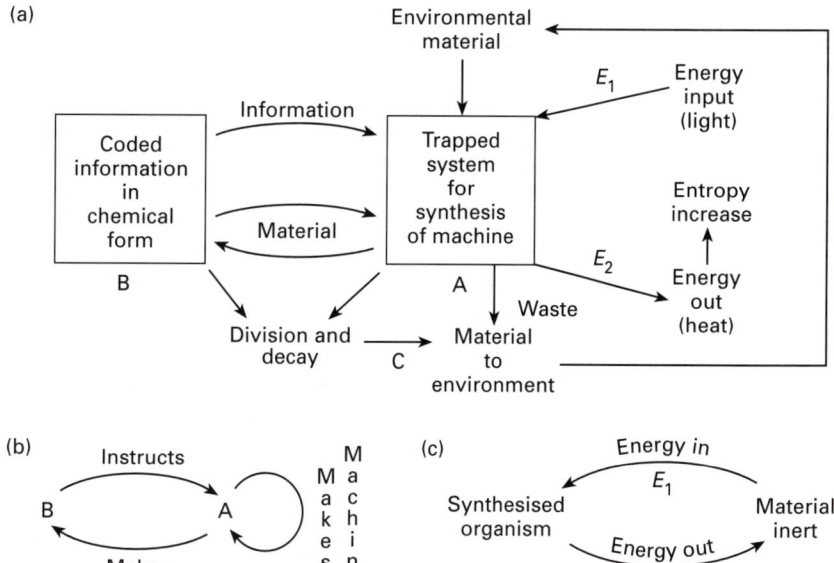

Fig. 10.2 To generate a living system, (a) it is necessary to have sources of energy and material, a coded programme (B) and a machine (A). The essence of the machine is that it provides routes to synthesis (as irreversible reactions) since synthesis has to absorb energy, E_1, and decay generates it, E_2, see (c). (b) The code and the machine evolve together.

nation of the fundamental variables *composition*, in terms of chemical 'components', their *distribution in space* and the *time dependence* of this distribution. In this way, organisms will be seen in the same chemical light as non-living kinetic systems, described in Chapter 8, no matter in what ways they differ from them and although their chemicals are selected for functional purpose, not stability (see Section 8.6). We shall not be much concerned with the origin of life (see references in further reading) but with the development of its functional activity (Table 10.1).

Table 10.1 Functional activities of a living system

(a) Operates so as to control the exchange of materials and energy with the environment
(b) Synthesises its own 'components' from materials available from its environment and self-assembles them
(c) Extracts or absorbs energy from the environment and converts it into useful work of many kinds
(d) Performs controlled synthesis limited by an information centre with machinery
(e) Catalyses selected reactions using energy that support its activities
(f) Informs the synthetic machinery, including the information centre itself, to ensure the correct use and reproducible replication of cellular materials
(g) Controls its activity by feedback mechanisms to preserve its organisation or to adapt it to external changes
(h) Reproduces its genotype through replication
(i) Allows phenotypic variation due to the changes of the environment
(j) Evolves with environmental changes or by chance

10.2   The atomic elements of life

As stated, our initial step is one of simple elementary analysis. In Table 10.2 we list those elements that are known to be functional in *every* living organism. (The amounts of the elements are also shown by percentages in the table.) Clearly, it is a mistake to concentrate on just C, N, O and H; there are about 15–20 essential elements without which life would not exist. In addition, several living organisms require some of the other elements shown in Fig. 10.3, where the essential elements listed above are also included. The

Table 10.2 Elements in all organisms

Non-metals	H, C, N, O, P, S, Cl, (Se), Si
Metals	Na, K, Mg, Ca, Fe and/or Mn, Zn
Possible additions	Co, Mo

Average composition [weight % and atomic ratio to hydrogen (10) in parentheses] of some elements in organisms

	Plants	Animals
Oxygen	79.0 (4.9)	65.0 (4.1)
Hydrogen	10.0 (10.0)	10.0 (10.0)
Carbon	3.0 (0.25)	18.0 (1.5)
Nitrogen	0.3 (0.02)	3.0 (0.21)
Calcium	0.1 (2.5×10^{-3})	2.0 (5×10^{-1})
Chlorine	0.07 (2×10^{-3})	0.2 (5.6×10^{-3})
Sulphur	0.01 (3×10^{-4})	0.3 (9×10^{-3})
Potassium	0.3 (7.7×10^{-3})	0.4 (1.0×10^{-2})
Sodium	0.03 (1.3×10^{-3})	0.2 (8.7×10^{-3})
Magnesium	0.08 (3.3×10^{-3})	0.05 (2.1×10^{-3})
Iron	0.02 (3.6×10^{-4})	0.004 (7.1×10^{-5})
Manganese	0.12 (2.1×10^{-3})	1×10^{-5} (1.8×10^{-7})
Silicon	0.15 (5.3×10^{-3})	Low

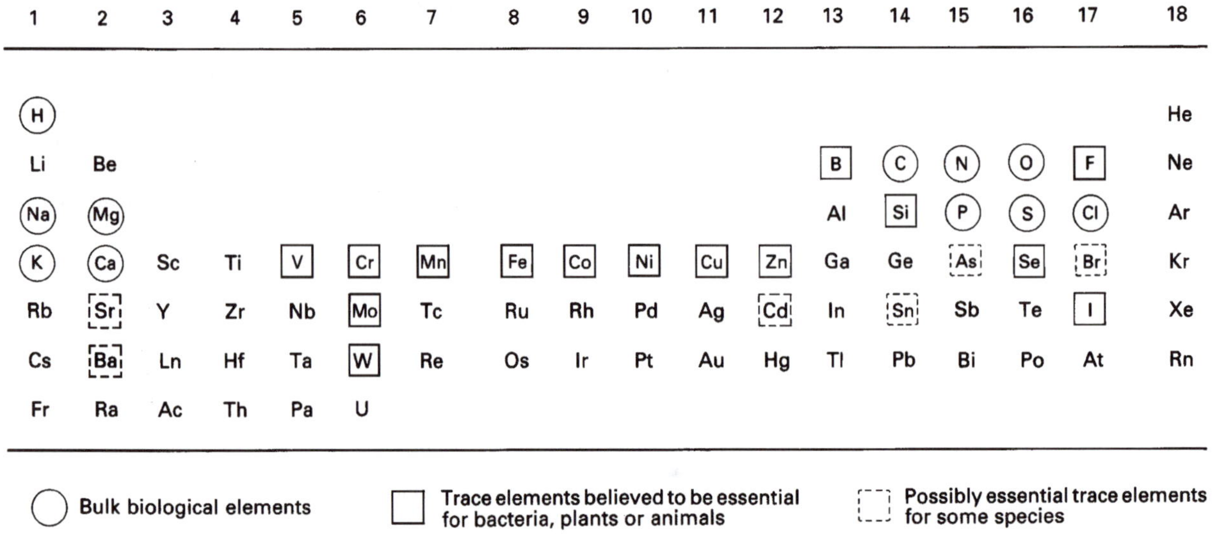

Fig. 10.3 The periodic table showing elements of functional value in organisms.

remaining elements in the periodic table have never been found to be specifically associated with any essential function in life, although they may occasionally be present in some organisms perhaps due to parallel chemical properties with essential elements (e.g. Nb alongside V). This is not to say that they could not be essential in some, as yet unknown, way.

10.3   The link to availability of elements

A quick glance back to Chapters 5, 6 and 9 shows that the elements in organisms are those that are or were *available* to a greater or lesser degree in (sea) water or the atmosphere, rather than those that are or were *abundant* on Earth. Some may be difficult to obtain, for example if they are present as insoluble species such as Fe is today in Fe_2O_3. It is also shown that one element at least, from almost every group of the periodic table is present, usually the first and/or the second element of each group. These lighter elements are often the most available. Noticeable missing exceptions are the elements of groups 3, 4 and 13 (all but the non-metal boron) as well as the inert gases of group 18. Some elements of groups 3, 4 and 13 are abundant but not very readily available due to the insolubility of their naturally occurring compounds (oxides), although there may also be other reasons for their absence which we discuss later. Group 18 elements do not combine under ordinary physiological conditions.

The intriguing feature is that the incorporation of elements from some 14 groups in the periodic table implies an almost complete involvement in organisms of the full variety of chemistry that can possibly be generated by the nature of the periodic table. The chemistry of elements within groups is, of course, similar (but not identical). This suggests that living systems have found the best of all possible uses for the available elements, employing their particular properties as expected from the position they occupy in the periodic table, for the diversity of functional tasks organisms need to undertake. Note that the *fundamental units of composition* are now *restricted* to some 20–30 rather than to the full list of 90 stable elements on Earth (see Chapter 1). We can, in effect, plot *survival strength* against composition, the amount of an element in an organism, as a landscape diagram (Fig. 10.4) much as we plotted thermodynamic stability (ΔG) against composition for inanimate chemical systems. The further fact that the availability of elements has changed with time (see Chapter 9) will be shown to have had a major influence on evolution. Availability is also variable with geographical location, as we have stressed in Chapter 9.

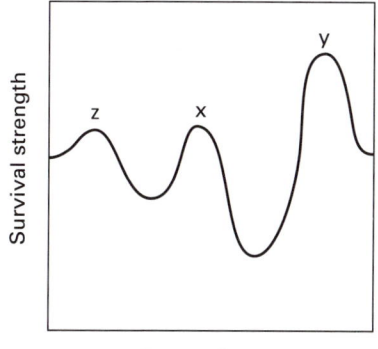

Fig. 10.4 Survival strength plotted against composition, for a given element. For carbon, for example, X might correspond to animal, Y to plant and Z to fungal species. The height to depth distance could be small or large when the regions of compositional survival can be limited or considerable. (See also Fig. 8.23(b)).

10.4   The chemicals in organisms

10.4.1 Water and organic compounds

In the discussion of the units that limited the possible composition of inanimate systems that could form on Earth (i.e. their compositional variability), we found it to be operationally valuable to examine not just elements but their chemical combinations which did not exchange elements, the so-called components. We attempt to follow the same logic here by looking next at observed essential molecules rather than elements in living organisms, but before we ask about exchange reactions. For this reason in this chapter we write 'components' in inverted commas until we do or do not show that exchange is absent or not.

The elements present in organisms are overwhelmingly but not absolutely limited to *molecules*. Of these, the outstanding one is water, H_2O. Life's chemistry is, in effect, largely aqueous solution chemistry, which restricts the kinds of reactions that are possible. In Chapters 6 and 9 we have discussed the presence in aqueous solution of many elements, mainly as small ions, but in living organisms we must concentrate to a larger degree on the chemistry of small organic, H, C, N, O, compounds and their larger molecules. Next to H_2O they are the most common combination of elements within organisms and often include S and P. We have already described some of them in Chapters 3, 4 and 8. Here it is useful to look at them as a hierarchy of non-metal 'organic' molecules starting from combinations of C and H. These two elements give molecular 'components' (units) in organisms mainly of the kind C_nH_m, where n lies in the range 16–20 and m in the range 26–42. They are fatty substances, but they usually have terminal hydrophilic groups containing

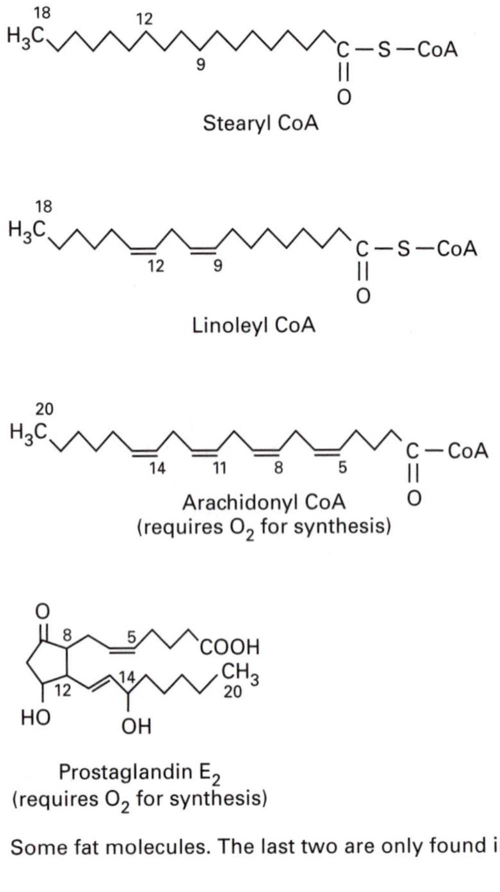

Fig. 10.5 Some fat molecules. The last two are only found in aerobes.

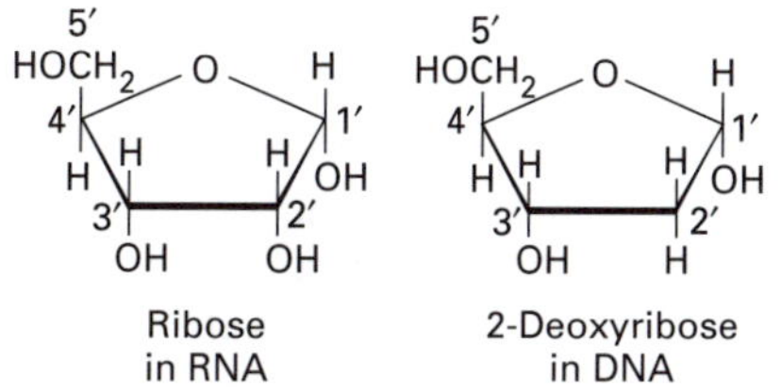
Ribose
in RNA
2-Deoxyribose
in DNA

Fig. 10.6 Sugars of nucleic acids: ribose in RNA and 2-deoxyribose in DNA.

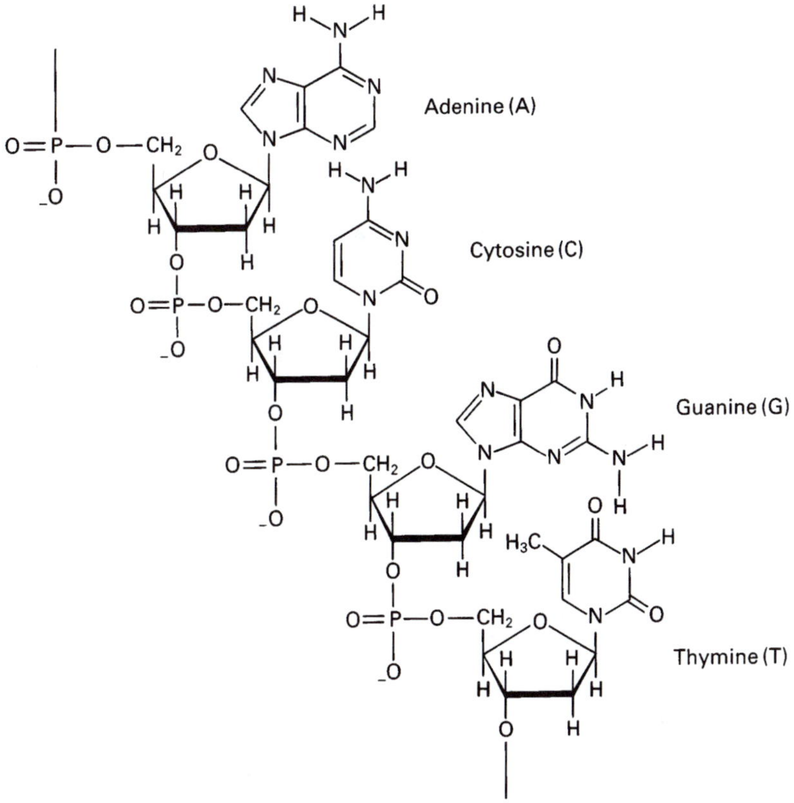
Adenine (A)
Cytosine (C)
Guanine (G)
Thymine (T)

Fig. 10.8 A short string of single-strand DNA giving the formulae of four bases. In RNA, thymine (**T**) is replaced by uracil (**U**) and deoxyribose is replaced by ribose (see Fig. 10.6).

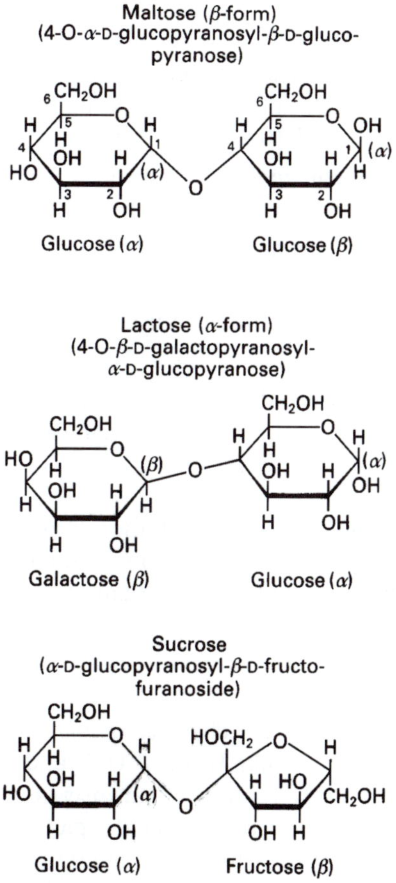
Maltose (β-form)
(4-O-α-D-glucopyranosyl-β-D-gluco-pyranose)
Glucose (α)
Glucose (β)
Lactose (α-form)
(4-O-β-D-galactopyranosyl-α-D-glucopyranose)
Galactose (β)
Glucose (α)
Sucrose
(α-D-glucopyranosyl-β-D-fructo-furanoside)
Glucose (α)
Fructose (β)

Fig. 10.7 A simple saccharide, glucose, in three disaccharides, showing an ether link to itself and two other monosaccharides.

oxygen (as shown in Figs 4.10 and 6.3) and are frequently negatively charged. Hydrogen may or may not saturate the binding ability of all the carbon in the chains (Fig. 10.5). We shall come to their value in organisms later, but notice that they are of very low solubility in water (Section 6.3) where they form micelles or vesicles (see Fig. 6.3 and Section 7.8).

The next unit we shall consider is a three-element combination of C, H and O, such as $C_6H_{12}O_6$ which is the composition of some sugars. The simplest member of the group is called formaldehyde, CH_2O, better written HCHO. If such CH_2O units are joined together they give pentoses $(CH_2O)_5$ (Fig. 10.6), or hexoses $(CH_2O)_6$ (Fig. 10.7). These units can also be linked directly through OH groups (by elimination of H_2O) to give polysaccharides, or linked through phosphate to give the backbone of DNA (see Fig. 10.8). Sugars, like fats, are obvious sources for building other larger molecules, such as cellulose, and they, like fats, also give energy when oxidised (burnt) in reaction sequences such as those shown in Section 2.5.6

$$\text{Fats} \quad (CH_2)_n + O_2 \rightarrow CO_2 + H_2O + \text{energy}$$
$$\text{Sugars} \quad (CH_2O)_n + O_2 \rightarrow CO_2 + H_2O + \text{energy}$$

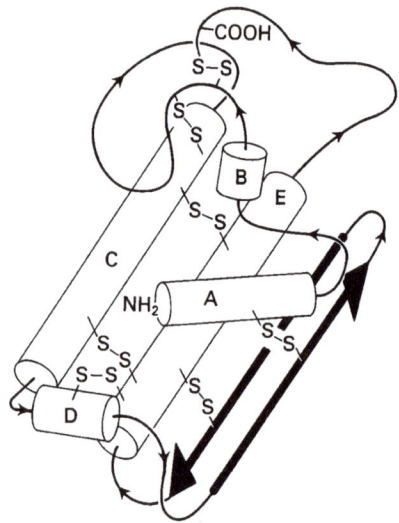

Fig. 10.9 An outline structure of a protein, here phospholipase A$_2$, showing α helical runs of amino acids as cylinders, A to E, and anti-parallel β sheet runs as heavy black arrows. Disulphide cross-links are shown, and runs of no secondary structure appear as thin lines. This structure is relatively immobile and binds calcium in a constrained loop (compare with Fig. 8.10).

The next combination of elements in 'components', still starting with carbon, involves C, H, N and O. Such combinations obviously have much greater variety and lead to a vast range of small molecules, but in biology the small molecules are mainly those related to a major family of polymers, proteins (Fig. 10.9) which are formed from the constitutive amino acids, in which carbon is reduced relative to CO_2 (Table 10.3) and to the genetic coded molecules (DNA and RNA), namely the purine and pyrimidine bases (see Fig. 10.8). Some proteins have S and P incorporated, and all DNA and RNA have P (phosphate) as part of their backbone. There are thousands of different proteins and RNAs in each living cell, but only one *central*, specific and distinctive DNA molecule (double-stranded). (N.B. Organelles in cells have their own DNA.)

There is clearly a huge range of organic molecules in any organism, but we know it is not a random set. There are five obvious limitations imposed by most cells on this range.

(a) The final polymers produced, a very limited set, are derived from a limited range of monomers, and hence their special synthetic pathways have many common chemicals.

(b) Degradative pathways will only yield energy that is readily usable by the cell provided they are carefully chosen. Optimal energy yields limit possible pathways drastically, for example in the glycolytic pathway and the citric acid cycle, see Figs 10.17 and 10.23.

(c) Certain limited pathways of special small molecules and ions, especially protons and phosphates, are needed in energy capture.

(d) Some large molecules, DNA (and RNA), are needed as codes and DNA itself is directly reproduced.

(e) The 'components' are for the most part organic anions so as to maintain solubility in water.

In addition to the essential common pathways, there is, of course, a vast number of lesser secondary reaction sequences used by one organism or another. In an overall sense the chemicals are also selected so as to serve the organism, that is, they are *functionally* selected.

However, this list in no way restricts the *concentration* of these organic chemicals. Before describing the way in which concentration limitations are brought about we turn to the other elements found in organisms and their restrictions.

10.4.2 Mineral elements in organisms

Glancing at Table 10.1 we see that there is a wide diversity of *inorganic* elements present in all organisms. Some behave like H, C, N, O, P and S, in that they form a limited set of covalently bound, apparently non-exchanging molecules, when they too, like the organic elements, give rise to possible 'components'. This group of elements includes selenium and some halogens. At the other extreme are elements binding, if at all, through fast-exchanging ionic interactions, and which form 'components' mainly as bare ions since they equilibrate rapidly from all bound associations. Here we include Na^+, K^+, Mg^{2+}, Ca^{2+} and other elements of groups 1

Table 10.3 The natural protein amino acids

Name	3-letter symbol	Symbol	Side-chain	Character	Metal binding
Aspartic acid	Asp	D		Acid (polar)	Mg, Ca
Glutamic acid	Glu	E		Acid (polar)	Mg, Ca
Tyrosine	Tyr	Y		Neutral (non-polar)	Fe
Alanine	Ala	A		Neutral (non-polar)	
Asparagine	Asn	N		Neutral (polar)	
Cysteine	Cys	C		Neutral (non-polar)	Cu, Zn, Ni, Fe
Glutamine	Gln	Q		Neutral (polar)	
Serine	Ser	S		Neutral (polar)	
Threonine	Thr	T		Neutral (polar)	Mg
Histidine	His	H		Basic (polar)	Cu, Zn, Mn, Fe, Ni
Arginine	Arg	R		Basic (polar)	
Lysine	Lys	K		Basic (polar)	
Glycine	Gly	G		Non-polar hydrophobic	
Isoleucine	Ile	I		Non-polar hydrophobic	

Table 10.3 (*continued*)

Name	3-letter symbol	Symbol	Side-chain	Character	Metal binding
Leucine	Leu	L		Non-polar hydrophobic	
Methionine	Met	M		Non-polar hydrophobic	Cu, Fe
Phenylalanine	Phe	F		Non-polar hydrophobic	
Proline	Pro	P		Non-polar hydrophobic	
Tryptophan	Trp	W		Non-polar hydrophobic	
Valine	Val	V		Non-polar hydrophobic	

and 2, and Cl^- from group 17. There is an intermediate group of elements, the transition metals, that form some rather slowly exchanging molecules and some non-exchanging associations in partially covalent chelates. We refer in particular to Mn, Fe, Co, Ni, Cu, Zn and Mo (W). Inspection of cells shows that these 'components' are again a selected set of elements often found in long-living complexes, such as haem and vitamin B_{12}. Even so, at first sight the complexity of 'component' composition in a cell is very great indeed (Table 10.4). We shall now correct that impression of a con-

Table 10.4 Examples of apparent 'components' in cells

Organic	Ions*	Complex ions
All substrates, co-enzymes polymers, fats, etc.	Na^+, K^+ Mg^{2+}, Ca^{2+}, Zn^{2+} and several transition metal ions, e.g. Fe^{2+}, Mn^{2+}	Haems (Fe), Fe, Fe_nS_n Vitamin B_{12} (Co) (Mo)-cofactors F-430 (Ni)

* As we shall show, these simple elementary units, together with DNA and some basic small molecules, e.g. H_2O, may be the only true components (Section 10.21).

fusing set of 'components' by turning attention to the limited chemical nature of life rather than just to its elemental and molecular composition.

10.5   Observed limitations on cellular 'component' composition

In any given steady state of any particular living cell the DNA, the proteins and the RNAs are found to be in closely fixed concentration *ratios*, together with a relatively fixed amount of molecules of fats and polysaccharides. Much though the constancy of organic chemical composition is obviously extremely difficult to achieve, i.e. maintaining all the concentrations and hence the ratios of the four different kinds of large molecules fixed opposite a particular DNA, this composition is, together with the DNA itself, one major characteristic of a given species.

In order to build and maintain these ratios, all synthesis of all the large and small molecules must be correlated, which implies that starting from the environmental supply of H, C, N, O, S and P in their available forms the organisms of Fig. 10.1, each in its own way, must control all the small as well as the large organic molecule concentrations co-operatively! Therefore, we must search for a method of internal co-operative control over cellular rates of uptake, synthesis and rejection of chemicals to ensure the fixed turnover of molecule content (see Section 8.9). This brings about a dramatic reduction of possible compositional variables, as we shall discuss in section 10.20. There is, therefore, a rough and ready parallel in this steady-state control with equilibrium control. (In systems containing molecules in equilibrium we had to be very careful with the definition of a component and a similar problem arises here, see Section 10.20.) Note too, that a preferred stability of a compound of a given composition was due, in the non-living condensed state, to a *co-operative interaction* of atoms, ions or molecules. The interlinking of amounts of different chemicals as seen in living cells will cause us to think again about the elemental compositional variables and the *kinetic* co-operativity between them, and hence reconsider the definition of a 'component' in a living steady state (see Section 10.20).

As stated, it is not just the amounts of all these organic molecules that are controlled, but also the internal nature and cellular *concentration* of other substances such as salts and their ions, since the rates of synthesis and degradation of the organic materials depend on these factors (Fig. 10.10). Organisms are based on cells, that is, on an aqueous internal volume, the *cytoplasm*, contained within a membrane made from a fat material, which is itself composed of molecules that are very insoluble in water (see Fig. 7.13). As far as the internal composition is concerned, it must now be of low total charge so as to preserve approximate electrical neutrality. While many of the molecules of C, H, N, O, P and S contain easily ionisable groups, e.g. –COOH, –SH, –OPO$_3$H$_2$, and become negatively charged, this charge must be largely neutralized. Again, the cell must not have a summed concentration of soluble chemicals significantly stronger than that of the external fluids, so that it is prevented from bursting due to inflow of water (see Section 7.10). (Water movement will always attempt to

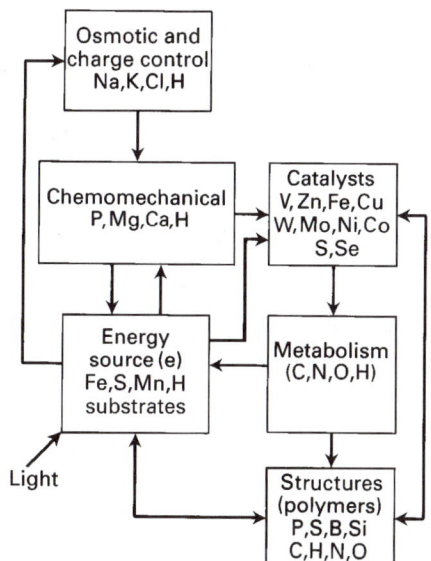

Fig. 10.10 The interrelationship of the functions of the elements is given here to stress the intimate connection between them. This diagram is only an aid to thinking about the ways in which energy and metabolism have to be linked, frequently using inorganic chemical functions.

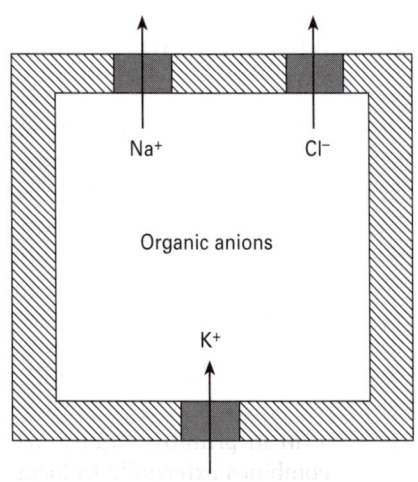

Ionic balance

equalise the osmotic pressure inside and outside cells.) It is obviously not possible to use organic chemicals alone inside cells to balance osmotic pressure and electrolyte out of balance across the cell membrane, while allowing equilibration with a high outside salt (NaCl) solution, such as the sea.

In most cells, therefore, the electroneutrality and the osmotic pressure are managed by exporting Na^+Cl^- (Na^+ and Cl^- ions), the most concentrated salt of the sea, and taking in K^+ ions, which are relatively dilute in the sea (see aside). In some other cells this process is also assisted by the rejection of much salt, NaCl, and the synthesis of special small neutral organic molecules. Since the internal content of C, H, N, O, P and S molecules is closely fixed, generating a given osmotic concentration and negative charge, so must the final content of K^+, Na^+ and Cl^- be fixed. Analysis shows that this is true. If the organism has many cells and lives out of water, as in the case of humans, then its body must also carry a balanced extracellular fluid of constant Na^+, K^+ and Cl^- concentrations to complement the cells' inner contents. Here, NaCl predominates. As everyone knows, humans die without the correct salt (sodium and potassium chloride) intake and every hospital is equipped to measure the mineral content as well as the organic content of blood to check the well-being of patients through their salt status.

Now, inside the cell it is necessary for chemical changes to take place at fixed speeds and directions in chemical local bond space (and often in bulk space) if all syntheses and degradations are going to yield constant ratios of organic molecules and a final functional structure. The elements that catalyse chemical change or bring anions together to assist change are very largely metal ions held by proteins. Na^+, K^+ and Cl^- are not useful in this respect since they bind weakly and are ineffective catalysts (see Section 8.5.1). Thus, a cell needs a range of metal ions for binding and carrying out selective catalysis, including probably Mg^{2+}, Mn^{2+}, Fe^{2+} and Mo^{VI}, as well as, possibly, Co^{3+}, Ni^{2+} and, in some organisms, Cu^{2+}. To these we have to add one non-metal catalytic element which could well be included amongst organic chemicals—selenium. (Note that organic groupings in proteins can also provide catalytic power but not for the activation of simple molecules such as H_2O, N_2, CO, CH_4 and so on, which cannot be held strongly enough in such catalytic sites.) Since all these elements act as homogenous catalysts their concentrations must be fixed too, for otherwise the ratios of different C, H, N, O, P and S compounds could not be fixed. This is managed by connections between all the elements as shown in Fig. 10.10.

In summary, there is an analytical content of about 20 elements, all required in fixed ratios in every cell we know of, but the exact contents in cells of each species, and within a species in each differentiated cell, can be different. How can a cell manage this analytical feat? Undoubtedly by having a very strictly *organised* internal (and, in higher multicellular organisms, external) chemical activity, which must include control over uptake pathways of material, reactions and energy used for both synthesis and degradation, as well as over movement of ions and molecules. The catalyst elements are all incorporated in proteins, making enzymes, which allows the metal ion-containing and all other catalysts to not only activate molecules but also guide them very selectively down certain chemical paths, i.e. to rearrange atoms specifically in local bond space. In the case of some catalysts they must also guide them in bulk physical paths, for example using the so-called 'pumps' (see Chapter 8). The guidance is supplied by the surfaces of the proteins. Clearly, proteins made to become metalloproteins are produced in proportion to the metal uptake.

Note that it is often said that carbon is the essential element of life; this is a judgement based on bulk amount which ignores its combined form with equally essential H, O and N. On the other hand, a trace element which is essential for life, such as Fe or Mg, can be seen to be of equal consequence when one considers life as a process, not just as a structure. An illustrative and strange example is that virtually no nitrogen, from N_2 or NO_3^-, can be fixed by life without molybdenum. We could well say molybdenum is the essential element of life since without it no proteins or DNA would exist. However, molybdenum is not very available, and therefore, in a landscape diagram of survival, we should plot against functional value x element composition, not just element composition (Fig. 10.4).*

In the above we have not mentioned calcium. As far as is known, calcium was rejected by all cells, at least to around 10^{-5}M in primitive organisms (and later to 10^{-7}M), where often the calcium combines externally to form (amorphous) calcium carbonate minerals. Since calcium binds easily and relatively strongly to negatively charged groups, aggregation of organic anions inside cells was and is avoided in this way. Outside cells, calcium acts to strengthen structure by cross-linking the 'walls' of cells or the connections between all cells, or later by forming shells (calcium carbonate) and bones (calcium phosphate). This is the origin of even primitive coral reefs. We shall show how evolution took advantage of this calcium gradient, as well as the gradients of Na^+, Cl^-, K^+ and H^+ across membranes to generate signals in more advanced cells.

Finally we shall from time to time in Chapters 11–14 comment on the total concentration of small and large molecules in solution. Some 20% of the cell volume is occupied and this is far from a dilute solution. Therefore, we must expect *localised kinetic effects*, especially close to membranes. We shall draw particular attention at first to the cytoplasmic spatial distributions common to all cells [see Fig. 11.1(b)], then to the way filaments occupy the cytoplasm of eukaryotes [see Fig. 12.1(b)], and finally to the space-filling in the very extended but highly convoluted structures of membranes of mitochondria and chloroplasts (see Fig. 13.10). Similar problems exist in the discussion of the free space in the Golgi (Fig. 12.8) and the reticula (Fig. 12.4) where membranes are also stacked. Inside membranes themselves, only two-dimensional diffusion is possible and requires quite separate mathematical analysis. In Chapter 8 we deliberately avoided the complications of rates of reactions in cells since there is no fully convincing treatment as yet. A similar problem arises with the external fluids of multicellular organisms when cells are in close contact.

---

*The functional effectiveness usually shows an optimal value below which an organism, even in isolation, struggles to survive, but above which it becomes disadvantaged by the element or 'component'.

All of the above features of concentrated solutions are common to all cells, though cells differ greatly in physical structure, that is in the way space is divided. In the next sections we consider only further common features of all cells and we look at specific organisms in Chapters 11–15.

10.6   Required cellular control

As stated, a living cell is essentially an enclosed reaction space (to control reactions and avoid losses) as shown in Fig. 10.11. At its very simplest it requires a single outer membrane. This membrane must admit the required chemicals, in essence the elements mentioned above, in simple forms, to give via small molecules and synthetic activity the large complex molecules on which life depends—as described, mainly fats, saccharides, proteins and polynucleotides. These two last large molecules form into self-assembled particles which can synthesise molecules of their own kinds, proteins and nucleic acids. This process is observed to be not a random synthesis but a succession of controlled pathways of reactions in steady states, leading to

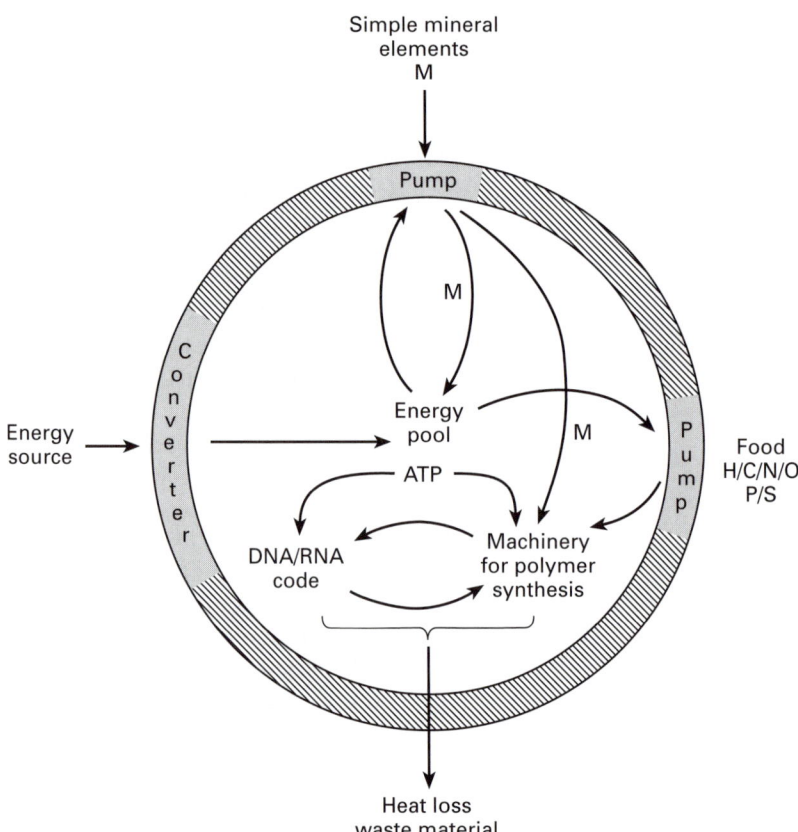

Fig. 10.11 An idealised cell requiring elements for metabolism, i.e. food (H/C/N/O/P/S) simple minerals (M) and energy within an enclosed space.

growth and reproduction. To see how such steady states are brought about we look next at the controls, one at a time.

As we have seen, the synthetic apparatus must have control over the following.

1. A source of basic elements and basic small molecules, i.e. a *suitable environment* with *available* raw materials. The environment is geographically variable and has changed with time as explained in Chapter 9.

2. A source of *energy*, i.e. constant input of energy and output of degraded thermal energy (heat) to move materials in and out of cells (doing work) and to maintain synthesis (uphill reactions, i.e. requiring energy) (see Fig. 10.12). Energy sources vary geographically and with time.

3. A *centrally organised unit* which carries the *information* so as to control what is to be reproduced, but itself being finally multiplied (DNA). This information has also to be transferred.

Furthermore,

4. Since the organization centre and many of the units for synthesis of proteins, pumping in and out of material, etc. are placed away from one another in space, a *system of messengers* is required, which must go between the centre and these units and also between the internal reaction pathways. There must also be molecules to carry small units of C, H, N, O, S and P compounds which are to be used to build the polymers and to share these small units amongst the pathways. These carriers may also generate modes of controlling the rates at which the machinery works, i.e. they may transfer information. Some of them must also transfer energy. These molecules are grouped under the heading of *co-enzymes* in Table 10.5.

5. As there is turnover of chemicals there must also be a steady output of waste chemicals. During growth the input exceeds the output, of course.

A cell, therefore, has controlled networks of transport of material, energy and information, which extends to the equipment in the outer membrane or even to the contained extracellular fluids of advanced organisms, as well as to the inner machinery for the movement and use of all

Table 10.5 Some coenzymes

Co-enzymes	Function
Nucleotide triphosphates	Distribution of energy and phosphate
Nucleotide sulphated diphosphate	Distribution of sulphate
Co-enzyme A*	Distribution of CH_3CO-
Nicotinamides* (NAD)	Distribution of hydride
Flavins*	Transfer of H• radicals and electrons
Haems	Transfer of electrons
Glutamine	Distribution of $-NH_2$
Adenosyl methionine	Distribution of CH_3-
Pyridoxal phosphate*	Transfer of $-NH_2$
Quinones	Transfer of $-H$ and e

* Several of the above are also known as vitamins.

the chemical components. Only if this overall organisation holds good will a cell self-assemble and self-reproduce in a succession of particular steady states. Any organism is then a co-operative flow system operating under extremely limited variables.

While we have described the major elements required by living organisms we have not yet mentioned the sources of energy. We will examine this next.

10.7   Free energy sources: the capability of cells to do work

The source of the free energy for life returns us to Chapters 2–6. There we saw that there were three major sources of energy: (1) chemical reactions of components in stationary states, i.e. energy stored in chemical bonds; (2) compartments out of equilibrium with respect to temperature or fields due to mass or charge (note that this condition allows radiation transfer); (3) chemicals out of physical exchange equilibrium in gradients of pressure or concentration between compartments. The primitive Earth readily supplied the sources considered under (1) by several possible methods, for example, (a) by forcing chemicals from the core and mantle up into the crust and oceans. Typical examples are Fe/S compounds and H_2S which can give

$$FeS + H_2S \rightarrow FeS_2 + H_2^* + energy$$

and (b) by the action of radiation (lightning) on CO_2, N_2 and H_2O, giving a diversity of organic molecules, but note especially NO, nitric oxide (see Section 9.11.3). The organic chemicals produced accidentally or by life are also sources of energy (food) for scavenging organisms.

One example of the second kind of source of energy, (2), is a temperature gradient such as that of the sun surface at about 6000K as opposed to the Earth at 300K (Fig. 10.12). The third kind, (3), is illustrated by the gradients of ions across cell membranes mentioned above, which, due to links with cellular activity, are a valuable source of energy. They provide both a concentration and a charge gradient or field, ψ (see Section 7.10).

Notice that as Earth has evolved the possible energy sources have changed, so that the earlier sulphide/sulphur and hydrogen/sulphur chemical reactions have been largely replaced by hydrogen/dioxygen reactions based on the prior use of light (Section 9.12 and see below). We discuss the relevance of these environmental changes further in Chapters 11–14. As described below, modern life is due to electron withdrawal from H_2O (oxidation of water) which gives O_2 plus combined H due to the sun's action in a process called photosynthesis (see aside). Thus O_2 plus reduced organic matter (oil, gas, sugars) became a major source of energy, but it was not available originally.

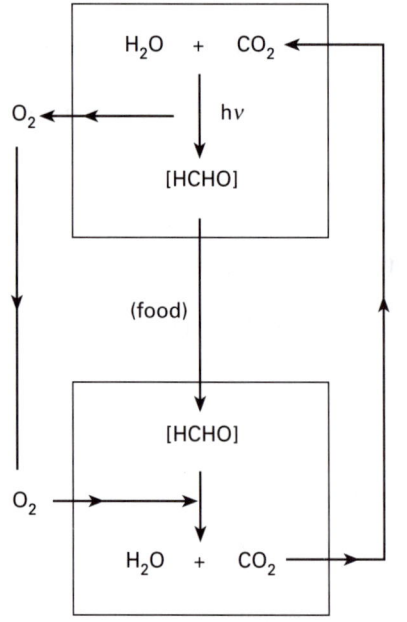

The C, H, O, cycle:
[HCHO] represents sugars.

---

*In this chapter and Chapters 11 and 12 we write H_2 as available hydrogen which may well be bound in organic co-enzymes, e.g. NADH.

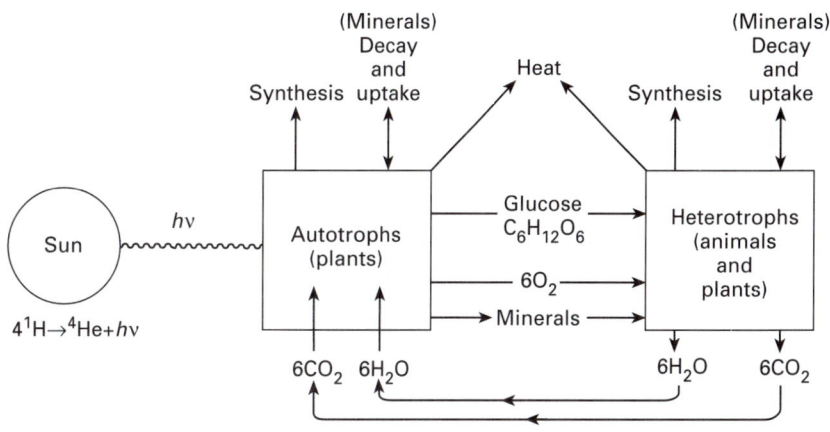

Fig. 10.12 The main routes of energy in the biosphere within chemicals.

10.8   The capture of energy by cells

Electron/proton flows

(a) Electron transfer hops in proteins

Distance (Å)	Rate (s^{-1})
3	10^{12}
10	10^{7}
15	10^{3}

(b) Proton hops in proteins or water

The hop distance is 0.2Å, which together with group rotation gives a rate of $10^{10}s^{-1}$ over 3Å.

The activities of pumping and sorting out of elements, as well as the synthesis of molecules, require the above energy supplies to be captured and used. This means that the organism (*every* organism) must have an energy capture system and a store for that energy. As stated, the energy supply is for the most part from the sun to photosynthesising organisms, such as *plants*, and then energy can be obtained from combustion with dioxygen of these synthesised plant materials or from combustion of materials from other animals by *animals*. The development of the systems for light capture are shown in Table 10.6.

The trapping of light energy needs, of course, photoabsorbing pigments (Fig. 10.13), but in addition these pigments must be organised so as to convert light to a usable form of energy which can assist metabolism. The step required uses light to ionise some chemical, i.e. to create an electrical charge separation by exciting (i.e. moving uphill) electrons away from the photoactive site. The electronic charge separation, flow, can be used to make an electric potential or a concentration gradient of protons (see aside). This requires an acceptor for the electrons excited and the migration of new electrons to the photoactive site, provided by an electron donor (H_2O) which is then oxidised through a manganese-based catalyst (see Fig. 10.13). The proton or charge gradient can be used directly in exchange

Table 10.6 Development of photosystems (PS) in anaerobic bacteria

Time (years $\times 10^9$)	System
3.5	Outer membrane (bacteriorhodopsin) (see Fig. 8.10)
3.5–3.0	Thylakoids in cyanobacteria, PSI (see Fig. 10.20)
3.0–	Thylakoids in cyanobacteria, PSI + PS II, (see Fig. 10.13)*
2.5	Chloroplast incorporation

* Photosystems (PS) are in two parts (see Fig. 10.13).

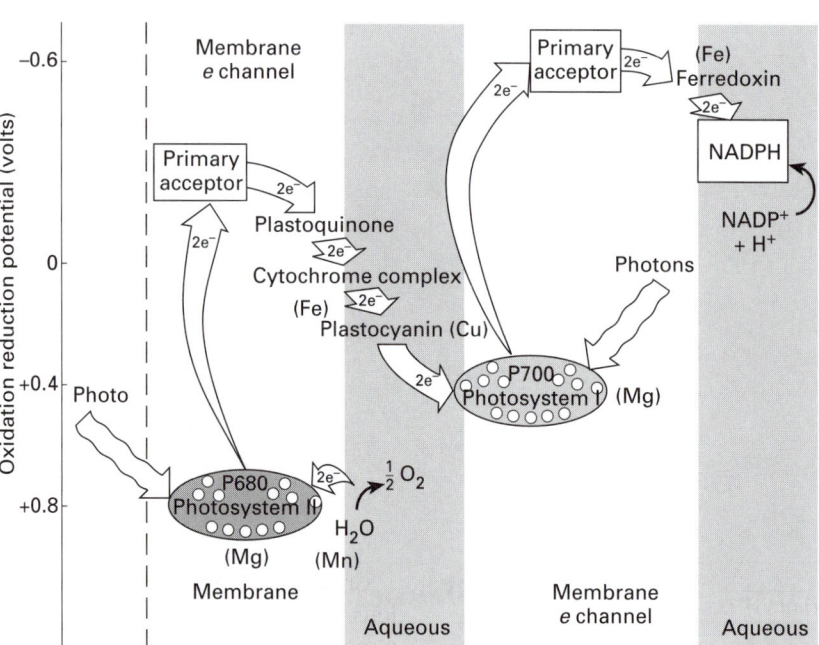

Fig. 10.13 When photosystem II is activated by absorbing photons, electrons are passed along an electron acceptor chain and are eventually donated to photosystem I and finally to NAPD+. Photosystem II is responsible for the photolytic dissociation of water and the production of atmospheric oxygen. This pathway is sometimes referred to as the Z scheme because of its zigzag route, as depicted here, but the two arms are in fact remote in space. (After Villee *et al.*, with permission, ref. 109).

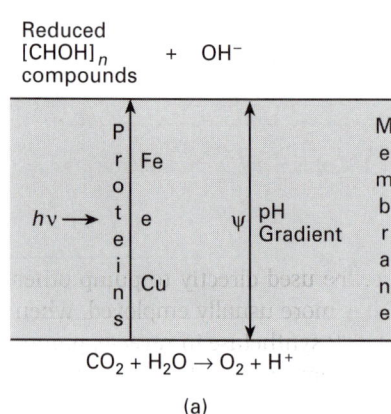

(a)

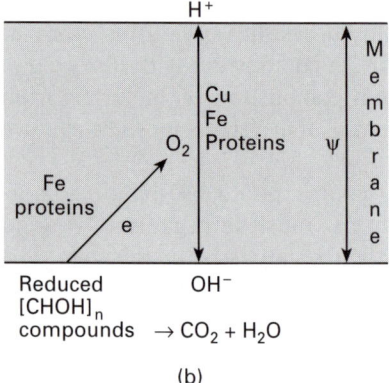

(b)

Fig. 10.14 Two examples of electronic conduction in biological matrices. (a) In photosynthesis dioxygen plus a set of reduced carbon compounds, $[CHOH]_n$, together with pH and potential (Ψ) gradients, are generated by the action of light. The light forces a flow of charge, electrons and protons, through the membrane using Fe and Cu proteins. (b) In respiration the dioxygen and reduced carbon compounds $[CHOH]_n$ are used to create, across another membrane, a pH and potential gradient while remaking CO_2 and H_2O. Electrons and protons are again moved within the membrane using Fe and Cu protein 'wires'.

reactions across membranes to give a gradient of any other ion or molecule. Alternatively, the energy of the gradient may be transferred (transduced) to a covalent, thermodynamically unstable compound (somewhat less capable of doing work since there are always heat losses) such as a pyrophosphate (ATP), a reaction catalysed by the enzyme ATP-synthase [see Fig. 10.15(a)], or again the reducing equivalents (bound H) which are generated by the action of light can be used to carry out uphill, energetically speaking, reductions, such as trapping CO_2 as formaldehyde (HCHO) and sugars ($C_6H_{12}O_6$). The burning of this reduced organic material with dioxygen, the major source of the energy of animals, leads to ATP production in a similar way (Fig. 10.14), utilising proton and electron gradients. (Note both electrons and protons flow in membranes while electrons do not flow in water.)

In effect, what life does is to transfer radiative energy from the sun to stored anti-entropic energy (concentration gradients) and/or potential energy of charge gradients, and then to free energy in chemical bonds (ATP or carbohydrates) (see Sections 4.5 and 5.9). The system of transfer, largely to pyrophosphate derivatives, implies a very limited way of distributing energy which had to be the case since the energy was demanded by a very special set of reactions and their catalysts. We see the further development of contained energy in compounds in the biochemical pathways of the cell (see Section 8.9.2). Observe once again the functional value of the chemistry of elements such as hydrogen and phosphorus in energy transduction. We turn immediately to the molecular machinery for connecting ATP to ion gradients.

10.8.1 Molecular machines for energy transduction

The question which now arises is how a cell can transfer the energy of the ion gradient of protons to ATP, or the reverse, i.e. how can the pumping of ions in or out of a cell using ATP be achieved? The molecular machine for ATP synthesis is shown in Fig. 10.15. Protons, while passing through a helical bundle, c, from a high [H^+] or a high potential across a membrane, cause the bundle to rotate. This rotation forces changes of conformation upon a synthesising enzyme $\alpha\beta$ with the aid of a rod-like structure, $\gamma\epsilon$. The conformational changes force the ADP + P to undergo transformation to ATP and to be released into the cytoplasm. To prevent slippage, the $\alpha\beta$ units are held by a stator, ba. The switch is then (see Chapter 2)

gradient energy → mechanical energy → chemical energy

Gradients of hydrogen ions can, of course, be used directly to pump other ions or molecules out of cells. However, ATP is more usually employed, when a molecular machine somewhat like that of ATP synthetase in reverse, pumps ions or molecules across membranes. A schematic for this is shown in Fig. 10.16. There are general structural features of these machines made from proteins. They are organised so as to provide at any one time both static platforms and moving rods or similarly mutually adjustable parts in space. As shown in Fig. 10.16 (and see Section 8.13), all mechanical machines are of

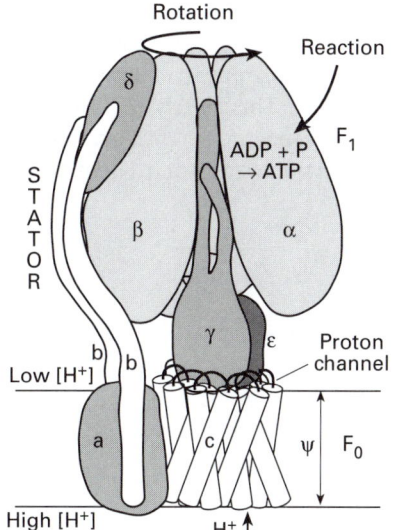

Fig. 10.15 The ATP synthetase machine. Energised protons entering at the bottom, F_o, drive the rotation of the central region so as to force ADP + P to form ATP in the F_1 region (from Junge *et al.* 1997). Trends in Biochemical Science, *22*, 420–423.

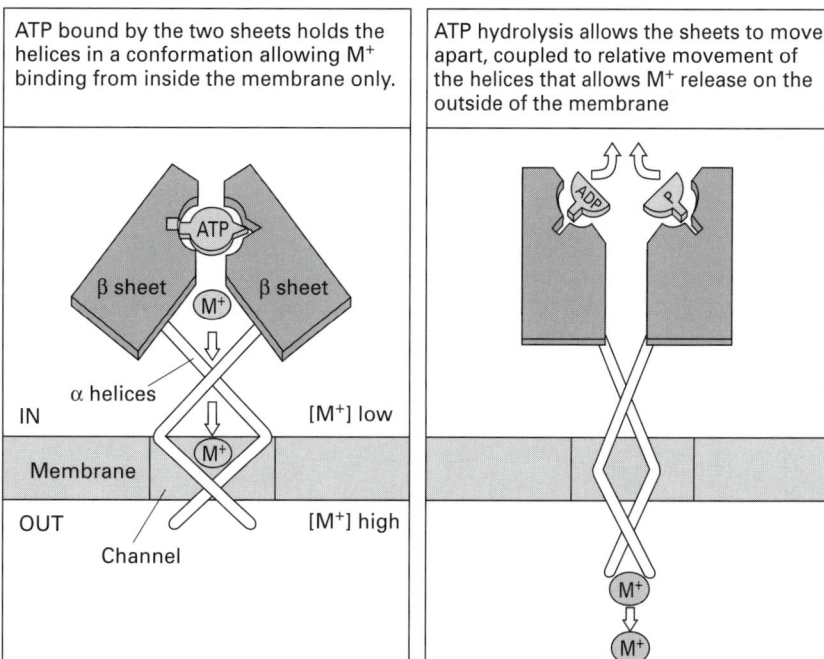

Fig. 10.16 A stripped down version of the way in which the dynamics of helices connect to reactions on β sheets in pumps and ATP synthetase (from ref. 98).

this kind. In cells the dominating static structures are membranes made of lipids and β sheet protein folds upon hinges that act as the enzymes (Fig. 10.16). The major moving parts in cells are protein α helices (see Section 3.6 and Figs 10.15 and 10.16). We need to see next how this biological cellular machinery works as a whole, i.e. how its different structural parts coordinate their functions, and how it is controlled. This clearly requires communication systems, which we shall examine in section 10.10. First, we give another example of energy capture by conversion from such chemicals as sugar to pyruvate while making ATP.

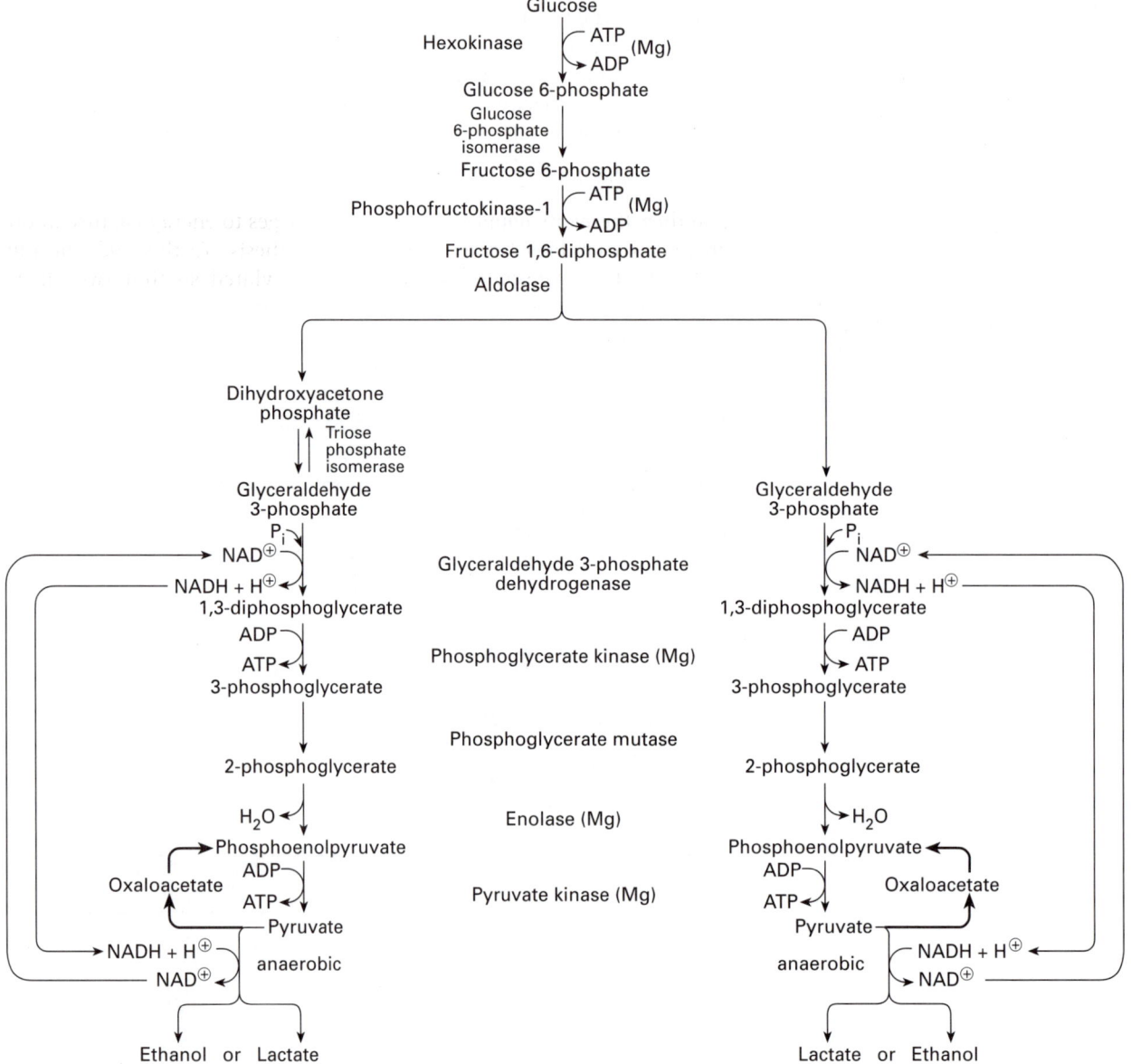

Fig. 10.17 Pathway of glycolysis and gluconeogenesis. The distinctive reactions of gluconeogenesis are denoted by heavy arrows.

10.8.2 The anaerobic conversion of sugar to give energy

The simplest product from the reaction of CO or CO_2 and H_2 is HCHO, formaldehyde. The polymerisation of formaldehyde gives sugars, $[CH \cdot OH]_n$, e.g. glucose. Such reactions are carried out by many organisms and these sugars then become available to other organisms as food. For example, all plants synthesise many sugars and both bacteria and animals use the sugars as food, giving energy. Now, this can only come about if the transformations of glucose, for example, are downhill, giving more stable carbon compounds, such as pyruvate (Figs. 10.17 and 10.18). Note that each step has a 'chosen' catalyst, an enzyme, that recognises one reactant only and transforms it into one product only. Thus a pathway is defined by catalytic control. But why this pathway? If we work our way through the glycolytic transformation shown in Fig. 10.17, we see that it is a particular way of bringing oxygen atoms on to one carbon atom and hydrogen on to another, i.e. in pyruvate $CH_3COCOOH + H_2$ (bound in NADH) starting from an even distribution in glucose, $[CHOH]_6$. This reaction path provides a release of energy (see Figs 10.17 and 10.18). Organisms must not lose this energy and they, therefore, couple the chemical changes to energy capture in other chemicals useful for energy transfer to synthesis. To this end, the initial sugar, $C_6H_{12}O_6$ (glucose), is first phosphorylated so that two carbons become CH(O–phosphate) in fructose diphosphate. This costs the energy of two ATP molecules. The phosphates are now on a half-reduced carbon atom. The molecule is then split to two molecules of glyceraldehyde–3-PO_4 (see Fig. 10.17). After this stage, in a series of steps, inorganic phosphate is introduced and 1,3-diphosphoglycerate produced as 2H are removed by NAD^+. There are now a total of four phosphates bound on oxidised C/H/O frameworks. The phosphate on the oxidised 1-carbon, a carboxylic acid group, can now be transferred to ATP. The phosphate groups remaining on the 3-carbons are transferred on the oxidised phosphoglycerates to the 2-carbons, when water is removed to give the two phospho-enoyl pyruvates

$$CH_2 = C \begin{array}{c} COOH \\ \\ OPO_3^{2-} \end{array}$$

The now *energised* phosphates are then removed to give ATP and the total of 4H that have been removed earlier are returned later from NADH to take two molecules of pyruvate back to the initial empirical composition $(CH_2O)_n$, as lactate $CH_3CHOH \cdot CO_2H$. The sequence of reactions is not simple degradation but a *functionally valuable* set of transformations to gain net energy stored in 2ATP.

As stated above, ATP, a thermodynamically unstable but kinetically stationary molecule in the absence of a suitable catalyst, is the energy carrier in living organisms for synthesis. Immediately we see that if we load a cell with ATP it can reverse the sequence in Fig. 10.17 from pyruvate so as to make glucose and then glycogen, but this pathway in cells is somewhat different from that of glycolysis. Of course, it is now essential that the cell runs towards pyruvate when it needs energy (glycolysis), and to glycogen when it needs to store sugar (energy) (gluconeogenesis), so that it can use

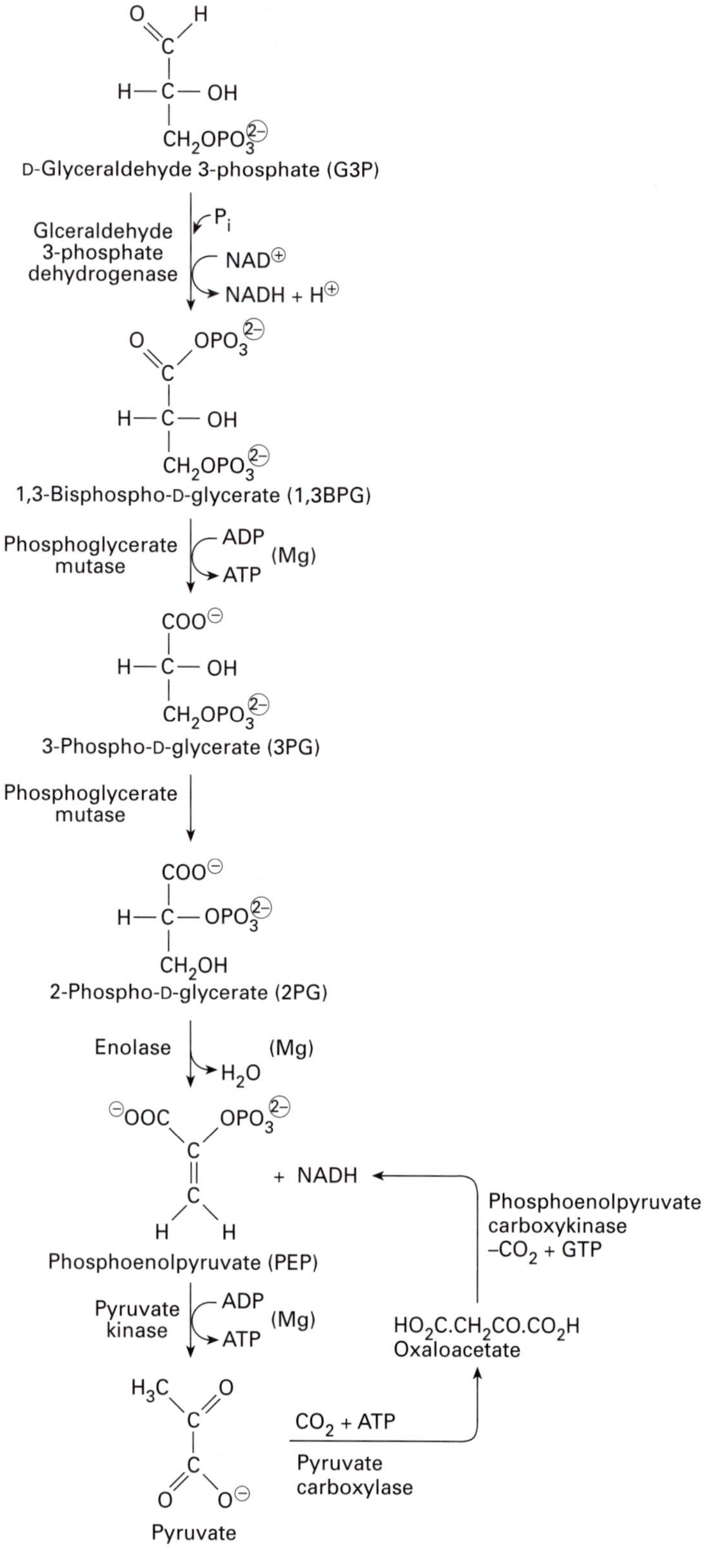

Fig. 10.18 The chemicals in the later steps of Fig. 10.17. In these reactions only Mg^{2+} is used as a metal ion catalyst. In steps that follow, e.g. reduction of pyruvate to give compounds leading to some amino acid syntheses, Fe^{2+} is required.

the energy later and not waste it. A sugar is a better, more kinetically stable, long-term energy store than ATP, which is more useful in energy transfer.

Now that we have seen ways in which ATP is made it is important to note that it has to act in many pathways just as do other co-enzymes (see Chapter 8). These co-enzymes, together with ions such as Mg^{2+}, Ca^{2+}, H^+, act as controls as well as distributing agents in communication networks.

10.9   Communication systems

10.9.1 Networks using the electron: models in man's equipment

Before discussing control at cell level we return, to illustrate the idea, to examples of man-made and industrial equipment.

Taking the risk of over-simplification in the description of feedback control systems we showed a door bell circuit in Fig. 8.12.

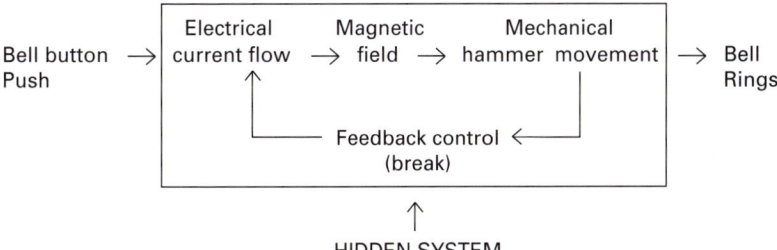

Here there is negative feedback information from the output to the input. Such circuits can be joined together and generalised in computer diagrams such as that represented in Fig. 8.14. [Note how the circuit of the door bell produces oscillations or 'waves' of activity. Waves are also often observed in living cells, e.g. of calcium ion input, (see Section 8.4.5.)]

When series of feedback electronic circuits are connected, then the circuitry as a whole has only a small range of permitted currents (flows) so that the individual units can be said to act together, co-operatively, towards a given purpose. This is part of the design of industrial control circuitry. For example, out-station computers can control the operations of a factory by monitoring many activities and feed information to the central computer, which then issues instructions to maintain steady activity. It helps us to look upon lowly organisms in this way since they have the purpose of producing repeatedly one kind of cell (DNA, membranes, protein machinery) from simple chemical and energy inputs. All activities of such a cell can be said to be DNA programmed. Of course, we are not suggesting that man's experience-based reasoning and activity responses can be described simply in terms of computer operations, but that these may be very useful in helping us to see how a bacterial cell works. Furthermore, we must be aware that a cell self-assembles while a computer does not.

In these examples of man-made controls the current carrier is the electron and the current-carrying material is a metal conductor made from Cu or Fe, or a semiconductor made from Si, GaAs, etc., in wires. The production of metals in wires appears to be beyond the ability of living systems today (but this need not be so forever!)

10.9.2 Feedback Pathway control in organisms

We now need to show how a parallel feedback organisation can operate even in free aqueous solution to control cellular chemistry. We do so by looking again at one pair of selected pathways, glycolysis and gluconeogenesis, and their controls as an example. The pathway from pyruvate compared with the reverse, that from glucose, is controlled by one or two different enzymes (Fig. 10.17), which are activated or inhibited according to need by feedback from small molecules, including ATP, in the pathway. This is a type of feedback control (here by ATP) and directly parallels the electronic circuit of the door bell, using chemicals in place of the electron. In fact, ATP acts partly like an electron in a bell circuit, opening and closing activity according to its concentration (compare with a current). In a similar way, any small molecule member of a pathway can be used in feedback (or feed-forward) to control that pathway, or to control a related one.

Finally, note that some of the molecules in the glycolytic pathway carry hydrogen or fragments of C/H/O compounds needed in many other pathways, e.g. NADH (nicotinamide co-enzyme), and at the end of the pathway, co-enzyme A carries *acetyl* (which links to the synthesis of fats) (see Fig. 10.23). These three co-enzymes, ATP, NAD and co-enzyme A, therefore, diffuse to and from different pathways, making a material network, as in a computer. Again, intermediates in these pathways link to the syntheses of amino acids and DNA (RNA) bases, which also have feedback controls (see Fig. 10.23). Finally, the energy carriers, ATP and other nucleotide triphosphates (NTP), which are to be seen as derivatives of pyrophosphate, are the common source of energy between pathways, and of pumps for moving ions and molecules in and out of cells and for syntheses (see Fig. 10.11). All the concentrations of the chemicals in these pathways are therefore seen to be interrelated by feedback and feed-forward of material and energy (see Table 10.7).

We leave the reader to pursue, in biochemistry textbooks, this topic of the co-existing co-operative synthesis and degradation of all biological chemicals in 'designed' functionally useful pathways dependent on controls linked to supply of material and energy. It will be seen that all individual pathways of cellular metabolism are also interlinked by a variety of small chemicals in a network. The obvious conclusion is that metabolism is a highly co-operative activity acting in the interests of cell survival. It is partly chemical rearrangement of atoms, but it is associated with energy

Table 10.7 Carriers of energy, material and information

Energy	Materials	Information (feedback)
Pyrophosphates	Co-enzymes,	Many substrate concentrations
Acid anhydrides	e.g. NADH,	Co-enzyme concentrations
Thio-acids	$FADH_2$, CoA	Energy carriers
		Ion gradients

capture. There are, however, other messengers in these systems which are not directly part of metabolism and which we shall examine next. They are often involved not so much in enzyme-linked controls of chemical reactions, microscopic movement in space (see Section 8.5.1), as in physical transfer controls across membranes (bulk space movement).

10.9.3 Networks using messengers other than organic molecules

The essence of the above kinds of circuit can be reduced to

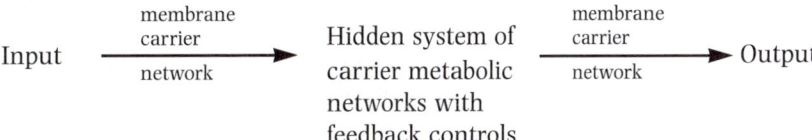

The choice of network rests, therefore, in the choice of carrier material and of the material for its transfer. A further possibility, different from inanimate electronics and the transfer of organic chemicals much used by organisms, is ion movement. For example, inorganic ions, especially H^+, Na^+, K^+, Ca^{2+}, Cl^- and $H_2PO_4^-$, which flow in aqueous solution, can be used in electrolytic circuits structured in a form analogous to tubes (or wires). Nerves, for example, use Na^+ and K^+. However, these are but examples of an infinite variety of possible message carriers and materials in biological solutions (see Chapters 11–15). In principle, any molecular ion, including organic molecules soluble in water, could be used, e.g. adrenaline, but the flows can take place even in an organic fluid as long as the message carrier is soluble in this medium or is provided with a channel. Clearly, the direction of flow needs to be controlled, as in an electronic circuit where it can be managed by an applied difference of potential: valves, resistors, condensors and transistors. A variety of devices in cellular electrolytic circuits match these devices (see Section 8.4.3) and they can be linked to feedback controls. Receptors for the messenger ions are also needed and are found to be present, some on the outside of cells, which generate an internal message, and some on the inside, when the messenger penetrates into the cell directly to an enzyme or to the DNA or RNA via protein receptors. The receptors acting on DNA are called *transcription factors* as they help to control polymer (DNA, RNA and protein) synthesis (see Fig. 7.12).

There is, of course, a question of the suitability or functional value of a chemical as a messenger. The further the distance it has to travel the better it is if the messenger does not interact (bind) to the material in which it moves. For this reason Na^+, K^+ and Cl^- are better messengers than Mg^{2+}, Ca^{2+} or HPO_4^{2-}, which bind more strongly. Here we meet again the problem of the natural selection of chemical elements for their functional value, see aside. We shall find that Ca^{2+} is a major carrier only over short distances, while Na^+, K^+ and Cl^- work over long distances. A chemical messenger of this kind can be fast or slow acting. Very slow messenger ions have still stronger binding constants, e.g. Fe^{2+} or Zn^{2+}. Organic molecules can also be

Early uses of K^+
(1) In Telomere of DNA
(2) In some Kinases
(3) Osmotic Control

chosen to be fast or slow, depending on the reactivity and positioning of enzymes, since they are metabolised and this restricts in one sense and increases in another their use. Now, a messenger is only useful if its arrival is noticed, thus we need next to examine how messages are converted into actions.

10.10   Conversion of messages to actions

In the door bell circuit, the electron current (the messenger) creates a magnetic field locally which operates the mechanical action of the bell and breaks the circuit. In living systems the equivalent to a magnetic field is needed to create mechanical change in molecular devices using, for example, ion currents. In principle, and sometimes in practice, the change of an electric (electrolytic) potential itself is sufficient. Such a current is called a 'gating' current. In these 'gatings' changes in Na^+ and K^+ concentrations are often involved in adjusting potentials. Alternatively, the current carrier (an ion) or a messenger molecule can bind to a protein and change its conformation (compare with the striker of the bell). Now the fastest ion current carriers are Na^+, K^+ and Cl^-, but they are not able to modify protein conformations since they bind too weakly. Thus, their gated message is used to release second messengers, e.g. Ca^{2+} ions, which bind more effectively but not too strongly to avoid a permanent effect. If a slower message system is wanted, then strongly binding molecules or ions (Zn^{2+} or Fe^{2+}) can be used and these may directly affect their protein receptors—many organic hormones act in this way. Here we see two different elementary functions combined, diffusion and binding, and of course they compromise one another. The binding of the messenger then acts as an allosteric switch by changing the structure of catalysts from active to inactive states (see Section 8.6.4).

10.11   The functional capability of elements

The value of the elements as messengers depends in part on their ability to move unimpeded, as in the above example. A very different type of value of the elements is their ability to form structures which, if they are to be included in mechanical devices, must have some dynamic parts, e.g. gates. Still other values are related to the ability to allow movement of electronic charge, to catalyse reactions or to bring about mechanical change. In some cases the roles are quite specific; for example, we have shown for certain phosphorus compounds the special value of carrying energy for dehydration reactions. Thus, the elements of the periodic table can also be described by functional value, as shown in Table 10.8. We now see why the making of a living organism requires the selection of chemical elements from some dozen (at least) of the 18 groups of the periodic table if it is to function with a high degree of effectiveness. It is only effective organisms (or indeed man's effective industrial plants which also utilise an appropriate

Table 10.8 Functional value of the elements

Information (electrical) transfer and storage	Mechanical transmission	Acid–base catalysis	Redox catalysis	Structural role (excluding the organic polymers)	Chemical energy transmission and storage
Na^+, K^+, Cl^-, H^+, Mg^{2+}, Ca^{2+}, HPO_4^{2-}	Mg^{2+}, Ca^{2+}, HPO_4^{2-}	Non-metals and divalent and trivalent ions, e.g. H^+, (N) (S) Zn^{2+}, Fe^{3+}, Mg^{2+}	Transitional metal ions and some non-metals, e.g. Cu, Fe, Mn, Co, Mo, Se, S	Si, B, P, S, Ca, Mg, (Zn)	P, S, (C)

N.B. The functional value of the elements arises partly from the chemistry and partly from the energy input to that chemistry. The energy input can be to the gradient of the free concentration of an element, to a chemical bond of given kinetic stability or to the synthesis of a polymer in a compartment that acts as a trap.

selection of elements) that will succeed in 'survival of the fittest' terms. Put simply, for organisms, of the available elements only C, H, N, O polymers can make the machinery, only Na^+, K^+, Cl^- can be fast current carriers, only Mg^{2+}, Ca^{2+} and HPO_4^{2-} can assist relatively quick mechanical change and only transition metals and sulphur and selenium can be used in electronic conduction or oxidation–reduction catalysts, while ions such as Zn^{2+} are the most effective acid catalysts, free from the complication of redox reactions. This list of more than a dozen elements has the makings of all biological organisation that we know about. Any cellular organisation has to have selective uptake and catalysed transfer of these elements.

The way elements are used clearly depends, and depended, critically on their availability, and often on the management of their use in cases where there are multiple possibilities. We shall need, therefore, to look at the *evolution* of organisation in living systems in different geographical regions and as the Earth changed with time, and with it the availability of elements. Before that, let us describe the structure of cells in a little more detail and then the nature of the central code to and from which messages must go. Finally we consider the total organisation of cells, especially seeing them now as a way in which available chemicals and energy are arranged and moved in space and time.

10.12 Complex molecular structures within cells

In Chapters 2–4 we showed that molecular structure, even up to the size of isolated proteins, came about from consideration of the energetics of internal binding. We observed that co-operative interactions of metal atoms, ions or molecular units of any size led to a symmetrical packing of the units in crystals that retain little motion. A biological assembly is very differently structured in that it is composed of many different units and has little if any symmetry. Some examples were given in Section 10.8, Figs 10.13 and 10.15. Thus, the ribosome has a multitude of proteins and more than one RNA (see Fig. 6.8), yet it has a fixed structure with no overall symmetry. Again, it is a dynamic 'structure' for the production of proteins—it is a machine. In Chapter 7 we contrasted these 'structures' with those of viruses, where packing of one or two proteins led to highly symmetrical *static* structures with no activity. Local 'structure' in active particles must be such that it can switch between more than one (similar) structure; thus, highly co-operative single states (crystals) are to be avoided. These 'structures' are then related to function in an *organisation*, such that as small units are added to and subtracted from them they can cycle. A simple reversible cycle of a protein is illustrated by haemoglobin uptake and release of dioxygen. A more complex but parallel cycle is that of the ribosome as it takes up amino acids and later releases proteins but needs also to transfer energy into the peptide links. Thus energy is used in order to make the 'components' of the machines of the cell. In such processes heat is released, so that some part of the energy is inevitably lost as entropy gain of the surroundings (see Sections 5.4.2 and 8.10).

The ways, in principle, in which the pieces of 'structures' are built for functional use were described in Chapter 7, but the details of a complex machinery such as that of the ribosome are not yet known. In Section 10.8 we have already described the essence of one machine for a relatively simple process—energy transduction in membranes—and indicated the common features of all biological machines. The aspects to stress in all cases are that there have to be multiple states of the 'structure' as it cycles *functionally*, and that because the machines are made from several proteins they have little or no true symmetry, unlike ordered crystals.

Typically, every machine in a cell has a set of shapes or 'structures' based on the conformations of the proteins in it. A cell is then a hierarchy of dynamic structures starting from those of the smallest molecules and leading up, through folded polymers, to their packing in aqueous complexes or in lipids in membranes, and then to the assembly of these complexes in mutual association. There is, then, *dynamic organisation rather than order*. The restrictions that generate organisation are not the same as those that generate order, but the structures involved in both cases are equally in an improbable or low 'entropy' state. The more complicated the cell and the more cells there are in an organisation the more improbable it becomes and the lower its 'entropy'. There is, however, an overall gain of entropy during the activity of the combination of cell and environment, since the environment constantly receives heat from metabolic syntheses—*cells are dissipative systems*.

The question arises as to whether we should use the word 'structure' opposite a protein in a cell where it is a time-dependent unit. Perhaps it would be better to refer to it as a unit of organisation, not of order, *when in a cell*, since order implies relatively rigid structure. Note too that the order appears to indicate a linear mathematical relationship between properties and sequence but organisation does not (see Section 5.8.2). We shall also have to think again about the nature of the polymers, proteins, which have available to them many conformations that are perhaps better thought of individually as microscopic states of a macroscopic molecular system (Section 5.4.1). Now, in order that a cell should be reproducible it must have a 'planned' activity and a replication of many key features. We turn to the source of such activity.

10.13   The need for a control code (DNA/RNA): inherited information

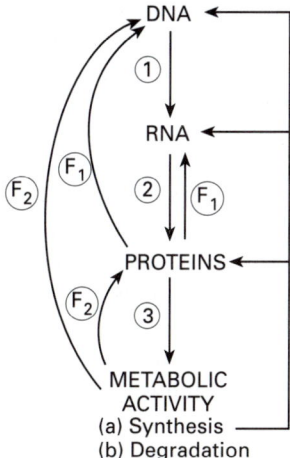

(1) Transcription:
    plus RNA polymerases

(2) Translation in ribosomes

(3) Enzyme activity + pumping in
    of minerals, etc. and energy
    capture

(F₁) Feedback by transcription factors
    (proteins)

(F₂) Feedback and controls from
    small molecules to DNA and
    enzymes

Fig. 10.19 The overall relationship of
DNA to RNA to proteins and their feed-
forward and feedback connections (F).

Once all the 'components' have been put in place in the simplest cell, which consisted, in large part, of a single outer membrane of proteins, some with catalytic activity, and a set of small molecules and ions, it was always possible that it could capture energy, repairing and making itself, and dividing, in principle. However, such activity could only be haphazard and inefficient and the cell could not be expected to reproduce itself accurately. Therefore, for life, although this primary machine-like activity was one objective that had to be achieved, a second one was needed to ensure survival as well as defined reproduction. The need was to have and to pass on coded information, otherwise the waste of effort (energy) in generating systems of long survival would have become enormous. While 'life' may have started from the basis of random production of polymers in cells with a very limited survival capability, this does not correspond to what we see as life today, which has controlled reproduction in well-defined forms. Sustained, faithful reproduction is ensured through a coded information molecule, DNA, which controls the process. To function, DNA has to connect to machinery of all kinds in the cell since it is not directly a part of the machinery. The transmission is through transcription to RNA, and then from RNA, by translation, to proteins (see Fig. 10.19). Communication between the different machines for 'transcription' (using RNA polymerases), for 'translation' (in ribosomes) and for reproduction of DNA (using DNA polymerases) is therefore necessary so that these machines are capable of working connectedly and leading to balanced production of proteins, RNA and DNA, and all other cellular parts such as membranes (fats) and walls (polysaccharides). Simultaneously, energy must be captured and small molecular materials must be generated from basic inorganic environmental substances, and all must be connected by a messenger network of small units (the co-enzymes; see Tables 10.5 and 10.7) and the metal ions described above.

Of course, the networks of large and small messengers (information carriers) have to be interactive with the reproduction of the central information-carrying centre. Thus, even DNA polymerisation is sensitive to the cell environment in that, for example, K^+, Na^+, Mg^{2+} and energy levels must all be interrelated with the protein machinery needed for DNA synthesis (see Fig. 10.11). Obviously, this synthesis of DNA requires that of nucleotides, which have needs of their own for other elements, proteins and extra energy. Only by sensing the complexity of the apparatus common to every cell do we have a feel for what life is. We shall see in the next chapters how both DNA and its messenger circuits developed with the evolving complexity of organisms. In fact, DNA does not show a mathematical linearity between its sequence and its functional properties. DNA, even as a single molecule, is complex (see Section 5.8.5).

In fact, controlled reproduction and repair is just a further step in the evolution of 'fitness for survival' and it co-exists with improvement in the machinery for fitness. The latter had to be 'discovered' by trial and error methods by the cell and again represented in coded form in the DNA reproduction centre (see Section 10.18). (In this book we cannot give a detailed account of the genetic code in DNA and the reader must seek this information in standard textbooks. Our task is to indicate the fundamentals of the logical structure of biological reactions.)

The way to control activity tightly is to give specific instructions—to pass information to the machine so that it only produces valuable products. It is necessary, however, to understand that, in principle, the control centre is not essential to the machine once it is up and running and not open to change or damage. In essence, then, DNA is a spare parts code for the distribution of proteins (parts) in the steady state and, of course, for parts in development in a *timed* sequence. It is only a blue print, that is it *represents* the coded part of the total machinery of life, but is not an essential feature of living or a life form itself, although it is the source of all information. Viruses made from DNA or RNA are not alive *per se* and red blood cells of animals survive for many days without DNA. (The same applies to a factory; once it is made, the plan can be thrown away and a computer incorporates the control and operates the factory continuously.)

10.14 The disposition in space of ions and molecules: compartmental control

In the above we have considered the simplest and most primitive cell enclosed by a single membrane. An obvious problem of control involving so many different atomic (ionic) and molecular units is the complexity of the system. A considerable gain can be made by reducing the number of types of units that are mixed together. Here, compartments offer a great advantage. *In effect, one of the great steps forward in evolution was the generation of 'isolated' parts of space within cells.* The earliest example we know about of compartmental structures is the thylakoid in cyanobacteria developed from invaginations (see Fig. 10.20) where energy capture is separated from other chemical processes.

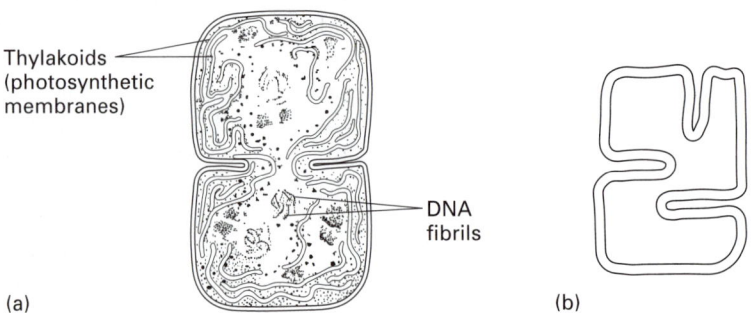

Thylakoids (photosynthetic membranes)

DNA fibrils

(a)

(b)

Fig. 10.20 (a) Thylakoids in blue–green algae (cyanobacteria). Note also the DNA fibrils in the cytoplasm, not enclosed in a membrane. Thylakoids and mitochondria are multinucleated and have large vesicle network structures in three dimensions. Presumably the thylakoids developed from invaginations of the outer membrane, see (b).

Later in evolution there were four further developments: one was radial division of space, as found in its simplest form in the double membranes of certain bacteria giving a cytoplasm and a periplasm (see Fig. 11.2). These organisms are included in the class prokaryotes. The second was the development of many other internal compartments, which gave the unicellular organisms called eukaryotes, where any resemblance to radial symmetry is lost and in particular the control code (DNA) is separated in the nuclear compartment. Some of these cells captured bacteria by endocytosis and after incorporation modified them to give organelles for energy capture (see Chapter 12). The major ones are the mitochondria and the chloroplasts, see aside. (A third may be the flagella motor but the evidence for this is still not definitive.) Other compartments formed, including peroxyzomes, the endoplasmic reticula, vacuoles and many storage vesicles, all with separate chemical activities. Much of this diversity of spatial discrimination became extremely useful in that it not only separated required internal activities in vesicles but it allowed removal of poisons or even ejection (exocytosis) of poisons and active catalysts (enzymes) to act on other species by emptying the vesicles outside the cell. It was important for the perfection of this organisation that the different compartments were located in space and in communication with each other and the major cellular activity. The spatial organisation was achieved by creating networks of filaments, at first inside cells and then outside, while communication came about through the use of new messengers.

In the *third* development, where activities are segregated in organs, the vesicles came to be of use in the storage of messenger chemicals for communication systems between organs. Messages must now pass between cells to co-ordinate the activity of the whole. This development of multicellular organisms is relatively recent (some 700 million years ago) and was followed by the last stage, a *fourth* development, *that of the nervous system and the brain*. The peculiarity here is that the brain is a new store of information unrelated to that in DNA. In fact, the information stored in the brain is not in a coded linear sequence, as in DNA, but in storage gradients of charge, as in a computer, or chemicals. An appreciation of the functioning of the brain is not yet within our grasp, but in contrast with the sequential code of DNA, this store of information is in a three-dimensional net. The brain has allowed control over materials and energy in space to extend to regions (compartments) outside organisms. We shall try to show how all this evolution depended upon realisation of the full value of the chemistry of particular elements distributed in space while flowing in time.

Mitochondria and chloroplasts

These organelles and some vesicles are long weaving compartments which may well not act as homogeneous bodies

10.15 External shapes of organisms

Shapes of organisms are not just products of amounts of chemicals in systems at equilibrium (see Chapter 7); they are functional. The principles of construction are not different from those of shapes of internal units of cells, such as ribosomes described in Section 10.12. The simplest shapes generate virtually fixed *protective* outer structures. Only when they are built

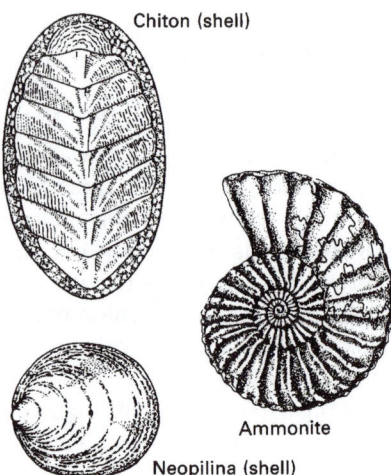

Fig. 10.21 The forms of some mono-valve shellfish showing different variations on conical shape and illustrating, to the right, the logarithmic spiral. Here the shells are dominant in the appearance of the animal, but the control over the open end of the shell rests internally and it grows two-dimensionally. Other features of shape are DNA related and species selective, see Fig. 14.7.

from repeated single units do they have symmetry, as is seen in the outer structure of viruses. In such cases there are no dynamic features. An excellent additional example is the conical shape of limpet shells made from $CaCO_3$ building blocks. The opening of the cone grows as an expanding circle, while the organism grows roughly as an expanding sphere or elipse (see Figs 10.21 and 14.7).

Many more or less rigid structures have some limited left/right or up/down symmetry, e.g. the rib cage of humans. Of course, the rigid shape is for strength. Here, shape and growth pattern have some secondary dependence on DNA, but they arise mainly from stresses induced by the growth of soft parts. D'Arcy Thompson in his book (see ref. 58 in Further reading) was able to look at many patterns in nature and show mathematically that they arose from considerations of growth stresses and did not need specific instruction from the DNA code. It is sufficient to generate the stress in the soft tissue for many hard parts to grow as observed. Support for such an argument comes from the fact that external environmental stresses imposed during growth can determine the shape of a shell or bone. This description does not apply to many of the dynamic functional soft parts of an organism, e.g. an eye. Here, we must observe that proteins generate the moving active parts and that these proteins, as they are synthesised, assemble in shapes that are dynamic. Another example is the formation of a muscle. Here we see the assembly of several proteins in a repeating pattern where the whole is functionally active. The DNA decides the shape of each protein unit, which it did not do in the above case of $CaCO_3$ crystals, and the proteins self-assemble into a pattern that is open to several shapes, as in a ribosome. In principle, the whole organism is also open to construction in this way, e.g. a worm or a caterpillar. These organisms only have soft parts that repeat along a rough cylinder. The shape is then controlled by an association of contractile units for motion while containing an inner tube for digestion. We shall see that the tubular shape has advantages in many, many organisms (plants and animals) and is clearly related to survival. Now, it is not just the outside of organisms that have a shape, and we also need to examine the internal shapes of cells.

10.15.1 Internal shapes of cells

The biological activity of all the above units is contained by membranes. Membranes of synthetic vesicles made from lipids are roughly spherical since the tension in them is the same everywhere. Biological membranes are made from proteins and lipids, hence where the proteins come together curvature surface tension is different from elsewhere. Biological membranes have shape according to the nature of their proteins. However, the membranes are held by proteins attached to them *along* the lipid bilayers as well as *in* them. The way in which the proteins cluster affects the shape. Mitochondria, chloroplasts and some bacterial membranes have well-defined shapes opposite differently energised states of their proteins. They are functional shapes in that they generate localised activities connected to the cytoplasm.

10.16 Growth, reproduction, development and evolution

So far we have treated a living cell or even an organism as being in a *steady state* of activity, which has a fixed appearance although all its chemicals constantly turn over. Life is, of course, *reproductive* and all cells go through *growth*, which is looked upon as increase in size. All cells also die and it is now known that the death of cells (apoptosis) and organisms is programmed. Again, those organisms that have multicompartments all *develop* and differentiate from simpler single cells (gametes). Thus, the steady state apparently seen in adult organisms does not provide a full description of any living body, no matter what its complexity. In this book we cannot go into any detail of single cell cycles or of development, but we give a brief section of each to direct attention to further difficulties with a full description of living cells. Before doing so, it may help the reader to have the nature of metabolic steady-state cycles clearly separated from reproductive cycles (see Section 10.16.1).

Finally, we must stress that life forms have *evolved* continuously with time, so that a steady-state analysis is again inadequate to explain change in life forms.

10.16.1 Cycles of reactions and cell cycles

Cycles of reactions in cells are not related to, for example, cycles of a wheel, which, in the absence of friction, continues at a steady rate once started, in what we call a stationary state, with no loss or gain of energy or matter—the whole cycle has no time-dependent character. A cycle in a cellular reaction refers to a recurrent identical mode of synthesis or degradation in which one part of the starting material is constantly replaced and processed in a succession of steps which returns a second part of the starting material to the initial step. For example, in a material cycle, such as that in Fig. 10.22, B can be looked on as a carrier of A until A is transformed into C. During the cycle, energy may be extracted, or introduced, and extra chemicals may be added to complete the transformation of A to C. Metabolic cycles are often held in vesicles or organelles to keep the concentration levels of intermediates high but with a low total capacity of compounds in the cycle. An example is the citric acid cycle (see Section 10.20) in mitochondria.

The total *cell cycle* effectively has many interrelated reaction cycles producing the necessary complement for viable reproduction. In one schematic cell cycle similar to Fig. 10.22, A is just the source of H, C, N, O, P necessary to make DNA, while B is DNA and so is the product C. The cycle produces two DNA from every starting one and they separate.

Given the nature of developmental cycles it is obvious that, unlike in stationary cyclic states, there is additional coupling with *time*. We return to this topic in Chapter 16 under sequential change.

10.16.2 Growth and the cell cycle

Growth, in its simplest form, just leads to reproduction of identical cells following an increase in size, so that it applies to a colony as well as to a

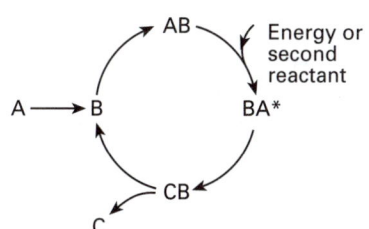

Fig. 10.22 A scheme for conversion of A to C using energy and based on cycling carriers, B. For a true cycle see Fig. 8.2. In the citric acid cycle A is acetate, B is oxaloacetate, C is CO_2 + H_2O and the second reactant is O_2 (see Fig. 10.23).

multicellular organism. For any single cell it means that the process of cell division has to be understood, which is very different from considerations of a cell as a simple steady-state reactor in a fixed volume. To this end many of the products of reaction have to be retained in order to duplicate the contents of the cell after one division, before the next division takes place. For cell growth without a coded set of instructions the repeated division could in principle be controlled by the surface: volume ratio, given that at some critical value of this ratio the cell divides. Cells, as we know them, are DNA coded and usually divide on doubling the DNA. However, this is a necessary but insufficient condition for reproduction since many varieties of cells can be multinucleated. It appears that cell reproduction is brought about by the growth in the cell of an apparatus for dividing the cell, which requires DNA replication after rough duplication of all the contents has occurred. Frequently, cells under-supplied with essential factors do not divide. As far as this book is concerned, therefore, we treat cell division as an automatic response to growth in a given cell when accumulated products give duplication of *all* material in that cell. We must then assume, of course, that in a complicated organism the growth of all compartments (organs) is synchronised by message systems in a cycle. In the simplest case, all of the above structures are contained within a single cell's membrane, which is often shaped, but the shape of a complex organism is not fixed during the growth.

In more advanced cells the proteins attached to the membrane are under tension from the cell filaments which criss-cross the cell. Shape now depends on the machinery of the cell as a whole, as it uses energy to maintain stress. Note how even amoeba change outer membrane shape, and that change of shape is fundamental to locomotion of all animals, even bacterial flagella.

10.17 Development

Much of the above describes a steady state of a cell, but an organism, and indeed many cells, develop. This adds an extra time-dependent feature to life, on top of the consideration of flow of materials and energy in the steady state or in a state of expansion through multiplication. Simple growth could be taken within a steady-state system, but a switch in pattern, metamorphosis for example, is quite different. The machinery changes with time no matter that the DNA is constant. Thus, one and the same DNA represents a caterpillar and a butterfly. The messengers, too, are constantly adjusted. We shall find that development, like growth, requires a feedback of information as to the nature of the present state of a cell in order to instruct it to enter its next steady state. The time of maintenance of all steady states increases, of necessity, with complexity, so that more evolved forms show a gradual increase in the periods of development and adult life. We associate such modifications with changes in the messages to DNA, so that the part of the DNA expressed can be different. In the simplest cells this is achieved by phosphorylation and dephosphorylation, which controls the successive appearance and disappearance of certain proteins, cyclins. The cyclins manage each stage of the cell cycle. In advanced multicellular organisms a similar mechanism is operative, but now cell cycles are also managed by messages between cells using so-called hormones, such as thyroxine and certain oligopeptides (see Chapter 14). How is the changing of DNA itself related to evolution of new organisms? What circumstances force evolution to occur and yet act at the level of DNA? (See Molecular Strategies in Biological Evolution. ed. L. H. Caporale. Annals New York Acad. Sci. *870* (1999).)

10.18   Speciation amongst living systems: evolution

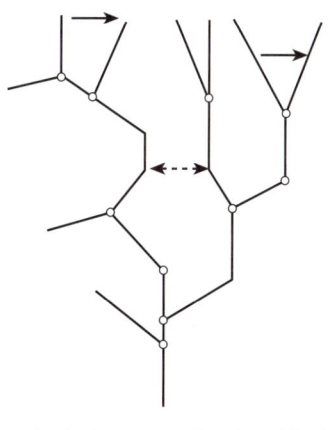

◄‑‑► incorporation (symbiosis)
⟶ transfer

Our thoughts concerning evolution have been guided by classifications of living systems into species (Fig. 10.1). We may see this as progressive diversification of organisms forming an evolutionary tree. One development is certainly in physical organisational complexity (shapes and structures), but another is in chemical sophistication, both of which may well be dependent on the changing environment. In fact, evolution is not quite like this simple picture; it may well spread out from a central point to diversity within simple and complicated forms, but there is much interaction and exchange between branches (see Fig. 10.1 and see aside). Before going into any detail it is necessary to be clear about the nature of species.

We may well believe that all life has a common ancestor some 4×10^9 years ago (see Fig. 10.1). However, this is not proven and it is possible to consider that life's machinery* developed in several forms before a common code of information transfer arose and, of necessity, imposed itself on a variety of machines. No matter which idea is correct, we know that today there exists a huge variety of so-called *species* of living organisms. Since DNA represents the code for a given cell there is then the temptation to say that species arise just because of chance changes in DNA, called mutations. However, mutations occur in simple steps and it is not conceivable that single base changes would lead to anything but a rich continuous variety of *individuals* within a population, one species.

The problem may be overcome in principle if, by accident, different groups of one original population become isolated under different physical conditions, when the mutations produced in isolation could lead to organisms unable to interbreed with organisms developing in a different environment, i.e. they would become different, although related species. However, this explanation does not appear to be very satisfactory either, particularly for the description of early speciation when prokaryote bacteria lived in the sea, since different populations would always mix. Moreover, gene transfer is common amongst bacteria (and not only amongst bacteria!) so that speciation would not be persistent. An alternative explanation could be that *only certain different kinds of efficient complex machinery, not a con-*

*By machinery we mean not only structural components but also their connections and operational conditions. A biological machine is a dynamic not a static concept involving continuous flow of material and energy.

tinuous variety, have considerable survival strength. They might employ different elements or different amounts of the same elements and different energy flows, either using the same or not the same environment. If the sets of machinery are not compatible, mixing them together, no matter what the role of the DNA could be, would not yield a living organism. It is this possibility that we wish to study. It implies that DNA had to evolve into the codes for each successful machine, e.g. a cat and a dog. Species of bisexual organisms, by definition, are not able to interbreed and we may say that the DNA, the codes in the different organisms, represent machines that are sufficiently different so that combination is impossible. Evolution would then be a development of new machinery as the environment changed, which would become most efficient in certain forms, isolated species, and which would come to have the appropriate DNA for it.

The question arises as to why this change would be coupled to changes in DNA, see aside. We explore this possibility later. The alternative mode of evolution is for entirely random searching by continuous adjustment of the code until a new viable system is reached and preserved by natural selection. Here the environment is not felt as a positive pressure on the system.

In either eventuality the DNA and the machinery would accumulate around, say, two good survival regions of small variation, and now back interaction, cross-breeding, would disappear as intermediate forms would not be viable in terms of efficiency of machines, whether or not the differences in DNA were large or small. We can represent this situation in a landscape diagram, such as that of Fig. 10.4. We must now examine the logic of such biological machines in the same terms as the logic we used for the consideration of non-living chemicals. There we plotted stability (ΔG) against variables, such as element composition. Here, changes in variables will lead to species in terms of the maxima in similar survival landscapes. What are the units and variables of organisms throughout evolution?

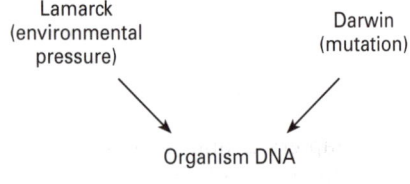

Lamarck (environmental pressure)

Darwin (mutation)

Organism DNA

10.19 Units and variables in a biological steady-state organization

We have described the units, the genuine components within the chemical variable *composition*, and the physical variables, *temperature*, *pressure* (volume), *size* and *external fields*, which are the factors controlling stationary inanimate objects within a given compartment (Chapters 4–7). We went on to show that the increase in numbers of compartments increased the number of variables (Chapter 7) so that literally billions of speciated chemical systems are expected to exist and are indeed found on Earth (see Chapter 9). Eventually, the removal of any equilibrium constraint means that chemical speciation on Earth, or through the synthetic efforts of man, has virtually infinite variety.

Turning to living systems, we have already stressed certain peculiarities of them. They are not stationary but developing flow systems, they cannot be in true temperature equilibrium with their environment since they need energy to maintain them, and they are broken down into species. What then are their units and variables and, in particular, how can we look at man against this background? How does the inclusion of *information* (DNA) alter our analysis? Furthermore, how does the inclusion of the additional central machinery of the brain within individuals of higher organisms, whose contents are not inherited but slowly accumulated in the 'form' of long- or short-term 'memory', affect our analysis? Before we decide upon the *variables* in particular living organisms we must look more closely at the different forms of life that have arisen (see Chapters 11–14), but certain general features of the chemical units in life have already been described. We discuss these next under the title of 'components'. Notice that we have always written 'components' in inverted commas in this chapter when referring to the molecular content of organisms and will continue to do so in subsequent chapters. We now explain the problem.

10.20 Are there 'components' in living organisms?

The study of compositional variation in Chapters 2 and 5 allowed us to define components operationally as those units that could be varied in concentration independently from all other units, since exchange of elements was excluded. They were then defined by element composition, energy content and the barriers to change, i.e. space co-ordinates, local or bulk. All equilibria then reduced variability since they removed the independence of components eventually to the level of the fundamental chemical elements and charge.

Now, the study of living organisms shows us that, over very considerable periods of time, while in steady state, all H, C, O, N atoms of the vast majority of the molecular constructs present in cells, probably all except DNA, are exchanging between different molecules, at *particular rates*, through controlled pathways of degradation and synthesis (Figs 10.23 and 10.24). Leaving aside DNA, the fact that these molecules are in such fixed exchange means that all their *relative concentrations* are not variables in this steady state but are fixed, much though they may fluctuate slightly. *Thus, a living steady state does not have components as variables.* (Note again that the fixed relative concentrations of such molecules are *not* now related to the free energy differences, ΔG, corresponding to the transformation of them to other forms, and are not, therefore, related to any equilibrium consideration. Furthermore, they are not restrained by barriers which keep chemicals in stationary states, i.e. as true components. In fact, the concentrations have evolved in an overall network of flow to optimise *overall* cellular function, using energy and material input from outside to generate this chemical flow.) Since all the elements in these molecular 'components' of life exchange, it would appear that the *variable of composition in organisms, which could dominate speciation in biology, is reduced to the number of elements involved* (compare with Section 10.5 and see aside).

Thus the ratios of concentrations are decided by the co-operative functional use of *elements* (not molecular components), including catalysts, in the whole machinery of the cell, while energy passes through it. Hence, one and the same element *composition* may give rise to several different

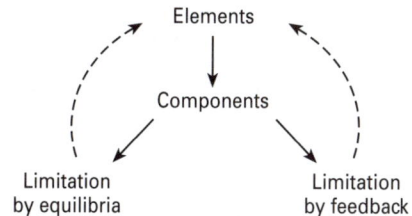

Restrictions on components

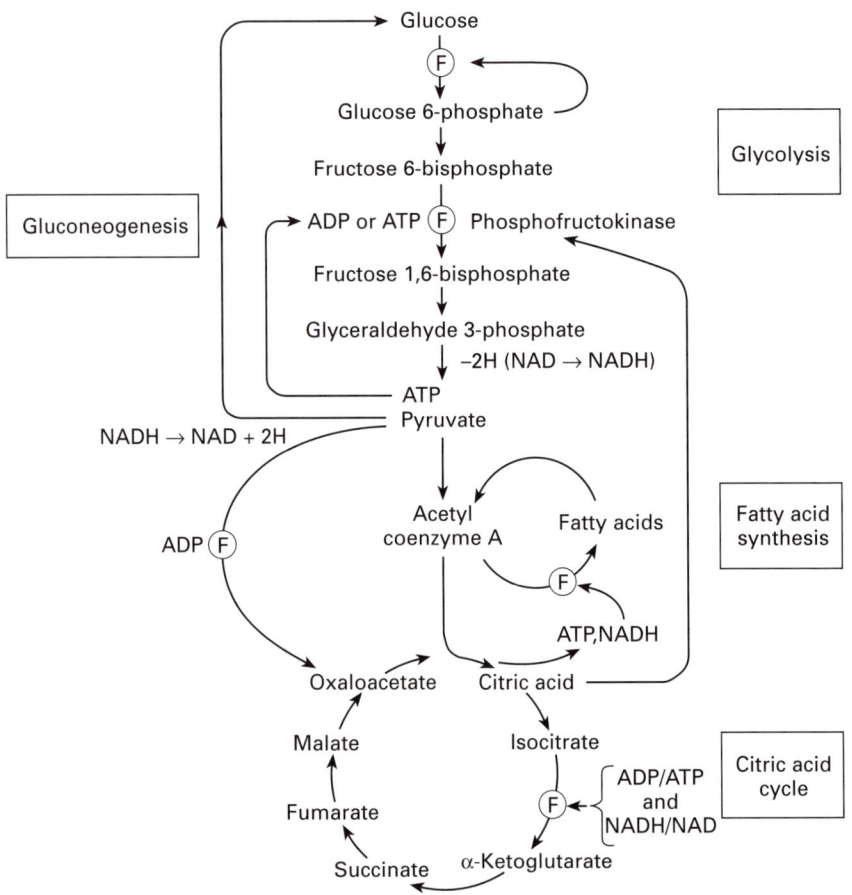

Fig. 10.23 The whole of the metabolic activity of a cell is linked as in this flow chart which shows feedback control sites, (F), of glycolysis, fatty acid synthesis and the citric acid cycle. Steps most affected can be acted upon in a negative or positive manner. Note that feedback controls glycolysis and the citric acid cycle at crucial early steps in the pathways; this increases the efficiency of the pathways and prevents excessive build-up of intermediates. Note also that the compounds inhibiting or activating enzymes are often the energy-carrying or chemical-carrying compounds themselves—ATP, ADP, NAD⁺, and NADH—that is co-enzymes. The incorporation of nitrogen from ammonia leads from pyruvate or later substrates to amino acids and nucleotide bases. Feedback operates here too.

chemical combinations, that is, different steady states exist due to energy flow differences. For example, it could be that CH_3–CHO, together with HO–CH=CH_2, could be found in different steady-state ratios due to different energy flow, while, through metabolic activity, the molecules exchange C, H, and O between themselves and even with the contents of the whole system at particular rates. This contrasts directly with the use of these two chemicals as components of a non-exchanging mixture, as discussed in Section 5.11.2. Note again that the ratios that develop are not equilibrium values: they are decided by energy input and catalysts. If we look upon a cell as such an holistic steady state and co-operative dynamic activity, then there could also be a discontinuous variety of compositional flow systems

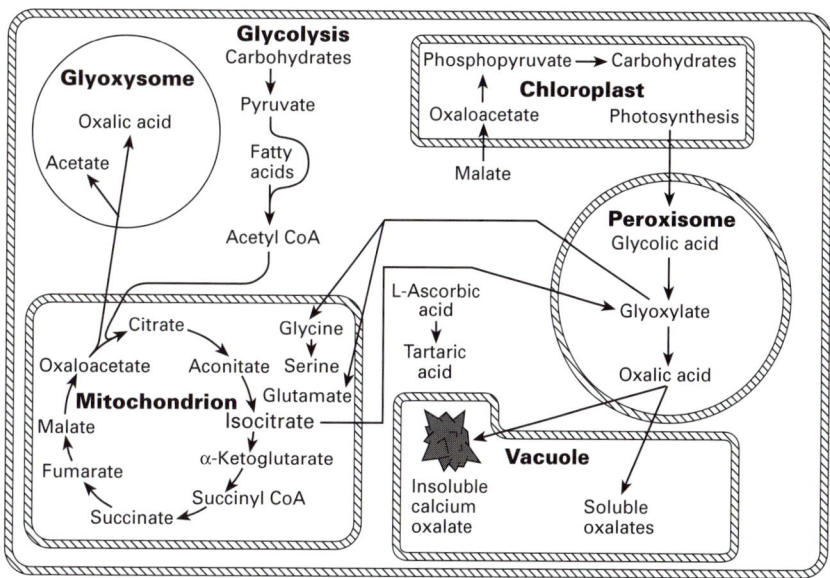

Fig. 10.24 Summary of metabolic pathways in plants illustrating that, as well as the chemical pathway connections shown in Fig. 10.23, many links are made across physical membrane barriers.

of elements plus energy throughput, which could have given rise to successful machines where rate of energy input together with selected element input dominate. *Such an approach allows us to consider survival of steady states in terms of these two variables, element composition and energy throughput.* (Developmental states may require us also to consider rate of change of element composition.)

What we are proposing is that once the variables, energy input and chemical element uptake, are fixed, the *steady state* will be maintained no matter that all the elements within the hidden machinery of the cell exchange. The products of such a cell could be externally generated heat, mechanical work and a discharge of elements in any combination such that it equalled the intake. Such a cell is maintained, we consider, by the intense feedback activity internal to it. Quite correctly, it can be stated that as such it could not reproduce. To reproduce requires *developmental state growth*, when the elements and energy entering do not equal those leaving but are retained in the activity of replication. As we stated in Section 10.18, we wish to separate out the survival value of the machinery itself and that of reproduction. Many cells are not engaged in reproduction, even during the whole of their life, but we must also turn to those that are. Again, all cells appear to develop towards death and this also requires additional consideration from the analysis of steady states. However, the inevitable conclusion of this section is that the only *true units of cellular activity are the chemical elements*, some 20–30.*

*With some justification it might be argued that we should also include very basic chemical compounds, especially H_2O, although even H_2O, like CO_2, undergoes transformation in cells. Thus, we shall use the simplification that the chemical elements are the variables of composition.

We have seen in Section 10.13 that DNA is the coded molecule that secures reproduction of proteins, although historically the same objective could have been reached using another coded molecule, e.g. RNA. Such coded molecules do not exchange atoms with the metabolic components, nor with the growth, the developmental or even the reproductive machinery of the cell. They are unchanging in a given cell. Thus the DNA is a true *component* of a cell. However, by itself it only represents the reproducibility of a cell's code. Without elements and energy from the environment the cell cannot exist. Thus, the DNA codes for survival strength in terms of chemical element and energy intake. Unfortunately, we cannot read the code in this way, so that DNA is generally used as a labelling description of a living system although it is impossible to appreciate the appearance of any particular species from a DNA sequence. In principle, we are nearer to an appreciation of species difference if we examine the active element input by analysis plus the energy input. However, we must accept that the DNA code is an extremely valuable way of classifying species. Given that it too is open to feedback and feed-forward interaction with the cell, it could be said to represent, in coded form, both the element composition of the cell and all the variables of energy flux at any time. In this sense DNA is not a molecule similar to any other molecule since it represents an holistic activity, i.e. it represents all the composition and energy variables that characterise the cell!

10.21   Linear and non-linear systems

When we analysed inanimate systems in Chapters 2–4 we concluded that once we knew their fundamental units and variables we could describe their properties. Some properties also had simple linear dependence upon one another, for example the pressure and the volume of units, assumed to behave as a *perfect gas*, were related to temperature by the equation

$$pV = RT$$

Using the kinetic theory of gases, these properties were linearly related to the fundamental units and their space variables. However, some properties such as transition temperatures of *real* materials, e.g. melting and boiling points, which are indicative of the phase changes discussed in Chapters 5–7, did not conform to such simple equations. They arose when bulk co-operativity of the constitutive units came in balance with the statistical tendencies, which oppose such co-operativity. In such cases the reductive explanation of behaviour in simple linear terms of the relationship between fundamental variables, composition and dynamic time-independent occupation of space, gave way to mathematical equations which are complex or non-linear, e.g. the van der Waals equation (see Section 5.8.2). We then used a set of derived thermodynamic variables when we could again obtain linear relationships such as $\Delta G = \Delta H - T\Delta S$, which were not readily reduced to fundamentals. In Chapter 8 we introduced time and, with it, momentum, which we could analyse again in linear terms for ideal gases.

(However, the flow of a liquid system was found to be non-ideal.) Moreover, when we analysed reaction systems which had feedback-catalysed mechanisms (in Chapter 8) we saw that they too were non-linear.

Even greater complexity arises in systems such as living organisms that contain multiply linked feedback pathways of energy and material. To be alive is then another property of systems which cannot be explained by simple reductive analysis, that is, as a linear property of its units and fundamental space and time variables. We might say that life is again an *emergent* 'bulk' property, see aside the result of an immense set of interacting patterns of flow. When we look at the simplicity of DNA as a linear code, the idea of differences between organisms seems to be a simple matter of changes in linear base sequences. It is not! DNA codes for a variety of living machinery, compositional and energy possibilities, which generate particular substances (protein molecules). The proteins interact with other constituents of the cell and the environment. Even a protein molecule itself has non-linear properties relative to its sequence. The protein activities are also all interactive in reaction patterns and feedback, both on and between themselves and on the DNA. Thus, only certain DNAs have the representational character required for life. The hidden activity described in Section 10.9 is then a complex, non-linear interaction between variables. The simplest observables remain the elements available from the environment, energy fluxes, external temperature and pressure and any acting fields of the environment. Hence, in the following chapters we shall proceed by considering not the DNA, the code, but the environmental variables and their evolution, and correlate them with the evolution of living forms as shown in Fig. 10.1. We shall use as a derived variable the observed survival value of an organism, symbolised as ΔL, which can be looked upon as a variable parallel to ΔG, the free energy of an inanimate system for time-independent systems. Both generate speciation. Note that a survival value like ΔG would give an indication of the probability of observing a system.

Emergent properties

Phase changes
Dynamic patterns
Life itself
Biological development
(Metamorphosis)

10.22   Conclusion

We wish to describe the long journey from formless inanimate matter to man, while explaining the logic of all related processes, including our suggestion that chemical change of the environment allowed evolution of life to occur. To do so we need to analyse the relationship between chemical element content and speciation amongst living organisms in terms of the variables described in Chapters 1–8, where they were mainly applied to inanimate systems. In this chapter we have shown that this programme is not unlike that of analysing non-living, equilibrium systems, as we have done in Chapters 2–7, in terms of their units and variables, in order to discover limits of existence of phases within specified composition boundaries. When studying those limits in these earlier chapters we observed that it was the ordering co-operativity between atoms (metals), ions (salts) or covalent molecules opposing the disorder due to statistical probability that caused separation of *condensed* (liquid or solid) matter in phases. As well as examining such chemical variables as composition we found that we also had to consider physical variables such as temperature, pressure, total amount of material and fields affecting the system. We concluded that through such analysis we could appreciate the breakdown of mixtures of chemicals into limited chemical composition and physical zones with shape, which corresponds to a speciation of chemicals. It is against this background of *chemical co-operativity*, which we say gives rise to *chemical speciation* and complex properties of inanimate substances, that we ask: is the breakdown of biological activity into shaped species due to *kinetic co-operative organisation* in the chemical reactions of organisms, again opposing disorder, due to statistical probability, but generating organisation?

We are immediately involved with the extra fundamental variable, time, and derived variables that include it. Of course, this co-operativity is dependent upon the availability of elements in the environment. Now, the kinetic co-operativity we observe in living species requires an energy influx too, which affects the variables such as composition, temperature, pressure and the other variables of static systems, and has a particular value for a particular survival strength of a given living system, which is complex. We concluded that any *coded information*, such as the *genetic code, DNA*, much though it is necessary, *is far too simple to provide a cause of the emergent biological diversity, even if it is a representation of it, and it is the requirements of the successful survival of systems of molecular machinery that separate and generate species.* (Compare the presentation of variables in chemical species in terms of molecular formulae, *components*, and the appearance of substances in physical states which are emergent properties of bulk systems.) Evolution may then occur through changes in the environment, which affects the variables.

We have no wish to quarrel with the advocates of other factors that have had a strong influence on the ways in which evolution can occur. Thus, adaptation to change of climate, to geographical separation or any other external factor can influence both chemical availability and energy input, to which organisms must adapt or die. Chance mutation or rearrangement of genes, amongst other 'forces', can obviously assist evolution to find good survival scenarios. They represent one way of *searching* the variables, which we did not have to consider in the description of inanimate chemical species. However, we shall insist that any such evolution has to be related to chemical and energy supply changes and to the advantages or disadvantages they bring to the possible machineries of life's steady or developmental states.

To appreciate more fully the above simple outline of the general chemistry and physics of biological systems we must ask how the necessary organisation works in more detail. The first point to make is that each cell is a flow system of chemicals and energy which for long periods can approximate to a *steady state*. Remember that such a state is not to be likened to a thermodynamically stable condition nor to a stationary state (see Table 5.3). It is a new way of combining the variables: the spatial distribution of material and energy and their *time dependence*. In this steady state of flow all chemicals have pathways of change in two senses—the local reorganisation of elements in molecules, e.g. the glycolytic pathway, and the bulk reorganisation of chemicals in space. In an organisation such as a cell we need to understand what flows (chemicals and energy), what causes and restricts flow to these pathways (which energies and which catalysts), but, outstandingly, how the patterns of flow are knitted together so as to make an holistic organism with a particular functioning emergent property, life. The subject matter of patterns of flow, i.e. guided trajectories of chemical and energy change and physical direction, is part of kinetics. It is biological kinetics in particular kinds of cells, i.e. the rates of change of the biological elements in combinations and in space distribution, which we must study before we elaborate further upon the nature of living organisms and their evolution. Clearly, it is limited by the availability of elements, not their abundance. It is this study that must reveal features of essential biological organisation and how it has evolved, even if we cannot say how it originated. In this critical investigation we are not primarily concerned with growth, development and reproduction.

The evolution of living systems can obviously be examined in different ways. The most obvious, which we can follow, are changes of shape, size and compartments of organisms, while looking at functional development (see Tables 10.9 and 10.10). The major switches are from single compartment cells of very small cylindrical shapes (prokaryotes), to internally

Table 10.9 Changes in function

	Organisms	Function
(1)	First true cells (prokaryotes)	Overwhelmingly, capture of material and energy for reproduction
(2)	Earliest eukaryotes	Greater stress on survival, otherwise as for (1)
(3)	Multicellular eukaryotes	Increased stress on survival relative to reproduction
(4)	Mankind	Stress on self-satisfaction and survival beginning to dominate reproductive drive

Table 10.10 Major changes of metabolism

Time ($\times 10^9$ years)	Major metabolic small molecules and ions	
	Small Molecules	Ions
4.0–3.0	CO_2, N_2, CH_4, H_2S, NH_3? HCN, H_2O, NO?	Fe, Mg, Mn, (Ni), (Mo), (W)
3.0–2.5	CO_2, N_2, H_2S/SO_4^{2-}, NO? H_2O	Fe, Mg, Mn, (Ni), Ca, (Mo), (W)
2.5–2.0	CO_2, N_2, SO_4^{2-}, (NO_3^-), H_2O, (O_2)	Fe, Mg, Mn, Ca, (Mo), (Zn), (Cu)
2.0–0.0	CO_2, N_2, SO_4^{2-}, NO_3^-, H_2O O_2	Fe, Mg, Mn, Ca, (Mo) Zn, Cu

divided or compartmentalised, somewhat larger, single cells (simple eukaryotes) and then to multicellular organisms. This is the way in which we divide the next four chapters, always looking at the change in the use of energy and physical conditions and chemical elements which must have helped to bring about, and may have forced, the change. We believe that these two kinds of change are to a large degree coincidental and co-operative. Thus, we shall look for the strong effect of environmental change upon evolution. Chapter 15 describes man's self-conscious realisation of the nature of chemical systems, so that he can and does develop activities closely parallel to life—his industrial society. This will close the account of the development of matter to man up to the present.

Finally, we stress again that the variables in rate constants of feedback circuits cannot be related to one another in any simple linear mathematical fashion, but belong to the realm of complex, non-linear systems. Thus, biological speciation, like condensed phase formation in chemistry, does not generate properties reducible to formulation in terms recognisable just from the analytical content. Evolution of species cannot be brought about by removing one chemical and replacing it by another, since we are dealing with an holistic (complex) value: survival. All such simple changes to a pre-existing organism, which has survival strength in a given environment, are deleterious. On a landscape diagram, such as that of Fig. 10.4, change to improved survival in a given environment is via initial descent, lowering of survival strength involving several small changes but resulting in the later possibility of climbing to a new height by further changes from a point of bifurcation. However, in making such changes it is essential that:

(1) the features of great strength are maintained, e.g. the basic code and the reductive machinery of the synthetic apparatus;

(2) energy and material capture and their means of distribution are ensured;

(3) connections between any new development and the pre-existing systems are devised so as to make (1) and (2) kinetically effective, compatible and co-operative.

We shall look for these features of complex systems as well as the utilisation of the variables already described as we examine organisms from the most simple to man. In the final chapter we shall attempt to pull together the description of the transition from matter to man in terms of the units and variables that we have uncovered.

11

Early life: anaerobic prokaryotes

Double, double, toil and trouble;
Fire burn and cauldron bubble.

W. Shakespeare, *Macbeth* (1564–1616)

11.1   Introduction

In Chapter 9 we have given an account of the early crust of Earth, of its sea and of its atmosphere. We have also described there the development of these physically separate zones up to the present day. In Chapter 10 we described the essential features of all living systems, in general terms of cellular content and activity. The fundamental ultimate units of the two systems, inanimate and animate, were the same—the chemical elements—but while abundances restricted their composition in inanimate systems, *availability* was a major restriction on life. Again, the variables causing change, apart from composition, imposed on both systems, were the same, related to available free energy, including its dependence on temperature

and pressure. Now, in the case of inanimate systems we concerned ourselves mostly with the material and the energy associated with equilibrium
or stationary states, but in living systems we are concerned overwhelmingly with material and energy flow in steady and developmental states. As
discussed in Chapter 10, this is to say that living systems are dependent
upon the additional variable *time*.*

In this and the next chapters (12–15) we wish to describe the nature of
the earliest life and then the changes that have taken place in living organisms with time, how, in part, they have caused the changes of Earth's
surface and how, in part, Earth's surface changes have affected life. We
begin with a description of what we take to be the first organisms, single
cells of anaerobic prokaryotes or bacteria, possibly the so-called eubacteria

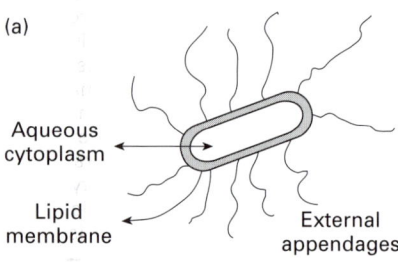

(a)

Aqueous
cytoplasm

Lipid
membrane

External
appendages

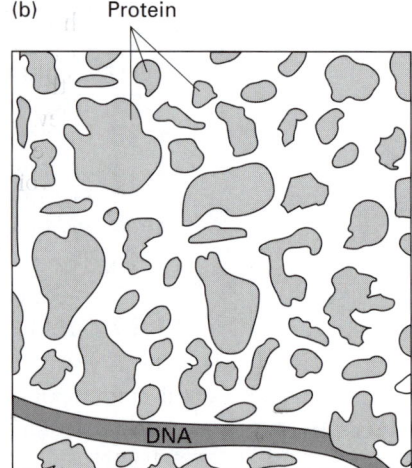

(b)   Protein

DNA

Fig. 11.1 (a) An outline of the most primitive kind of cell we can observe. It is
10^{-4} cm long and 2×10^{-5} cm wide. There may well be an outer wall-like structure
surrounding the membrane. The two most primitive forms are the eubacteria and
the archaebacteria which differ considerably in chemical detail but not in essential
physical form. (b) The density of proteins in a bacterial cell filling $\geqslant$ 20% of the
volume (x 1 000 000). [After Goodsell, D. S. (1998). *The machinery of life*, Springer-
Verlag, New York.]

---

*The Earth, of course, also depends on this variable, but the time-scale is much larger (see
Chapter 9).

Table 11.1 Common features of all living cells

(a) Physical properties
1. A cell is contained by an outer membrane.
2. A cell requires continued input of energy and materials, which it has to capture.
3. Cell activity is dependent on physical transfer of chemicals across membranes.
4. A cell must maintain electrical potential and osmotic pressure at constant low levels.

(b) Major chemical elements and compounds
5. The major metabolism is of H, C, N, O, S and P in small molecules, many of which are synthesised into four classes of polymer: nucleic acids (DNA, RNA), proteins and polysaccharides, as well as lipids for membranes.
6. Carriers of fragments, co-enzymes, aid the transfer and activity of segments of organic molecules, atoms and electrons.
7. Reactions are catalysed by certain proteins, enzymes, many of which contain co-enzymes and/or metal ions.
8. Energy is built first into gradients of inorganic ions and pyrophosphates, and then used in synthesis; see (5).
9. The electrolyte background is largely maintained by K^+, while Cl^- and Na^+ ions are largely removed from cells.
10. The divalent cation Ca^{2+} is largely removed from cell cytoplasm while Mg^{2+} assists in cell activity.
11. The major catalytic trace elements are Mn and Fe, though Co, Ni, (Cu), Zn, Mo, (W) and Se all play a role in virtually all organisms.
12. The centre of information is DNA.
13. The transmission of information is from DNA to RNA to proteins.
14. Activity is then linked by feedback messages to and from DNA, to and from cytoplasmic activity by proteins, metabolites and some ions. The whole is organised.

or the archaebacteria (Fig. 11.1). Subsequently, we shall attempt a short description of how life may have begun; although there are many suggestions in the literature they are all very speculative and we will limit ourselves to a few general remarks (see Further reading). Before we start this description we list the common features of all cells as described in Chapter 10 so as to avoid repetition (Table 11.1).

11.2 The earliest cells

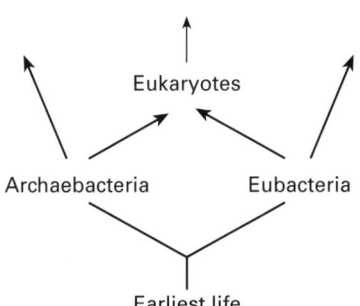

In Section 10.4 we have described in a general way the nature and chemical content of all living organisms (see Tables 10.1 and 10.4). Here, we must particularise as to the content of the earliest organisms—some form of bacteria—in the *reducing anaerobic environment* as it was 4×10^9 years ago (see Aside). We have to judge this chemistry from our knowledge of the strictly anaerobic bacteria of today, some of which are listed in Table 11.2 (compare with Table 11.3). (We shall use the past tense in our account although we clearly make an assessment of early life based on knowledge of today's extant organisms.)

In essence these bacteria had but one containing membrane with, in some cases, a strong surrounding wall [Fig 11.1(a)]. Thus, we shall concern ourselves initially with only three compartments: an inner aqueous solution (the cytoplasm), a surrounding membrane which is fluid and can be compared to a viscous non-aqueous solvent and an outer

Table 11.2 Some early anaerobic bacteria

Strict anaerobic species	Examples of specific characteristics
Bacteroides	Use CO_2 and H_2S reactions
Clostridia	Use CO_2, fix N_2
Methanogenic bacteria (archae)	Produce methane as an energy source
Nitrobacteria	Use oxides of nitrogen, especially note NO
Sulpholobus (archae)	Use sulphur as an energy source: thermophile and acidophile
Eubacteria (Anaerobic?)	
Sulphate users (Desulphovibrio)	Not strictly anaerobic, use sulphate as an energy source
Rhodobacteria (purple non-sulphur)	Phototrophic (light user) (use reduced carbon as energy source)
Chromatia (purple sulphur)	Phototrophic (light user), use H_2S as a source of energy and H_2
Azobacter, Rhizobium	Can fix N_2
Lactobacillus	Ferment sugars. Do not need iron

Archaebacteria differ from eubacteria in their ribosomal RNA and in their DNA to such an extent that they are now considered to belong to different kingdoms of life (see Fig. 10.1). The lipid (membrane) components are distinct; for example, in archaebacteria they include long-chain *ethers* not found in eubacteria and their cell walls do not contain peptidoglycan (a glucose peptide polymer). They also incorporate more primitive enzymes and co-enzymes, such as co-enzyme M (see Table 11.6), particular pathways, such as the so-called pyroglycolytic pathway, and use certain metals more extensively than eubacteria, namely W, Co and Ni.

Table 11.3 Redox potential range of activity of some organisms

Examples of species		Redox potential range (volts)	O_2 reactivity
Strict anaerobes	Many	–0.5 to 0.0	Poison
Sulphate users	Many (desulphovibrio)	–0.5 to 0.2 (?)	Detoxification to H_2O
Nitrate users (precursors of mitochondria)	Many (denitrificans)	–0.5 to 0.5	Facultative energy capture
Aerobes	Very many	–0.5 to 0.8	O_2 requiring

aqueous compartment (the environment). Spatial complexity of single cells was then introduced through either an increase, internally, of membrane complexity or of separate membrane-contained compartments, or by the construction of a second outer membrane leaving an aqueous controlled space, effectively a new compartment, called the *periplasm*, between two organic solvent domains (see Fig. 11.2).

What was in the cytoplasm [see Fig. 11.1(b)]? In these early prokaryotes this compartment already contained the instruction tape, the DNA, which was accompanied by a range of intercommunicating machineries for carrying out chemical and mechanical tasks. For much of the lifetime of the primitive bacterial cell, the DNA was freely accessible so that it could be

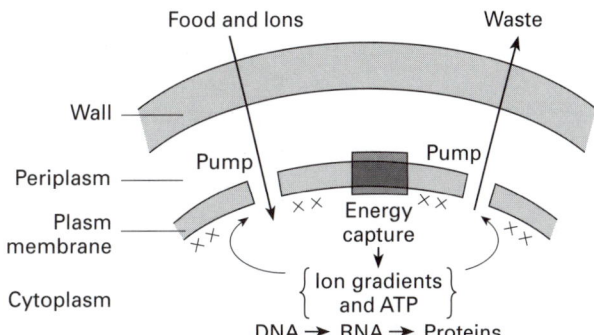

Fig. 11.2 The input and output pumps of cells connected to energy from external sources such as light or unstable chemicals. We show a gap between the inner membrane and the outermost structure which was not present in Fig. 11.1. It was probably a secondary development. Typical examples are shown in an Aside.

read rapidly, producing proteins via RNA. Much of the metabolism of the cell occurred in this well-packed compartment, including all the syntheses of proteins, fats, polysaccharides and both RNA and DNA. In order to do this, very many of the proteins (enzymes) had catalytic sites. The non-catalytic proteins served to give structure, sometimes, for example during division, as filaments criss-crossing a cell, but for long periods there was little such internal structure while always there were protein supports (x) beneath the lipid membrane (Fig. 11.2). The tensions in the cell protein structures and in the membrane were adjustable in order to force the cell to divide at a given stage of development. Finally, some proteins were bound to DNA and RNA in a flexible manner, so that their conformation changed on change of cellular constituents, which acted as molecular instructions to DNA expression. Thus, precise expression of proteins depended on the chemical condition of the cell. The adjustable proteins are called *transcription factors* (see Section 10.9.3) and responded to the cell's internal messenger chemicals, which arose mainly from the metabolic pathways of the cell. Much activity was constantly maintained, but *the major target of activity was to reproduce the DNA and its associated machinery*, whereupon division occurred.

All the activities required a controlled distribution of energy utilising nucleotide triphosphates (NTP). Within the membrane there were, therefore, yet other proteins, which acted as energy capture devices or as pumps for uptake or rejection of specific chemicals (Fig. 11.2), (see aside). The whole functioned as an organisation (see Chapter 10) so that each and every activity was linked and timed; for example, the pumps were connected to metabolism. Controls consisted of messenger molecules or ions which passed between the metabolic and pump systems, the proteins (many of which were enzymes) and the DNA (RNA). To some limited degree the controls responded to the nature of the external environment, so that the cell could take advantage of or protect itself from external conditions. At some stage some early cells developed a motor so that they could swim in a selected direction. We stress again that: (1) the whole purpose of this cellular life was devoted to chemical reproduction; and that

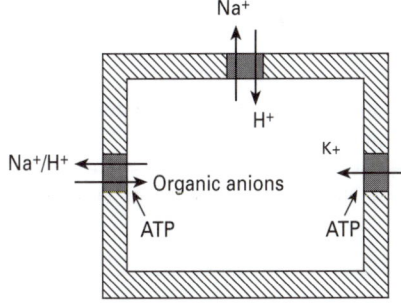

Primitive exchangers and ATP-pumps

(2) the chemistry in it was reductive, i.e. toward $(CHOH)_n$ from CO_2, to NH_3 from N_2 and to H_2S from S_n, and there was virtually no O_2 present. This is the description we can give with some degree of certainty concerning early life since it is that of anaerobic bacteria to this day. The surprising feature is its very high degree of sophistication, which means that we are far from understanding how life began. It is for this reason that in this book we pay little regard to the origin of life and concentrate on reasonable models for these earliest living systems (Table 11.2), turning then to evolution.

11.3   The earliest available elements and their uptake

Against this background and the sure knowledge that life arose in a *reducing medium* (see Figs 6.15. and 9.2 and Table 11.3) we can ask not just which elements were available then, but in what form did they occur and how were they employed? We shall use again the known composition of today's anaerobic prokaryotes as a guide (Table 11.4). We return to the description of the early sea (see Chapter 9), exposed to CO_2, H_2S, N_2 and probably also to some CH_4 and NH_3 atmospheric gases, and containing high concentrations of some elements from dissolved salts, for example, Na^+, Cl^-, K^+, Ca^{2+}, Mg^{2+}, some HPO_4^{2-} and MoS_4^{2-} (WS_4^{2-}), some HS^-, HCO_3^-, NH_4^+, and HSe^-, and less considerable quantities of Fe^{2+} and Mn^{2+}, but little Co^{2+} and Ni^{2+}, less Zn^{2+} and even less Cu^{2+} in any form due to the relative insolubility of the sulphides of these metals (see Sections 6.5, 9.3 and 9.5). Many other elements could only have been available in very small amounts. Given this environment, the earliest cells used the elements as shown in Table 11.4, but apparently they also avoided certain elements such as aluminium, titanium and fluorine.

Table 11.4 Major functions of elements in primitive cells

Elements	Major functions
H, C, N, O, P, S	Formation of polymers from H_2O, CO_2, NH_3, HPO_4^{2-} and S^{2-}
Na^+, K^+, Cl^-	Electrolytic and osmotic balance
Mg^{2+}	Mild catalysis (phosphate compounds). Structural in RNA, DNA, etc.
Mn^{2+}, (Zn^{2+})	Some stronger acid–base catalysis
V, Fe^{2+}, Co^{2+}, Ni^{2+}, Se, Mo(W)	Redox catalysts, often devoted to reduction
Ca^{2+}, Si	Strengthening of outer structures

Note the absence of Cu, Br, I, the low content of Zn, and the reduced state of iron, sulphur and selenium.

11.4   Functions of elements in primitive life

Such a list demands us to show why even the simplest cells had to be so complicated in composition, by giving details of the specific major functions

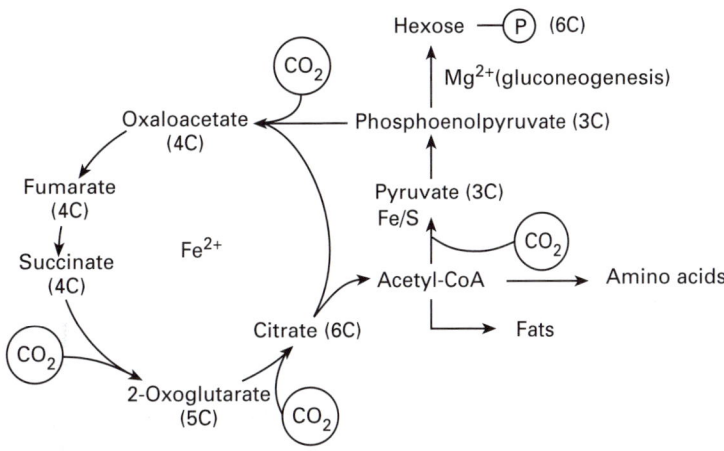

Fig. 11.3 The reductive citric acid cycle for autotrophic CO_2 fixation in *Thermoproteus neutrophilus*, an archaebacterium.

of the individual elements. We start with the smaller organic molecules made from a combination of H, C, N and O, with some S and P in particular cases, which had to be soluble in water for the most part or have a mixed character so as to form membranes. They, therefore, contained charged or polar groups (see Section 6.3). Some examples are given in Figs 6.1 and 6.3.

An outstanding feature of these combinations of C, H, N, O, S and P chemicals is that they are not a random set, but each chemical is related to others via a step or series of steps, so that special synthetic and degradative chemical pathways were already present (see Table 11.5). Major pathways had a strict rationale, one of which we describe as an illustration—glycolysis. In this case (see Fig. 10.17) the purpose was and is today twofold: (1) to change a sugar storage polymer into valuable intermediates for the synthesis of proteins, DNA and so on; and (2) to supply energy as ATP for their syntheses. As stated before, the reaction system can go in reverse (gluconeogenesis) by a slightly different pathway (see Fig. 10.17).

Now, this same logical construction applies to the reasoning for the presence of pathways of fat synthesis, CO_2 incorporation (reversed citrate cycle;

Table 11.5 Major primitive metabolic pathways of H, C, N, O

Pathway	Function
Glycolysis	Energy production from glucose, small metabolites for other pathways
Gluconeogenesis	Synthesis of glucose
Fatty acid synthesis	Synthesis of fatty acid esters for membranes
Fatty acid degradation	Source of energy and metabolites for other pathways
Peptide bond synthesis	Protein synthesis (on ribosomes)
Nucleotide condensations	RNA and DNA synthesis
Reverse citric acid cycle	Incorporation of carbon dioxide

Table 11.6 Some primitive co-enzymes and co-factors

Co-enzyme	Function
NADH, NADPH*	Mobile carriers of H^-
Flavin	Fixed carrier of H and e
Quinone	Mobile carrier of $2H^+$ and 2e (in membranes)
Nucleotide triphosphates	Mobile carrier of phosphate and energy
Glutamine	Mobile carrier of $-NH_2$
Haem (Fe)	Fixed carrier of e
Vitamin B_{12} (Co)	Fixed carrier of CH_3-
Factor F_{430} (Ni)*	Fixed carrier of H and CO
Co-enzyme M*	Mobile carrier of CH_3CO-
(Co-enzyme A)	Mobile carrier of CH_3CO-

* Mainly in archaebacteria; the other co-enzymes and co-factors occur in many organisms but not all synthesize them; see Table 10.5. See aside for the early history of porphyrin.

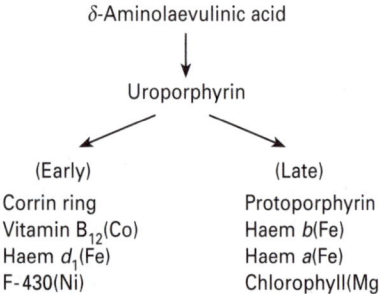

δ-Aminolaevulinic acid

Uroporphyrin

(Early)
Corrin ring
Vitamin B_{12}(Co)
Haem d_1(Fe)
F-430(Ni)

(Late)
Protoporphyrin
Haem b(Fe)
Haem a(Fe)
Chlorophyll(Mg)

Fig. 11.3), amino acid and nucleotide synthesis and their degradations and so on (Table 11.5). Moreover, since most of the pathways used C, H, N and O fragments, for example $-H$, $-CH_3$, $-COCH_3$, $-NH_2$, these groups had to be carried and shared between pathways in order to keep all activities in balance. As stated before, the carriers of fragments are called co-enzymes (Table 11.6). The primitive mobile co-enzymes acted additionally, as co-enzymes do today, to control pathway rates by feedback (see Section 8.6.4). Each pathway also used or produced energy, and this commodity was shared since the chemical form of energy (ATP) travelled between the pathways and also adjusted pathway rates differentially by feedback and feed-forward control. To maintain the medium of constant ionic quality, i.e. to ensure the supply from the environment of chemical elements and of energy, the ATP was also connected to uptake systems, pumps (and channels). Some chemicals, such as H^+ and Ca^{2+}, were also exchanged across membranes. We refer the reader to a text on biochemistry for all the details, but we have to insist on one feature: all the internal organic compounds of any primitive organism, even in the most primitive cells we know about, were linked, so that there had to be linked feedback control of every pathway or pump in every organism. This is a fundamental feature of a living steady state (see Figs 11.3–11.5; see also Section 10.9.2) which

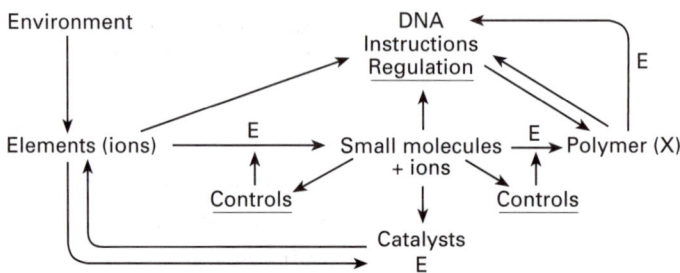

Fig. 11.4 The polymers (X) acting on DNA are transcription factors activated frequently by small molecules or ions with a wide variety of time constants. E is an enzyme. Instructions are passed to a polymer (X) or directly to DNA. Energy requirements not shown.

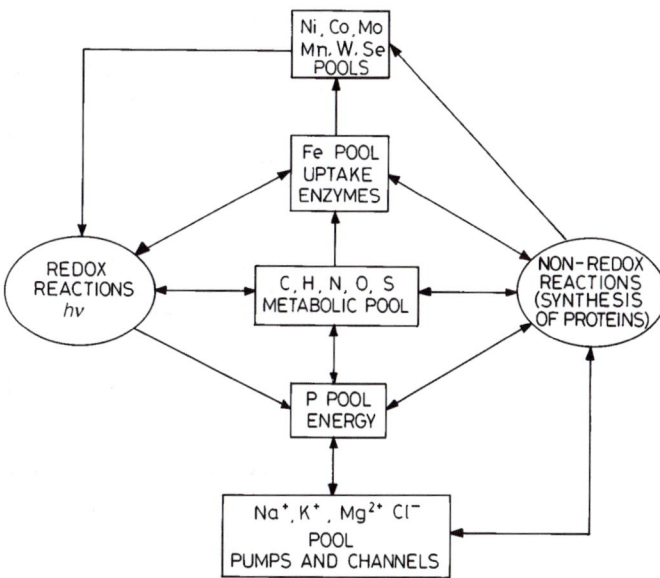

Fig. 11.5 A simple view of the necessary interaction between a minimum of 15 elements to maintain a primitive cell. Calcium is assumed to be of little consequence and is expelled from the primitive cell (see Chapter 12). Discussion of copper and zinc is deliberately omitted until Chapters 13 and 14, respectively.

also requires several trace elements as catalysts. The steady state can, therefore, be characterised by a given content of elements as well as by H, C, N, O, P and S compounds, as shown in Table 11.7 and Fig. 11.5. The concentrations shown in Table 11.7 are, of course, approximate values and could vary somewhat from primitive species to species, but they could not vary by much. The overall primitive picture of metabolism for synthesis is given in Table 11.8.

Table 11.7 Concentration (mol l^{-1}) of some ions and molecules inside and outside primitive cells

Ion	Inside concentration	Outside concentration
Na^+	10^{-3}	10^{-1}
K^+	10^{-1}	10^{-2}
Cl^-	10^{-3}	10^{-1}
Mg^{2+}	10^{-3}	10^{-2}
Ca^{2+}	10^{-5}	$<10^{-2}$
Mn^{2+}	10^{-7}	10^{-6}
Fe^{2+}*	10^{-6}	10^{-6}
Ni^{2+}*	10^{-14}	$<10^{-9}$
Zn^{2+}*	10^{-15}	$<10^{-12}$
ATP^{4-}	$\sim10^{-3}$	zero
H_2S	$\sim10^{-3}$	$\sim10^{-3}$
HPO_4^{2-}	$\sim10^{-3}$	$\sim10^{-5}$
HCO_3^-	$\sim10^{-3}$	$\sim10^{-3}$

* Limited by the presence of sulphide; see Fig. 6.5.

Table 11.8 Overall polymer synthesis scheme

CO_2 (in H_2O) or CH_4 $\xrightarrow{\text{energy}}$ HCHO $\xrightarrow{\text{polymerisation}}$ $[HCOH]_n$
The polymers are polysaccharides

$2HCHO \rightarrow CH_3CO_2H$ $\qquad N_2 \rightarrow NH_3$
$CH_3COOH + CH_3COOH \rightarrow CH_3\cdot CO\cdot CH_2COOH$
Acids and keto-acids + $NH_3 \rightarrow$ amides + amino acids
Condensation of amino acids $\rightarrow$ proteins

$HCHO + NH_3 \rightarrow$ purines and pyrimidines (bases)
Condensation of bases, sugars and phosphate $\rightarrow$ nucleotides
Nucleotides $\rightarrow$ RNA + DNA

Note that the reactions require reduction of CO_2 and N_2. To this day, the above reactions take place in some anaerobic bacterial cells and require:

(1) reducing media linked to the H_2S/S_n couple or the $RSH/(RS)_2$ couple;

(2) reactions *directed* by catalysts and controls;

(3) energisation;

(4) control over amounts of every chemical synthesised or taken into the cell.

11.5   Primitive energy sources

All this pumping in and out of ions and the synthesis of a variety of small and large covalent molecules required energy, which had to come either from the sun or from unstable element combinations of the Earth. As stated in Section 10.7, one such chemical which was available on the primitive Earth was iron sulphide which can undergo the reaction

$$FeS + H_2S \rightarrow \underset{\text{pyrite}}{FeS_2} + H_2 + \text{energy}$$

The reaction, when suitably linked to some organic chemical transformations, also produced reducing H_2 in bound form for the initiation of carbon chemistry. In short, we can write

$$H_2S \rightarrow S + \text{bound hydrogen } (H_2)$$
$$\text{bound hydrogen } (H_2) + CO \rightarrow HCHO$$
$$\text{(formaldehyde)}$$

(This is a reduction of carbon from oxidation state II to zero.) The bound hydrogen is invariably found first in a reduced pyridine nucleotide (NADH), a co-enzyme that connects to several different redox pathways and is in all cells to this day.

From formaldehyde and ammonia almost every required organic chemical was synthesised. Alternatively, the reducing H_2 could have been reacted with any oxidising agent present, such as free sulphur, to give energy

$$H_2 + S_n \rightarrow H_2S + \text{energy}$$

The problem with schemes for initiating syntheses or capturing energy according to the above reactions is that all required, and to this day require, catalysts. The only effective catalysts (see aside) for such reactions are transition metals such as nickel, iron and cobalt for synthesis, and an agent such as iron for energy capture utilising electron transfer. The synthesis can be written

$$
\begin{array}{ccc}
(H)_2 & CO & \\
| & | & \\
Ni \cdots Ni & \rightarrow & Ni \cdots Ni + HCHO \\
(H)_2 & CO & \\
| & | & \\
or \quad Ni \cdots Fe & \rightarrow & Ni \cdots Fe + HCHO
\end{array}
$$

The energy capture requires more complex steps and takes place across a membrane. The general scheme is given in Fig. 11.6. The reaction of FeS with H_2S to give (bound) H_2 can only supply energy if the reactions are differentially placed in space to create a hydrogen or charge gradient (see Chapter 7).

Another possible early source of chemical energy, (see aside), which could also have led to the development of membrane proton or charge gradients, was the formation of oxides of nitrogen. For example, lightning could have generated quite considerable amounts of nitric oxide from the photoreaction of $N_2 + H_2O$

$$N_2 + 2H_2O \rightarrow 2NO + 2H_2$$

The NO generated would not have been reduced rapidly since it can bind to Fe^{2+} ions and sulphides, and is then temporally stabilised. This would give a water-soluble reservoir of a strong oxidising agent. It is unlikely that

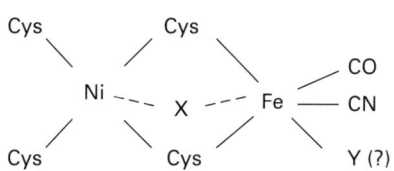

Ni/Fe hydrogenase centre

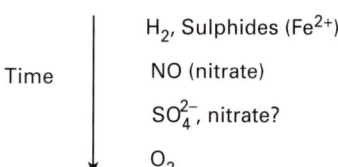

Development of chemical energy sources

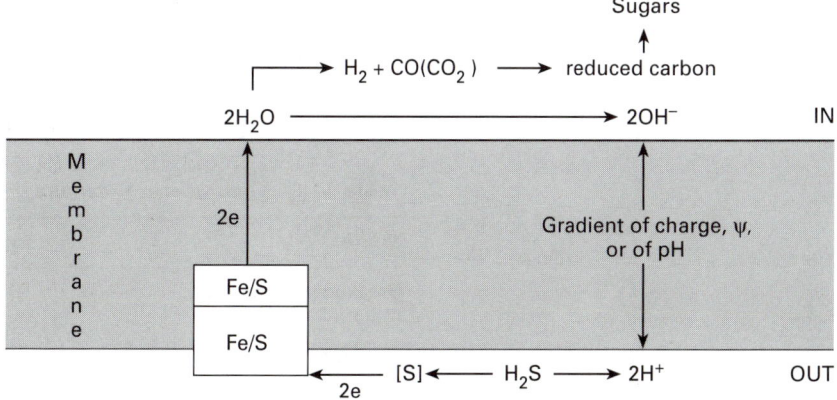

Fig. 11.6 A simplified description of electron flow and its connection to carbon oxide reductions using a membrane. It is the presence of a transition metal (Fe) that is essential. In fact, the electron transfer usually meets the proton transfer pathway in the membrane (see refs. 167–173). NO rather than H_2O could have been the oxidising agent being reduced to NH_3.

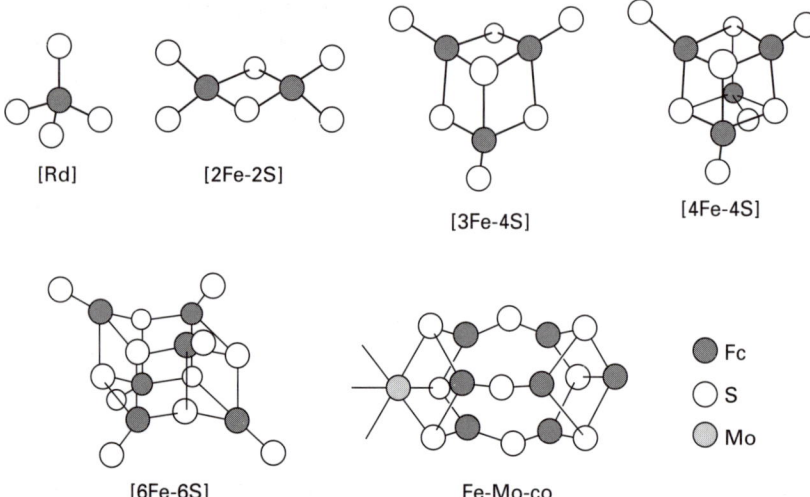

Fig. 11.7 Structures of iron–sulphur clusters found in proteins.

further oxidation to NO_2^- or NO_3^- would have happened until there was O_2 in the atmosphere. The bound NO would then have been an electron acceptor and could have been linked to the production of proton or charge gradients (Fig. 11.6). It could also have been reduced to give a supply of ammonia. Thus, there are many ways in which energy could have entered chemical cycles. We insist here that the initial enzymes using NO were either based on Fe^{2+}, or RS^-, or both, in Fe/S clusters (Fig. 11.7). These centres were present in association with cells and their membranes in very early life forms. (The present-day reactions of copper enzymes in nitrogen oxide metabolism will be discussed in Chapter 13 since we believe they all arose later.)

Apart from chemical sources of energy, light could also have been used in these early organisms to supply energy, by generating a photochemical gradient of electrons and protons. The several possible ways of using pigments are listed in Table 10.6 and discussed in Section 10.8. Note, however, that the synthesis of such organic pigments is rather complicated.

There is also the often-suggested idea that light gave rise to the major amino acids and nucleotides directly outside cells, seemingly obviating the need for an organised energy supply. We do not see how a cell could or can

Table 11.9 Primitive energy capture giving ATP

Mode	Energy captured
Breakdown of organic material	Glycolysis → ATP
$FeS/H_2S \rightarrow FeS_2 + H_2$	Charge (e/H^+) separation
Light	Charge (e/H^+) separation
Charge separation	Formation of ATP
Lightning	NO absorbed and used as oxidant giving charge separation
Light (direct)	Variety of chemicals in stationary states

operate without prior creation of an energy trap of limited volume which looked after its need to pump in or to synthesise complex molecules. It is difficult to imagine all such molecules free in the ocean, somehow becoming used in the absence of cells. All the modes of primitive energy capture are listed in Table 11.9. As we have mentioned in Section 10.8, the energised proton led to the synthesis of ATP and in turn the ATP could energise gradients or assist uphill chemical reactions.

The hydrogen ion gradient generated in these processes was also used in part in exchange reactions to move other ions across membranes, e.g. Na^+, K^+, Cl^-, Ca^{2+}, Mg^{2+} or even complex molecules (food), against a concentration gradient, either into (K^+, Mg^{2+}) or out of (Na^+, Ca^{2+}, Cl^-) cells. Any gradient can then be coupled to drive a gradient of another ion or of food molecules.

In the preceding chapter we have shown that the production of a H^+/OH^-, or a charge gradient as in Fig. 11.6 could be followed by the synthesis of pyrophosphate, observed as pyrophosphate derivatives such as NTP and, particularly, ATP

$$H^+ \text{ Gradient} + 2 \text{ phosphates} \rightarrow \text{pyrophosphate}$$

The great value of pyrophosphate, as stressed in Chapters 9 and 10, returns us to a description of organic condensation reactions. Here we note that all inorganic element pumps were also connected to this common source of energy. They were each controllable by a feedback switch-off when the appropriate level of incorporation of each element concerned had been reached (Fig. 8.13). We now see, not only the need for the earliest cell we know about to obtain the content of some 15–20 elements, but also the need to manage their levels while reproducing their complicated organic compounds (Figs 11.4, 11.5 and 10.10). To illustrate the sophistication of the special use of elements, i.e. the reasons why we think they were selected, see Table 11.4. Apart from the use of metals as carriers of fragments, we must not forget the related non-metal chemistry, especially the use of sulphur in early life.

11.6   The use of thio-acids and thiols: redox buffering

Synthesis requires intermediates to carry fragments other than H to reaction sites and so bring them into juxtaposition. The route which was favoured in the earliest bacteria, and still exists to some degree today, was to carry organic fragments on organic sulphur compounds. The fragment carried can be an acyl or a methyl group

$$R–S–COCH_3 + XH \rightarrow RSH + X–COCH_3$$
$$R–S^+(Me)_2 + X^- \rightarrow XMe + R–S–Me$$

Thio-acyl groups could also have acted as dehydration agents.

$$R–S–COCH_3 + H_2O \rightarrow RSH + CH_3COOH$$

Some of these sulphur-containing co-enzymes are rather different from those of today's aerobic life, e.g. co-enzyme M (Table 11.6).

The further fundamental value of thiol chemistry in providing relatively fast exchange kinetics has been mentioned in Chapter 8. A particularly important case is that of hydrogen exchange and the maintenance of redox buffering. Primitive cells had to maintain a steady state of many reaction intermediates, as we have stressed many times. Now the overall reactions were towards the reduction of C and N, whence it was necessary to keep the cell cytoplasm with a steady-state buffer of reducing equivalents. This was done, and in part is done today, by the controlled use of glutathione (see formula) in the reaction

$$2H - carrier + R\text{--}S\text{--}S\text{--}R \rightleftarrows 2RSH + 2 \ carrier$$

Glutathione was the redox buffer for hydrogen transfer. Today this activity remains, but it is aided by ascorbate.

11.6.1 Sulphate-using bacteria

We have stated that strictly anaerobic chemistry was the only form of metabolism open to the earliest life forms. Here, anaerobic implies that there is no involvement of dioxygen, which does not indicate that there is no use of oxidative chemistry that does not require dioxygen. From chemical considerations we have seen that the oxidation of sulphur

$$H_2S + FeS \rightarrow FeS_2 + H_2 \ (bound)$$

is a favourable reaction and could have been the initial source of energy for life. Similarly, the production of NO from N_2 and H_2O due to lightning

$$N_2 + H_2O \xrightarrow{\ h\nu\ } NO + H_2 \ (bound)$$

followed by the use of NO

$$NO + H_2 \ (bound) \rightarrow N_2 + H_2O$$

was another possible energy source.

Geochemical records show that another oxidant, sulphate, appeared sometime before 3×10^9 years ago, as baryte ($BaSO_4$), and its source could have been the non-biological reaction

$$H_2S + 2O_2 \ (very \ low \ pressure) \rightarrow H_2SO_4$$

Once there was sulphate, no matter what its origin, living organisms could use it (see aside) since the reaction

$$SO_4^{2-} + 2HCHO \rightarrow 2CO_2 + 2H_2O + S^{2-}$$
$$2H^+ + S^{2-} + FeS \rightarrow FeS_2 + H_2$$

also gives considerable energy. Inspection of sulphur isotope ratios in very early sediments indicates that this transformation of sulphur by organisms

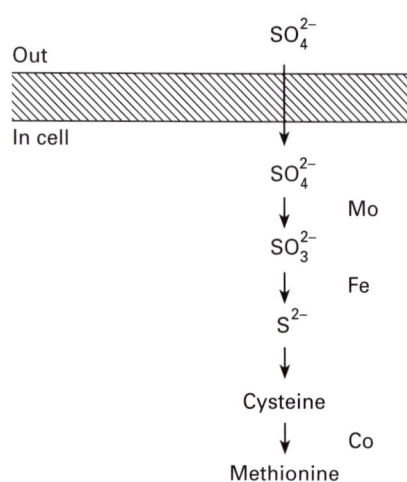

began about 2.5×10^9 years ago (see Fig. 9.13). Today, many prokaryotes living at the bottom of partially anaerobic lakes, or even parts of the ocean, use sulphate as an energy and sulphur source. Of course, most of life today uses sulphate as the chemical that provides sulphur generally. Note that in Table 11.2 we have included nitro- (not nitrato), sulphato- and photobacteria amongst the earliest life forms.

11.7 Early electronic devices for energy capture: the use of iron/sulphur compounds

Now, while the selection of phosphorus described in Section 11.5 and of sulphur described in Section 11.6 have been stressed, along with the synthesis from H, C, N, O of polymers, the electrical equipment for energy capture in Fig. 11.6 required a quite different element which had to pass electrons back and forth in a wire-like organisation. We observe, from a study of anaerobic bacteria today, that, in the initial reducing conditions, iron atoms were selected since iron: (1) was available as Fe^{2+}; (2) easily undergoes single redox steps, $Fe^2 \rightarrow Fe^{3+} + e$; and (3) binds quite strongly to an organic matrix and can be held in place, especially in primitive conditions, by sulphides, that is, in Fe/S complexes (see Fig. 11.7). Hence, sulphur is as important as iron in this process.

In these respects iron and sulphur are more valuable than the other elements that could carry out electron transfer steps. Thus, manganese was quite available but binds much less well, copper was not available since it was precipitated as sulphide, nickel and cobalt were less readily available since they are of limited abundance and, as for manganese, it is more difficult to place their one-electron redox potentials in the correct range for C/H redox reactions. A possible early example of the use of iron/sulphur compounds is given in Fig. 11.8 which describes the way Fe/S compounds

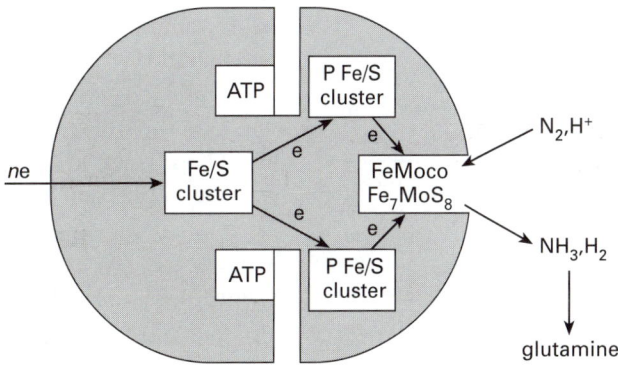

Fig. 11.8 An impression of the nature of nitrogenase with its different Fe/S clusters, its ATP reaction site and its active site for N_2 binding and reduction. The outline structure for both proteins is now known. The ATP acts on the first protein, which is in the form of a hinge, forcing a change on the Fe/S cluster so that electrons (and protons?) go to the FeMoco via P centres. [The structure is taken in all but details from Kim. J. and Rees, D. C. (1992). *Science*, **257**, 1677–81.] (See also Fig. 11.7).

Table 11.10 Catalysts in strict anaerobes

Process	Catalytic Centre
$H_2 \leftrightarrows H^+ + H^-$	Fe, Ni, S, Se
O-atom transfer	Mo or W pterin
H transfer	NAD, NADP, flavin, Ni, Se
e transfer	Fe/S, Fe, haem
CH_3 transfer	Vitamin B_{12}, (CO), folates, R_2–S–CH_3^+
CH_3CO transfer	RS^- in Coenzyme A
CH_3–S transfer	Ni (F430), Co-enzyme M

are used in electron transfer in nitrogen fixation (see also Table 11.10). However, as we note in the next section, such transition metal ions have other advantages.

11.8   Remarkable early enzymes: catalysts in anaerobes

A feature of the chemistry of the earliest organisms that we can be fairly certain about is the catalytic sophistication using different elements in reaction steps (Table 11.10 and see also Table 11.6). Here we can list but a few.

1. Cobalt was used in vitamin B_{12} (Fig. 11.9) for the making and breaking of Co–C bonds in two ways:

   (a) $Co^{3+} - CH_3 \rightarrow Co^{3+} + CH_3^-$, e.g. in methylation

   (b) for breaking C–H bonds in the synthesis of DNA from RNA precursors, using free radicals

$$Co^{3+} - CH_2\text{-}R \rightarrow Co^{2+\bullet} + CH_2^\bullet - R$$

2. Nickel was used in Factor F-430 (Fig. 11.9) and elsewhere in hydrogen (H_2) and possibly CO catalysts making Ni–H and Ni–CO bonds, e.g. in the reaction $H_2 + CO \rightarrow HCHO$

3. Molybdenum (or tungsten) was used in pterin-like co-factors (Fig. 11.9) in the transfer of oxygen (and perhaps sulphur) atoms via Mo=O bonds, e.g. in the reaction $CH_3CHO \rightarrow CH_3COOH$. Molybdenum was also used in nitrogen fixation (see Fig. 11.8).

4. Zinc was, to some degree, used to activate water for hydrolysis reactions, as described in Section 8.5.3, e.g. in peptide hydrolysis, but also in syntheses (Table 11.11).

5. Selenium was used with nickel or tungsten to transfer hydrogen atoms, e.g. in hydrogenases.

6. Iron in haem (Fig. 11.9) was used in reactions of NO, and especially of electrons.

Figure 11.9 gives the structural formulae of several of the metal co-factors. The reasons for the well defined uses of Co and Ni are not obvious.

Table 11.11 Early zinc enzymes used in synthesis

Synthetic step	Zinc enzymes
RNA	RNA polymerase
DNA	Reverse transcriptase
Viral synthesis	Terminal dNT transferase
Transfer RNA	tRNA synthetase
Essential amino acids	Dehydroquinate synthase
Essential nucleotides	Aspartate transcarbamylase

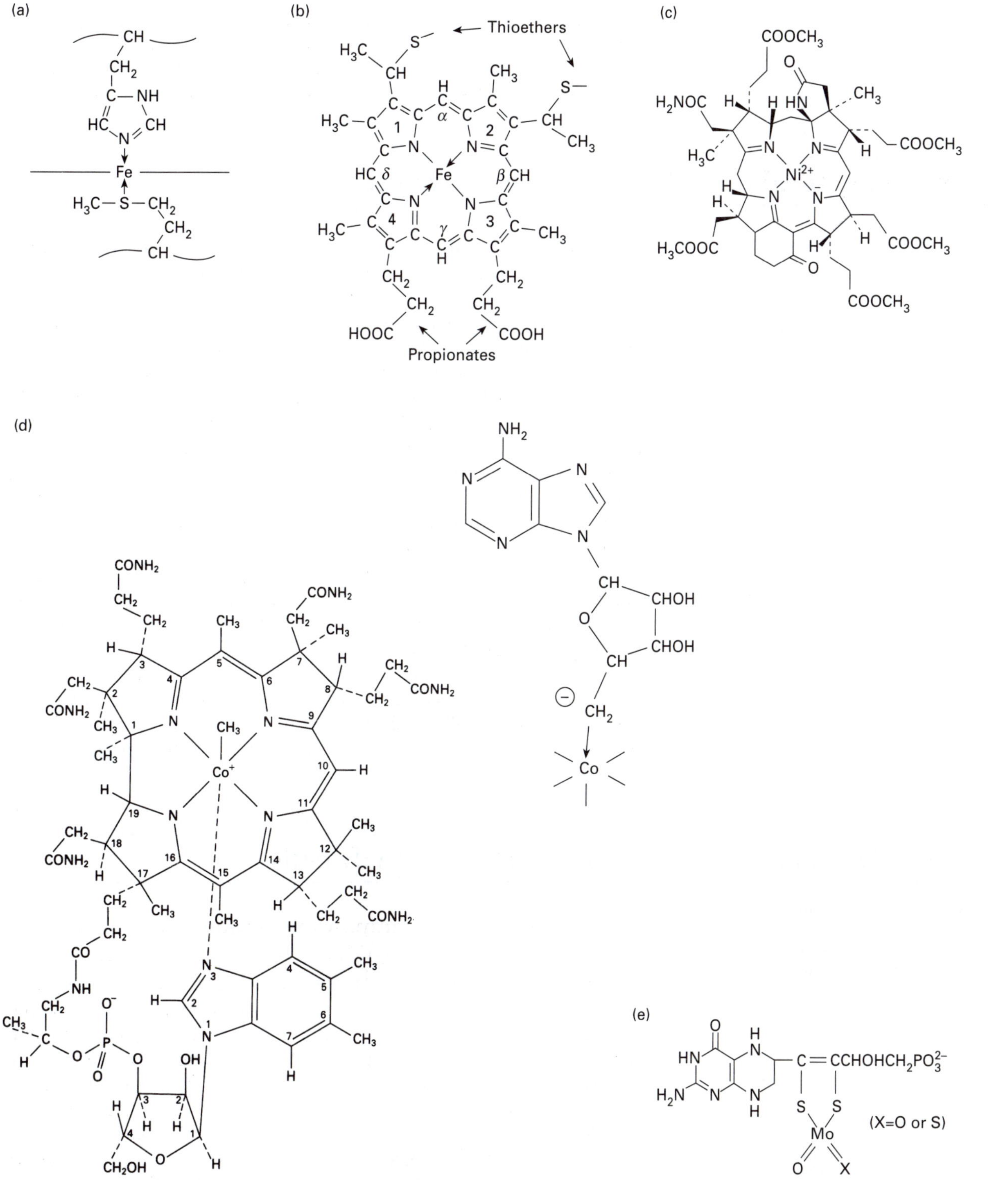

Fig. 11.9 (a) The axial ligands, methionine and histidine, to the iron prophyrin in cytochrome *c*. (b) The formulae of protoporphyrin bound here through thioether links, as in cytochrome *c*. (c) The structure of nickel-containing co-enzyme F-430. (d) The structure of methyl co-enzyme B_{12}, which contains cobalt. Other forms of the co-enzyme have a 5′-CH_2–deoxyadenosyl attachment instead of the methyl group. (e) The formula of one molybdenum-containing co-enzyme: there is a similar tungsten co-enzyme.

These features must be seen against the gentler and more general use of Na, K, Mg, Ca, Fe and Mn, as simple ions. All these elements had developed a special use more than 3×10^9 years ago. In later chapters (12–14) we shall see how some further sophisticated uses of elements evolved.

11.9   Structural organisation in primitive cells

It must never be forgotten that the structure of the cellular organisation was provided by fatty compounds for membranes and proteins for almost every other activity. The fatty compounds, lipids, are already diverse in the earliest known life forms (Table 6.2 and Fig. 10.5). Proteins are extremely versatile in their ability to form platforms for catalytic atoms such as iron, but also in their ability to form surfaces (which themselves can be catalytic) and which recognise specific organic molecules (see Chapter 8). With or without an assembly of individual protein catalysts, a series of reactions can be managed based on a different protein for each step in a pathway. Thus, pathways had, even in primitive prokaryotes, a recognisable trajectory, perhaps in space but decidedly in a timed sequence. This allowed control at various steps. Since many anions were involved, Mg^{2+} was also frequently utilised as the best of all chemical element partners to aid association and as a gentle catalyst. In fact, virtually all phosphate transfer, including reactions such as those of pyrophosphate ($P_2O_7^{4-}$ as ATP), required Mg^{2+}, so that the associated unit Mg^{2+}/organic anion was frequently the reactive species.

In Section 8.13 and in Fig. 10.15 we have shown that certain types of protein are also ideal molecules for incorporation into machines that do the mechanical work in cells, and the accessory rod-like and particularly the required helical structures necessary were present in the earliest cells we are able to describe.

11.10   Dynamic chemical organisation in early cells

11.10.1 Protein production: transcription and translation

As just shown, even the most primitive organisation required considerable structuring of chemical equipment for energy capture and synthesis, but it also required modes of transferring material and instructions, demands and commands, from a central governing system. The organisation also had to be self-assembling and reproducible. In the above sections we have described the analytical bits which are found to be required, listing these requirements down to the level of the most basic of all—the elements themselves. Before looking at the governing centre, we can examine the earliest working chemical machinery and its message operations as we presume they must have existed. We note again that in order to make DNA, RNA, proteins, sugars and fats there needed to be several different and separate, but co-ordinated, metabolic pathways. Secondly, in order that a cell should operate the amounts of materials supplied to different RNA, fat, protein and polysaccharide syntheses had to be controlled. Thirdly, energy had to be introduced into all the pathways of synthesis at a certain rate. To keep all these selected different activities running smoothly to a timetable, each required the production of proteins.

In essence, the central feature of each and every activity was then the presence of proteins. There had to be, therefore, two necessary controls over: (1) protein production and (2) activity of the proteins, and these two had to be linked. Protein production was through the translation from RNA transcribed from DNA. As stated, this synthesis is controlled by: (1) material and energy supply; and (2) by feedback (see Section 10.9 and 10.13). The direct control of RNA synthesis at the DNA level is by proteins themselves, transcription factors (TF), which also require synthesis and which can be activated or deactivated in their function by the binding of small molecules or ion units of the cell cytoplasm (see aside). We have then (see Fig. 10.18).

Early regulatory genes

Gene	Element regulation
Fnr	Fe^{2+} and see Ni^{2+}
Arc	Phosphorylation
Nif	Nitrogen
Narc	Na^+/H^+ exchange

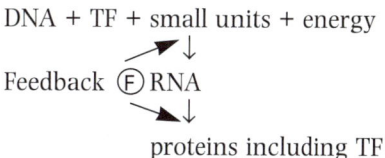

$$DNA + TF + \text{small units} + \text{energy}$$

Feedback $\enspace \textcircled{F} \enspace RNA$

proteins including TF

The small units are those essential to cellular activity, including some substrate molecules from pathways, some catalytic ions such as Fe^{2+}, S^{2-}, and Mg^{2+}, and some units derived from co-enzymes, such as phosphate and cyclic-AMP. *The essence of the control over selected protein production is that it is a function of proteins linked to DNA and its machinery.* The second feature of activity in the overall network is the link to small molecule reactions (see Table 11.12) and to the environment by pumps and channels made from proteins.

As an illustration of the complexity of the genetic structure we can examine the so called *nif* genes for nitrogen fixation (Fig. 11.10) although this particular enzyme may not have arisen in *strictly* anaerobic organisms. The enzyme itself consists of the several parts shown in Fig. 11.8, but additionally it requires two reducing proteins shown shaded to the left

Table 11.12 Primitive messengers

Messenger	Functional control upon
Mobile co-enzymes	Distribution of metabolic fragments H^-, CH_3^-, $-COCH_3$, etc.
Nucleotide triphosphates	Distribution of energy
Fe^{2+}, $2RS^-/(RS)_2$	Distribution of electrons Redox state balance
Some simple substrates (feedback)	Metabolic products, e.g. glutamine, nucleotide bases, amino acids, and upon gene expression
Phosphorylation of proteins, Fe^{2+} (Mn^{2+})	Gene expression

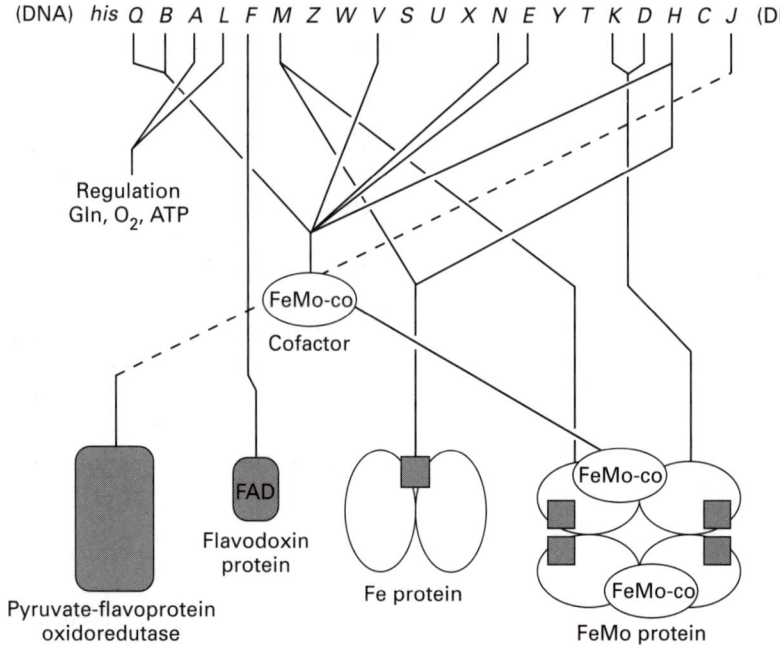

Fig. 11.10 *Nif* genes required for nitrogen fixation, as arranged in *K. pneumoniae*, and their respective gene products. The regulation by product, glutamine (Gln), (dioxygen) and ATP (energy) is shown. This genetic structure became fully necessary only after the advent of O_2 (see Chapter 13). All genes except *A* and *L* generate RNA for the production of proteins directly or for the production of a synthesis protein unit for FeMo–co (Fig. 11.9).

in Fig. 11.10. Notice that apart from the large number of genes required for the synthetic steps there are four genes related to regulation over syntheses (*Q, B, A, L*) which have feedback connection to the level of product, that is glutamine, the carrier of the NH_3 produced, to the level of dioxygen, which is a poison for nitrogen fixation, and to the energy level of the cell, ATP. ATP, O_2 and glutamine link to all other activities of the cell. Of course, the protection from O_2 was only required after a somewhat aerobic atmosphere arose (see also Section 13.4.1). A very important point is that the numbers of protein molecules of a given kind are very variable and not related, therefore, to the fact that they are equally represented as single coded stretches of DNA. This statement is true of all genetic information and we return to its significance in Chapter 16.

11.10.2 The early chemical messengers and protein activity controls

As outlined in Chapter 10, messages using chemicals are propagated in two ways: first, an ion gradient across a membrane is used; secondly, a chemical is changed from one form to another. In both cases change is imposed on the system by the binding of the newly introduced charge (or by the accompanying effect of electrostatic field changes) or by the binding of the introduced molecular form to proteins. In order that the message system can be recharged, the messenger must be removed. Examples of the two schemes are

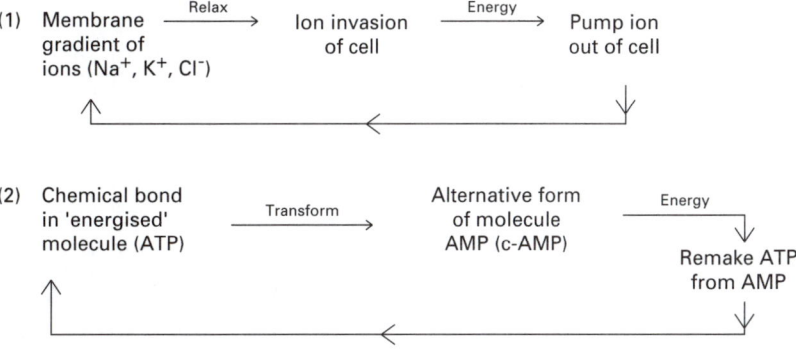

Under scheme (1) the ions used were initially very simple, e.g. Na^+, K^+ and Cl^-. In the most primitive cells more sophisticated gradients were little used for inside/outside messages since the main concern was management inside the cell just to produce DNA. Inside the cell the major messenger molecule was ATP, which was partly converted into cyclic-adenosine monophosphate (c-AMP), a second messenger, or used to phosphorylate a protein or substrate. This change in messenger concentration then activated metabolism. Other internal controls were based on organic substrates or, especially, a thiol–dithiol redox buffer and the standing levels of free Fe^{2+} or Mn^{2+} (Fig. 11.11). It does not appear that ions such as Ca^{2+} and Zn^{2+} had any great influence on primitive anaerobic cellular life as messengers. It may well be that phosphate, sulphur and iron messenger systems were closely interconnected see aside. Note that energy (ATP) is always also coupled into both of the above messenger systems and protein production. The development of new messengers is a major feature of evolution (see Chapters 12–15). (N.B. We do not imply that all messengers in today's prokaryotes are the same as those in primitive cells.)

Naturally enough, management of the available elements had to be tightly coupled to protein production to keep the whole cell in homeostatic balance. This was achieved by the common, exchanging, small units (see Fig. 11.11).

Early chemical messengers

Messenger	Control function
H^+	pH Buffering
K^+	Osmosis
Phosphate	Energy Transfer
Fe^{2+}	Redox Balance

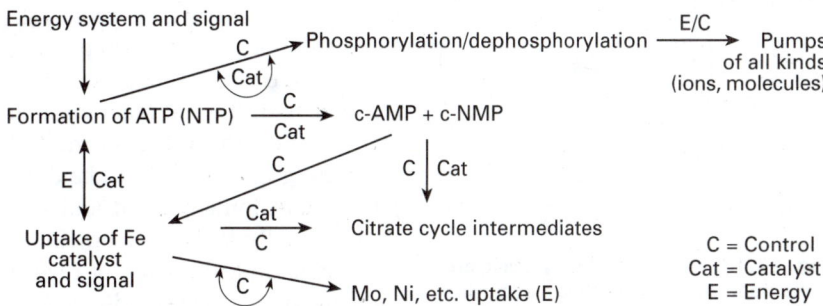

Fig. 11.11 The multiple interactions shown as feedback controls (C) and catalytic connections acting on enzymes (Cat) or pumps (P) which also need energy (E).

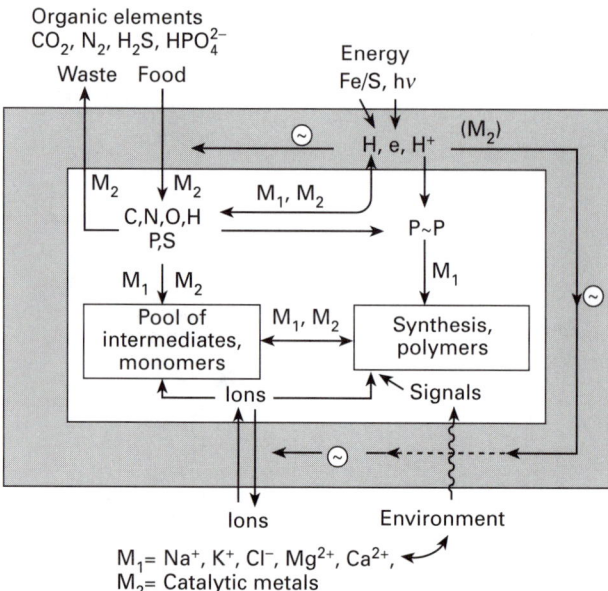

Organic elements
CO_2, N_2, H_2S, HPO_4^{2-}

Waste   Food

Energy
Fe/S, $h\nu$

Fig. 11.12 A schematic picture of the roles of different elements in a primitive cell and a low redox potential environment. O_2 and SO_4^{2-} were introduced later. ~ is a charge or concentration gradient.

We must now remember that the survival value of such primitive cells lies partly in the machinery of the cell as described, but also in the ability to reproduce quickly, since these cells have very little protection, and are therefore vulnerable and short-lived. A view of the whole cell is given in Fig. 11.12. The remarkable feature of this system is that it could self-assemble as well as reproduce cyclically very quickly—in one hour.

11.11   The prokaryote cell cycle

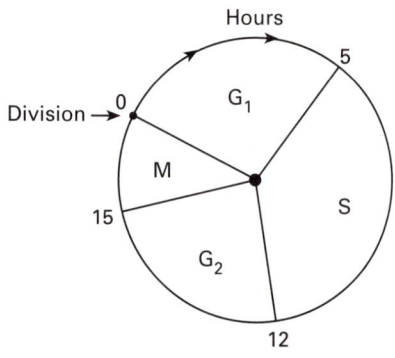

Fig. 11.13 The cell cycle: G_1, basic proteins duplicated; S, synthesis of nucleotides and DNA synthesis started; G_2, machinery of division made, DNA synthesis finished; M, mitosis to division.

As stated, the objective of the earliest cells was in large part simple reproduction, which required repeated development in a cycle. (Remember that we regard living systems as those that can self-make themselves and reproduce, while extracting materials and energy from the environment, through a form of organised chemistry—abiotic chemical cycles undoubtedly predated them). Reproduction means a doubling of the DNA content, together with the cell machinery and membrane, during the life cycle. The cycle is then the route to multiplication through division (Fig. 11.13). Everything in a cell had to be duplicated, including a number of enzymes, pumps, fats, ionic content and structural components, as well as the DNA (the code). By looking upon the genetic code alone, the DNA, as the self-preserving unit, the sense of regeneration of activity is lost. It is the *activity* (now including the reproduction of the code) that had to be generated time and time again. Here we include the apparatus for energy capture as well as for material intake and incorporation. The cell cycle concerned, therefore, all the 15–20 essential elements for a living species. Only if the right amounts of ions and of proteins, fats and so on were obtained or made, was it possible to reproduce by self-assembly a cell with fidelity. As far as we can extrapolate from today's examples, the cell cycle was largely governed by phosphorylation by enzymes called kinases, dephosphorylation by phosphatases, and destruction of proteins by proteases. Remember that energy flow was equally essential for this primitive cellular reproduction, so the coupling of the cell cycle to phosphate of ATP was equally essential.

In essence it is believed that the cell cycle proceeds in the following way. We suppose that the cell has an adequate supply of energy and this is seen by the level of ATP in the cell. We also assume it has an adequate supply of chemicals, 15–20 elements. The cell is then observed to pass through the stages shown in Fig. 11.13. The early stages, G and S, require synthesis of most of the proteins of the cell and then of DNA. Each of the stages is governed by the timed appearance and hydrolysis of certain protein molecules, which allow it to pass into the next phase. Thus, at a given stage of synthetic activity, a protein, X_1, is produced with a kinase and, given that enough ATP is present, X_1 is phosphorylated. The X_1P molecule alters the synthetic apparatus through binding to DNA, but the stimulated synthesis generates a phosphatase and a protease which destroys X_1P and the first specific kinases while producing new kinases and a protein, X_2, so that the next step in the cycle of phosphorylation/dephosphorylation occurs. The procedure is repeated several times using phosphorylated proteins X_nP until the production of the last signal molecules in sequence brings about cell division. Note that all the essential steps are governed by phosphate signalling, strengthening the belief that this was the most primitive method of general message transmission. As stated, it is hardly surprising that this is so when phosphate is so closely linked to energy transfer using NTP (nucleotide triphosphates).

We conclude by noting that a cell 'cycle' is not a dynamic overall steady state but a *developmental activity* in that the cycle doubles the number of cells. We return to the problems that this raises in Chapter 16. It is truly remarkable that such sophisticated development is present in the earliest forms of life we can postulate.

11.12 Differentiation of prokaryotes

There is a question now of the chemical internal tolerances inside cells in relation to the external environment within which a primitive cellular system of activity could be maintained. We refer back to Chapter 5 where, in a discussion of thermodynamic stability, we showed that some materials, e.g. water, had a precise composition and fixed physical transition points based on values of ratios of chemical elements and on moderately strong bond co-operativity at a given pressure. However, in other combinations, such as alloys and many minerals, the chemical composition could vary over a range and with it the physical properties also varied. We asked there how the chemicals observed were related to the change of free energy with element composition. The parallel question we ask here is: did a primitive cell need to maintain composition very exactly? Now, analysis shows that only cells of particular composition *ranges* (of chemical elements) were formed. We must ask why this is the case.

We observed a further feature of inanimate systems in Chapter 7, which was that for ordered systems of limited size the shape was also fixed at equilibrium. It is a characteristic of a cell of a given kind, to this day, that the structure that encloses the dynamic organisation is contained in space in a fixed way, that is, it is shaped both as a whole and even at the sub-cellular microscopic way in its particles and proteins. Taking the observation that a given shape represents a given composition, the composition of a cell had to be rather tightly, though probably not exactly, kept from the beginning of life. There had to be some latitude since a cell which can manage a variety of foodstuffs has a great advantage. To some degree, therefore, the primitive reading of the DNA code had to vary so as to produce the right proteins, enzymes, etc., opposite the environmental food supply. Basing discussion on the properties of bacteria today, we feel confident in asserting that early cells had this capability, but this in no way affected the code itself—only the reading of it. Thus, an early bacterial cell, like today's bacterial cell, was adaptable to food supply and hence could make, to some degree, switches of internal composition and could then have a region of existence over certain somewhat limited ranges of chemical composition (see Figs 10.4 and 11.14). The ranges may not be continuous in composition, so that the cell may undergo the equivalent of a 'phase' change. While noting this point we must remember that some 15–20 elements were involved, which probably allowed variation in some of them, e.g. of Ca^{2+} and Cl^- concentration, but maybe not all. A particular primitive example of adaptation to the environment is the ability of some cells to become dormant in a spore when food is scarce, but we shall see more sophisticated changes in Chapter 13.

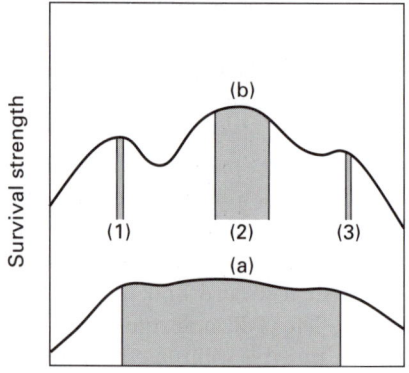

Fig. 11.14 Viability of primitive organisms may be a relatively slowly changing function of composition (a) giving rise to difficulty in the definition of species. Compare the difficulty of characterising non-stoichiometric compounds. Alternatively, the viability may have a more rapid change with composition (b) when species are better defined. Shaded areas indicate regions in which organisms are found. Compare plots of free energy, ΔG, against composition in Chapter 5.

11.13   Sporulation

Many bacterial cells can differentiate into a dormant state—called a sporulated state—in which the bacterium becomes a small cytoplasmic zone with its DNA screened by a heavy coat. Sporulation comes about when food reserves are reduced below a certain level. The cell enters an inactive condition, waiting to be aroused at some later time by heat, light (i.e. energy) or a new food supply. In one sense a spore is not living as it has no detectable flow of energy or material. One element which is then pressed into service is calcium. In effect, the loss of energy due to deprivation of food from the active cell allows the calcium level in the cell to rise. The knowledge of calcium entry is transmitted (via phosphorylation?) to the DNA which switches on synthesis of materials to make the coat of the spore. Thus, a spore is a particular differentiated state of a cell of quite different elemental composition. Many spores have coats made from calcium complexes.

We give this example to stress the fact that primitive cells may well have been able to adjust to adverse as well as favourable circumstances. We shall look again at the role of calcium in cell death (apoptosis) in Section 14.12. In sporulation a cells is in a death-like coma, but its DNA has not changed!

11.14   Mutation and evolution

In order to develop persistent variety of chemical content and physical form (the two go together) which is observed amongst bacteria, there had to be a change in the DNA code and/or the reading of the code. As stated above, the code itself was and is entirely conservative, only open to chance mutation and/or rearrangement. While most such changes must be damaging, it is generally held that now and then a change, or rather a series of changes, may be cumulatively advantageous so that an organism could develop, which is not recognisably the same as its earlier precursor, and natural selection leads to the predominance of its descendants, forming a new species. (We shall see that this also relates to the inevitable problem of cancer; Section 14.12.2.) We can observe too that when food and energy are limiting the different organisms will compete. In the most rough and ready way this leads to the survival of the fittest (in the prevailing conditions) where fitness has to be judged on many different grounds. Since the environment is variable from place to place the general advantage is not clear-cut. This is especially true when we consider the fact that any two organisms, including primitive ones, may differ in chemical element requirements. Now we must ask to what degree were such organisms able to mix their DNA and form intermediate types of organism. This is the problem of speciation of living forms.

11.15   Prokaryote speciation: variability

What is the definition of a *species* of life? At first, scientists based the definition of a species on physical characteristics, such as shape, but there does not seem to be grounds, in principle, for anything but continuous gradation of size and shape in kinetic schemes. Thus, no separate species would arise. Neither is it convincingly obvious to state that the DNA of two organisms differs so greatly that the organisms cannot interbreed. Since DNA can, in principle, be continuously varied by changing the bases one at a time, all intermediate organisms should appear. This is not observed. The question arises then as to why DNA itself is not found to be continuously varied so that separate species would not exist. A tentative answer to the problem of speciation must be sought, therefore, by considering the *survival value* of different chemical machineries.

It is highly probable, of course, that a given species can always show minor variation around a mean, this mean representing the most effective cell type and a particular set of reactions in a feedback relationship. The cells of this species will have a somewhat variable chemical composition within a limited range. In parallel with it there could be a rather different cell type with another set of reactions in feedback relationship, but now this second cell may have a considerably different chemical composition. This being the case, we may draw a figure showing survival value against chemical composition (see Fig. 10.4) which has maxima and minima since the element compositions near the minima do not yield either a cell that reproduces (quickly) or survives (strongly) (see Fig. 11.14). As we stated, viability (fitness, at this level) is a product of reproductive speed and life-time, cell survival. Observable organisms will exist close to each maxima but not near the minima. Hence, the separation of species. The diagram has a relationship to plots of stability of chemicals in equilibrium against chemical composition (see Fig. 5.19.)

Further thought leads to the idea that for some kinds of chemical machinery a wide range of composition may be permitted without species demarcations [line (a) in Fig. 11.14]. The variations within early bacteria (and still today) may be of this kind, so that bacterial may be very difficult to classify within species. Although bacteria have certainly been classified (see Table 11.2), this approach may have greater significance in the laboratory, where chemical control over environment is exerted, than in conditions in the wild. In effect, bacteria were and still are now able to transfer and pick up quite considerable lengths of DNA to and from different strains and even between apparently identified 'separate species'. Thus, some extra-genetic DNA load, and hence some metabolic load, was and is not a problem in these cells. In many cases the extra DNA is not located in the main DNA but in *plasmids*, separate short circular DNA units (Fig. 11.15) which replicate independently. This proves to be a very effective way of transmitting 'resistance' to poisons (note the resistance to antibiotics and to rare elements which are becoming a serious problem) from bacterial strain to strain through the movement of phage, compare viruses transfer from one cell to another, which may have helped such transfer of genes.

In the above we have spoken of *competition* for resources. In a later chapter we shall ask if evolution could also develop through *co-operation*, e.g. *symbiosis*, given that one type of chemical machinery can do a task very different from another and thus complement it. By sharing activities, co-existence could be more powerful in assisting survival than if competition dominates.

All the above seems to indicate that DNA is an *opportunistic code*, not a commanding one, so that DNA changes reflect evolution of the sophistication of machinery responding to environmental changes. How can this be?

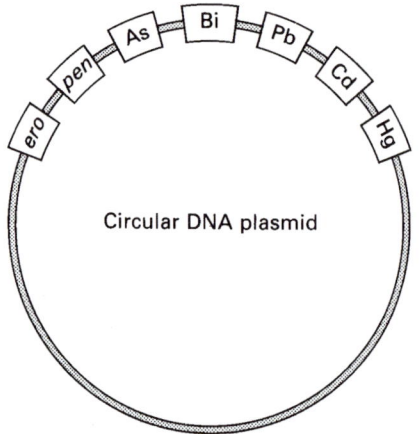

Circular DNA plasmid

Fig. 11.15 A series of resistance genes built on to a bacterial plasmid. *ero* is erythromicin; *pen* is penicillin.

11.16 The beginnings of life and reproduction: a speculative approach

We shall now approach the problem of the beginnings of life from the point of view that the most fundamental requirement is persistence of the chemical machinery, i.e. survival value, so as to allow time for the organisms to reproduce. A major point is, of course, confinement. We shall assume, therefore, that fatty molecules (see Section 6.3) arose in the reducing sea to give *enclosed space*, 'cells'.* A second essential requirement is energy to drive reactions and the machinery. We assume that the initial enclosed space *trapped energy* from an external source (see Sections 11.5 and 11.7). Such a system can do repetitive chemistry, such as making complex molecules and moving ions and molecules across its membrane. At first, several kinds of action could lead to more long-lasting 'cells'; for example, especially those that reduced osmotic pressure relative to the sea, i.e. removed Na^+ and Cl^-, accumulated those organic chemicals that protected the membrane, e.g. polypeptides, and pumped in selected elements, such as K^+, Mg^{2+} and Fe^{2+}. These more stable vesicular chemical factories would have a longer life span and were therefore able to develop kinetic stability further. One can imagine that they found ways to capture better catalysts, so that production capability and protective systems for the 'cells' improved. Eventually, such a cellular system 'learnt' to limit the variety of its internal products by producing roughly only those molecules that were useful to the cell (thermodynamic or kinetic stability may have driven this process), so that it increased in size, divided and multiplied. This crude method was bettered by the development of templates for the production of their own required materials.

Today it is often thought that these were molecules of RNA, but there may have been mineral precursors such as zeolites and other silicates (see aside), which were not required to be reproduced since they were part of the environment. This is far from necessary since RNA and proteins could have been separately generated at first and have then led to the preferred RNA/protein co-operative production we now see in ribosomes. Only at this stage is coded reproduction a very strong beneficial development. We refer to the bibliography for further discussion, but note that the kinetic stability of this system already demands utilisation of some 15 elements, minimally.

There is a further problem with the discussion of the origin of life. While the living systems we discuss today exist largely at about one atmosphere pressure and at 25 °C, the conditions when life started could have been very different. The pressure locally could have been over 100 atmospheres and the temperature over 100 °C in deep trenches of the ocean. Such conditions would have aided reactions, both of synthesis and degradation, but many organic chemicals of today's life do not persist for long under these circumstances.† However, many organisms are now known that are able to live in such extreme conditions, for example in the deep-sea chimneys, in hot springs and even underground to a depth of perhaps 5 km beneath the surface. They are called *extremophiles*, and many scientists think that some of them may resemble more closely the universal ancestor of all living species (see Fig. 11.1). Many are archaebacteria (see Table 11.2). It is easy to formulate possible schemes for their beginning but there are far too few good, relevant, chemical experiments in such conditions to give confidence in understanding. For these reasons other authors prefer to suggest that life

Clay minerals as possible DNA precursors

Kaolinite
Montmorillonite
Glauconite
Cacoxenite
$[Fe_8(PO_4)_6O_2(OH)_4(H_2O)_8]$

---

*This does not exclude that some abiotic chemistry could have developed out of closed compartments, using templates (clays, metal surfaces, etc.).
†See Amand, J. P. and Shock, E. L. Science, *281*, 1659–62, (1998).

originated in milder conditions (or even at quite low temperatures) and that extremophiles were latecomers that adapted to unusual conditions.*

11.17   Units and variables in primitive cellular organisation

Much as the Greeks asked which are the elements (units) and qualities (variables) of all observable objects (see Chapter 1), and much as a similar question was asked of equilibrium situations after the ideas of reversible thermodynamics were established in the last century (see Chapters 2–7), we can now ask which are the units and variables of *primitive* organisms, cells. The units, of course, should be the *independent* chemical 'components' made from the *elements*. We have established that the elements here are some of the available elements in the conditions of the Earth, some 15–20, not the total number of 90 naturally occurring elements of the periodic table. They are shown in Figs 10.3 and 10.10 and listed in Table 11.4.

Now, the true units of variability are not seemingly these elements by themselves but, as in mineral chemistry, they should be the *independent* 'components' of the cell machinery made from these elements. In a steady state, however, the number of independent components is not the number of observable different chemical compounds in the cell, since virtually all are controlled by the fact that the cell must have a *fixed relative concentration of all its compounds*. Just as we found that equilibrium exchange between compounds (in thermodynamic systems) reduced apparent variability, i.e. the number of compounds present did not give the number of independent components in thermodynamics, so we see now that feedback interrelationships between compounds (in the living cells) strongly reduces chemical content variability in steady states. In fact, we could well conclude that in *a given cellular steady state* all compounds, substrates, proteins, fats, sugars, nucleotides and element ions, in whatever form they occur in the cytoplasm of one *species*, are in feedback relationships except for one—DNA (and H_2O to a certain extent). Since this includes the uptake of all the elements across the cell membrane, linked by feedback to the energy of pumps, it could be concluded further that *the units of composition are just the chemical elements taken in. They are the only independent variables of composition applicable to a simple primitive organism in a steady state* (see Section 10.19), apart from the DNA sequence.

Now, we must stress that the final composition of the cell is not fixed, as it would be by chemical equilibration of these 15–20 elements, since the content is dependent upon the energy used to manipulate each element. Thus, *rate of energy input* is another variable since this controls the patterns of flow (or even the choice of reaction pathway). A viable machine will only develop for certain *selected* combinations of elements and rate of energy input, when the whole chemistry and physics (even the shape) of a cell may well be defined. The total variables of the cellular system reduce,

*See Woese, C. Proc. Natl. Acad. Science U.S.A. *95*, 6854 (1998).

Table 11.13 The units and variables of the simplest cell

Units	Variables	Restrictions
Elements (restricted by availability) Charge	Composition External T and p Fields External energy supply	DNA (feed-forward and feedback) and Controls on enzymes (see aside)

Limitation of variables in cells by controls

(1) Enzymes: feedback by subtrates in pathway
(2) Enzymes: feedback by energy (ATP)
(3) Pumps: feedback by pumped element
(4) Channels: feedback by diffusing element.

DNA represents a dynamic state:

(1) element uptake/output
(2) element use
(3) energy intake/output

therefore, to the composition in terms of units (the available elements), to the rate of energy input applied to the system and to the external variables, temperature and pressure, plus fields, (especially gravity) (Table 11.13). Temperature and pressure, like composition, are part of the nature of an environmental niche, and so is gravity. However, the *internal* temperature and pressure are related to the external temperature and pressure through the free energy flux of the cell. In other words, the cell interior is not at equilibrium with the environment in these respects. It constantly generates heat, as we know from experience in fermenting organic material. Thus, the internal temperature and pressure are not simple variables in primitive cells, though they can be measured and reflect the variable 'energy flux' (see also Sections 5.6 and 8.15). We state then that, based on this thinking, only certain (selected) combinations of units, or ratios of combinations of the variables, as yet unknown, generate a stable (viable) steady state of a particular organism, and these combinations and ratios of combinations are those that *have to be coded* in a DNA so as to control and reproduce the viable organisation (Fig. 11.12). In other words, the DNA is a *guarantee through coded instructions of a particular viable steady state*. It is not the only fundamental feature, but it reflects restrictions on variables.

DNA is, then, a restriction (see aside), like a balanced set of ΔG values in an inanimate equilibrium system (Section 5.9). We do not know how many different self-sustaining machines, cellular 'species', could be produced in this way, but since the rate of energy input used, even if it is fixed, can be dissipated using about 20 different element variables of composition, it could be a very large number. Together, then, the chemical composition variation of a niche that controls element availability and the variation of the rate of energy uptake are the two dominant variables. Thus, we may consider that a multidimensional plot of, say, 23 variables (20 elements + energy applied + external T and p) gives rise to the possible viable species of bacteria. This is an incommensurate number, but it is limited by the kinetic patterns securing survival and represented by DNA restrictions. Note that we are not yet able to give a full explanation of this viability. We just observe that, if this description is correct, while it is possible to characterise an organism by its DNA, it must also be the case that an alternative characterisation could be the composition in terms of the chemical elements in it, together with the rate of energy throughput, if this could be accurately measured. Compare the problem with the use of elements plus energy instead of components in chemistry. We repeat that it is not obvious why a particular DNA corresponds to a particular viable machinery of a cell's activity. We have to be aware, too, that different DNA could represent different viable organisms using the same selection of elements and different or even closely the same energy intake but different networks.[*]

Finally, we stress that this logic leads us to conclude that a major factor limiting the possibility of variety amongst species was and is *element availability*, since a variety of energy sources have always been present. Temperature and pressure have varied somewhat and this must have been a secondary cause of evolution. When life began it was the reducing atmosphere that limited cell species and hence the ability to evolve new species. As the Earth evolved it was inevitable that element availability changed and so life evolved too (see Chapters 12–14).

While describing the compositional variables it should be observed that, in cells, different elements carry different weightings not just dependent on analytical composition or concentration. Thus, a trace element such as iron is as important as a bulk element such as nitrogen. The weight of iron should, therefore, be represented by its concentration in the cell multiplied by its effective functional value, f_v. Vague though this notion is, an appreciation of weighting factors allows us to see the equal importance of the 10–15 trace elements in the composition, relative to the 5–6 bulk elements. It is then foolish to assert that carbon is the element of life when it is one of some 20 elements, each with a special function and without which no life would exist.

In the next chapters we must analyse the way in which this system has evolved, keeping our eyes especially on the available elements, the available energy input, the control over temperature and pressure, i.e. all the controls belonging to the restrictions in the parallel evolution of machinery and of its coded DNA. It is unfortunate that we cannot go directly from DNA to any real understanding of a cell. DNA is an index of a system not the system itself, as we shall show more clearly in Chapter 16. However, we have now made our first steps from inanimate matter towards man, describing real biological systems.

---

[*]A way of seeing this possibility is to consider optically active isomers, Figs 3.16 and 8.4. The two isomers, *laevo* and *dextro*, have the same chemical composition and equal potential energy. They differ only in the relative way the system of atoms are disposed in space. For a large number of optically active centres joined together, a large number of possible components arise. The parallel with an identical set of atoms exposed to an identical energy supply is that the number of different ways of directing their flow can be devised by instructions.

12

The development of anaerobic organisation: from prokaryotes to eukaryotes

All things by immortal power
Near or far
Hiddenly
To each other linked are

Francis Thompson, *The mistress of vision* (1859–1907)

12.1   Introduction

The proposed anaerobic primitive prokaryote described in the previous chapter represents the simplest form of living organisation that we know of. As we have seen, space is divided in these organisms into three compartments: inside (cytoplasm), membrane and outside, Fig. 11.1. The cell as a whole is small and often cylindrical in shape. The chemical pathways within its cytoplasm are limited by the small range of redox reactions permitted under reducing conditions. The large molecule chemical partners are the least complex of those in any organism: the DNA is small, about one million bases, and the variety of proteins limited in number to about 3000. The messenger and transfer systems are often based on the exchange of small molecules made from H, C, N, O, S and P, sometimes co-enzymes and sometimes substrates, with P dominant in energy partitioning through ubiquitous ATP and other nucleotide triphosphates (NTPs). Alternatively, messages are based on metal ions such as Mn^{2+} or Fe^{2+}. The uptake and rejection of elements depends on simple pumps and energy capture rests in the membrane. Some of these elements are quite heavy and function as catalysts only, for example molybdenum. The local internal environment is not stringently regulated and a not inconsiderable adaptation to external circumstances is often observed. The definition of species is not very clear as the DNA accepts considerable insertions or increases, in part due to the presence of mini DNA in plasmids.* As evolution progressed most of these features of metabolism in the cytoplasm have remained, but organisation became more complex and restricted, with DNA especially enclosed in a compartment, the nucleus. However, bacteria (even archaebacteria and other anaerobic prokaryotes) did not disappear. At the same time, the use of chemical elements in the new organisms, eukaryotes, changed somewhat. Note the changing use of elements can mean the introduction of a new element into the composition or a change of energy flow involving the previously used elements, or both. These are changes in variables, as is increase in the number of compartments, which we describe first (see Table 12.1).

Table 12.1 The probable development of eukaryotes from bacteria

1. Invagination of bacterial membrane
2. Appearance of separate vesicles associated with permanent filamentous cell structure and the reticulum
3. Capture of organelles
4. Nuclear membrane linked to filamentous structure and surrounded by the reticulum.

N.B. The first eukaryotes may have evolved from archaebacteria while the organelles came from eubacteria [Fig. 10.1(b)].

12.2   Compartmental changes

The first step in the introduction of this complexity possibly took place in the absence of dioxygen, which caused numerous *chemical* changes, and we shall, therefore, consider the *initial* evolution of anaerobic eukaryotes not

---

*Free independently replicating circles of DNA, usually obtained by infection from virus or from another prokaryote (see Fig. 11.15).

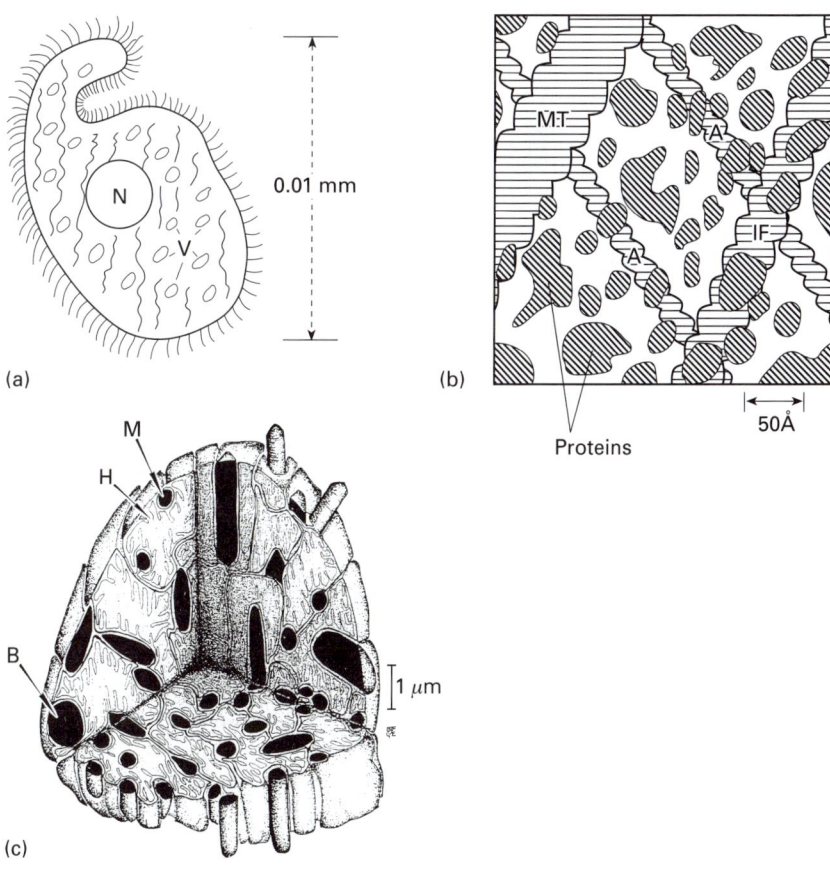

Fig. 12.1 (a) An anaerobic ciliated protozoa with a nucleus (N) and many small vesicles (V). The protozoa can digest bacteria and they are able to swim in a directed way. Note the internal filaments. (b) The three kinds of filament in a cell: microtubular (MT), actin (A) and intermediate (IF) (x 10⁶) (*see aside*). The aqueous solution is > 70% of the volume. [After Goodsell, D. S. (1998). *The machinery of life*. Springer-Verlag, New York.] (c) A tripartite symbiotic consortium between methanogenic bacteria (M), eubacteria (B) and hydrogenosomes (H), recently discovered as a permanent intracellular complex in an anaerobic ciliate (*Cylidium porcatum*). The whole complex is about 1/100th of a millimetre from top to bottom.

Sizes of polymers

Polymer	*Size (Å)*
Protein (small)	20 (radius)
Protein (large)	>200 (radius)
Fatty Acid	15–20 (length)
Filaments, MT	100 (width)
Filaments, A	30 (width)

so much in terms of new oxygen chemistry as of new organisation. (We shall look again at this evolutionary change in Chapter 13.) There was an early such step even in one prokaryote: the formation of a single separate compartment for light capture, a thylakoid in blue–green algae, now called cyanobacteria (Fig. 10.20). This is a development from the localised invaginated activity seen in some photobacteria. They are believed to have been responsible for the later release of dioxygen into the atmosphere (see Section 13.1). There is also the appearance of hydrogenosomes in archaea, which are strict anaerobes.

The new organisation seen in the eukaryotes involves two major changes in the cell's structure (see Table 12.1). The first, as stated, is an increase in the number of compartments due to new membrane structures within the cell cytoplasm which gave rise to vesicles [Fig. 12.1(a) and (c)]. In

Table 12.2 The organelles of eukaryotes

Organelle	Function
Mitochondria	Energy (ATP) supply from O_2 reactions (or first from NO- reactions). Citric acid cycle
Chloroplasts	Energy capture, hydrogen supply and O_2 generation from light
Kinetosomes(?)	The motor units of undulipodia (and nuclear division?) (See Fig. 12.2)
Hydrogenosomes	Release of H_2 in anaerobes derived from archaebacteria and used to supply energy [See Fig. 12.1(b)]
Peroxyzomes (?)	Long-chain fatty acid oxidation

particular, a new membrane structure enclosed the nucleus, and this attribute is used as a distinguishing feature from the prokaryotes. The new membranes are made from lipids similar to those of the cell membrane. The second new structural feature was the introduction of protein filaments holding the vesicles in particular parts of space and connecting with the underside of the major cell cytoplasmic membrane [see Fig. 12.1(a) and (b)]. This allowed the cells to become much larger and the outer membrane to be flexible, when cells became of particular shapes.

Two types of special larger internal vesicles, called organelles, are observed in most eukaryotes: the mitochondrion in many of them and the chloroplast in those that are photosynthesising (see Fig. 10.17 and Table 12.2).* Mitochondria today are dioxygen (aerobic) organelles degrading C/H/O compounds, while chloroplasts are light-harvesting, carbohydrate-synthesising and frequently dioxygen-producing organelles. We refer to mitochondria and the associated dioxygen chemistry mainly in Chapter 13, since we believe they arose later than the most primitive eukaryotes. (N.B. If mitochondria developed from nitric oxide-utilising energy-capture devices in bacteria not using dioxygen, then the organelle may be much older *in anaerobes*.) It is accepted today that these 'organelles' originated from symbiotic association with small bacteria captured and retained inside the original large eukaryote anaerobic cell and then much modified. Such capture processes can occur in anaerobic eukaryotes as is shown in Fig. 12.1(c), where we also show the presence of hydrogenosomes. This capture itself is dependent of the presence of a flexible membrane and a set of filaments internal to the eukaryote. We return to the value of organelles later, but remember that in this chapter we do not wish, as yet, to include the consideration of the major changes of the chemistry of living organisms produced by changes in the environment, especially due to the presence there of dioxygen.

In Chapter 11 we have described the development of methods of propelling simple cells using the beating of flagellae activated by bodies called kinetosomes. It is believed by some authors, refs. 195 and 196, that the appearance

---

*Some authors include peroxysomes and hydrogenosomes within organelles. We shall not discuss these vesicular systems in that way.

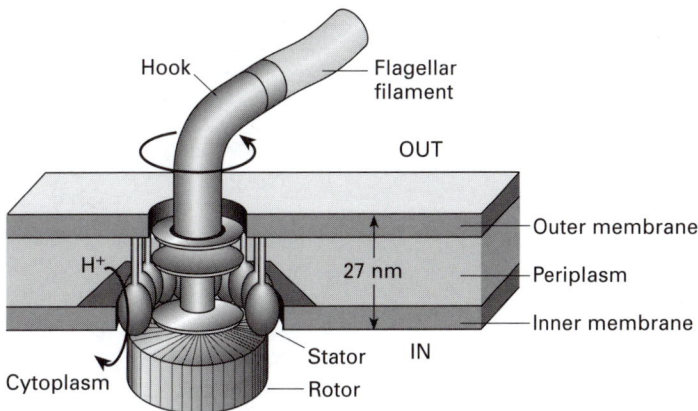

Fig. 12.2 Flagellar motor rotor–stator architecture (schematic) depicting location of the transmembrane ring particle and cytoplasmic motor modules, thought to be rotor and stator components, respectively, as denoted, within the flagellar base. [After Khan, S. (1997). *Biochimica et Biophysica Acta*, **1332**, 91.]

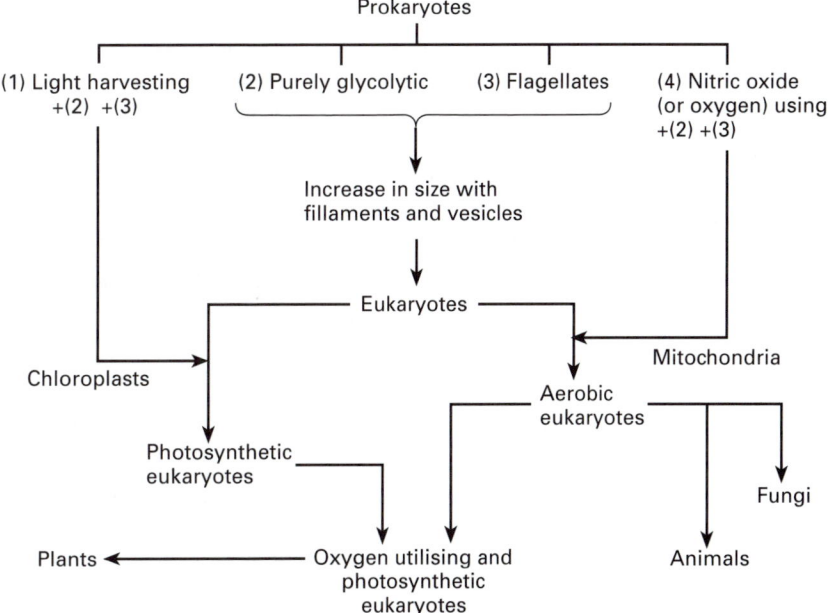

Fig. 12.3 A scheme to show the interdependence of the origin of many eukaryotes upon genetic crossing and combination with early prokaryote species.

of similar apparatus in larger anaerobic eukaryotes was also due to the capture of bacteria holding kinetosomes (see Fig. 12.2 and 12.4). If this is the case, then there is a third organelle to be added to the above two. The kinetosomes could have become part of the machinery of nuclear cell division. We have, then, the possible development sequence shown in Fig. 12.3. The full complexity of a eukaryotic cell is shown in Fig. 12.4 and Table 12.3 lists differentiating features from the simpler prokaryotes.

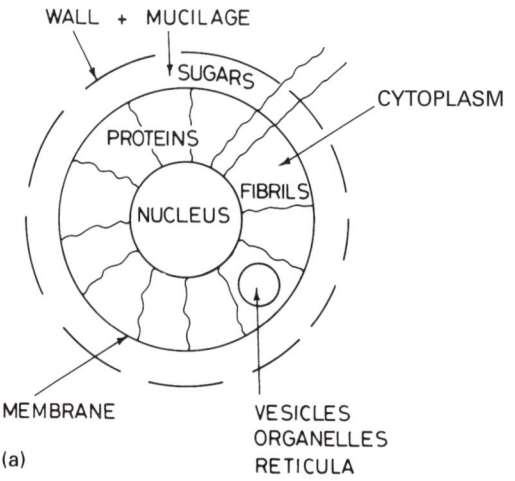

WALL + MUCILAGE

SUGARS

CYTOPLASM

PROTEINS

FIBRILS

NUCLEUS

MEMBRANE

VESICLES
ORGANELLES
RETICULA

(a)

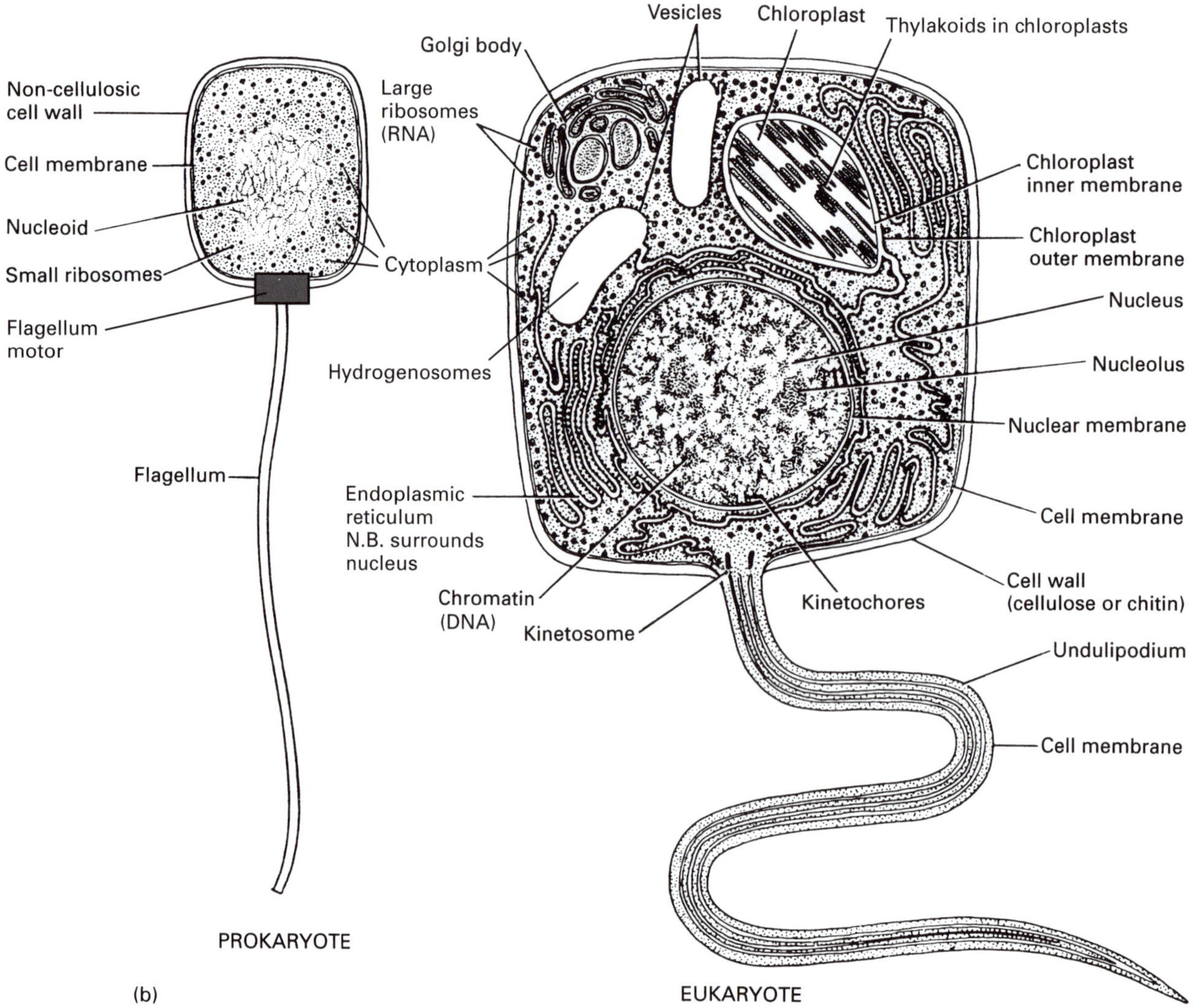

Non-cellulosic
cell wall

Cell membrane

Nucleoid

Small ribosomes

Flagellum
motor

Flagellum

Golgi body

Large
ribosomes
(RNA)

Cytoplasm

Hydrogenosomes

Endoplasmic
reticulum
N.B. surrounds
nucleus

Chromatin
(DNA)

Kinetosome

Vesicles

Chloroplast

Thylakoids in chloroplasts

Chloroplast
inner membrane

Chloroplast
outer membrane

Nucleus

Nucleolus

Nuclear membrane

Cell membrane

Cell wall
(cellulose or chitin)

Undulipodium

Cell membrane

Kinetochores

PROKARYOTE

(b)

EUKARYOTE

Fig. 12.4 (a) A schematic diagram of a eukaryote. (b) Typical prokaryotic and eukaryotic cells, based on electron microscopy. Not every prokaryote or eukaryote has every feature shown here and observe the change from a flagellum to a kinetosome motor. N.B. The mitochondria were not present until the dioxygen level was considerable, that is, until aerobic conditions, unless they used NO originally. (ref. 129 in Further reading.)

Table 12.3 Main differences between prokaryote and eukaryote cells

Prokaryotes	Eukaryotes
Small cells (1–10 μm)	Mostly large cells (10–100 μm)
Nucleus, not membrane-enclosed (DNA)	Membrane-enclosed nucleus associated with proteins—chromosomes
Cell division by fission in equal parts; no filaments, centrioles or microtubules. Sexual systems rare. Fast reproduction rate	Cell division mostly by mitosis; filaments, centrioles and microtubules present; sexual systems common. Slower reproduction rate
Mostly strict anaerobes (but also facultative anaerobes and aerobes)	Mostly aerobes (but many anaerobes)
No mitochondria	Mitochondria present in aerobes
Simple bacterial flagellae composed of flagellin (a protein)	Complex (9+2) undulipodia composed of tubulin and other proteins
Various patterns of aerobic and anaerobic photosynthesis (end-products such as sulphur, sulphate and dioxygen). Enzymes for photosynthesis in thylakoids close to the cell membrane	All photosynthetic species have dioxygen as final product (released to the environment). Enzymes for photosynthesis packed in membrane-enclosed chloroplasts
Absence of calcium-binding proteins in cytoplasm	Presence of calcium-binding proteins in cytoplasm
Phosphate signalling; calcium expelled (poison)	Calcium signalling interacting with phosphate in anaerobic unicellular organisms; evolved systems in multicellular aerobic organisms

Adapted from ref. 129 in further reading.

12.3   The value of extra compartments

A feature of the new compartments in single cell organisms is that they can be used to remove some chemical reactions or products of reactions from the cytoplasm. Yeasts are the best-known examples and they can function in anaerobic conditions. A compartment can carry out reaction steps that would normally (in prokaryotes) increase risk of damage had they been allowed in the same homogeneous solution as the DNA of the nucleus. A vesicle can also act as a storage region for elements or special compounds to be used later, when required.

Taking the last example, even some bacterial cells managed to store iron in a *protein* (not a lipid) vesicle such as ferritin, an $Fe(OH)_3$ precipitate, which we have not considered as a separate compartment since iron exchanges readily from it. Elements other than iron, especially calcium, can also be stored in many of the vesicular structures, called reticula, which are lipid-based vesicles or tubes that criss-cross a cell (see Figs 12.4 and 12.8 and Table 12.4). The calcium in these vesicular structures does not exchange freely with the cytoplasm. Vesicles can also hold a catalyst for hydrolysis (digestion) away from the cytoplasmic proteins which it could attack. Such catalysts (enzymes) can also be ejected from the vesicles into the environment, when required, in order to break down large

Table 12.4 Chemical elements concentrated in compartments

Compartment	Element concentrated
Thylakoid	Manganese for photosynthesis (Ca^{2+}, Cl^-)
Mitochondria (NO using?)	Iron for redox reactions and haem synthesis (Ca^{2+})
Golgi	Manganese for glycosylation
Vesicles endoplasmic reticulum	Calcium for triggering
Exocytic vesicles	Poisonous elements

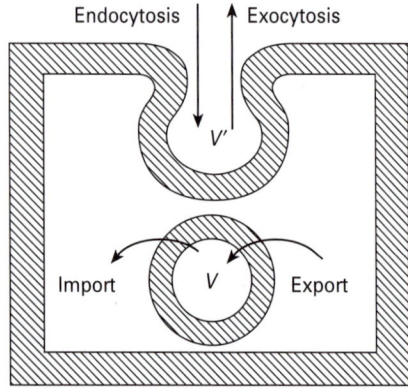

Vesicle uptake and release mechanisms:
V vesicle
V′ vesicle coalesced with membrane

molecules and hence provide food. The enzymes are nucleases (to break down DNA or RNA), proteases (to break down proteins) and saccharases (to break down polysaccharides).

As far as poisonous chemicals are concerned, vesicles can be used again, first to store and then to release them into the environment, especially to attack foreign organisms (see aside). An example is the release of molecules that can form channels in membranes of bacteria, so killing them. They act in much the same way as certain modern synthetic antibiotics in that, on forming a channel in the membrane of the cells of the competing (enemy) organism, these poisons allow sodium or chloride ions to enter those cells, increasing their internal osmotic pressure and so causing the enemy cell to burst. The timing of release of such vesicular chemicals was, of necessity, due to the sensing of nearby targets, e.g. bacteria. Thus, the eukaryote has better control over its reaction paths and its timing of release and uptake of chemicals, and better protection than the prokaryote. It does not necessarily have a greater diversity of kinds of reaction, since, effectively, we are still describing anaerobic metabolism in all compartments, but the variety of catalysts for a process such as hydrolysis may well have increased considerably in the different compartments. Note again that as far as anaerobic eukaryotes are concerned, and in the account we give in this chapter, we do not suppose that the availability of elements had changed, though the use of some elements did change.

We have already described the use of sunlight in some prokaryotes as a source of energy. The reactions utilise one-electron steps which could be a source of free radical species, an obvious risk to DNA. The photosystems of some bacteria are usually distributed in the underside of the cell's outer membrane, where, in effect, they do present a risk to DNA; but note the development of algal thylakoids (Fig. 10.20). The capture and modification of such photosynthetic bacteria by eukaryotes in the form of chloroplasts allowed the free radical-producing reactions of photosynthesis to be safely housed. The chloroplast is an organelle with part of its own DNA and synthetic machinery retained; as stated above, it is perhaps as well to think of this incorporation as the first example [or the second (after NO-using mitochondria) see Fig. 12.4] of true symbiosis between organisms. The filaments of the eukaryote allowed it to position the new energy-capturing organelle. Of course, the eukaryote cell has to donate material to the organelle to keep it functional.

In the eukaryote cell, its own DNA could remain in a largely dormant condition, only looking after replacement of proteins needed to ensure the steady state, since the function of the cell became more a matter of survival rather than just of reproduction, which is now much slower (see Section 12.14). The DNA in the organelles could (and can) act to some extent independently from the main DNA.

12.4   The development of internal filaments and shape

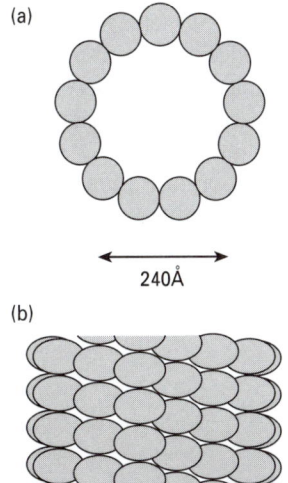

(a)

240Å

(b)

Helical pattern of tubulin subunits in a microtubule: (a) cross-sectional view showing the arrangement of thirteen protofilaments; (b) surface lattice of α and β subunits.

A prokaryote cell has little flexibility and, on the whole, maintains its simple, approximately cylindrical shape (Fig. 11.1). Its membrane is stabilised by walls and under-pinning scaffolds of proteins. A eukaryote cell is different in that its body plan is not confined by a wall and its overall shape is adjustable to allow it to take advantage of the asymmetry of its environment, e.g. of environmental fields (see Section 7.4). Thus it may have a greater range of shapes, (see Figs 12.1(c) and 13.15) and sensing devices that affect the cell structure. The cell membrane itself is now supported by permanent internal radial filaments, as well as those running underneath it. The internal filaments are contractile devices which act in given directions. The contraction is based on a system of actomyosin proteins which evolved into muscles of all kinds [Fig. 12.1(b)]. They can then be used to extend or contract sections of the cell membrane (forming so-called pseudopods) [see Fig. 12.1(a)] to allow local formation of attachment. These proteins are parts of the mechanical machinery which now acts not just at division. They also hold all vesicles in positions in the cell, so that compartments are not placed homogeneously. Later, vesicles were able to travel along filaments, called tubulins (see aside), to and from the central regions of a cell, driven by special 'motor' proteins such as kinesin and dynein.

As mentioned, the filamentous structures of the eukaryote can be activated so as to reject vesicles and release poisons to the environment. This rejection is called exocytosis. The reverse, endocytosis, is also possible—the formation of vesicles from the outer membrane—involving an ability to capture large particles, or even small organisms such as bacteria, before digesting them in other vesicles called lysozomes, which contain digestive enzymes and other chemicals. These movements, dependent on the activity of the filaments, can be looked upon as a very primitive precursor of muscular activity. For the activity, new or modified proteins are required and, as for the introduction of all the other novel features of eukaryotes, an enlarged DNA was essential. There was clearly a requirement for all these extra features of the eukaryote cell since it was very much larger than the prokaryote cell. We stress that both the introduction of vesicles and of filaments, which can be energised, are tantamount to the introduction of new variables, allowing a vast increase in the diversity of organisms using the same chemical elements.

12.5   Changes in organisation and communication

In Chapter 11 we explained how the reactions of prokaryotes were orchestrated. For efficient reproduction a coded molecule, DNA (RNA), a plan and an apparatus for interpreting the code and translating it into active protein units was required. The making of these units needed supplies of energy and material involving many reaction pathways. Moreover, the cell had to maintain osmotic and electrical balance by moving simple ions in or out of cells. To bring this diversity of activity into a co-operative system leading to reproduction, each pathway of synthesis and degradation, and of uptake and rejection, had to be interconnected in a selected feedback network of internal reactions. The communication network had three different functions: (1) transporting fragments such as –H, –CH_3, –$COCH_3$, –NH_2 and so on, between pathways; (2) carrying energy, e.g. ATP; and (3) carrying information, using messengers, from one pathway to another. Co-enzymes formed a vital part of all actions and they were also linked to feedback messages ensuring the homeostasis of the concentrations of simple ions such as H^+, Mg^{2+}, Fe^{2+} and HPO_4^{2-}. A major part, integrating both energy and material distribution, was played by phosphate compounds (see Section 11.10.2).

The nature of evolution implies that all of this organisation had to be maintained in the eukaryotes, but the development of the outer membrane as an adjustable structure in the eukaryote demanded additional signalling

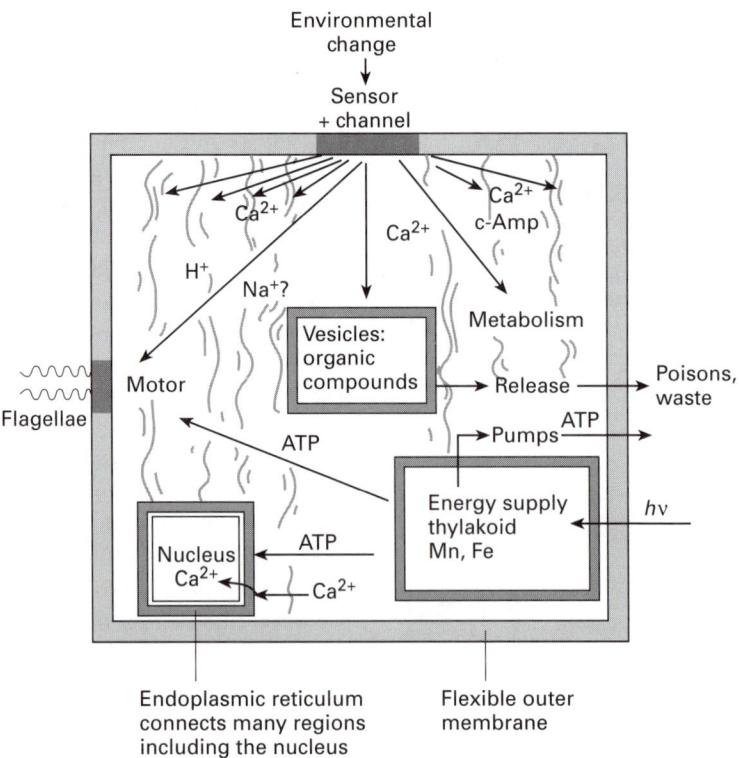

Fig. 12.5 The most primitive connection between an external event (message), filaments, vesicles, metabolism and mechanical motor reactions using calcium, energy and chemical release from different compartments. Compare the reactions of plants and animals in Chapter 13. Note the involvement of Ca^{2+} and more primitive messengers, c-AMP, and energy transfer systems, ATP.

(see Fig. 12.5) which had to extend to new signals for the synthesis of new proteins while the old organisation had to remain intact but connected to any novel features. Clearly, the two organisations of the cytoplasm and of control over cell shape by filaments had to be knitted together. Furthermore, the greater sensitivity to environmental fields made it very useful to have mechanisms of sensing the environment, which required messages to pass through the outer membrane and to be amplified internally, since the cell is now quite large.

12.6   The major new messenger: calcium ions

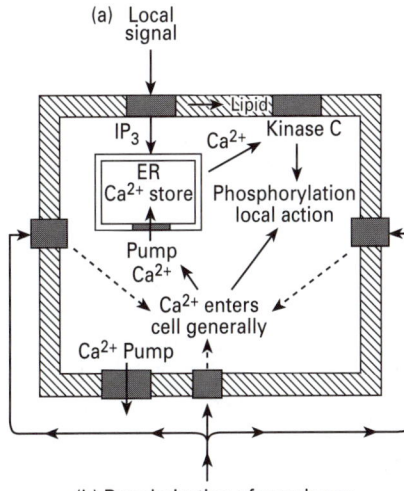

Fig. 12.6 The two kinds of calcium signal entering a cell due to: (a) a localised receptor for transmitters, see also Fig. 12.5; and (b) a depolarising pulse which is delocalised. ER, endoplasmic reticulum. The use of IP$_3$ (inositol triphosphate) as shown may have come later in evolution. The further effects of calcium are shown in Fig. 12.7.

The new vesicular compartments in the cell differed from the cytoplasm in many respects. Normally they were not provided with a condensing agent such as ATP, so that, generally speaking, they did little or no synthesis of polymers. An exception is the Golgi apparatus (see Fig. 12.8) in which polysaccharides are made using imported nucleotide triphosphates. As stated above, a major function of the vesicles, at least before oxygen was available, was the storage of chemicals and ions for subsequent use or rejection. The vesicular solutions resembled, to some degree, the external environment, since their membrane pumps acted to move ions such as Cl$^-$, Ca^{2+} and H$^+$ into them while leaving the cytoplasm richer in Mg^{2+}, K$^+$ and HPO$_4^{2-}$. They also contained selected proteins which had to be directed into particular vesicles. Many of these proteins are required to assist storage or are enzymes. By giving the vesicles pumps and channels as well as certain proteins for binding ions such as Ca^{2+}, the vesicles came to have a *buffered* content of ions very different from that of the cytoplasm. In effect, the vesicles were loaded electrolyte condensers of ions, especially H$^+$ and Ca^{2+}, relative to the current-carrying (ion-carrying) water of the cytoplasm. As such they became a very useful means of amplifying any message received at the outer surface of the cell.

As an example, consider again the scheme of a major new message transmission of eukaryotes for sensing an external event, interactive with the cell surface, and which generates a small molecule, or particularly an initial ion release (generally) into the cytoplasm (Fig. 12.6 and see section 11.10.2). This messenger molecule or ion could then go directly to the (old prokaryote) phosphate signalling system internal to the cytoplasm (Fig. 12.5) and/or, now more usefully, it could go to a receptor on the membrane of a vesicle to open a message channel for stored ions (Fig. 12.6). Thus, the message would be strongly amplified locally. Here we draw attention to the major messenger which came to be used—calcium ions. We are now turning to novel uses of chemical elements in supposedly anaerobic eukaryotes. As stated, this represents an introduction of a new variable since the calcium ion is energised in a compartment. We next look more closely at this development of the use of calcium through the use of newly evolved proteins.

12.7   Calcium ions and signalling

It is probable that the earliest organisms, prokaryotes, had to lower the calcium concentration in the cell to around 10^{-5} M from the external 10^{-3} M environment. The lowering was achieved by $2Na^+/Ca^{2+}$ exchange backed up by a Na^+ pump energised by ATP. At this level the calcium does not precipitate with 10^{-3} M phosphate or with the carbonate which is present as bicarbonate at about the same concentration (see Table 11.5 and Section 6.5). If precipitates were formed they could be retained or perhaps rejected by the cell. It might be thought that this energised gradient could have been used in an electrolytic signalling device immediately; however, the admittance of calcium down this gradient, 10^{-3} M outside and 10^{-5} M inside, would not have been a very useful triggering mechanism to alert the cell of the change of its environment for the following reason.

To act as a trigger in a cell it is realistic to assume that whatever the basal level of calcium, here 10^{-5} M, a pulsed increase of concentration of

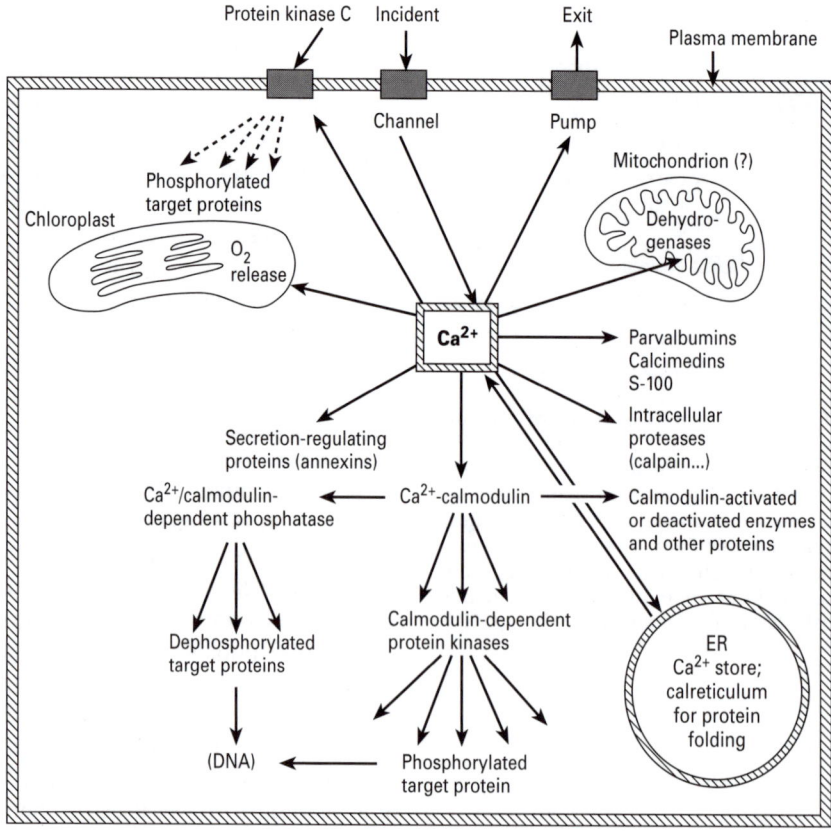

Fig. 12.7 Condensed overview of the interaction of Ca^{2+} with intracellular proteins. The Ca^{2+} can be free or released from a vesicle, e.g. the ER, endoplasmic reticulum (Figs 12.5 and 12.6). The ER surrounds the nucleus and calcium signalling to it may be differential.

at least 10-fold would be required. There are two unsatisfactory features of this possible change using the extracellular source of 10^{-3} M calcium and an intracellular concentration of 10^{-5} M. First, it would raise calcium internally close to the precipitation concentration of calcium salts and possibly cause the binding of calcium to many membrane and DNA phosphates, which are by nature anionic, and other organic anions. Secondly, it requires considerable time for a channel to let in sufficient calcium to raise the concentration from 10^{-5} M to 10^{-4} M using a 10^{-3} M external solution. A slow response is not a very valuable device for a signalling or triggering device. Now, eukaryotes are known to have not just $2Na^+/Ca^{2+}$ exchangers but ATP-driven Ca^{2+} pumps, which are certainly absent from most prokaryotes, even today. (It is likely that primitive prokaryotes did not have such pumps since they could serve little purpose.) Once these outward calcium pumps established a cytoplasmic concentration of 10^{-7} M, as is observed in eukaryotes, the value of the calcium gradient became quite different. Now the pulse of warning or excitation needed only be to a cytoplasmic level of 10^{-6} M. Of course, this requires receptors for calcium ions in cells and, in fact, wholly adequate, novel Ca-binding proteins such as annexins and calmodulins, see aside, are found in abundance in eukaryotes (Fig. 12.7) but not in prokaryotes. To create advances in organisation it was essential to increase lifetime by improving not just structure but signalling such as this, so that dangers can be avoided over prolonged periods and new advantageous environments found.

The extra Ca^{2+} ions that enter a cytoplasm at 10^{-7} M calcium cause events of several kinds, generating advances according to the nature of the eukaryote cell and the signal received (Fig. 12.7). For example:

(1) cell movement can be initiated (Fig. 12.5) (see also the functioning of the muscle in Chapter 13);

(2) cell shape can be changed, as in many simple eukaryotes (Fig. 12.5);

(3) contents of vesicles can be released to the environment, e.g. to poison or to help to capture other organisms by calcium ion manipulation of filament tension and of membranes (Fig. 12.6);

(4) The entering calcium can switch on kinases (phosphorylating enzymes) or phosphatases (dephosphorylating enzymes) and is, therefore, connected to the pre-existing (in prokaryotes) phosphate signalling system of the cytoplasm. Here the use of calcium-binding proteins is essential, (Fig. 12.7). This gives an outside/inside network to add to the inside one of prokaryotes. It can be extended so that it affects differentiation, perhaps to the extreme of causing cell death (apoptosis) after injury (see the previous account in Section 11.13 of sporulation).

(5) Energy-producing reactions can be switched on locally.

The eukaryote cell became, then, a very flexible organisation able to adjust its metabolism and energy use to its surroundings without involving much protein production, i.e. without involving DNA. Hence, when required, energy could be used in bursts in response to danger or discovery of food. Note that messengers such as calcium can act *locally* around the

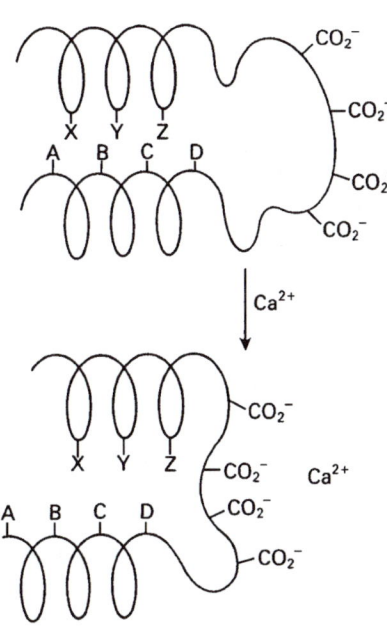

A schematic representation of calcium triggering of a subdomain (EF hand) of a calcium-binding protein. The site connects to helices which are adjusted relative to one another by the calcium binding.

Table 12.5 The changes introduced for calcium signalling

	Change	Location
1.	Compartmental concentration of Ca^{2+}	Endoplasmic reticulum (chloroplast and mitochondria). Other vesicles, Golgi
2.	Calcium channels	Membranes (cell membranes and vesicle membranes)
3.	Calcium receptors	Membranes of vesicles. Many enzymes (kinases). Filaments for contraction (calmodulins, annexins, troponin)
4.	Calcium pumps	Membranes (driven by ATP)
5.	Links to phosphate signals	Kinases and phosphatases with attached Ca^{2+} trigger proteins
6.	Calcium stores	Calsequestrin in reticula
7.	Nucleus (separation) (calcium concentration?)	Calcium signals to the nucleus from the surrounding endoplasmic reticulum (Fig. 12.5)

cell and, as stated, their messages can be amplified by the local storage of this ion held in the reticulum or vesicles close to the receptors of the environmental change. This again saves energy since not all the cell is involved. The full extent of the required changes from a primitive prokaryote to a calcium-utilising eukaryote are listed in Table 12.5. (The multitude of calcium current-based responses can be likened to switching on electric current in modern electronic equipment; see Chapter 16.)

12.8 Other activities in vesicles and organelles

Elements in vesicles

Mn^{2+}	Chloroplasts
Ni^{2+}	Hydrogenosomes Vacuoles
H^+	Lysosomes Vacuoles
SO_4^{2-}	Golgi

As well as the Ca^{2+}, H^+ and Na^+ concentrated in vesicles, other elements were predominantly exported into these compartments. Examples are manganese and nickel (see aside). Manganese came to be associated with glycoprotein formation in the Golgi apparatus, while nickel became the preferred catalytic metal in the enzyme urease, though it is uncertain if this enzyme is to be found in eukaryotic anaerobes. In the organelles (chloroplasts and mitochondria) certain elements were also strongly imported. Thus, manganese is concentrated in chloroplasts, while iron was pumped into mitochondria. (Do not forget that mitochondria could have evolved before O_2 was available.) Some tasks previously performed in the cytoplasm were now found in these organelles. A major activity was the synthesis of porphyrins, both for haem and chlorophyll. In a sense, the eukaryote was, therefore, symbiotic with (captured) prokaryotes. A major metabolic route largely using iron catalysts is the reverse citric acid cycle shown in Fig. 11.3, and this cycle may now have been held in 'captured' mitochondria (NO-using) while photosynthesis took place in the chloroplasts. Energy in the form of ATP was generated in these organelles and passed to the cytoplasm. It can be seen that the organisation of the cell had become more complex and had to be correlated in many ways. Part of the correlation between organelles and cytoplasm was achieved by calcium messages, as described, as well as by phosphate compound and substrate exchanges, but part undoubtedly involved more extensive use of novel phosphate compounds and other metal ions such as Mn^{2+} and Fe^{2+}.

12.9 The structure of the new vesicular compartments

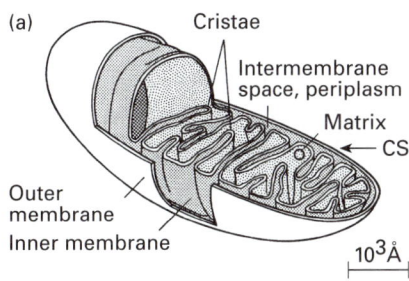

(a)

Cristae

Intermembrane space, periplasm

Matrix

← CS

Outer membrane

Inner membrane

10^3Å

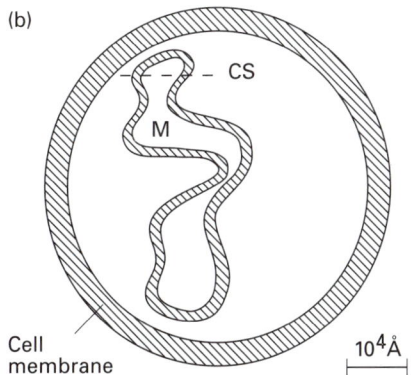

(b)

– – CS

M

Cell membrane

10^4Å

(a) Old impression of mitochondria (M) based on cross section (CS) shown also in (b) a modern view of this weaving organelle in a cell. See Perkins *et al.* (1997) J. Structural Biol. *119*, 260–272.

The endoplasmic reticulum (or ER) was recognised very quickly as being a continuous or almost continuous, thin, tube-like body which appeared to wander over all the cytoplasm turning around the nucleus (Fig. 12.4). Other vesicular bodies, such as lysosomes in which digestion occurs, are seen as very small, roughly spherical bodies, and yet others, such as vacuoles, as much larger three-dimensional vesicles, especially in plant cells. Problems have arisen with the description of mitochondria and chloroplasts. When cells are examined in cross-section these organelles appear as many separate, roughly elliptical, small bodies with an invaginated or lamellar inner membrane (Figs 12.4 and 12.7 and see Fig. 10.20). However, serial cross-sections of cells show that, in fact, many of the supposed separated small mitochondria are parts of tubes that twist and turn through large areas of the cell, see aside. Chloroplasts are known to have a similar very extensive extension to all parts of cells.

All these organelles and ER structures have large ion concentration and electric field gradients across their membranes, but a major question now arises concerning the spatial localisation of their activities. There is little doubt that local ER regions of eukaryote cells respond to calcium differentially since calcium has been tracked as it enters different regions of a cell from outside and interacts locally with different parts of the continuous endoplasmic reticulum. The response of mitochondria (assuming they existed) and chloroplasts to chemical concentration of nitric oxide, or to electric or redox potential changes, or to light-generated proton gradients, may well be similarly localised. It makes good sense for a message, a chemical substance or an energisation, to be local in a large cell, since with its localised sensors and response activities it becomes more sensitive.

12.10 Difficulties of many compartments

The first and most obvious difficulty in the construction of cells with many compartments lies in the need for more complicated synthesis of new membranes, proteins and so on. It is apparent that many vesicles requiring these syntheses arise from a special new unit in eukaryotes—the Golgi apparatus (Fig. 12.8). This apparatus puts together specific proteins (often modified—glycosylated or sulphated) from the cytoplasm and delivers them in particular vesicles to selected zones in the cell. Each type of vesicle has its own complement of proteins, and often of small molecules, as well as of free ionic elements. Obviously, all these separated activities require many extra syntheses and a considerably larger DNA, thus the eukaryote cell again has, of necessity, a structure of much greater complexity.

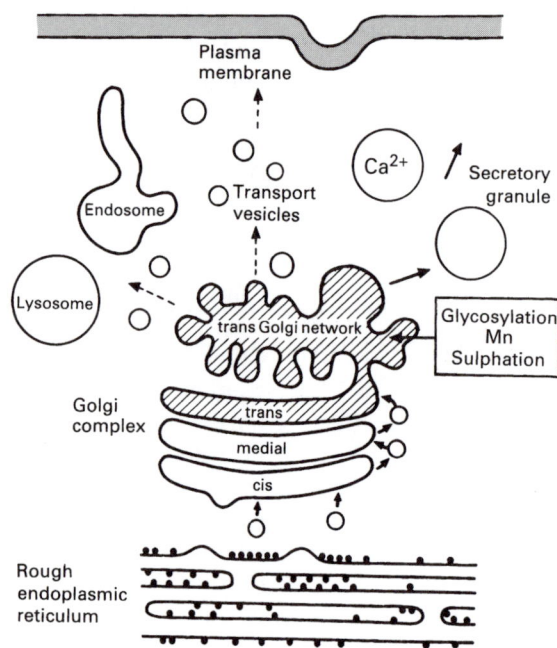

Fig. 12.8 A representation of the Golgi apparatus, which is supplied with proteins by ribosomes of the endoplasmic reticulum for processing and then transfer to specific functional compartments.

12.11   Folding and positioning of proteins

Protein processing

Synthesis (ribosome)
↓
Membrane transport
to vesicle
↓
Chaperone folding
↓
Export

Now, this transport of proteins raises quite new structural problems. Whereas in the cytoplasm of prokaryotes we can see that a folding pathway has to be found for proteins produced as linear sequences in ribosomes in the cytoplasm, a new difficulty in the eukaryote is that some of the proteins have to cross membranes. To do so they have so-called *leader sequences*, which are recognised by membrane apparatuses that then assist them in being transferred. But now they have difficulty in folding. A solution to this problem is to devise a folding machinery—special proteins called 'chaperones'—which assist folding. Some of them are calcium-bound proteins. It is now known that such steps occur in the Golgi apparatus and in the endoplasmic reticulum after the protein has been synthesised by ribosomes resting on the vesicles' membranes. Folding and positioning of proteins became a very complicated process with the evolution of eukaryotes (see aside).

12.12   The extracellular matrices of eukaryotes

Given that the outer membrane had become more flexible and that the eukaryotes were able to use calcium to relay information about their external environment to the interior, it goes without saying that their outer surfaces developed in two ways. First, they developed proteins for sensing the environment and relaying the information to the interior through new membrane channel proteins. Secondly, the character of the 'wall' around the cell came to be adaptable in some cells but open to mineralisation in others. It is likely that the outer polymers of these cells contained large polysaccharide extensions, a type of glycoprotein made in the Golgi apparatus.

12.13   Mineralisation of eukaryotes

Whereas mineralisation of prokaryote colonies seems to occur in a rough matrix in such bodies as stromatolites, made from silicious material and calcium carbonate in an almost amorphous material, the development of vesicles with high calcium content and the cell's ability to put glycoproteins

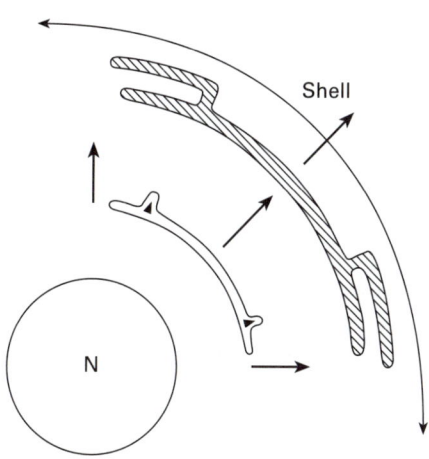

Development of coccolith, see Fig. 12.9(b). N, nucleus

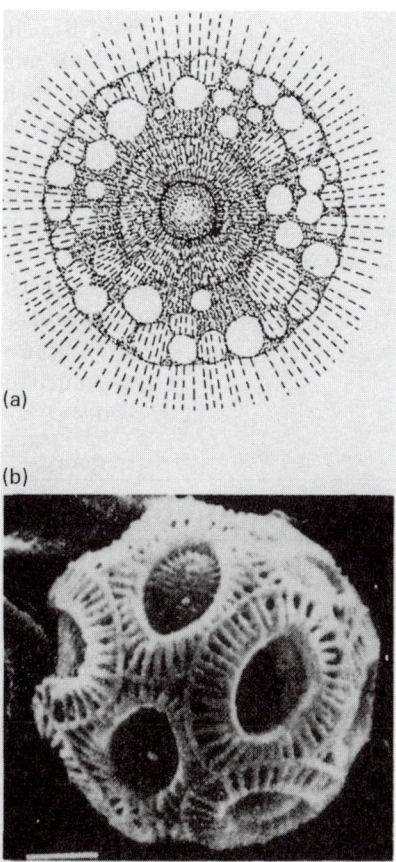

(a)

(b)

Fig. 12.9 (a) An example of a radiolarian. (b) The structure of the complete shell of a coccolithophorid, *Emiliania huxleyi* (scale bar, 1 μm). They are both aerobic organisms but serve to illustrate outer shell constructions.

into the vesicles allowed eukaryotes to make highly organised spicules, or other crystalline units, and even shells (Fig. 12.9 and see Figs 10.21 and 14.7). The shells are made of units, e.g. coccoliths or opals, often crystalline calcium carbonate (animal cells) or amorphous silica (plant cells), which can be exported from the vesicles into an extracellular matrix by exocytosis. The whole forms an wonderful protective layer or outer shell through which food is sieved (see Fig. 12.9).

12.14 Lifetimes of eukaryotes

A little thought will show that the making of a eukaryote cell will take much longer than that of a prokaryote. The doubling time of bacteria can be much less than one hour while that of a eukaryote single cell organism may be many hours or days. There is obviously a balance of advantage between a fast reproduction rate and a greater survival chance of highly effective individuals, and the result is a co-existence of bacteria (prokaryotes) and eukaryotes. The balance is only partially competitive since it pays the eukaryote (which can consume bacteria) to allow bacteria to make some of the chemicals essential for its form of life. For example, we have listed some co-enzymes, the most essential small molecules, in Table 11.6. Not all of them are made by most eukaryotes, so that they depend on digestion of bacteria for their existence. The big surprises in the evolutionary process are that, in general, eukaryote cells are unable to be self-sufficient through their own gene products (see aside), even in energy capture, relying on captured bacteria which became organelles (chloroplasts and mitochondria), nor are they able to convert dinitrogen gas into ammonia. Perhaps they were not even able to move around before they engulfed a bacterial motor. We have, therefore, to see evolution as a developing *ecosystem* in which mutual assistance is at least as strong as competition. Individual competition in the strictest sense is only between similar or even very similar organisms, e.g. higher animals.

Chemical dependency on lower organisms

(1) Essential co-enzymes
(2) Energy (organelles)
(3) Nitrogen (plants)
(4) Minerals (animals)

12.15 Environmental change

We do not know the environment in which life began. There is the suggestion that the ocean was somewhat acidic, pH < 6, i.e. considerably lower than the present pH = 8.6 (see Section 9.3). (Notice that the interior of cells is at about pH = 7.4.) In this chapter we are also assuming that the sea remained essentially highly reducing over the first one billion years, but it is likely that as volcanic activity fell supplies of energy from beneath the crust, say iron sulphides, diminished and photosynthesis, using $H_2S \rightarrow Sn$, became the dominant energy source (Table 12.6 and see Table 11.9). This could well have been one cause of the initial development of thylakoid vesicles (see Fig. 10.20). Another possibility is that life came to depend more on nitric oxide (NO), which could explain incorporation of mitochondria before the advent of dioxygen. Of course, the temperature fell

Table 12.6 Stages in the environment of early life

I.   Initial environment of life (4×10^9 years ago)
     (1) Acidic pH < 6
     (2) Highly reduced due to H_2, CH_4, NH_3 and eruption of minerals such as sulphides into the ocean.

II. First environmental changes (say 3.5×10^9 years ago)
     (1) Neutralisation of acidity to about pH = 7 or higher.
         Lowering of calcium and other hard metals.
     (2) Reduced eruptions and loss of H_2 and CH_4 from the atmosphere.

III. Advent of dioxygen (see Chapter 13)

Changes of element use

Oxidation	Fe only → Fe, Cu
Superoxide dismutation	Fe, Mn → Cu, Zn
Nitrate reduction	W, Mo → Mo
Methylation	Co(B_{12}) → Zn
RNA/DNA synthesis	Fe, Co → Fe
Hydrogenation	Ni, Fe → (Lost)

from close to 100 °C to around 40 °C, and then lower. How did these changes affect element availability and how did this availability affect organisms?

The progressive change in acidity to a more alkaline solution (due to the erosion and dissolution of oxides and carbonates, basic rocks) would have increased the availability of certain elements, e.g. molybdenum, and decreased that of others, e.g. calcium (easier precipitation as carbonate and phosphate). It is possible that this explains the virtual disappearance of tungsten (W) chemistry and its replacement by Mo in eukaryotes. It is certainly the case that any decrease in external calcium with increasing alkalinity would help the lowering of internal calcium and so enable its use as a messenger (see Section 12.7 above). As we have explained, the use of vesicles effectively lowered the cytoplasmic concentration of not only calcium but also of manganese, concentrated in vesicles, particularly thylakoids. The higher manganese contents of thylakoids may well have initiated O_2 release since the oxidation of water in photosynthesis involves a manganese-dependent enzyme. Further changes may have occurred, for example nickel may already have largely disappeared from use as dihydrogen and methane gas became unavailable [although nickel has persisted in the enzyme urease], and cobalt in the co-enzyme B_{12} may well have replaced Fe/S proteins in some enzymes such as ribonucelotide reductase, the enzyme for making DNA units from RNA units. However, no clear identification of these steps (see aside) in an evolutionary sequence has been established. As stated in Section 10.18, by postulating a relationship between evolution and environmental change we introduce two possibilities: (1) evolution depended on random mutation of DNA which eventually proved to be an advantage in a specific environment; or (2) evolution depended on the effect of the environment on DNA, causing stresses and eventually 'directing' mutation. We return to this point later.

12.16   The variables of eukaryotes

While describing the variables of bacteria we discussed only the availability of chemical elements and the supply of energy (Fig. 12.10). With these variables (plus temperature and pressure) we discussed the speciation of prokaryotes in terms of a single aqueous steady-state compartment, the

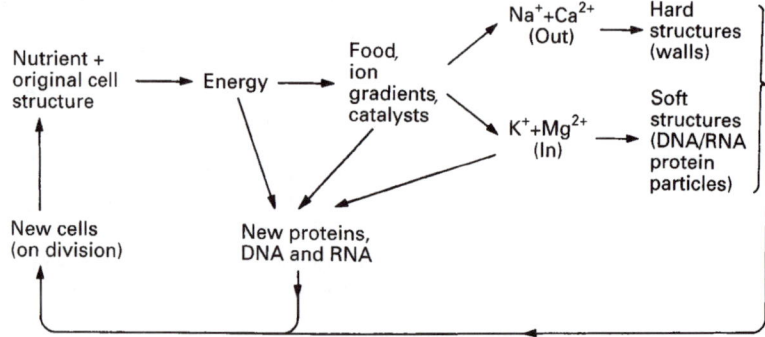

Fig. 12.10 A diagram showing the partitioning of selected elements inside and outside a cell, all of which are essential for the cell's stability. For eukaryotes 'outside' a cell includes vesicles and reticula.

cytoplasm. Now, eukaryotes have more than one compartment. If these compartments were in exchange equilibrium then they would not introduce new variables or indeed new advantages. However, they are only in a steady-state relationship with the cytoplasmic compartment, and as such can be differentially energised. There are, then, new variables for each new compartment: its separated rate of energy use and its element content. The increase in variables in this case parallels the increase we described for non-living systems when isolated compartments, as opposed to balanced phases, developed (see Section 7.12). A further increase in variables results from the way energy is applied, not just across a membrane but to the shape it has. Here it is the new filamentous structure which is involved since, once again, this is a new energised system.

Furthermore, the simple description of the environment as a source of energy and material must also be changed since the environment now includes a greatly increased multiplicity of different organisms which compete for materials (and energy) with one another. To protect any organism a more complex system of sensors and message transmission is, therefore, required. In a sense we can liken this change of variables to the introduction of fields in Chapter 7. We may suppose that the increase in variables in different organisms would lead to survival advantages of different organisations in the different environmental zones, but simultaneously internal conditions become more stringent and the range of viability over chemical composition decreased (see Figs 10.4 and 11.14). This may well apply to temperature and pressure too, so that new species of many kinds could arise which are limited by the environment to a greater degree than was true of prokaryotes.

12.17 Speciation

When we discussed prokaryotes we had difficulty in defining species since their DNA exchanges between organisms relatively readily. Our best description allowed for a considerable variation in chemical element composition or in energy utilisation (and possibly temperature and pressure). Thus, there was a relationship between viability and survival against composition, with broad maxima and minima in the 'landscape' plots (Fig. 11.14). It may be that true maxima did not exist, so that there was an almost continuous gradation between "species". Notice that DNA by itself says nothing about these landscape plots, although a particular DNA is associated with each of the maxima, of course.

Turning to eukaryotes, the plots of viability against variables are likely to be much more steeply changing. The more pronounced locality of maxima results from the need to refine the systems due to the multiplicity of cross-connections. This is a major problem of complexity of organisms—the more complex, the greater the refinement necessary and the clearer the identity, but the narrower the range of viability (see Fig. 13.15). The larger DNA is now more closely differentiated, so that the machinery, the shape of the organism and the DNA define the organism clearly. A consequence of the constraints on variation required to make a viable eukaryote machinery is that the organism has a closely defined speciation. There is an analogy here to the manner in which increased co-operativity between chemical components led to compounds of narrowly fixed composition while low co-operativity allowed broad zones of composition within a phase, (see Section 5.9).

12.18   Summary

The Eukaryotes

(1) Internal compartments
(2) Internal filaments
(3) Calcium became the new messenger to and from the cytoplasm to and from the internal compartments and the external environment

At this stage in evolution, when there was still virtually no dioxygen in the air, the major change we are postulating is a discovery of new compartmental structures, both of membranes and filaments, and particularly a completely new use for one element, calcium, which had been treated very largely as a poison by *primitive* prokaryotes, see aside. (N.B. We do not exclude the use of calcium signalling in some *modern* bacteria.) It may well be that this development allowed essential control quite generally over events separated in the cytoplasm, in vesicles, or due to external factors, and the linking of them back to the cytoplasm in an energy-dependent manner. However, calcium changes could also have become involved in a part of the cell cycle at this time. The value of calcium as a carrier of current and a trigger of events is not unlike the value of the electron in man's circuits and in his many appliances. As repeatedly stated, all these variables, composition and compartmentalisation, are increased by the variety of ways of applying energy to them.

In the description of change at this stage (anaerobic eukaryotes) we can imagine that chloroplasts had been incorporated and that light was used as a source of ATP. Today we know that calcium levels affect even this photochemistry. It was this activity, already present in the blue–green algae (or cyanobacteria) (which are prokaryotes) that increasingly led to the production of dioxygen. It was the greed for hydrogen (from H_2O rather than H_2S) that led to this dioxygen production, which was, of course, a multi-headed disaster for anaerobes, but eventually resulted in the evolution of multi-

Table 12.7 Size distribution of proteins of *Escherichia coli* and *Saccharomyces cerevisiae*

	Escherichia coli (prokaryote)	*Saccharomyces cerevisiae* (eukaryote)
Total number of proteins	4277	5809
Soluble	2787	4116
Transmembrane	1490	1693
Average length of all proteins (amino acids)	317	496
Soluble	302	475
Transmembrane	346	548
Number of proteins longer than 500 residues	588	2208
Soluble	340	1445
Transmembrane	248	763

Table modified from Netzer, W. J. and Hartl, F. V. (1998). *Trends in Biochemical Sciences*, **23**, 68–73.

cellular organisms based on *aerobic* eukaryotes (see Chapter 13). To our knowledge, there are no real multicellular prokaryotes or multicellular *anaerobic* organisms, only colonies.

In the next chapters we shall consider how environmental evolution has affected the variables in the new organisms. We shall find again that the restriction on the composition and energy variables leads to an even closer definition of species, while we also observe an enlarged DNA (see Table 12.7) and a larger specific shape. However, the number of possible species will increase as the number of different available elements changes, as the energy flow changes and as the number of compartments changes.

13

The coming of dioxygen: unicellular organisms

... In those days, again, many species must have died out altogether and failed to reproduce their kind. Every species that you now see drawing the breath of the world survived either by cunning or by prowess or by speed ...

Lucretius, *De rerum Natura* (99–55 BC), (transl. R. E. Lathan) (Penguin Books, 1951, p. 196).

13.1   Introduction

The changes in living organisms, especially multicellular organisms, since the strong build up of dioxygen from about 10% of the present atmospheric level, somewhere between a billion and seven hundred million years ago, were relatively rapid (Figs 13.1 and 13.2). These changes were preceded by the evolution of single cell *aerobic* bacteria and eukaryotes, from one to two billion years earlier, which started initially at much lower dioxygen pressures. The object of this chapter is to give the development of single cell, aerobic, living organisms a logical systematic *chemical* background based at first on the effects of low levels of dioxygen. We shall assume that the basic structure of eukaryotic cells changed but slightly on

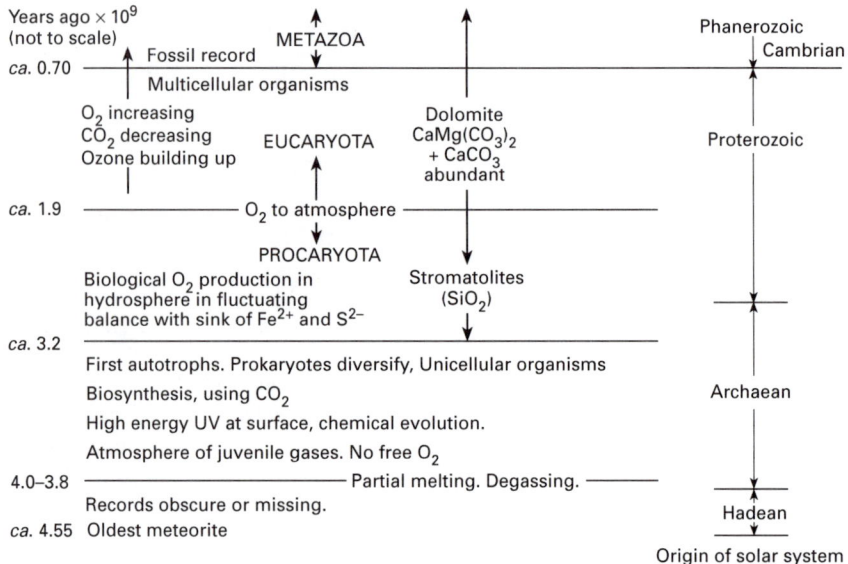

Fig. 13.1 Biospheric, lithospheric and atmospheric evolution the primitive Earth.

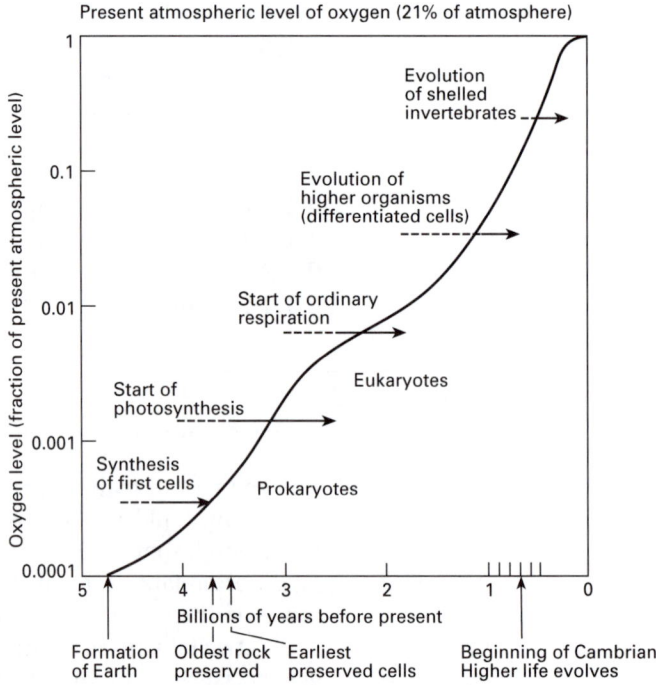

Fig. 13.2 One hypothesis of the evolution of oxygen in the atmosphere in relation to the origin of life and the evolution of higher organisms. (After ref. 81 in Further reading).

going from the anaerobic to this aerobic atmosphere (see Figs 13.1 and 13.2). We shall consider that it was not just random shuffling of DNA that decided the evolutionary change but that the new chemistry, which became possible through changes of the environment, forced a pattern of change on cells and generated local changes in the coded molecule DNA due to localised random mutational searching. We shall show that the emergence of dioxygen caused major environmental changes over a long period of time (seen in geological records) but it is unfortunate that our inevitable reliance on fossil records means that we know very little about the gradual introduction of dioxygen-using species over the one billion years before some one billion years ago. We can only search, therefore, for a probable reason for change of organisms coincident with and, we shall say, dependent upon, the ever-increasing oxidation of many elements in the environment (see aside). In effect, this is an examination of the possible manipulation of organisational variations by chemical compositional variables. To explore this evolution we reintroduce the changes in chemical element availability.

Redox potentials after dioxygen in atmosphere

Couple	Potential range (mV)
Fe^{3+}/Fe^{2+}	−500 to +400
FeO_2^+/Fe^{3+}	~ +800
Cu^{2+}/Cu^+	+200 to +700
Mn^{3+}/Mn^{2+}	> +300

13.2   Changes in redox chemistry

We have shown in Chapters 6 and 9 that a switch on Earth from a methane, ammonia and hydrogen sulphide (H_2S) atmosphere to a dioxygen atmosphere while carbon dioxide levels fell, changed the chemistry of the available elements gradually but dramatically and introduced several new elements into the sea, hence making them available also. In essence, as shown in Chapter 9, this change altered the surface layers of Earth, that is, its atmosphere, the sea and the top layers of the crust. In Fig. 13.3 and Table 13.1, we record again the major changes of non-metals and metals. The changes were extremely injurious to anaerobic life, of course, not just because they removed easily available sources of elements such as nitrogen, sulphur and iron, but also because they slowly introduced much higher concentrations of several metal elements, such as zinc and copper, previously held in insoluble sulphide minerals. For anaerobic life, higher concentrations of such metals were and are very deleterious. There was

Table 13.1 Major element availability changes with the advent of dioxygen*

Non-metals

$CO_2\downarrow$	$H_2\downarrow$ $N/O\uparrow$	$O_2\uparrow$ $S/O\uparrow$ $Se/O\uparrow$	$Cl^-(\uparrow)$ $Br^-(\uparrow)$ $I^-(\uparrow)$			

Metals

$V/O\uparrow$		–	$Fe/O\downarrow$	–	$Cu\uparrow$	$Zn\uparrow$
–	$Mo/O\uparrow$					

* $\uparrow$ Indicates increases in availability. The formulation $N/O\uparrow$ indicates an increased presence of nitrogen oxides (CH_4 and NH_3 disappeared). Note that S and Se changed from base anions to oxyanions. ($\uparrow$) for the halides represents changes in reactivity only to give covalently bound halogens.

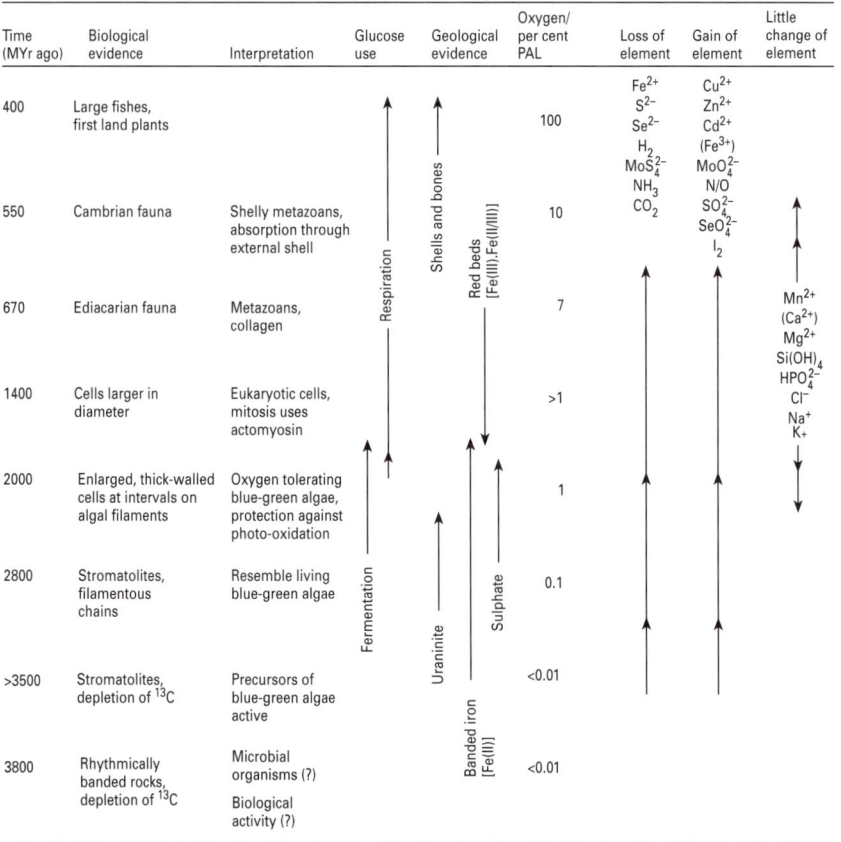

Time (MYr ago)	Biological evidence	Interpretation	Glucose use	Geological evidence	Oxygen/ per cent PAL	Loss of element	Gain of element	Little change of element
400	Large fishes, first land plants				100	Fe²⁺ S²⁻ Se²⁻ H₂ MoS₄²⁻ NH₃ CO₂	Cu²⁺ Zn²⁺ Cd²⁺ (Fe³⁺) MoO₄²⁻ N/O SO₄²⁻ SeO₄²⁻ I₂	
550	Cambrian fauna	Shelly metazoans, absorption through external shell			10			
670	Ediacarian fauna	Metazoans, collagen			7			Mn²⁺ (Ca²⁺) Mg²⁺ Si(OH)₄ HPO₄²⁻ Cl⁻ Na⁺ K⁺
1400	Cells larger in diameter	Eukaryotic cells, mitosis uses actomyosin			>1			
2000	Enlarged, thick-walled cells at intervals on algal filaments	Oxygen tolerating blue-green algae, protection against photo-oxidation			1			
2800	Stromatolites, filamentous chains	Resemble living blue-green algae			0.1			
>3500	Stromatolites, depletion of ¹³C	Precursors of blue-green algae active			<0.01			
3800	Rhythmically banded rocks, depletion of ¹³C	Microbial organisms (?) Biological activity (?)			<0.01			

Fig. 13.3 Evolution of the biosphere and the atmosphere—a summary of the biological and geological evidence that suggests how oxygen levels in the atmosphere may have progressed towards their present-day values and how the availability of chemical elements changed. PAL = present atmosphere levels. [Based on Cloud, P. (1983). *Scientific American*, **249** (3), 132.]

also a change in the speciation of many elements, such as iron, selenium and molybdenum, so that new methods of scavenging for, incorporating and using them had to be devised. Living organisms still demanded a local reducing atmosphere in the cytoplasm of every cell, whence the change to an oxidising atmosphere also required the immediate development of protection of the cytoplasm. As we have stressed in Chapter 12, on the change from prokaryote to eukaryote organisms we see the pressure for change exercised upon life by the new environmental chemicals that are novel to life and therefore poisonous, as the switching

$$\text{Poison} \xrightarrow{\text{demands}} \text{Protection} \xrightarrow{\text{becomes}} \text{(Signal)} \xrightarrow{\text{becomes}} \text{Incorporated use}$$

Our earlier illustration in Chapter 11 concerned the switch from poison to use for the elements calcium (mostly) and manganese. In this chapter we shall be particularly interested in the decrease of iron and sulphide and the introduction of increased levels of the elements copper and zinc, apart from dioxygen itself. *The changes of life's organic chemistry, we shall state, were totally dependent upon these changes of inorganic elements.*

13.3   The photochemistry of water

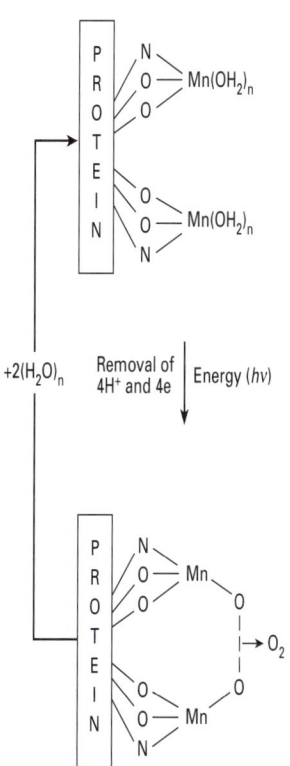

The switch from the use of H_2S to that of H_2O (which was responsible for dioxygen release) as a source of hydrogen depended on the skilful manipulation of the chemistry of manganese in cyanobacteria (blue–green algae). Manganese has the special quality amongst the transition metal ions that it is held available in the divalent ionic state, Mn^{2+}, in reducing conditions, and is not, therefore, a readily usable redox catalyst. It has low affinity for sulphide and, as Mn^{2+}, is held by side-chains of proteins using O– or N–, not S–, donors. Due to this limited co-ordination chemistry and the size of this cation it can be selectively incorporated into special proteins in cells. In this situation it can be energised locally to higher oxidation states, Mn^{3+} or Mn^{4+}, but this requires considerable energy in the removal of electrons. The fortuitous discovery by some prokaryote cells that light could bring about this energised condition of manganese changed life on Earth. The scheme is, in essence, simple.

This photochemical decomposition of water is possible using other elements such as iron, but within the context of early cellular life such reactions were probably prevented by the presence of sulphide, which binds to and interacts strongly with most available transition elements including iron but, as stated, not manganese. Moreover, iron in early life was used in reductive systems.

[As a further pointer to the chemistry of manganese which became available in an oxidising atmosphere, note that manganese as Mn^{2+} is readily oxidised to Mn^{4+} (MnO_2) in alkaline aerobic conditions, so that further new oxidation chemistry of manganese, not photochemically driven, became available once the dioxygen concentrations of the atmosphere increased.]

13.4   The possible new aerobic biological chemistry in prokaryotes

We shall start the discussion of aerobic life with an examination of the effect of dioxygen on *prokaryotes*. Dioxygen is a powerful oxidising agent, but it is not very soluble in water, neither is it very aggressive. On reduction towards water, however, it passes through stages of increasingly aggressive power—superoxide (O_2^-), hydrogen peroxide (H_2O_2) and hydroxyl radical ($OH^{\bullet}$). The intermediate chemicals, $O_2^{\bullet-}$ and H_2O_2, are soluble in water. They are produced by reaction of dioxygen with many reactive reduced bio-organic species, such as dihydroflavins and quinols, as well as by reaction with many reduced transition metal compounds, especially iron sulphides, all of which were found in anaerobes (see Chapter 12) and some of which were present in their environment. Thus $O_2^{\bullet-}$ and H_2O_2 would have been the first great threat to anaerobic life as dioxygen built up in the air. What was required for protection was, therefore, a catalytic means of removing these intermediates so that dioxygen cycled back to water (futile, even wasteful, cycling). We then have

$$H_2O \xrightarrow{\quad h\nu \quad} H \text{ for reduction + excreted } O_2$$

$$\text{some } O_2 \rightarrow O_2^{\bullet-} + H_2O_2 \text{ (poisons)}$$

$$\text{but } \quad O_2^{\bullet-} \rightarrow O_2 \uparrow + H_2O_2, \text{ catalysed by the enzyme superoxide}$$
$$\text{dismutase (protection)}$$

$$\text{and } \quad H_2O_2 \rightarrow O_2 \uparrow + H_2O, \text{ catalysed by the enzyme catalase}$$
$$\text{(protection)}$$

Of course, this protection did not prevent the build up of O_2 since the incorporation of H from H_2O and the generation of O_2 was much greater than the destruction through reduced side-products, $O_2^{\bullet-}$ and H_2O_2. The presence of the above enzymes, catalase and superoxide dismutase, is found to this day in all O_2-tolerant prokaryotes, even when they can only tolerate very low O_2 pressures. They are either manganese or iron enzymes, where the ability of both the elements to be involved in redox cycles is put to use. As stated, even organisms appearing early in evolution, which were only able to survive low pressures of dioxygen and were thought at one time to be strict anaerobes, e.g. sulphate bacteria, have these detoxifying enzymes. (Some modern aerobic prokaryotes also have a copper–zinc superoxide dismutase which became the common detoxifying enzyme of eukaryotes.)

Now, as noted in Fig. 13.4, sulphide and ferrous iron were the first reducing chemicals to be removed from the sea by dioxygen, being oxidised to sulphate and ferric ion, which precipitates as $Fe(OH)_3$. Major requirements for anaerobes after gaining protection from H_2O_2 and $O_2^{\bullet-}$ were, therefore, ways of reducing sulphate to obtain sulphur (and so gaining energy) and of capturing iron from Fe^{3+} sources.

The capture of sulphur from sulphate used and still uses in aerobic prokaryotes a modified enzyme of the strict anaerobic organisms—a molybdenum enzyme which is still useful even in eukaryotes for the reaction

$$CH_3COOH \rightleftarrows CH_3CHO + O \text{ and } O \rightarrow 2H_2O$$

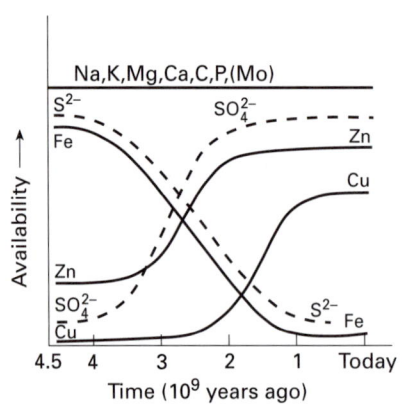

Fig. 13.4 The changing availability of some elements with time and as redox potentials rose due to dioxygen pressure increases. Note that it is solubility as well as standard redox potential that control availability. N.B. $NH_3 \rightarrow N_2$ as $S^{2-} \rightarrow SO_4^{2-}$.

namely organic acid reductases or (the reverse) aldehyde oxidases. This *flavin-using* enzyme became modified to reduce sulphate and later nitrate.

$$SO_4^{2-} \rightleftarrows SO_3^{2-} + O \text{ and } O \rightarrow H_2O$$

$$NO_3^- \rightleftarrows NO_2^- + O \text{ and } O \rightarrow H_2O$$

The O atom is carried on the molybdenum in all cases. [Note that molybdenum is very useful in oxygen atom transfer reactions at low redox potentials due to the almost equal stability of oxidation states IV, V and VI, which are easily interconnected (see Section 6.12.2).] Sulphite, SO_3^{2-}, was then reduced to sulphide on a haem/Fe/S combined catalyst.

Ferric iron was also captured by using the reducing power of the cytoplasm once bacteria had ways of manipulating extracellular iron chemistry. One simple device was for surface enzymes of the outer membrane to reduce the iron, now Fe^{3+} in the environment, and then to pump in the Fe^{2+} formed. This is a relatively small adjustment of the activity of anaerobic prokaryotes.

Iron, as Fe^{3+}, was later captured by *aerobic* bacteria (and some eukaryotes) by their sending out of chelating organic compounds from cells to scavenge for the ferric iron in their environment (Fig. 13.5). These organic compounds could be easily moved into cells and the iron then reduced to give Fe^{2+}. However, this chemistry required the synthesis of the chelating agent by new enzymes generated by the DNA and its machinery, a process which was only needed when the organism ran short of iron. The message to the DNA demanding production of the chelating agent was the lack of iron, while the presence of iron shut down the message. The regulating protein messenger, a transcription factor acting at the DNA level, is called FUR, ferric uptake regulatory protein. Figure 13.5 shows the way in which it operates. FUR also controls the production of superoxide dismutase, so that it controls protection from dioxygen side-products while allowing production of enzymes for the synthesis of chelating agents for iron uptake. The enzymes for production of the chelating agents often require molecular oxygen (see below). [Note that the gene which controls production of FUR is cyclic-AMP dependent, so that messages for iron metabolism were connected to those related to phosphate controls (see Section 11.11).]

As stated above, the superoxide dismutases which appeared first are relatively simple iron or manganese enzymes. They are still found in all prokaryotes and organelles but not in (aerobic) eukaryotes, where they have been replaced by copper/zinc superoxide dismutases. The catalases are haem enzymes which again do not use new metal ions or co-enzymes but employ an old co-enzyme (haem is common in anaerobes) in a new way. (There are some examples in bacteria of manganese-containing catalases, but notice that the earliest forms of all the systems we have mentioned do not use copper in any of these reactions.)

A further damaging effect of the presence of dioxygen was the removal of available nitrogen, i.e. of ammonia, and its replacement by the inert, insoluble gas, dinitrogen. It may be that only at this stage of evolution was a new enzyme needed for the reduction $N_2 \rightarrow NH_3$ (see Section 11.7). This

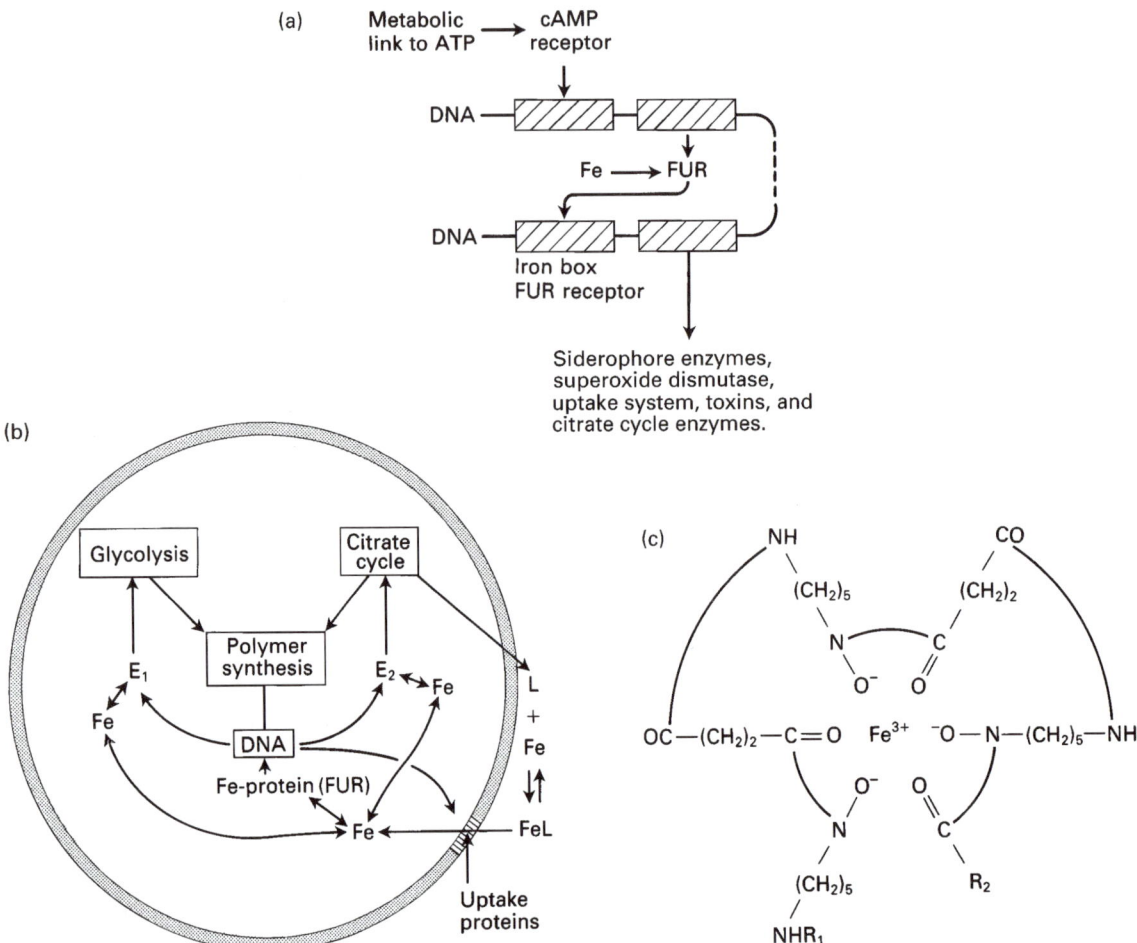

Fig. 13.5 (a) The protein FUR regulates the iron-box gene so as to produce the range of proteins shown. FUR itself is produced under the control of a gene regulated by c-AMP. Thus phosphate and iron are connected through the genes as well as in metabolism. (b) The scavenger system for iron in bacteria. A siderophone scavenger chelate, L, is shown in (c).

enzyme, nitrogenase, is extremely complicated. In effect, apart from the use of iron/sulphur centres, it requires a novel molybdenum/iron or a vanadium/iron, mixed-metal sulphur centre (Fig. 13.6) and much energy, ATP. Its production is under the control of a multiplicity of genes, some for synthesis of the novel co-factor and some related to the levels of carbon and nitrogen metabolites (Fig. 11.10). Furthermore, the enzyme requires to be protected from dioxygen (Section 11.10.1). Curiously, it is only found in prokaryotes. Thus eukaryotes came to depend on bacteria, not only internally (as modified organelles) but also externally as symbionts for supplies of nitrogen. This symbiosis is a major feature of evolution, as important as competition between organisms.

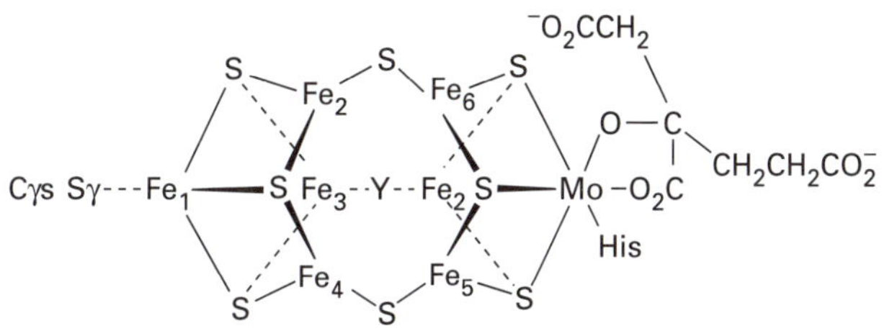

Fig. 13.6 The iron–molybdenum–sulphur cluster of nitrogenase (see Fig. 11.8). [Structure taken from Kim, J. and Rees, D. C. (1992). *Science*, **257**, 1677–81.]

13.4.1 The first direct uses of dioxygen in prokaryotes

While we consider that at first dioxygen was treated as a poison and protection from it arose through the arrival of detoxification, by removal of its even more dangerous products, $O_2^{\cdot-}$ and H_2O_2, the very fact that it damaged certain reaction sites of enzymes would have led to the increased production of these very enzymes. This is a direct consequence of feedback balance which maintains cell homeostasis. The most vulnerable sites do not so much bind O_2 as react with it and are therefore very sensitive to this chemical at very low levels. They were likely to have been Fe/S centres (see Figs 13.6 and 11.7) and flavins. While Fe/S centres oxidised in this way became the source of signals of the presence of O_2, the so-called FNR proteins (see below), and later led to adjustment of metabolism to utilise dioxygen (see Section 13.7), the flavins, which are non-diffusible, evolved so as to be able to use O_2 in enzymes called flavin oxidases. A list is given in Table 13.2.

Outstandingly, one of these new flavin enzymes is the squalene mono-oxygenase which generates cholesterol (Fig. 13.7). The precursor enzymes of the mono-oxygenase carry out non-oxidative ring closure only. The initial production of cholesterol could have been as a protection for the flavin enzyme against dioxygen, but by altering the utility of membranes cholesterol could have assisted a major change in evolution, namely the transition to eukaryotes, still following the switching scheme

Poison → Protection → Messenger or useful metabolite

Table 13.2 Flavin mono-oxygenases

Enzymes	Function
Phenol mono-oxygenase	Detoxification
Amine mono-oxygenase	Detoxification
Squalene mono-oxygenase	Cholesterol synthesis
Aldehyde oxidases	Production of acids
Sulphite oxidases	Production of sulphate
Amino-*N*-hydroxylase	Production of siderophores

N.B. They are often associated with Fe/S, not haem, electron transfer centres.

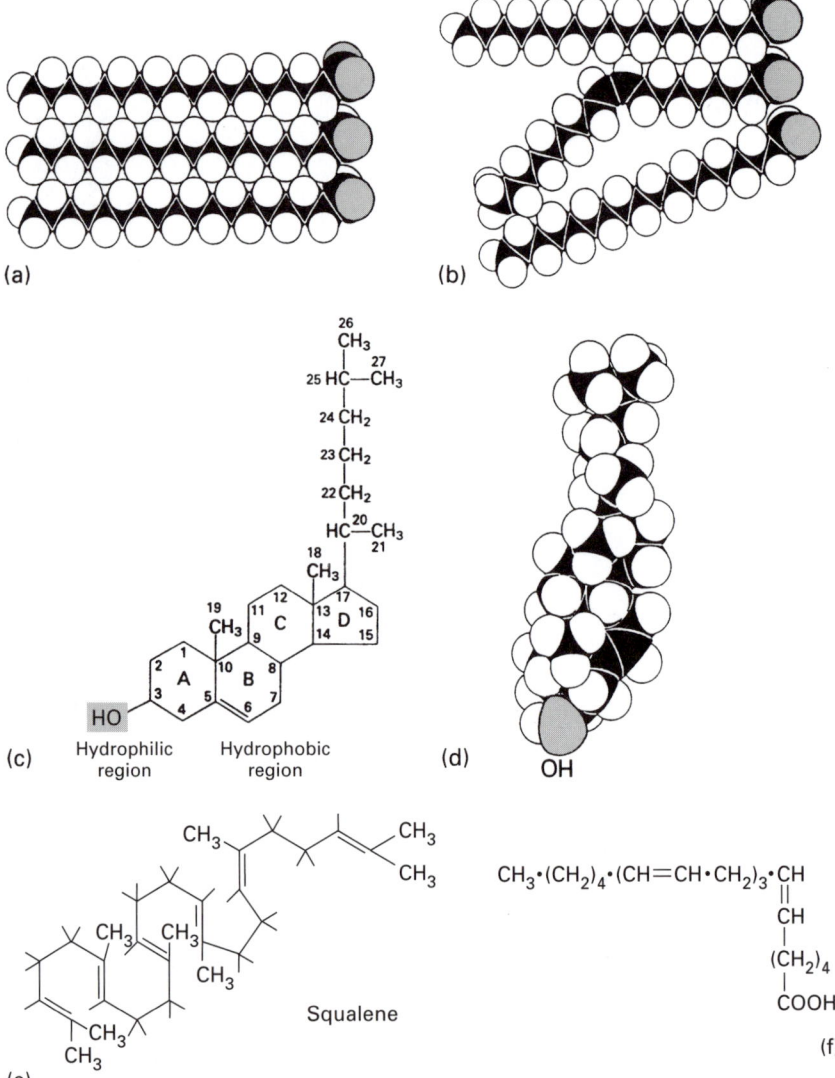

Fig. 13.7 (a) The highly ordered packing of fatty acid chains is disrupted by the presence of *cis* double bonds or by cholesterol. These space-filling models show the packing of (a) three molecules of stearate (C_{18}, saturated) and (b) a molecule of oleate (C_{18}, unsaturated) between two molecules of stearate. (c) Formula of cholesterol. (d) Space-filling model of cholesterol. (e) Formula of squalene. (f) Arachidonic acid, precursor of the prostaglandins.

This is an alternative cause of the evolutionary sequence to that given in Chapter 12 (see Table 13.1 and see Section 13.8). Other flavin oxidases are used in detoxification, i.e. protection.

We then see the appearance of dioxygen as acting as a stress on flavin-based reductive enzymes. The stress is relieved in part by removal of the products $O_2^{\cdot-}$ and H_2O_2 and in part by converting flavin into a co-factor for useful oxidation using, perhaps, H_2O_2 and then O_2.

During the development of these new or modified ways of handling elements, many of which were already required in considerable quantities in anaerobic life for synthesising organic compounds, bacteria learnt to be flexible, in that many of them gained the faculty of switching between anaerobic and aerobic metabolism. This switch requires a message to the DNA leading to the production of new sets of enzymes according to circumstances; the messenger has been recognised, as mentioned above, as the second very sensitive centre to O_2 damage, an iron/sulphur protein called FNR. [It is called FNR (ferrous nitrate reductase) due to the first observed feature of its activity.] It is not known precisely how FNR switches the metabolism, but it is thought that an iron/sulphur centre is responsible through redox changes which allow dissociation of iron, or of an Fe/S cluster, from its protein, where the oxidised condition switches on dioxygen-dependent metabolism. This adaptability is present in some eukaryote species but, as we stressed in Chapter 12, the complexity of eukaryotes limits their variability. Note that FNR senses dioxygen but also the levels of iron and sulphur.

There were, of course, advantages in the presence of dioxygen. A major one was that the reaction of dioxygen with reduced organic matter is a very fine source of energy. Thus, the oxidation of the fragment $CH_3CO–$ (acetyl) in the citric acid cycle, reversed from that in anaerobes, followed by an electron transfer chain (see references 97 and 98) gave rise to much greater amounts of energy than glycolysis

$$CH_3COOH + 2O_2 \rightleftarrows 2CO_2 + 2H_2O + 36ATP \text{ (oxidative phosphorylation)}$$

$$C_6H_{12}O_6 \rightleftarrows CH_3COCOOH + 2ATP \text{ (glycolysis)}$$

The first bacteria to carry out this reaction sequence may have been the result of a development of their NO-using energy capture systems (see Sections 11.5 and 12.15). (The bacteria able to carry out the use of dioxygen as an energy source became the mitochondria of eukaryotes.) Provided that some organisms, such as algae, generated dioxygen using light, others (as well as themselves) could just use the rejected dioxygen. It was this advance which paved the way for the enzymatic apparatus giving sources of energy in today's plant and animal life.

Further advantages developed slowly, undoubtedly, and a variety of new organic chemicals were produced using C/H/N intermediates by reaction with dioxygen, hydrogen peroxide or superoxide. One type of reaction was the introduction of oxygen atoms in aromatic compounds, e.g. benzene which provides phenol, a very useful compound for further synthesis.

$$2C_6H_6 + O_2 \rightarrow 2C_6H_5OH$$

Another useful organic intermediate is the epoxide produced from the reaction of O_2 with double bond-containing compounds, e.g.

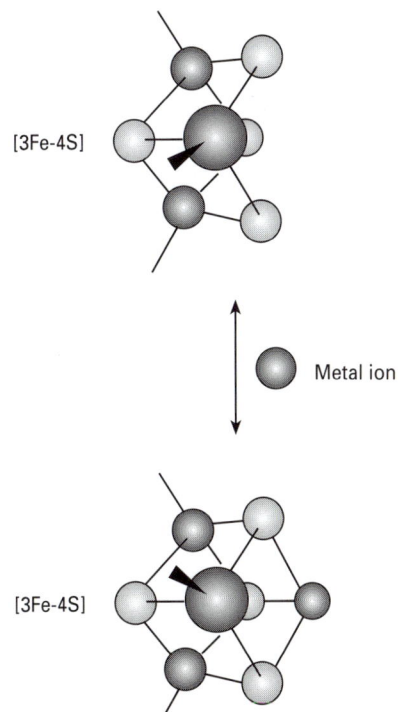

[3Fe-4S]

Metal ion

[3Fe-4S]

The Fnr Fe-cluster gene switch

Epoxides are used in many syntheses in all organisms and aerobic prokaryotes produced numerous novel compounds in this way, including penicillins. The enzymes employed use either haem or simple iron active sites, but very rarely copper, as catalysts.

In fact, the diversity of the organic chemistry of small molecules, inside and outside single prokaryote cells (and even of larger molecules outside cells), increased remarkably through the use of new catalytic oxidative processing. Since this development is vastly increased again in multicellular organism we describe it in the next chapter. However, the internal cytoplasmic basic chemistry leading to proteins, DNA and RNA, fats and saccharides had to remain virtually unaffected and reductive. The conclusion is that there were in evolution, and are at present, conserved regions of DNA and adjustable regions. Moreover, the adjustable regions, e.g. Fe/S and flavin genes, were those that were sensitive to the new agent, dioxygen. A possible explanation is that damage to proteins leads to increase in their production, which itself exposes DNA to change. Thus, DNA responds indirectly by local, not global, searching of environmental changes.

13.5   Catalysts of aerobic prokaryotes

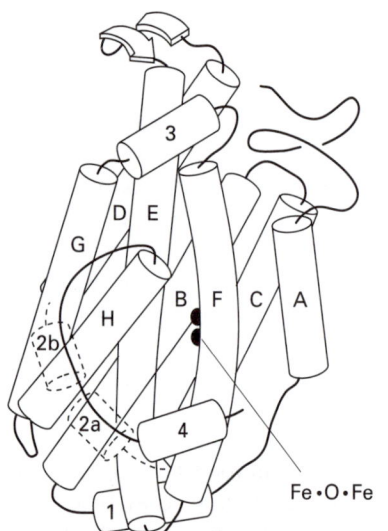

Fe·O·Fe

Ribonucleotide reductase

The most important observation concerning the catalysts of reactions in the cytoplasm of aerobic prokaryotes, compared with their anaerobic precursors, is that there was little modification. As stated above, the major pathways of the reactions of simple metabolites had to be kept essentially unchanged. There was, then, the requirement to keep making Fe/S clusters, flavins, quinones and so on. There were, however, some changes, even in the catalysts for these reactions. Notably, the reaction used to produce DNA from RNA, previously based on reduction using cobalt (vitamin B_{12} derivative) or an iron/sulphur centre in anaerobes, came to use an Fe–O–Fe (or a Mn–O–Mn) centre in aerobes. These iron (or manganese) centres are no longer based on sulphur co-ordination chemistry. Their formation depends today on the oxidation of iron by dioxygen. In fact, more and more frequently, organisms developed iron bound to nitrogen and oxygen donor centres of proteins, rather than sulphur, which allowed easier use of these iron centres as catalysts for the reactions of O_2. The iron was often held in binuclear (as in ribonucleotide reductase, see aside) as well as mononuclear centres (Table 13.3).

Table 13.3 Some metallo-oxidases in prokaryotes

Enzyme	Reaction centre
Methane oxidase	Fe–O–Fe
Ribonucleotide reductase	Fe–O–Fe
Penicillin synthetase	Fe
Cytochrome oxidases	Haem, (copper)
Ascorbate oxidase	Copper or haem
Cytochrome P-450	Haem

N.B. Most of the above enzymes reduce O_2 directly to H_2O, in contrast with one-electron extracellular oxidases (see Chapter 14), which produce radicals.

Table 13.4 Some new specific metal ion catalyses after the advent of O_2

Small molecule reactant	Metal ion	Examples
N_2	Mo(Fe)	Nitrogenase
NO_3^-	Mo	Nitrate reductase
SO_4^{2-}	Mo	Sulphate reductase
$O_2 \rightarrow H_2O$, NO, N_2O	Fe	Cytochrome oxidase*
Oxygen insertion (high redox potential)	Fe	Cytochrome oxidase*
SO_3^{2-}, NO_2^-	Fe	Reductase
$H_2O \rightarrow O_2$	Mn	Oxygen-generating system of plants
H_2O_2/Cl^-, Br^-, I^-	Fe(Se)(V)	Catalase*, peroxidase*
$H_2O/urea$, CH_3CO^-	Ni	Urease

* The porphyrin in these and other haem-containing enzymes is quite variable.

Cunningly, the O_2 inside the cell was reduced in effective two-electron steps so as to incorporate atoms of oxygen in organic molecules or to reduce it to water and avoiding the formation of dangerous free radicals. It was possible to use haem enzymes for this direct reaction of O_2 to H_2O too, e.g. cytochrome P-450. In these developments the chemistry and uses of iron porphyrin as well as of iron itself evolved considerably (see Table 13.4). The variety of porphyrins increased, especially in enzymes using dioxygen in metabolism or oxidative phosphorylation, while the ligands to the haem iron also increased in diversity, now using thiolate and phenolate (see Fig. 11.9).

It was not just iron that could be used in new ways, but also molybdenum (as described above), and, due to the removal of sulphide oxidised to sulphate, it became possible to use zinc extensively in hydrolytic enzymes and to some degree in transcription factors (see Section 13.10). Particularly, and probably considerably later, a novel element also became available for organisms: copper. Before considering the effect of these two catalytic elements upon bacteria, the general point must be made that while developing new, especially oxidative, catalysts, the reactions had to be selective so as to avoid non-specific damage. This is achieved by enzymes through the use of the high selectivity of their surfaces towards substrates, but also by the disposition of enzymes using oxygen or hydrogen peroxide *outside* the cytoplasm, often in the periplasm of prokaryotes (see Fig. 11.2), or on outer surfaces of the cytoplasmic membrane. Note that this corresponds to the use of an extra physical variable, compartmentalisation, common in eukaryotes.

13.6 The outer (periplasmic) space of bacteria

While the cytoplasm had to be maintained at a low redox potential by the presence of thiols, e.g. glutathione and newly evolved enols such as ascorbate, such homeostasis was no longer possible outside the cell where O_2 was in contact with metal ions. For cells to be able to use dioxygen easily they therefore had to do so in a separate compartment. The first such

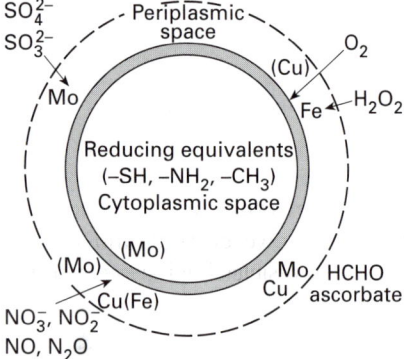

Fig. 13.8 The distribution of elements in the periplasmic compartment of aerobic bacteria.

compartment is the periplasmic space of bacteria (see Section 11.2 and Fig. 13.8). There, outside the cell, thiols became readily oxidised to disulphide bridges, –S–S–, which had a new functional value in the stabilisation, by cross-linking of extracellular proteins. In this outer region of 'high' redox potential it is also very difficult to maintain a standing concentration of free Fe^{2+} for catalytic purposes. Free iron came to be very low in concentration (10^{-17} M) due to precipitation of $Fe(OH)_3$. The use of iron in this outer aqueous medium then depended on keeping the iron tightly bound in a chelate trap, for example within a porphyrin ring, where the ring is cross-linked by oxidised thiols to the protein. Undoubtedly some iron porphyrin proteins (all cytochromes c) were therefore placed outside cells at an early stage after the advent of dioxygen. They were used to protect cells not only from H_2O_2 but also, even earlier, to prevent NO and CO getting inside cells. Additionally, again mainly in the periplasm, bacteria came to use H_2O_2, with the aid of iron porphyrin enzymes called peroxidases, to oxidise a variety of organic chemicals. Molybdenum is another metal which found a new use in enzymes in the periplasm, mainly in the reduction of nitrate and sulphate.

Somewhat later, in all probability, the nature of some of the catalytic processes changed due to the increasing availability of copper in the environment. Copper is very strongly retained by proteins in the absence of sulphide and is normally redox active in one-electron steps in the high potential range +0.3 to +0.8 volts, which is not the range of Fe^{2+}/Fe^{3+} activity. Now, inside the cytoplasm, such high redox potentials of *one-electron* couples, if at all exposed, are dangerous. Thus, while high potential one-electron reactions have functional value, e.g. in free radical polymerisation, they are very rare in the cytoplasm. In this reaction, copper, together with haem iron, which can catalyse such free radical reactions, became most useful on the outer surface of cells, generally, and especially in the energy capture from high redox potential reactions in both photochemical reactions and the use of O_2 itself.

Processes in the periplasm, which also came to be dependent on copper enzymes, now included the reduction of nitrogen oxides and oxidised compounds of sulphur (Fig. 13.8). The largest effect of copper, however, was not on prokaryote chemistry but on that of eukaryotes, leading eventually, we believe, to the production of multicellular organisms (see Chapter 14).

The new redox chemistry in the periplasm is summarized in Table 13.5. However, the use of separate compartments, including those involved with dioxygen chemistry, developed much more considerably in eukaryotes.

Table 13.5 Substrates of metalloenzymes in periplasm

Copper enzyme	Molybdenum enzyme	Haem iron enzyme
NO	NO_3^-	e (cytochrome c)
N_2O	SO_4^{2-}	
O_2	Dimethylsulphoxide	
Cytochrome oxidase	Aldehyde oxidase	Cytochrome oxidase (O_2)
(outer face of membrane)		(outer face of membrane)
(NO_2^-)		NO_2^-

13.7   The eukaryote cell and dioxygen

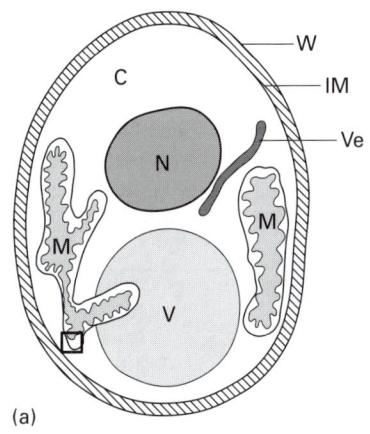

(a)

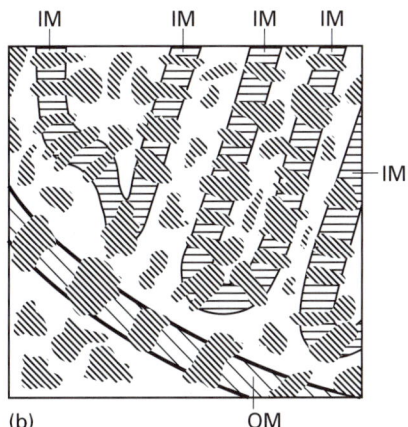

(b)

Fig. 13.10 (a) A typical eukaryote yeast cell, (x 25,000) is 20 times larger than a bacterial cell [see Fig. 11.1 (a)]. Vacuoles (V) may occupy a considerable volume, as do the weaving mitochondria (M) (shown) or chloroplasts (not shown). The protein density is much as in bacteria [Fig. 11.1 (b)]. Other features shown are the wall (W), the inner membrane (IM), a vesicle (Ve) and the nucleus (N). (b) A schematic diagram of a small section of a mitochondrion showing inner (IM) and outer (OM) membranes. The mitochondrion section is x 10^6 and is shown as an inset in (a). A thylakoid of a chloroplast is similarly constructed and indicates local coupling. [After Goodsell, D. S. (1998). *The machinery of life.* Springer-Verlag, New York.]

A great advantage of the eukaryotes over prokaryotes lies in the utilisation of compartments, allowing an increase in variables (see Chapter 12). Given that the major reductive chemistry of the cytoplasm had to be maintained, the prokaryote's overwhelming concern had to be protection from oxidative damage once the O_2 levels rose. The eukaryote, however, could much more readily *utilise* such oxidative chemistry, firstly in vesicles held away from the cytoplasm (Fig. 13.9) and secondly, since it could put to greater use these vesicles to export chemicals, e.g. enzymes, it could begin to use the immediate world outside the cell more extensively. Thus, we find in the first case that in single cell eukaryotes, oxidation occurs in vesicles such as the reticulae and the peroxisomes, and in the periplasmic space of mitochondria (captured prokaryotes acting as organelles) (see Table 13.6 and Fig. 13.10). Often the oxidised products were also stored in these vesicles. For example, the Golgi apparatus, an extended reticulum (see Fig. 12.8) is now used to handle oxidised sulphur, as sulphate, and metals such as copper. In the second case, the outer cellular polymers of cells could be made in new ways by treating them with dioxygen in the presence of exported enzymes from vesicles. Just because the eukaryote cell is prevented from bursting by

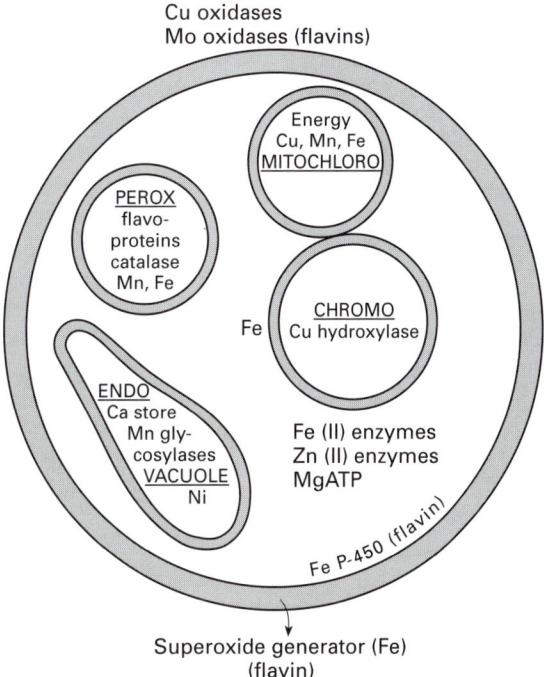

Fig. 13.9 A schematic indication of some of the different membrane-separated compartments in an advanced cell. PEROX is a peroxisome; MITOCHLORO is either a mitochondrion or a chloroplast; CHROMO is a vesicle of, say, the adrenal granule; ENDO is a reticulum, for example the endoplasmic reticulum. Other compartments are lysosomes, vacuoles and so on. Localised metal concentrations are shown.

Table 13.6 Compartments and reactions in aerobic eukaryotes

Compartment	Reactions	Elements
Cytoplasm	Synthesis of DNA, RNA proteins Glycolysis	Mg, Zn, K Ca, Mg
Mitochondria	Oxidative phosphorylation Citric acid cycle	Fe, (Cu), (Ca), (Mg)
Chloroplasts	Photophosphorylation Carbon assimilation	Mg (chlorin), Fe(Cu) Mg
Golgi apparatus	Glycosylation Sulphation	Mn
Endoplasmic reticulum	$2RS^- \rightarrow R–S–S–R$	(?) (Ca)
Vacuoles	Urea hydrolysis	Ni, (Mn)
Peroxisomes	Oxidative degradation	Fe, Mn

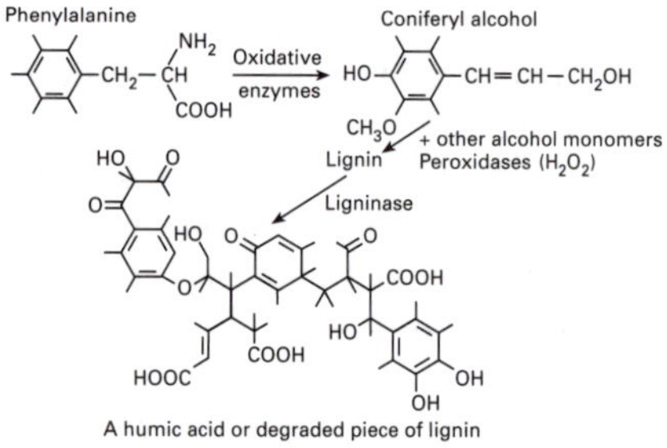

Fig. 13.11 Lignin is made from coniferyl alcohol by oxidation. Lignin is a very complex material of molecular weight > 10 000, often described as a statistical polymer of oxyphenyl–propane units. It is remarkably stable and is insoluble even in hot H_2SO_4 (70%). This matrix is laid down in plant cells as their primary walls harden into wood, and frequently forms an internal secondary wall. Its oxidative degradation gives humic acids, important constituents of soils. The structure of a humic acid is shown.

internal scaffolding (filaments) and not by a rigid wall, it could use new oxidised and external coats very differently from bacteria. In particular, lignins (Fig. 13.11) began to be used externally. These polymers are the product of oxidative cross-linking. (Compare the cross-linking of cell walls of bacteria using glycopeptides.)

The first organisms to develop extensive cross-linking of extracellular polymers may well have been the large algal constructs of such plants as seaweeds. (Organisms like seaweeds may contain many cells in colonies but they are not all true multicellular organisms.) The first oxidative catalysts for the cross-linking and breaking of cross-links, both needed to allow development of large structures, were probably haem enzymes: peroxidases. They often contain manganese as well and are a simple development from bacterial peroxidases. Essentially they now use H_2O_2 to generate organic free radicals in precursors of lignin which then form cross-links. It is still true today that the external stiffening of plants, e.g. woody materials, depends on such haem enzymes to cross-link these lignin polymers made from short peptides and aromatic alcohols (see Fig. 13.11). Similar cross-linking is found in advanced fungi (multicellular organisms) which also release enzymes (manganese-using peroxidases) to break down plant lignins. These processes are not important in later multicellular insects or animals generally, which base their connective tissue on proteins, chitins and collagens (see Section 14.5). They do use the same process of external oxidative cross-linking for their stabilisation however, but employing different means.

Since within vesicles proteins can be modified before export, the proteins outside eukaryotes differed not only in their stabilisation by –S–S– bridges but also in their glycosylated surfaces which could include sulphates. These surfaces could be used to bind cells to one another in colonies or to bind minerals (see Section 13.13).

All such development of new chemistry in vesicles and outside the cells demanded the development of new signalling devices since, even while the anaerobic message circuitry of phosphates, iron and substrates in anaerobic prokaryotes and of these plus calcium in anaerobic eukaryotes had to be maintained, management had to be extended to oxidations even across membranes. Before describing these signalling devices we need to describe the development of eukaryote membranes from those of prokaryotes.

13.8   Changes in plasma membranes

As described in Section 13.4.1, a major new chemical which was produced, eventually in quantity, by the reactions of dioxygen, was the sterol cholesterol (Fig. 13.7 and Table 13.2). It is a lipid and was incorporated into membranes. The effect was to produce membranes of greater strength, thus removing further the need for stabilisation by outer cell walls. We have suggested that it was produced first in certain prokaryotes (see Section 13.4.1), but it is not used in their membranes, in contrast with its extensive incorporation in eukaryote lipids. It is even suggested by some authors that the switch from prokaryote to eukaryote life forms was brought about by this change in membrane structure since it removed the constraint of building the strong outer cell walls. We leave this point as open for discussion, though in Chapter 12 we have deliberately described eukaryote compartment development without involving dioxygen. A further advance in membrane chemistry was the production of new unsaturated lipids by Fe enzymes using dioxygen. The different lipids used in membranes are listed in Table 13.7.

Table 13.7 Lipid (membrane) components (fats)

Prokaryotes (Archaebacteria)	Bacteria (cyanobacteria)	Eukaryotes (O_2)
Saturated	Saturated	Saturated
Some unsaturated	Some unsaturated	Some unsaturated
Ethers in membranes	Polyunsaturated*	Polyunsaturated*
	Sterols*	Sterols (cholesterol)
		Later sterol hormones*,[†]
		IP_3? Arachidonate?
	Haem	Prostaglandins*,[†]
	(Chlorophyll)	Halogen compounds

Source: Mead, J. F., Allin-Slater, R. B., Houston, D. H., and Popjak, G. (1986). *Lipids: chemistry, biochemistry and nutrition*. Plenum Press, New York.
* Require dioxygen desaturases (flavin + Fe_n/S_n) or oxygenases (Fe or haem) (see Fig. 10.5).
[†] Only found in multicellular eukaryotes (see Fig. 10.5).

13.9   New signalling in eukaryotes

We have already mentioned new signalling devices in simple prokaryotes based upon iron and sulphide levels, which were changed by dioxygen. They were the FNR protein for the anaerobic/aerobic switch and the FUR protein for iron uptake control. As stated above, they were linked to phosphate control systems. A similar set of controls were seen to be employed over the new genes for the metabolism of dinitrogen. Message systems parallel to these are found in eukaryotes. Such linking of different signalling systems so that the whole cellular activity is integrated, is an essential part

Evolution of the uses of calcium

(1) Wall cross-linking (prokaryotes)
(2) Side/inside communication (eukaryotes)
(3) Vesicles release (eukaryotes)
(4) Organ connections (multi-cellular organisms)
(5) Brain (higher animals)

of each evolutionary change. The major change we are describing in this chapter is oxidation due to the presence of dioxygen and we need to show not only the new signalling systems that evolved in eukaryotes, but also how they were connected to the pre-existing ones of prokaryotes.

In Chapter 12 we have seen how the evolution of calcium signalling in eukaryotes (see aside) was coupled to phosphate signals in these organisms. The use of dioxygen to produce new organic molecules therefore

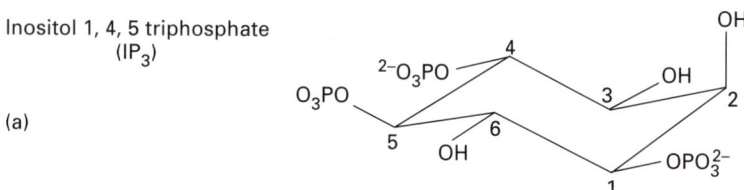

(a)

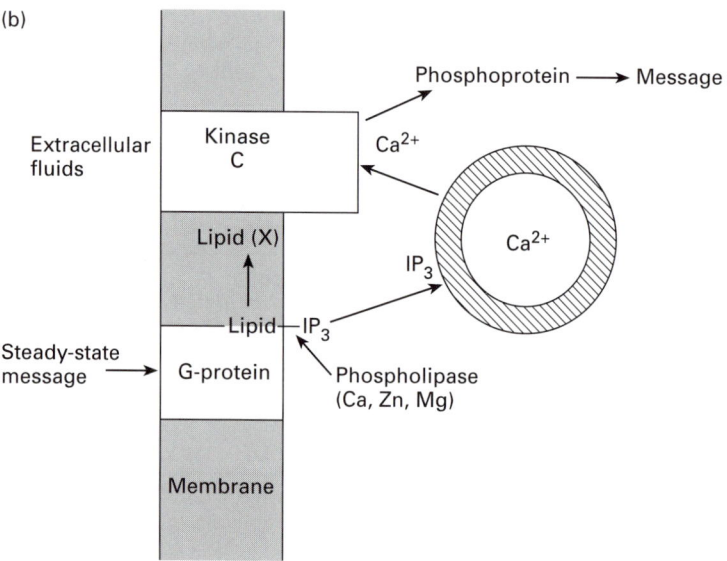

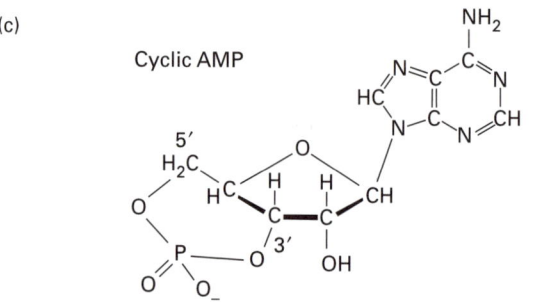

Fig. 13.12 (a) Inositol triphosphate, IP_3, and (b) some of its effects after release from phospholipid inside a cell, together with arachidonic acid, X, which gives rise to the prostaglandins in higher organisms. (c) c-AMP is used as an internal messenger and now and then as an external messenger, e.g. by slime molds to generate a colony not far distant from a true multicell organisation.

demanded links to this Ca^{2+}/HPO_4^{2-} network too. It is not certain how this signalling arose, that is, whether it appeared in unicellular or multicellular species, but it came to use the compound inositol 1,4,5-triphosphate (IP_3) (Fig. 13.12). IP_3 is bound to membrane lipids, usually containing arachidonic acid (see Fig. 10.5) a newly produced unsaturated fatty acid. On a signal from outside the cell, IP_3 is released internally and the lipid is released in the membrane. While IP_3 then activates invasion of Ca^{2+} from vesicle (ER) stores, the arachidonic acid lipid activates membrane enzymes such asso-called G-proteins, enzymes utilising guanosine phosphates. The further connections of calcium to phosphate signals are those shown in Figs 12.5–12.7. Arachidonic acid also became an internal signal since it is *oxidised* by iron enzymes to prostaglandins (see Section 14.6). The relationship of the iron and dioxygen of the Ca/P signalling network is seen to be complete when it is remembered that arachidonic acid lipids are produced by an Fe/O_2-dependent desaturases, a dehydrogenase. The connection between aerobic eukaryote oxidative reactions and (anaerobic) hydrolytic reactions in signalling is intense.

A major problem in the discussion of further novel message systems to and from the oxidative vesicles in eukaryotes is once again that it is not known which part of them arose in single cell as opposed to multicellular systems. An example is the control over the use of H_2O_2. This oxidising agent can be managed by the level of catalase activity, which removes it. Plants and maybe simple eukaryotes produce, by oxidative metabolism, salicylate, as a control messenger, which can block catalase and hence increase the level of H_2O_2. This aggressive chemical is then used to protect the cells from prokaryote invaders (see Section 14.12). Later, in Chapter 14, we shall see that other hormone-like oxidised substances, such as adrenaline, are similarly produced to assist the protection of multicellular organisms.

Other signalling also had to develop, related to the new uses of zinc and copper, which we describe next.

13.10   The value of zinc in aerobic metabolism

Until the rise of dioxygen in the atmosphere, free zinc was of relatively low concentration in the sea owing to the presence of sulphide (Fig. 13.4). Removal of sulphide increased the availability of zinc and reduced the availability of iron to extremely low levels, so that in place of *iron/organic sulphide* chemistry it was possible to develop *zinc/organic sulphide* combinations. These zinc thiolate proteins are mainly intracellular. What is found is that, apart from the limited use of zinc in anaerobes, e.g. as an acid – base catalyst bound by very strong non-exchanging N/O bonds to proteins such as proteases and phosphatases, zinc now occurs in thiolate (organic sulphide) complexes in some enzymes and becomes of major value in proteins as part of transcription systems (Table 13.8). It is also stored in a thiolate-dependent buffer protein, metallothionein, to maintain zinc's homeostatic level.

Table 13.8 Some novel zinc proteins in eukaryotes

Protein	Function
Zinc-fingers	Transcription factor (Zn)
Metallothionein	Zn and Cu buffer
Kinase C	Kinase (Zn)
Calcineurin	Phosphatase (Zn)
Superoxide dismutase	Detoxification (Cu and Zn)
Zinc ATP-ase (pump)	Export of zinc

The combination of metallothionein with other zinc thiolate proteins used as transcription factors allowed zinc to become a new regulatory element in addition to substrate (C, H, N, O compounds), phosphorus (phosphates), sulphur, calcium and iron. In particular, it could be used in association with the *newly oxidised organic chemicals, such as sterols*. Sterols, thyroxine and retinoic acid, produced by oxygen-utilising oxidation, all have zinc protein transcription receptors whose particular conformation led to the designation 'zinc-fingers'. As we stated above, we do not know at what stage in the evolution of eukaryotes these 'zinc-fingers' became so important. Some similar proteins are even found in certain aerobic prokaryotes. For this reason, the general extended use of zinc in eukaryotes is shown later, in Fig. 14.8, while describing multicellular organisms. Note immediately, however, that new circuitry involving zinc and new catalysis involving either zinc or copper had again to be knitted into the existing networks of substrate, phosphate, calcium, dioxygen and iron signalling even in single cell aerobic eukaryotes. This became possible through the evolution of proteins simultaneously sensitive to the levels of more than one of these elements, such as the zinc-requiring kinases, phosphatases, S-100 proteins and so on, which also require, variously, calcium and/or phosphate and/or iron, e.g. in the protein transcription factor, calcineurin.

13.11   The value of copper to eukaryotes

We have stated that copper was probably the latest of the elements to be used in the evolution of complicated organisms (see Fig. 13.4). Its sulphide is the most insoluble of those of the common biological metals, and therefore liberation from sulphide by oxidation to sulphate could not have been an early change for it (see Section 6.5). As stated above, its use is developed even in periplasmic spaces of the energy-generating organelles (mito-

Table 13.9 Some copper proteins in unicellular aerobes

Nitric oxide reductase (P)	Dinitrification (Cu)
Nitrous oxide reductase (P)	Dinitrification (Cu)
Ascorbate oxidase (P)	Redox buffer (Cu)
Cytochrome oxidase (P.M)	Energy capture from O_2 (Cu)
Copper ATP-ase pumps (P.M)	Export of copper

P, periplasmic; P.M, periplasmic face of outer membrane.

chondria and chloroplasts) and of prokaryotes, but its greatest value is not seen until we examine *animal* multicellular organisms. It may well be, in fact, that the earlier adaptation of copper chemistry by eukaryote colony-forming cells lead to the later evolution of animal multicellular organisms. We will explain why we believe this may be the case in Chapter 14, where we also describe the homeostasis of this metal. A summary of the uses of copper in single cell organisms is given Table 13.9, and see aside.

One particular, and as far as we know unique, development of copper chemistry in eukaryotes generally, and involving dioxygen chemistry, is the switch from manganese- or iron-dependent enzymes for the removal of the dangerous chemical superoxide, which can appear by accident in the cytoplasm, to a copper–zinc enzyme. We describe the change as unique since we know of no other copper enzyme in the cytoplasm. Possible explanations for the switch are as follows.

(1) free Mn^{2+} or Fe^{2+} are in themselves quite considerable mutagens; while prokaryotes can risk mutations, they are increasingly damaging to eukaryotes and both manganous and ferrous enzymes are known to release free metal.

(2) In eukaryotes, free Mn^{2+} is stored to a greater and greater extent in vesicles, e.g. vacuoles.

(3) The availability of iron had been seriously reduced by the increase of dioxygen.

(4) The levels of copper and zinc had risen, and while free copper would be dangerous in the cell, in this enzyme it is held very tightly, deeply inside the protein.

This is a very telling example of the way in which evolution has adapted to environmental change. To protect cytoplasmic polymers from any free copper, the element is pumped out of cells into vesicles by a copper pump closely related to the calcium pump, and held at very low levels by metallothionein buffering.

Peculiarities of copper chemistry

(1) Stable low oxidation state
(2) High affinity for sulphides (thiolates)
(3) Square (Cu^{2+}) or linear (Cu^+) binding
(4) O_2-binding (Cu^+)

13.12 Changes in non-metals

The increase in available dioxygen also led to a change in the availability of some non-metals and a change in the speciation of others. We have already discussed some changes in carbon (see Table 13.10), nitrogen, hydrogen and sulphur chemistry.

Perhaps the most intriguing example of non-metal oxidative chemistry is that of nitrogen, which requires us to return for a while to the discussion of the oxidative chemistry of bacteria, but not now involving dioxygen. We shall develop this theme in the next chapter, but it is very interesting to notice that certain modern bacteria can handle nitrate, using it as a source of nitrogen and of energy. Clearly, there is a problem as to when this metabolism arose (see Fig. 13.13). We have suggested that use of NO itself may have been much developed before eukaryote organisms arose.

Table 13.10 Typical new organic products due to oxidation

Class of compound	Example
Hydroxylated aliphatic	Hydroxyproline, cholesterol
Hydroxylated aromatic	Adrenaline, serotonin
	Many phenols
Carboxylated compounds	γ-Carboxyglutamate
Unsaturated compounds (dehydrogenation)	Arachidonic acid
Epoxides	Penicillins

N.B. A vast range of organic chemicals result from carbon oxidation, i.e. formation of –C–O–bonds, including small molecules, consider the alkaloids, and polymers. They are generally included under secondary metabolism. Many, including some of the above, may only be present in multicellular organisms and properly belong in Chapter 14.

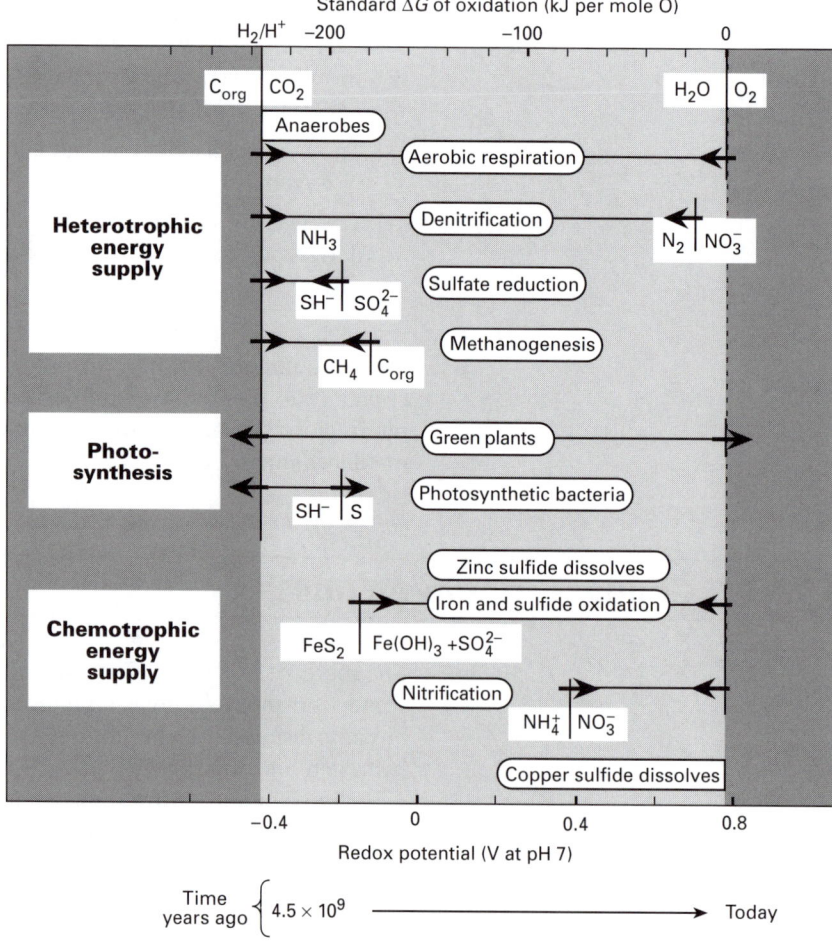

Fig. 13.13 The redox potential ranges of certain organisms and simple reactions allowing concentration changes of greater than 10^{20} in element speciation. [From Cox, P. A. (1995). *The elements on earth*. Oxford University Press, Oxford.]

However, the presence of dioxygen led to the production from NO of NO_2 and then of nitrite and nitrate.

Nitrate reduction to nitrite utilises the same molybdenum enzyme as sulphate reduction to sulphite, even in eukaryotes. Now, nitrite is a much more powerful and reactive oxidant than nitrate and is quite an aggressive chemical. Bacteria have two modes for its further reduction: one uses copper and the other iron. Following the inorganic trends in evolution, we believe that the mechanisms using copper were relatively late adaptations. They generally take place in the periplasmic space (see Table 13.5 and Fig. 13.8). Now, this dating apparently conflicts with the fact that dioxygen-using mitochondria, in eukaryotes, are related to nitrite-reducing bacteria that use copper enzymes. [A clear reason for this association is the finding of a special dinuclear (Cu_2) centre in N_2O reduction in both the nitrite-reducing bacteria and in cytochrome oxidase of all mitochondria in all organisms.] The implication in evolutionary progression may well be that mitochondria evolved not from nitrite- or dioxygen-reducing bacteria but from NO (nitric oxide)-utilising bacteria which may well have evolved very early on indeed (see Section 11.5). The nitric oxide (NO) reductase from some bacterial sources uses non-haem iron, not copper, but this protein is very similar in its primary sequence to the cytochrome oxidase of mitochondria, where copper now replaces this iron. The suggested evolution of mitochondria and nitrite-using bacteria from NO-metabolising prokaryotes is shown in Table 13.11 and Fig. 13.14.

Table 13.11 Possible development of nitrogen oxide-using bacteria and of mitochondria

Strict anaerobe	NO-using bacteria (no Cu)	
Aerobes	Mitochondria and bacteria using O_2, Cu replaces Fe of NO reductases in cytochrome oxidase and new [Cu]$_2$ centre appears	NO_3^--using dentrifying bacteria. Cu used in several N/O-utilising enzymes, including new [Cu]$_2$ centre

(a)

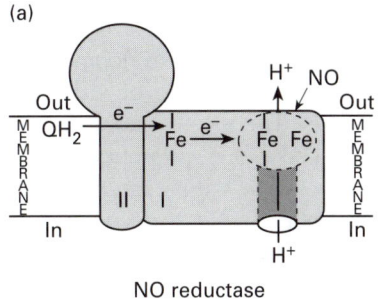

NO reductase

(b)

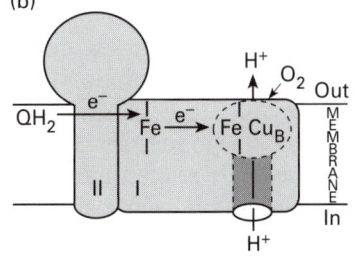

bo_3-type (*E. coli*)
ba_3-type (*Acetobacter aceti*)
aa_3-type (*Bacillus subtilis*)

(c)

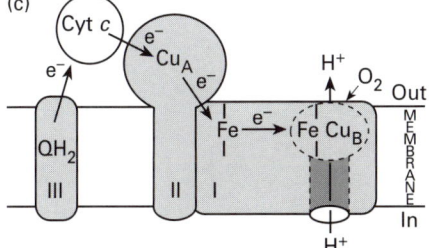

aa_3-type (cow, *P. denitrificans*, yeast, *Rb. sphaeroides*)
ba_3-type (*Thermus thermophilus*)

Fig. 13.14 The possible development of energy capture systems from (a) NO to (b) O_2 then (c) O_2 in three stages.

Selenium uses

Early	Late
–SeH, SeCH$_3$;	>SeO;
H$_2$ and reduced	metabolism
C metabolism	of RO$_2$H and RI

In aerobic eukaryotes, selenium became associated with oxygen chemistry and no longer with hydrogen chemistry, see aside. This element may have been used in this new way quite soon after the rise of dioxygen since it can act to remove peroxides, as indeed it does in many aerobic multicellular eukaryotes. An element that requires a similarly increased redox potential to develop its 'organic' chemistry is iodine. In prokaryotes it is not used, but it is incorporated into organic compounds already in algae, by peroxidation. Interestingly, bound iodine is removed by a selenium enzyme, suggesting that the use of both was generated at a similar time in evolution. They have similar redox potentials. Selenium has become peculiarly important in the protective systems of higher animals (see Section 14.12.1).

The hardest non-metals to oxidise are bromine and chlorine, but it is known that aerobic bacteria and some eukaryotes can make organochlorine and organobromine compounds, though organic nitro compounds are not known. While these halogen compounds can be used as drugs (antibiotics, insecticides, etc.) in higher organisms, there is always a risk since this halogen covalent chemistry is associated with aggressive attack, even on DNA. There is little or no covalent halogen chemistry in man (except for iodine). The iodine-containing hormones (thyroxines) appear to be a relatively late development in evolution.

A development of non-metal chemistry of great importance which we have mentioned already was the oxidation of pairs of thiolates to form –S–S–units outside the cytoplasm. The reaction, which was used in the cytoplasm of prokaryotes to act as a redox buffer (Section 11.6), became important externally in the stabilisation of proteins by the formation of protein disulphide bridges (see Fig. 10.9). The process, which is carried out in the Golgi-derived compartments of eukaryotes, has to be selective between many specific thiolates within each protein. This is managed by a protein, disulphide isomerase, which removes incorrect bridges. This protein, in the vesicle compartments of the cell, is dependent upon the high concentration of calcium there and is another example of co-operative action of elements, here a metal ion and a non-metal. Later, the –S–S– bridges and calcium ions became additional 'cross-linking' agents assisting the stability of the extracellular matrix of eukaryotes, as well as the internal folds of individual exported proteins (see Chapter 14).

As also mentioned earlier, one of the most interesting developments in the case of oxidised sulphur (sulphate) is the production of sulphated saccharide units to be positioned outside cells. This use of sulphation in eukaryotes is due to the new co-enzyme APS (adenosine 5′-phosphosulphate), an energised carrier of sulphate, again in the Golgi. We presume that the control over sulphate metabolism in the Golgi is due to feedback and feed-forward communication using APS in much the way that other adenine nucleotide co-enzymes are employed in the cytoplasm. Sulphation of saccharides is largely developed in multicellular systems. The sulphate transport protein is closely similar to the iron transport protein transferrin, and perhaps they evolved together.

13.13   Mineralisation and rejection of elements

Evolution of minerals

(1) Low organisation of external crystallites (Algae)
(2) Organisation of small external crystallites (Coccoliths)
(3) External shells
(4) Internal stores e.g. oxalates in plants
(5) Internal skeletons e.g. bones

Some improvements in mineralisation in eukaryotes were described in Section 12.13. Further development (see aside) was possible through the use of oxidation of the side-chains of external proteins and polysaccharides, which allow better binding to mineral surfaces. Such surfaces include those of lignins and, later in multicellular organisms, of chitins and proteins (see Table 14.9). Once again the full value of mineralisation appears in multicellular organisms, which we describe in the next chapter. All external mineralisation is dependent, of course, on outward pumping of elements, which is also a way of rejecting poisonous elements.

We have stressed that certain elements available from the earliest times are not used biologically, including aluminium and titanium and, in general, heavier elements of groups 3, 4, 13 and 14. They are effectively treated as if they were poisons. Now, the removal of sulphide by dioxygen made a new range of heavy elements more available, including Cd, Pb, Hg, Sb and As. These elements are again treated by many aerobes, both prokaryotes and eukaryotes, as poisons and there has been little or no adaptation to their presence. The elements are positively rejected. The rejection system for Cd, excess Zn, and indeed Cu uses the buffer protein for zinc, metallothionein, and then an ATP-dependent pump; that for rejection of arsenic uses a modified system for phosphate transfer, and also an ATP-dependent pump.

Today, when many prokaryotes are exposed to new elements, additional modes for their rejection have appeared. Bacteria have evolved rejection systems against, for example, Hg*, Sb and Bi, in addition to the above. This is a further illustration of the ability of prokaryotes to modify their chemistry when the proteins for new activities are often found coded in small DNA units outside the main genes. These small DNA units (plasmids), (see Fig. 11.15), can be transferred between species. Thus evolution does not require random mutation of all the DNA to find solutions to new chemical problems. DNA can change also by transfer as well as by local mutation.

13.14   Summary

This chapter has stressed early evolutionary changes in chemistry after the advent of dioxygen, whereas Chapter 11 put emphasis on the evolutionary changes of compartments, a physical development. The major changes for unicellular prokaryote and eukaryote organisms resulting from the coming of dioxygen (see Table 13.12) were:

(1) new possibilities for energy capture;

(2) new processes for the capture of iron (and sulphur);

(3) new uses of elements such as zinc and copper;

---

*The Hg pump is closely related to the pump for Cu^+—both have a linear $2RS^-$ binding site.

Table 13.12 New element biochemistry after the advent of dioxygen

Element	Biochemistry
Copper	Most oxidases outside higher cells, connective tissue finalization, production of some hormones, dioxygen carrier, N/O metabolism
Molybdenum	Two electron reactions outside cells, NO_3^-, SO_4^{2-}, aldehyde metabolism
Manganese	Higher oxidation state reactions in vesicles, organelles and outside cells, lignin oxidation (note especially plants); O_2 production
Nickel	Virtually disappears from higher organisms
Vanadium	New haloperoxidases outside cells
Calcium	Calmodulin systems, signalling; general value outside cells
Zinc	Zinc fingers connect to hormones produced by oxidative metabolism
Selenium	Detoxification from peroxides, de-iodination?
Halogens	New carbon–halogen chemistry, poisons, hormones
Iron	Vast range of especially membrane bound and/or vesicular oxidases; peroxidases for the production of hydroxylated and halogenated secondary metabolites; dioxygen carrier and store

(4) new demands for protection from reduced O_2, $O_2^{\cdot-}$ and H_2O_2;

(5) new synthesis of organic poisons for aggressive protection.

New developments of eukaryotes alone were:

(6) improved inner membranes using cholesterol;

(7) improved cell outer membranes using new polymers;

(8) novel signalling systems;

(9) new outer surfaces and mineralised structures.

These changes, giving new 'component' variables, based on elements, and chemical pathways based on new energy uses, accrued together with the development of vesicles and filaments in eukaryotes. Frequently, this new oxidative chemistry and the new 'component' chemicals were placed in these vesicles, which themselves increased in number. Since all this activity was held behind compartmental barriers, the variables of composition were greatly increased by the advent of dioxygen. This includes both the new inorganic elements and the new, mainly organic, 'components'. The increase in materials, 'components', can also be thought of as increased energy throughput in new modes of chemical element interaction.

If we allow that a plot of either material (element) composition or energy input against survival value takes the form of a landscape or contour diagram, (see Figs 10.4 and 11.14) then we must concede that it may well be that for aerobic eukaryotes survival has once more a more sharply featured landscape, as in Fig. 13.15. Therefore, we stress, as at the end of Chapter 12, that many new, well-defined species could appear as a result of a changed environment. This follows inevitably from the principles we have described in Chapters 1–10 concerning increase in diversity from increase in derived variables. Meanwhile the ancient anaerobic species, both prokaryotes and eukaryotes, remained in the environmental zones deprived of dioxygen. Thus, we see evolution as a continuous adaptation to developing variables, dependent on the composition of the environ-

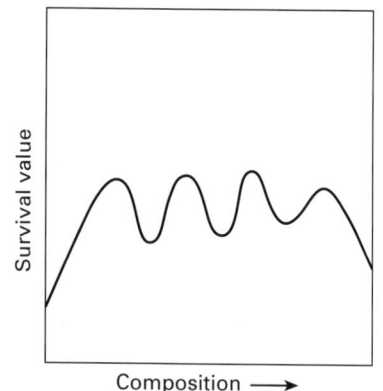

Fig. 13.15 A landscape diagram, survival value versus element composition or energy input per compartment, for eukaryotes with several compartments. Each *new* element, or compartment, or energy input requires a separate *x*-axis, multiplying the possible number of species relative to the prokaryotes of Chapter 11. Note the more marked definition of species than in Fig. 10.4.

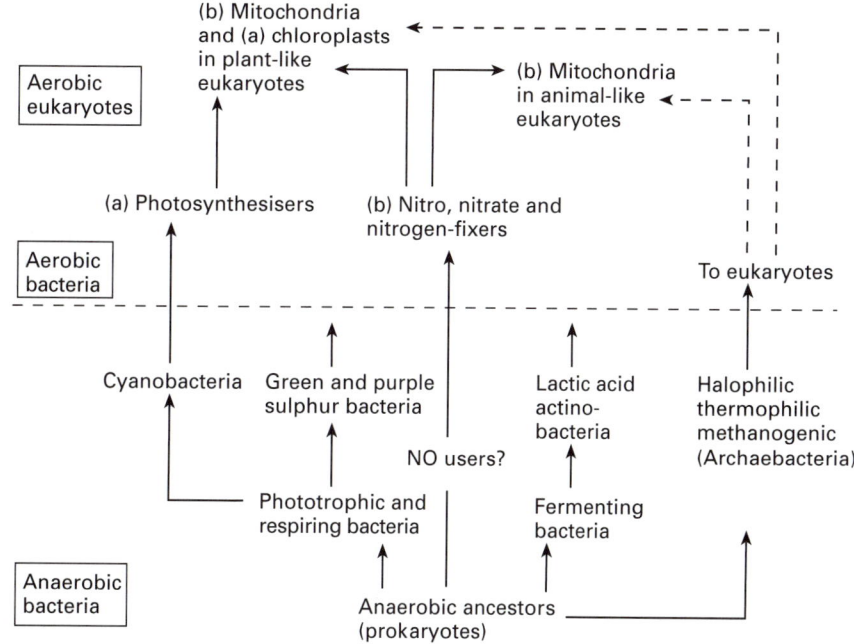

Fig. 13.16 The development of aerobic eukaryotic organisms from the earliest anaerobic ancestors, all single cells.

ment, while there was always some improvement in those organisms that were left in the environment of 3.5×10^9 years ago.

Variety is also dependent on the fact that the environment is graded between extremes of old and new and locally from place to place, by temperature and geochemical weathering. The huge variety of eukaryote possibilities that evolved is illustrated in Fig. 13.16. Changes in DNA represent the coding of these new possibilities and a protective restriction on the various activities so that they remain co-ordinated. Increased complexity of a living system has a risk in that the multiplicity of controls, all of which can develop errors, had to increase. An alleviation from complications then rested in symbiosis, where many simple tasks were performed only in prokaryotes, products from which were incorporated in eukaryotes. In this way ecosystems developed. (The extreme feature of symbiosis is, of course, the use of organelles.)

In the next chapter, we shall see the further advance of both physical (compartmental) and chemical evolution jointly, in the coming of multi-cellular organisms.

14

The coming of multicellular organisms

Evolution is a change from an indefinitive incoherent homogeneity to a definite coherent heterogeneity.

Herbert Spencer, *First principles*, Ch 16. (1850)

14.1   Introduction

Given the time when multicellular organisms first appeared (Table 14.1), less than a billion years ago, it is likely that a considerable level of dioxygen in the atmosphere was necessary to generate both their external structure

Table 14.1 The geological time-scale of life after the advent of dioxygen*

Era	Period	Epoch	Distinctive features	Years before Present
Cenozoic	Quaternary	Recent	Modern humans	11 000
		Pleistocene	Early humans	1 700 000
	Tertiary	Pliocene	Large carnivores	5 000 000
		Miocene	First abundant grazing animals	23 000 000
		Oligocene	Large running mammals	38 000 000
		Eocene	Many modern types of mammals	54 000 000
		Palaeocene	First placental mammals	65 000 000
Mesozoic	Cretaceous		First flowering plants; extinction of dinosaurs and ammonites at end of period	135 000 000
	Jurassic		First birds and mammals; dinosaurs and ammonites abundant	192 000 000
	Triassic		First dinosaurs; abundant cycads and conifers	223 000 000
Palaeozoic	Permian		Extinction of many kinds of marine animals, including trilobites	280 000 000
	Carboniferous	Pennsylvanian	Great coal-forming forests; conifers; first reptiles	321 000 000
		Mississippian	Sharks and amphibians abundant; large primitive trees and ferns	345 000 000
	Devonian		First amphibians and ammonites; fishes abundant	405 000 000
	Silurian		First terrestrial plants and animals	
	Ordovician		First fishes; invertebrates dominant	495 000 000
	Cambrian		First abundant record of marine life; trilobites dominant, followed by massive extinction at end of period	570 000 000
	Precambrian		Fossils extremely rare, consisting of primitive aquatic plants	700 000 000

* Source: Brown, J. H., and Gibson, A. C. (1983). *Biogeography*. Mosby, St Louis; see also Ricklefs, R. E. (1993). *The economy of nature—a textbook in basic ecology* (3rd edn). W. H. Freeman & Co, New York.

Event	Billion of years ago	Manifestation	Oxygen (%)	Events and consequences
Full diversity of life forms present	0.35	Large fishes and land animals; primitive trees 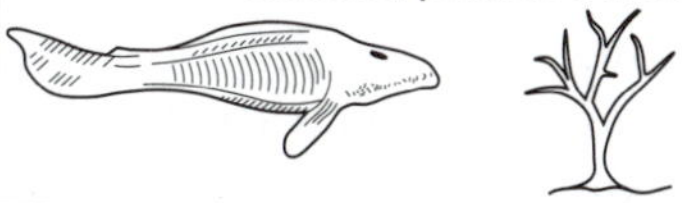	20	Complete ozone screen, atmosphere same as today. Large fishes, land plants. Intra body mineralization.
Shelled metazoans and early land plants appear	0.55	Hard-shelled animals 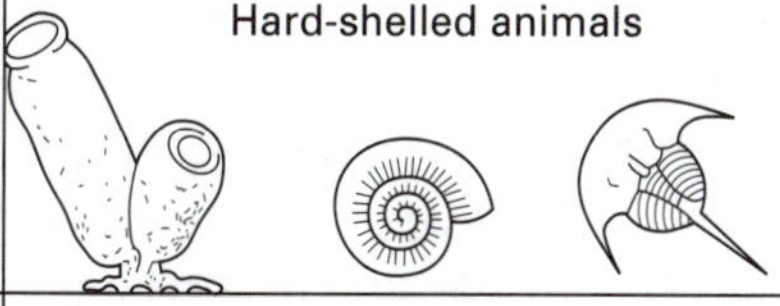	10	Diversity evident in fossil record: mineralized extracellular structures.
Metazoans appear	0.70	Soft-bodied multicellular animals and sea plants 	7	Fossils and tracks. Accumulation of oxygen and ozone. Connective tissue. Differentiated cells.
Eukaryotic cells well established	1.5	Large nucleated cells 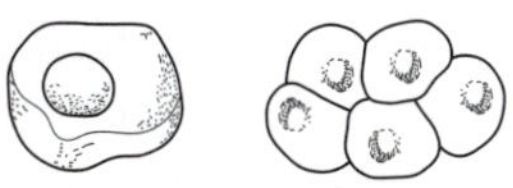	1–2	Mitosis, meiosis, genetic recombination, aerobic respiration. Organelles incorporated. O_2 fully used.

Fig. 14.1 The last four steps in the evolution of life from about 1.5 billion to 350 million years ago. The thick line indicates the appearance of multicellular life. Adapted from Postlethwait, J. H. and Hopson (1989) *The Nature of Life* McGraw-Hill, New York.

(Fig. 14.1) and some new internal features of their cells. The special internal features are not present in single cell aerobic eukaryotes, but clearly multicellular organisms are derived from these organisms. Anaerobic eukaryotes and any kind of prokaryote are not known to have become truly multicellular, which is quite a different construction from colony formation, and these organisms undoubtedly underwent almost optimal development before the dioxygen levels rose.

The novel features of multicellular organisms are as follows.

1. *Differentiated cells*, that is, different kinds of cells with the same DNA and with a great degree of permanence of their special functions, which develop from a single initial cell. Effectively, each group of such cells forms a new 'compartment' of the whole organism, which introduces new variables (see Chapter 7).

2. The differentiated cells may be irreversible in later life in advanced animals, but are reversible in simple organisms. All plant cells can dedifferentiate, and so can cells of such animals as planaria and hydra. [It now appears (in 1998) that even some animal cells can revert to dedifferentiated 'stem' cells and so allow cloning, e.g. of the sheep Dolly.]

3. The differentiated cells are assembled in a 'planned' disposition in which many more or less identical cells may be separately *organised in organs with boundaries*. The cells and organs are separated by connective tissue and bathed by the extracellular fluids of the organism. They are all positioned within the whole, which is limited by an outer 'skin' in animals or an outer 'bark' in plants.

4. In final form, such an assembly has a more or less permanent structural scaffold outside each cell (*connective tissue*) as well as inside each cell (filaments) to which the *connective tissue* is joined so as to maintain the positions of the cells and organs in the organism. This tissue can be energised and its energisation is a further new variable, allowing dynamic activity.

5. Since the organs of differentiated cells cannot survive as an autonomous system, they must function as a whole. They, therefore, have to communicate with one another using *messengers in the extracellular fluids* of the whole organism. These new messengers are often organic molecules (and Zn^{2+}) stored prior to release in new kinds of vesicular compartments.

6. The *extracellular fluids* of the cells are no longer part of the environment but become a controlled, flowing 'body' fluid compartment of special composition containing special proteins and other chemicals. The extracellular fluids provide a new variable in that they correspond to a particular controlled 'environment.'

7. The whole organism must constantly obtain material and energy from the environment and reject waste to the environment. A tubular form (Fig. 14.2 and see Figs 14.3 and 14.4) is then frequent but far from universal. The tubular form in animals contains the digestive system.

8. In addition to the genetic DNA of the instructional organising centre of each differentiated cell, the whole organism developed, later in the evolution of animals, an organising apparatus to instruct through a memory bank and to control *phenotypic* response. The apparatus is called a *nervous*

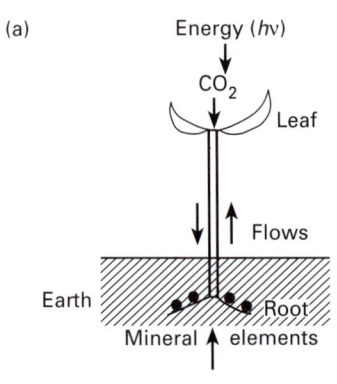

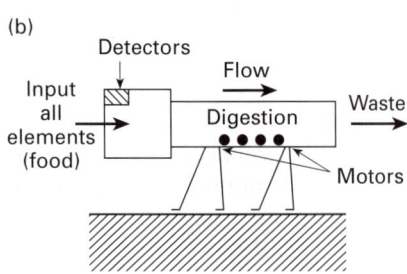

Fig. 14.2 (a) The basic plant structure receives energy from the sun and carbon from the air. All other elements (15–25) come from the Earth as inorganic minerals or from bacteria ($N_2 \rightarrow NH_3$), shown as filled circles. (b) The basic animal structure. Motors are connected to arms, legs, wings, fins, etc. The animal acquires 15–25 elements and energy from food (plants) and requires a multitude of assisting bacteria.

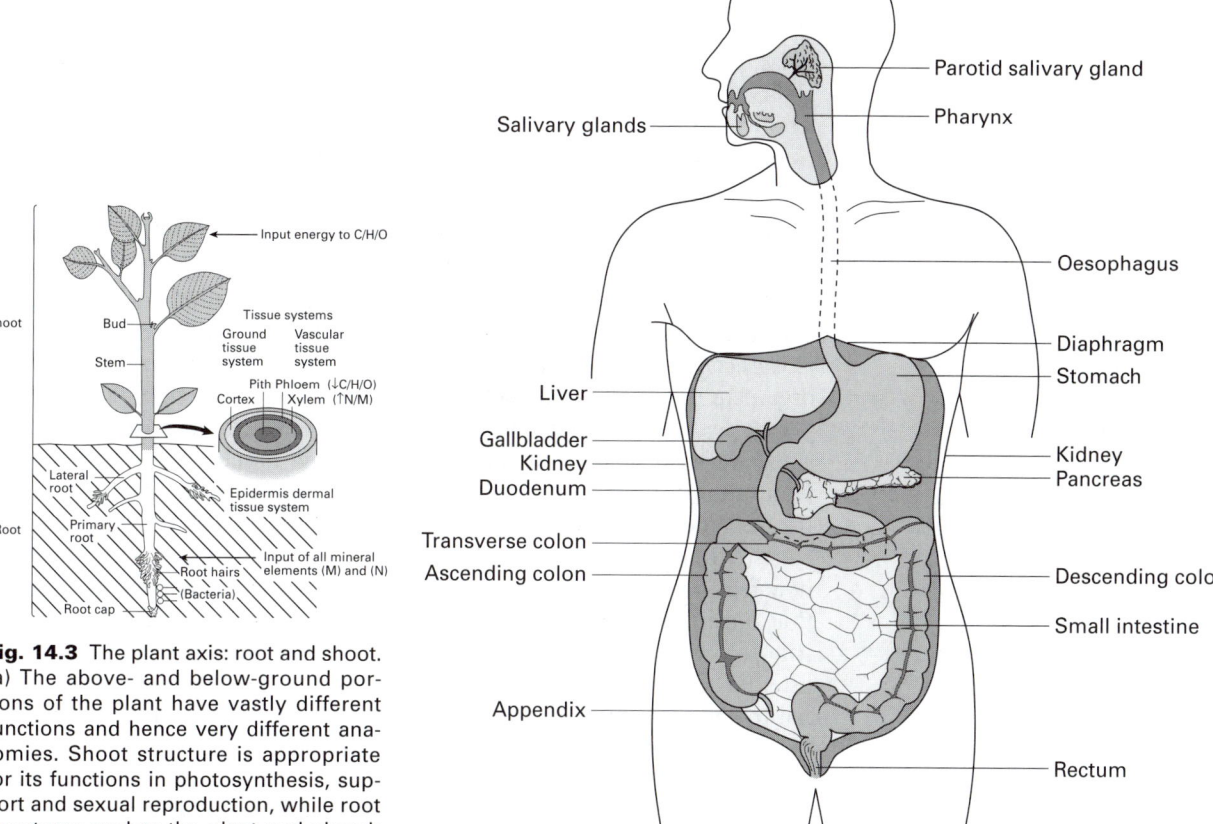

Fig. 14.3 The plant axis: root and shoot. (a) The above- and below-ground portions of the plant have vastly different functions and hence very different anatomies. Shoot structure is appropriate for its functions in photosynthesis, support and sexual reproduction, while root structures anchor the plant and absorb water and minerals. Plants have three tissue systems, each of which carries out a different function.

Fig. 14.4 Diagram of the human body showing the parts of the digestive system. The liver, which normally covers part of the stomach and duodenum, has been folded back to reveal these and the gall bladder on its under surface. Many animals have similar features.

system and in particular it has an organ called the *brain* in later animals, which allows the organism to have planned, co-ordinated responses. We turn to the nervous system and the brain in higher animals in Chapter 15. This development of a brain introduces a quite novel variable into living systems based on the individual.

9. The nervous system of animals communicates via sensors with the environment, but these *sensors* are large-scale apparatuses (not molecular constructs as in single cell eukaryotes) within organs: the eye, the ear, the nose and so on.

10. Reproduction requires the special creation of gametes in separate organs and a way of bringing male and female cells together.

11. Given that reproduction is based initially on a single cell, the whole—many millions of cells—has to go through extensive *development*, which can only occur over days, months or even many years depending on

the organism (see Section 10.16 and 13.1). Clearly, as multicellular structure develops the properties of cells must change before reaching a final adult form.

12. Mineralisation, external or internal, is very common, hence our ability to follow evolution over the last 700 million years can be based on a fossil record.

13. Multicellular organisms have evolved both internal *protective immune systems*, linking intracellular activity with the extracellular fluids, and *repair mechanisms* in the extracellular fluids. Intracellular repair, e.g. to DNA, is common to all organisms, even prokaryotes.

14. Apart from these complexities we shall find that no multicellular organism has an existence independent from the assistance of *symbiotic prokaryotes* and sometimes of unicellular eukaryotes, so that a multicellular organism becomes, even inside itself, a balanced ecosystem.

Clearly, we must introduce considerable simplification in the discussion since we wish to refer in this chapter to new functions of chemical elements, the ultimate units which, with extra divisions of space and new uses of energy, we believe gave rise to all multicellular living species. Of course, the number of variables has increased dramatically with the above new features. We, therefore, describe first the new compartments (organs) that have arisen in multicellular organisms and the way elements are distributed within them. Two of the other more important novel features listed above are the extracellular matrices and the message systems external to cells. We describe these in Section 14.5 and 14.6 while considering how they could have arisen in evolution. Note again that much of the internal *reductive* metabolism already seen in the prokaryote and unicellular eukaryote remains intact in the cytoplasm of all the cells of all the species. Finally, we have to become more and more concerned about developmental as opposed to steady states in these multicellular organisms.

14.2 Element distribution in extracellular fluids

In Chapters 12 and 13 we observed the advantages that could be gained by separating functions in different compartments, that is, in vesicles of single cell eukaryotes. The compartments then contained different elements, especially trace elements for catalysis or signalling. This evolution of organisation increased in multicellular organisms with the formation of organs, i.e. large-scale compartments, which can be likened to vesicle compartments in eukaryote cells, where the whole animal or plant body has become the equivalent of a single cell. Additionally, as stated above, the multicellular organism has, as a separate new compartment, a controlled extracellular fluid. This fluid contains considerable amounts of organic chemicals, including proteins and even cells such as erythrocytes. Analysis shows that the element distribution in this *controlled* extracellular fluid is strikingly different from the intracellular fluid (Table 14.2) particularly in Cl^-, Na^+, K^+, Mg^{2+}, H^+, HCO_3^- and HPO_4^- concentrations. The control over the fluid is managed by inwardly and outwardly directed pumps differently

Table 14.2 General features of element distribution in extracellular fluids

Elements not concentrated, i.e. evenly distributed	Elements concentrated in cytoplasm of cells	Elements concentrated outside cells or in vesicles
Mg^{2+}*, H, O	K^+, HPO_4^{2-}, Fe, Co, Zn	Na^+, Cl^-, Ca^{2+}, Cu
	C, N, P, S, Se (in organic molecules)	Mn, Ni, Si, (Mo)

* The reference is to ions of elements where charges are shown, i.e. not to magnesium in chlorophyll. Most elements are concentrated in combined forms.

organised in different organ cells. It varies in different plants and animals, although all extracellular fluids have some resemblance (not close) to the sea. Bathing the organs in very strictly controlled extracellular solutions is of very great advantage for signalling between organs and cells. The management of these fluids has become tighter and tighter in evolution, especially in animals, using in some cases the internal precipitation of salts, e.g. bone, which can act as buffers, and by controlling the body temperature so that solubility products are kept strictly constant. (Control of the body temperature of animals removes one variable, of course, in a given species, and might then be used highly selectively to differentiate species.)

Organs used to manage extracellular fluids are the root uptake systems of plants and the digestive intake plus the kidney excretion systems of animals. In higher animals, dioxygen is distributed, as well as simple chemicals such as glucose, via circulation of these body fluids, e.g. the bloodstream, and is carefully controlled. Since plants and animals are very different in their organisation we shall describe separately the essential features of the distributions in plants and animals of elements in organs.

14.3   Outline structure of plants and their element distributions in organs

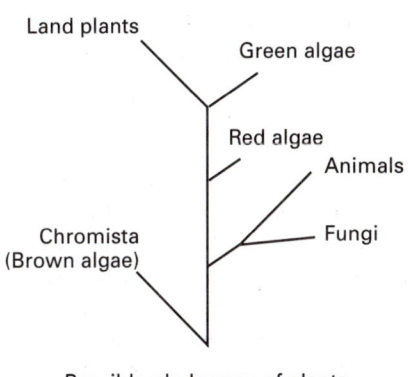

Possible phylogony of plants

It is a remarkable feature of plants that from a single cell in the sea, which could utilise light energy, there developed not just large colonies of unicellular organisms, which do the same thing, but huge plants, first in the sea and then on land (ultimately trees) that evolved organs such as leaves, (see aside) to capture light and carbon high above the major sources of mineral nutrients (soils or the sea bed), and roots, which absorbed the inorganic nutrients from these sources. These organisms could live *between* air and wet land or between sea and sea bed. Linking the two organs there evolved massive structures, stems, trunks and branches, allowing the connection between root and leaf to be further and further extended (Fig. 14.3). Finally, in due season the plant generates the reproductive organs associated with spores or flowering and the scattering of seed. It is startling that this growth is periodic, annual, allowing the rejection of the light-collecting leaf when there is low light intensity in autumn and simultaneously rejecting unwanted chemicals, which had been stored in the leaves, later to be degraded in the soil or in the sea. Meanwhile, during the periods of quiescence or growth, the stem or the outer fringes of the cellulosic structures (wood) of many plants are protected by a skin or bark, full of material poisonous to many invaders—for example many phenols.

In this book we cannot direct attention to all the chemicals and structural compartments involved in these processes which give co-operative survival capability to the organism; instead, we simplify grossly and look largely at the chemical elements involved with some particular functions and which are, therefore, found in particular organs. The most obvious example is the concentration of magnesium in the chlorophyll of green leaves. Equally we note that it is to these leaves, in vesicles (vacuoles), that excess or unwanted (poisonous?) elements are transported, e.g. nickel. The cellulose wood structure is effectively dead skeleton, but the extracellular fluid which constantly moves up and down the plant (see Fig. 14.2) has wound-healing copper and iron enzymes in it, such as laccases and peroxidases. Damage anywhere to the plant is sealed off by the action of these enzymes once the affected tissue is exposed to O_2 or its reduction product, H_2O_2. All of us recognise the browning of cut plant surfaces and we know too of the curious forms of galls that protect a plant after, for example, an invasive act by an egg-laying insect.

Elsewhere, manganese is concentrated in chloroplasts of the leaf cells for the generation of combined hydrogen from water and the rejection of dioxygen. It is also used in haem–iron enzymes, called ligninases, to cross-link and break cross-links of lignins (see Fig. 13.11), though these enzymes are most often found in fungi. Calcium, sodium and chloride are pumped out of the cytoplasm of cells to extracellular fluids or vesicles while potassium and free magnesium are retained, but the distributions of these elements are not identical in root, seed, stem and leaf cells, though generally the first group of elements are high in the vacuoles and reticula. The separation of elements allows, of course, differential synthetic activity. The seed accumulates proteins rich in N, S and P while the polysaccharides form most other structures. A plant is, therefore, a graduated system from top to bottom (and also in the reverse sense) of a variety of elements. Carbon/hydrogen/oxygen compounds from photosynthetic activity go from leaf to root through a vascular tissue called the phloem, while inorganic elements and nitrogen go from root to leaf through a separate system, the xylem (Fig. 14.3). Thus, the plant has to select and distribute in its organs, in quantitative proportions, at least the following elements from the soil: B, (F), Na, Mg, (Al), Si, P, S, Cl, K, Ca, (V), Mn, Fe, Co, Ni, Cu, Zn, Se and Mo (apart from obtaining C, N, H and O), while not taking up unwanted elements, or at least, if they are taken up, hiding these poisons in vesicles, say the large vacuoles of the leaf cells. Since every plant species differs in its chemical nature at a fundamental level, the species must accumulate these elements in different proportions as well as differently in different organs. The differences in ability to absorb energy must also be noted, so that the flow of both fundamental materials (elements) and of energy distinguish one plant species from another. We trust that the reader will see the vast numbers of possible values of these variables, allowing millions of species of plants.

The different distribution of elements carries over to the symbiotic organisms on which plant life depends, for the nature of multicellular organisms is not solved once we relate each species to its own DNA. There is no independence of a multicellular species in this sense. For example, despite the sophistication of the different organs of higher plants it appears that they have not evolved a way of fixing nitrogen directly from N_2 gas. This activity is left to symbiotic bacteria attached to roots of some plant species (legumes) which have a particular catalyst—nitrogenase—containing molybdenum (or vanadium) and an independent DNA. In return, these bacteria take carbohydrates from the host plants. Plants may also get N from N-compounds in the soil, e.g. nitrates.

To maintain all the activity, not just in steady states but throughout in the progression of development seen in all plants, a tight control is required over not only metabolic activities in each organ but also over uptake of elements and synthesis. The activities in each organ must be correlated, and for this purpose plants, like animals, require messengers that go between cells of the organs. They are described in Section 14.6.

Plants, we see, have an essential requirement for only very few elements or compounds that they do not take in or make themselves. Let us contrast this system with that existing in animals.

14.4 Outline structure of animals and their element distributions in organs

An animal is very unlike a plant in that it depends on digestion of plants for its source of all materials. Thus, an animal does not absorb even H, C, N or O from the primary small molecules H_2O, CO_2 and N_2. Neither does the animal take in directly any significant amounts of minerals other than from water itself. Finally, an animal takes energy not from light or the environment directly, but from the photosynthesised plant products, C/H/N/O compounds (food) and dioxygen of the air during respiration. No matter how this system developed, it has evolved around a tubular digestive tract (Fig. 14.4) and see aside. The ingested material is broken down into usable small compounds or even to elements, and passed for synthetic activity and energy production purposes to all cells. A major organ for the monitoring of this supply is the liver and a major organ for monitoring the waste products is the kidney (see Fig. 14.4). The evolution of these two organs (and others) from a single original combined system is a fascinating topic which we cannot describe here.

As the organisation grew in size, flow to all cells required the extracellular fluids to be circulated systematically, carrying dioxygen and other chemicals (Table 14.3). This is achieved by the use of a macroscopic pump and additional tubes, which are effectively organs—the heart, arteries and veins. The fluid, the blood in humans and other vertebrates, is obviously rich in the dioxygen-carrying haem–iron of the erythrocyte cells in it. In

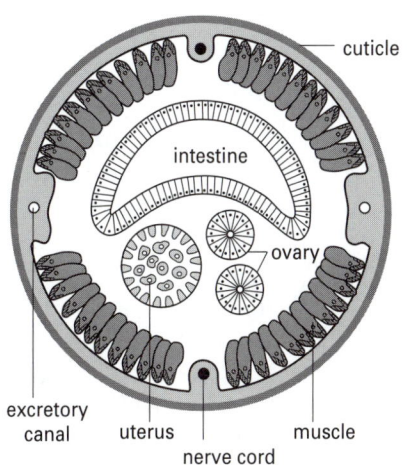

Cross-section of nematode worm

Table 14.3 Novel flow in multicellular organisms

Flow	Example of Function
General extracellular fluid of animals	Bloodstream containing special cells, erythrocytes. Lymph containing special cells for protection
Cerebrospinal fluid of animals	Brain and spinal cord chemical communication in animals
Plant plasma	Movement of chemicals up and down stems (see Fig. 14.3) and note phloem and xylem

N.B. A major problem for an aberrant growth, a cancer, is the need for blood flow to be redirected to it.

some invertebrates, e.g. some arthropods and molluscs, copper proteins carry the dioxygen. Of course, the whole animal needed sensing devices to find food and avoid danger. Thus there developed sense organs—eyes, ears, tongue and nose in higher animals for light, sound, taste and smell, as well as nerves for electrolytic conduction and touch. Obviously, this complexity of different organs had to be organised. From the earliest stages the sense organs used nerves to contact the muscles, e.g. in worms, but the organising ability of the whole could develop further through connections in the blood stream using different messages preparing the organs of digestion and motion for action. In more advanced animals the nerves were linked, therefore, to glandular organs (glands) excreting special chemical messengers to the pre-existing, more primitive organs, (see Fig. 15.6). As complexity grew so did the need for greater organisation. This led to the development of the brain as an integrator of electrolytic (nervous), and chemical (glandular or hormonal) activities. Throughout all these developments of behaviour it was essential to maintain chemical element balance differently distributed in the different organs.

Within animal organs, i.e. in the differentiated cells, as within plant organs, there are very different distributions of elements, especially those stored and particular catalysts. Table 14.4 lists some features of advanced animal organs. Notable are the storage and functional use of iron in the blood, liver and kidney, and of zinc in certain parts of the male reproductive tract and the pancreas. The authors believe that far too little detailed analytical work has been carried out to give a proper appreciation of this selection of elements to allow functional value of organ compartments in either plants or animals to be understood. The reader will see that given this increase in the number of compartments, hence of variables, a vast number of animal species is possible, as is observed.

In passing we note that the brain, as a combination of glandular and nerve organs, has an extracellular fluid of its own, the cerebrospinal fluid (CSF). We return to its description in Chapter 15 given that the isolation of the brain is one step further in the evolution of organisation, whereby it became almost a multiorgan system (the brain) within a multiorgan system (the animal) since it acquired its own environment. Note that it does not

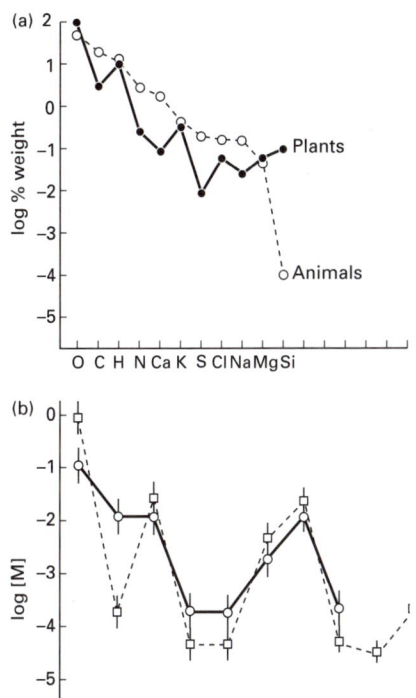

Fig. 14.5 (a) Concentration (logarithm of average weight percentage) of major chemical elements in plants and animals (including mineral deposits). [Data from Calvin, M. (1969). *Chemical evolution*. Clarendon Press, Oxford.] (b) Estimates of the total trace elements in plants (circles) and animals (squares) including precipitates. [M], per cent dry weight.

Table 14.4 Qualitative distribution of elements in advanced animal organs

Element	Element content in organs		
	High	**Medium**	**Low**
Iron	Liver, spleen, blood	Kidney, heart, muscle, brain	Skeleton
Copper	Liver	Skin, brain, kidney	Skeleton, some glands
Molybdenum	Liver	Kidney, spleen	Lung, brain, muscle
Cobalt	Liver	Kidney, heart	Pancreas
Nickel	Kidney, lung	Liver, heart	Muscle
Manganese	Liver, bones	Kidney, pancreas	Brain, spleen, muscle
Zinc	Prostate, muscle, liver, kidney, pancreas	Heart	Brain, lung, spleen
Cadmium	Kidney	Liver	Muscle
Selenium	Kidney	Liver, heart	Muscle
Silicon	Lung, lymph nodes	Liver	

Data from Underwood, E. J. (1977). *Trace elements in human and animal nutrition* (4th edn). Academic Press, New York.

have a separate DNA but it is in itself a separate information depository, based on nerve storage memory systems.

We can now compare the very different element distributions in plants and animals referring all concentrations to the water content which is around 80–90% of the wet weight of all species (Fig. 14.5). One big difference is the higher nitrogen content of animals since they build structure with proteins, not polysaccharides. Proteins are required for movement. Again, the animal usually has higher calcium content used in structures such as shells and bones. The extracellular fluids of plants are more acidic than those of animals, pH = 5 as opposed to pH = 7, which prevents calcium salt precipitation but allows that of silica. Hence, plants are often richer in silicon. They are richer too in manganese, required for photosynthesis, and are relatively low in iron. Plants and animals separated quite early in evolution and no intermediate species are known, e.g. advanced animals that are facultative photosynthesisers. Why is this? There is also a separation of both groups of species from fungi, which are known to concentrate elements differently again, e.g. vanadium in some species, but they have been little studied in this respect.

14.5 The extracellular matrices of plants and animals

During the discussion of aerobic eukaryotes we indicated that in their outer layer structures old and new polymers were generated by *oxidative*

Table 14.5 Major classes of extracellular structures

	Organisms	Major structure	Cross-link	Function
Organic	Plants	Cellulose, lignin	Phenols, alcohols	Relatively rigid exterior (wood)
	Animals	Protein/cellulose, chitin	Phenols	Relatively rigid exterior
		Protein, collagen	Lysine (Amino-acid)	Relatively mobile polymer (ligaments)
Inorganic	Plants	Silica	Cellulose	Rigidifies (opals)
	Animals	Calcium phosphates Calcium carbonates	Proteins	Rigidifies (bones) Rigidifies (shells)

reactions. These new polymers are similar to the lignins, (see Fig. 13.11), chitins (a polymer of *N*-acetyl glucosamine) and proteinaceous filaments such as collagens (animals only) found in the extracellular matrices of multicellular organisms (Table 14.5). The further great value of oxidative reactions, after particular polymers (made inside cells and modified in vesicles, e.g. the Golgi) have been assembled outside cells, is that they can be used to cross-link these external polymer matrices (Table 14.5). It is this cross-linking via various free radical oxidative reactions that allowed semi-rigid extracellular matrices to be made. They could then also be connected to cells and bones as in the muscles of animals (Fig. 14.6). The matrices permitted cells to be assembled (semi-permanently at least) relative to one another in an organisation. Of course, this requires catalytic activity outside the cell in the extracellular fluid. Most intriguingly, the oxidative enzymes used are either haem (largely in plants) or copper (in animals)

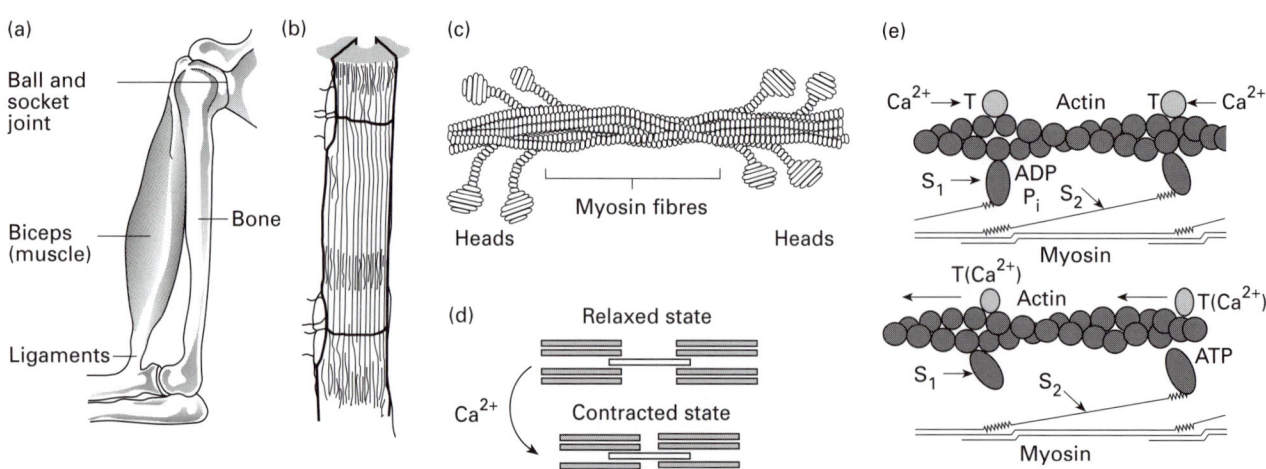

Fig. 14.6 (a) Diagrammatic representation of the frog sartorious muscle. In fact, several cells (b) coalesce to form the muscle units. The cells contain (c) fibrous proteins which form working filaments (d). The muscles link to bone and extracellular collagen (tendons) to form a mobile frame. (e) Diagrammatic representation of the tilting myosin head cross-bridge mechanism, producing a relative sliding motion between actin and myosin filaments. T is a troponin complex activated by calcium in certain muscles. S_1 and S_2 are myosin heads.

Table 14.6 Extracellular oxidases

Organism	Protein	Functional metals
Plants	Ligninase, phenol oxidases	Haem (Fe), Mn Cu
Animals	Chitinase	Cu
	Lysine and proline oxidases	Cu and Fe
	Tyrosinase (phenol oxidase)	Cu

enzymes (Table 14.6). (Remember, copper became available only after dioxygen concentrations rose and removed sulphide, and that free iron, but not haem, became of low availability at that time, especially outside cells; see Sections 13.4 and 13.5.) The use of haem or copper here is associated with two facts: (i) they are good one-electron redox systems, (Fe^{2+}/Fe^{3+}) and Cu^+/Cu^{2+}), at moderately high potential, whence they can produce free radicals; (ii) they could be rejected from cells. The form of iron rejected is in haem. To the authors' knowledge, copper is very generally pumped out of the cytoplasm into vesicles or the extracellular fluids of all cells. The implication is that to bring about multicellular organisation a cell had to evolve to provide a controlled supply and disposition of haem iron and/or copper outside cells.

The haem enzymes involved in the synthesis of the extracellular matrices, especially in plants, often use H_2O_2, not O_2, and saccharides, not proteins (but note lignins: Fig. 13.11), and undoubtedly such systems evolved first. There is evidence that later the extracytoplasmic oxidative activities of copper and iron became co-ordinated. Some animal organisms using chitin also cross-link the external extracellular polymers with zinc co-ordinated to phenols. Such chitin is common in the unmineralised structures of insects and some invertebrate worms, especially in their teeth. Zinc too became more available with the removal of sulphide.

The extracellular matrices of multicellular organisms cannot be permanent constructs since such organisms have to grow and develop. The matrix must be constantly reshaped, especially during growth, and therefore this matrix must be constantly broken down. The breakdown involves either the oxidative destruction (e.g. of lignin) or the hydrolytic destruction (e.g. of collagen) by enzymes. (Chitin appears to be very difficult to break down and often forms a fixed plastic external shell in insects.) The enzymes that developed for these purposes are haem oxidative enzymes (sometimes assisted by manganese) using dioxygen or hydrogen peroxide, especially in lower organisms and plants to break down lignin,* and zinc hydrolytic

---

*As stated in Chapter 13, lignins are particularly resistant to destruction but can be degraded by micro-organisms and fungi, originating the humic acids of soils. In the paper industry, they are solubilised by treatment with disodium bisulphite, originating a waste liquor with sulphite derivatives which is a major source of pollution of rivers (millions of tonnes of such effluent are discharged annually in them).

Table 14.7 Zinc and filamentous structures

Filamentous structure	Role of zinc
Collagens	Zinc collagenases
Proteoglycans	Zinc stromelysin
Denatured collagens	Zinc gelatinase
(Chromosomal proteins)	Zinc structural role?
Keratins	Zinc cross-links

Table 14.8 Zinc in degradation processes

Degradative process	Zinc enzyme
Pancreatic juice action	Carboxypeptidase
Venom haemorrhagic action	Zn proteases
Extracellular digestion (yeast)	Zn aminopeptidase
Extracellular digestion (bacteria)	Thermolysin
Breakdown of DNA and RNA	3' Nucleotidase
Peptide hormones	Peptidases 24.11
Connective tissue degradation	See Table 14.7

enzymes (collagenases, elastases and so on) to break down proteins in animals (Table 14.7). These enzymes are usually stored in vesicles and injected into the extracellular fluid on demand. Once again we stress that the increased availability of zinc due to the oxidation of zinc sulphide by the increase of dioxygen made this aggressive hydrolytic action more possible. In passing, we note the general increase in the value of zinc in digestive processes (Table 14.8) and in message systems (Section 14.6).

While the extracellular matrix is obviously not permanent in growing organisms, it is often so in the fully fledged adult. Thus, some plants have permanent ligninised cellulose structures (wood) and some animals metamorphose, e.g. caterpillars, into organisms with unchangeable 'skins' (cuticles) of chitin which even enclose their bodies, e.g. adult insects. Alternatively, the extracellular matrix may grow by accretion as in the logarithmic growth cone of many a shell, when each layer of the shell polymers is permanent in 'dead' $CaCO_3$ mineralised tissues (Fig. 14.7). The organism here moves continuously as it grows larger to live in a new extracellular matrix, and so forms a conical shell. All of these features are external constructions around the whole organism where the problem of growth is very different from that in the interior.

Due to the (oxidised and therefore charged) nature of the outer polymers of the cells of multicellular organisms it was possible to incorporate calcium ions, as on the outside of prokaryotes, and even calcium salts, mainly carbonates, as on external surfaces of single cell and multicellular eukaryotes. It was also possible to build *internal* extracellular calcium phosphate skeletons such as bone in vertebrates (Fig. 14.6), again using some partially oxidised connective tissue, e.g. collagen. Such skeletons also became a remarkably useful buffer for the extracellular fluids, controlling calcium, phosphate and pH (see refs 108 and 109 for a description of bone). The calcium-binding proteins frequently contained oxidised amino acid residues (see Table 14.14). Plants, with a more acidic extracellular matrix, often make silica opals to strengthen their internal extracellular matrices, while the primitive sponges colonies use silica externally. An interesting feature of bone (and hydrous silica?) is that it conducts protons. The resulting piezoelectric effect of stress on the bone gives rise to asymmetry in an electric field in the bone. Thus bone development follows stresses due to growth. Calcium carbonate is not able to do this.

Before ending this section on the extracellular matrices, the essential role in evolution of the changed inorganic environment of life has to be

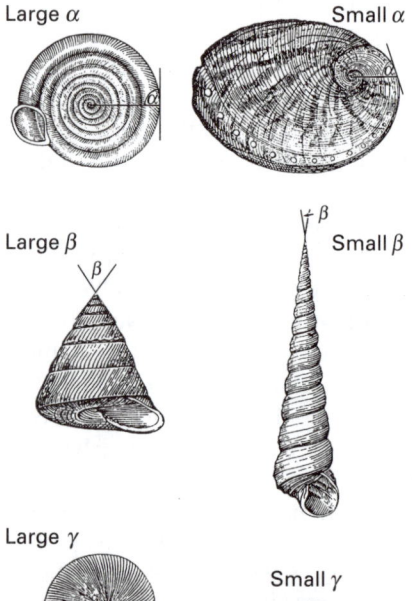

Fig. 14.7 Various gastropod shells showing the effect of the variation of different growth angles. In the top row the shells have large and small spiral angles (α); in the middle row they have large and small enveloping angles of the conical ends (β); in the bottom row there are large and small angles of retardation (γ) which govern the extent to which the whorls overlap. Some parts of shape are DNA controlled, e.g. α, β and γ, but the general expansion is due to body size. (From J. C. Chenu, see ref. 58).

Table 14.9 The successive replacement of Fe/Mn by Cu/Zn catalysts

	Prokaryotes (organelles)		Eukaryotes (unicellular)		Multicellular organisms		
					Plants		Animals
Energy capture	Outer membrane Fe (earliest)	→	Organelles, Fe/Cu	→	Organelles Fe(Cu)Mn	→	Organelles Fe/Cu
Protection in cytoplasm	SOD (Fe/Mn)	→	SOD (Cu/Zn)	→	SOD Cu/Zn	→	SOD Cu/Zn
Extracellular cross-links	–		–		Cross-linking lignin, Mn/Fe		Cross-linking collagen, chitin matrix (Fe/Cu)
Messengers; inorganic	Fe/Mn/P		+ Ca (Zn?)		+ Zn		+ Zn
Messengers; organic	P compounds, substrates		P compounds, substrates		+ Organics oxidised by Fe(Mn)(Cu)		+ Organics oxidised by Fe, Cu (adrenaline)

stressed. In our opinion, *it was the generation of markedly increased copper and zinc, together with the loss of iron supplies (Table 14.9) which forced (or allowed) the evolution of multicellular plant and animal life.* This is not an independent process since it required the development of messengers external to cells in a contained extracellular fluid. They too are very dependent upon copper, zinc and iron (haem) (Section 14.6).

Of course, we have to stress that apart from the general uses of elements in plants and animals, each species is different, not only in variation of element content, relatively small, but also in the energisation of the elements in the compartments and particularly in chemical 'components'. Thus, organisms differ in their nucleotide, protein, saccharide and lipid constitution. However, much as we divide man's industrial transport vehicles today into organised classes—cars, boats, aeroplanes, trains, etc.— and subdivide every class into types with non-exchangeable similar parts, it is not the parallel of the diversity of present-day life with this diversity that interests us so much in this book, as the parallel with the general change in complexity of the evolution of organisation. In industry this is from the whole group of early disorganised movements of powerless simple carts and crafts to modern energised complex vehicles. The analogy, then, is with a change in level of organisation from unicellular to multicellular systems. Further general changes were essential to manage this new level of organisation.

14.6　Cell–cell messenger systems

The new message system that was required in multicellular organisms had to link cells and organs into an holistic unit. Two solutions of this problem are found in multicellular organisms. The first is the use of long, thin, cylindrical extensions of one cell reaching to another—a physical network of organic 'wiring', i.e. nerve axons in animals (see Fig. 15.5).

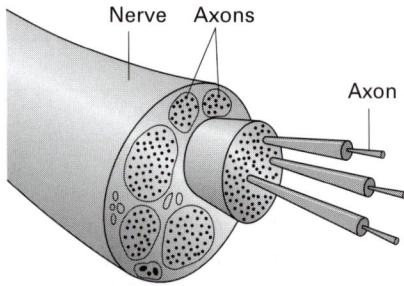

Structure of a nerve

The cylindrical axonal extension of a cell (see aside) can only grow and become useful, of course, within a (relatively) fixed extracellular matrix. It uses, as new messengers, the ions Na^+, K^+ and Cl^-, which are not employed in quite this way in unicellular organisms, i.e. as current carriers along the extensions (axons) of the nerve cells. In fact, a new Na^+/K^+ pump (ATP-ase) evolved to meet this need. The 'nerve' extensions meet at *synapses*, where there is release of organic messenger molecules from vesicles to stimulate neighbouring cells. The stimulation is through calcium invasion at the synapse, as described in the next chapter (see Fig. 15.5). In the case of automatic nerves the network is fixed, as is the relationship between nerves and muscles, but the nerves of brain cells alter their morphology with activity and in response to experience, to form a 'flexible' memory. Thus the brain is *plastic*, in which property it differs from all other organs, and is constantly changing with experience. (Actually it also dies differently and slowly.) We describe it in more detail in Chapter 15. Note immediately the newly energised and organised use of simple ions, Na^+, K^+, Cl^- and Ca^{2+}.

The second new message system involves the ejection from controlling cells (of secretory organs, glands) of organic chemicals into the extracellular fluids. These chemicals are recognised at some distance by target cells and organs (Fig. 14.4). This system is used by both animals and plants, and we treat animals first. An example is the effect of adrenaline, from the adrenal gland, on muscles in animals. These external organic molecules (see Table 14.10), are additional to messengers of the internal cytoplasmic systems of muscle and other cells, namely phosphate compounds, H^+, free Fe^{2+} and some substrates (originating in prokaryotes), as well as to the use of Ca^{2+} ions as a connector between the environment and the cytoplasm (originating in unicellular eukaryotes, see Fig 12.7 and 13.12, but used in more complex ways in multicellular systems), and now even additional to the use of Na^+, K^+ and Cl^- in electrolytic nerve signals (see Table 14.10). These signalling molecules, such as adrenaline, peptides and sterols, control the homeostasis of adult organisms as well as the switch in their activities using receptors on receiving cells, e.g. in

Table 14.10 Some new signals in multicellular organisms

Molecules or ions	Function
Na^+, K^+	Nerve transmission;
Ca^{2+}	signal to intracellular systems generally activated by many messengers in extracellular fluids
Acetylcholine	Ca^{2+} release, junctions of nerves
Adrenaline	Signalling, e.g. to muscles
5-Hydroxytryptamine (serotonin)	Trigger of c-AMP cascade
Peptides	Activation of G-proteins
Sterols	Control of differentiation
Prostaglandins	'Hormones' in cells
Nitric oxide*	Relaxing factor for muscles
Gibberellic acid	Plant hormones for growth
Ethylene	

* NO may have been functional earlier in evolution.

Table 14.11 Zinc proteins related to peptide hormonal action

Hormone	Mode of association with hormones
Insulin	Zinc associated with hormone storage
Angiotensin	Zinc in angiotensin-converting enzyme
Enkephalin	Zinc in enzyme enkephalinase
Several peptides	General hydrolysis by endopeptidase 24.11
Neurotensin	Degradation by zinc-dependent enzyme
Other peptides	Amino and carboxypeptidases (Zn) at synapses

development. At all stages they must interact with the control networks that pre-existed in eukaryotes (see Section 13.9).

We shall see, in Section 14.9, how the circuitry of, say, peptides in extracellular fluids is intermeshed with that of calcium, ferrous ions and phosphate compounds, which affect the inside of all cells. Note the involvement of zinc in the production and destruction of these peptides (Table 14.11). Again, we must see how sterols, which can cross membranes directly, also affect, but now in new ways, the expression of genes inside cells, which is essential to development (see Section 14.9). Furthermore, these new organic chemical messengers must be prepared and handled by new synthetic (and degradative) paths which do not damage primary metabolic pathways. This is frequently managed by limiting their syntheses to vesicles using new enzymes often employing the newly available catalytic elements, copper and zinc. In animals, zinc is especially associated with release of peptides from vesicles and their destruction in the extracellular fluids (Fig. 14.8). Some of the major new organic messenger molecules made and/or destroyed in these ways are given in Table 14.12. Haem enzyme reactions are not normally used in vesicles for the purpose of producing messengers, e.g. sterols, but are used in the outer membrane or in the extracellular fluids, especially of plants.

Notice that some of the peptide messengers are made by oxidation of their terminal carboxyl glycine units, X-CONHCH$_2$COOH, to give amidated

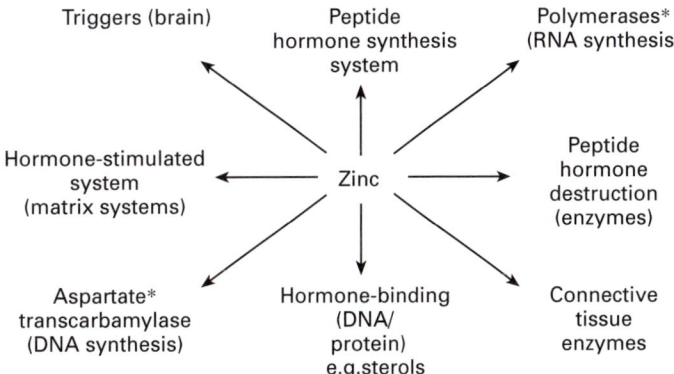

Fig. 14.8 The extensive connections of zinc to multicellular controls and regulations. Zinc links to sterol-like hormones, to peptide hormones, to the external matrix and to a vast list of enzyme and control activities in cells. *Older enzymes, see Tables 11.11 and 13.8.

Table 14.12 Examples of catalysts of messenger reactions in organisms

	Messengers	Enzymes	Producing group	Removing group
Plants	Ethylene		Haem (Fe)	Haem (Fe)
Animals	Adrenaline	Oxidases	Fe, Cu	Cu
	Peptides	Peptidases	(Zn)	Zn
	Sterols	Cytochrome P-450	Haem (Fe)	Haem (Fe)
	Thyroxine	Peroxidase	Haem (Fe)	Se
	Nitric oxide	Oxidases	Haem (Fe)	Haem (Fe)
	Peptides	Oxidases/peptidases	Cu	Zn

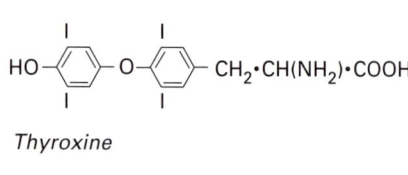

Thyroxine

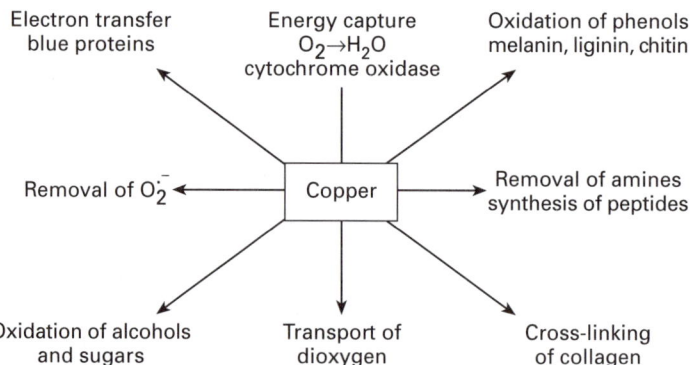

Fig. 14.9 A diagram showing the various activities in which copper became involved after the advent of dioxygen.

peptides, X-CONH$_2$, also in vesicles. The enzyme used for this purpose is a copper oxidase (using O$_2$). Copper is also used in the response to some of these same peptides; e.g. the peptide melatonin activates the production of melanin through the action of a second copper oxidase in the receptor cell. In our bodies this is seen as the browning of skin. In other messenger systems copper acts in synthesis, for example in the production of adrenaline in a vesicle (see Table 14.12), but iron is required as well, outside the vesicle.

While the functions of copper (see Fig. 14.9) and zinc in the synthesis and degradation of the organic messengers are vitally important, especially in animals, there were also developments, probably earlier, of iron (haem) enzymes separately, and in all organisms, not just outside cells. Iron, as well as being a one-electron transfer agent, is a particularly suitable element for oxygen atom transfer at high redox potential. Thus, it can carry out reactions such as

$$Fe^{(II)} + O_2 \rightarrow Fe^{(II)} \cdot O_2$$
$$Fe^{(II)}O_2 + \text{bound } H \rightarrow Fe^{(II)}O + H_2O$$
$$Fe^{(II)}O + RH \rightarrow Fe^{(II)} + ROH \text{ (atom transfer)}$$

There are two families of such iron catalysts in all organisms: one is the P-450 cytochrome oxidases and the other is the non-haem oxidases. The first uses a haem complex and the second a simple iron complex or an iron dimer, where iron is bound directly to N,O donors of proteins. Since both catalysts transfer atoms of oxygen they do not release radicals which would attack essential molecules in the cell, so that they can be used to generate many oxidised organic molecules (messengers) such as sterols, prostaglandins and so on, inside cells on membranes or even in the cytoplasm (see Table 14.10). In this way further new message systems, especially sterols, were devised. Again, they are used more in animals than in plants.

A final messenger in animals, which is dependent on haem enzymes, is nitric oxide. It is produced by a haem enzyme oxidising arginine, a haem protein is its receptor, and it is probably destroyed by another haem enzyme. There seems to be a common theme here in that, frequently, production and destruction use the same catalytic elements. The nitric oxide synthetase is dependent on calcium, thus linking the new and the older signalling systems.

In plants it is found that peroxidases rather than oxidases are often used to produce hormones, even by free radical reactions (Table 14.8). Free radicals can easily cause cancerous growth, but this is a much greater risk to animals than to plants since cells do not migrate in the latter. None of these enzymes is inside a vesicle. There are also a few haem *peroxidases* which are used to produce messenger hormones in animals, but these are not radical reactions. An example is the oxidative chemistry of the halide ions (here, iodide) with incorporation into particular organic molecules (thyronines). Thyroxine (tetraiodothyronine) is an example of many related brominated and chlorinated organic molecules found in aerobic prokaryotes and eukaryotes, such as algae and sponges. We consider that all such compounds were made originally as poisons.

As already stated the removal of thyroxine is by de-iodination using a selenium enzyme. Here we are reminded again that with the coming of dioxygen it was not just metals that changed function. Selenium in anaerobes is a useful hydride transfer agent for reductions, but in the presence of oxygen it is a useful oxygen atom transfer agent for peroxidases, being activated to destroy peroxides and remove iodine from thyroxine. Selenium peroxidases appear to take on an ever-increasing role in higher, especially animal, organisms.

14.7 Dioxygen, multicellular organisms and new metabolic products: summary

We must always remember that the primary metabolism of the cytoplasm could not be changed; the basic anaerobic chemistry of early life had to be preserved. The necessary essential central activity is the production, in a reductive environment, of proteins, nucleic acids, lipids and saccharides. The introduction of oxidative chemistry could only be added-on chemistry, secondary metabolism (Table 14.13), from which the cytoplasm had to be protected. It occurs largely, then, in the extracellular liquids or in vesicles. Most frequently, the new pathways start from existing products of the

Table 14.13 Suggested sequence of development of oxidative enzymes

Time (years ago)*	Enzymes
3.5×10^9	Development of proteins for handling nitric oxide?[†]
3.0×10^9; prokaryotes	Production of dioxygen by *manganese* enzymes
	Removal of H_2O_2 by haem or vanadium enzymes
	Removal of O_2^- by iron and/or manganese enzymes
2.5×10^9; eukaryotes	Oxidation of squalene to give cholesterol
	Use of copper to remove superoxide in dismutases
	Utilisation of O_2 and especially H_2O_2 by haem/copper enzymes, e.g.
	(a) cytochrome oxidases for energy capture
	(b) peroxidases, often with the assistance of manganese (production of chitin and iodination reactions)
	Utilisation of vanadium and selenium in peroxide chemistry
	Use of O_2 by iron enzymes (non-haem)
1.5×10^9; plants, multicellular	Use of copper outside cells in oxidases
1.0×10^9; animals, multicellular	Development of messenger systems based on (or derived from) oxidised or halogenated organic molecules, e.g. NO, thyroxine, sterols

* The years in this column are speculative.

[†]N.B. There is the possibility that the use of NO predates that of O_2 in the evolution of oxidative means of generating energy (see Sections 11.5 and 12.15).

HO—⬡—CH(OH)·CH₂·NH·CH₃

Adrenaline: animal hormone

Hydroxy tryptophan: animal hormone

Indoleacetic acid: plant hormone

Table 14.14 Some oxidised compounds in multicellular organisms (see aside)

Starting material	Product	Function	Catalyst
Tyrosine	Dopa	Messenger	Cu/Fe
	Melanin	Protection	Cu/Fe
Tryptophan	Serotonin	Messenger	Fe
Proline	Hydroxyproline*	Structure	Fe
Lysine	Hydroxylysine,* etc.	Structure	Cu/Fe
Glutamate	Carboxyglutamate (Gla)	Structure	Flavin
Arginine	NO	Messenger	Fe
Cysteine	–S–S–	Homeostasis, Structure	Se
Glycine	Amidated peptides	Messenger	Cu
Cholesterol	Hydroxysterols	Hormones	Fe
Lipids	Prostaglandins	Hormones	Fe

* These amino acids support bone formation on collagen.

cytoplasmic reaction systems. Typically, these are amino acids, even in proteins which are made for extracellular use. Some of these products are listed in Table 14.14. Another source of new materials is the unsaturated fats and terpenes: note especially the 5-carbon unit isoprene. Reference should also be made to the details of secondary metabolism in the usual biochemical texts.

Now, many of the products of these reactions are stored in vesicles and became molecules for signalling, e.g. modified amino acids, or were released directly from membranes, e.g. sterols. They then add considerably to the chemical composition variables. Do not forget that they are not only different in chemical composition but they are differentially energised (concentrated), so that the local free energy change (ΔG) is an important combined variable to take into account (Chapter 5). Table 14.13 gives a possible history of oxidation, leading from protection to the production of cholesterol and then to extensive use of messengers. The development of hydrolytic products such as peptides grew at the same time, and here and there oxidative and hydrolytic paths were used together. While stressing the extracellular and vesicular chemistry of dioxygen it would be unwise to neglect some of its oxidative reactions in the cytoplasm. However, all such reactions should, if possible, lead directly from O_2 or H_2O_2 to H_2O. The reader is referred to books on enzymology which detail reaction mechanisms of cytoplasmic oxidases such as cytochrome P-450 (see ref. 105). The internal enzymes are based on haem, not on copper.

Before proceeding further with this description of the novelty of multicellular organisms we must stress heavily one point: *all of the above developments in evolution arose from the changed availability and distribution of (inorganic) elements.* This was the major continuous change in variables with time, and paralleled the increase in physical compartmental variables.

14.8   New receptor systems: zinc-finger proteins

Receptors for the new external organic messengers were essential and had to be made from different kinds of proteins so as not to be confused with those for internal messengers. The parallel is with the appearance of protein receptors for calcium in single cell eukaryotes. There are two different types of messenger and receptor. First, in membranes, receptor proteins were introduced, e.g. the G-proteins (see Fig. 13.11) for messengers that cannot pass through membranes. The use of this system is amplified in multicellular organisms and so we describe it again here. The membrane receptors act to create a new second messenger, now within the cell, by releasing parts of the molecules of the membrane, e.g. fatty acyl groups or inositol phosphates, where the latter often serve further to activate internal calcium stores (see Section 13.9). (The calcium receptors are those found in unicellular eukaryotes). Alternatively, the membrane receptor can act as a channel which opens on contact with the messenger molecule allowing a second messenger to flow through it, e.g. calcium. In the cytoplasm, the free receptor proteins for the entering message (calcium) may act so as to change tension in the cell filaments (as in muscles) [Fig. 14.6(b)] or to activate enzymes which control metabolism, e.g. kinases (phosphorylating enzymes). In association with the nucleus, a number of these internal (phosphorylated) protein receptors, while binding the messenger or not, may work as transcription factors, altering the reading of DNA (RNA) and eventually controlling the production of new proteins. Some of these latter proteins are then exported. The new receptors of this first kind that have developed most in the evolution of multicellular organisms are linked to both calcium and phosphate internal transmission so as to connect to the

(a)

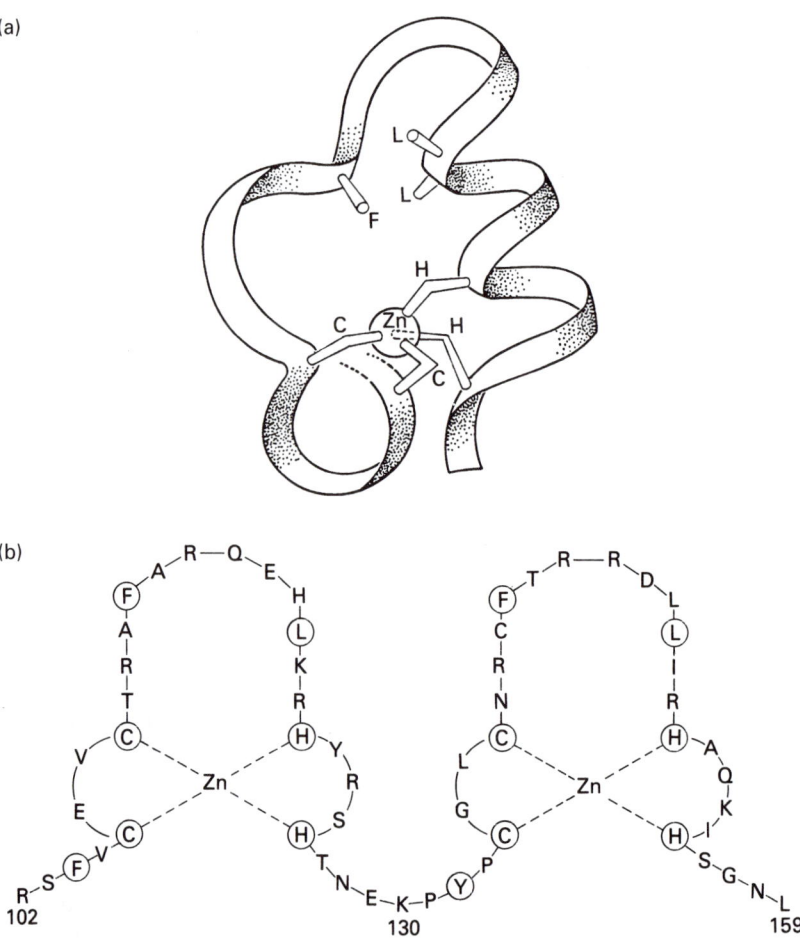

Fig. 14.10 (a) The structure of one kind of zinc-finger as determined by NMR methods. Here, certain amino acid residues are shown: L, leucine; F, phenylalanine; H, histidine; C, cysteine. [After R. E. Klevit and P. E. Wright.] (b) An example of a usual sequence around zinc sites in two zinc-fingers. The circled amino acids are conserved and are either ligands or part of the hydrophobic core.

already-present eukaryote Ca^{2+}/HPO_4^{2-} network. A new range of proteins had to be made for this last purpose, e.g. novel calmodulins.

Certain of these receptors are more complex and have more than one activating agency, which allows cross-communication between different circuits of carriers of information. Thus, some phosphatases, e.g. calcineurin, respond to Ca^{2+}, phosphorylation, Fe^{2+} and Zn^{2+}, and very frequently other receptors are at least responsive to two of the many message-carrying ions or molecules.

Intriguingly, the second group of novel multicellular hormone messenger molecules, those that pass directly through membranes to reach the DNA (see above) are often, if not always, mediated by an extension of the DNA-linked receptors called 'zinc-finger' proteins (Fig. 14.10). (The zinc-fingers are also used in cytoplasmic controls.) Zinc ions with sterols, thyroxine, prostaglandins and related oxidised hydrophobic molecules, see aside thus

Messenger to zinc finger protein

H_3C CH_3 CH_3 CH_3
CH=CH·C=CH·CH=CH·C=CH·COOH
CH_3

Vitamin A or Retinoic acid

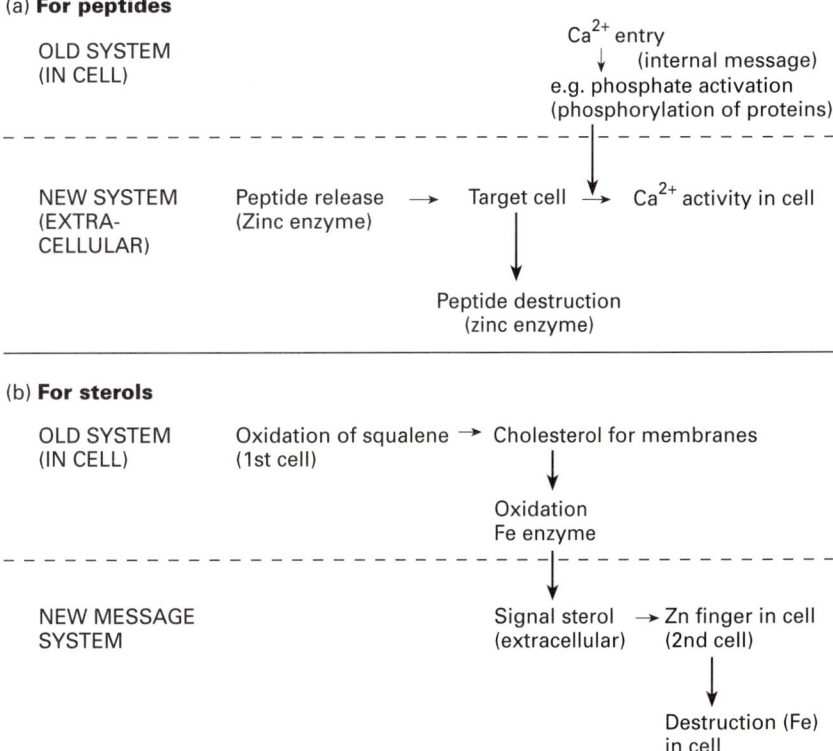

Fig. 14.11 The connection between the old single cell and new multicell systems for use of zinc and iron (Fe), where Fe is usually part of haem.

providing a new long-term message system for the cell *growth control* so necessary in multicellular eukaryotes. Note that zinc-fingers are found to some extent in single cell eukaryotes (see Section 13.10) but rarely, if ever, in prokaryotes. Their increasing use in multicellular species is a feature of evolution (Fig. 14.11) much as is the use of calcium and haem proteins. Zinc, initially a possible poison, and then used in functions, such as digestion, and then as a messenger, has therefore shown a continuous evolution. A representation of many of these connections is shown in Fig. 14.12.

14.9 Interdependence of new and old pathways

We now summarise the continuous evolution of life which can only be based on the introduction of novelty against the background of the most primitive organisation that we see in ancient prokaryotes. The major metabolic pathways to polymers of amino acids, nucleotides, fats and saccharides persist in all cells. They are based upon anaerobic chemistry. The messengers which help to maintain and limit this metabolism are also unchanged through three to four billion years. They are the common substrates, co-enzymes and ions described in Chapter 11. The switch to

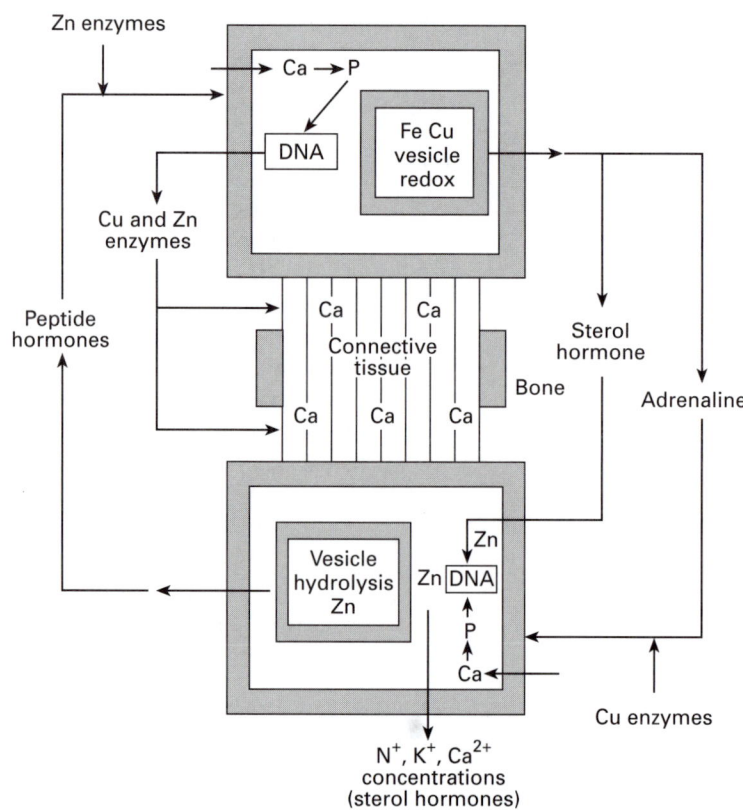

Fig. 14.12 A schematic diagram illustrating the connections between cells of multi-cellular organisms based on inorganic and organic chemicals but stressing the links between different elements.

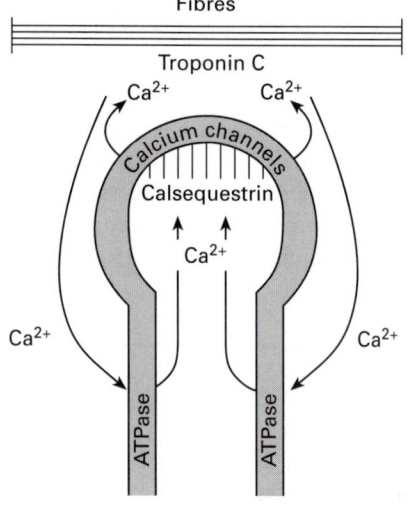

The flow system for calcium at the tip of a sarcoplasmic reticulum of a muscle cell. The calcium channels are placed very close to the contractile fibres and here calcium is stored in a calcium-binding protein, calsequestrin. The calcium is removed by pumps (ATPase).

eukaryotic life forms, probably still anaerobic, did not change the organisation except for one major feature. The introduction of filaments and separate (vesicle) compartments into single cells with enlargement of cell properties came with a new communication network based on calcium ion flow (see aside). This flow characterises both a new steady state and new triggering responses allowing the cell better knowledge of and response to its environment. Calcium became a major player in eukaryote life but it was dove-tailed into the earlier message systems based on, in particular, phosphate compounds (see Section 11.10.2). The next step in evolution, the coming of dioxygen and the new series of reactions it introduced, demanded further additions to the controls. In large part, the oxidative chemistry was kept in compartments, such as in membranes, organelles, vesicles or outside the cell, so that the major essential reactions of the cytoplasm are still those seen in the earliest prokaryotes. Controls over production of these chemicals rested with management of new catalysts based on iron, zinc, copper and so on, all of which were themselves managed by ATP-linked pumps. Moreover, release of the new chemicals was and is managed by calcium signals from the environment. Thus, old and new systems were integrated.

The further step which we are analysing in this chapter is the extension from single cell to multicell organisation. Here, cells in organs act and grow in a mutually dependent manner. Novelty does not rest in the primary reactions of the cytoplasm of any of the cells but in the secondary metabolic products and the communication network, often using them, between cells. Thus, we observe not just modified uses of Na^+/K^+ and Ca^{2+} gradients, but circulating organic compounds going between cells. We have given many of these in Table 14.10. We stress here that these communication modes must link to those originating in the earliest prokaryotes and eukaryotes. Hence we note that many of the organic transmitters and hormones are released in calcium and phosphate compound-dependent processes and many generate signals in cells utilising these same two more primitive messenger systems. The interdependence of calcium and these new organic message systems, often based on oxidation, is seen in that calcium levels in blood have become connected to sterols (vitamin D_3, see aside). The overall consequences of this feedback network of the new and old entities is the controlled pattern of growth of an organism. At present we tend to see the interdependencies of the earliest metabolic reactions and those introduced step by step in evolution in bits and pieces, but it is becoming necessary to appreciate the overall intermeshing of the many flows of chemicals that generate the total organisation.

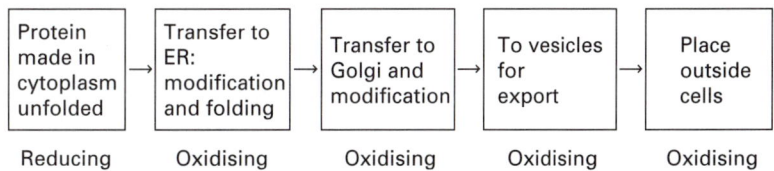

Vitamin D$_3$/Cholecalciferol

14.10 Intracellular vesicular transfer

When cells began to manage extracellular fluid compartments it was necessary to develop larger internal vesicles, namely the endoplasmic reticulum (ER) and the Golgi, to process proteins for external use. The exporting steps became

Protein made in cytoplasm unfolded	→	Transfer to ER: modification and folding	→	Transfer to Golgi and modification	→	To vesicles for export	→	Place outside cells
Reducing		Oxidising		Oxidising		Oxidising		Oxidising

The steps of modification included in the ER, Golgi and vesicles are:

(1) storage of small molecule transmitters (and zinc) as well as calcium;

(2) oxidation of 2SH to –S–S– bridges in proteins;

(3) attachment of sugars, glycosylation of proteins;

(4) folding of proteins with the help of, say, calcium-dependent folding assistants, so-called 'chaperones', e.g. calreticulum and calnexin.

The protein, in its correctly folded form, is then transferred through several stages to vesicles, with or without small molecules and ions. The vesicles themselves can be moved along tubulin polymers or by actinomyosin filaments to an appropriate position in the cell. Finally, their contents may be rejected to the extracellular fluids either for structural, protective or response purposes. (Mineralisation can be achieved in this way.)

It is particularly instructive that some extracellular copper and iron proteins are folded and transferred as *glycosylated* proteins. The implication is that copper(I) or Fe^{2+} pumps put the copper or iron into the ER or Golgi and that folding there is assisted by –S–S– bridges and glycosylation. It is probable that many zinc proteins for export are handled in a similar way. Thus, the ER is extended from a calcium store in early eukaryotes to a device for protein distribution outside the cell.

14.11   Elements falling into disfavour

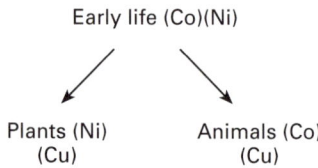

Early life (Co)(Ni)

Plants (Ni)   Animals (Co)
(Cu)   (Cu)

A peculiarity in the development of multicellular organisms in an aerobic atmosphere is that while new elements, such as copper and halogens, have come to be used, one or two elements are now rejected. The major example is nickel. As far as the authors know, there is no protein coded by human DNA that binds nickel, so that nickel is an essential element in a curiously indirect sense (see below). Again, the dependence on cobalt is very small in humans (vitamin B_{12} is a vitamin needed in the smallest amounts), yet cobalt in B_{12} compounds is functionally very important in anaerobes. Even so, the existence of humans depends on nickel and cobalt since these elements are essential for the symbiotic organisms within the human body.

14.12   Internal risks for multicellular organisms

Risks for prokaryotes that are common to all single cells, are:

(1) breakdown of osmotic barriers;

(2) blockage of uptake mechanisms (poisons);

(3) blockage of metabolic steps (poisons);

(4) attack on DNA (poisons);

(5) failure of food supply;

(6) invasion by viruses.

In a sense, some of these are obvious and apply to any single cell. The very diversity of the bacterial population (wild-type) is its main protection, but we have seen that there are devices for nullifying poisons and there exists the possibility of lying dormant (sporulation) until more food arrives (see Section 11.13). The best protection for preservation of these species, however, is rapid division and dispersal.

A eukaryote cell faces graver additional problems and hence, as stressed in Chapter 13, protection is increased by: (1) warning signals, and (2) aggressive responses, but all the time its longer life increases the risk of predation. The eukaryote also has a less flexible range of living conditions (see Section 13.14).

When we come to multicellular organisms, with their own circulatory fluids, the new risks are greater still:

(1) invasion of the organism by unicellular (or even small multicellular) organisms;

(2) internal blockage to flow—haemorrhage;

(3) failure of specific organs due to further increase in complexity and lifetime;

(4) accumulated damage—ageing;

(5) genetic errors generating incorrect cells—e.g. cancer, aberrant growth of cells, particularly in animals where cells can travel.

We do not need to list the huge range of diseases known in humans, but undoubtedly parallel lists could be drawn up for all advanced animals and plants. Such is life. The saving factors are in part old and in part new:

(1) repair mechanisms from the molecular level (DNA) to whole organs;

(2) protective devices of great sophistication:

    (a) the immune system (acting against foreign material);

    (b) the blood clotting and declotting mechanisms;

    (c) regrowth of tissue through *locally* permitted de-differentiation;

    (d) new enzymes to control radicals, e.g. ascorbate oxidase.

To understand this battery of protective devices stemming from new and enlarged DNA is to appreciate how gene evolution matches the new problems, but the success of humans does not rest there. Humans have, like other animals but to a larger degree, a conscious appreciation of the environment, whence the risks to life can be assessed and humans can learn (through their brains) to overcome these risks. The opportunity for man then lies in a quite novel evolutionary development unconnected with genes: to employ brain power to manipulate the environment rather than

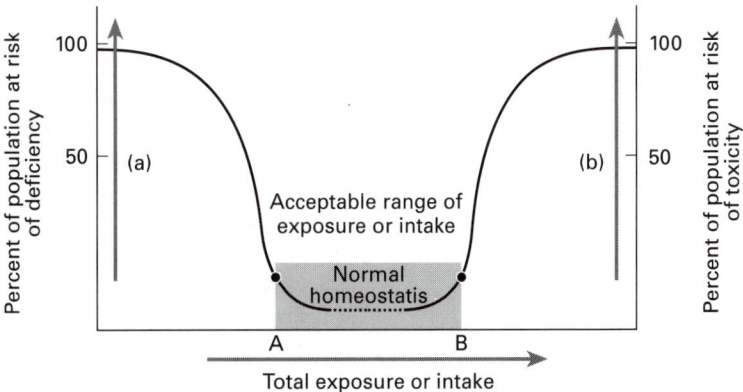

Fig. 14.13 Per cent of population subjected to (a) deficiency and (b) toxicity effects according to exposure or intake of the essential trace elements.

to wait for genes to find new modes of protection. In the next chapter we describe this evolutionary chemistry *external to man*, which involves using a greater variety of elements. Of course, it brings new risks as any extension of organisation is bound to do. In fact, even all new uses of elements, though advantageous, introduce risk at high internal levels (Fig. 14.13), as do all activities. In the next sections we indicate some special internal risks to life which have arisen since the advent of dioxygen. [In passing we note that programmed cell death (apoptosis) is apparently linked to excessive entry of calcium, and the immune system can give rise to auto-immune diseases].

14.12.1 Risks of, and protection from, dioxygen reactions

By itself dioxygen is not a risk since it is not very reactive. However, as stated in Section 13.4, in contact with biological systems, which are reducing by their very nature and contain metal ion redox catalysts, dioxygen generates increasingly reactive and damaging species

$$O_2 \quad \rightarrow \quad O_2^{\bullet -} \quad \rightarrow \quad H_2O_2 \quad \rightarrow \quad OH^{\bullet}$$

dioxygen   superoxide   peroxide   hydroxyl radicals

The immediate necessity, therefore, was to protect against the two chemicals superoxide and peroxide, as described in Section 13.4. As we noted earlier, initially, the superoxide was removed in prokaryotes by Fe(Mn) superoxide dismutases, but later in eukaryotes, the same reaction utilised a Cu/Zn enzyme (see Table 14.9). (Note again the increased use of these two elements.) All organisms remove H_2O_2 using catalases.

However, H_2O_2 can also be put to use and we find a range of H_2O_2 peroxidases in many organisms as haem, vanadium or selenium proteins. One value of some of these peroxidases, as stated in Section 14.5, is that their oxidation of organic compounds can be used to generate cross-linking protection and stabilise structures *extracellularly*, or to produce hormones or protective devices in extracellular fluids or vesicles. A second use is in direct attack on an enemy. Thus, deliberately, when under attack, plants especially block catalases and use H_2O_2 defensively.

A different use of reduction of O_2 is found in animals, where special cells produce $O_2^{\bullet -}$, not H_2O_2 directly, to attack invaginated foreign organisms. The cells are called myelocytes and are thought to be able to use oxidation of even chloride to give very aggressive chemicals such as ClO^-. Again, protection against foreign chemicals is afforded by O_2 giving rise to hydroxylation employing the enzyme cytochrome P-450; the hydroxylated molecules are made water soluble by conjugation to sugars and are then excreted. Most of these enzymes in plants and animals are based on haem (iron) catalysts. Algae and fungi also use peroxidases based on vanadium, which are usually haloperoxidases. Vanadium, as vanadate, became available by oxidation of VS_2 much earlier than did copper from CuS.

We believe that the sequence of protective and then useful catalysts in O_2 reduction probably followed in a time sequence

Haem (Fe)	Haem (Fe)	Haem (Fe)
Fe or Mn	Fe or Mn	Fe or Mn
	V, Se	V, Se
		Cu
Earliest use of O_2	Early eukaryote use of O_2 (note especially plants)	Dominant modern use of O_2 (note especially animals)

It is notable that not only do animals avoid the use of haem but they also avoid manganese in the step $Mn^{2+} \rightarrow Mn^{3+}$ in extracellular oxidative systems, both of which readily produce radicals, a step which is currently used in plant life. Very curiously, while plants and animals use ascorbate as a free radical-trapping agent outside (and in plants inside) cells, humans cannot synthesise this chemical, which has become a required vitamin. Note that ascorbate oxidase is either a copper (animals and plants) or a haem enzyme (plants).

14.12.2 Risk of invasion by lower organisms: sensing self and protection

The very fact that a plant or an animal had become large and rich in materials and energy, which could be used to sustain any form of life, prokaryote or eukaryote, naturally led its whole body to be invaded by lower and much smaller organisms. The parallel is with the invasion (and later incorporation) of bacteria into eukaryotes (as organelles). This invasion has its advantages and its disadvantages. Microbial life, once trapped, can be used to advantage, for example in the supply of nitrogen to plants by certain bacteria, and the human body also plays host to vast numbers of unicellular organisms that sustain it. At the same time, many invaders are only self-interested and must be removed. These invaders, now including both viruses and bacteria and even protozoa and animals (worms), are classed under the diseases they cause. The plants and animals had to combat them chemically, but to do so they had to sense friend and self from foe. Both plants and animals, then, have responses that recognise and kill, or at least contain, the invading viruses, bacteria or lower organisms. Table 14.15

Development of "immune" systems from external poisons

Plants	Animals
H_2O_2 release	(O_2^- release)
Salicylate	Myelocytes
[Internal]	Macrophages Lymphocytes (Antibodies) [All internal]

Table 14.15 Examples of protection against invading organisms

Mode of attack	System
Hydrolysis	Antibody recognition; invagination and phagocytosis; lysozomal degradation (white cells of animal blood)
Oxidation	Superoxide and peroxide generation; (capture by special cells: leucocytes)
Encapsulation	Extension of external matrix (galls in plants)

lists some of the protective devices and we refer the reader to books on the immune system for details (see refs. 108 and 109).

Internal sensing must distinguish self from non-self, but it cannot be done perfectly. Consider a self-cell which loses control over growth, i.e. a cancer. This 'disease' is the result of a multiple mutation, not of an invasion. Internal sensing frequently fails to detect such changes. To a large degree this had to be the case, since evolution itself depends on the failure to note and constrain damage to DNA, and on the ability to accept internally foreign cells, starting with the acceptance of bacteria in eukaryotes giving rise to organelles, and the development of various kinds of tight symbiosis, and including even the acceptance of the fetus by the animal mother. Advantage and risk are intimately mixed. (Note that in the human lymph the cells are constantly inspected by lymphocytes, some of which recognise 'rogue' cancer cells and others which kill foreign organisms. A very interesting protein in protection from aberration is called p53, which is a zinc protein and which on undergoing mutation appears to allow cancer to develop.)

14.13   Growth and development patterns

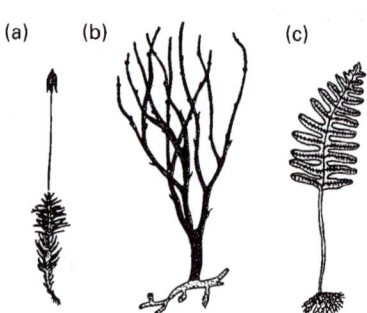

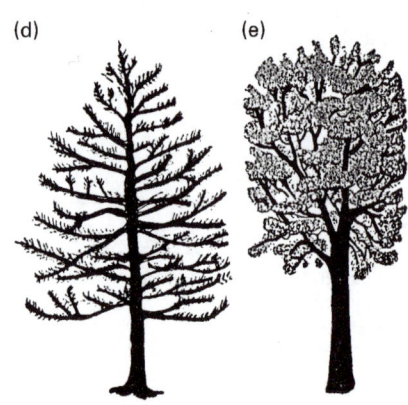

Development of plants:
(a) Moss (b) Psilopsid (c) Fern
(d) Conifer (e) Flowering plant.

An increasing problem as the lifetime of organisms was extended was that the message system not only had to respond as quickly as possible to threats or to immediate advantage but also had to drive the modification of the organism during growth and development, over periods of days, then months, then years. In many species the messenger levels are, therefore, geared to day and night, the yearly cycle or even moon cycles, as well as to the state of growth. The day/night cycle is controlled by light in both animals and plants, while the yearly cycle is controlled largely by temperature. All kinds of cyclic features are then found in multicellular organisms, for example yearly rings in trees and in the gravity apparatus (the otoconia) of fish. It is now well recognised that enzymes can be switched on and off by relatively small changes in temperature, which allows cyclic concentration of messengers to control many activities. Harder to appreciate are the long-term changes during the growth of multicellular organisms (see aside) including the fixing of the sizes of organs. Here it is most likely that cell–cell contact and such properties as membrane curvature cause changes in membrane activities. All such constraints can be formulated in a mathematical way so as to give limits to size and shape or to sudden changes of patterns of growth associated with particular size and shape. It is likely that all these phenomena are associated with critical concentration levels of messenger (hormone and transmitter) molecules. They are, of course, linked to the concentrations of simple inorganic elements. It is undoubtedly the case that changes in such elements as calcium stimulate quite remarkable changes in cells, e.g. in fertilisation and apoptosis.

An example of a switch in morphology which can be induced by (artificial) increase in the level of a messenger is the effect of thyroxine (containing iodine) on the axolotl, the larval form of certain salamanders.

This organism normally stops development at the tadpole-like stage, but the addition of thyroxine to the waters in which it lives causes it to develop into a normal amphibian. There are, of course, many known switches in hormone levels such as those associated with morphic changes in animals, for example, human beings. We return to this topic in Section 14.17, but note that in this book we are concerned largely with steady states of organisms and not developmental patterns. We refer the reader elsewhere for treatment of these problems (see refs 108, 109, 151, 154, 156).

14.13.1 Shapes and multicellular life

It is a feature of co-operative binding that it leads to preferred shape. We have discussed this property in Chapter 7 and again in Section 10.15. Whereas shape is usually locked into *thermodynamic stability* in inanimate systems, i.e. ΔG values, and changes with the conventional variables, the shape of an organism is linked to *survival value*. The implication is that flow patterns of chemicals in a large organism must be not simply the best chemical survival kits but have to be compromised to match the physical functional value in given surroundings. For example, animals on land hold their heads high, while fish hold them horizontal to their bodies. Rather than extend this discussion we draw attention to refs 58, 180, 188 concerning the values of shapes of living organisms. Note particularly the

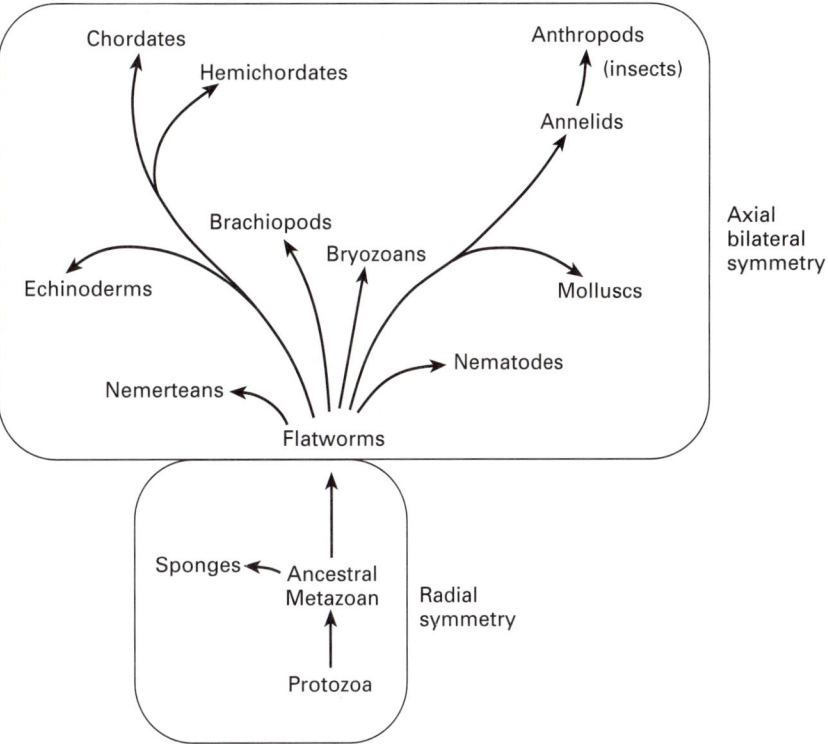

Fig. 14.14 Distribution of symmetry within the animal kingdom and the development from ancestral worms to modern animals. [After Barnes, R. D. (1987). *Invertebrate zoology*. Saunders College Publishing, Philadelphia.]

increasing influence of Ca^{2+} in gradients on morphogenesis. An evolutionary scheme for animals is given in Fig. 14.14. Our view is that the shape of an organism and its physical chemistry (internal patterns of flow) are part of a unity for survival value, ΔL, much as the free energy of a crystal derives from the shape and the composition generating stability (ΔG). Do not forget that shape is a consequence of the chemical activity internal to the organism.

There is a curious relationship in plant morphology to a set of numbers. These numbers, the Fibonacci series (ref 242), give a pattern of leaves or petals in space. The suggestion is that such a patterning derives from a restrictive relationship between how a system can grow and the spatial packing to secure optimal advantage—exposure to light or air. The whole can be given a complex mathematical logic.

14.13.2 Cell death: apoptosis

The idea that a cell is programmed for development is not new, but that it should have a programmed death has only recently become accepted. It now appears that each cell contains genetic information related to its life span and that this span is different for cells in different organs. Red blood cells in humans last for some 30 days and are constantly replaced. Each cell of the brain can last tens of years and is not often replaced upon death. The death rate also appears to be related to the species—long-lived species, such as elephants, have long-lived cells. The triggering of death is a protection against mutation of offspring cells which are more exposed to risk than the parent stem cells. The actual triggering of death may itself arise through the accumulation of damage which eventually allows excessive calcium into the cell. This calcium switches on proteases internal to the cytoplasm. However, the phenomenon of cell death, like other activities with timed rhythms, such as day/night changes, is likely to be genetically complicated and is ill-understood at present.

14.14 Summary of multicellular organisation

The description of multicellular life is far from complete. At one level we see that organs now communicate with one another (Fig. 14.12). The communication involves a multiplicity of elements (Na, K, Mg, Ca, P, Zn, Cu, Fe and Mn, as well as organic molecules) and a multiplicity of compartments. Thus, in any part of the organism there is temporary homeostasis of all these elements. At a second level we have to examine the way the organs (organised cells of a somewhat similar kind) also communicate internally, largely through organic chemicals and Na^+, K^+ and Ca^{2+} ions. The organs (glands) of the endocrine system of the human body are shown in Fig. 15.4. Clearly the description of activity of a whole organism in a steady state is difficult. Furthermore, we have to connect this system to others for discontinuous action, e.g. the digestive tract and the muscles, as well as to the brain. Finally, we have to see that this set of interactions, like the cell 'cycle', is developmental and not really a steady-state activity. The task is only in its infancy. Superimposed upon this activity of a given organism is its ability to reproduce and to die. We insist that to understand these complexities of living forms we must concentrate upon the change of variables, external *availability of elements* distributed within the altered internal confined *space* of organisms and its continuous *time dependence*.

14.15   The variety of multicellular life

There are undoubtedly many prokaryote species and, equally, there are vast numbers of single cell eukaryotes. The most obvious feature of life around us, however, is the variety of multicellular organisms, especially plants and animals (Figs 14.15 and 14.16), which relates to the large size of their DNA. A further feature of this variety is its regional distribution. Nowhere is growth and diversity of plant (and animal) life so varied as in a tropical forest. Even small changes of temperature and humidity especially, though soil quality too is a very important factor, change (select) the plant species that are able to survive (see Fig. 14.18). In all probability, therefore, the restriction on variety and the varieties that succeed are due in part to the influence of such factors as temperature and the availability of water, which then affect the availability of elements. The suggestion that arises is that multiplicity of species is in part due to the rate of energy input—little grows north of the Arctic Circle. If this analysis is correct, then, given an adequate soil which is not necessarily dependent on geographical region and temperature, the variety of plant life is closely determined by photosynthesis. Energy flow (as well as temperature, since higher temperatures allow access to the uses of energy through lower barriers to reaction) and

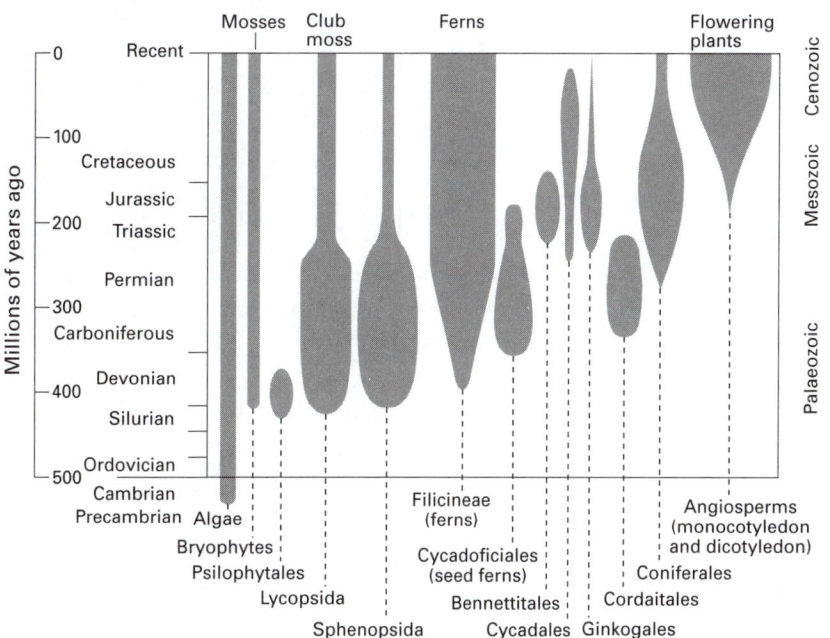

Fig. 14.15 The history of the plant kingdom. The width of the shaded area gives the magnitude of the population. At the left time *periods* are shown, and on the right the *eras*.

element availability are therefore the major variables for multicellular life, much as they were the major variables analysed in living systems in earlier chapters. Just as the huge variety of organic chemicals is not due to the diversity of elements used to make up empirical compositions but to the variety of stationary energised spatial arrangements, which a small number of elements can have when used in large assemblies, so, variety of living species is not due to possible empirical chemical composition variation alone, but to the diversity of rates of *directed* energy and patterns of material (element) flow and usage.

14.16 Variables of multicellular organisms: Increased complexity

It is useful to have a summarize here the new variables of multicellular organisms that we have mentioned in this chapter and which give rise to their enormous diversity. It is these variables that caused the explosion in the number of species while within each species the handing on of DNA maintains each line but with opportunity to uncover new functioning machinery (see Section 16.10). The size of the DNA means, of course, that relatively small differences allow separation of species, since even small changes may represent huge changes to internal complexity. The complexity arises from the following new variables:

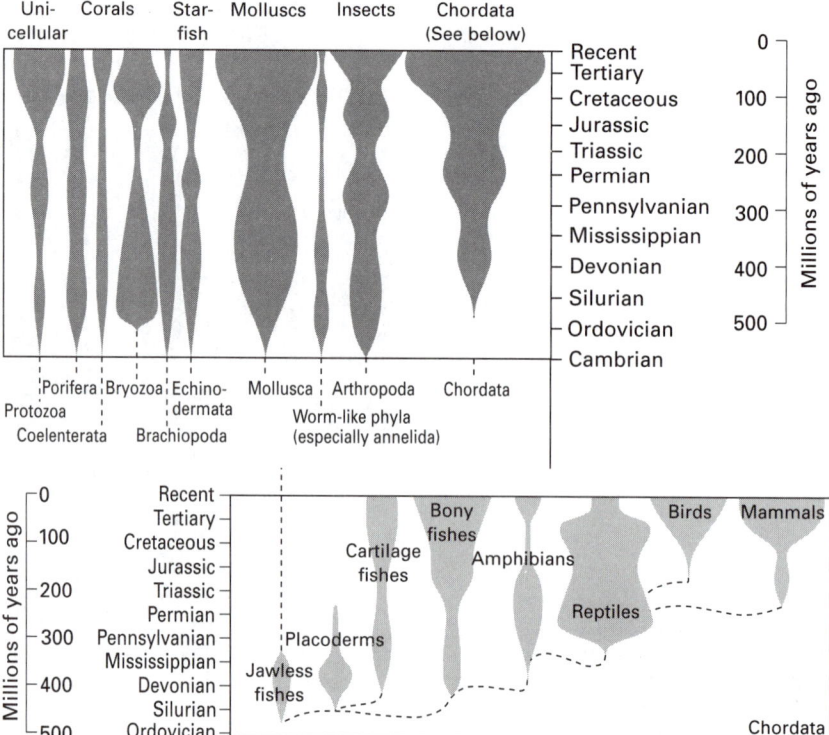

Fig. 14.16 The history of the animal kingdom. The width of the shaded area gives the magnitude of the population. The invertebrates are shown above, the vertebrates below.

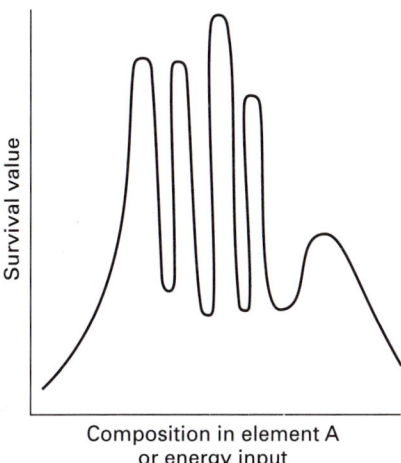

Fig. 14.17 A plot of survival value against a variable such as chemical composition of the environment or energy input from the environment. This is a landscape diagram. Note the sharp variation.

(1) new elementary chemicals, mainly in vesicles and in extracellular fluids;

(2) new polymers in external matrices, such as structures, receptors and protective devices; organ/organ organisation;

(3) new intracellular and extracellular flow pathways;

(4) new energisation of all these systems—maintained synthesis, tension and concentration gradients;

(5) new mineralisation.

The increase in complexity, i.e. in the number of variables related to composition, energisation and flow rates, has demanded, in any one species, tighter and tighter controls. Some of these we have outlined under risks, together with the greater complexity introduced to combat them, under protection (see Section 14.12). The most important observations are, however, that factors such as the composition of extracellular fluids (in particular), the internal temperature of the organism and the control over organic and inorganic chemicals present in the body are critical for survival of the multicellular organisms, which are very sensitive to changes in any of these variables. Plotted on a landscape diagram we see now a more rugged contour (Fig. 14.17) than for earlier, simpler organisms. An example, the differential growth of tree species with conditions, is shown in Fig. 14.18. The implication is that different species may differ only very very slightly in certain characteristics while differing considerably in others.

This being the case, we should not be too surprised that organ transplants using organs from a different but related species can frequently work well. The most obvious case is that of the grafting of trees, where the root is from one species and the higher reaches of the stem from another. The supply of chemicals to and from the root of two different species is sufficiently similar so that it does not change the higher reaches of the hybrid plant. However, new shoots from the roots or below the graft are

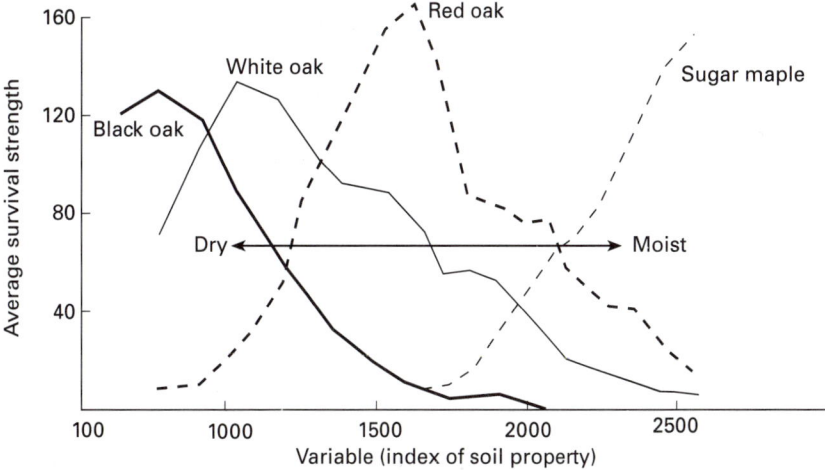

Fig. 14.18 Relative importance of several species of trees in forest communities of south-western Wisconsin arranged along a continuum index. Soil moisture, exchangeable calcium and pH increase to the right on the abscissa. [After Curtis, J. T. and McIntosh, R. P. (1951). *Ecology*, **32**, 476–96.]

Present day

Orangutan Gorilla Chimpanzees Human

10 million
years ago

Ancestral
hominoid

Possible evolutionary relationships: The
chimpanzee has 98% the same DNA as
humans.

'sports', true to the lower reaches. Thus, the root sends only some cellular materials and growth signals appropriate to the upper reaches of both plants above the level of the graft. Presumably, selective extra signals come from exposure to light. Meanwhile, the higher reaches of the plant send the energy and basic C/H/O material which they alone collect for the purposes of either the lower region or themselves. The two species must be sufficiently similar for the graft to take, of course. Rejection is also the fate of many organ transplants in higher animals, which have a more stringent limitation on variables through the immune system, but increasingly it is proving possible to use organs from an animal such as the pig in transplants into humans (and we are not very far from controlling the mechanisms that cause rejection). The overall implication is that at least some organs from different species are sufficiently similar (see Aside) that they can be exchanged and function is only somewhat impaired. Most importantly, the energy and material can come from a source governed by one DNA and utilised by a receiving organ governed by another very similar DNA. Machineries can be added on to one another without changing DNA in any cell. The purity of the DNA does not decide whether the body can live or not. However, the whole assembly cannot now reproduce itself. To do that the DNA itself would have to be adjusted in the germ cells.

Another example of combined 'organisms' is that involved in doubling or tripling the number of chromosomes, called polyploidy, usually seen in angiosperm plant species, and which has been a major technique in the breeding of plants for agricultural purposes. Polyploidal stem cells have DNA from two or more different but closely related species and they can generate a new species, e.g. hybrid maize.

What this shows is that living machines will accept parts from other very similar machines but they then produce a mixed organism that cannot reproduce. Mixing the DNA, however, has to generate a new species that reproduces or fails. Survival of a given organism is different from reproduction of a species: consider a horse/donkey cross—a mule.

As we have stated, the complexity of multicellular organisms brings with it much increased risks and thence the need for much improved protection. The difficulty with this increase in refinement is that it lacks flexibility, so that, as stated, landscape diagrams (Fig. 14.17) where we plot survival value against a variable, are likely to show much steeper contour changes. Three features follow: (1) an alteration in the variables external to the organism is likely to be extremely deleterious; (2) greater survival is achieved by increased dependence on simpler organisms for essential chemicals— symbiosis—which reduces the complexity encoded in one DNA; and (3) advantage will accrue to those organisms that can improve their own environment, e.g. humans (see Chapter 15). We look next at environmental changes during the last one billion years on Earth.

14.17   Survival and environmental fluctuations

The survival of the fittest, that is of those organisms that occupy peaks in landscape diagrams, is easily understood if the conditions of the so-called competition are fixed and, of course, especially where there are large differences in relative fitness. However, when the differences in fitness are small and fluctuations in the environment are considerable and rapid, then fitness is also a fluctuating factor. Obvious properties of the environmental fluctuation are day and night temperatures, daily rainfall and wind, seasonal changes and sometimes chemical variation, say in tidal zones. In such circumstances it is to be expected that many species co-exist. Around us it is not survival of one or two species that we observe but that of vast numbers of very similar species in the same environment—plants in a meadow, insects in a wood, fish in the sea. The question facing the survival of the fittest is 'Fittest for what?' Generally speaking, then, we also expect extinctions of species to occur only due to more persistent temperature changes or chemical variations in the environment.

One example is the dependence on levels of dioxygen in the atmosphere that is critical to the diversity of multicellular life. If the level falls then many species may disappear and on return of the level of dioxygen there is no certainty that the same species will reappear. The fossil record indicates many mass extinctions of organisms and it has been suggested that at least one of these was caused by a loss of dioxygen over a period of time around 250 million years ago [see Wignall, P. B. and Twitchell, R. J. (1996). *Science*, *226*, 1155]. Remember that loss of dioxygen has a major influence on the availability of many elements.

Another possible cause of extinctions is the dual effect of physical bombardment of Earth by meteorites of unusual chemical composition. A meteorite (or a giant volcano) can also cause a huge dust storm, with resultant years of winter, and it may also introduce unusual elements, e.g. iridium. Plants, as well as many animals, are easily lost if the energy supply is seriously reduced—for example in a nuclear winter.

Some authors, however, argue that mass extinctions may not be the result of calamitous events, but are driven naturally by the internal dynamics of the global ecosystem, i.e. the complex interactions between the communities of species and their environment. The global ecosystem may be considered to be in a 'steady state' of 'self-organised criticality' which can very fragile and respond non-linearly to even small changes of, say, temperature or oxygen levels, amplifying dramatically their effects. The extent of the response is, therefore, determined by the internal dynamics of the system, and not so much by the magnitude of any external event.

14.18   Analogies between static and dynamic changes of state

In Chapter 5 we saw that there were co-operative switches in static states of systems at equilibrium which could not be described by the properties of single or pairs of molecules. Amongst these are changes of state including melting, boiling, allotropy and changes of physical properties, such as the co-operative switches in metal/insulator transitions, superconductor/conductor transitions and the onset of ferromagnetism. These are all 'emergent' properties of bulk systems. In the discussion of such changes there can be no involvement of the variable time. A general name for such changes of state is 'phase changes'. These changes do not obey a simple linear connection between the variable, say temperature, and the property concerned, although they are all time-independent changes of spatial arrangement *at equilibrium*. The switches are related to the balance of order and disorder, and we showed in Chapter 7 that these co-operative order $\rightleftarrows$ disorder switches occur also for all structured polymers such as proteins. It is then obvious that just as non-catalytic proteins can switch states (allosteric changes; compare with allotropic crystals) so can enzymes. However, enzymes have the added property of altering material and energy transfer and then rates of change which in turn are linked with flow patterns in space in cells. Even in a single compartment the pattern produced has a non-linear relationship to its variables since it is dependent on feedback, so that compartments are clearly complex systems.

Now, in living systems, as we consider multiplying compartments from prokaryotes to eukaryotes to multicellular life, in effect there is an *increase in complexity associated with co-operativity* whenever new (feedback) interactions between compartments appear and new emergent properties can arise, such as metamorphosis. In living systems all these new properties depend equally on spatial distributions of material, as for phase changes, and on timed energy and material input. Thus, switches in co-operativity can occur in growth patterns in multicellular life. On isolating a single protein, such properties of the time dependence will be lost. In effect, a flow system in a co-operative organisation has quite new features compared with a bulk, static, inanimate system. The parallel with inanimate systems is that their complexity is due to the statistics of spatial and energy distributions, while in dynamic systems there is this complexity and that of the added variable of the time distribution of both material and energy in patterns (organisations). Therefore, surprising as it is that cells adjust to new reactivities and new patterns as they develop into larger and larger assemblies, there are parallels with inanimate static systems (Table 14.16 and see aside). Both are features of the complexity of relationships between properties and variables. Perhaps we should not just sit back and be amazed that a caterpillar metamorphoses into a butterfly. After all, water vapour can become ice. The DNA has not changed, the amounts of energy entering and leaving may not be very different, neither are the basic elementary chemicals of the two forms. It is just that when a certain critical level of activity is reached new controls are activated and the system of

Cooperative phenomena

Critical condition

Phase change (equilibrium)   Metamorphosis (flow balance)

Allotropy (Allostery)   *Homeosis*

Table 14.16 Morphic changes in inanimate and animate matter

Materials	Emergent properties
Inanimate materials (generally)	Phase changes: solid $\rightleftarrows$ liquid $\rightleftarrows$ gas; Solid $\rightleftarrows$ solid (allotropy)
Proteins (DNA, RNA)	'Phase' changes (folding); allosteric switches
Animals, plants	Switch of flow pattern; metamorphism
All living cells	Reproduction, differentiation

flows is critically altered so that a new steady-state pattern of energy and material becomes dominant. This is a characteristic of *complex dissipative systems* far from equilibrium. Thus, in landscape diagrams, changes in flux, plotted on the *x*-axis, may give rise to closely associated species or metamorphic forms, just as temperature on the *x*-axis can give rise to changes of state or allotropic forms with inanimate materials.

Finally, since we have observed that species are not independent, this overall system is also time dependent and complex, and it may exhibit new emergent properties, such as population dynamics of an ecosystem. The reason for this behaviour is due to the nature of feedback connections between the different organisms comprising the ecosystem. Thus, the whole system is *complex* in its mathematical, non-linear relationship between variables (and since it is energised it is also always at risk).

14.19   Summary

The major general features of multicellular organisms were outlined in Section 14.1. In essence, complexity of cell organisation could increase once cells became attached to one another in fixed spatial patterns through newly devised extracellular matrices. This advance, we propose, was due to the use of oxidative cross-linking of extracellular filaments employing novel haem and then copper enzymes. By itself this development did not allow organised intercellular activities—messenger systems going between cells were essential. Once again they are often dependent directly on oxidative systems employing, most frequently, haem iron and copper enzymes to produce molecules such as sterols and adrenaline. Indirectly, oxidation of sulphides had also increased the availability of zinc. In particular, we noted its widespread application to degradation of connective tissue and peptide messengers, and in receptors (transcription factors) for the newly developed sterol and related hormonal messengers. Its use increases too in the evolutionary progression

Prokaryote $\rightarrow$ Single cell eukaryote $\rightarrow$ Multicell eukaryote

In parallel with these developments in the diversity of (extracellular) polymers, of messengers derived from organic molecules and of trace

Table 14.17 The evolution of pumps

Pump	Organism
$Na^+/Ca^{2+}/H^+$ exchangers Ca^{2+} ATP-ase	Primitive prokaryotes Eukaryotes (anaerobic?)
Cu^+ ATP-ase Zn^{2+} ATP-ase	Evolved with increase in dioxygen
Cd^{2+} ATP-ase Na^+/K^+ ATP-ase	Nerves in multicellular animals

N.B. These devices were initially part of the machinery for removal of elements such as Na^+, Ca^{2+}, Zn^{2+}, Cd^{2+} and Cu^+, which were poisons. Later they became particularly associated with messenger systems.

element such as iron, copper and zinc, there was an increase in the variety of uses of other trace elements, such as selenium and iodine, and of bulk elements such as sodium, potassium and calcium, for example in electrolytic message systems. In particular, the combination allowed the development of nerves and then brains. Many of these changes are associated with increased ability to energise movement of elements (Table 14.17). Now, while the new uses of copper and iron, in particular, derive from their variety of oxidation states (and the same is true for the non-metals selenium and iodine, and even for carbon), the new uses of the bulk metals, Na^+, K^+ and Ca^+, arose from the energy input to their gradients. We noted how these novel systems were grafted on to the pre-existing message linkages based on substrates, organic phosphates, iron and hydrogen, all of which were modified.

All of this new extracellular chemistry allowed the development of new compartments in the form of organs. Variability, then, is due not just to the multiplicity and different uses of elements in single cells but to the variety of ways in which organs can be assembled and then put to functional use. Obvious examples are the different development of 'limbs' in fish, birds and mammals. Referring back to Chapter 7 and the discussion of variables in non-living systems it is easy to appreciate that the potential for variability has again increased rapidly in the order (with complexity see aside):

$$\text{Prokaryote} \rightarrow \text{Single cell eukaryote} \rightarrow \text{Multicell eukaryote}$$

The way in which species can be distinguished also increases in this order (Table 14.18). We shall consider in Chapter 15 how this potential for variability has increased once again in humans due to the new possibilities allowed by the development of the brain, but the variability is now curiously extended to within what we define as a single species, i.e. to the level of the individual—see Chapter 16.

Now, complex systems are likely to be highly susceptible to variation in conditions, whence, internally, a multicellular organism is protected in a variety of ways. (Of all species, humans are possibly the best protected from environmental stress through their own constructions; see Chapter 16.) We have illustrated the survival strength on a plot against element content or energy flow in landscape diagrams, by sharp peaks with very deep valleys (Fig. 14.17) and ever closer relationships. This treatment has allowed us to maintain the connection between what we observe and

Increase of complexity and of emergent properties

Proteins/DNA

↓

Prokaryote cells

↓

Eukaryote cells

↓

Multicellular organisms

↓

Animals with brains

Table 14.18 Characteristics of species

Organism	Characteristics
Prokaryotes	Shape, metabolism, element composition, DNA
Eukaryotes, single cell	Shape, metabolism, internal construction, element composition, DNA
Multicellular organisms	Shape, metabolism, reproduction, internal construction, element composition, DNA, average behavioural pattern
Higher animals and humans	Shape, metabolism, reproduction, internal construction, element composition, DNA, brain, individual behaviour pattern

the fundamental variables: chemical element composition, spatial distribution, temperature and pressure, but spatial distribution now includes time. However, the complexity is such that just as we had to use operational sums of variables in thermodynamics, such as ΔG, we now needed to use non-mathematical qualitative concepts such as 'survival value' for organisms.

Another feature of multicellular life is that it is more easily sustained in co-operation with lower unicellular organisms than in direct competition with them. Thus, symbiosis is extraordinarily common, probably general, to all higher plant and animal life. This forces us to consider all living systems as an ecosystem. The way in which to develop ecosystems one step further will be apparent in the next two chapters. While the life of each individual species is an emergent *property* of complex interacting patterns of flow of materials and energy in an organism, to be sustained and to evolve living *forms* must interact with one another and with the environment. The new emergent property, then, is an ecosystem.

Returning to the progression of the use of the four fundamental variables outlined in Chapters 1 and 2, as we move from matter to man we see that life has gradually come to use the *distribution and flow* of mass and charge in space and time in more and more effective (for survival) ways. The remaining variables for exploration are quite clearly a wider use of chemical elements (mass and charge); an increase in organisation to larger and larger regions of space; and a more subtle use of the rate of movement of mass and charge in space. The last is, of course, a more sophisticated use of time dependencies and an easier example is electronic conduction in metals. It is humans who have uncovered these possibilities by rational thought and not through changes in DNA (see Chapter 15). In all this evolution we are keeping strictly to a programme based on physical–chemical systems (Table 14.18).

15

The evolution of man and his chemistry

Time present and time past
Are both perhaps present in time future
And time future contained in time present
          T. S. Eliot; *Burnt norton, in Four quartets* (1944)

15.1   Introduction

In this chapter we describe the latest great step in evolution, that to human beings (Table 15.1 and Fig. 15.1) which stresses the increase in differentiation based on the number of cell types. In fair part humans are animals, but they have two very distinctive features. First, the brain has developed in such a way as to make them self-conscious and able to develop skills such as the use of complicated tools and languages. Secondly, in the last 300 years they have come to understand the physics and chemistry of the observable world around them. Together, communication skill and management of chemicals have led towards a controlled, though as yet limited, ability to manage their *external* environment. Of course, in many other essentials, such as chemical composition (Fig. 15.2) human beings remain

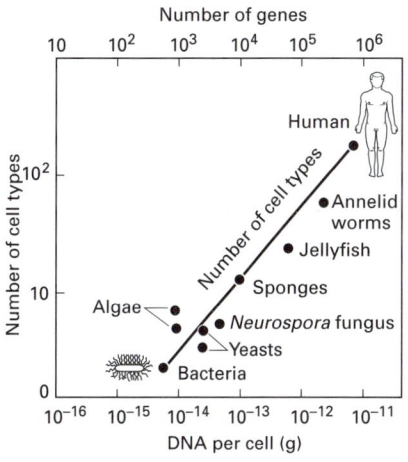

Fig. 15.1 The number of cell types in an organism seems to be related mathematically to the number of genes in the organism. In this diagram the number of genes is assumed to be proportional to the amount of DNA in a cell. The actual number of cell types in various organisms appears to rise accordingly as the amount of DNA increases. [Adapted from Kauffman, S. A. (1991). *Scientific American*, August 1991, 64–70.]

Table 15.1 Evolution of nerve systems

Years ago	Evolved species
700 000 000	Multicellular organisms, earliest animals (?)
600 000 000	Shells and skeletons
?600 000 000	Worms without organised nerves [see Fig. 15.3(a).]
?<500 000 000	Worms with centrally organised nerves [see Fig. 15.3(b)]
	Vertebrates
~450 000 000	Fish
~350 000 000	Reptiles
~200 000 000	Mammals
4 000 000 (?)	Bipedal 'animals'
1 000 000	*Homo erectus* in Asia
400 000	*Homo sapiens* in Africa
40 000	Cultured humans: self-conscious man ?
300	'Understanding' of our surroundings
↓	and full ability to manipulate the ' environment to advantage'
Today and tomorrow	

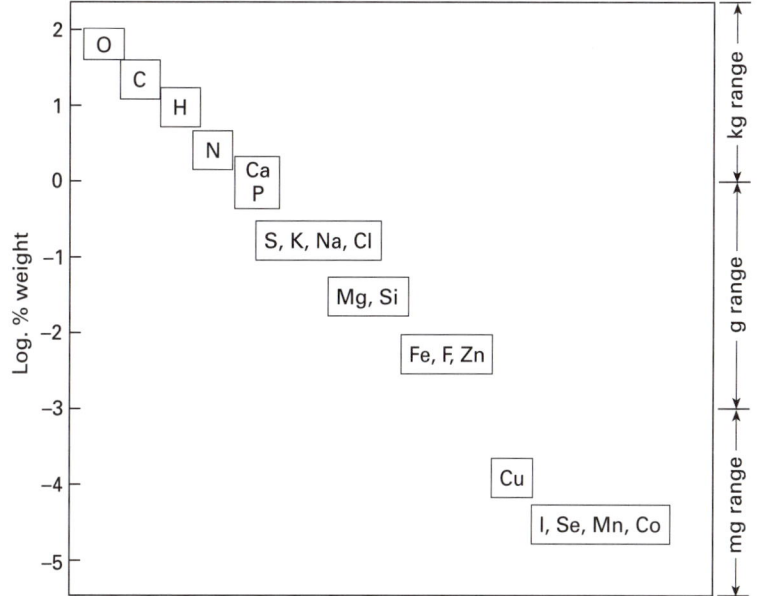

Fig. 15.2 The elements in the human body by weight per cent.

as animals. Again, the manner of construction of their shape and most of their organs for sustaining life is little different from that of literally thousands of other species, as described in Chapter 14. The human body is devised to assist movement (nerves, muscles and bones), to aid digestion (a series of organs, from intake of food to exit of waste), to provide a circulation system for transport of chemicals (the heart, arteries, veins and air streams) and to sense and react to the environment (sense organs and a neuronal system which includes a brain). However, in some way, human brains are very different from those of other animals and therefore we start this chapter with an outline description of the brain as an organ, which we left out of Chapter 14.

The second part of this chapter will be devoted to recent developments in man's chemistry, since this is a fully innovative advance in the ability of an organism to handle the environmental chemistry and physics external to itself. Finally, we ask: Have these two developments changed the derived variables we have associated so far with living organisms? There is no possible change in the fundamental units or variables, of course.

15.2   The evolution of the brain

Nervous systems are present in very primitive animals and may have developed some 600 million years ago (see Table 15.1). The value of a nerve message, as opposed to a chemical message in an internal stream of the body fluids, is its speed, about 10^3 times faster. Nerves are of greatly increased value if their activity becomes co-ordinated so that a message from one part of the organism is relayed to several other parts at a consid-

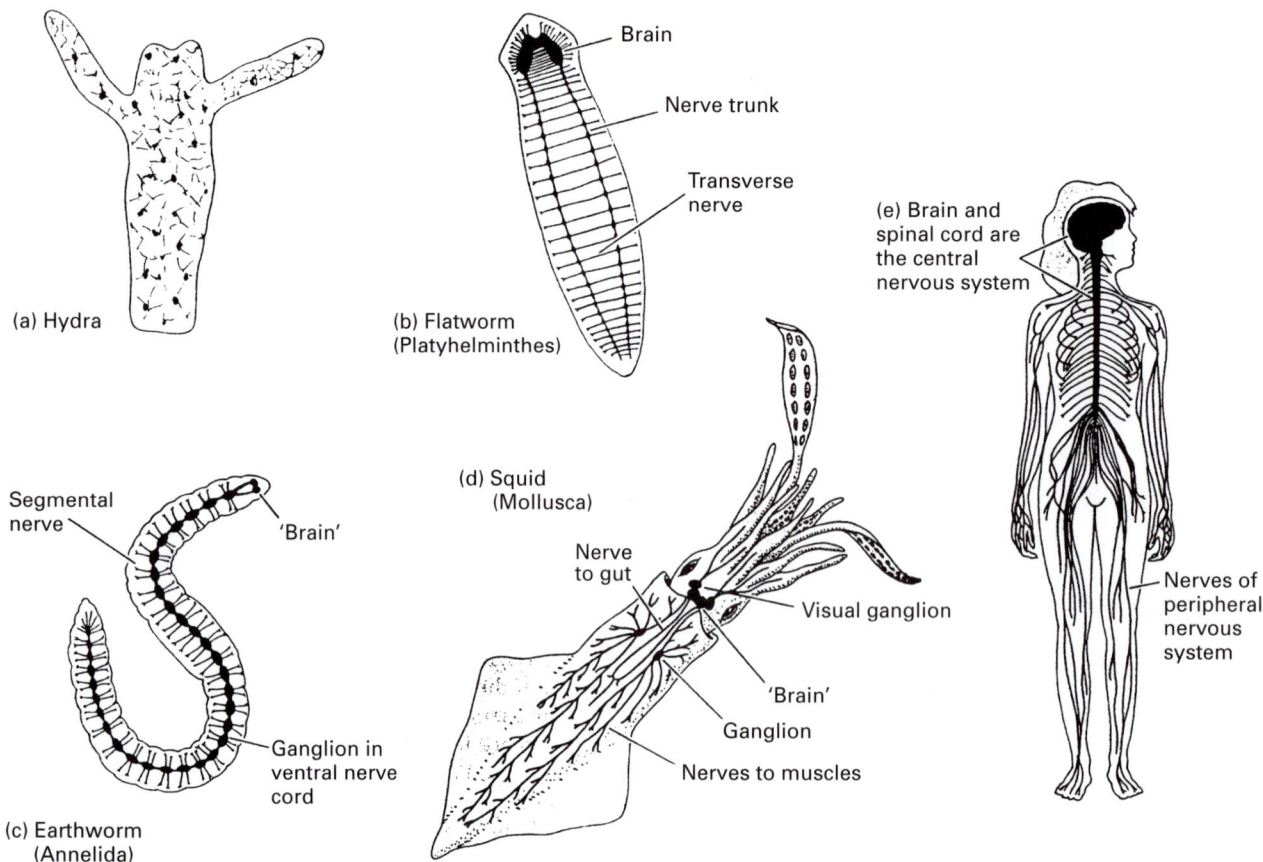

Fig. 15.3 A comparison of the brains of some very primitive organisms with that of man.

erable distance. This co-ordination became centralised in what we recognise as a brain even in animals such as some early worms (nematodes) which have very few nerves (Fig. 15.3). Undoubtedly, initially the response from nerve to nerve via the 'brain' relay, say from sensing by eye or touch to muscle, was automatic. This electrolytic message is also automatically relaxed so as to be ready for further transmission. Even at this stage this novel nerve message system had to be linked to all other internal cell message systems inevitably inherited though evolution and based on such chemicals as phosphate, iron, calcium, zinc and a multitude of organic compounds (see Chapter 14) which use the body fluids for transmission. The integration occurred locally at the level of the receiving cell. Subsequently, certain of these chemical messenger-releasing organs (glands), which also contain receptor systems, were linked into the central co-ordinating system, or brain (Fig. 15.4).

The glands continued to release chemicals to adjust distant cells through the blood, as before, but now in connection with the nerve system (Fig. 15.5). Such chemical messengers (hormones) are slow in action and so could be used to prepare the body in such a way that it will respond quickly to later nerve input. Loosely, we may consider that many organic hormones maintain, or awaken slowly, distant local cells in organs, while

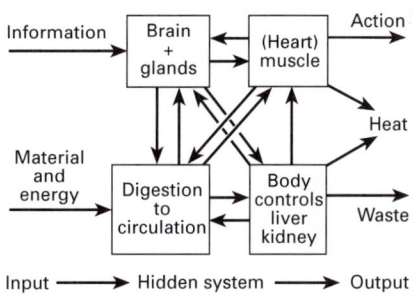

Fig. 15.5 The network of feedback through both nerve and chemical messages gives the hidden system a steady state and an ability to respond intermittently (see Section 7.14). Material transport is also required in the hidden system.

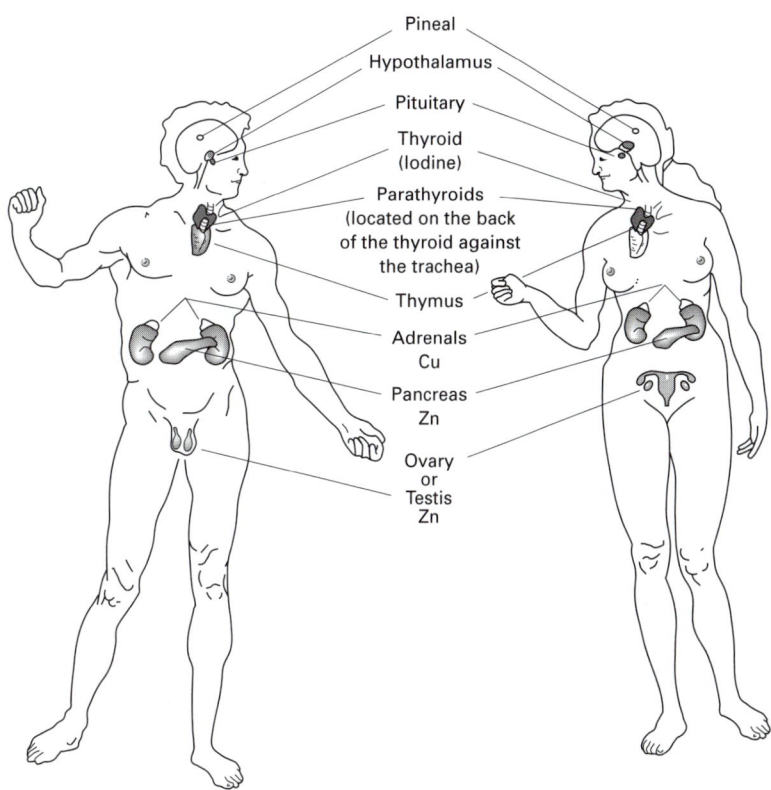

Fig. 15.4 The endocrine glands of man with some specially required elements indicated. (Adapted from ref. 108 in Further reading.) Note that organs such as liver are rich in certain elements, for example, iron.

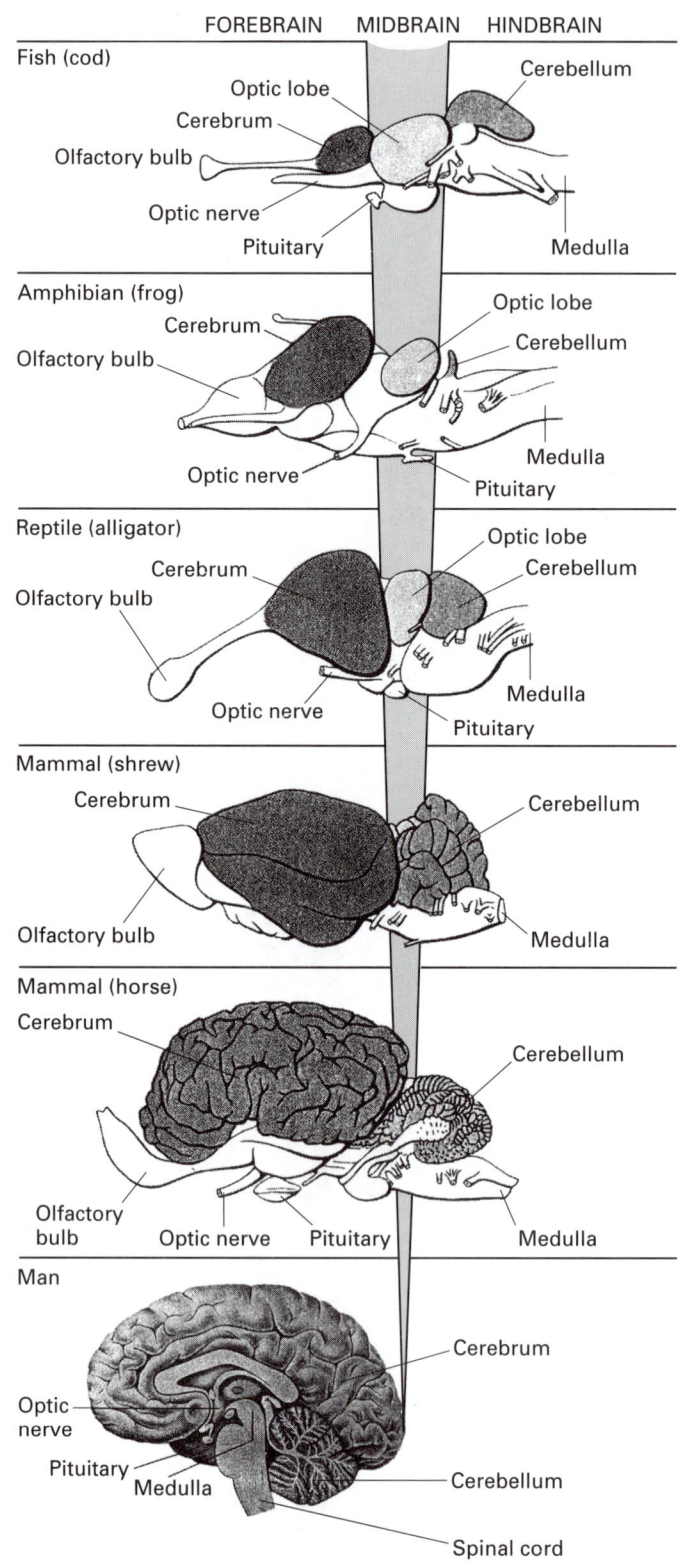

Fig. 15.6 Evolutionary change in relative size of the midbrain and forebrain in vertebrates shows a marked decrease, and that of the forebrain a very considerable increase. The brains are aligned to show the development of the cerebrum. The human brain is shown in more detail in Fig. 15.7. [Modified from Romer, A. S. (1962). *The vertebrate body*, Saunders, and Simpson, G. G., Pittendrigh, C. S., and Tiffany, L. H. (1957). *Life: an introduction to biology*. Harcourt Brace Jovanovich]

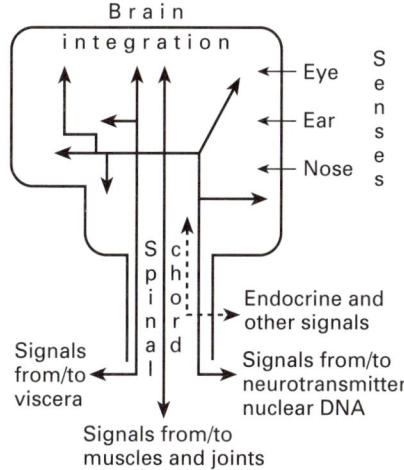

Fig. 15.8 A schematic diagram of the brain and spinal chord in constant exchange condition with the body and the environment (through the senses). The mental and physical circumstances are interpreted by inherited and experience-based connectivities in the brain such that every individual is different. The scientific method used throughout this book depends upon the observing instruments all 'seeing' in the same way. This is not a necessary feature of human perception. (After A. R. Damasio, ref. 213).

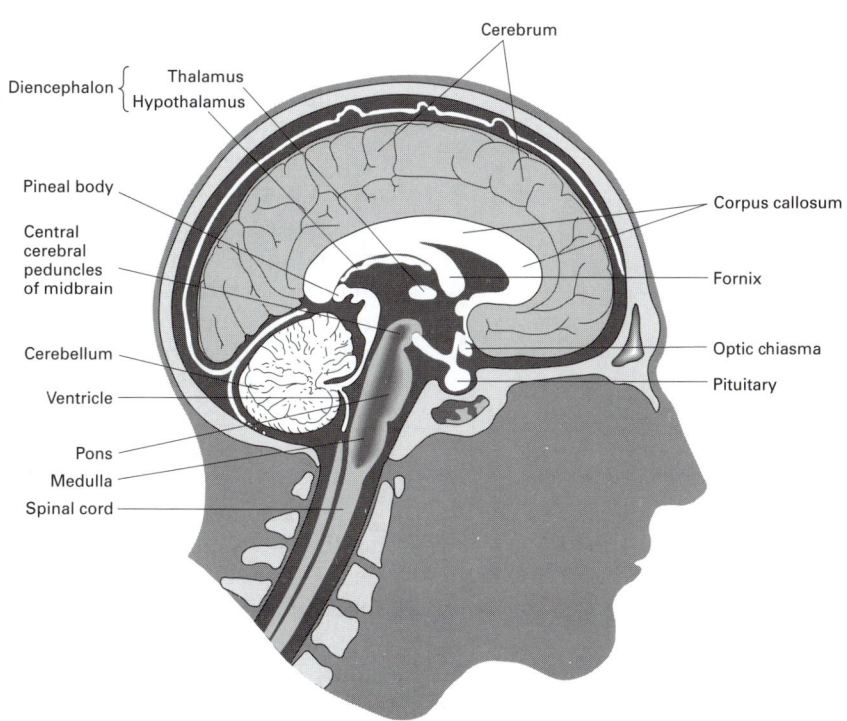

Fig. 15.7 A midsagittal section through the human brain. Note that in this type of section half of the brain is cut away so that structures normally covered by the cerebrum are exposed.

nerve pulses activate these cells quickly. The hormones that maintain general steady states in many organs are, for example, sterols, while through levels of awareness certain other hormones, e.g. adrenaline, change and heighten the readiness for action of muscles, but it is nerve messages that trigger muscle contraction. Hormone levels also control the growth and development of the whole body; this activity would appear to rest with chemical releases from the glands alone, not with nervous activity (see Chapter 14). Thus, the hormonal system is not very different in humans from that in earlier animals.

An efficient co-ordination of nerve and chemical signalling demanded evolution parallel in time and spatially in close proximity with such sense organs as eyes, mouth, ears and nose (see Fig. 15.8). There then developed a localised region of the body, the head, in which sensing and both electrical (nerve) and chemical messages (glands) were organised. The head is the leading region of the animal and this construction is of obvious good sense. A picture of the primitive brain then shows it to have a quite large region occupied by glands and a smaller section devoted to central nerve connections, all placed near sense organs (Fig. 15.6). The cells of this region are sometimes divided into white matter lying within the grey layer of the cerebral cortex. The brain system, at the initial stages of this evolution, is little more than a relay centre for effective automatic response and to think of it as much of a memory store of highly particularised events may well be incorrect. At the same time we can all observe that birds, insects and more lowly animals find their way to and from objectives.

Before going further, note that the bringing together of nerves to a locality gained improved organisation, by leading the nerve wires *together* through a specialised region of the body where there could be no interference with a message. The terminal nerve regions approaching the brain were encased in the *spinal cord* (Fig. 15.7, compare with Fig. 15.8) and this cord and the brain itself later became surrounded by a special new environmental extracellular fluid, the cerebrospinal fluid (CSF). Connection between the general extracellular fluids and the brain fluid (CSF) was prevented by a membranous structure called the blood–brain barrier. This new extracellular fluid has a novel ionic and chemical composition (see Table 15.2) which has yet to be understood in terms of brain function. Curiously, it contains very few proteins and a lowered potassium, calcium and glucose, but higher magnesium, chloride and lactate, compared with blood.

A further, quite early evolutionary step in the development of the brain was that of reinforcement of connectivity between certain nerves and others, but not all others, through cells totally internal to the brain. The reinforcement of particularised connections, say visual responses to an event, within a nerve network in the brain, is due to *synapse* growth and contact (Fig. 15.9) and leads to a selected memory or recall, either when the event is met again or even when it is accidentally triggered. In this way, there is, in the brain, a latent image (an information content) of an object or event. Internally, the image may have less relationship to what we call objective 'reality', since it is individually coded in electrical and chemical language that can be faulty due to errors of perception (see below). Thus, an image can be 'seen' ('imagined') by humans even by calling its name. For example, you can 'see' a cat when someone says 'cat'. We are sure a dog can too and a small bird surely sees an image of a hawk when it hears one cry. A few images are merely reinforced and elaborated from inherited knowledge, e.g. the image of the hawk in the small bird's brain,

Table 15.2 Elements in rat brain CSF and organs (meq per kg H_2O)

	Plasma	CSF
Na	148	152
K	5.3	3.36
Ca	6.14	2.2
Mg	1.44	1.77
Cl	106	130
Glucose	7.2	5.4
Pyruvate	0.17	0.18
Lactate	0.7	2.0
Proteins (mg per 100 ml)	6500	25

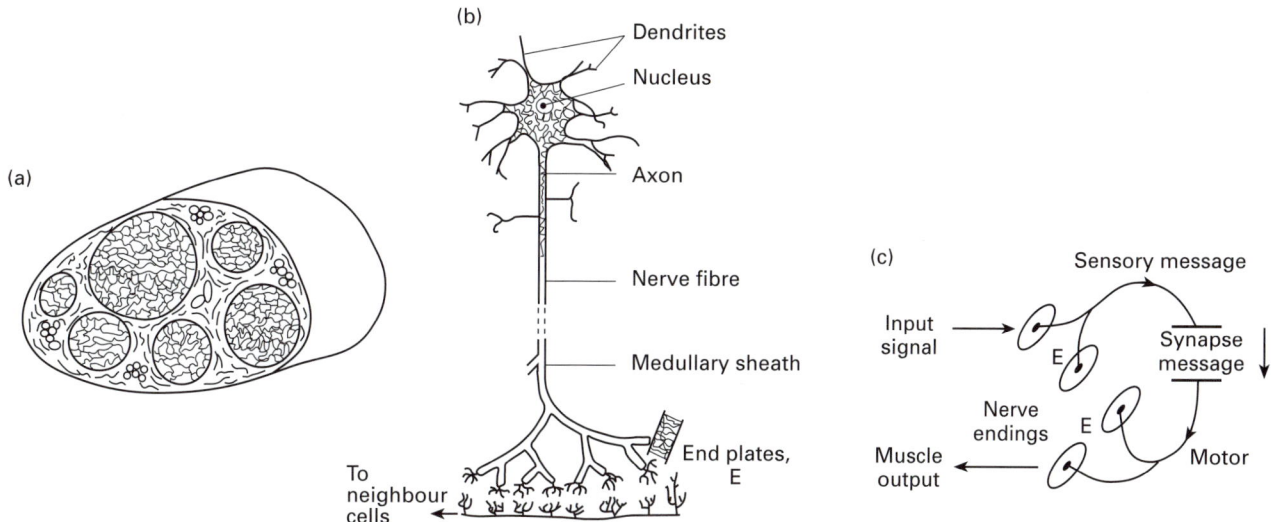

Fig. 15.9 (a) Cross-section of a nerve trunk with individual nerve fibres grouped into six separate sections. (b) A nerve cell or neuron. (c) Sensory and motor nerves connected via a synapse.

Table 15.3 Development of cerebrum in brain during evolution

	Shark	Frog	Alligator	Pigeon	Rabbit	Human
% Cerebrum	<5	10	20	30	50	80

N.B. The brain increases in size, keeping other regions of the body of fixed relative size, approximately.

but the vast majority of images settle in the brain just from experience after birth. This increased capacity for constructing images is a very powerful development in the evolution of animals (see Fig. 15.6). It occurs in the region of the brain called the *cerebrum*, which has developed to be a dominant feature of the brain of humans (Fig. 15.7), and see Fig. 15.6 and Table 15.3. Intriguingly, man appears to be born with extremely poor use of this system. In fact, a human baby is a peculiarly helpless animal since the neuronal network in the brain corresponds only to very basic reactions, but when the interaction with the external world begins, the networking (wiring) of the nerve system develops quickly, especially during the childhood years (see Fig. 15.10). [The complexity of the organisation of the human brain is given in Table 15.4, but we shall not refer to these details which are given simply to lead the reader to the appropriate references (see Further reading).]

15.3 The brain, DNA and information

The obvious question to ask next is 'what causes such an image?' In effect, what has been created is a *memory* or a persistent *information store*, visu-

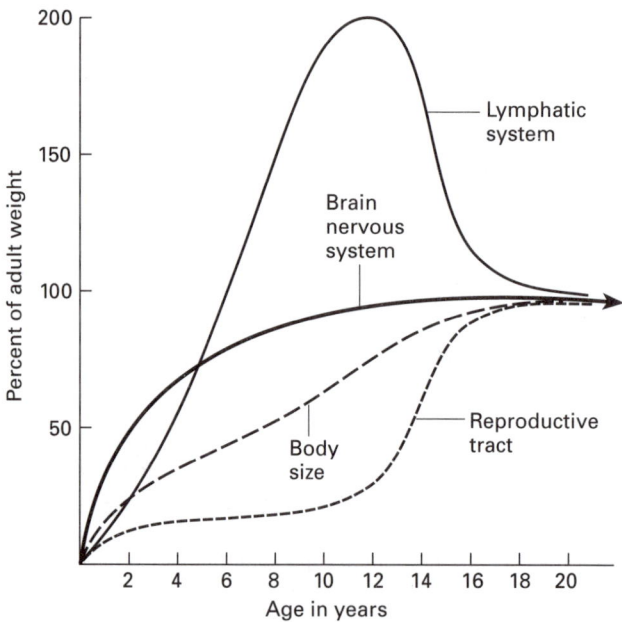

Fig. 15.10 Relative rates of growth of several different organ systems during human development. Note the brain and the lymphatic system.

alised, heard, tasted or smelled by sense organs, and now recreated (in 'virtual reality') by what we call, in common terms, the 'imagination'. This information is of a totally new kind, completely different from that in DNA, and relates not to internal but to external goings on. Thus, it is not of necessity in any way DNA based, though DNA is responsible for laying down a nerve construct (the fetal brain), but with very little information content. The environmentally dependent growth of synapses became a new electrically and chemically coded form of information, where the code is a three-dimensional distribution of not quite permanent chemical and physical gradients or fields, and is largely taught or experience based, not inherited. We may contrast it with DNA information in many different ways.

(1) DNA is linear, the brain is three-dimensional.

(2) DNA is conservative, the brain is adaptive.

(3) DNA is based on the ordering by chemical bonds of four chemicals, the nucleotide bases, while the brain is based locally on perhaps 50 transmitters, as well as on their concentrations in compartments, and upon fields; there may be 10^{15} different storage zones for the information-carrying chemicals in any one brain.

(4) DNA is inherited; the connectivity in the brain is very partially inherited, but overwhelmingly it is experience based, i.e. based on personal history.

(5) Information about the external world is passed on ('inherited') more and more successfully through spoken language, books, films, records

Table 15.4 Divisions of the human brain

Division	Description	Functions
Medulla	Most inferior portion of the brain stem; continuous with spinal cord. Its white matter consists of nerve tracts passing between spinal cord and various parts of the brain.	Contains vital centres (within its reticula formation) that regulate heartbeat, respiration and blood pressure.
Pons	Consists mainly of nerve tracts passing between the medula and other parts of the brain	Serves as a link to connect and integrate various parts of the brain; helps regulate respiration.
Midbrain	Just superior to the pons; contains red nucleus.	Superior colliculi mediate visual reflexes; inferior colliculi mediate auditory reflexes. Red nucleus integrates information regarding muscle tone.
Diencephalon thalamus	Located on each side of the third ventricle; consists of two masses of grey matter partly covered by white matter.	Main relay centre conducting information between spinal cord and cerebrum.
Hypothalamus	Forms ventral floor of third ventricle. The pituitary stalk connects pituitary gland to hypothalamus.	Contains centres for control of body temperature, appetite and fluid balance.
Cerebellum	Second largest part of the brain; consists of two lateral cerebellar hemispheres; superior to the fourth ventricle.	Responsible for smooth, co-ordinated movement; maintains posture and muscle tone; helps maintain balance.
Cerebrum	Largest, most prominent part of brain. Longitudinal fissure divides cerebrum into right and left hemispheres.	Centre of intellect; memory, language and consciousness; receives and interprets sensory information from all of the organs; controls motor functions.
Cerebral cortex	Convoluted, outer layer of grey matter functionally divided into three areas;	
	(1) motor areas	(1) Control voluntary movement
	(2) sensory areas	(2) Receive incoming sensory information from eyes, ears, touch and pressure receptors
	(3) association areas	(3) Responsible for thought, learning, language, judgement and personality; store memories.

and computers, developed by the brain and which are, effectively, extensions of the brain, but more long-lasting.

The more primitive brain images spark responses such as the barking of a dog at a danger, and are still partly DNA based, but even a dog's ability to distinguish its owner from a stranger has little to do with DNA.

The brain is a special organ in further ways.

1. After a certain period of growth of an animal, the number of neurons of the brain does not increase greatly. In fact, the majority of the neurons in the human brain are already formed even some months before birth. Hence, they can only die (hundreds of thousands per day). Brain cells are not replaced. This is in contrast with, for example, the behaviour of red blood cells, which are constantly replaced (life time 120 days) and liver

cells, which are replaced when lost. Synapse development in the brain is very fast in the childhood and youth years and continues more slowly throughout life.

2. The morphology of almost all organs is DNA based and is fixed, and during growth there may be slight changes only. The brain is a set of cells that develop (and lose!) unique patterns of connectivity with time and according to the experience of each individual. The internal structuring of the brain is the only structure not based on DNA.

The characteristics of the brain make it different in kind from all other organs. The huge development of the brain in mankind then makes human very special animals. As stated, it is just the peculiar development of the cerebrum that effectively separates humans from all other species. We consider in this book that the development of the brain is such that we must see it as a very recent and major change in evolution since it has even given humans the power to control evolution deliberately (at least in part). We are not suggesting that there is any discontinuity between humans and animals, but in a very short period of evolution one organ, the cerebrum, has grown differentially to such an extent that the change can be compared with, say, the steps in evolution that led from single cells in colonies and then to organised, multicellar life. The increased organisation in compartments seen in humans is complemented and vastly increased by the organisation of the environment by humans, which has been enabled by their brains. Darwinian evolution will have a lower importance from now on, being replaced by planned, systematic and controlled change.

15.4   Special chemical features of the brain

We have already mentioned that the extracellular fluids of the brain differ chemically from those of the main body. Just as we had to consider the cells of different organs in the main body of an organism as being of different chemical composition for a purpose, so we must examine the composition of the brain (Table 15.5) and of different regions of the body

Table 15.5 Averaged values of elements in soft tissues* of human (μg/g dry weight)

Element	Liver	Kidney	Brain	Muscle
Mg	500	500	500	500
Ca	200	400	400	100
Mn	4	3	0.5	0.2
Fe	500	300	150	150
Co	0.2	0.1	0.1	<0.1
Cu	20	12	5(20)	<3
Zn	150	200	50†	200
Mo	3	1	0.1	0.1
Se	3	2	1	0.2
I	0.2	0.04	0.02	0.01

* Most tissues do not differ greatly in composition from those of other animals.
† Zinc is found concentrated in some neurons in vesicles and may be a transmitter.

Table 15.6 The ionic composition during development of the human brain

	Electrolyte composition (meq kg ^{1})				
	Human brain at given age				
	Prenatal age (weeks)				
Electrolyte	**13–14**	**20–22**	**Newborn**	**Adult**	**Senescent**
Na$^+$	97.5	91.7	80.9	55.2	Rises
K$^+$	49.6	52.0	58.2	84.6	Falls
Cl$^-$	72.1	72.6	66.1	40.5	Rises
Mg^{2+}		8.4	7.9	11.4	Unknown
Ca^{2+}		4.9	4.8	4.0	Rises
P	57.0	52.2	54.0	109.0	Falls

Table 15.8 Signal compounds between cells in the brain

Fast signals

Excitatory (+)/inhibitory (–) amino acids
Glutamate (+)
Glycine (–)
GABA (γ-aminobutyric acid) (–)

Monoamines
Noradrenaline
Dopamine
Serotonin
Acetylcholine

Inorganic signals
Nitric oxide
Carbon monoxide?
Na$^+$, K$^+$, Ca^{2+}, (Zn^{2+})

Slow signals
Neuropeptides (large number >50)
Substance P
Cholecystokinin
Corticotrophin-releasing factor

Alzheimer's Disease

Abnormal concentrations of Cu and Zn in protein deposits in the brains of patients with Alzheimer disease are thought to aggravate this condition (See Huang, X. *et al.* (1999) Biochemistry **38** 7609–7616).

comparatively and separately. In Table 15.6 the general change in brain electrolytes with age is also recorded. (Figure 15.10 outlines the way the nervous system as a whole develops with age.) In particular, copper and zinc vary two-fold in different parts of the brain and like other elements changes are differentially age related (Table 15.7). The extraordinary nature of the brain and its changes in physics and chemistry, from initial beginnings to final states, are, as yet, poorly understood. Here we stress that this development is a remarkable switch in evolution, since no other organ has this kind of fast chemical development. Some information on organic chemical messengers between cells in the brain is included in Table 15.8. (Note also the inorganic messengers and the presence of nitric oxide.) The significance of the chemical composition and the use of elements in the brain is an area for intense future study. However, what is required above all else is an appreciation of how the brain 'system' is based

Table 15.7 Trace element distribution in the compartments of the brain (μg/g dry weight)* and see aside

	Compartment of brain						
	1	**2**	**3**	**4**	**5**	**6**	**7**
(a) 4-days old							
Mo	0.14	0.12	0.09	0.10	0.12	0.15	0.12
Rb	8	7	13	8	10	10	8
Zn	46	29	52	38	47	61	47
Cu	9	6	12	8	9	8	7
Mn	1.9	2.1	2.8	2.5	2.1	2.1	1.1
(b) 20-days old							
Mo	0.14						
Rb	8	8	12	8	11	12	10
Zn	45	28	56	37	60	86	76
Cu	12	6	14	8	12	13	10
Mn	1.3	1.0	1.8	1.1	0.7	1.1	0.8

* Numbers have been rounded. From Saito, T., Itoh, T., Fujumura, M., and Saito, K. (1995). *Brain Research*, **695**, 240–44.
Distribution of trace elements in seven brain regions of the rat: 1, cerebellum; 2, medulla and pons; 3, hypothalamus; 4, midbrain and thalamus; 5, striatum; 6, hippocampus; 7, cerebrum.

not just on compositional variables but also on the inter relationship of these variables with those of energised and constrained flow, so that we appreciate how it functions. It is undoubtedly different from a computer since it is built upon a different 'knowledge' base: human, senses, not machine code.

15.5   Nerves and brain functions

Whereas the autonomous nervous system, which helps to maintain a steady state within the internal environment of the human body, e.g. to maintain the temperature or to regulate the rate of the heartbeat, is readily understood today as just an example of an external event generating an automatic response in, say, a muscle, and is common to all animals, the voluntary action of the brain is not at all easily explained. At one level, the simplest, we may believe, as stated before, that it records messages through growth, locally, of axons and then of connections (synapses), so that it is a spatial physicochemical map record with recall (a memory) of individual phenotypic experience. This has great and novel value in that the response to events is largely learnt (reflective) and is not automatic. All such activity could be analysed in terms of messages, their reception and their registration (as growth and connections), making the brain into a three-dimensional map of sense experience and controls, but it is clearly different in different individuals. This description leaves us short of an explanation of the human development of self-consciousness, which allows the evaluation of possible or even impossible foreseeable future events to be contemplated on the basis of past experience and imagination and then made the cause of present action (see the quotation at the beginning of this chapter). It is the huge development of this capability within the cerebrum that separates *homo sapiens* from all other creatures. Instead of passively living through and recording past events, and then causing the body to repeat them, the brain seemingly creates, almost simultaneously, several images of possible future events and then selects actions by individual judgement of weighted expectations. Of course, as already stated, the individual record of past experience, memory, can be faulty due to errors of perception—senses may deceive us, so that these images are not necessarily related to any objective consensual 'reality' (that is 'reality' to a certain extent observable with instruments). Therefore, the action of any individual to a given circumstance will always contain a degree of unpredictability related not to a *real* previous history but a *perceived* personal history.

At the present time, then, there is no satisfactory explanation of this self-conscious activity, but it has to be an holistic concept referring to a further new *emergent property* derived from the extremely complex networking of neurons and links to the senses, called self-consciousness. Thus the brain analyses the activity of the whole organism, including nerves and glands, in an integrated way connected to its history and simultaneously to its organic, DNA-based, well-being. Clearly, the development of the human brain is an enormous evolutionary event, equivalent to the building up of a temporary second information centre quite different from DNA and coded with the individual's internal chemistry in a novel fashion. As far as humans are concerned, we may well say that it contains a *second code*, far more complicated than that of DNA. (It has to be repeated that a part of this activity is based on misconceptions, as judged when analysed by instruments, but it is 'reality' as interpreted by each individual's brain.)

The fact that the brain allows consideration of future events made it possible for humankind to a plan purposefully and therefore to develop ways of manipulating the environment to ensure provision of future needs—warmth, food and shelter at first. Later, this activity extended to the development and use of tools and machinery and then of the chemical element units of all materials. We see this as a movement towards the last possible major step of evolving control over all compartments, still within natural biological development (see Aside). The first step, described in Chapter 11, was control over the cytoplasm in prokaryotes. The second step, described in Chapters 12 and 13, was the ability to respond to the environment through increased internal organisation in eukaryotes. The third step, in multicellular organisms (Chapter 14), allowed a controlled environment for much of the cellular system within a body carried with the organism. The final step is that deliberately undertaken by man which, at least in principle, allows control over the whole external environment. (It is this that changes our whole passive, Darwinian attitude to evolution.) Man has also found external ways to store and pass on his knowledge using devices that are external to himself.

There are two ways in which the knowledge store has become further enlarged and effectively inherited. The first is by dividing and sharing responsibility for activity. This development means that each individual (by agreement or compulsion) can be used separately to store particular memories and useful abilities. This shared communal insight is already found in primitive animal organisations, e.g. that of ants or bees, but at a very low level. In its highest form it demands developed communication skills between individuals, such as visual or sound signalling, and then language in these communication modes. It is probable that some basic grammar of signalling can be inherited, i.e. DNA based, but it is insignificant compared with a fully developed, learnt skill. The inheritance is directly from parent (or teacher) to offspring, repeated time and time again.

The second development is through the recording of information, so that the knowledge gained in one lifetime by an individual is, in effect, 'inherited' by external means. Here we refer to the printed, recorded or computerised storage of knowledge. This extension of the brain is not a new brain, but a utilisation of technology to avoid wasted effort in repetition. It is this huge extension of information by man which has allowed his gradual cultural, scientific, technological and practical evolution over about 40 000 years. Thus, he has extended his internal organisation to the organisation of everything around himself in a continuous upgrading. It is the knowledge of the physics, chemistry and biology of all inanimate and animate objects allied to the capacity to 'think' conceptually and mathematically that gives humans an extraordinary organising power today. This knowledge is general and cumulative and can be passed on to future generations so that it evolves without repetition of the experiments or experiences that

Development of compartments

1 Chemical pattern + codes
2(a) Internal compartments
2(b) Internal/external messages
3 Differentiation of cells
4 Body organisation, organs
5 Physical communication in societies by individuals

led to it. However, knowledge also develops differently, in part, within each individual, so that there is an independence of each one based upon one's own self-consciousness—one's phenotypic, not just one's genotopic brain—not generalised from experience and science-based reasoning. In this way the individual can no longer be treated as a statistical object who has to conform to general rules since he has a new phenotypic development—a novel form of handling information and acquiring variable 'instruction' which entitles him to a new individual freedom. The variety of human beings no longer corresponds to the idea of uniformity applicable to the word 'species', for this defines only his DNA information and hence the reproduced inherited characters, but nothing of the ability to manipulate the environment in a thoughtful cultural way. Shakespeare is not obviously one of us in that we can all do what he could do! Man-made machines, like lowly organisms, are strictly reproducible, but adult individual men and women are not.

The above account completes all that we can do in this book to describe the chemical variables and physical functions of the brain of humans. It completes the description of the evolution of internal physical–chemical systems but not the extension to the environment. Much though the brain of man has some special chemical features, the major advance in *chemistry* and *physics* that most obviously characterises man, apart from his culture, is the chemical activity external to himself. This extension of life's activity into the larger environment has brought with it a totally new evolution of chemistry on Earth. It provides another reason for considering mankind as a living form quite different from any of life's predecessors.

15.5.1 Classical experience and modern science

Here we must reflect momentarily on the change in attitudes to the objects around us that the brain has allowed mankind to develop. The classical approach to observable external objects (see Chapter 1) was based on study using our animal-like senses. There are two peculiarities in the very nature of these observations; they are to some degree idiosyncratic, i.e. personal, and they are limited by the purpose for which senses were developed—survival. Thus, they are overlaid by value to the organism, which needs food and avoids danger. Knowledge is then highly selected. With the coming of instruments for measurement, observation detached from personal reactions and benefit became validated strictly by precise reproducibility. We can distinguish the two approaches as largely subjective and largely objective. In the next sections we illustrate how brain power has been employed to develop *uses* from objective studies of objects around us. We shall see clearly how great benefits have arisen, but at a price. Only in Chapter 16 shall we attempt to integrate subjective and objective experience.

15.6   Man's purposeful chemistry and physics

There are broad general distinctions between man's industrial chemistry and both the general equilibrium chemistry and geological chemistry

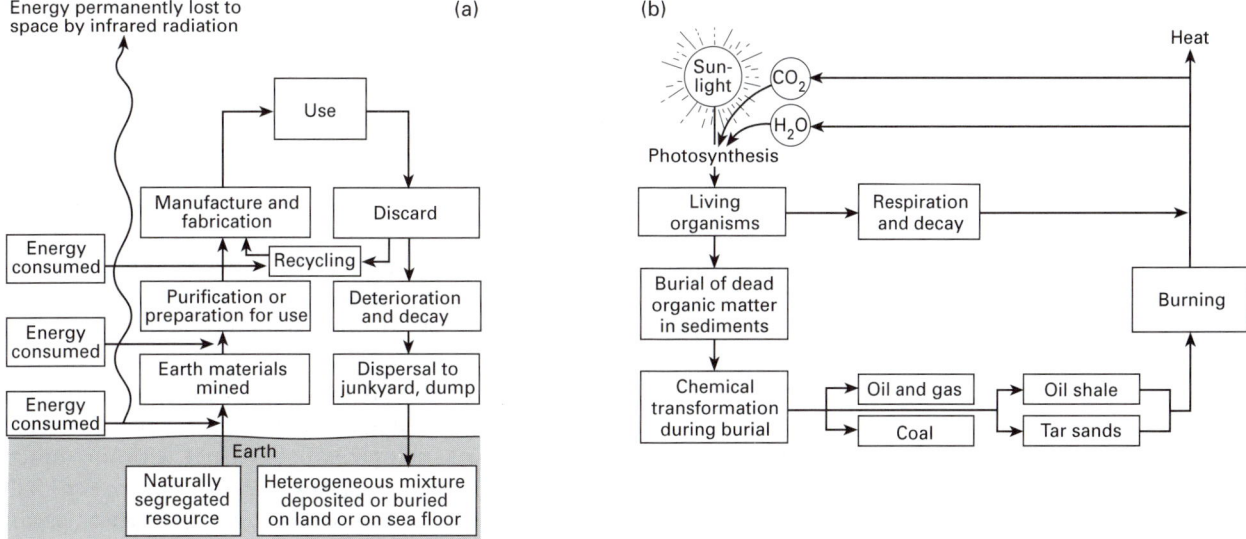

Fig. 15.11 (a) The cycle of mining, preparation, use and discarding of useful Earth materials is one in which a naturally segregated resource, such as an iron ore body, is mined, processed and fabricated and then returned to Earth in a dispersed, heterogeneous mixture. The cycling is slow, for example, iron oxide to iron oxide. [From Press, J. and Siever, R. (1986). *Earth*, (4th edn). W. H. Freeman, New York, with permission.] (b) Photosynthesis produces organic matter that is buried and transformed, and so becomes a fossilised product of photosynthesis—a fossil fuel. The fossil fuel has been very slowly recycled, but increasingly rapidly owing to man.

described in Chapters 5–9 and the biological chemistry described in Chapters 10–11 [compare Fig. 15.11(a) and (b)]. Man's objective is to process chemical elements in a deliberate way, making new combinations (new components) to further his own well-being as an organised, self-conscious species. He, therefore, has developed new ways of increasing energy input to obtain new stationary combinations of elements in compounds. In this, man has extended the *functional use* of chemical elements from internal biological needs, common to all organisms, to external applications. Much though this could be seen as merely an advancement of (natural) selection of the elements from earlier biological actions, it involves, in fact, a quite distinct change in chemical processing over that achievable within living systems. Moreover, it is deliberate (planned) not accidental. In this chemical processing there are four major changes, the fourth of which is just being developed.

1. The chemical elements that can be used are not restricted in a simple way by availability, but rather by abundance. Thus, man can now use some 100 chemical elements, including even some that are man-made, not just some 20–30, as used in organisms. (In minerals all the naturally occurring 90 elements are found but in restricted combinations.)

2. The modes of processing of the elements in controlled ways are no longer limited by conditions such as those of low temperature (270–370 °C), low pressure (1–10 atm) and aqueous solvents. Man now

has the means to operate degradations and syntheses from very low up to very high temperatures (10^4 K), at pressures exceeding 10^3atm and in solvents of very low or very high polarity.

3. The flow of energy into external systems has been vastly altered.

4. Man can manipulate the way organisms work via genetic alterations.

It is conventional to describe these activities under the headings of inorganic industrial (or mineral) chemistry, organic (or molecular) industrial chemistry and genetic engineering. We shall follow this approach, noting all the time that the activity is opposite foreseeable future demands, not just present needs, and is based on an understanding of the underlying principles of chemistry and physics.

There is a further distinction between man's chemistry and biological chemistry. Man *deliberately* generates materials as new, effectively isolated, components for *specified* functions, neither in equilibrium with any others nor in steady-state exchange. His chemistry *has no cyclic features* especially in the short term of 10–100 years. Let us see how all this activity began, starting from the distant past of some 10^5–10^4 years ago, noting how small a fraction of evolutionary time this is (see Table 15.1).

15.7   The use of energy in inorganic or mineral chemistry

Possibly the first chemistry we can recognise as being due to man is the deliberate manipulation of clays, silicate minerals, to give pots and bricks. All that was required, then, was an appropriate selection of a hydrated silicate mineral and the application of heat. The heat drives condensation. We represent a hydrated silicate by $MHSiO_3$, when heating gives

$$2MHSiO_3 \xrightarrow{\text{heat}} M_2Si_2O_5 + H_2O$$

Wet clay   Dry clay   Water

This use of temperature to create a polymerised mineral is virtually irreversible, so that many of the pots, bricks and so forth are recognisably similar to, but not identical with, geological minerals, e.g. fossils. Over time, man extended this chemistry to the production of cement, mortar, plaster, glass and concrete. The new manufactured minerals did not at first compete with minerals such as marble, sandstone and granite for the building of massive structures, but today they are the major construction materials in buildings, roads, pipes and so on. Notice that the condensation reactions are parallel to those used by living organisms to make proteins and DNA, for example.

Not long after the development of new minerals by the use of heat, man discovered metallurgical chemistry—ways of extracting metals from ores (see Section 5.9.2). The activity required higher temperatures, thus greater use of energy. A few metals are found native, but the vast majority are in oxides or sulphides. These are processed by reductive removal of oxygen and oxidative removal of sulphur, employing largely natural reserves of reduced biological organic matter—gas, oil, and coal.

$$FeO + C \xrightarrow{1000K} Fe + CO$$

$$FeS + O_2 \longrightarrow Fe + SO_2$$

The metals, once cooled to 300K, are in a stationary condition (to a first approximation) resistant to re-oxidation in the air, but this does occur slowly, as in the case of iron, since the oxide formed is thermodynamically favoured

$$2Fe + O_2 \rightarrow 2FeO$$

Only the so-called *noble* metals, such as gold and platinum, resist air oxidation. Once again, the new 'minerals', metals and alloys have been used by man in diverse ways in construction, not only of buildings, railways, pipes and so on, but also of machinery. Some uses of metals are listed in Table 15.9. They include the development of message systems and transport systems, so that there is, today, a huge, controlled movement of material in an organised manner. The transport is by land, sea or air, using a variety of vehicles. The controls are largely based on electronic or electromagnetic wave transmission linked to computers. All the controlled machinery has been made possible by the introduction of new components through chemical and physical processing of extreme sophistication. Externally there is much feedback control of activity, although this is not self regenerating. Notice the parallel with the internal structures of living organisms, especially of higher animals.

Table 15.9 History of element usage: the use of natural materials (e.g. wood, stone) without processing is excluded*

Application	Era of first major usage of element or its compounds			
	Prehistoric (before 2500 BC)	**Pre-industrial (2500 BC–AD 1750)**	**Industrial (AD 1750–1940)**	**High-technology (after AD 1940)**
Metals: vessels, tools, coins, weapons, etc.	Cu, Au	Fe, Zn, Ag, Sn, Pb	Al, Ni, Mo, W	Zr, Nb
Metals: construction, transport	–	–	Al, Cr, Mn, Fe	Be, Mg, Ti
Fuels and explosives	C	N, S	H	U, Pu, Th, H
Glass, ceramics, refractories	Na, Al, Si, Ca (clays)	Pb	Mg, Zr, Th	Li, B, La–Lu
Pigments and dyeing	–	Al, Fe, Co, Cu, Cd, Hg, Pb	Ni, Zn, As, Se	Ti
Pharmaceutical	C, N, O, H (plants)	S, As, Sb, Hg	Bi, Br, I, Ra	Li, Pt
Fertilisers and pesticides			N, K, P, Cl, As, S, B, Br, Hg, Cu	Sn
Industrial chemicals and catalysts	–	–	C, N, F, Na, S, Cl, K, Hg, Pt	Ar, Rh, Ba, La–Lu, Re
Electrical and electronics	–	–	Fe, Cu, In, Pb, W	Si, Ga, Ge, As, Li, Cd, Cd, Ni, Se, Ta, Ir
Household goods and chemicals	–	–	C, N, Na, Cl	B, P, Br, Sn, C, H, N, O, F

* After Cox (1995), ref. in Further reading. Note each column adds to the previous ones.

15.8   Machinery and energy

The new materials allowed man to design ways of creating useful objects such as pots and weapons and then machines (with which he could cook, fight or do constructive work). At first these ways of increasing the ability to do constructive work, i.e. transform the chemical or physical environment constructively, used extremely simple devices (wheels, pulleys, levers, etc.) and were driven by man's own energy. Once the principles became understood, engineers set about the task of increasing the ability of machines to carry out chemical or physical processes and to couple the action to engines. The role of man moved then to control of the machines. Obvious later developments are the steam engine, the petrol engine and all kinds of power stations coupled mechanically or electrically to thousands of different functional machines. Finally, man is moving control to 'brain-like' machines, computers. It is obvious what man has achieved all around us, since within a modern town almost every object we see is machine-made. It is important to observe that this use of energy and material is quite novel in evolution. It is a self-conscious, purposeful activity to increase control over the external environment by man. Unless one attributes the natural world around us to a purposeful God there is no other equivalent *purposeful activity* we know of in the universe. The activity is not absolutely novel to man as we can see its beginnings in the activities of fish, birds and insects, but the scale and novelty associated with thoughtful action is unknown in species other than human beings. As stated, the activity of the machinery and its products were at first designed to make life more secure. Thus, necessities could be made and supplied. Before amplifying this point we must see that self-consciousness can be used to increase internal as well as external well-being, through medical rather than engineering practice.

15.9   Organic chemistry and medicine: the development of protection

The development of a large-scale chemistry of non-metals was not started until about 150 years ago. One difficulty was that the combinations of elements such as H, C, N, O, P and S in stationary states are not very secure against oxidation (they are thermodynamically unstable to reaction with O_2 to give CO_2, H_2O and other oxides), while the inorganic minerals are stable, or at least are stable under the high temperature conditions at which they are prepared, and are resistant to change at low temperature. Furthermore, identification of organic compounds, except the most simple, requires quite sophisticated analytical equipment. Urea, the first organic compound to be fully characterised, was only identified in 1836. Once the principles of synthesis and practices of analysis were available the subject developed rapidly, so that chemists became able to isolate and examine biological molecules and to synthesise molecules not made naturally.

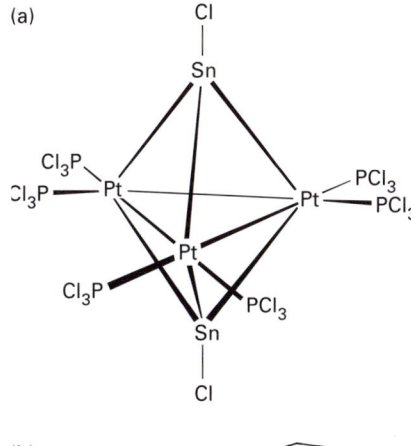

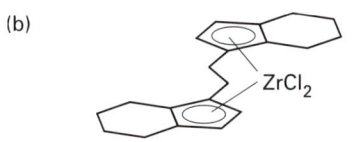

Fig. 15.12 (a) An organometallic cluster compound with Pt–Pt and Pt–Sn bonds and P and Cl donor ligands. (b) A zirconium organometallic compound used as a catalyst in polymerisation reactions.

The methods of organic chemistry today utilise the following strategy:

(1) identify an objective, small molecule or polymer;

(2) purify suitable starting materials;

(3) devise a scheme of intermediates and selective catalysts;

(4) test on small scale;

(5) build a factory for production.

This organic chemistry uses energy in a more subtle way than was illustrated for inorganic chemistry. The treatments are gentle—low temperature and pressure for larger molecule syntheses—and kinetic (pathway) control is then of the essence. Energy has to be introduced by such tricks as water removal by energised drying agents, the use of photochemistry and the addition of fragments in unstable combinations. Reactions can be carried out in specially devised solvents and atmospheres. These methods extend to organometallic chemistry, where organic moieties made from C, H, N and O and sometimes P and S, are linked to metal elements (see Fig. 15.12) to be used as catalysts. Note the parallel functional value of elements in both biological and man's activities (Table 15.10).

The great advances in organic chemistry are in the fields of medicine and agriculture, to which we must add plastics (so important in the modern life style). Man is now heavily protected, by clothing, and cure of diseases is highly advanced. His food supply from plants and animals is equally well looked after. Knowledge of the way in which biological chemistry works is not complete, but it does allow management by interference to the benefit of organisms. Unlike the engineering of machines, this activity introduces no new machinery but tampers with existing systems.

In the last thirty years, however, engineering of biological-type machines has become a real possibility (see Section 15.11).

Table 15.10 Functional utilisation of elements

Biological systems	Man
Mineralised structures	**Minerals and metals**
$CaCO_3$, $SrSO_4$,	$CaSO_4$, CaO, cements
SiO_2, Fe_2O_3,	and silicates (concrete),
$Ca_2(OH)PO_4$	Fe, Cu, Al, Zn, Mg and alloys
Organic structures	
Proteins, polysaccharides,	Plastics,
C/N/O	C/N/O
Catalysts	
Enzymes and metalloenzymes,	Metal complexes;
Mo, Mn, Fe, Co, Ni, Cu, Zn	very many, including heavy
	elements Pt, Zr, Pb;
	surfaces of solids
Message carriers	
Ions, organic molecules:	Electrons, light
	(wires and pipes)
Na^+, K^+, Ca^{2+}, etc.	Cu, Fe, Ge, Si, etc.
(water in tubes)	

15.10 Elements and chemicals in the diet and in medicine

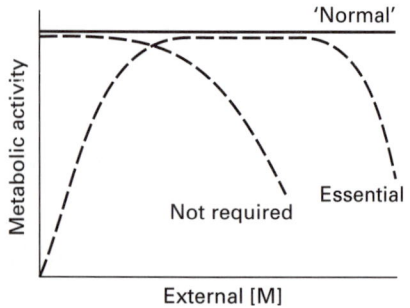

Fig. 15.13 Diagram of the effect of external concentration of M, [M], on metabolism. ('Normal' corresponds to optimal activity.) Essential elements can be toxic at high concentrations; 'not required' elements can be poisons at low concentrations, but may be therapeutic.

While we have stressed the great advantages of the complexity of *homo sapiens* based upon his mental skill we must also point to his fundamental chemical weakness. As stated earlier, as animals developed complexity of activity they increased dependence on other species for essential chemicals. Some of the needs of humans are listed in Table 15.11 under vitamins and minerals, but there are many more, such as the essential amino acids and fats. The human animal is a very poor chemical factory and depends on a vast host of other organisms, from bacteria to plants and probably even to animals, to sustain life. It may well be that this reduction in the complexity

Table 15.11 Some essential dietary requirements of humans

Vitamins		Minerals and Trace Elements	
Vitamin A (as β-carotene)	2500 i.u.	Calcium	100 mg
(Retinol equivalent)	(750 μg)	Iron	12 mg
Vitamin D_2	400 i.u.	Copper	2 mg
	(10 μg)	Phosphorus	77 mg
Vitamin E	10 mg	Magnesium	30 mg
Vitamin B_1	1.2 mg	Potassium	4 mg
Vitamin B_2	1.6 mg	Zinc	15 mg
Vitamin B_6	2 mg	Iodine	140 μg
Vitamin B_{12}	3 μg	Manganese	3 mg
Vitamin C	60 mg	Selenium	50 μg
Biotin	100 μg	Chromium	200 μg
Nicotinamide	18 mg	Molybdenum	250 μg
Pantothenic acid	4 mg		
Folic acid	400 μg		

Formulated in line with the UK Dietary Reference Values (DRVs, 1991) and the US Recommended Dietary Allowances (RDAs, 1989).

Table 15.12 Medical uses of unusual elements in inorganic drugs

Element	Medical use
Lithium	Hyperactivity drug
Boron	Neutron capture therapy
Fluorine	Tooth protection
Aluminium	Anti-acid
Chlorine	In several antibiotics
Copper	Anti-inflammatory
Zinc	Wound healing
Platinum	Anti-cancer
Gold	Arthritis; anti-inflammatory
Bismuth	Drugs for gastric diseases
Arsenic	Anti-syphilis (now obsolete)
Antimony	Anti-schistosomiasis
Barium	X-ray diagnosis (gastrointestinal tract)
Mercury	Antiseptics, diuretics (now obsolete)
Selenium	Anti-seborrhoeic
Tin	Anti-boil and anti-acne drugs
Gadolinium Manganese	Magnetic resonance imaging

Table 15.13 Pesticides (plants)

Types of pesticide	Examples	
First generation (inorganic or organometallic substances)	Sulphur Lead Copper Arsenic Mercury	salts
Second generation (organic substances		
(a) Chlorinated hydrocarbons	DDT, Chlorin, Dichlorin, Mirex	
(b) Organophosphates	Malathion, Parathion	
(c) Carbamates	Sevin, Temik	

of his internal chemistry allowed the development of the brain. In any event, man himself is an ecosystem which is dependent internally on micro-organisms and also dependent externally upon the ecosystem in his environment. However, man, through his brain, knows of his limitations and therefore attempts to correct them using external manufacture of many required foodstuffs, including vitamins. The development of chemicals is also for protection against other organisms or even mismanagement of his internal chemistry. There is now a huge pharmaceutical industry and while most of its chemicals are organic there is a developing interest in other types of chemical medicine involving inorganic elements.

While mismanagement by industrial chemistry introduces risks, careful use of even extremely toxic elements can be used in medicine or agriculture to great advantage. In Fig. 15.13 we show that all elements (and components) are associated with risks at high levels (see Fig. 14.13). However, those that are not used by organisms may well carry risk at extremely low dose. This risk can be turned into an advantage if it can be shown that the poisonous effect of the element (component) is greatest against an invasion of unwanted organisms or even against cancer cells. Examples are given in Tables 15.12 and 15.13.

15.11   Genetic engineering

A differing approach to the future, at the moment of limited use, is to consider the units and variables of reproducing living systems and to ask how they could be changed. Small changes involve manipulation of plants and animals for man's benefit by altering their genes. This can be and has been done but still only to a limited extent. A wider issue is to ask about the kind of machinery that is living and then to enquire as to whether or not living material could be brought about in a fundamentally different way. The simplest model is a self-assembling computer, but we must add that it would have to find its own energy and materials to put into a flow system. As the understanding of robotics and biological systems increases we have to face the possibility of creating 'living' machinery of a new kind with new sensory images. It can be designed for a useful purpose but, of course, it must not be made self-conscious! Then it would challenge us! Note that this possible future development is quite different from and much more challenging than cloning, which merely manipulates existing biological chemistry. Genetic analysis and genetic engineering can also lead to a deeper appreciation of the factors that underlie our very being. As stated earlier, we must not think that there is a simple linear relationship between coded data (DNA) and the properties of an organism, which are complex.

15.12   Industrial organisation

As knowledge grew, the possibility of producing chemicals on a vast scale arose. At first, metallurgical processes were developed. The industry required ever-more complex transport of material from different sources—coal and ores—and, of course, with transport there grew the necessity for a communication network. There is a rough parallel with biological evolution. At first any activity, any form of life or industry, was all located in one

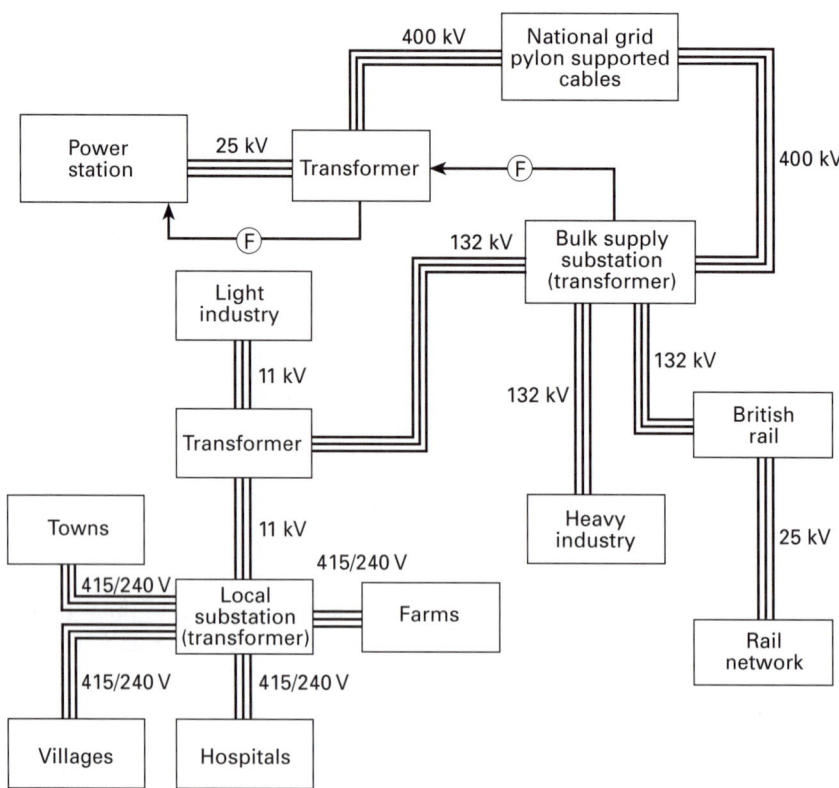

Fig. 15.14 Connections used in the British National Grid power supply. The voltages used in the National Grid are very high and the current is relatively low so that less energy is lost as heat. When the electricity reaches its destination the voltage is reduced by a transformer. Controls are involved at all points. Ⓕ is feedback.

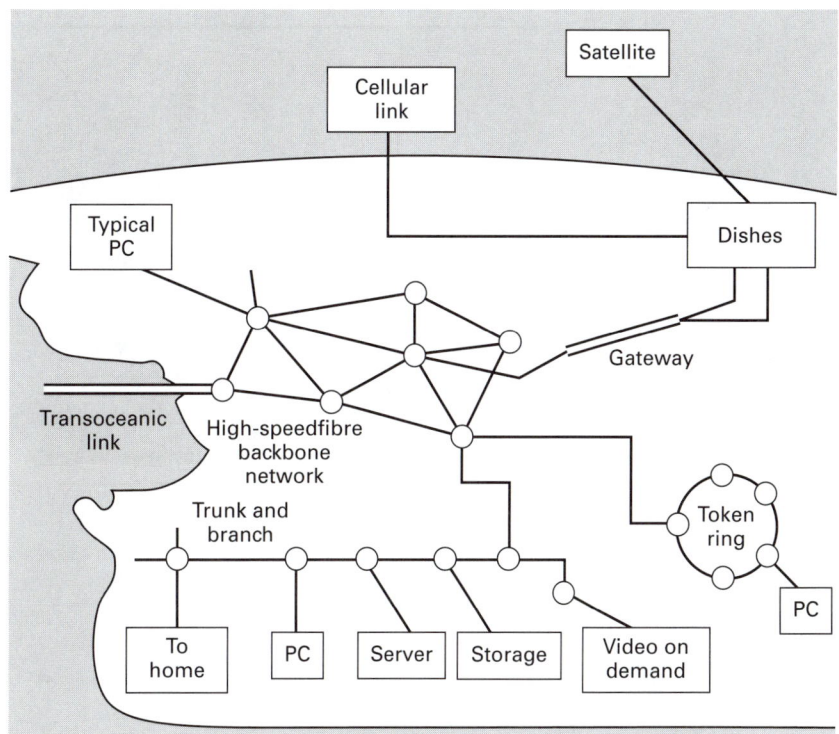

Fig. 15.15 The internet is a network of computer networks. Information travels around the world over different routes, including fibre optic cables, satellite links, radio transmission and traditional telephone lines. [Adapted from *New Scientist* (1994).]

place, compare one local cell (a prokaryote), or spatial zone. Subsequently, industry, like cellular operations, became divided into compartmental activities in different spatial zones. Material then had to be transported from zone to zone. In both cases there followed a need to control and inform each compartment of the activities of every other one. Networking of messages is then a common feature of advanced biological organisms and of industrial production. We illustrate some examples in Figs 15.14–15.16.

Much though the two activities, biological and industrial, have common features, they do not use the same chemicals in most cases. The major turnover of elements is also very different; for example, the use of minerals and plastics is common to the building of biological and man-made frameworks but the development of metals for both frameworks and communication is due to man alone. The catalysts are somewhat similar in life and industry in homogeneous but not in heterogeneous systems. The modes of energising systems are also quite different in that man has evolved huge coal-, oil- and gas-burning equipment and has developed power from water flows, dams, thermoelectric plants and even controlled nuclear reactions. However, man has failed, as yet, to harness the energy of the sun, as many forms of life do. The industrial communication network is based on electronics using metals or near-metals, such as silicon and

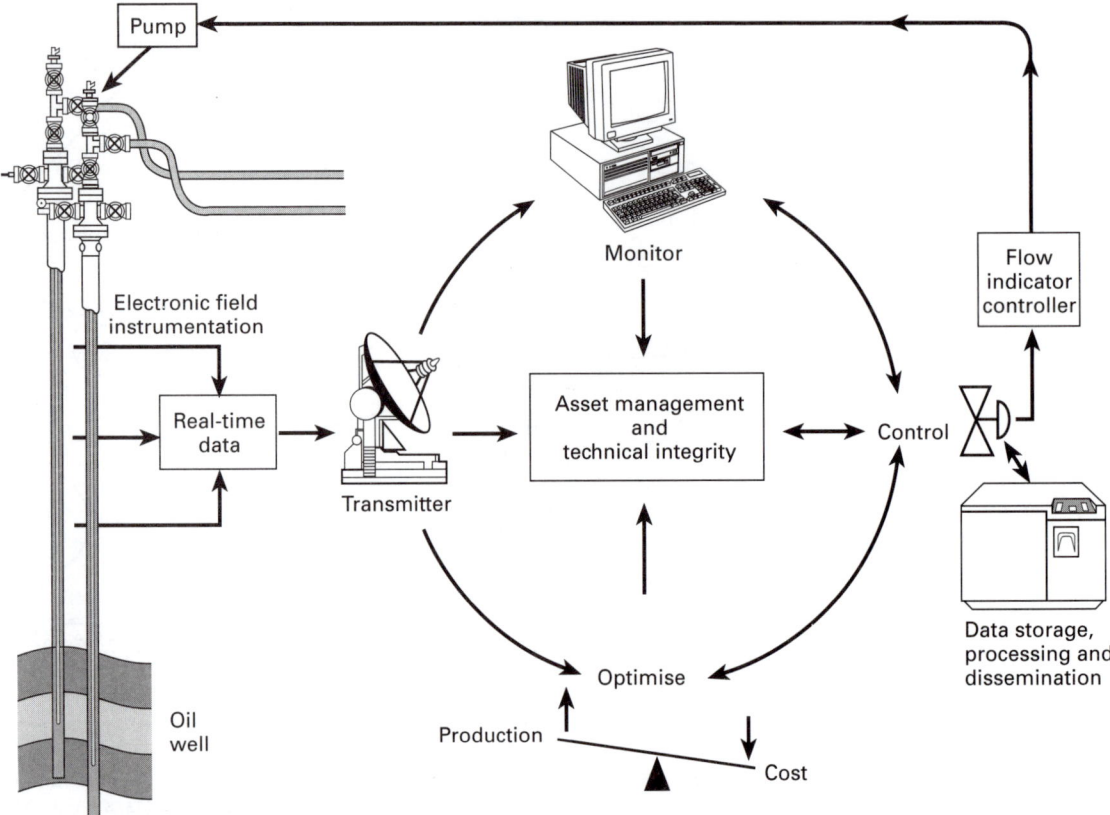

Fig. 15.16 The management scheme of an oil-pumping system showing feedback responses.

germanium, not electrolytic conduction, as in nerves, with inorganic or organic messengers. We have computers, but no manufactured brains yet.

As was the case with every step forward in evolution (see Chapters 11–14) it is essential to look at the gains and losses. Anaerobes lost territorial regions with the coming of aerobes; plants lost some control over land with the coming of animals; all animals lost much of their control with the advent of self-conscious man. Can man himself be replaced? Alternatively, can man set up a chemical system to his own disadvantage much as did anaerobic, photosynthetic, oxygen-producing organisms by the very act of producing the pollutant for them, dioxygen? Once again we must be aware that the nature of living organisms in societies is likely to be given by complex connections between properties and variables.

15.13   The essence of industry

The essential features of industrial activity can now be given in diagrammatic form, much as was used to describe living organisms

Materials $\rightarrow$ | Hidden Flow System Factory | $\rightarrow$ Products

Fuel (energy) $\rightarrow$ | | $\rightarrow$ Heat

The hidden flow systems included feedback controls over steps in the processing, such as over transport of chemicals and energy, their input and output, introduction of catalysts at specific times and management of conditions (temperature and pressure). There is one major feature quite unlike those in many biological cellular processes: the products do not degrade readily to give back initial materials (see Table 15.18 and Fig. 10.2). There is a parallel here with two particular sets of biological reactions some $3–2 \times 10^9$ years ago—the overall production of oxygen and reduced carbon compounds such as coal, oil and gas

$$2CO_2 + 2H_2O \rightarrow 3O_2 \uparrow + 2[CH_2]_n \downarrow$$

and the production of calcium carbonates

$$Ca^{2+} + CO_3^{2-} \rightarrow CaCO_3 \downarrow$$

Thus, some elements are only cycled very slowly by organisms over long periods of time. For the anaerobic systems that existed at the time, the new chemicals, especially dioxygen, were pollutants. They could not recycle them. This was not the result of a 'demand' for dioxygen but of an 'accident' associated with a 'demand' for hydrogen (from water).

In contrast, in most of man's industry the wish is to produce deliberately long-lasting material objects and to avoid decay. *This means that man produces new real components (Section 2.4) not biological 'components' (Section 10.19).* Decay of man-made minerals, for example concrete and steel, is a relatively slow trickle compared with the massive decay of biological material. In the short term, perhaps over the one hundred years from 1850–1950, there was no grave disadvantage, since the scale of man's operations was relatively small, but in the long term it is inevitable that problems will arise as more and more products are generated in slowly decaying states. There is, in effect, the general result.

Products $\rightarrow$ Waste $\rightarrow$ Decay $\rightarrow$ Pollution (not use)

In other words, pollution automatically increases with production unless a deliberate effort is made to clean-up. Examples are given in Tables 15.14–15.17 and Fig. 15.17.

In Table 15.18 we stress some chemical elements which are more prominent in industry than in living organisms. The comparison of the two groups gives some impression of the potential risks of possible contamination by presently non-biological elements. There is, in addition, a variety of hazards from components (combination of elements) (see Table 15.13), which are carefully examined by health and safety authorities. Once again, note that, as components, they do not equilibrate their elements with the environment and are not limited by feedback and so remain as risks for living organisms.

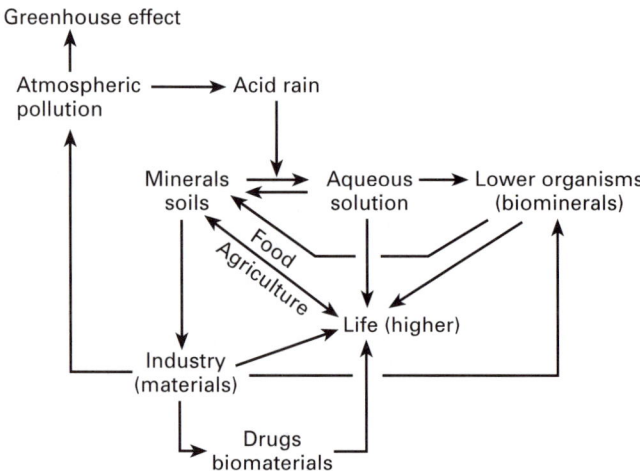

Fig. 15.17 A generalised scheme of the way in which man's industry and his environment interact with living systems.

Table 15.14 Global problems

Problem	Source (agent)	Effects
Greenhouse effect	Carbon dioxide, carbon monoxide, methane, water vapour, chlorofluoro-carbons, etc.	Global warming and associated effects
Acid rain	Sulphur dioxide, nitrogen oxide, carbon dioxide, dimethyl sulphide (after oxidation)	Acidification of rivers and lakes: solubilisation of rocks and minerals (e.g. with aluminium): destruction of vegetation
Ozone hole (in the stratosphere)	Atomic chlorine derived from chlorofluorocarbons (freons), methyl chloride and other products	Skin diseases; possibility of increased incidence of skin cancer
Photochemical smog	Carbon monoxide and nitrogen oxides (from exhausts), and hydrocarbons (terpenes)	Respiratory affections and diseases
Radioactivity	Nuclear power stations (release of radioactive isotopes)	Mutations (cancer)

15.14　Summary of the nature of man and his activities

This chapter has been divided into two main parts. In the first part we have attempted to describe what is special about man internally. We could not find any novel feature of man's basic physical and chemical construction, with one exception—the brain. For the rest of man's organs we could see only small evolutionary changes relative to other mammals. If anything, man's overall chemistry is more, not less, dependent on external supplies and suppliers. For example, very few organisms require vitamin C,

Table 15.5 Atmospheric pollution (local and regional)

Pollutant	Source	Effect
Sulphur dioxide (SO_2)	Burning of fossil fuels	Respiratory diseases; acid rain, release of aluminium
Hydrogen fluoride (HF)	Glass factories	Destruction of grazing fields and pinewoods
Hydrogen sulphide (H_2S); ammonia (NH_3) methane (CH_4)	Bacterial activity	Poisons
Nitrogen oxides (NO_x)	Burning of fuels and of vegetation	Photochemical smog
Lead tetraethyl ($Pb(CH_3)_4$)	Combustion of petrol (exhaust)	Destruction of vegetation (in zones close to transport systems)
Dust and smoke (with heavy metals)	Industry: traffic	Respiratory diseases; toxicity of heavy metals

Table 15.16 Aquatic pollution

Source	Nature	Effects
Domestic sewage	Excrements; detergents; soap; paper	Pathogenic micro-organisms, skin and intestinal infections
Industrial effluents	Oxidisable organic materials; foams; heavy metals	Toxicity for plants and animals; deoxygenation of rivers, lakes and estuaries; death of fish and vegetation
Agriculture and cattle breeding effluents (rain and irrigation water)	Manure; fertilisers (nitrate and phosphate); insecticides, pesticides, herbicides, etc.	Toxicity for fish and vegetation; eutrophication
Transport of crude oil (in the sea)	Heavy hydrocarbons	'Black tides'; death of birds and sea animals
Storage of toxic and radioactive wastes	Variable	Risks of toxicity and radioactivity at dangerous levels

Table 15.17 Soil pollution

Source	Nature	Effects
Fungicides, herbicides, pesticides and insecticides	Especially chlorinated and phosphorylated compounds	Poisoning of animals; metabolic deficiencies (e.g. absence of eggs, too thin egg shells, dead embryos, etc.)
General household and industrial rubbish	Variable including trace metals	Crop limitations

Table 15.18 Novel elements important in man's activities

Activity	Some novel elements*
Metallurgy	Al, Cd, Pb, U, Cr
Construction	Al, Si (cements), Pb (pipes)
Medicine	Pt, Bi, Au, see Table 15.10
Electronics	Si, Ge, Cu
Nuclear energy	B, U, Pu

* Elements not prominently used in living systems.

but man does and as a species is obviously dependent on plants and animals for food and on a considerable range of lowly organisms for digestion and internal cleansing. The big jump in evolution to man was, then, the rapid increase in his brain, especially the cerebrum. This allowed an unparallel freedom of action by the individual, since his phenotypic development is so large compared with his genotypic expression that it has come to play a dominant role.

The outstanding feature of the brain, not seen in the other organs, is the continuous acquirement of knowledge of surroundings, which improves amazingly during the first five years of life. It is this that allowed the development of self-awareness and understanding. We have looked for, but so far have failed to find, a chemical reason for this difference. All the chemical 'components' of the brain are found elsewhere in the body and in other animals. There are, however, some clear distinctions in the concentrations of these chemicals from those in other body organs, not just in cells but in the special extracellular fluid, the cerebrospinal fluid, that surrounds them. The future will surely reveal some underlying peculiarity in the way the brain works, which differentiates its physical–chemical cellular growth pattern from that of all other organs. As stated, it reflects phenotype more than genotype, individuals as much as species. Thus, it is the peculiarities of physical development, the making of circuitry, that we must appreciate.

Now, it might have been the case that this brain power would have led to a contemplative animal, but in fact it has led to an extremely active one. This has been particularly noticeable in man's ability to manipulate both the chemical and the physical environment outside, and even inside, his own body. It has meant that a living organism, man, can exploit the value of all 90 naturally occurring chemical elements on Earth (and even some synthesised ones), where earlier life could only manage internally some 20–30 elements (see Tables 15.19 and 15.20). The mode of usage has also been developed, so that a vast realm of new chemicals and materials have been produced. These chemicals have been employed in the construction industry, in the manufacturing of machines of very diverse kinds, in medicine, in agriculture and even in the control of genetic codes. Physics knowledge, together with new chemicals, has also allowed access to vast new sources of energy. Thus, for the first time in evolution, the environment outside the organism has become closely linked to the purposeful activity of the organism itself. This, together with the phenotypic development of the brain, has led to individual expression, to cultural development and to the organisation of societies. A revolution in the nature of

Table 15.19 Changes of variables and derived variables on Earth

Time (years)	Components and compartments	Temperature (K)
5×10^9	90 elements equilibrated	1000–5000
4.5×10^9	Core and mantle equilibration; crust continuously changes	300–400
3.5×10^9– $+1.0 \times 10^9$	As above plus release of O_2; set of 20–30 elements + large number of compartments gives life	273–373
1.0×10^9– 4×10^4	As above; little change in chemistry: large changes in compartments in life	273–373
4×10^4– today	As above plus large change in man's industrial chemistry; 90 elements not equilibrated; large increase in external compartments	100–10 000

Table 15.20 Man's expanded variables in industry over living organisms

Units	Increase in the number of elements and component chemicals used
Composition	Adjustment of energy input. Increase in components
Compartments	Building and incorporation of compartments entirely *external* to an organism, e.g. in the environment (industry)
Energy	Increased utilisation of resources—fossil fuels, sun, nuclear energy, etc.
Message systems	Telephone, radio, television, E-mail, fax
Transport	Huge variety of vehicles

evolution have taken place. With all this novelty there is now the need for the flexible management of these human organisations, societies, which have evolved, so that they are self-sustaining both with respect to the external physical and chemical resources and conditions but also to the internal individual and collective motivations. In the last chapter, where we summarise the contents of this book, we shall look at these problems in somewhat more detail. In the process of development from matter to man, matter has fallen under the control of man.

16

Survey and conclusions

The next great era of awakening of human intellect may well produce a method of understanding the qualitative content of equations. Today we cannot see that the water flow equations contain such things as the barber pole structure of turbulence that one sees by rotating cylinders. Today we cannot see whether Schrödinger's equation contains frogs, musical composers, or morality—or whether it does not. We cannot say whether something beyond it like God is needed, or not. And so we can all hold strong opinions either way.

R. Feynman (1991)

16.1   Introduction

We have titled this book '*Bringing chemistry to life*' and subtitled it '*From matter to man*'. An alternative subtitle would be 'The physical–chemical evolution of natural systems'. In essence what we wished to show is that the development of the universe, all the way from the initial 'matter' to man on Earth, is to be understood in terms of the ways in which natural ordered or organised systems arise spontaneously. (The arrival of man has introduced quite different inputs recently.) We have shown that to do this in a logical way we have to analyse first the nature of all materials, the basic units from which all matter is constructed. This corresponds to the description of an obvious variable—composition. Knowing the nature of matter we can ask what other variables, inevitably associated with space and time dimensions, in which matter resides, have affected the way in which systems involving the units of matter (elements or chemicals) can behave. To do this we found that it was convenient to separate systems which had different restrictions on the variables. The major distinction was between systems assumed to be *unchanging** and in which order and disorder were balanced in stable or stationary, time-independent states, and *changing* systems, which could evolve as flowing organisation (such as life) against an ever-increasing disorder. The second set of systems, evolving organisations, are time dependent. We could then ask how both types of system arose. In both cases, as the complexity of the observed systems increased we found that new *emergent properties* appeared which forced us to move from fundamental to operationally useful derived variables. In this summary we examine first the time-independent systems, analysing the ways in which the spatial distribution of matter can be shown to be related to the properties of its constitutive units. (N.B. Spatial distribution here may be dynamic as well as static, but it must be unchanging when averaged over time.) Before doing so it is necessary to show briefly how matter itself arose.

An explanatory history of everything can be said to originate from the Big Bang. This was the moment when space and time began (an anthropocentric view), some 15×10^9 years ago, corresponding to the creation of an enormous quantity of localised energy and outward momentum, which led, and still leads, to expansion of *our* universe. *Space* and *time* created simultaneously with energy and momentum must, therefore, be two of the fundamental variables in all our discussions. It was the energy created which evolved into matter as we know it. After about one second, material particles—protons and neutrons derived from quarks, electrons, neutrinos and other exotic species—arose from the initial energy source (see Table 16.1). Free neutrons, however, are unstable and decay into protons and electrons, which some one million years later, when the temperature cooled down to 3000K, formed hydrogen atoms, some deuterium, trace amounts of lithium and about 10% of helium (relative to hydrogen). These two elements, H and He, which still make up about 99% of all known matter in the universe, then associated, in part due to gravitational forces, to form stars and galaxies of stars. Elements heavier than helium—only about 1% of the known matter in the universe—were synthesised later by

The origin of energy, matter, space and time

---

*In fact, all things in the universe change but some systems change so slowly that we can discuss their properties as if they were unchanging (see Chapters 1–7).

Table 16.1 The formation of our universe

Time	Average T (K)	Situation
0	(Infinite?)	Creation
10^{-43} s	10^{32}	Inflation
10^{-34} s	10^{27}	Hot Big Bang
10^{-3} s	10^{12}	Quarks and exotic particles fill universe
1 s	10^{10}	Electrons, Neutrons and protons
3 min	10^{9}	D, He, (Li)
10^{5}–10^{6} years	3×10^{3}	Decoupling matter/radiation
10^{6}–10^{10} years	$<10^{3}$	Atoms and galaxies, compounds
1.05×10^{10} years	$<10^{3}$	Earth
1.15×10^{10} years		Life on Earth
1.5×10^{10} years*	3	Today's universe (man)

* Recent data telescope suggest that the universe is not quite so old, but this is still an open problem. N.B. The temperature is very different locally.

Chemical composition: atomic mass and charge

successive nuclear reactions in giant stars, as discussed in Chapters 1 and 8 (and see Table 16.1). It is at this subsequent stage of the evolution of the universe that we find our chemical units of the composition of matter (atoms and charges) which are central to the discussion of chemistry and biology. Atoms and charges provide, therefore, the third fundamental variable (in the physicochemical systems of interest to us), *composition*, additional to space and time, since only certain (less than 100) types of atoms and two types of charge (positive and negative) were allowed. Thus, in purely descriptive terms, we may say the universe proceeded, on expanding, from the creation of a small number of basic units to their (inevitable) association through different types of forces (see Table 2.2) but we must observe that the formation of giant stars, where the chemical elements were born, represents only a part of the evolution of the universe from an initial homogeneous equilibrated system of energy (radiation) *and* matter, to one of great energy and material inhomogeneity, with hot and cold, occupied and "unoccupied" spaces, all of which are part of a whole (the universe) that is expanding with time. For our purposes, however, whatever the particular nature of this process, what we state again is that today the *fundamental units* of all chemical and biological materials are atoms and charges and the *fundamental variables* are composition, space and time. While the changes in the state of energy and matter developed, the original outward momentum evolved also into turbulent momentum, which in some regions of the universe became organised into changing patterns of flow. This requires us to introduce the final fundamental variable, *time*. We turn first to those objects around us here on Earth which are, to all intents and purposes, unchanging.

Turbulence and flow

16.2 From fundamental to derived and operational variables of unchanging systems

The initial decoupling of matter and energy, the first loss of equilibrium, probably occurred some one million years after the Big-Bang and several

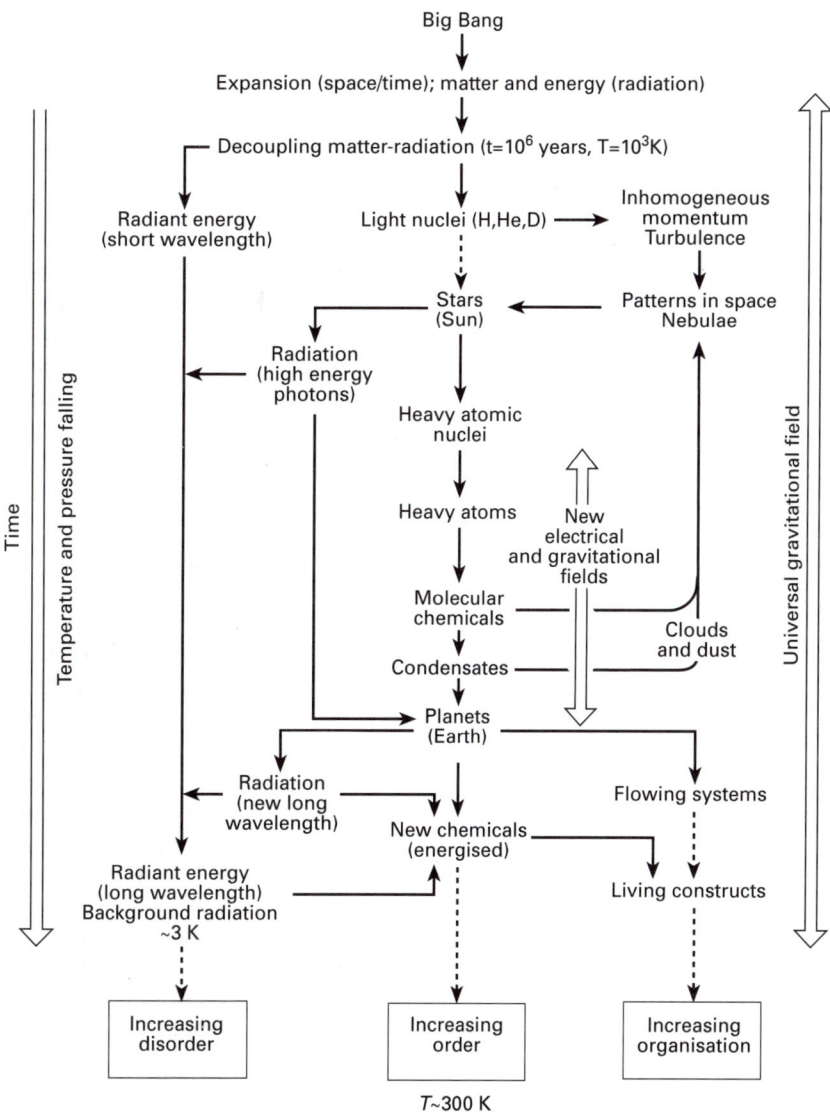

Fig. 16.1 An outline scheme given as an indication of the development of objects around us, and some of the underlying interactions and energies since the beginning of time.

Radiant energy: photons

million years before the formation of galaxies and stars and then of elements, compounds, condensates, planets, etc., a persistent process that has continued ever since (Fig. 16.1). We stress again that, at the stage of Earth's formation, the fundamental underlying, unchanging units of chemical and biological systems had evolved (see Table 2.2). [To the list of fundamental units we should perhaps add photons (radiation) but these are best left largely to one side in our discussion of static chemical systems while remembering that radiation (temperature) is also not homogeneous in the universe.]

Newtonian and quantum
mechanics

Now, the *fundamental* units and variables that we have identified are not the most *useful* quantities for the description of what we observe, especially here on Earth. In fact, we find that atoms and charges have come together in particular ways in space to form static materials, chemicals, molecular or bulk, which are described by structures, due to forces (hence potential energies) of interaction between them. As we have seen in Chapter 2, the dominant internal interactions (forces) were gravitational and electrostatic (attractions) and repulsive exclusion. [We frequently use the *energy* of interactions instead of forces, but note that both derive from two of the fundamental variables—composition (atoms and charges) and space (distance between pairs of units).] In effect, these interactions and their cyclic dynamic counterparts are best described in terms of sums of pairs of units, and this was of the essence of the early descriptions, included under Newtonian mechanics, which allow a continuous, smooth variation of interaction with distance and of certain characteristics of dynamic properties in stationary conditions (see below). This treatment remains valid today for bulk objects, but, as we have seen, it fails for atomic interactions and has been replaced by quantum mechanics. Quantum treatments show that the dynamic–space interactions between electrons in atom give true discontinuous stationary states (not losing energy). They lead to favoured states of atomic combination, such as H_2O and $NaCl$ but not, for example, to H_4O and $NaCl_2$, with defined structures. Hence, we separated bulk interactions, in terms of continuous field energies, from localised, discontinuous chemical bonds and their energies used to describe atom associations.

Chemical components

Many of these atomic assemblies formed stable or (more commonly) energised stationary chemicals, which are isolated and more or less permanent. For this reason, composition in terms of atoms, their spatially defined locations and their potential energies is a clumsy way in which to describe chemicals, since in chemistry and biological systems we find that in many operations most of these associations of atoms and charges remain unaltered, e.g. on mixing water and alcohol the identity of the two chemicals is not affected. The *operationally useful*, derived description of the composition of chemical systems is then in terms of non-exchanging combinations of these two variables, i.e. of a combination of atomic composition and specific space location of atoms and charges (or composition and potential energy), often in molecular compounds and generally called *components*. The subject matter of chemistry and biochemistry is then written in terms of components, not atoms and potential energies. (We considered whether this is a truly useful approach to living organisms in Chapter 10–14.) We showed also that chemicals which exchange atoms in equilibrium are not separate components since their variables of composition are not independent. The molecular units, while they are very small, are looked upon as having fixed structures, but we found that we needed to look at larger polymeric molecules in a different way (see Section 5.8.5 and 16.4.1).

Initially, that is, shortly after the Big Bang, all the precursor particles were also in motion, that is, their space co-ordinates changed with time. As stated, they had an overall outward movement due to the expansion of the universe which we described by the variables *momentum (mv)* and *directed*

Pressure and temperature

kinetic energy of flow $(1/2m\ v^2)$. Since all this movement is coherent it can be described by the properties of each particle separately. As mentioned above, there was, however, a loss of homogeneity in the universe, so that, as well as outward flow, turbulence developed locally. As far as individual particles are concerned, each one within a system came to have a new individual momentum and kinetic energy in some extra direction other than along the outward line of flow. In effect, when a large number of such particles exist together, the collisions that are generated between them give rise to an overall distribution of motion and energy quite different from that associated with outward flow. They now have an additional random motion and random kinetic energy which does not give the whole assembly any additional flow in any direction. It was then no longer possible to handle the properties of this system by reference to the properties of single or pairs of particles except by crude averaging, as in the kinetic theory of gases, which failed to explain many properties. We describe this (localised) randomised momentum *distribution* by a measurable *pressure* and its (localised) randomised kinetic energy *distribution* by a measurable *temperature*. These are *bulk* properties (derived variables) and, unlike flow, are time independent with respect to spatial distribution. (As all such systems of particles or bodies are also emitting and absorbing energy as radiation, a fixed temperature is also characterised by an equal fixed emission and absorption of radiation.)

We can apply the same description to atomic *or* component units, so that both can be given statistical properties (temperature and pressure) when in a system of many units. We may, therefore, use a set of more complex *derived* variables, all related to the fundamental ones (Table 16.2), to describe unchanging systems on the Earth in terms of component composition, temperature and pressure. (We leave flow to one side for the moment and return to it later.) In this way we could develop equilibrium thermodynamics in an operationally convenient way for a physically homogeneous system.

Table 16.2 Fundamental and derived (and other operationally useful) units and variables for pair-wise interactions

	Fundamental	**Derived (operationally useful)**
Units	Mass and charge (photons)	Components
Variables	% Composition (fundamental units)	% Composition (components)
	Space	Potential energy
	Time	Directed momentum
		Directed kinetic energy
		Temperature and Pressure (random kinetic energy and random momentum)

See Chapters 2–4 for consequential properties.

Equilibrium thermodynamic functions

An observation which we have yet to explain in this summary is that all chemicals (components) from animate or inanimate matter can occur in three states of matter: solid, liquid and gas. Clearly, the value of the internal potential energy of interaction between units of the components is greatest (more negative) in the solid and least in the gaseous state, while the liquid state is intermediate in nature. The states are, therefore, less ordered in the sequence solid, liquid and then gas. However, the random kinetic energy is greatest for gaseous, liquid and then solid states, that is in the reverse sequence, and of course a gas has a pressure due to the mechanical momentum while a solid does not. While the most ordered condition, the solid, is found at the lowest temperature and highest external pressure, the least ordered, the gas, is found at higher temperature and lower external pressure. Now, we have seen that disorder, especially, is a statistical feature of a large number of particles and cannot be reduced to a linear sum of pair-wise interactions. We again needed new *derived* variables to quantify the probability that any bulk system could be observed in a given state. We evaluated, therefore, a statistical measure of the disorder called the *entropy (S)* which is a useful variable, related quantitatively but not linearly to temperature and pressure and connected to the probability of a given physical state of a bulk material, independent from its internal energies due to binding between atoms. From the entropy of states we could calculate an entropy *bias* in energy terms in favour of change from an ordered to a disordered state, $T\Delta S$, which competes with the binding in the ordered condensed state expressed by the heat content or enthalpy change, ΔH, at constant pressure. The derived thermodynamic variables are given in Table 16.3. At certain temperatures and pressures, those corresponding to the melting and boiling points, the different physical states come into balance, whence disorder and order are in equilibrium, and new physical states form.

Phase equilibria

This formation of *phases* is then an *emergent property* resulting from the balance of variables. The presence of boundaries between different states in balance, *phase equilibria*, is a restriction on variance in the sense that the variables temperature and pressure cannot be adjusted independently when two phases of one material are present. Note that a phase, by definition, allows both equilibrated transfer of energy and material across the phase boundaries. At equilibrium, order and disorder remain in balance in every phase across their *phase* boundaries. We can now describe an unchanging *static* system in terms of the derived variables component composition,

Emergent properties

Table 16.3 Thermodynamic functions (variables) for bulk systems (changes at constant pressure)

Composition	Components (C)
Compositional variable	ΔU
Compositional variable and work of expansion	$\Delta H = \Delta U + p\,\Delta V$
Statistical variable (hidden energy)	$T\Delta S$
Maximum available energy for change	$\Delta G = \Delta H - T\Delta S$
Temperature and pressure	T, p (concentration)

See Chapters 5 and 6.

Table 16.4 Restrictions on thermodynamic variables

(1) The gas law, $pV = RT$. (Internally equilibrated velocity)
(2) Physical equilibria across phase boundaries
(3) Chemical equilibria between units (of apparent components)

See Chapters 5 and 6. (1) + (2) give the phase rule when combined with component (C) composition.

temperature and pressure, and/or in terms of the derived composite 'equilibrium thermodynamic' bulk variables heat content, enthalpy, (ΔH) and entropy energy ($T\Delta S$), which are more useful for the description of change. (Note again that the appearance of each new phase above one acts as a constraint on variance) The heat content at a given temperature and pressure here is also related to the bulk substance and is not just related to fixed binding energies, for example in molecules.

We also showed that chemicals in equilibrium were not separate components and just as the number of degrees of freedom, variance, was limited by phase boundaries across which units equilibrate (Table 16.4), so chemical equilibria reduced compositional variance. Thermodynamic equilibrium is the condition to which all systems aspire and is one of reduced variance. If a system is not in equilibrium it can obviously change and will then give out heat. The heat increases entropy, whence all systems move towards a state of maximum entropy at equilibrium.

Now, equilibrium exchange does not characterise the condition across most boundaries which limit objects around us. In effect, the transition of the universe from a single homogeneous system of energy and matter in one phase to a separated system of heterogeneous 'compartments', as seen in the uneven distribution of mass and energy in nebulae, stars and other bodies, represents a loss of equilibrium, which gives rise to a vast increase in the number of possible systems (systems with different values of the derived variables) since each compartment is independent in space and time to some degree. Despite limited energy exchange, each separate compartment could then evolve additionally in its own way since it could have its own composition of chemical elements and energy. This break-up of a single system is an almost inevitable consequence of rapid expansion with cooling and was repeated on a smaller scale as planets evolved in the neighbourhood of stars. Thus, the sun is at a very different temperature from Earth, the Earth has zones at different temperatures and pressures and the chemical composition varies from zone to zone without balanced exchange, e.g. between sea and land.

While we can look upon many local static (unchanging, that is time independent) zones as being internally at equilibrium, they interact with one another through physical fields, gravitational and electrical. The sizes of the independent '*compartments*', and the fields they exert generate new variables for static systems (Table 16.5). (Notice that we always include stable and all kinds of dynamic stationary conditions within static systems. Hence our planetary system is 'static' in the sense that over a sufficiently long period of time it is not distinguishable from its initial condition; this is

| Degrees of freedom (variance) |

| Divisions of space: compartments |

Table 16.5 Variables due to compartmental separation

The variables are those of Table 16.2 increased by:

(1) one, in total number of compositional variables per compartment to describe amount, not % composition, of each compartment. Each compartment is independent;

(2) the number of independent compartments;

(3) fields due to gravity and electrostatic potentials acting on each compartment.

N.B. 'Shape' is derived from all the variables of compartmental separation and the internal variables of Table 16.3. Size is defined by the full composition (see Chapter 7).

due to the fact it includes repeated and, to a first approximation, exactly fixed cycling, i.e. with fixed angular momentum, and does not involve collisions.) Table 16.3 lists the way scientists have found it useful to describe systems using these further operationally useful, derived variables, all ultimately related to combinations of the fundamental units and variables (see Fig. 16.2). Unfortunately the relationship becomes mathematically so complex, see section 16.4, that it is not practical or useful to apply strict reductionism.

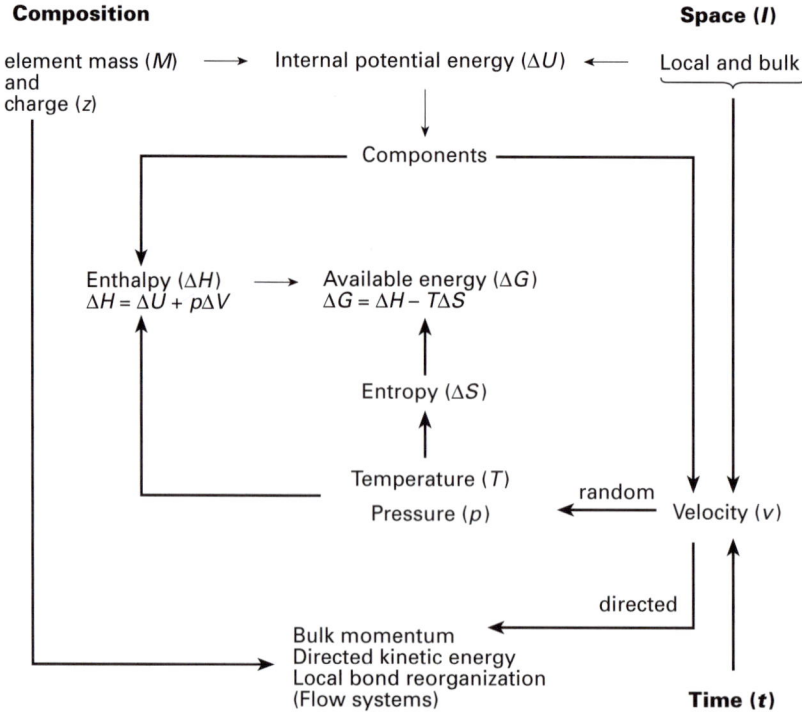

Fig. 16.2 Schematic description of the links between fundamental units and variables (heavy type) and derived variables used in equilibrium thermodynamics for operational convenience. Symbols are in brackets.

16.3   The capacity for change

Now, we wish to discuss change rather than fixed systems since our universe has continually evolved. We need to note two very different aspects. First, we can ask what is the maximum capacity for change going from one static condition of two or more systems to another, with gain or loss of energy. Secondly, we can ask what is the rate of change of these two systems. The second aspect concerns non-cyclical motion, i.e. flow of units, energy, mass or charge, in space, even very locally, to give novel objects or conditions. Clearly, the capacity of a system for change is the simpler and has to be characterised in some relative way by properties of the different unchanging states of the two systems, that is, the initial and the final conditions of composition, temperature, pressure, ΔH, ΔS, size of compartments and any fields present, all in local, i.e. internal balance, in each state.

We have seen in Chapter 5 that any such capacity for change of condition of a given system of interest in chemistry and biology on Earth could be discussed most relevantly in the presence of an environment, the second system, from which heat (energy) could be taken or supplied without material transfer and with which, we postulated, the system of concern had to remain in energy balance, being, therefore, a so-called *closed* system. We discovered, in fact, that capacity for change then had a peculiar feature. In large part because change of a system invariably imposed itself on the environment either as an alteration of pressure (volume) or temperature, the corresponding energy had two parts: the first was just internal, often overwhelmingly linked to change of binding, and was reversible, but the second, transferred as heat, was in fair part external (environmental) and as it turns out irreversible. Note that the total energy, as well as the total mass, is fixed in these changes. In any such change from one local set of equilibria to another, then, we found that changing variables—composition, temperature and pressure or exchanging material between compartments within it—involved loss of energy to the environment due to the heat released. It is the maximum loss of heat, which generates maximum entropy increase, which we consider to be the attractor condition of the changes.

We are very concerned with these transfers of energy since they also allow one system to change and to do work on any other (the environment being only one of these), which is the source of evolution. However, that part of the energy involving the change of temperature and pressure (or volume) is in large part concerned with increasing the statistical probability of the system plus the environment, that is ΔS_{total}, and it is obviously impossible to reverse this change effectively. This energy cannot be used to do work. The maximum amount of energy available for useful work of a change, called free energy change (ΔG) was shown to be

$$\Delta G = \Delta H - T\Delta S$$

at constant pressure, where $T\Delta S$ is the energy 'hidden' or lost in changes of T and v, and ΔH is the heat content of enthalpy change, which is equal to

Free energy: available energy for change

the change of internal energy of the system, ΔU, if the volume is also constant. As stated above, all of these thermodynamic quantities are variables of bulk systems and it becomes increasingly difficult to relate them quantitatively to the initial set of fundamental units and variables. We therefore used the sum of thermodynamic systems variables, ΔH and $T\Delta S$, i.e. ΔG, to allow us to analyse the maximum capacity of a system to bring about change. Examination of ΔG made it possible for us to appreciate both chemical changes and physical changes of state, but not their rates, through the examination of two sides of any equilibrium, e.g. the change from $2H_2O$ to $2H_2 + O_2$, or from ice to steam. The analysis of the free energy changes, ΔG, also allowed us to develop phase diagrams, in which composition of phases varies, using ΔG plotted against T, or p or composition. Of course, it is also possible to analyse ΔG in terms of ΔH and $T\Delta S$. It was through such analysis that we learnt how to appreciate the occurrence of particular chemicals in stationary states, i.e. why particular speciation is observed in chemicals as well as in physical states. Many geochemical as well as chemical observations could then be rationalised.*

We could extend this analysis of ΔG to systems in which space was further divided, so that there were static compartments which interacted with one another through the fields they generated or by exchange of energy. As stated, the size of a compartment and the field strength at a point are further variables. Within each compartment local equilibrium was assumed to be established. Sizes of compartments had to be defined, where-upon their shapes followed (Table 16.5). (Note that shape is often used to define a chemical or a physical species.) Thus, the equilibrium thermodynamic treatment gave us an appreciation of many aspects of the formation of our planet, but it left unresolved the way in which the planet has changed with time as opposed to the ability it had for change.

Now, as we have stated, starting from a given state and allowing all possible changes of composition, temperature and pressure (or volume), any system will release energy as far as is possible to the environment until the system and the environment with which it is in contact are in complete balance and stable, when $\Delta G = 0$. The ability to change is then zero and, as stated above, the entropy is at a maximum. The whole system can do no more work. This description of change is applicable to the whole universe as well as to local chemical (and biochemical) activities. Thus, we can draw a diagram (Fig. 16.3) of the entropy increase in the universe relative to the maximum entropy which evolution towards equilibrium would have given during time. At any one moment the difference between the maximum

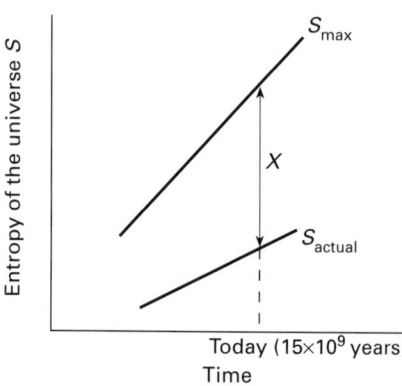

Fig. 16.3 Schematic time dependence of the entropy of the expanding universe. S_{max} corresponds to the equilibrium value (the maximum), and S_{actual} to the value that occurred at a given time. X is the difference today.

*During the course of this discussion we made use of certain plots which are called landscape diagrams. In these plots we connect the observed probability or the predicted probability of observing a given system against a variable such as temperature, pressure or composition. For example, we plotted abundance of elements against their proton composition (atomic number) or free energy ($-\Delta G$) against composition of phases. Later, we will apply the same treatment to changing systems, when the most obvious feature, the probability of observing them, is their survival strength (or survival value).

possible entropy and the actual entropy corresponds to the free energy still available for work which has arisen since large parts of the universe are in compartments not at equilibrium with one another. Now, as we have pointed out, such a universe is always changing since its distribution of energy and material is very inhomogeneous. Hence, today, the free energy difference, see X in Fig. 16.3, lies in part in the differences between the inhomogeneous 'compartments' into which the universe has fallen, e.g. the sun and the Earth.

On a cosmic scale, evolution locally has not followed a path to equilibrium for two reasons. First, it met rapid cooling and local turbulence which prevented even energy and material distribution; thus compartments were formed. Secondly, it met barriers to nuclear synthesis in the giant stars which prevented rapid equilibration of atomic nuclei.* We can see, in essence, how the interactions of the variables mass and charge, within composition, and their distribution in space with time have created today a particular universe that is *far from equilibrium*, full of a variety of compartments—stars, black holes, planets and living organisms—and therefore very difficult to analyse. It is very open to the creation of new variety due to the absence of this equilibrium. In effect, it is this failure to establish equilibrium and maximise entropy which drives local as well as universal continuous change, i.e. evolution, since the separated mass systems have remained in contact through both energy and matter exchange. They struggle towards equilibrium, the ultimate 'attractor' state, to gain entropy overall, but the far from equilibrium conditions gave, and still give, new possibilities for movement in space and in time, and hence all flows (except universal expansion) stem from it. Universal expansion without loss of internal equilibrium is the original flow or momentum defining time (and space). We could look on this as a continuous change of gravitational interaction. This is a subject of considerable debate, particularly since the total amount of matter in the universe is not known. As we shall see below, another such directed flow, of a quite a different nature, is life.

Before proceeding to flow, we wish to stress a mathematical feature which is very important for our appreciation of the predictive strength of any analysis such as that we are making of static systems even. Do not forget, we make a constant effort to be reductive in science, but it may well be that this leads us to approximations or simplifications that cannot be sustained. The important feature to remember is the level of the variables to which the discussion can be safely reduced. For example, two different levels of static systems are: (1) the fundamental variables composition and spatial distribution; (2) the derived statistical thermodynamic variables ΔG, ΔH, p, v, T and ΔS. (All quantities here are time averaged, see Tables 16.2 and 16.3.) Once we describe a *bulk* system by statistical thermodynamics some ability to reduce the variables to the fundamental level is lost. We explain this further in the next section.

States far from equilibrium

---

*Equilibration would have led to the dominance of iron, the most stable atomic nucleus.

16.4   Linear and non-linear systems: the mathematical consequences of description in terms of bulk derived rather than fundamental particle variables

Linear and non-linear relationships

Before we proceed to the analysis of systems undergoing change we need to underline the switch which we made in Chapters 5–7 from the descriptions in Chapters 1–4. In the earliest chapters we used, as fundamental variables, composition (even if this included components rather than atoms and charges) and spatial distribution averaged over time where necessary. This allowed us a description of derived variables such as gravitational and electrostatic potential energies, directed momentum and kinetic energy, temperature and pressure corresponding to random kinetic energy and random momentum. We stressed that the descriptions of all these variables were *linearly* related to atoms, charges or molecular components interacting in pairs and *with averaged* speeds (see Section 2.5.4), but it was noted that this treatment failed to describe the so-called *emergent properties* (properties of wholes, not of parts) of bulk systems, such as phase changes. They were describable only by changes in bulk thermodynamic properties under the umbrella-derived variables ΔH and $T\Delta S$, which themselves could be combined linearly (see Table 16.3). However, it was apparent that the relationship of these derived variables to the fundamental variables associated with composition and spatial distribution was *non-linear*. Thus, there is a difficulty with ultimate reductionism, since we cannot describe such systems deterministically (unit by unit) but only in statistical, probabilistic distribution terms. The treatment of individual systems required additional

Table 16.6 The relationship between the properties of bulk substances and their constitutive units

Units	Property and mathematical analysis
1. Units themselves give	Composition (linear)
2. Units interacting in pairs treated by	Potential energy; Newtonian mechanics and quantum mechanics (linear)
3. Averaged speeds for units related to	Temperature and pressure; kinetic theory of gases (linear)
4. Statistical analysis of single systems, in bulk, related to	Emergent properties; phase changes (order); temperature and pressure (non-linear)
5. Interacting compartments related to	Fields (linear)
6. Polymeric molecules (statistical analysis) giving rise to	Emergent properties (folded state/ random coil changes) (non-linear)
7. Flowing systems of all kinds of units treated by	Fluxes giving organisation (non-linear)

N.B. Points 1–3 are easily related to fundamental variables, but 4–6 are not and are better described by the derived variables of thermodynamics. Point 7 requires irreversible thermodynamics.

variables when there were more systems present which were restricted in their ability to exchange energy and material. Besides the internal variables corresponding to each system, we had then to introduce external variables corresponding to the fields arising from the non-equilibrium conditions. (N.B. There is even a problem of prediction, except in terms of probabilities, at the fundamental level—the uncertainty principle.) All the relationships of units and variables are brought together in Table 16.6.

16.4.1 Variables applicable to single polymeric molecules

Emergent properties of polymers

We also showed that this difficulty of relating the properties of a collection of atoms or molecules to fundamental linear relationships between variables occurs in the analysis of *single polymeric molecules*. Composition is here described by the sequence of units in a single polymer molecule and we might have expected a linear relationship between sequence and the properties of the polymer, now looking at each monomer in the polymer as a separate unit. The expectation fails since the random form of the single polymer molecule has to be described by configurational and motional entropy that is equivalent to the spatial and temperature-dependent entropy of a bulk gas. The polymer molecule may then show *emergent properties* parallel to those of an atomic or molecular ensemble, e.g. a melting point, allosteric changes (compare phase changes), etc. It is essential that this *non-linearity* between sequence and polymer properties is recognised since it affects our attitudes to the treatment of biological polymers such as DNA, RNA and proteins, and then our approach to living organisms, which exist only through the properties of polymers. In this sense *molecular biology* is a misnomer and refers to extracted static structural biological components which have been studied as if they were rigid units (in crystals). A living organism contains a multitude of such systems (polymer molecules) made even more complex by polymer–polymer selective association properties (and also by constant turnover). The degree of complexity of such time-independent polymer systems becomes greater as they are placed into separate compartments, as in cells, when fields are again a variable. Perhaps we should call the real study of life *'systems' biology*. It is through systems that emergent properties such as life can arise. A quite different source of non-linearity arises once time dependence is treated, and it leads to the possibility of evolution (see the following sections).

Before we turn to time-dependent properties we must stress the remarkable achievement of modern analysis. Almost all physical and chemical properties, including chemical speciation and phases, *are now qualitatively explicable*, at least in terms of known units and variables. There is an underlying mystery and a complication in complexity, but reduction has been extremely successful.

16.5   Rate of change

While we can comfortably describe local static features of the universe at any one moment in the above terms of composition (mass, charge) and energy (space), and describe capacity for change in terms of thermodynamic (equilibrium) functions (ΔH, ΔS and ΔG), when we consider the *rates* of change we must obviously include a new variable—*time*—which we measure in units that can ultimately be related to the overall expansion of the universe or to a universal constant such as the velocity of radiation (light) (see Chapter 8).

Now, while the capability to create change is locked in the operational variable ΔG, the rate of change is locked into the rate of transfer of matter or energy which is limited by physical and/or chemical barriers between compartments as well as ΔG differences, plus or minus, and allowed by obtaining an activation energy due either to thermal collisions or to radiant energy transfer, some of which may become stored.

From our point of view in this book we became interested, therefore, in the flows of energy *and* material from one compartment to another, especially within systems on Earth, which are new variables related to timed activities of the compartments. The overall energy flow, for example radiation from sun to Earth, is in one direction, of course, in these systems. The sun's radiation is of high energy quanta which are absorbed by the Earth, held for some time and re-emitted as low energy quanta, while only a small part of the energy is retained temporarily. Approximately every high energy quantum from the sun gives, eventually, 30 low energy quanta emitted by the Earth. Thus, flow in Earth-bound systems is entropy driven in terms of the increase in the number of photons. The process is part of the drive to overcome the entropy deficit of the universe due to non-equilibrium (Fig. 16.3) and it is, therefore, irreversible. (It can be thought of as an effort to equalise temperature everywhere.)

Overall, the surface of Earth came to a temperature *steady state* some 4×10^9 years ago due to this dynamic process of absorption of energy from the sun and radiation of this energy and of energy transfer from the interior of the Earth, but in which other (changing) factors must also be taken into account, such as the luminosity of the sun itself, which has increased progressively, and the composition of the atmosphere, in which the concentration of CO_2 (a greenhouse gas) has decreased. Frequently, we ignore such effects and assume that the Earth is in a fixed stationary state. However, there is always a second amount of energy which is being retained for some time and this is the major source of the capability of systems on Earth to change and drive change. The changes are overwhelmingly on Earth's surface. The energy retained by this surface is not converted just into random kinetic energy shown by its temperature (see Fig. 16.4), but is held in extremely important activities such as the creation of unstable atomic associations in cyclic chemical potential energy traps, for example

$$\text{light (radiation from the sun)} + CO_2 + 2H_2O \rightarrow CH_4 + 2O_2$$

which is followed, after an interval, by

$$CH_4 + 2O_2 \rightarrow CO_2 + 2H_2O + \text{heat (radiation from the Earth)}$$

The best cyclic traps of this kind are organic molecules because of their kinetic stability in stationary states. [These traps are of atoms or charges

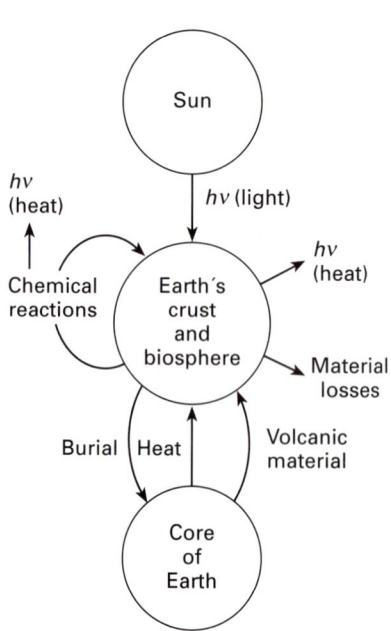

Flow of energy and material

Fig. 16.4 The energy and material transfers to and from the Earth's compartments and life.

moved locally in space to form thermodynamically unstable but kinetically persistent (molecular) structures with shapes, giving rise to new chemical components.] Given the Earth/sun relationship and the Earth's atmosphere, the steady-state production of organic chemicals was certain to come about, sooner or later. Remember, however, that all such components are unstable and temporary so that they revert cyclically to the initial, more stable state. As stated, the whole process is entropy driven. Now, there is good reason why such a driven process could evolve new chemical features over long periods of time.

As well as generating chemically energised systems, the energy of the sun causes steady-state, *directed kinetic energy*, i.e. mass flow. This is seen in the circulation of water and air all around the Earth. It was the combination of energisation of chemicals, not just organic, and of bulk transfer that allowed new features, specifically life, to appear in flow systems, using vesicles as compartments which became cells. By random experimentation the time of survival of specific flow patterns would increase so that they would dominate kinetically less stable fluctuations. The flow patterns would then evolve. (Note that material and energy flow into and out of cells is *twice irreversible* in the sense that entropy is always increasing due to heat losses to the environment and that it proceeds at a given rate).

What we see is that the newly introduced variable *time*, here due to the local lack of equilibrium between the sun and the Earth, introduces the variable *flow* (or, more precisely, *flux*) of radiation, which becomes more complicated as this flow drives three others: (1) flow of energy into and out of energised traps (molecular components and structures such as membranes); (2) the *local flow* of atoms in and out of component chemicals; and (3) flow of units in *bulk space*. The overall process is, therefore

Materials of Earth
↓
Light (radiation from the sun) → Chemicals in flow → Heat (radiation
↓ from the Earth)
Materials on earth

The creation and evolution of *life* is but one example of the outcomes of this scheme;* in effect, as we have shown, it is one *emergent property* from a particular kind of interacting set of flowing 'units', but some external conditions must be fulfilled (see Table 16.7) and some form of *organisation* of the flows is obviously required.

We need to know how these flows became organised in particular patterns, i.e. how they became restricted, and to what extent the derived and operational thermodynamic variables we have uncovered above underlie flow within living systems. Asking the question differently: are flow patterns describable by the same variables, ΔH, ΔS and ΔG, as was true for the equilibria of changes of physical and chemical states in static systems?

Steady states and developmental states

---

*Consider the circulation of water through clouds, rivers and the seas.

Table 16.7 Probable requirements for the beginning of life and its evolution

Beginning of Life
1. A planet with a temperature between 0 and 100°C.
2. Available water on the planet.
3. A planet of the correct size to keep an atmosphere of CO_2, N_2 and H_2O, but not of H_2 and He.
4. A sufficiency of available elements to provide functional survival—perhaps 15–20 in total.
5. A source of radiation powerful enough to generate chemical energy and physical flow (compare with the sun).
6. An ability to make fatty molecules and polymers.

Evolution of Life
1. A planet which could hold a temperature within the range 0–100 °C for 5×10^9 years.
2. A size of planet that would keep an atmosphere for 5×10^9 years.
3. A solid–liquid surface, compartmentalised, to provide local environments and diversity of conditions.
4. A changing atmosphere which the activity of life could utilise advantageously.

Obviously the answer is *no* since those variables are equilibrium thermo-dynamic functions of state and, as such, depend only on the initial and final states, not on the path followed. On the other hand, living organisms are not equilibrium systems, nor simple energised stationary systems. They are not even steady-state systems, where we make the total energy and the material contents entering and leaving the system equal and do not allow it to develop. On the contrary, they are *far from equilibrium, open developmental systems which evolve irreversibly along series of paths*.

We shall now summarise the properties of living organisms in a little more detail so as to appreciate their fundamental features and the difficulties that have arisen in this book with the attempted reduction of a description of them to simple relationships between a given set of units and their variables.

| Developmental systems |

16.6   Living systems: patterns of activity and pathways

As we have seen in Chapter 10 (and see Section 8.10) the apparent violation of the second law of thermodynamics by the increasing organisation of living organisms was explained when it was realised that they are far from equilibrium open systems in dynamic developmental states maintained by continuous fluxes of matter and energy, and that the overall system plus environmental entropy does, in fact, increase continuously as required. The entropy change is due to heat going to the environment. What we have observed is that such flow systems can generate patterns of activity.

This is not fundamentally different from what we see to some degree in inanimate systems. It is not usual to look on these systems as developmental, but many of them are obviously in steady states. For example, the radiation absorbed from the sun can lift water against gravity, which later descends again and the energy released is dissipated as heat. During this

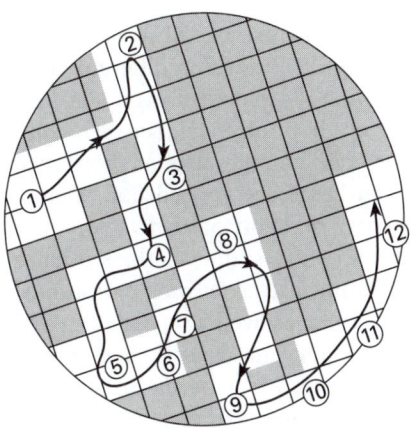

Fig. 16.5 A pathway, 1–12, is defined by barriers to bulk or local flow along its sides (gravitational or electrostatic) and the flow along the path is reduced (controlled) in particular directions by smaller barriers, crossed, for example, with the aid of catalysts or pumps along the path. A living cell has bulk and local controlled barriers to flow.

period the water forms particular patterns of only some permanence and without instructions, in clouds and rivers. Global winds are similar patterns of air. Of course, these *physical patterns* (Fig. 16.5) are subject to fluctuations and external perturbations, and can develop or collapse to complete disorder, but they keep reforming. There is, then, a certain instability in the steady state as there is in any energised and ordered static stationary state, but now we must also take into account the instability associated with the incoming material going to the outgoing material. These instabilities in time could, in principle, take on any pathway, i.e. any sequence of bulk space changes with time, but only a specific set of changes gives the observed patterns at a given time. There is, therefore, a new improbability associated with time, that is with the rate and direction of flow, as well as with spatial order internal to a flowing liquid on a pathway. The implication is that spatial flow pathways of bulk materials arise through restrictions on random momentum in both bulk space, (energy) and time.

When we looked at biological systems we found that there were not just bulk flow patterns, as above, but patterns of chemical local bond *reaction*

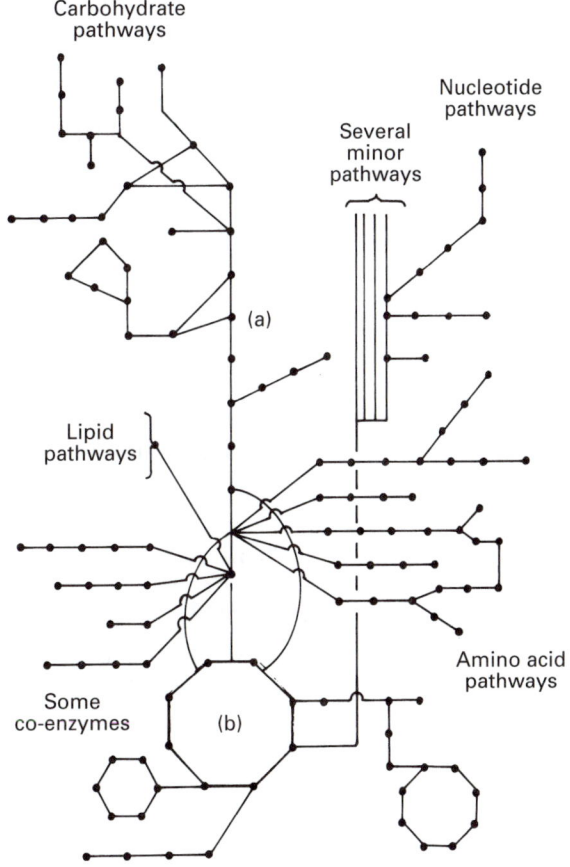

Fig. 16.6 A deliberate attempt to make a cellular metabolic reaction sequence chart look like a computer circuit. Enzymatic steps are dots: (a) is the glycolysis pathway; and (b) the citric acid cycle. (After Kauffman, see refs. 122 and 149).

pathways (Fig. 16.6). They are a very strong feature of living organisms and represent additional, new, statistically unlikely events different from ordering of atoms, i.e. life is an improbable *timed sequence* of events in *local bond as well as in bulk space*. The pathways are both physical, between compartments, and chemical, along particular bond changes, e.g. as those seen in glycolysis (see Fig. 10.17). (Organic syntheses are carried out by man are not very dissimilar.) The fact that such reaction pathways are very reproducible implies that there must be some controls, now of the sequence of chemical as well as physical events, so that *instructions* for maintained continuity giving ordering in space and time as well as providing the required energies for the changes, were necessary. That is, biological organisation, like any other kind of long term organisation, needs instructions (for example, from coded information, see later) to avoid collapse and turbulence, and these instructions must generate restricted bulk and local fields for the passage of chemicals. Such physical–chemical organisation is a constraint much as is order in static systems, and we need to differentiate ordered flow from turbulence just as we differentiated order (ΔH) from disorder ($T\Delta S$). (Of course, as stated, in the discussion of flow, the thermodynamic functions are insufficient since they are pathway independent.) Furthermore, the instructions in biological systems must also provide for *growth*, *development* and *reproduction*, and in these aspects they differ from any other natural process. In the first instance it is easier to consider their instantaneous steady state rather than development of any kind. Thus we need to describe how pathways of all kinds can arise.

Generally, bulk flow patterns in inanimate systems derive from the containment of movement in space due to barriers. To give examples, water on Earth circulates physically through sea → clouds → rain → rivers → sea finding easy *gravitational routes*, while organic chemicals in man's hands go through preferred routes of synthesis and degradation by finding the lowest energy of chemical activation of reaction steps. The latter processes use local electrostatic field gradients where flow is directed by catalysts imposed by the human agent. It so happened that biological systems developed both physical containment possibilities, e.g. membranes, as well as chemical catalysed pathways, e.g. the citric acid cycle. The constraints in cells, therefore, must be exceedingly complicated (Table 16.8) and they are

Table 16.8 Instructions giving patterns

Flowing system	Pattern	Instruction (constraint)
Atmospheric gases	Clouds	Thermal gradient
Water on Earth	Rivers	Gravity gradient
Chemicals in life	Organisms	Electric and chemical potential gradients; catalysts
		Energy input (gradient)
		All represented by DNA*

* Included here are catalysts and feedback connections.

self-generated! In Chapters 10–15, when describing the chemical and physical flow of *components* such as water and 'organic' compounds in organisms, we showed that this was due to the necessary presence of selected polymeric molecules which give rise to locally shaped fields, protein and enzyme surfaces, with active catalytic groups, especially metal elements, which alone can manipulate large energy barriers acting as catalysts as well as physical diffusion controls. While we may never be sure how the pathways leading to the major chemical components of life appeared, it is certain that they owed their origin to the creation of lipids and polymers and the incorporation of available inorganic metal as well as non-metal elements (see Section 16.7). It is useful to remind the reader of the constraints on flow introduced by these chemicals and catalytic elements. While in Chapter 8 we concentrated on the kinetics of individual steps, note that in Chapter 10 and here it is combinations of steps in sequence that are of the essence of the system.

16.7 Selection of materials in living systems

All organisms require a basic set of elements, some 20 or so. Before they could be accumulated and organised there had to be certain special types of molecules forming a compartment. We describe these chemical requirements in this section.

| The formation of vesicles: surface fields |

16.7.1 Creation of lipids and polymers

The synthesis of lipids was essential for life to begin since it is these molecules, insoluble in water, which can form membranes and so create local space in *compartments* with their own properties. They form barriers to bulk diffusion. Given the reducing nature of the atmosphere when Earth formed we presume that oily fat molecules (lipids) were created in large numbers due to reactions of the kind

$$2CO_2 + 2H_2S \xrightarrow{\ h\nu\ } 2HCHO + S + SO_2$$

Polymerisation of formaldehyde then gave sugars and, on reduction, linear fatty molecules and then vesicles.

The formation of polymers, proteins, RNA, DNA and polysaccharides is obviously more difficult since they are required to arise with selected sequences of functional value within the vesicles. Even the initial formation in quantity of random polymers from their monomers is not easily achieved. However improbable the events were, it was the generation of the selected polymer sequences trapped in vesicles which gave rise to the *surface fields* of the polymers. It was these selected surface fields, with incorporation of suitable inorganic elements which increased fields locally, that generated the physical and chemically catalytic pathways for the small molecules and energy to synthesise more fats and polymers. We can see this pathway creation as parallel to the creation of directed water flow by gravitational fields along valleys between hills. Note how such pathways evolved from the formation of molecules to that of compartments, and then with catalysts in compartments. The events are extremely improbable both in the final creation and in the timing of the flows, especially since they were self-produced. The need to keep the system informed for it to proceed to duplication clearly requires, additionally, a set of instructions and reading machinery, further improbable events. Before turning to them we asked about the acquisition of the necessary inorganic elements to give the protein-catalysed pathways extra field strength in the control of pathways.

16.7.2 Selection of elements

Essential elements

			H			
		B	C	N	O	
Na	Mg	•	Si	P	S	Cl
K	Ca	•	•	V	•	→
→	Mn	Fe	Co	Ni	Cu	Zn
		Se	Mo	I		

The survival strength of cells of all kinds in a compartment arises in part from control of osmotic and electrical pressures, and requires specific uses of energy in the transport of certain major elements, Na, K, Mg, Ca, Cl as well as C, H, N, O, S and P. A further group of some 10 minor elements, Mn, Fe, Co, Ni, Ca, Zn, Se, Mo (W), I, is required to bring about appropriate catalysed syntheses (see aside). We showed in Chapters 10–15 that, again and again, the elements were placed in sequences in particles, e.g. in nitrogen-fixation proteins, and/or in membranes (energy capture). Thus, survival requires first selection of elements by *functional value with energetic efficiency*. (Table 10.1 shows the basic minimum primitive selection, and that the elements chosen were those which were available.) Secondly, the elements had to be positioned so as to be functionally useful to a sustained particular flow (Fig. 16.7). The very fact that life was required to handle and organise the uptake, transport and utilisation of some 15–20 elements, at particular concentration levels (and some or all of these levels vary in a timed sequence) and process them in pathways, introduces the extreme improbability of the beginning of life, ever occurring (see Table 16.7 and Fig. 16.7). That it did occur might have been an accident, the probability of which is very difficult to calculate, though many authors,

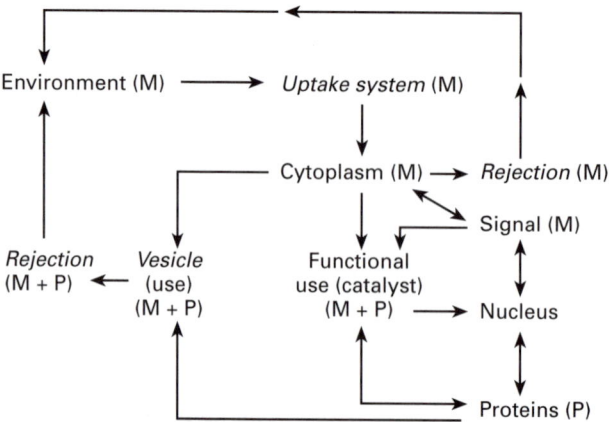

Fig. 16.7 A very simplified scheme for the passage of elements through cycles of incorporation in the environment and in life. Connections (in italics) include membranes.

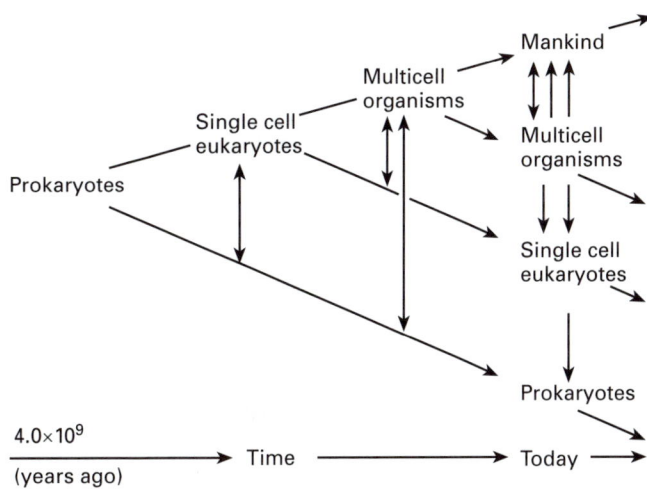

Fig. 16.8 The evolution of the essential ecosystems. A schematic parallel between static and flowing systems indicating emergent properties.

and not necessarily creationists, think that it was a cosmic imperative at some moment in time, see for example ref. 211 in Further reading.

With time, *survival* was ensured in more sophisticated ways by developing tighter and tighter organisation of the polymers and elements, i.e. more and more constrained pathways (see Chapter 12). The functional value of the basic set of 15–20 elements remained much as at the beginning within the cytoplasm of cells, but extended to new roles as space and timed use of space was split into more and more compartments,for example the internal vesicles or bodies of many cells in organs (see Chapters 11–15). The elements used were always restricted to those that could develop a function in a selected region of space *without excessive use of energy*. Once again the activity became increasingly improbable. At this time also a total ecological activity developed which had a loose organisation, where species of longer survival were helped by more primitive species (Fig. 16.8). (Much crossing over of genes and symbiosis occurred.) The environment itself was affected by these chemical changes in organisms. Biological organisation is not just adaptive in physical space over time; it adapts also to such chemical changes of the environment (Table 16.9).

We noted that quite generally a substantial use of elements at any one time can generate waste, which affects the availability of other elements. The best known case is the release of dioxygen following the photochemical oxidation of water as a source of hydrogen. It was the change in *availability of certain elements* following the release of oxygen as waste that helped to drive the evolution of organisation further and further towards increased complexity, to secure survival capability of the complete set of energised chemicals in the following ecosystem. We may look upon this as a change in the possible compositional variables of energised systems on Earth. However, at all stages of evolution, the cell, using containment in compartments and controlled transport and catalysis, works as a unity, meaning that all activities of the elements in the cell act co-operatively. The

Table 16.9 Changes in environment and life

Environment changes	Changes in life
Initial conditions (see Chapter 11)	Origin of prokaryotes; Anaerobic life (photosynthetic?)
Rise of pH, fall of free calcium and CO_2 (Chapter 12)	Unicellular anaerobic eukaryotes
Initial rise of O_2; changing availability of Fe , Cu , Zn etc. (Chapter 13)	Unicellular aerobic eukaryotes
'Final' rise of O_2; further changes in availability (Chapter 14)	Multicellular aerobic eukaryotes
Control of internal environment (Chapter 15)	Specialised organs (e.g. brains) in multicellular aerobic eukaryotes
Control of external environment (Chapter 15)	Self-conscious animals (man) and genetic procedures

Environmental changes

probability of finding new pathways and marrying them with older ways is again extremely improbable. The fact that it occurred requires further analysis of the variables and restrictions on them. Thus, not only do we need to appreciate the value of elements in single pathways, but we need to see that the pathways had to be linked. Furthermore, we had to show how cells become linked in tight organisation in multi-cellular systems and, more loosely, in ecosystems.

16.8 Feedback control

Here we struck a further interesting problem in the analysis of living organisms: if all the chemicals in a system are on a pathway they cannot be independent and if all the pathways are co-operatively interlocked they are not independent either. This includes the use of energy as well as material. Variance is then extremely diminished in that the description of components is reduced to that of the elements of composition, which is similar to equilibrium restrictions of static chemical and physical balance (see the phase rule), and in that the organisation is persistent and reproducible provided the external conditions do not vary. Under these conditions the number of independent chemical variables is small. Now, the interlocking of pathways and the control of catalysts (a space–time-dependent or sequential process), as we described in Section 8.6.4 and many times in Chapter 10, requires feedback control, i.e. appropriate messages. [Messengers (from instruction centres) are at the heart of the control of biological systems and like catalysts act to restrict pathways (see Figs 16.6 and 16.9) since, in fact, the messengers control the catalysts (proteins + elements) and their amounts.]

Several questions then arise.

(1) Feedback-controlled systems are, by their very nature, non-linear complex organisations and their analysis requires systems dynamics concepts and

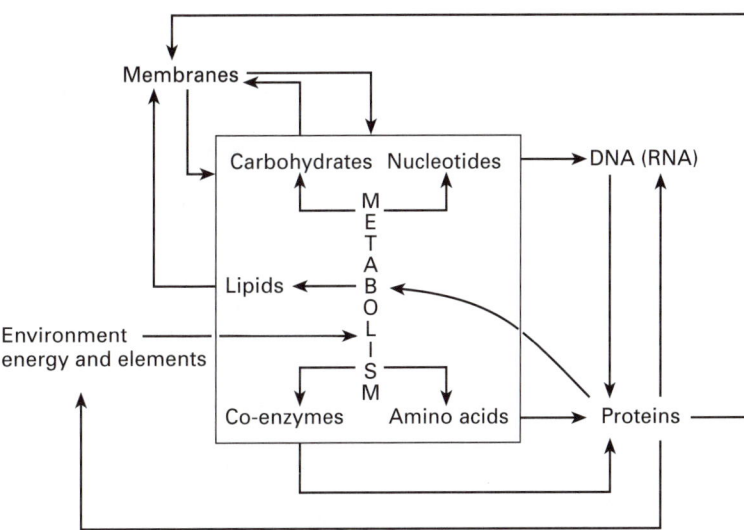

Fig. 16.9 An indication of the interlocking feedback connection of the metabolism, shown in Fig. 16.6, to the major cellular activities at the level of polymers of several kinds.

mathematical tools. As stated above, we need operational derived variables equivalent to ΔG, ΔH and $T\Delta S$ to describe flow in fixed pathways.

(2) These feedback systems are inherently unstable (remember that they are iterative and any small deviation is self-amplifying) and live 'on the edge of chaos', that is, they can undergo entropic collapse.

Pathway interdepencies

(3) Any living system is characterisable by its pathways, but these pathways are based on constructions which represent a restriction on variability (see Table 16.8),but see the brain.

(4) Since a living system is self-assembling and reproducible the constructions must be imposed upon it by instructions. We found that these instructions are in a controlling coded molecule, DNA, which does not participate in the flows but influences them.

Finally, the ecosystem as a whole is co-operative and has similar feedback features, so it too has the same features as those of single cells and more complicated organisms. Extending the argument to the involvement of the

Table 16.10 Development of environmental interaction

Prokaryote	Little sensitivity to environment
Unicellular eukaryote	Increased sensitivity to environment. Signalling due to calcium controlled motion
Multicellular eukaryote	Increased sensitivity to environment; electrolytic, chemical signals; controlled movement; *internal environment* controlled (Gaia possible)
Mankind	As for multicellular organisms but increasing control over *external* environment; man-made signalling and constructions (science)

environment (Table 16.10) had led to the discussion of Gaia (see Section 9.14). We turn first to the nature of DNA.

16.9   Self-organisation and DNA

From the above we appear to hit a dichotomy. We have stated that an organism can only be described by its *complex* activity using the language of *systems*, based on instructions from DNA, yet, it is often said that DNA, *a single molecule of unique linear sequence*, represents an organism—hence the reductive term 'molecular biology'. Apparently, the two descriptions are not compatible. To remove the problem we shall revise the description of DNA, no longer seeing it in a living organism as just a molecular spatial sequence but also as a timing device which visits a continuous series of states (see Fig. 16.5 and Sections 16.4.1 and 16.10). This set of states refers both to the spatial distribution of the continuously varying, controlled, time-dependent conformations of the DNA, and to the time dependence of the material and energy flow to and from it. We must look more closely at DNA to show that it is not linearly related to an organism's features but it is a non-linear functional unit of organisation in a biological system due to its information content. This is a possible emergent property of a polymer, by which description we imply that it is not reducible in a simple linear way to its sequence.

16.10   Information in living systems: DNA

The code: DNA dichotomy

To survive for long and certainly to reproduce any self-organised organisation, biological or not, needs an active *master plan*, which relates to *a set of instructions*, and this in turn derives from *information* contained in or put into the system concerned.* The more internally disordered in space–time a system can be, the more instructions are required to organise it and to make it operate. Hence, organisation needs not only a master plan (information), but also a set of outgoing and incoming messengers so that, while the action has some flexibility it constantly constrains and is itself constrained by the system. It also requires energy. How can such a system be described by variables which must be related to tuned application of instructions?

The obvious feature of DNA is that it is a coded molecule which can be represented by a sequence of symbols corresponding to the four organic

---

Information, which is stored, is by definition no more than knowledge. By itself it is passive. Specific action requires compulsory instruction based on information, that is commands. While DNA can be said to carry information, in itself it has no power to instruct or to command, which requires in addition reading machinery, message carriers, active agents (enzymes) and energy, for activity. DNA is not simply related to RNA and proteins. There is a complicated dependent relationship to DNA hidden in the words 'transcription' (DNA → RNA) and 'translation' (RNA → proteins).The words suggest linear relationships, but the chemical transformations which arise are much more complex, as shown in this section.

bases. Likewise, a chemical is represented by a set of symbols chosen amongst the naturally occurring chemical elements. The fact that the bases appear in a non-repeating sequence while the chemical has repeating units and an ordered structure is not relevant in the first stage of the description: both are ordered in sequence. However, a crystal array, for example of sodium chloride (NaCl), does not *contain* information in the same sense as does a molecule of DNA since any reading machine examining the structure can take only a single simple message from NaCl, while the DNA message, being a non-repeating long sequence, contains much more information. Compare a single word with a chapter of a book.

Let us examine the nature of DNA, but note first that the chemical information in a sodium chloride crystal, written NaCl, is not affected if we write it as ClNa. The formula is written in an arbitrary way. It is just common salt. Thus, the information here is not dependent on a directional sense of the reading of the symbols. DNA is very different apart from its non-repeating sequence, in that the code (*information*), is sequential, say to be read only from left to right, and is divided in sections. It then becomes important to know where to start reading and where to stop. The code has to contain spatial and time controls which a machine must read and obey, but without the machine, which can read the code in a special way, DNA is useless. The essence of an information sequence is that it has time as well as space co-ordinates; this, we noted above, is also the essence of organisation. *Hence, DNA, plus a reading machine represents organisation.*

To illustrate the reading problem consider the 'structure' of three letters in a triangle. The message

P

T   A

is ambiguous in that it can be read as PAT, TAP or APT, while there are also nonsensical forms (in the English language), TPA and PTA; however, ATP means something to every biochemist. Clearly, a coded message has to have an instruction as to where to start and where to stop and which way to read in time, that is to say, it is organised. The above triangle readings are pieces of structured information which belong to a code with no such symmetry (top, bottom, left and right) and must be given a time axis in each sequence if we are to get the right message. Words are also in (time) sequence. Very few messages read the same backwards as forwards. An example is Napoleon's dictum: line (1) below

ABLE	WAS	I	ERE	I	SAW	ELBA	(1)
Able	was	I	ere	I	saw	Elba.	(2)

But here, if we put in capital letters and a final full stop, the time sequence of reading is clear, as in line (2). To give further examples, each sentence in this book tells us where to start—at a capital letter after a stop—and where to stop—at a stop. In English we know we must read left to right and down the page. The three-base per amino acid DNA code is not symmetric as in (1) but reads as in (2). Reading a coded message then, needs instruction as to the rules of reading so as to get information from *timed* sequential

Reading DNA: (information)

reading. DNA is not, therefore, an independent information molecule for giving instructions. It needs a reading machine, which like all machines needs energy to drive it, and executive units (enzymes) which relay back to it so as to produce correctly from environmental material.

It follows from this lack of time symmetry in the way DNA information is read by a machine that it is separate from the spatial and thermal thermodynamic consideration of 'ORDER' relative to a 'DISORDER'. Time reversal in the reading is not meaningful whereas thermodynamic entropy itself has no time dimension.* DNA represents then a *time sequence*, as well as a *structure sequence* in the conventional sense, and codes for *sequences* in other polymers and their properties (not just static structures) and for the whole timed sequence of activities of cells. The DNA sequence is constructed so as to be transcribed and then translated (by RNA) while, in time, it receives feedback, reading instructions from proteins themselves and other messengers. DNA is, therefore, a somewhat curious creation in a cell having a relationship to it similar to the relationship between the egg and the chicken. Again, DNA is read in a time sequence which does not necessarily follow the space sequence. For example genes X..Y..Z can lie in sequence in a chromosome but may be read Y..Z..X if the protein product of Z alone directs the reading of X. This is not unusual; after all, when reading a book we can be told to stop and turn to a page either further into or back in the book to see a table or figure.

The fact that one synthesis of proteins follows another means that the first prepares space (or opportunity) for the second. There is then a *timed-ordered developmental* activity imposed upon a *spaced-ordered* sequence of genes. The analogy is with a selected reading of the triangle of letters, APT, shown above. The same applies to RNA. In fact, the whole of the cell 'cycle' is linked to time. (Note that this is not a true cycle such as in a dynamic stationary state, but a time sequence, a sequence of events; life does not repeat, it proceeds.)

Proteins in cells are then made in a time sequence, but the *rate* at which any one protein is produced also depends on the state of the cell at any one moment hence proteins are not produced quantitatively in direct relation to the DNA code. Hence, the DNA code senses the flow of energy and material in the cell in a non-linear feedback relationship with the cell contents. An obvious result is the different flows of different protein molecules in succession in the cell 'cycle', where these flows are connected to the timed activity of proteins in the cell at any one moment so that syntheses follow a unique set of patterns in time and space (see Fig. 16.5). The spatial timed sequence in all cells develops towards division, each step preparing for the next. Thus, all the units of a cell move in time along specified space pathways, a typical feature of patterns of organisation. The variables for different types of cell are then very restricted by the DNA information, but the instructions, derived from the restricted reading of the DNA, are linked

> DNA is a single molecule (concentration independent)

> All molecules (substrates and proteins) and ions are concentration dependent

---

*Time is more closely linked to the entropy deficit of the universe and therefore must flow one way, just as entropy is said to increase in the universe. Of course, production of entropy is time dependent.

Configurational entropy of polymers (DNA)

to the whole organisation in the cell. The survival of such a cell is then an emergent property only represented in part by its DNA sequence.

In Chapter 5 we had to describe separately (1) a structured state such as a crystal of sodium chloride at 0K, when we could represent it by a unique ordered condition and formula represented by NaCl, and (2) a gas of any kind which had to be described by a bulk concept, a huge variety of microstates within a macrostate. The danger is that DNA is looked upon as a *singular* structure (compare with NaCl) and not a huge variety of possible structures with many sequences and conformations. However, DNA, through its conformational entropy has some of the characteristics of a gas (see Section 5.8.5). The variety of possible interpretations of a single molecule of DNA is therefore due to the possibility of reading its many conformations in many different ways. Thus it has a space–time-dependent statistical character. (In fact, it is read in a selected way to produce a particular organism.) Of course, since it is an extended polymer, not folded, it also has an energy state dependence of its microstates. The parallel thermodynamic picture for DNA is then a very complicated *system* such as a partially ordered gas. The functional interpretation of its intrinsic variety of microstates is constrained, however, by the ability of machinery to bind to it, to obtain energy and to read out from it only certain more structured molecules (RNA and then proteins) at certain times. These proteins produce small molecules and pump in elements and the *concentrations* of substrates and elements then act so as to restrict or demand use of DNA. In one sense the partial pressure (concentration) of all units in the cell constrains it in a particular organised state. Much as a gas has a condensed liquid or solid state due to ΔH, so this binding to and reading of DNA constrains its readout to give the *sequence* significance opposite the cellular system it helps to produce, *an organisation not an ordered system*. While states of order of chemicals are described as liquid or solid we have no classifying words for states of organisation (see Fig. 16.12).

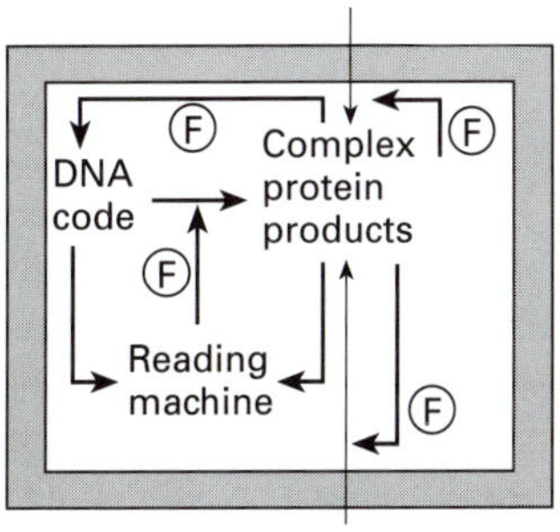

Consider a self-assembled organisation with its machinery controlling the flow of energy and material through a sequence of steps (paths), as in Fig. 16.10, and which is in a steady state. There are two parts to this system one of which can be represented by the thermodynamic characterisation of the cycling state as if it were an averaged stationary state of chemical elements over a cell cycle. Its stability relative to standard states of elements is then given, at constant pressure, by

$$\Delta G = \Delta H - T\Delta S$$

where ΔH is the order energy parameter restricting the disorder of space/energy states, ΔS. We have used ΔG as a measure of the likelihood of observing a stationary system throughout this book. Note that ΔG in a cell, an average value of all its chemicals over a cell cycle, is very highly positive relative to commonly observed states of elements in stable compounds, and many of the chemicals seen here as stationary are constantly formed, degraded and reformed, e.g. DNA, RNA, proteins, fatty substances and sugars. However, a ΔG as low as possible is advantageous to maintain many ordered regions of the cell, as well as the possibility of a reasonable life-time for each chemical, and a high efficiency per cell cycle.

Fig. 16.10 The network which produces the emergent activity, life, here related to the code, DNA, and its connections. (F) is feedback. The relationships between the parts are non-linear, and are not easily related to molecular characteristics. The cell is a very complex system. [In fact, all the polymers of biological systems, including DNA, are to some greater or lesser degree complex, in that they have properties which can only be described statistically—they have configurational entropy (see Section 7.6). This is reflected in the fact that they have melting points. Thus, the reading of DNA or the functioning of proteins is not simply related to temperature or pressure. These statistical properties are independent from concentration terms or time dependencies.]

A different treatment is needed to include, additionally, the flow through of material and energy. We can imagine that there are very many possible ways in which these flows can occur, i.e. many paths, dependent upon the space and energy distribution *of barriers to the overall flow*, even though the composition remains the same when measured in terms of elements. The probability of a *given* cellular path being observed can be related to a second entropy per cell cycle.

$$\Delta S_t \propto \log P_t$$

where P_t is a probability expression similar to that for the thermodynamic entropy, representing the large number of possible flows with time, t, dependencies and will have a similar temperature dependence for its energy equivalence since barriers restrict paths. Now the restriction to a particular set of pathways is imposed by instructions, activated information (I). I is an organising parameter, not an order parameter, and can be treated as opposing the ΔS_t, or pathway randomness, of the system. It forces the production of only certain element combinations and energy flow in sequences. Very roughly we can write the term ($\Delta S_t - I$) for the combined probability of the flow. The overall ("energetic") probability of observing the steady state is then related to ΔL_{cell}, where ΔL_{cell} is an energy representing survival fitness of an open flow system, including composition and energy throughput per cell cycle

$$\Delta L_{cell} = \Delta H - T\Delta S - T(\Delta S_t - I)$$

It must be remembered that change, ΔG ($\Delta H - T\Delta S$), here is not the free energy of a conventional static condition but an averaged value for a cell cycle, that is, of a system in which all chemicals are continuously changing—*that is* they are in a flow. Now, if the flow stopped the whole would rapidly degrade, giving out heat energy and increasing the entropy of the environment. To maintain this in a far from equilibrium steady state, energy must be put into the system and degraded continuously. This energy is included in ΔH. We can draw a diagram to illustrate this (Fig. 16.11) where we note that the driving force of the system is the entropy gain of the environment. The above equation for ΔL_{cell} represents the average for a cell cycle energy. Now we expect (ΔS_t-I) to be close to zero so that there are few wasted pathways of chemicals. The optimal conditions for survival then are that ΔL_{cell} should be at as *negative* a value as possible.* There is then an efficiency demand implying that energy throughput should be as small as possible. Thus, we can compare ΔL to ΔG of *equilibrium* thermodynamics, but now for an out-of-equilibrium open flow system—a living cell.

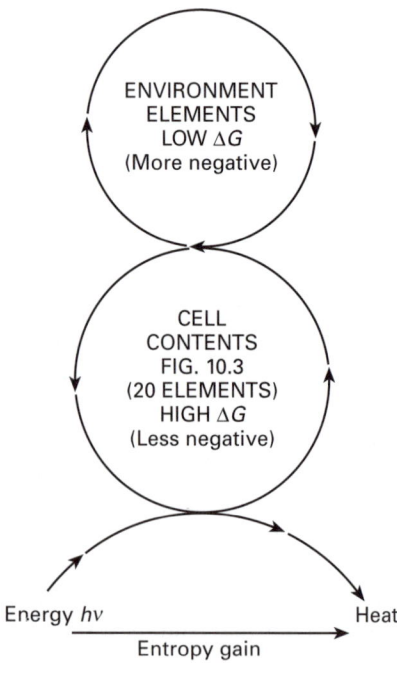

Fig. 16.11 The entropic drive which underlies the cycle of chemicals in life, from and to the environment. The elevated steady state ΔG of life is an averaged value over a cell cycle.

ENVIRONMENT
ELEMENTS
LOW ΔG
(More negative)

CELL
CONTENTS
FIG. 10.3
(20 ELEMENTS)
HIGH ΔG
(Less negative)

Energy $h\nu$ Heat

Entropy gain

---

*Note that a more *negative* ΔG means a higher stability of a static system. Here the more negative ΔL the higher the survival strength of a flow system, a cell. While ΔG is truly quantitative, ΔL is used to give a qualitative impression.

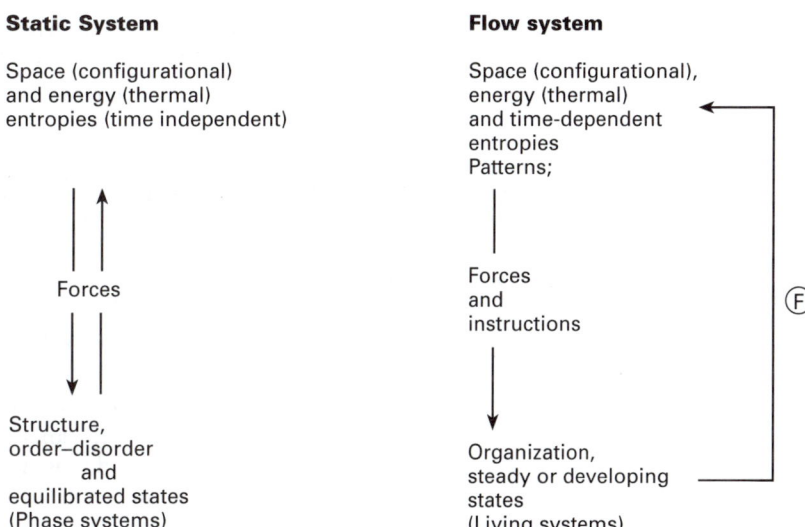

Fig. 16.12 An attempted parallel between the variables of static and flow systems and their emergent properties.

16.11   Landscape diagrams of survival strength

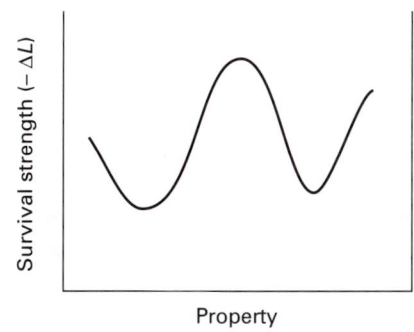

The analysis in the previous section leads us to plot, in Chapters 10–15, survival strength, now labelled ΔL_{cell}, against appropriate variables, see aside much as we plotted ΔG against variables for static systems (see Figs. 5.19–5.22). The variables must again be composition of the environment (based on the available elements, which changed with time to a large extent), temperature and pressure, again of the environment (which also changed to some degree with time), divisions into compartments (also developing in time) and energy and material intake rates. Thus, we can obtain a qualitative picture of opportunistic evolution against environmental and random development of element content. The fact that the environmental changes are often due to biological evolution means that overall development of Earth, until man arrived, was a combination of physical ordering and disordering and biological organisation. The plots of survival strength changed in character with time over billions of years. Gaia is one possible result, but we must look at man's activities to see how it can fail.

However, although the concept of survival strength is simple in theory, in practical terms it is difficult to represent this survival value quantitatively on the basis of statistical probability, though it is obviously of a very low numerical value. In effect, a survival landscape diagram has extra variables besides those included in ΔS_t and extra constraints besides ΔH, which are the coded instructions, included in I. Clearly, in the future, we need to appreciate in greater depth the information and the probability distribution of flows, but at present the exercise is exceedingly difficult since we must match the thermodynamic description of static systems with the non-linear flux descriptions of cells, (see Fig. 16.12). For this reason it is more common to represent survival of populations against a combined set of environmental and internal characteristics (see Fig. 16.10 and see Fig. 14.17). At this stage we abandon our analytical description of the nature of cellular life in terms of variables, and move to its evolution.

16.12   Survival strategies and evolution of complexity

Evolutionary developments could hardly have been haphazard, much though organisation may have begun by chance. In this book we have related changes in evolution to two specific systematic developments—the increase in the number of compartments and the change in the availability of chemical elements. However, the persistence of all modes of life depended and depends, in fact, on instructions relayed from coded information. As stated above, this information today is in the DNA molecule, which represents all known survival systems in coded form (but see our considerations above). This means that DNA also had to change *following*, or coincidental with, the change of the environment in order that evolution, as we see it, could occur, even while much cytoplasmic chemistry is fixed throughout evolution. How was this possible? How can the environment drive evolution? Let us look briefly at survival strategies of organisms and then turn to the DNA itself.

From what we have said, and experience confirms this, organisms can adopt different strategies for survival of their own kind. The simplest systems use the fastest rate of reproduction and have a short lifetime since they lack the skill of searching the environment for food and are poorly protected. Increase in these two abilities, to search for food and protection, increases lifetime but reduces reproductive rate since the organisms are larger, more complex (they have more compartments) and their life-cycle is longer. We can therefore write, schematically,

$$\text{Survival strength} = \text{Lifetime} \times \text{Reproduction rate}$$

Clearly, and additionally as shown in Chapters 14 and 15, the more complex organisms gain by allowing the simpler organisms to produce their required common materials and then consuming them; they become predators and initiate what we call a food-web. Equally, the simpler systems gain if the complex ones search and find food for them and ensure better shelter. *Symbiosis* is the result—it has high survival value and is a major reason for the evolution of living species in ecosystems (see Fig. 16.8).

All of this, however, does not answer the fundamental question: how have biological systems adapted so as to follow and to gain seeming advantage from the changes of the environment listed in Table 16.9? In large part this is a matter of understanding the relationship of DNA, the instruction code, to the machinery of the cell. It is this under-standing which is required as we progress in the analysis of the physical–chemical evolution of natural systems as it is applicable in biological organisation.

16.12.1 Environmental changes and DNA responses

Clearly, from our inspection of cells from Chapters 10–15 it is also information about the environment which must be sensed by the DNA. Now, in

Table 16.11 Evolution of chemistry of hydrogen peroxide employing peroxidases and catalases

Organism	Function
'Anaerobic' sulphate bacteria	Detoxify O_2 using NADH/flavins to produce H_2O_2 and O_2. Removal of these by catalase and superoxide dismutase (Fe and Mn)
Early plants, animals and fungi	Use of H_2O_2 to protect cell by cross-linking cell walls. Peroxidases (Fe) help to make stable multicellular organisms. Protection against other organisms by generating H_2O_2 from O_2 by flavoproteins. Use O_2 in long electron transfer chain to NADH to give energy
Later plants and animals	Use of H_2O_2 to generate hormonal message systems, e.g. thyroxine produced in animals, and general control of plant hormones by peroxidases (Fe)

N.B. Protection against excess H_2O_2 is managed by catalases and selenium peroxidases. The reactions are all linked to the handling of a set of oxygen- and hydrogen peroxide-sensitive co-factors.

Effects of Stress on DNA

1. Mutation of genes
2. Gene duplication
3. Gene transposition
4. Gene incorporation
5. Recombination

See Molecular Strategies in Evolution. ed. L. H. Caporale Annals New York Acad. Sci. 870 (1999).

Poison
↓
Messenger
↓
Purposeful use

principle, environmental change is harmful to a cell. The cell responds, due to feedback, by attempting to repair those machine parts, proteins, that are damaged. Thus damage causes increased reading (transcription) of local regions of DNA, and as regions of DNA pass through single-stranded states they become open to mutation and other changes, see aside. Damage to a cell thus causes damage locally to DNA. There will then arise, by local chance mutations, proteins modified to handle the damaging agent related to those proteins damaged. The damaging agent becomes a messenger for transforming DNA in subsequently surviving generations. These new cells will have the ability to sense the poison and counteract it, which is equivalent to treating it as a messenger. It is but a small step for the poison to become incorporated and made functionally useful. The developed organism is better adapted to the environment and it can also then evolve new features. The roles of Ca^{2+} and O_2 in the environment illustrate this progression, from Chapters 12 to 15. If this is correct, evolution is driven within DNA by environmental change in a localised Darwinian manner. A possible example is given in Table 16.11, where we note that living organisations are following chemical environmental change—increase in atmospheric dioxygen.

Returning to our equation for the survival value of a cell system we see that in a curious way DNA is only a part of the problem, in that it provides information, I, to reduce the ΔS_t, but while alone it can represent species, it cannot explain why they exist. It is the *interaction* of the information-containing molecule with the chemical machinery of the cell that generates the survival value and the interaction with the environment that helps DNA to evolve. Put in other words, life is an emergent property of information-carrying molecules together with reaction flows of chemicals and energy in a system not describable by a property of a single molecule.

16.13   Further evolution of complexity

As we have stressed in this book, evolution of living organisms is characterised by two major features acting holistically: (1) increase in the number of compartments; and (2) increase or change in the availability of chemical elements and energy and their use. Both are *represented* by increase in the complexity, even the length (but not linearly) of DNA. Complexity increased again later with the introduction of a brain which could collect information independently from DNA. Man has increased this complexity by further increasing the association of compartments (now external to himself) and the use of additional chemical elements and sources of energy (see Chapter 15). Before describing how this last organisational activity is coded we turn to the peculiarity of man—his brain—since here life has evolved a new approach to survival strength.

16.14   The brain: a second code

The evolution of the nervous system and the brain has changed the nature of information held in living systems. Until this development all information was stored in DNA; once it was possible to retain knowledge of the *outside world* in a separate organ, a nerve centre or brain, then the two forms of information could come into co-operative (or conflicting) use. The DNA has, as we have described, information in a compulsory programme; the brain is in large part a plastic, coded memory which remains only partially DNA dependent, in that DNA contains the gene combinations for making the various kinds of differentiated neurons and generates all neuronal wiring, i.e. all the possibilities for neuron connections. At first, in primitive animals, neurons only linked muscle cells to muscle cells and allowed *automatic responses* based on electrical as much as chemical signals which were faster than a response through DNA directly or through chemical messages alone, commanding, for example, general contraction (see Chapters 14 and 15). Then, connections of neurons to other neurons allowed crossed information to circulate—not only from senses to neuron but from neuron to neuron and neuron to muscle—eliciting co-ordinated motor responses to the environment. As multiplication and diversification proceeded, neurons became a kind of receiver/transmitter device, using *dendrites* to obtain one-way information from the outside world, still *through the senses*, and using the axons and junction *synapses* to respond chemically to the stimuli received. The response could be contraction, secretion, stimulation or inhibition of other neurons. Diversification was achieved by the use of different chemical messengers or distinct synapses. Once more from a simple property, now of single neurons, there arose an emergent property, a coded memory (stored information), see aside and then self-consciousness.

The individual human brain, for example, has more than 10×10^9 neurons interconnected by an average of 10 000 junctions *per* neuron, and

Senses
↓
Brain
(A second code)

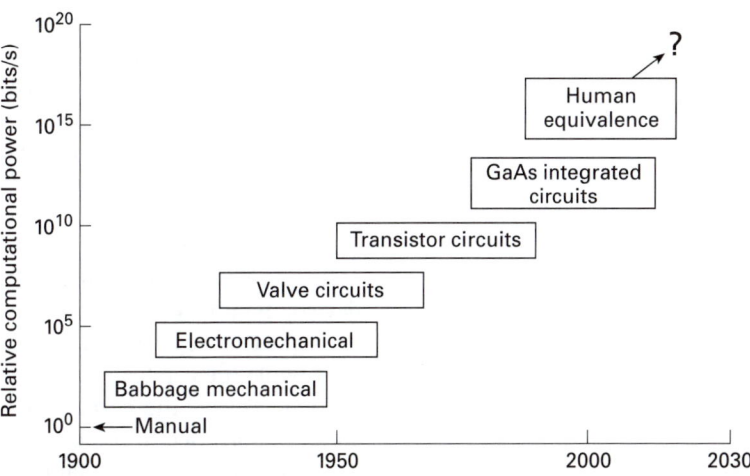

Fig. 16.13 The evolution with time of computer power. Also shown is an estimate of the computer power, at much lower speed, of a human brain.

some 50 kinds of synapses (a network allowing computing power probably superior to all the world's computers put together; see Fig. 16.13). It is, however, important to note that a fair number of the neurons in the human brain are already made before birth, but with few connections (just those necessary to ensure the basic body functions). Multiplication of neurons then proceeds, perhaps up to the age of 3–5 years but more slowly. What increases vastly, particularly in the early childhood years is the neuronal networking (cross-linking), with all the mentioned junctions and types of synapses, as well as the development of the secretions (hormones) from specialised glands. The connections are, at first, based on sense perceptions, not on analysed data.

As we have noted, the sight, hearing, taste and olfactory organs are also located in the head, and the brain is split into different, but co-ordinated, compartments, of which the cortex, with six distinct cell layers surrounding the more ancient brain parts, increased 60-fold from lowest mammals to chimpanzees and a further three-fold to humans. All of this must be looked upon against the background of the general increase in brain size, which tripled in a few million years. It is impossible to escape the conclusion that it was these developments, taken together, that were responsible for mankind's largely increased intellectual capability. The brain became able to collect and respond to coded messages from members of its own species. These codes became spoken words and sign languages, which helped co-ordinated behaviour in societies (see section 16.17), and cultural development, (see Fig. 16.17). As sophistication of co-ordination developed, so did the ability to predetermine advantageous activity. As stated, only the general lines of such development were encoded in the genome; all the rest arose in the course of evolution through constant interaction with the environment using the brain. The evolution of man's society is very largely due to acceptance of ways of living not related to constraints within DNA. Evolution took a twist to favour those that were both the best adapted to the social environment concerned and that could make better use of it thoughtfully.

It is clear that the development of the brain was a major selective advantage; the fact that its responses increasingly became unrelated to genetic control generated man's ability to assess himself and his needs (in a very *subjective* way). The independence of the brain's ability to assess options is today strongly based on a *memory* of past events within a lifetime, that is, it corresponds to information stored, open to retrieval (largely automatic) and coded quite separately from DNA. In the future it will allow self-determined, by man, development of DNA itself by genetic manipulation.

Before proceeding with the analysis of the further consequences of development of the human brain, we can summarise the situation we have arrived at by saying that DNA evolved to manage internal body activities only while the cerebrum evolved to react to and to manage external events. In this respect, the cerebrum contains, for all purposes, a *second code* of equal importance to the genetic code for mankind.* We have to remember too that the endocrine glands play an integrating role to some degree, so that the two codes are interactive. Man's independence from his DNA has increased rapidly in the last 200 years as objectively collected information (and now inherited outside the body in books, tapes, etc) has come to dominate inherited and subjective information. Such is the impact of an understanding of the underlying rules of physical–chemical sciences. However, the older form of inheritance remains, stretching back all the way to the earliest DNA-based organisms.

16.15   Mankind's activities

Development of society: Imagination Language (New codes) Culture (Arts) Technology Organisation

After the evolution of the brain, the changes that came about in man's general phenotype development, which, as far as we know, are absent in other animals, were twofold. First, he began to be aware of himself and to communicate with others using, especially, language, a totally new structured communication system, in effect a new coded message system, and, secondly, he began to construct, through experiment, a 'scientific' knowledge of his surroundings. This knowledge has been used to build up an empirical objective (experiment-based) activity, which now relates not only to the controlled use of the environment but to the management of the behaviour of man himself in a *society*, a community of shared interests. Beginning with agricultural and farming activities, which were the basis of the first communities 10 000 years ago, the great step forward was the evolution of *inorganic chemistry* using fire, which allowed the making of utensils, from simple cooking pots to sophisticated ceramics and then porcelains, pigments and glasses, not to mention building materials (bricks and mortar) and then objects and tools made of metals—gold, copper, bronze, brasses, steel. Early technical achievements also led to the development of painting, music, writing and other art forms, related to ways and means of expressing inner self. Thus, science and art (culture generally),

---

*Memory is stored in developed concentration gradients and cell structures, not in sequences within molecules.

the attributes of mankind, developed slowly together. Simultaneously, a huge industry has appeared for the handling of inorganic materials and creating all manner of constructions and machines to assist mankind's survival and other needs. The details are described in Chapter 15 but the resulting change in 200 years to the surface of the Earth is almost as great as that of the previous four million. As stated man learnt to codify this knowledge in increasingly sophisticated spoken and written language so that experience was inherited (and is inheritable).

It was only 200 years ago that man began to examine his own (*organic*) chemistry. While there has evolved a quite new industry in medicine and agriculture using this knowledge, the most startling result has been understanding (still far from complete) of life itself, both in terms of its organic chemistry, extending to the gene level, and of its integrated inorganic chemistry, as described in this and our previous books. Here, the possibility of guiding the future evolution of life arises since the DNA code of present-day life is now known.

If we examine this evolution carefully, we see that it was again based on knowledge (coded information) and communication, and both increased exponentially over the years in quantity and complexity. There are also feedback loops here, which control the entire system, keeping it organised but allowing development, as in living systems (see Fig. 16.17), not internally now, but externally and used purposefully. Chemistry, of necessity, underlies all this past and further development.

16.15.1 The computer: a new external code

Computers and coded knowledge

There is one particular form of man's industrial evolution of data handling which deserves some additional comment here, since it has generated a fantastic extension to the brain itself: the development of more and more powerful computers (see Fig. 16.13). Once again it is instructive to see how this innovation has been based on a purposeful selection of elements, the units of materials, and charges. First, computers were constructed of wood and ordinary metals or alloys, as in the Babbage calculating machine, then of semiconductors like germanium, silicon (now and again), gallium arsenide (as in the Cray supercomputers) and on certain other materials, and possibly in the future even on proteins such as bacteriorhodopsin, which takes us back to carbon chemistry.

These are quite novel developments in that they do not make use of a liquid state, or of electrolytic conduction, or chemical signals and messengers, as in the brain. They are based especially on electronic conduction in solid-state materials, and the progress in this field derives from the rate of development of smaller and smaller circuits and faster and faster conduction with minimum production of heat. The physics (of chemicals) here plays a dominant role and so does organisation, both in the construction of the semiconductor units (chips, layers, films, etc.) and in the design of the circuits. We must see in this activity not just a *third* form of coded knowledge which can be used to control activity but, in essence, an eventual possible ability to function independently from man's will and a capacity to reproduce itself. At the moment this is not possible since computers cannot self-assemble, but the possibility is there.

16.16   Survival strength of mankind: species and individuals

In several chapters describing biological organisms we have looked at survival value against variables such as composition, energy and material flow rates, compartments, temperature and pressure. We have shown that the corresponding landscape plots divide effective organisms, which we observe, from less effective ones, which we postulate. In this way we defined species which we could frequently correlate with the shape and behaviour of the organisms (see Table 14.18). However, we considered that each species was represented in the controlled activities of its holistic machinery by the coded instructions in the DNA. Thus the genetic information was also a way to characterise an effective species. (We noted also that it was not possible to go directly from the coded information in DNA to the nature of the organism any more than it was possible to go from the formula H_2O to bulk collective properties such as the melting point of ice.)

Now, when we come to mankind, imposed upon the species differences which can be correlated with DNA, i.e. genetic differences, there are strong intellectual (phenotypic) differences in individuals since the individual brain builds, using a quite new code, a personality influenced by the memory of an individual's history. This ability to use the brain then allows behavioural patterns to distinguish each person, so in effect breaking down the human species into individuals. Hence, survival strength, now with a broader meaning, has a new parameter, rational power, which allows choice. Hence, we plot in Fig. 16.14 a new landscape diagram showing the selective advantage of brain in terms of compositional, or perhaps better the combined informational properties found in individuals, with environment factors and energy constraints. It must be appreciated that this variable is not based on genes alone but on the happen-stance of environmental exposure which can, in fact, be managed. Thus, inheritance is not just genetic but cultural and it is transferable down generations through education and recorded information in various forms. Will this cultural inheritance back-interact with DNA?

Moreover, the juxtaposition of the nervous system and the glands in the head region has a value in that the *sense perceptions* of the external world are linked both to the electrolytic (if you like computer-like) brain recording and response and the organic hormone responses of a more primitive DNA kind. It might be said that the psychological and the physiological aspects of any man or woman are linked. The outcome is not just an individual with objective outlook but one who has subjective (call it cultural and emotional) content. There is then no possibility of an individual having a totally objective response. It becomes difficult, therefore, to match views of such individuals with scientific (objective) thinking.

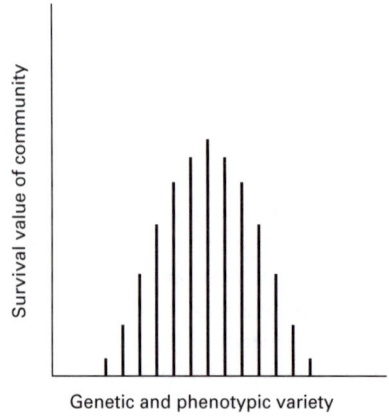

Fig. 16.14 A landscape plot showing the variety of individuals in a human population based on a particular measure of cultural difference or individual content of information in the brain. Individual differences are advantageous to a limit which concerns the collective strength of a society.

16.17   Subjectivity and objectivity

The breakdown of the description of living organisms from species to individuals forces us to reconsider attitudes to the survival of organisms

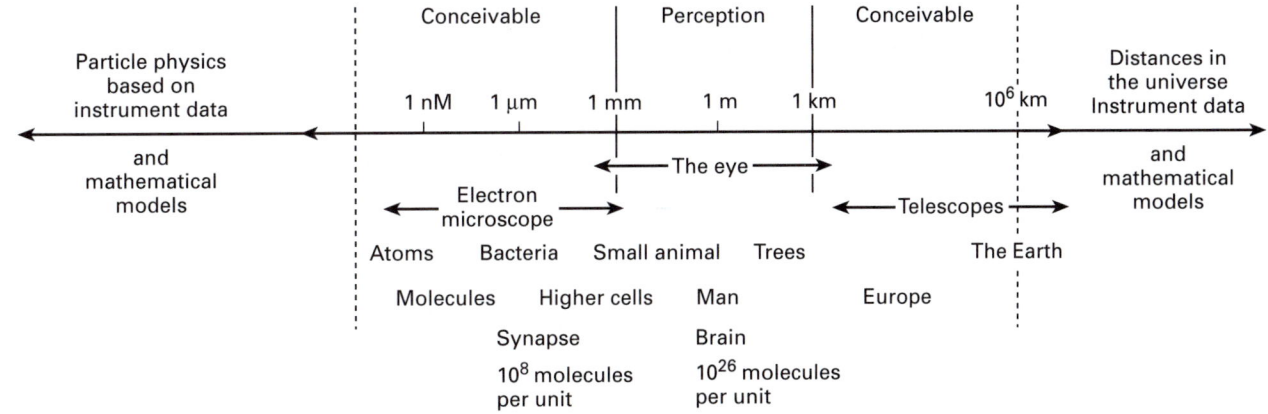

Fig. 16.15 The extent of the electromagnetic radiation spectrum open to man and two extensions of it to the very, very large and very, very small. The regions beyond the dashed lines are not open to conception by the senses. The figure is not to scale.

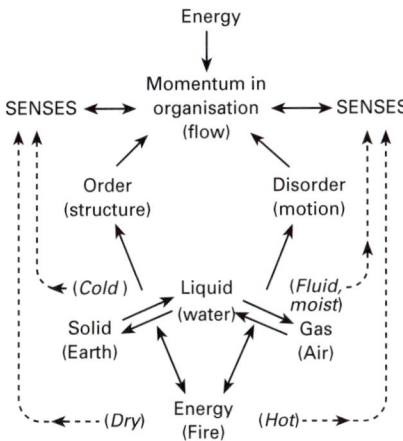

Fig. 16.16 This figure indicates how a return to consideration of the senses leads us back to a subjective, not an objective, description of all that is around us (see Fig. 1.1).

(compare Figs 14.16 and 16.14). No longer can we use solely objective assessments, such as we have applied throughout the book, in an effort to come to terms with what is around us. In that treatment we always analysed collectively on the basis of large numbers of individual units and we used, therefore, a generalised agreed account, more often than not based on measurements by instruments. The notion of a species is a bulk statistical consideration. A human with a very large computing power in his brain can make assessments as an individual, but these subjective evaluations do not use equipment of the same kind as that used by external measuring devices (see Fig. 16.15). Although the individuals ability to assess is based upon common general principles, the senses and experiences of each one differs from all others. Thus, whether we, as scientists, like it or not, there is a strong personal interference between objective (equipment-based) and subjective (individual) response to external events. This affects the ability to survive. We try to incorporate this in Fig. 16.16, which then returns us to what the senses actually measure. In a peculiar way we have to go back to diagrams more like those of Figs 1.1 or 1.2. This conclusion states that lying beside the examination of matter around us by objective means, there is, in each one of us, subjective examination. This subjective self may come to be trained to make rational 'scientific' observations and interpretations but it has, underlying everything, a biological character based on "senses", which may conflict with 'reason' (or the intellect) (see the quotation in Chapter 1). The nature of this biological system belongs to inheritance from 4×10^9 years ago and cannot be fundamentally changed. Hence, whatever we create using objective knowledge, no matter how obtained, has to be dove-tailed into this foundation—the nature of man as an individual. There is no doubt, however, that man's survival depends not on individuality alone but on his ability to accept constraints on that novelty. To survive, man has to accept organisation (see Fig. 16.17) which leads us to the subject of complexity in society.

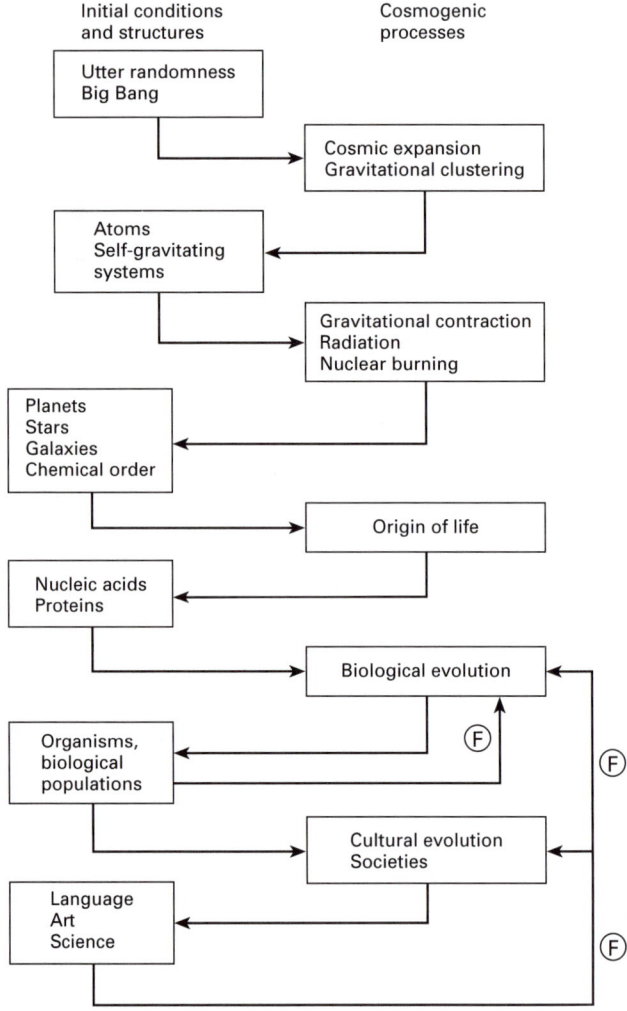

Fig. 16.17 The relationship and development of both order and organisation, the latter is underlined. Notice the feedback loops Ⓕ stemming from the introduction of the brain. (Adapted from Layzer, D. (1990). Cosmogenesis,Oxford Univ. Press).

16.17.1 The evolution of society: a new organisation

The very fact that mankind became a collection of individuals has forced upon him the need to create a new organisation—societies. It is not the purpose of this book to discuss such a development, in effect a new emergent property, but it is worth noting that it is built in order to increase survival. It also has within the organisation all the features we find in a single cell. These are:

(1) structured units;

(2) a command centre;

(3) message systems;

(4) feedback controls;

(5) protective apparatus; and even

(6) containment within boundaries at different levels.

We leave it as an exercise for the reader to pursue the parallels between biological and economic/political organisations.

16.18   Conclusion

Bringing chemistry
to life:
From matter to man

In writing this book we want to bring to the attention of our readers that the study of the progression from matter to man incorporates knowledge from all sciences, and their interaction with all of man's activities (Fig. 16.17). This knowledge is essential for the well-being of our own future. While the book details a physical–chemical evolution of all natural systems within a logical framework, utilising the idea of substances and objects made from units, which gives the variable *composition*, and the distribution of units in *space* and *time*, it is obvious that we meet increasing difficulties with the ever-increasing complexity of what has evolved. We could not in the end describe the most complex of all systems, a single man, in descriptive terms of units and variables, which had been our aim throughout the book. The emergent properties of life itself coupled to those of the brain of individuals defy such analysis. However, we are convinced that, despite the complexity, all systems on Earth are chemical, so that the constitutive units of all objects are well defined: hence our title 'Bringing Chemistry to Life'. It is the complexity of their interactions that has permitted man to evolve, now with an awareness of this chemical evolutionary inheritance, even if without detailed understanding, which is still challenging us. In this there lie two features which are, presently, to some degree at daggers drawn. Both are obvious. Mankind will be able to develop not only with the aid of the machines he has devised but by manipulating life itself. This, his objective scientific knowledge allows, but besides foreseeable disturbing developments that still belong to the realms of science fiction, e.g. self-assembling intelligent robots or cloned human beings, a more immediate problem we face is that our present wilful activities have imposed themselves on the materials external to ourselves at a pace faster than the rate of natural evolution. This generates risky consequences which may damage the chemistry internal to all living systems. In this case brains will have defeated genes instead of assisting them, and we, a very vulnerable and dependent species, will probably be the first victims of the resulting instability. A second problem is for scientists to see that respect for individuals severely limits the ability to apply science-derived ideas of organisation to human well-being. The essence of our science study has been to show how bulk properties emerge as the effects of large numbers of identical units are taken into account. Mankind cannot be handled in this way since individuals have eccentricity which is not discernibly related to anything, except perhaps their own history. Social studies can only be scientific in part.

The present generation is perhaps now more aware of the problems, but unfortunately it tends to blame science for many of the difficult situations with which we are already confronted. In fact, the problem lies not with our knowledge of systems (science) but with our inability as yet to apply knowledge considerately to the greatly improved evolved status of man relative to all previous forms of life, so as to secure an appropriate relationship with all that is around us. The solution lies not in the rejection of science, since only science can be used to solve many existing or future problems which some applications of science have brought to us. It is science that allows us to understand, generally, the nature of our environment. This we have shown. What is required is sound, scientifically informed, shared wisdom, together with appreciation of individuals in societies. What is further needed is a co-operative effort to bring once more 'organisation out of chaos', but the control to be imposed is now of a different nature from any in the past: it is a social code of behaviour in the form of internationally monitored agreed *legislation*, which can be realistically enforced against individual and many political and corporate group selfish interests. Perhaps then, if joint efforts succeed, a complement to this book can be written which will outline a new sustainable emergent system, based on individuals in a global society and in balance with the rest of the material universe.

Further reading

The main reference for the present volume is our previous book, written at a higher technical level: Williams, R. J. P. and Fraústo da Silva, J. J. R. (1997). *The natural selection of the chemical elements: the environment and life's chemistry*. Clarendon Press, Oxford.

Complementary reading for the several chapters is given below. Most of the titles refer to relatively simple, layman-type books, or to texts written at about the same level as the present volume. In a few cases, particularly commendable college or university textbooks are also indicated. For a more comprehensive list of references see our previous book. The Oxford Chemistry Primers (Oxford University Press) contain 100-page accounts of different areas of chemistry for undergraduates. There are some 70 volumes on almost every section of chemistry.

Chapter 1

1. Asimov, I. (1956). *A short history of chemistry*. Anchor Books, New York. Simple and concise.
2. Partington, J. R. (1960). *A short history of chemistry* (3rd edn), Harper Torchbooks. Harper and Brothers, New York. A compact version of the monumental treatise on the history of chemistry by the same author.
3. Brock, W. H. (1992). *The Fontana history of chemistry*. Fontana Press, London. A modern account of the history of chemistry up to recent days.
4. Puddephat, R. J. and Monaghan, P. K. (1986). *Periodic table of the elements*. (2nd ed). Oxford University Press, Oxford. A brief account of the atomic structure and properties of the elements, leading to the build up of the periodic table.
5. Rossotti, H. (1998). *Diverse atoms*. Oxford University Press, Oxford. An element by element introduction to chemistry in a simple catalogue-like format.
6. Atkins, P. (1996). *The periodic kingdom*. Phoenix, London. A light 'travel guide' to the features of the periodic table in terms of the structure and properties of the atoms of the chemical elements.
7. Cox, P. A. (1989). *The elements—their origin, abundance and distribution*. Oxford University Press, Oxford. A clear descriptive, non-technical account of the elements, which fulfils the aims of the title.
8. Cox. P. A. (1995). *The elements on earth—inorganic chemistry in the environment*. Oxford University Press, Oxford. A complement to the previous reference, by the same author, which provides a good overview of the roles played by the different chemical elements on Earth and constitutes an excellent reference work.
9. Pagels, H. R. (1992). *Perfect symmetry*. Penguin Group, London. An accurate presentation of cosmology for the layman where the synthesis of the heavy elements is clearly described.
10. Weinberg, S. (1983). *The first three minutes*. Fontana Press, London. A clear account of the synthesis of elements in the Big Bang.
11. Primo Levi (1986). *The periodic table*. Sphere Books, Ltd, London. An enchanting narrative by a famous Italian writer who happened to be a qualified industrial chemist too (a fact that saved his life in Auschwitz). It has little to do with the contents of this chapter, except the titles of its 21 chapters which are the names

of 21 chemical elements. But, according to the author, 'every element says something to someone' and in our intended holistic view of the world where art, culture, science and even religion are complementary approaches to the understanding of the whole, this book is almost a compulsory reference.

See also:

12. Norman, E. B. (1994). Stellar alchemy—the origin of the chemical elements. *Journal of Chemical education*, **71**, 813.
13. Viola, V. E. (1990). Formation of the chemical elements and the evolution of the universe. *Journal of Chemical Education*, **67**, 723 and see also *Journal of Chemical Education* (1994), **71**, 840.
14. Prantzos, N., Vangioni-Flan, E., and Cass, M. (eds), (1994). *Origin and evolution of the elements*. Cambridge University Press. Cambridge. A series of papers reflecting our current knowledge of the subject.
15. Hoffman, D. and Lee, D. M. (1999). Chemistry of the heaviest elements. *Journal of Chemical Education*. **76**, 311.

Chapter 2

16. *Science—a science foundation course*, S 102, Units 9 and 10 (1991). The Open University Press, Milton Keynes. A basic introduction to the concept of energy and its forms at a fairly elementary level.
17. Narlikov, J. V. (1996). *The lighter side of gravity* (2nd edn). Cambridge University Press, Cambridge. An up-to-date introduction to one basic force of nature in a clear, non-technical style. Recent advances and problems of cosmology are also discussed.
18. Physical Science Study Committee (1965). *Physics* (2nd edn). D. C. Heath and Company, Boston. A widely used undergraduate textbook which presents physics in a phenomenological basis, with many illustrative examples.
19. Asimov, I. (1996). *Understanding physics*, Vols. I–III. Walker and Company, New York. (Reprinted as a single volume in 1993 by Barnes and Noble Books, New York. A typical Asimov book full of historical information, written in a clear non-mathematical style. Like ref. 1, it is not a recent book, but classical physics has not changed much.

See also:

20. Haish, B., Rueda, A., and Puthoff, H. E. (1994). Beyond $E = mc^2$. *The Sciences*, Nov./Dec. 26–31. A speculative discussion on the possibility of mass, inertia and gravity arising from underlying electromagnetic processes.

Chapter 3

21. Kerwin, L. (1963). *Atomic physics—an introduction*. Holt, Rinehart and Winston, New York. Chapters 1 and 2. A simple introduction to atoms and atomic properties.
22. Atkins, P. W., Clugston, M. J., Frazer, M. J., and Jones, R. A. Y. (1988). *Chemistry—principles and applications*. Longman Group, London. Chapter 1. A textbook at undergraduate level providing an introduction to the topics discussed in this chapter.
23. Pimentel, G. C. and Spratley, R. D. (1970). *Chemical bonding clarified through quantum mechanics*. Holden-Day, Inc., San Francisco. A simple introduction to the theory of chemical bonding and molecular structure.

24. Shriver, D. F., Atkins, P. W. , and Langford, C. H., (1994). *Inorganic chemistry*. Oxford University Press, Oxford. Part I. An undergraduate textbook, at a slightly higher level than the previous reference, covering the topics discussed in his chapter and in Chapters 4 and 6.

25. Dekock, R. L. and Gray, H. B. (1989). *Chemical structure and bonding*. The Benjamin/Cummings Publ. Company, Inc., Menlo Park.

26. Gray, H. B. (1995). *Chemical bonds—an introduction to the atomic and molecular structure*. University Science Books, New York.

27. Lawrence, C., Rodger, A., Compton, R. (1996). *Foundations of physical chemistry*. Oxford University Press, Oxford.

Chapter 4

Most texts of general or inorganic chemistry cover the subjects discussed in this chapter, in greater or lesser depth. The choice is very much a matter of individual preference. We limit ourselves to indicate a few recent books and to suggest a number of other classic texts that still have much to recommend them.

28. Atkins, P. W., Clugston, M. J, Frazer, M. J., and Jones. R. A. Y. (1988). *Chemistry—principles and applications*. Longman Group. London. Mostly Chapters 1, 2, 5–8 and 15.

29. Huheey, J. E., Keiter. E. A., and Keiter, R. L. (1993). *Inorganic chemistry—principles of structure and reactivity* (4th edn). Harper Collins College Publishers, New York. Mostly chapters 2, 4, 5 and 8.

30. Shriver, D. F., Atkins, P. W., and Langford, G. H. (1994). *Inorganic chemistry*. (2nd edn). Oxford University Press, Oxford. Chapters 1–4, 9 and 18.

31. Phillips, C. S. G. and Williams, R. J. P. (1996). *Inorganic chemistry*. Oxford University Press, Oxford. Chapters 2–5.

32. Day, M. C., Jr and Selbin, J. (1969). *Theoretical inorganic chemistry* (2nd edn). Reinhold Book Corporation, New York. Chapters 3–6.

See also:

33. Pauling, L. (1960). *The nature of the chemical bond* (3rd edn). Cornell University Press, Ithaca, New York. Chapters 3, 7, and 11–13.

34. Mason, B. and Moore, C. B. (1982). *Principles of geochemistry*. J. Wiley and Sons, New York. A classic text with useful complementary reading for most of chapters 4–9 of this book. Phase diagrams for the formation of Earth's minerals are found in Chapter 5.

35. Ball, P. (1994). *Designing the molecular world—chemistry at the frontier*. Princeton University Press, Princeton, New Jersey. An overview on modern chemical problems.

Chapter 5

This chapter deals mainly with some basic principles of equilibrium thermodynamics. There are, literally, hundreds of texts on this fundamental branch of physics, from elementary introductions to advanced treatises, many of which could be recommended. The following are just a few references which we found useful to clarify concepts, but the selection is very much a matter of individual preference.

36. Atkins, P. W. (1998). *Physical chemistry* (6th edn). Oxford University Press, Oxford. Chapters 1–5, 7, 19, 20.

37. Moore, W. J. (1972). *Physical chemistry* (5th edn). Longman Group, Ltd, London. Chapters 1–3 and 5.

38. Maham, B. H. (1972). *Elementary chemical thermodynamics*. W. A. Benjamin Inc., New York.

39. Price, G. (1998). *Thermodynamics of chemical processes*, Oxford Chemistry Primers no. 56. Oxford University Press, Oxford.

40. Pimentel, G. C. and Spratley, R. D. (1970). *Understanding chemical thermodynamics*. Holden-Day Inc., San Francisco.

41. Bent, H. A. (1965). *The second law*. Oxford University Press, New York.

42. Craig, N. C. (1992). *Entropy analysis*. VCH Publishers, New York.

43. Atkins, P. W. (1994). *The second law—energy, chaos and form*. Scientific American Books Inc., New York.

44. Zemansky, M. W. and Dittman, R. H. (1981). *Heat and thermodynamics*. McGraw Hill Book Co., New York.

45. Denbigh, K. (1966). *The principles of chemical equilibrium*. Cambridge University Press, Cambridge. (At a more advanced level.)

46. Maczek, A. (1998). *Statistical thermodynamics*, Oxford Chemistry Primers no. 58. Oxford University Press, Oxford.

47. Johnson, D. A. (1968). *Some thermodynamic aspects of inorganic chemistry*. Cambridge University Press, Cambridge. General reference for this chapter and Chapters 6–8.

For applications to biology:

48. Harold, F. H. (1986). *The vital force: a study of bioenergetics*. W. H. Freeman and Company, New York. Also a reference for Chapters 10–15.

See also:

49. Adams, S. (1994). No way back. *New Scientist*, 22 October. (Inside Science number 75).

50. Wilson K. G. (1982). The renormalisation group and critical phenomena. *Nobel Prize Lectures*. Almqvist and Wiksell International, Stockholm. This lecture gives the mathematical explanation of phase transitions and illustrates the problems of dealing with complex phenomena in terms of variables. It is not easy but it is important.

Chapter 6

Detailed treatment of reactions in aqueous solution can be found in many standard texts; the following cover the subject from many complementary points of view.

51. Phillips, C. S. G. and Williams, R. J. P. (1996). *Inorganic chemistry*. Oxford University Press, Oxford. Chapters 7, 9, 14 and 32.

52. Fraústo da Silva, J. J. R. and Williams, R. J. P. (1997). *Biological chemistry of the elements—the inorganic chemistry of life* (reprint). Oxford University Press, Oxford. Chapters 1, 2, and 4.

53. Huheey, J. E. Keiter, E. A., and Keiter, R. C. (1993). *Inorganic chemistry*. Harper Collins College Publications, New York. Chapters 9–11 and 13.

54. Stumm, W. and Morgan, J. J. (1995). *Aquatic chemistry* (3rd edn). J. Wiley and Sons, New York. Chapters 2, 3 and 5–7.

55. Laitinen, H. A. (1960). *Chemical analysis*. McGraw Hill Book Co., New York. Chapters 1–3, 6, 7 and 15.

56. Martell, A. E. and Hancock, R. D. (1996). *Metal complexes in aqueous solution*. Plenum Publishing, New York.

See also:

57. Siegel, H. (ed.) (1984). *Metal ions in the environment*, Vol. 18. *Circulation of metals in the environment*. Marcell Dekker, New York.

Chapter 7

The topics discussed in this chapter are found dispersed in many texts on thermodynamics and general physical chemistry, although the approach is different from that used here. The reader can gain further insight by consulting the references indicated for Chapters 2 and 5, and the following ones.

58. D'Arcy Thompson (1961). *On growth and form*. Cambridge University Press, Cambridge.
59. Van Vlack, L. (1973). *Materials science for engineers*. Addison-Wesley, Reading (Mass).
60. *The Structure and Properties of Materials* (1964). J. Wiley and Sons, New York, especially Vol. 1. (ed. Moffat, W. G., Pearsall, G. W. , and Wulff, J.) *Structure*, and Vol. II (ed. Brophy, J. H., Rose, R. M., and Wulff, J.) *Thermodynamics of structure*.
61. Smith, C. A. and Wood, E. (1991). *Biological molecules*. Chapman and Hall, London.
62. Compton, R. G. and Sanders, G. H. W. (1996). *Electrode potentials*, Oxford Chemistry Primers no. 41. Oxford University Press, Oxford.
63. Bergethon, P. R. and Simons, E. R. (1990). *Biophysical chemistry*, Springer-Verlag, New York. Especially Chapters 7, 19 and 22. Also a reference for Chapters 5 and 6 in this volume.
64. Birkeland, P. W. (1984). *Soils and geomorphology*. Oxford University Press, New York.
65. Anderson, G. M. and Crerar, D. A. (1993). *Thermodynamics and geochemistry: The equilibrium model*. Oxford University Press, New York.
66. Lesk, A. M. (1991). *Protein architecture*. IRL Press, Oxford.
67. Diamond R., Keotzle, T. F., Prout, K., and Richardson, J. S. (1991). *Molecular structures in biology*. Oxford University Press, Oxford.

Chapter 8

The topics discussed in this chapter are only orthodox in part and so no simple appropriate reference deals with them all. Some parts were discussed in our previous books.

68. Fraústo da Silva, J. J. R. and Williams, R. J. P. (1991, reprinted 1997). *The biological chemistry of the elements—inorganic chemistry of life*. Oxford University Press, Oxford. Chapters 2–7.
69. Williams, R. J. P. and Fraústo da Silva, J. J. R. (1996, reprinted 1997). *The natural selection of the chemical elements—the environment and life's chemistry*. Oxford University Press, Oxford. Especially Chapter 7.

Classical approaches to chemical kinetics are found in the following widely used textbooks.

70. Atkins, P. W. (1998). *Physical chemistry* (6th edn). Oxford University Press, Oxford.

71. Laidler, K. J. (1987). *Chemical kinetics*. Harper and Row, New York.
72. Connors, K. A. (1992). *Chemical kinetics in solution*. VCH, New York.

See also:

73. Raines, R. T. and Hanse, D. E. (1988). An intuitive approach to steady-state kinetics. *Journal of Chemical Education*, **65**, 757.

On biological kinetics the following texts provides ample information.

74. Purves, W. K., Orians, G. H., Heller, H. C., and Sadava, D. (1998). *Life—the science of biology* (5th edn). Sinauer Associates, Inc. and W. H. Freeman and Company, Sunderland (Mass.) Chapters 6–8.
75. Roberts, D. B. (1977). *Enzyme kinetics*. Cambridge University Press, Cambridge.
76. Harrison, L. G. (1993). *Kinetic theory of living patterns*. Cambridge University Press, Cambridge.

See also:

77. Coveney, P. and Highfield, R. (1990). *The arrow of time*. W. H. Allen, London. Especially Chapters 6and 7.
78. Nicolis, G. and Prigogine, I. (1989). *Exploring complexity*. W. H. Freeman and Co., New York.
79. Kauffman S. (1995). *At home in the Universe*. Oxford University Press, Oxford.

Chapter 9

The following references provide extensive complementary information on the various topics discussed in this chapter—from the formation of the Earth, its structure and dynamics, to the origins of early life and the so-called Gaia hypothesis, i.e. the preservation of a global steady state due to the interaction between living and non-living systems in our planet.

80. Cox, P. A. (a) (1989). *The Elements—their origin, abundance and distribution*. Oxford University Press, Oxford. Chapters 4 and 5. (b) (1995). *The elements on earth*. Oxford University Press, Oxford. Chapters 1–4.
81. Press, F. and Siever, R. (1998). *Understanding Earth* (2nd edn). W. H. Freeman, New York. Chapters 1–3, 6, 7, 13, 14, 17, 20.
82. Brown, G. C., Hawkesworth, C. J., and Wilson, R. C. L. (eds) (1974). *Understanding the Earth*. Cambridge University Press, Cambridge. Chapters 2, 3, 8, 9, 22.
83. Mason, B. and Moore, C. B. (1982). *Principles of geochemistry* (4th edn). J. Wiley and Sons, New York. Chapters 1–3, 8–10.
84. Henderson, P. (1986). *Inorganic geochemistry*. Pergamon Press, Oxford. Chapters 2, 4, 10, 11.
85. Wayne, R. P. (1993). *Chemistry of Atmospheres* (2nd edn). Clarendon Press, Oxford. Chapters 1, 4, 5, 9.
86. Kasting, J. F. (1993). Earth's early atmosphere. *Science*, **259**, 920.

On prebiotic chemistry and the origin of life:

87. Cairns-Smith, A. G. (1982). *Genetic take-over and the mineral origins of life*. Cambridge University Press, Cambridge.
88. Cairns-Smith, A. G., Hall, S. J., and Russel, M. J. (1992). Origins of life and the evolution of the biosphere, **22**, 161.
89. Cairns-Smith, A. G. (1985). The first organisms. *Scientific American*, June 1985.

90. Wächtershäuser, G. (1990). Production of the first metabolic cycles. *Proceedings of the National Academy of Sciences USA*, **82**, 200.

91. Oró, J., Rewers, K., and Odom, D. (1982). Criteria for the emergence and evolution of life in the solar system. *Origins of Life*, **12**, 285.

92. Monastersky, R. (1998). The rise of life on Earth. *National Geographic*, **193** (3), 54.

On Gaia:

93. Lovelock, J. E (1979). *Gaia, a new look at life on Earth*. Oxford University Press, Oxford.

94. Lovelock, J. E (1988). *The ages of Gaia*. Oxford University Press, Oxford.

95. Joseph. L. E. (1990). *Gaia: the growth of an idea*. St. Martins Press, London.

96. Volk, T. (1997). *Gaia's body*. Copernicus, New York.

Chapters 10–16

The sequence of these chapters, the arrangement of the topics and the style of treatment is unusual in that we wish to present the evolution of living organisms as closely dependent and interactive with the evolution of the availability and utility of mainly inorganic as well as of organic elements. The background is, of course, the bio-inorganic chemistry of life and here a variety of texts could be recommended to complement the various aspects discussed. We indicate only a few, but the choice is in no way limited. These elements are the units of life and we go on to discuss the variables, space and time. A second novel stress is on the use of space within organisms, compartmentalisation, as it evolves to elaborate kinetic schemes. We deliberately avoid development through chance genetic DNA variation. The most comprehensive general references are our previous two books, see below; more specific references to particular topics are grouped together and presented in the approximate order of their appearance in these six chapters. The reader should keep in his mind that we are looking for a logical way in which to describe evolution in terms of chemical units and the variables composition, space and time, so that the connection with Chapters 1–9 is strong.

General references

97. Fraústo da Silva, J. J. R. and Williams, R. J. P. (1991). *The biological chemistry of the elements—the inorganic chemistry of life* (4th printing, 1997). Oxford University Press, Oxford.

98. Williams, R. J. P. and Fraústo da Silva, J. J. R. (1996). *The natural selection of the chemical elements—the environment and life's chemistry* (2nd printing, 1997). Oxford University Press, Oxford.

99. *The limits of reductionism in biology*. (1998). Ed. Novartis (Ciba) Foundation Symposium 213. John Wiley & Sons, Chichester.

100. Maynard-Smith, J. (1968). *Mathematical ideas in biology*. Cambridge University Press, Cambridge.

101. Goodsell, D. S. (1998). *The machinery of life*. Springer-Verlag, New York.

Organic chemistry and biochemistry

102. Hornby, M. and Peach, J. (1993). *Foundations of organic chemistry*. Oxford University Press, Oxford.

103. Holum, J. R. (1995). *Elements of general organic and biological chemistry* (4th edn). J. Wiley and Sons, New York.

104. Solomons, T. W. (1994). *Fundamentals of organic chemistry* (4th edn). J. Wiley and Sons, New York.
105. Stryer, L. (1995). *Biochemistry* (4th edn). W. H. Freeman, San Francisco.
106. Lehninger, A. L. (1975). *Biochemistry* (2nd edn). Worth Publishers, New York.
107. Metzler, D. E. (1977). *Biochemistry—the chemical reactions of living cells.* Academic Press, New York.

Biology

108. Purves, W. K., Orians, G. H. and Craig-Heller, H. (1998). *Life—the science of biology* (5th edn). Sinauer Associates/W. H. Freeman, San Francisco.
109. Villee, C. A., Solomon, E. P., Martin, C. G., Martin, D. W., Berg, L. R., and Davies, D. W. (1989). *Biology* (2nd edn). Saunders College Publishing, Philadelphia.

Origin of life (Chapters 10 and 11 and see Chapter 9)

110. Oparin, A. I (1957). *The origin of life on the Earth* (3rd edn). Oliver and Boyd, Edinburgh. For a recent comment see also Lazcano, A. (1997). Chemical evolution and the primitive soup: did Oparin get it all right? *Journal Theoretical Biology*, **184**, 219.
111. Bernal, J. D. (1967). *The origin of life.* Wedenfeld and Nicolas, London.
112. Kenyon, D. H. and Steinman, G. (1969). *Biochemical predestination.* McGraw Hill Book Company, New York.
113. Bock, G. R. and Goode, J. A. (ed.) (1997). *Evolution of hydrothermal ecosystems on Earth*, Ciba Foundation Symposium 202. John Wiley & sons, Chichester.
114. Calvin, M. (1969). *Chemical evolution.* Clarendon Press, Oxford.
115. Fox, S. N. and Dose, K. (1972). *Molecular evolution and the origin of life.* W. H. Freeman and Company, New York.
116. Miller, S. L. and Orgel, L. E. (1974). *The origins of life.* Prentice-Hall Inc., New York.
117. Day, W. (1981). *Genesis on planet earth.* Shiva Publishing Ltd., Nantwich, Cheshire.
118. Cairns-Smith, A. G. (1985). *Seven clues to the origin of life.* Cambridge University Press, Cambridge.
119. Shapiro, R. (1986). *Origins—a skeptic's guide to the creation of life on Earth.* Heinemann, New York.
120. Davies, P. (1998). *The fifth miracle—the search for the origin of life.* Allen Lane, Penguin Books, London.
121. Mason, S. (1991). *Chemical evolution.* Oxford University Press, Oxford.
122. Kauffman, S. (1995). *At home in the universe.* Oxford University Press, Oxford.
123. Eigen, M. (1996). *Steps towards life.* Oxford University Press, Oxford.
124. Smith, J. M. (1993). *The theory of evolution.* Cambridge University Press, Cambridge.
125. Wright, S. (1982). Macro-evolution-shifting balance theory. *Evolution*, **36**, 427.
126. Joyce, G. F. (1989). RNA evolution and the origins of life. *Nature*, **338**, 217.
127. Lazcano, A. and Miller, S. L. (1996). The origin and early evolution of life: prebiotic chemistry, the pre-RNA world and time. *Cell*, **85**, 793.
128. Baltscheffsky, H. (ed.) (1996). *Origin and evolution of energy conversion.* VCH Publishers, New York.

Primitive forms of life and fossil records

129. Margulis, L. and Schwartz, K. V. (1998). *Five kingdoms* (3rd edn). Freeman and Co., New York.
130. Woese, C. R., Kandler, O., and Wheelis (1993). Towards a natural system of organisms: proposal for the domains Archae, Bacteria and Eukaria. *Proceedings of the National Academy of Sciences USA*, **87**, 4576.

See also Morrell, V. (1997). *Science*, **276**, 699 and Kerr, R. A. *Science*, **276**, 703.

131. Pace, N. R. (1997). A molecular view of microbial diversity and the biosphere. *Science*, **276**, 734.

132. Madigan, M. T. and Marrs, B. L. (1997). Extremophiles. *Scientific American*, April, 66.

133. Danson, M. J., Hough, D. W., and Lunt, G. G. (ed.) (1991). *The archaebacteria*, Biochemical Society Symposium no. 58. Portland Press, London.

134. Levett, P. N. (ed.) (1991). *Anaerobic Bacteria*. IRL Press at Oxford University Press, Oxford.

135. Ballows, A. Truper, H. G., Dworking, M., Harder, W., and Schleifer, K. H. (ed.) (1992). *The prokariotes* (2nd edn). Springer-Verlag, Heidelberg.

136. Gross, M. (1998). *Life on the edge*. Plenum Press, New York.

137. Lipps J. H. (ed.) (1993). *Fossil prokaryotes and protists*. Blackwell Scientific, Oxford.

138. Harold, F. M. (1990). Morphogenesis in micro-organisms. *Microbiological Reviews*, **54**, 381.

139. Silver, S. and Misra, T. (1988). Bacterial element requirements. *Annual Reviews of Microbiology*, **42**, 717.

140. Silver, S. (1998). Genes for all metals—a bacterial view of the periodic table. *Journal of Industrial Microbiology and Biotechnology*, **20**, 1.

141. Benton, M. J. (ed.) (1993). *The fossil record* (2nd edn). Chapman and Hall, London.

142. Cavalier-Smith T. (1991). In *Evolution of life* (ed. Osawa S. and Honjo, T.). Springer-Verlag, Tokyo.

143. Thomson, K. S. (1988). *Morphogenesis and evolution*. Oxford University Press, New York.

144. White, D. (1995). *The physiology and biochemistry of prokaryotes*. Oxford University Press, Oxford.

145. Doolittle, W. F. (1999). Phylogenetic classification and the universal tree. *Science*, **284**, 2124.

Self-assembly

146. Hemsley, A. R., Collinson, M. E., Kavach, W. L., Vincent, B., and Williams, T. (1994). The role of the self-assembly in biological systems. *Philosophical Transactions of the Royal Society, London. (Series B)*, **345**, 1.

147. Peacocke, A. R. (1983). *The Physical chemistry of biological organisation*. Clarendon Press, Oxford.

148. Capra, F. (1996). *The web of life*. Harper-Collins Publishers, London.

149. Kauffman, S. (1993). *The origins of order*. Oxford University Press, Inc., New York. (At an advanced level.)

150. Eigen, M. (1996). *Steps towards life*. Oxford University Press, Oxford.

Cell organisation

151. Cooper, G. M. (1997). *The cell*. ASM Press, Washington D. C./Sinauer Associates, Sunderland, Mass.

152. Rensberger, B. (1996). *Life itself—explaining the realm of the living cell*. Oxford University Press, New York.

153. de Duve, C. (1991). *Blue-print for a cell—the nature and origin of life*. Neil Patterson, Burlington, N. C.

154. Lodish, H., Baltimore, D., Berk. A., Zipursky, S. L., Matsudaira, P., and Darnell, J. (1995). *Molecular cell biology*. (3rd edn). Scientific American Books, W. H. Freeman and Company, New York. (At a more advanced level.)

155. Goodsell, D. S. (1998). *The machinery of life*. Springer-Verlag, New York.

156. Doerfler, W. (ed.) (1992). *Molecular biology of the cell*. VCH, Weinheim, Germany.

The chromosome (DNA)

157. Heslop-Harrison, J. S. and Flavell, R. B. (1994). *The chromosome* Bios Scientific Publ., London.
158. Fitzsimons, D. W. and Wolstenholme, G. E. W. (ed.) (1975). *The structure and function of chromatin.* Ciba Foundation Symposium 28. Elsevier, Amsterdam.
158(a) Caporale, L. H. (ed.) (1999). *Molecular strategies in biological evolution.* Annals New York Acad. Sci. Vol. **870**, New York.

Biological inorganic chemistry

159. Lippard, S. J. and Berg, J. M. (1994), *Principles of bioinorganic chemistry.* University Science Books, Mill Valley.
160. Fraústo da Silva J. J. R. and Williams R. J. P. (1991). *The biological chemistry of the elements—the inorganic chemistry of life.* Clarendon Press, Oxford.
161. Kaim, W. and Schwederski, B. (1994). *Bioinorganic chemistry: (inorganic elements in the chemistry of life.* Wiley, Chichester.
162. Hughes, M. N. (1987), Coordination compounds in biology, in *Comprehensive coordination chemistry,* vol. 6 (ed. Wilkinson, G.). Pergaman Press, Oxford.
163. Fenton, D. E. (1995). *Biocoordination chemistry,* Oxford Chemistry Primers no. 25. Oxford University Press, Oxford.
164. Wilkins, P. C. and Wilkins, R. G. (1997). *Inorganic chemistry in biology,* Oxford Chemistry Primers no. 46. Oxford University Press, Oxford.
165. (1996) Bioinorganic enzymology. *Chemical Reviews,* **96** (7), 2237.

See also:

166. Berthon, G. (ed.) (1995). *Handbook of metal-ligand interactions in biological fluids (bioinorganic chemistry).* M. Dekker, New York.

Bioenergetics

167. Smith, C. and Wood, E. J. (1991). *Energy in biological systems.* Chapman and Hall, London.
168. Harold, F. M. (1986). *The vital force—a study of bioenergetics.* W. H. Freeman and Company, New York.
169. Cramer, W. A. and Knaff, D. B. (1990). *Energy transduction in biological membranes—a textbook of bioenergetics.* Springer-Verlag Inc., New York.
170. Lehninger, A. L. (1973). *Bioenergetics* (2nd edn.). W. A. Benjamin Inc. Menlo Park.
171. Williams R. J. P. (1961). Chains of catalysts. *Journal of Theoretical Biology,* **1**, 1.
172. Jones, C. W. (1981). *Biological energy conservation—oxidative phosphorylation* (2nd edn.). Chapman and Hall, London.
173. Rögner, M., Boekema, E. J., and Barber, J. (1996). How does photosystem 2 split water? *Trends in Biochemical Sciences,* **21**, 44.

Signalling and regulation

174. Evered, D. and Whelan, J. (ed.) (1986). *Calcium and the cell,* Ciba Foundation Symposium 122. J. Wiley & Sons, New York.
175. Rasmussen, H. (1988). The cycling of calcium as an intracellular messenger. *Scientific American,* **261** (4), 66.
176. Clapman, D. E. (1995). Calcium Signalling. *Cell,* **80**, 259.
177. Hunter, T. (1995). Protein kinases and phosphatases: the yin and yang of protein phosphorylation and signalling. *Cell,* **80**, 225.
178. Henneke, H. (1990). Regulation by metal-protein complexes. *Molecular Microbiology,* **4**, 1621.
179. Golbeter, A. (ed.) (1989). *Cell to cell signalling.* Academic Press, New York.
180. Thomson, K. S. (1988). *Morphogenesis and evolution.* Oxford University Press, Oxford.

181. Bradt, D. S. and Snyder, S. H. (1994). Nitric oxide: a physiological messenger molecule. *Annual Revenues in Biochemisty.*, **63**, 175.
182. Schmidt, H. H. W. and Walter, U. (1994). NO at work. *Cell*, **78**, 919.
183. Cohen P. and Klee, C. B. (ed.) (1988). Elsevier, *Calmodulin*. Amsterdam.
184. Fosket, D. E. (1994). *Plant growth and development—a molecular approach.* Academic Press, San Diego.
185. Ecker, J. R. (1995). The ethylene signal transduction pathway in plants. *Science*, **268**, 667.
186. Simons, P. (1992). *The action plant*. Blackwells, Oxford.
187. Barritt, G. J. (1992). *Communication within animal cells*, Oxford University Press Oxford.
188. Eckert, R., Randall, D., and Augustine, G. (1988). *Animal Physiology* (3rd edn). W. H. Freeman and Co., New York. Chapter 9.

Cell cycle

189. Murray, A. and Hunt, T. (1993). *The cell cycle: an introduction.* W. H. Freeman, New York.

See also refs 108, 109, and 154.

Biomineralisation

190. Williams, R. J. P. (1984). An introduction to biominerals and the role of organic molecules in their formation. *Philosophical Transactions of the Royal Society, London (Series B)*, 411.
191. Mann, S., Webb, J., and Williams, R. J. P. (ed). (1989). *Biomineralisation*. WCH, Weinheim, Germany. Especially Chapters 1, 2, 9 and 10.
192. Suga, S. and Nakahara, H. (ed.) (1991). *Mechanisms and phylogeny of mineralisation in biological systems*. Springer-Verlag, Tokyo.
193. Mann, S. (1993). Biomineralisation: the hard part of bioinorganic chemistry. *Journal of the Chemical Society, Dalton Transactions*, 1.
194. Crick, R. E. (ed.) (1986). *Origin, evolution and modern aspects of biomineralisation in plants and animals*. Plenum Press, New York.

Symbiosis

195. Margulis, L. (1981). *Symbiosis in cell evolution*. W. H. Freeman, San Francisco.
196. Margulis, L. and Sagan, D. (1987). *Microcosmos—four billion years of evolution from our microbial ancestors*. Allen and Unwin, London.
197. Sapp, J. (1994). *Evolution by association—a history of symbiosis*. Oxford University Press, Oxford.

Evolution and advanced forms of life

198. Knoll, A. H. (1991). End of the proterozoic eon. *Scientific American*, October issue, 42 and see also *Science*, **256**, 622 (1992).
199. Strickberger, M. W. (1995). *Evolution* (2nd edn). Jones and Bartlett, Boston.
200. Hopper, A. E. and Hart, N. H. (1985). *The foundations of animal development* (2nd edn). Oxford University Press, Oxford.
201. Maynard-Smith, J. and Szathmáry, E. (1995). *The major transitions in evolution.* W. H. Freeman and Company/Spektrum, New York.
202. Dawkins, R. (1986). *The blind watchmaker*. Longman, Harlow, and see also Dawkins, R. *The selfish gene* 2nd ed., (1989) Oxford Paperbacks, Oxford and Dawkins, R. *Climbing mount improbable* (1996) Penguin Books, London.

203. Bonner, J. T. (1988). *The evolution of complexity*. Princeton University Press, Princeton.

204. Gribbin, J. and Gribbin, M. (1993). *Being human*. J. M. Dent, London.

205. Harrison, G. A., Tanner, J. M., Pilbeam, D. R., and Baker, P. T. (1988). *Human biology*. Oxford University Press, Oxford.

206. Wood, D. W. (1974). *Principles of animal physiology*. Edward Arnold, London.

207. Sagan, C. and Druyan, A. (1992). *Shadows of forgotten ancestors*. Random House, New York.

208. Wilson, E. O. (1992). *The diversity of life*. Allen Lane The Penguin Press, London.

209. Raven, P. H., Evert, R. F., and Eichorn, S. E. (1992). *Biology of plants* (5th edn). Worth, New York.

210. Allègre, C. (1993) *Introduction à une histoire naturelle—du Big Bang à la disparition de l'homme*. Librairie Arthmème-Fayard, Paris.

211. de Duve, C. (1995). *Vital dust—Life as a cosmic imperative*. Basic Books/Harper Collins Publ., New York.

212. Fortey, R. (1997). *Life: an unauthorised biography*. Harper Collins Publ., London.

Brain and neural networks

213. Damasio, A. R. (1994). *Descartes' error*. G. P. Putnan's Sons, New York.

214. Calvin, W. H. (1996). *How brains think*. Weidenfeld and Nicholson, London.

215. Crick, F. (1994). *The astonishing hypothesis*. Simon and Schuster, Ltd, London.

216. Pinker S. (1997). *How the mind works*. Weidenfeld and Nicholson, London.

217. Churchland, P. and Sejnowski (1992). *The computational brain: models and methods on the frontiers of compuntional neuroscience*. Bradford Books, MIT Press, Cambridge.

218. Cairns-Smith, A. G. (1996). *Evolving the brain*. Cambridge University Press, Cambridge.

219. LeDoux, J. E. (1996). *The emotional brain: the mysterious underpinnings of emotional life*. Simon and Schuster, New York.

220. Changeux, J. -P. and Chavaillon, J. (1996). *Origins of the human brain*. Oxford University Press, Oxford.

221. Saito, T. and Saito, K. (1996) Determination of brain trace elements. *Tohoku Journal of Experimental Medicine*, **178**, 11.

222. *Journal of Theoretical Biology* (1994). Special issue on mind and matter **171**, 1.

223. *Scientific American*. Special issue devoted to mind and brain, September 1992.

See also ref. 234.

Technology and environment problems

224. Cardwell, D. (1994). *The Fontana history of technology*. Fontana Press, London.

225. Wayne, R. P. (1993). *Chemistry of the atmosphere* (2nd edn). Oxford University Press, Oxford. Chapters 1, 2, 4, and 5.

226. Baird, C. (1994). *Environmental chemistry*. W. H. Freeman and Company, New York.

227. Butcher, S. S. Charlson, R. J., Orians, G. H., and Wolfe, G. V. (ed.) (1992). *Global biogeochemical cycles*. Academic Press, London.

228. Lovelock, J. E. (1991). *Gaia—the practical science of planetary medicine*. Gaia Books, Edinburgh.

229. Gribbin, J. (1988). *The hole in the sky*. Bantam Press, London.

230. Gribbin J. (1990). *Hot-house Earth—the greenhouse effect and Gaia*. Bantam Press, London.

231. Hill, M. K. (1997). *Understanding environment pollution: a primer*. Cambridge University Press, Cambridge.

232. Schlesinger, W. H. (1997). *Biogeochemistry* (2nd edn.) Academic Press, New York.

Systems and irreversible thermodynamics. Complexity and dissipative structures

233. Bertalanffy, L. von (1950). The theory of open systems in physics and biology. *Science*, **111**, 23.
234. Mainzer, K. (1997). *Thinking in complexity*. (3rd ed.) Springer-Verlag, Berlin.
235. Prigogine, I. (1967). Dissipative structures in chemical systems. In *Fast reactions and primary processes in chemical kinetics*, (ed. S. Claessons). Interscience, New York.
236. Prigogine, I. and Glansdorff, D. (1971). *Thermodynamic theory of structure, stability and fluctuations*. J. Wiley and Sons, New York. (Advanced level.)
237. Prigogine, I. and Stengers, I. (1984). *Order out of chaos*. Bantam Press, New York.
238. Prigogine, I. (1996). *La fin des certitudes*. Odile Jacob, Paris.
239. Schuster, P. (1982). Irreversible thermodynamics—an overview. In *Biophysics* (ed. Hoppe, W., Lochman, W. H., Markl, H., and Ziegler, H.). Springer-Verlag, Berlin. Chapter 8.
240. Capra, Fritjof (1996). *The web of life*. Harper Collins Publishers, London.

See also:

241. Cohen, J. and Stewart, I. (1994). *The collapse of chaos*. Penguin Books Inc. USA, New York.
242. Goodwin, B. (1997). *How the leopard changed its spots*. The Guernsey Press, Guernsey, Channel Islands.
243. Bak, P. (1997). *How nature works*. Oxford University Press, Oxford.
244. Science (1999) *Beyond reductionism*. (A series of papers on complex systems) **284**, 79.

And see refs 36, 78, 79, 122, 149 and 150.

Information

245. Shannon, C. and Weaver, W. (1949). *The mathematical theory of communication*. University of Illinois Press, Urbana. (Advanced)
246. Weber, B. H., Depew, D. J., and Smith, J. D. (ed.) (1988). *Entropy information and evolution*. MIT Press, Cambridge.

Ecology

247. Ricklefs, R. E. (1993). *The economy of nature—a textbook of basic ecology*.
248. Wilson, E. O. (1992). *The diversity of life*. Allen Lane. The Penguin Press, London.

Cosmology

249. Kaufman III, J. (1991). *Universe* (3rd edition). W. H. Freeman and Co., New York, especially chapters 5, 28, 29 and "Afterwork".
250. Gribbin, J. (1994). *In the beginning—the birth of the living universe*. Penguin Books, London.
251. Layzer, D. (1990). *Cosmogenesis—the birth of order in the universe*. Oxford University Press, Oxford.
252. Rees, M. (1997). *Before the beginnings*. Simon and Schuster, London.
253. Smolin, L. (1998). *The life of the cosmos*. Phoenix. Orion Books Ltd. London.
254. Riordan, M. and Schramm, D. N. (1991). *The shadows of creation*. W. H. Freeman and Co., New York.

See also references 9 and 10.

Index